Geomorphology of Europe

Geomorphology of Europe

General Editor: Clifford Embleton

M

MACMILLAN PUBLISHERS
LONDON

Macmillan Reference Books

First published 1984 by
THE MACMILLAN PRESS LTD
London and Basingstoke
Associated Companies throughout the world.

Reprinted 1984
First published in paperback 1984

British Library Cataloguing in Publication Data

Embleton, C.
      Geomorphology of Europe.
      1. Geomorphology—Europe
      I. Title
      551.4'094   GB435

      ISBN 0-333-34638-6
      ISBN 0-333-37963-2 paperback

Phototypeset by Tradespools Ltd., Frome, Somerset
Printed in Hong Kong

CONTENTS

(The authors of chapters or sections
are shown in italics)

Preface

This book has been compiled and written by members of the Commission on Geomorphological Survey and Mapping of the International Geographical Union. Originally proposed in 1976 as a text to accompany and explain the *International Geomorphological Map of Europe* (1 : 2.5 million), the outline of the book was prepared by the Commission at a meeting in Nové Mesto, Czechoslovakia, in 1977. Since then, there has been a considerable delay in publishing the *International Map*; meanwhile, work on this book progressed and has overtaken the schedule for printing the Map. At the same time, the book has been enlarged and its aims have changed to some extent: it is no longer envisaged simply as an explanation of the Map but as a work in its own right, providing the first comprehensive survey of the geomorphology of Europe.

The editor is well aware of shortcomings in the work and of unevenness of treatment. The original contributions were in many cases written in languages other than English; the sections on the USSR were translated from the Russian by J. Demek, that on eastern Austria was translated from the German by B. Bauer, and those on France were translated by the editor himself. Translation of technical works causes many problems, for literal conversion into English usually produces a meaningless result unless the translator is closely conversant with the subject matter. Even then, obscurities often remain and are impossible to remove wholly unless direct contact between author, translator and editor can be achieved. Unfortunately this is not only very time-consuming, but has not always been possible. There have also been considerable problems of postal communication, especially between eastern and western Europe. I should like at this point to pay tribute to Dr J. Demek of the University of Brno, without whose assistance and guidance this book would never have been completed. His knowledge of eastern European languages other than his own, and of German and English has been of immense value, and the editor has had invaluable discussions with him on numerous visits to Brno to clear up various difficulties. Other areas that have presented problems have been in the standardization of place names (whose spelling is far from consistent even in the major atlases), on how far to preserve feature names in the original language (e.g., Giant Mountains or Krkonose?), in obtaining a representative selection of photographs and in deciding how many references to list.

The regional division of Europe employed in this book is based mainly on structure and to a less extent on relief. It should be recognized that it is an arbitrary division in many instances, and that the dividing lines are not hard and fast. For example, the separation of the Scandinavian mountains from the Fennoscandian Shield is in many ways quite unrealistic and arbitrary, yet there are major morphological differences between these two units. The divisions cut strongly across national frontiers, which has caused problems of continuity of treatment and of correlation.

The level of detail varies considerably from one region to another, and the lengths of individual sections are not in proportion to the area covered. This is inevitable, for far more is known about the geomorphology of some areas than of others, and some areas present a much greater variety of landform or of geomorphological history than others. No attempt has been made to standardize the contributions from different authors, although each author has been asked as far as possible and where appropriate to deal with certain basic topics such as relief, geology, structure, Tertiary and Quaternary landscape evolution, Pleistocene glaciation, coastal forms, etc. The several contributions reveal interesting differences of geomorphological thought and approach; in some the emphasis is on tectonics and tectonic history, in others on climatic geomorphology, while in others on the Quaternary. Some authors pay great attention to present-day processes of erosion; others feel that older events and structures are more significant in the present landscape.

This book, then, represents both an amalgamation and a compromise. It represents a considerable effort of international cooperation, and shows above all how much remains to be achieved in understanding the geomorphology of what is probably the world's most complex continent.

C. Embleton
Pedras da Rainha
Good Friday, 1983

Acknowledgements

First, I want to thank all the contributors to the book, who have dealt patiently with my queries and requests for information, but especially Jaromir Demek without whose help this book would never have been completed. Many other people have helped in the various stages of preparation of the material: I am particularly grateful to Mrs V. Dittrichová for help in translation at an early stage; to Roma Beaumont, Kathy Hopkirk and Gordon Reynell for the redrawing of all the maps and diagrams, including some of immense complexity; and to Clare Baynes, Helena Gardberg, Anne Rogers and Celia Vangen for typing the manuscript, sometimes in several successive versions and often from very difficult originals. Finally, I want to thank Dr. A. Ralph for her care and patience in seeing the book through the press.

Chapter 1
Structural and Tectonic Framework of the Continent of Europe

1.1 Introduction

The various chapters that comprise Part I are concerned with broad aspects of the relief and structural patterns, the evolution of these patterns through time and the formative processes controlling their evolution. At the outset, it should be noted that, while it is intended to be used as an independent guide to the geomorphology of Europe, this book also aims to complement and explain the *International Geomorphological Map of Europe*, published in 16 sheets on a scale of 1 : 2.5 million. Reference to individual sheets of the Map is made in this book; an understanding of the features depicted on the Map will be facilitated by consulting the relevant sections.

For the purposes of both the Book and the Map, Europe (*see* Fig. 1.1) is taken to extend from Iceland in the west to the Ural Mountains in the east. The southern limit is provided by the Mediterranean Sea. Turkey, other than the small part lying west of the Bosporus, is excluded. Covering a land area of roughly $10^7 \, km^2$, the continent includes a great diversity of relief and structural forms (*see* Fig. 1.2), including vast plains, plateaus and high mountain ranges. Elevations of the land range from over 4500 m (e.g., Mont Blanc, 4810 m) to a few metres below sea level in the Netherlands. The ages of the rocks exposed at the surface vary from the most recent sediments of the present-day to Precambrian rocks more than 2600 Ma old (*see* Fig. 1.3). Further contrasts in the landform arise from the diversity of climate: from Arctic to Mediterranean, from semiarid to humid; the shifting climatic patterns of the Cenozoic have powerfully influenced landform evolution, especially in connection with the repeated glaciation of the northern half of the continent in the Pleistocene. A separate chapter (*see* Chapter 2) is devoted to the submarine forms around Europe, which also display considerable variety, but about which less is known.

The authors of the three chapters in Part I are all Russian. Their viewpoint differs frequently from that of geologists and geomorphologists in western Europe: for example, in respect of the theory of plate tectonics and the importance of the structural control of landform. No attempt is made (nor would it be possible) to present any unified theory, acceptable to both eastern and western European countries, of the structural evolution of Europe. In Part I and elsewhere, considerable attention is devoted to morphostructure. This term, introduced by Gerasimov (1946), refers to a structural unit expressed in the relief, modelled by denudation and/or sedimentation. The degree of modelling depends on the tectonic activity of the area and on its climate. A hierarchical system of morpho-structures is recognized, ranging from the largest 'mega-morphostructures', such as the Fennoscandian Shield and Russian platform, to micromorphostructures, such as a fault-controlled valley perhaps only a few tens or hundreds of metres in size. The *International Geomorphological Map of Europe* incorporates three major elements in its conceptual framework: (1) the differentiation of relief classes according to relief amplitude, (2) the differentiation of relief according to morphostructural type and (3) the differentiation of so-called 'special relief forms' of mainly exogenic origin.

1.2 Structural Evolution in Europe

At the outset, it should be noted that there are differences of opinion on two fundamental questions. One concerns the interpretation of the Precambrian record in which there are many uncertainties, owing partly to inadequate data on absolute age determinations. The second relates to the particular hypothesis of global tectonics that is adopted.

The earliest stage in the geological history of Europe can be seen from Fig. 1.4. The distribution of Archaean and Proterozoic rocks (Lower–Middle Precambrian, Middle Precambrian and Upper Precambrian) shown in this diagram suggests that there are cores of Archaean rocks around the peripheries of which younger fold zones have been accreted. The Archaean cores can be recognized in the crystalline basement rocks of the north-western Scottish Highlands and Outer Hebrides, the Bohemian massif, the Balkans, the southern Carpathians and the Fennoscandian Shield – both in Scandinavia and the Kola Peninsula. As is seen from Fig. 1.4, the early Precambrian rocks also comprise separate portions of the basement of the eastern European platform buried under a sedimentary cover. It is assumed that in the Massif Central of France, and the Armorican and certain other massifs, late Precambrian rocks are found overlying the early Precambrian varieties. Similar assumptions, although without convincing evidence, are made from some other areas where the Precambrian rocks are common [see Matveevskaya (1975)]. Following numerous periods of folding in the Precambrian, the Archaean cores became joined together. All of these Precambrian complexes of different ages formed the basement complex of the western and eastern European platforms 2500–3000 Ma ago. Thus, by the beginning of the Phanerozoic, the basement of the European continent already existed. One school of thought believes that the early Precambrian formations originated on oceanic crust. This was a lengthy process for

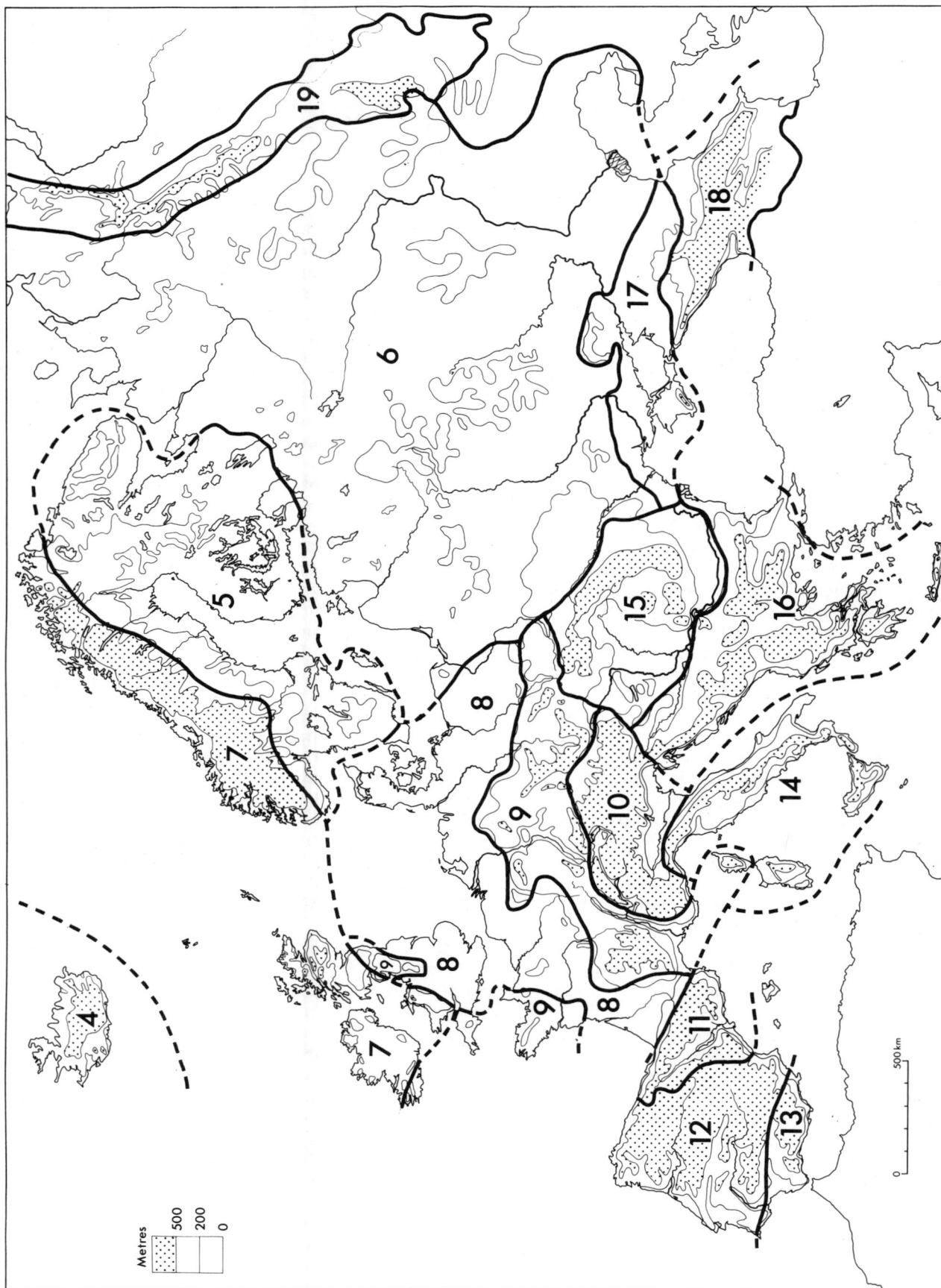

Fig. 1.1. Geomorphological regions of Europe. The numbers correspond to the chapters in this book.

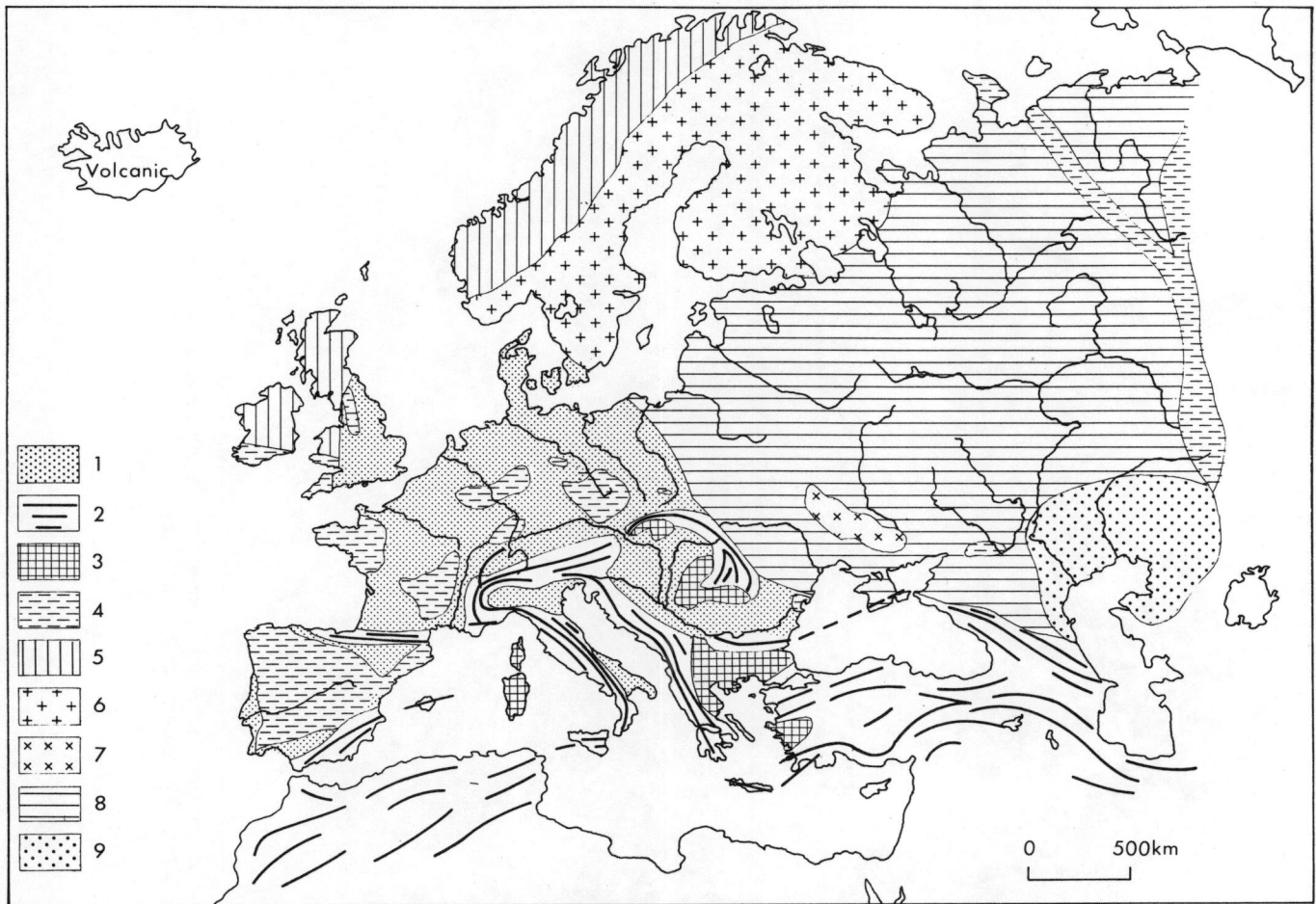

Fig. 1.2 The major morphostructural divisions of Europe: (1) lowlands and scarplands composed of Mesozoic and Tertiary sediments; (2) Alpine fold mountain belts; (3) Hercynian massifs involved in Alpine folding; (4) Hercynian massifs and fold belts; (5) Caledonian fold mountain belts; (6) Fennoscandian (Baltic) Shield; (7) Ukrainian Shield; (8) Russian platform; (9) Caspian lowlands.

the duration of the Precambrian era, three to four times greater than that of Phanerozoic. A similar picture may also be observed in some other ancient platform areas, such as North America and central Siberia. Characterizing the Kola and Pre-Dnepr megablocks of the eastern European platform as being the most ancient, Khain indicates that they '. . . are remnants of the most ancient continental crust which . . . evolved as a result of granitization of basic volcanic sequences, that is of the primary crust of the oceanic type' [see Khain (1977a) p.50].

The Precambrian history of the European subcontinent was not only marked by the accretion of younger fold zones around ancient cores (*see* Fig. 1.5) but also by the repeated fracturing of various Precambrian rock masses along deep-seated faults into large blocks (i.e. megablocks). Some of these megablocks were uplifted while others subsided, the subsiding blocks often becoming covered by the sea. Evidence of such subsidence is manifested locally in the form of horizontal or subhorizontal marine and metamorphosed marine sediments (e.g., the rocks of the middle Proterozoic mantle of Karelia). Within the megablocks, the direction of tectonic movement changed frequently over the course of the many

millions of years of Precambrian history. Therefore, the megablocks carved out of the Archaean and early Proterozoic rocks do not coincide with the ancient fold zones. By the end of the Upper Precambrian (Riphaean), however, some of the present-day structural outlines of Europe were becoming apparent.

The Ural geosyncline bounding the continent on the east was initiated in the Riphaean on the residual oceanic crust of the present-day western Siberian platform. Following an inherited but irregular pattern, the geosynclines of central Europe began to develop from the late Precambrian to the Palaeozoic. At the very end of the Riphaean (more precisely, at the boundary between the Riphaean and Cambrian: some 800 Ma ago) the Caledonian geosyncline of Scandinavia was formed. Finally, approximately the same age is attributed to the initiation of folding in the vast geosynclinal belt of the Tethys Ocean. Initiation of the present-day Alpine mountain system of Europe and north Africa also took place at that distant time. Many large platform structural forms (megablocks), such as graben–synclises and synclises, horst–anteclises and anteclises, were also formed then. These are expressed in the present-day relief both directly, as in the case of the

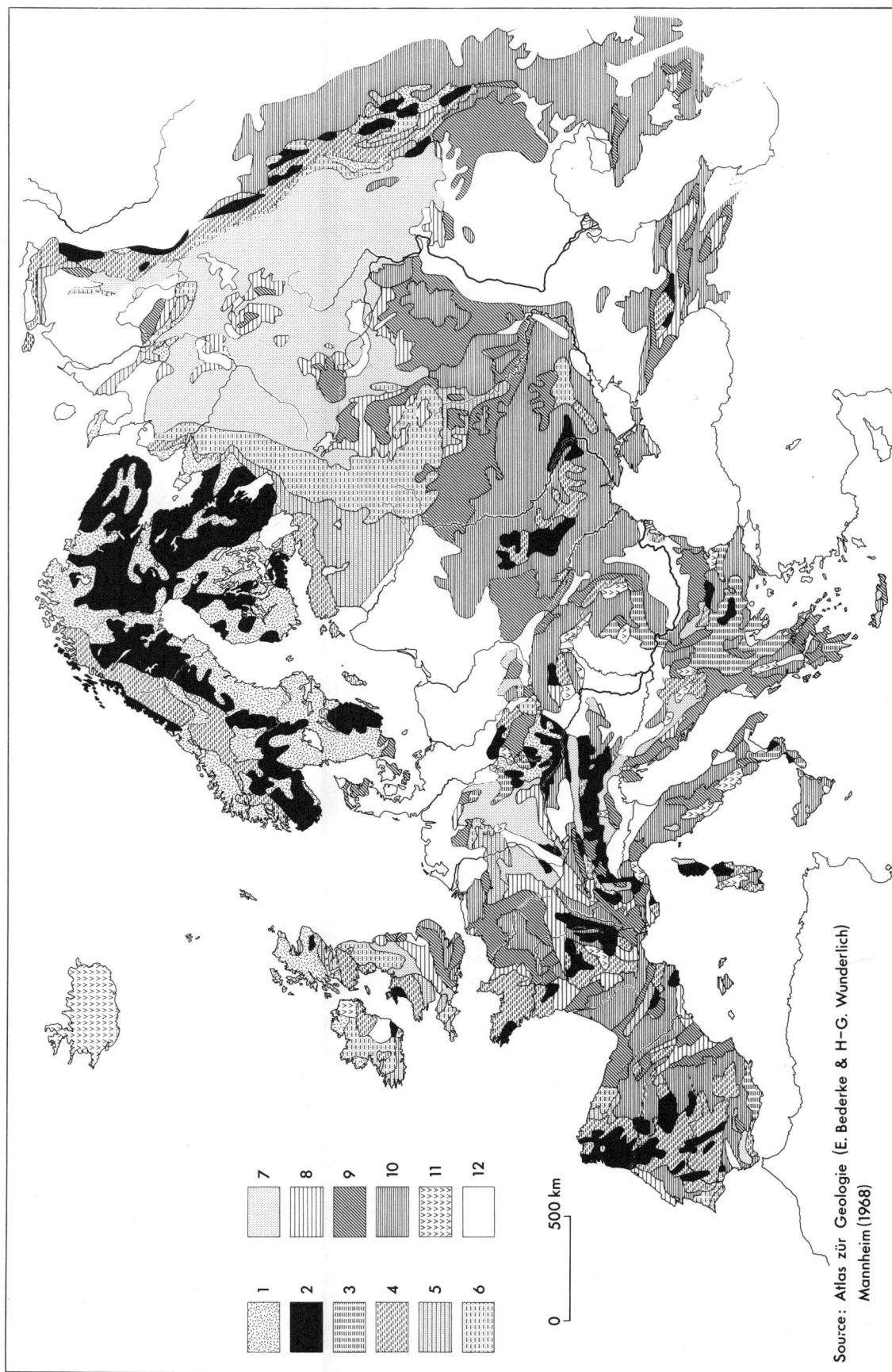

Source: Atlas zür Geologie (E. Bederke & H–G. Wunderlich)
Mannheim (1968)

Fig. 1.3 The geology of Europe [from *Atlas zür Geologie* (1968)]: (1) Archaean crystalline; (2) later Precambrian and Cambrian; (3) undifferentiated Lower Palaeozoic; (4) Ordovician and Silurian; (5) Devonian; (6) Carboniferous; (7) Triassic; (8) Jurassic; (9) Cretaceous; (10) Tertiary; (11) Neovolcanic; (12) Quaternary.

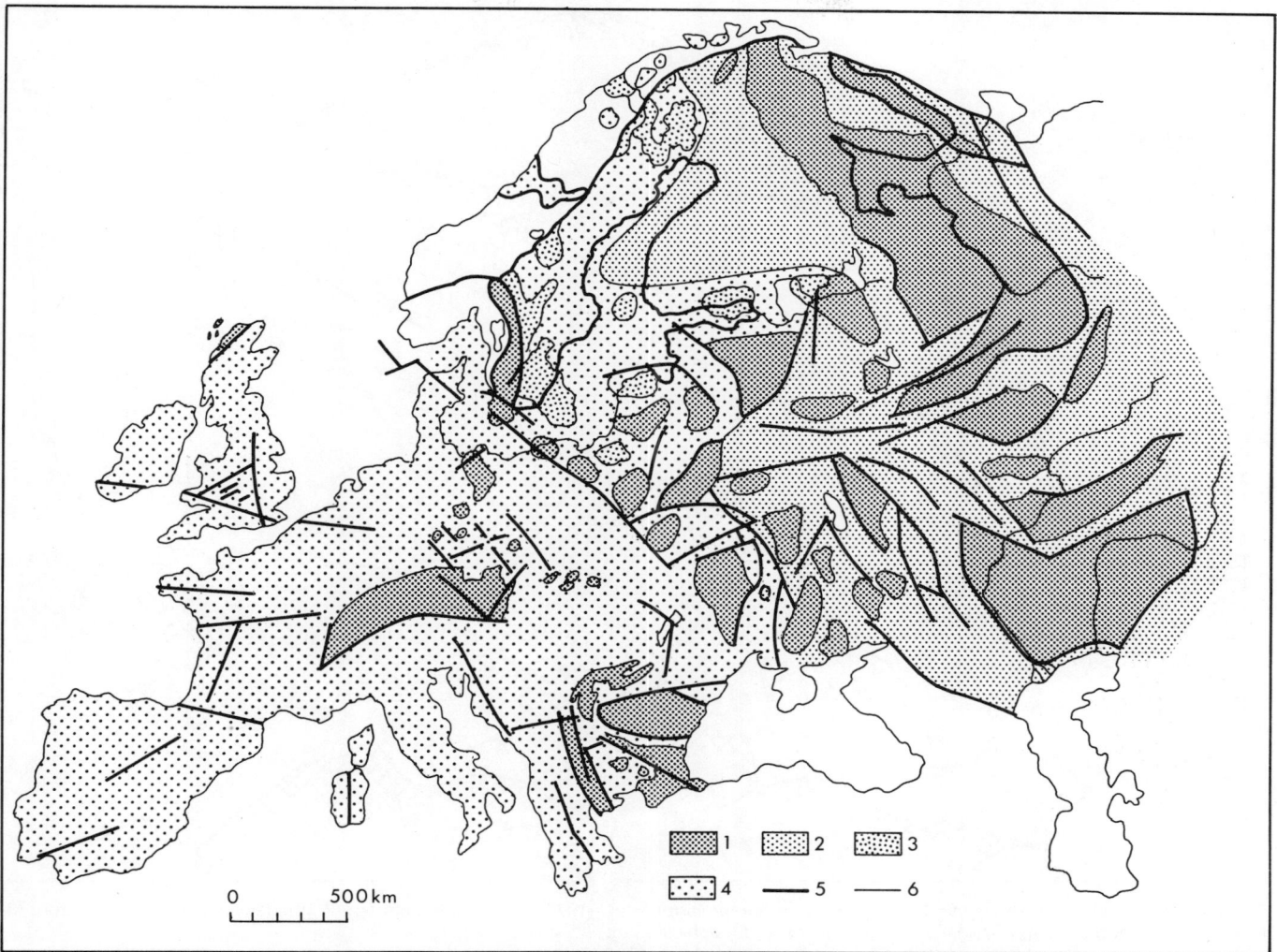

Fig. 1.4 Structure of the Precambrian basement [after Matveevskaya (1975)]: (1) crustal blocks consolidated in the Archaean (early Precambrian); (2) Karelides (middle Precambrian); (3) Rapakivi-type granites and granitoids of various ages; (4) Gotian–Dalslandian fold systems (late Precambrian); (5) boundaries of ancient blocks marked by major fractures; (6) other boundaries.

complex Baltic graben–syneclise, the Caspian syneclise and others, and in the form of inverted relief, such as the complex horst–anteclise of the central Russian elevated plain and the Volyn'–Podolsk plateau.

The Precambrian period was characterized by many orogenic episodes accompanied by metamorphism and complex magmatic processes (basic and acid magmatism), by the uplift and subsidence of megablocks and by frequent regional changes from marine regimes to continental. All these events led to the formation of the basement of the European subcontinent.

Despite the fact that the Phanerozoic covers a time interval three to four times less than that of the Precambrian, it has played an important role in the formation of the European subcontinent and its relief. The Phanerozoic history was at least as complex as the Precambrian, if not more so. This is not only because the Phanerozoic is a younger stage whose geological and geomorphological record is more fully and precisely documented, but also for the following reasons. It now seems likely that in the early

Precambrian (i.e. 3000–3500 Ma ago) at the outset of the formation of continental crust, the relief was less complex than it is today with a relatively low amplitude (including the ocean floors). This is evidenced by the rocks formed at that time and by the absence of deep-sea facies. Depressions were initiated in the oceanic crust (the Earth's primary crust) along the ancient faults as isometric, irregularly shaped and ring-shaped features. Fine detrital material that accumulated in these depressions gave rise to greywacke formations. Study of this detrital material has shown that it was derived from erosion of fissure lavas (i.e. from the primary basaltic crust). Apart from the depressions, there also existed rounded or irregular uplifts. Then there were uplifts in which the greywackes and basic lavas had already been granitized. Evidently, granitization was one of the first causes of the progressive differentiation of the relief: the density decrease of the crustal rocks caused the granitized portions to be uplifted. In other words, granitization disturbed the isostatic equilibrium of the Earth's crust, while subsequent uplifts

Fig. 1.5 Cratons and Precambrian folded complexes in Europe [after Khain (1977b)]: (1) ancient craton; (2) rifts (aulacogens) within craton; (3) reworked ancient massif; (4) Baikalian folded complexes; (5) ophiolite belt; (6) spilitic rocks.

restored it.

Apart from the rounded and isometric structures, linear structural forms began to develop in the newly evolving depressions. So-called protogeosynclines were formed, in which accumulated the products of erosion—not only of basic rocks but also of acidic varieties. Evidence for this is provided by the arkose sandstones found in these sequences. Acid volcanics are also observed in the depressions (Krivoy Rog, Saxonides, etc.). The rocks infilling these ancient depressions are crumpled into simple linear folds, often with gentle dips. The predominance of acid rocks and erosion products from crustal materials less dense than the basaltic crust accounted for the subsequent isostatic uplift of the folded structures. This stage of the Earth's development ended approximately 1700–2000 Ma ago but, like the previous stages, did not end everywhere simultaneously. In other words, its lower boundary is time-transgressive, varying in different areas over an interval of 300 Ma. The ancient cores of the shields in every continent are composed of Archaean folded zones that are restricted to the protogeosynclines and even to older depressions mentioned previously. All these zones should be thought of as being associated with deep-seated fault zones—a term introduced by Peive (1956a, b, 1960)

and by Rukhin (1959). The concept of a global framework of faults is one that has been developed in various countries [e.g., see Hobbs (1911), Sonder (1956), Belousov (1976)]. It is suggested that, at the earlier stages of the Earth's history, the faults were not so deep and their pattern was less distinct, the fault zone being confined to the thinner, more easily broken basaltic crust. The relief contrasts, as has been mentioned already, were negligible.

After the late Precambrian (i.e. Proterozoic, Riphaean) another stage of geological, tectonic and geomorphological development began—the next megachron—which continued throughout the entire Phanerozoic [see Yanshin (1966)]. This megachron was termed by Stille the 'Neogeikum', during which the same tectonic style was maintained. Faults penetrated to greater depths and the associated geosynclines became deeper, more extended and linear. This is borne out by the great thicknesses and the facies composition of the Riphaean and later geosynclinal formations. The continental crust continued to form in the geosynclines. Relief amplitude appeared to be increasing up to the present-day maximum. This increase has been caused by the very process of the formation of the continental crust and its increase in thickness and is intimately associated with maintaining the figure of the

Earth in dynamic equilibrium [see Bashenina (1967)].

Analysis of the present-day relief, crustal thickness and geological history of Europe shows that the longer the period of geosynclinal development (covering more than one tectonic cycle) and the more intense the postorogenic activation the greater is the thickness of the crust. The general elevation of the relief is and, probably was, directly associated, from a global point of view, with the crustal thickness.

Geosynclines of the new megachron began to develop at the outset on the oceanic crust along the peripheries of Europe. Thus, the former tendency—accretion by younger structural formations around ancient cores—was preserved during the new megachron. Development of the European geosynclines was extremely varied and intricate in space, history and time of completion. In the north-west of Scandinavia, geosynclinal development was restricted to the Lower Palaeozoic. In the Urals, the central European Hercynian blocks and central Spain (e.g., Sierra Guadarrama, etc.) this development continued up to the end of the Palaeozoic and the beginning of the Mesozoic. In the major part of Cantabria and Iberia, geosynclines developed even during the Mesozoic. In all the mountainous zones of the former Tethys Ocean (i.e. around the Mediterranean Sea, the Carpathians, Pontides, Taurus, Caucasus and Elbrus) geosynclinal development started in the Riphaean and ended only in the Cenozoic, locally even in the Holocene; the rocks infilling the marginal and intramontane basins appear to be deformed into folds, even locally forming the slopes of the mountains.

All the Tethyan mountain systems are extremely complex, which is likely to be due to the duration of their construction: for example, the Great Caucasus underwent a final inversion of relief as far back as the pre-Jurassic. Southwards and eastwards, the rocks and folded structures composing this mountain system become younger. No less complex is the history of other Alpine mountain belts with their fold–block and fold–block–thrust structural patterns and their great diversity of relief forms. The basement complex of the continent was subjected to frequent, although nonsynchronous, fracturing along reactivated faults and accompanied by differential subsidence. Subsidence was followed by marine transgressions and accumulation of marine sediments. These events led to the development of the thick sedimentary cover that rests on the basement over a major part of the continent. Because of the intricate arrangement of the zones of uplift and subsidence, and the frequent changes in the direction of movements, the sedimentary blanket is of varying thickness, age and composition. Many graben–syneclises and syneclises, horst–anteclises and anteclises were subjected to inversion. Some of them, however, retained a tendency towards uplift or subsidence over the entire megachron, giving rise to uplifted denudation plateaus and plains or to depositional lowland plains.

Another characteristic aspect of geosynclinal development in Europe was the inclusion of fragments of the basement complex in the geosynclines at various stages of their development. The geosynclines were initiated on the oceanic crust along the continental margins; fracturing of the continental crust allowed the separation of basement blocks which became incorporated in the orogenic zone. In the young mountain belts (i.e. the Alps, Caucasus and Pyrenees), these crystalline blocks occur up to the highest altitudes.

It can be seen then that the continent of Europe is still very active tectonically. Fracturing and complex differential movements occur even at the present time, not only in the mountainous areas but also in the lowlands. This is indicated—particularly in western Europe, which is more mobile—by active magmatism, by high seismicity and even by the fact that the Anglo–Parisian Basin was probably divided only during the Quaternary.

1.3 Endogenic Processes

1.3.1 General Relationships of Structure, Tectonics and Relief

At present, the structural–tectonic base is represented by the following major elements: (1) shields (e.g., Fennoscandian, Ukrainian); (2) platforms consisting of graben–syneclises and horst–anteclises, syneclises and anteclises of different ages which, from time to time, developed on the Precambrian basement; (3) Caledonian, Hercynian and Mesozoic orogenic zones; and (4) young Alpine forms. From this it follows that the chronological boundaries between the principal tectonic cycles, and within those cycles, are not always distinct.

Belousov (1976) suggested that these boundaries are roughly synchronous, basing his conclusion on the evidence of orogenic zones of different ages in different continents. It is hardly probable, however, that the structural history of Europe should be considered as one of alternating periods of tectonic activity and relative inactivity, followed by widespread denudation and levelling. Such a supposition is totally at variance with geological and geophysical data. It is true that throughout the tectonic history of the earth there have been regional 'relaxations' or 'intensifications' of tectonic activity, but there is no indication that tectonism is steadily diminishing or dying.

Although the platform and shield areas of Europe are, at present, less active tectonically than some of the younger orogenic zones, their low-lying character or their low amplitude of relief is not simply the result of prolonged denudation following cessation of tectonic activity. For example, the Russian platform—the largest in Europe—is believed to be a plain not just because it has been subjected to repeated phases of erosion but also because of the operation of isostatic compensation, along with other global megamorphostructures. The extensive western Siberian lowland is also a result of isostatic compensation, which caused the major part of the lowland to subside during the Mesozoic–Cenozoic era. The principle of isostatic compensation is a very important one in the morphostructural development of western Europe and especially the platform areas. The elevation of erosional plains is balanced by the subsidence of other areas where

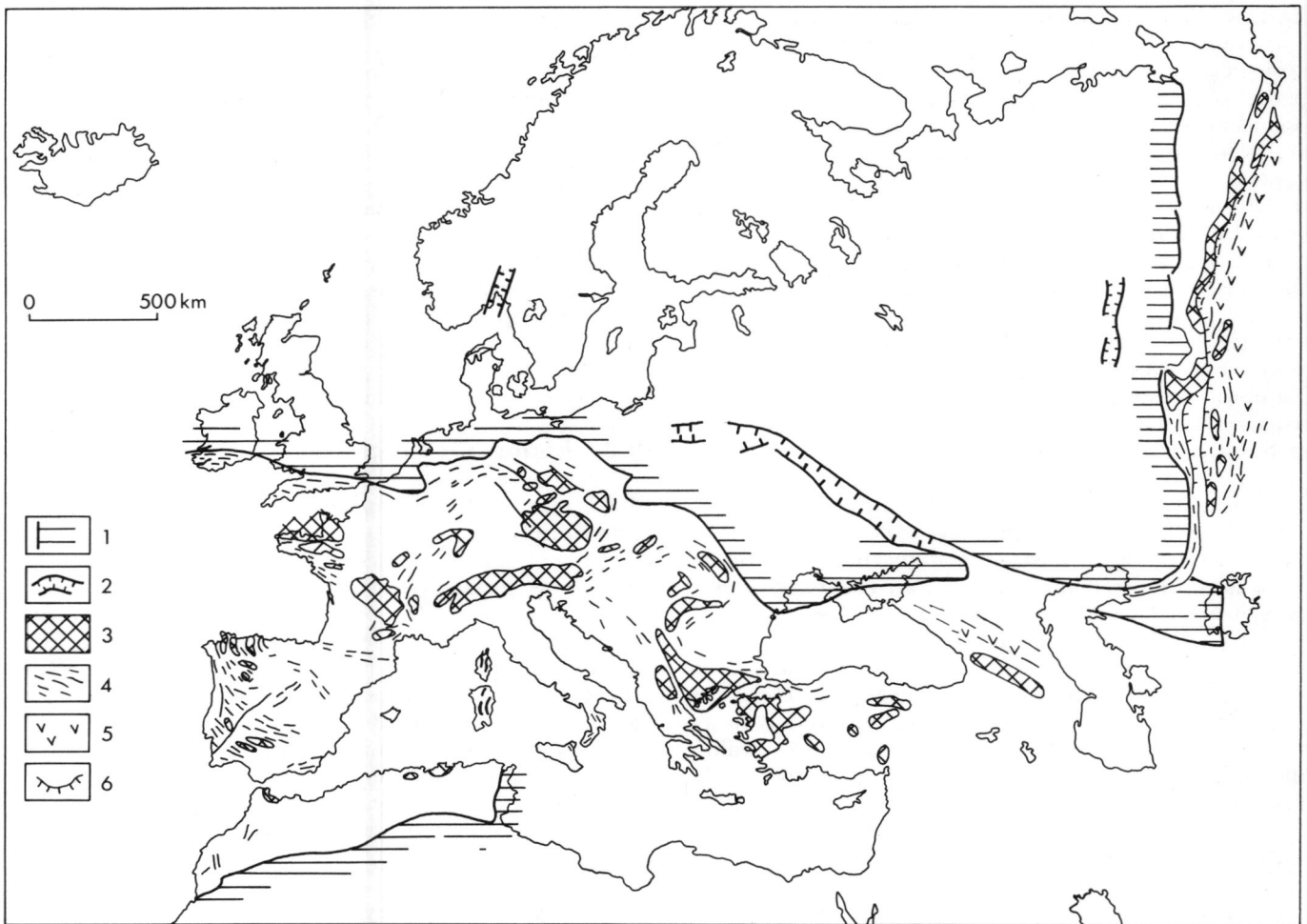

Fig. 1.6 Major Hercynian structures of Europe [after Khain (1977b)]: (1) pre-Hercynian platform; (2) Hercynian rifts (aulacogens); (3) pre-Hercynian reworked massifs; (4) Hercynian folded complexes; (5) basic volcanics; (6) thrusts and major faults.

sediments accumulate. Similar considerations apply to the behaviour of horst–anteclises or graben–syneclises throughout the entire Phanerozoic and also govern the formation of morphostructures in orogenic zones (e.g., the compensatory links between mountain systems and their marginal and intermontane depressions).

Geosynclinal development varied considerably between the orogenic zones. There was no further geosynclinal development in north-west Britain and Scandinavia after the Caledonian Epoch, nor was there any in central Europe or the Urals after the Hercynian. The Hercynides, however, require a special explanation (*see* Fig. 1.6). The Hercynian orogenic zones of the Urals, occurring under thick Mesozoic–Cenozoic sequences in western Siberia, extend as far as their eastern portion where the basement is older (Baikalian). If the younger geosynclines in every mountainous area are assumed to be initiated on the oceanic crust (i.e. from the ocean side) the idea of the accretion of the Hercynides of the Urals by younger geosynclines should be abandoned, since no oceanic crust existed at that time within the limits of western Siberia. The same also applies to the western European Hercynides.

1.3.2 Role of Faulting in the Formation of the Relief of Europe

All the major relief features of the Earth have been created by processes operating in the crust or mantle layers. These processes account for the irregularities of the Earth's relief and consist of folding, various disjunctive dislocations, vertical movements (i.e. uplift and subsidence) and various volcanic and seismic events at different depths. During the last decade, however, it has been established that, among the endogenic relief-forming factors, deep-seated faults of various orders (i.e. different extent, width and depth of origin) play a unique, major role throughout the geological history of the Earth. This was first shown by Karpinsky in 1887. Later many other scientists [e.g., Hobbs (1911), Cloos (1939), Stille (1924, 1936), Sonder (1956), Machatchek (1955), Peive (1956a, 1960), Khain (1964), Vardanyanz (1932), Dobrynin (1948), Borisov (1966)] recognized the importance of these faults.

It is only during the last 15 years, however, that geomorphologists, as well as structural geologists, have begun to emphasize the importance of faults as primary crustal features. They have also shown that the transition regions between oceans and continents are marked by deep-seated fracture zones which correspond to geosynclines [see Rukhin, (1959)]. The faults of the Canadian Shield are traceable in its relief and in its pattern of great rivers, such as the Saskatchewan, through a sedimentary cover 3 km thick or more. Geomorphological maps of Czechoslovakia by Demek and others, of France by Joly and Paskoff and of Italy by Sauro and others show the expression of deep-seated faults and other smaller tectonic dislocations in the relief. Soviet geomorphologists [e.g., Simonov and Bashenina] have prepared two sets of maps; one of neostructural forms and the other of geomorphology which show clear relationships. Specialized studies have been made by Bashenina (1967, 1973), Piotrovskiy (1972) and others of the role of faults and associated block tectonics in the evolution of the relief. These studies were undertaken mainly by geomorphologists working on mountain areas using aerial photographs and later using satellite imagery. As a result of these investigations, it has become evident that the major relief features of Europe, as of other continents, are determined to some degree by faults of different orders.

1.3.2.1 Zones of deep-seated faults and other subordinate tectonic dislocations

Fault zones are differentiated with respect to the depth of initiation, extent, width, manifestation of different tectonic processes operating in them and according to the role they play in the evolution of the relief [see Khain (1964), Bashenina (1973, 1977)]. The Soviet tectonists—Borisov, Khain and others—believe that the term 'deep-seated fault' is not strictly correct. These faults are weak zones penetrating to different depths in the Earth's crust and mantle. They '... are not a result of complication of structures, but represent primary features of the Earth's crustal structure, with respect to which many other features, as for example geosynclines, are secondary, derivative elements' [see Khain (1964) p.251].

Deep-seated faults are frequently associated with the Benioff zones. They dip at angles of up to 60° and are exposed at the Earth's surface in the areas of island arcs and deep-sea trenches. These faults are associated with earthquake foci concentrated at depths of a few kilometres to 100–200 km, more rarely being located at depths of 300–900 km (i.e. deep-focus earthquakes). The widths of the fault zones may be 200 km or more. Melting of the mantle and crustal material takes place in these fault zones. Magma and gas are extruded up associated vertical fractures of a lower order that intersect the oblique faults and give rise to the surface vulcanism of island arcs. The concept of oblique faults in the transition zone between continents and oceans was first introduced by Zavaritsky and Benioff, and was then further developed by a number of other geologists and seismologists. The major deep-seated faults were termed Benioff–Zavaritsky zones, probably without associating this term with Benioff's general concept. Since the latter is in dispute, it seems preferable to use the term deep-seated fault zone when associated with active geosynclinal zones.

Apart from the faults of transition zones, other large global faults of deep origin and great extent are known.

Many such faults that cross the Atlantic and Pacific Oceans have their own names in the published literature. Ocean faults extend to all continents. Their topographical expression on these continents is rather complex. They fail to maintain one and the same trend: their intersection with other faults, mostly perpendicular, causes them to shift somewhat and to change their direction. The same phenomenon may apply to the faults they intersect.

The faults of the Atlantic Ocean are clearly expressed in the relief of Europe, such as the Gibbs fault running from Newfoundland through the Porcupine Bank and the southern parts of Ireland and Great Britain, being distinctly represented in the relief of the latter, and the Gloria fault extending into Spain from the Azores, passing into the graben of the Guadalquivir River. There are also other large faults of a transcontinental type. For example, the major Ural fault running through the Aral Sea, Iranian highlands, and the north-western part of the Indian Ocean is traceable in Oman; transverse Transcaucasian faults are identified in the Arabian peninsula. Large mountainous massifs, complex grabens and young volcanoes (e.g., Elbrus, Kazbek, Italian volcanoes, the volcanic areas of the Massif Central, etc.) are associated with the intersection of faults of various trends. The largest mineral deposits are commonly localized at these intersections.

These major faults, together with others of lower orders, form a global network which divides the Earth's crust and mantle into blocks of various orders—from global megablocks to minor crustal blocks several kilometres in size. Many fault zones are long-standing features that have maintained their activity during several tectonic cycles: some Riphaean faults are still active.

1.3.2.2 Fault-induced tectonic processes

Tectonic processes operating during the entire period of the Earth's geological history are known to penetrate to the Earth's surface along the deep-seated fault zones. Folding, various magmatic events, secondary dislocations along lower-order faults, negative and positive vertical movements, horizontal movements and earthquakes took place in the geosynclines related to these faults. The deep-seated faults appear to act as conductors for all known deep processes.

The deep-seated faults account for the formation of a system of oceanic and continental rift zones (*see* Fig. 1.7). There is a lot of contradictory data on the problems of riftogenesis. Rift valleys expressed in the present-day relief are located mainly in fault zones not more than 70 km deep. Minor rift valleys, such as those of the southern Urals and the Ohže river valley, are restricted to crustal faults of lesser depths. The rift-forming process is commonly associated with a fault zone accompanied by some horizontal tension. Rift zones are recognized traditionally as consisting of one or more longitudinal, almost parallel depressions—narrow linear grabens showing a displacement *en échelon* along transverse faults. The sides of the troughs are represented by horst ranges and massifs which have also been formed as a result of tension. The latter is accompanied by the squeezing of a system of horsts to different heights [see Bashenina (1977)]. It is supposed

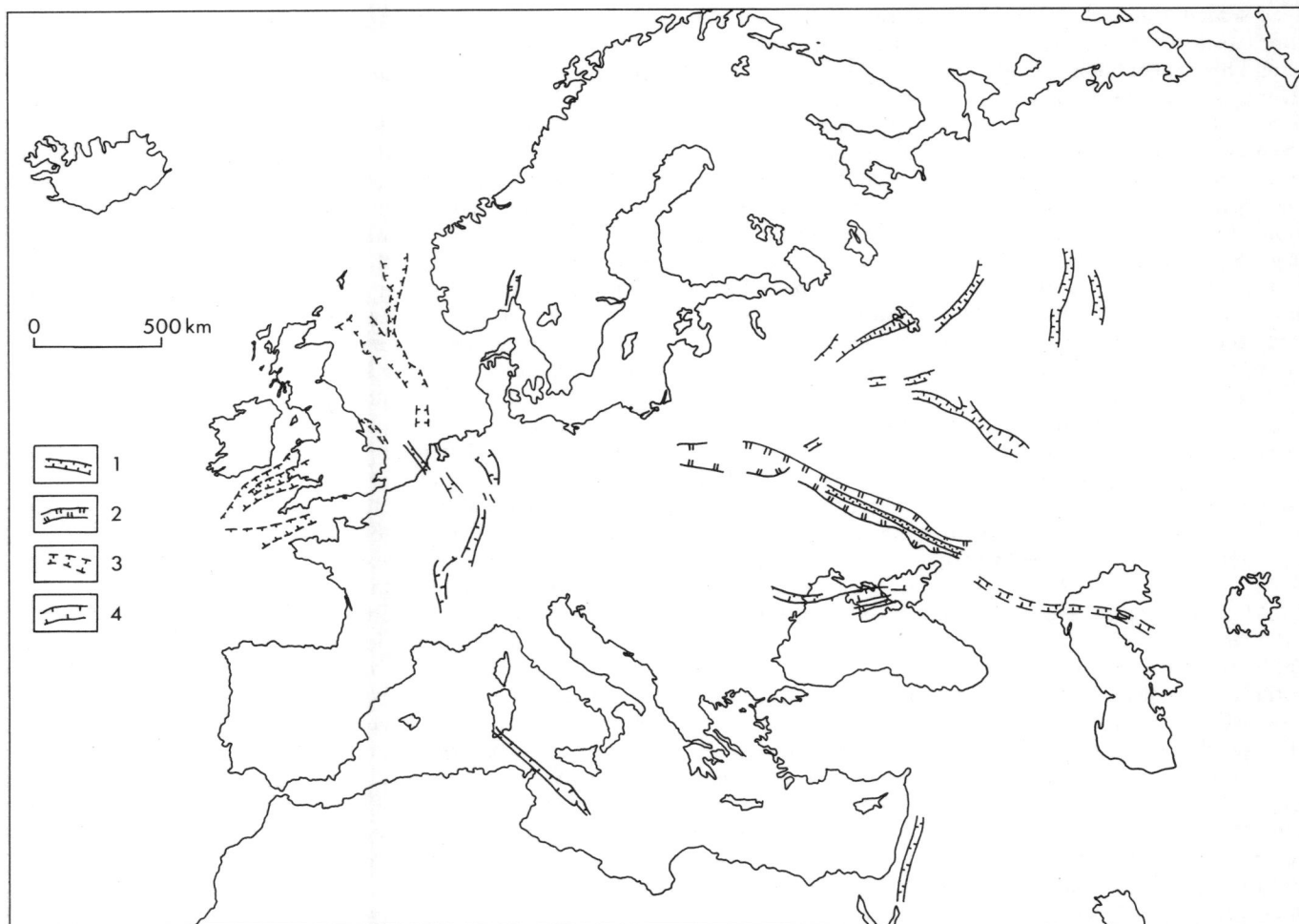

Fig. 1.7 Major rift systems of Europe [after Khain (1977b)]: (1) late Precambrian; (2) mid-Palaeozoic; (3) Mesozoic; (4) Cenozoic.

that continental riftogenesis occurs both on orogenic and platform megastructures; its age varies greatly. Formation of the European rift valleys has occurred in quite a different pattern. Like the valleys of the southern Urals, they do not represent rift zones akin to those of Africa–Arabia or Baikal. Illies (1975) indicated that the Oslo, Rhine and Rhône depressions were formed in Palaeozoic times and were frequently reactivated from the late Cretaceous onwards. Although the Oslo graben is called a Permian palaeorift, it is clearly manifested in the present-day relief.

Many crustal faults of lower orders are secondary features with respect to folding and vertical movements. They complicate the folded morphostructures and may be unique to every large morphostructure. They also complicate the global network.

1.4 Major Relief Features of Europe: Megamorphostructures and Morphostructures

The main megastructures (global structural forms) are quite distinctly expressed in the relief of Europe (*see* Fig. 1.8). The megastructures of platforms, shields and various orogenic zones differ from each other in the intensity, type and direction of the most recent movements, as well as in

their relationship with older tectonics. The structural outlines of the continent were determined by the beginning of the neotectonic period. This period saw the final completion of the relief of Europe.

1.4.1 Major Morphostructures and Megastructures of the Alpine Zone of Europe

The most important event during the neotectonic period was the completion of the major portion of the geosynclinal zone of the Tethys Ocean and the formation of mountain chains differing from each other in structure and relief. The modern relief of the Alpine orogenic zone results from all types of tectonic movement, among which folding has played an important role (*see* Fig. 1.9). The relief expresses folded structural features of various orders. Arched uplifts may be correlated with the so-called 'large folds' of Argand and Penck. However, these uplifts are often complicated by differential block movements which have taken place along faults. The major morphostructures of the Alpine system include the following features: (1) ranges and massifs of anticlinoria and horst–anticlinoria, (2) intermontane and marginal depressions of synclinoria and grabens, (3) massifs and ranges composed of crystalline rocks belonging to the ancient basement and

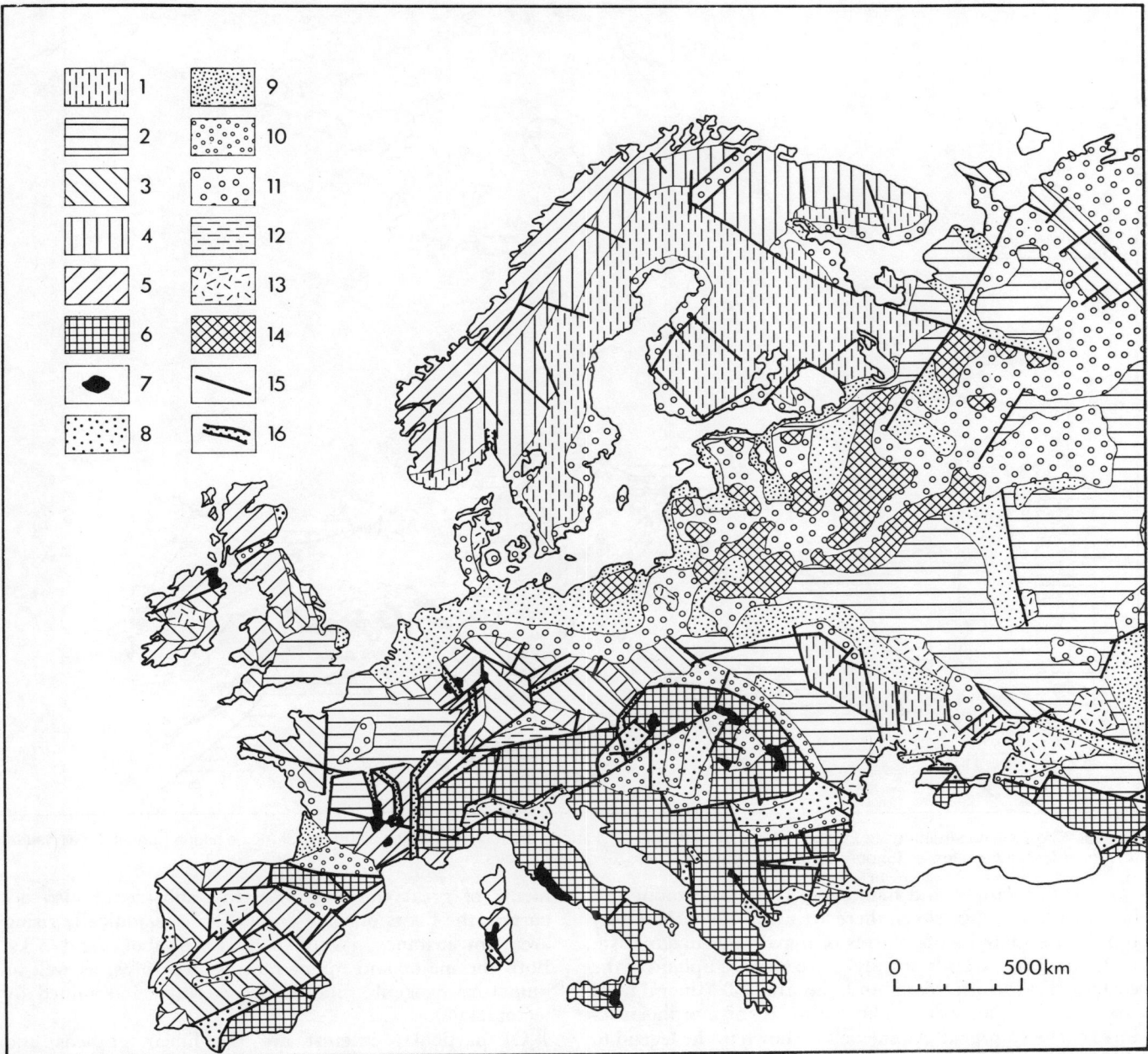

Fig. 1.8 Megamorphostructures of Europe [after Bashenina *et al.* (1979)]. (A) Erosional relief: (1) relatively stable shields; (2) relatively stable platforms; (3) epiplatform areas slightly affected by later tectonics; (4) shield areas affected by neotectonics; (5) epiplatform areas strongly affected by neotectonics; (6) young epigeosynclinal orogenic zones. (B) volcanic relief: (7) Cenozoic volcanics. (C) Depositional relief on a deep-seated basement: (8) marginal and intermontane depressions; (9) platform and shield depressions. (D) Depositional relief on a shallow basement: (10) marginal and intermontane depressions; (11) platform depressions. (E) Erosional–depositional relief on a shallow basement: (12) marginal and intermontane depressions; (13) platform depressions; (14) platform and shield rises; (15) faults; (16) rift valleys.

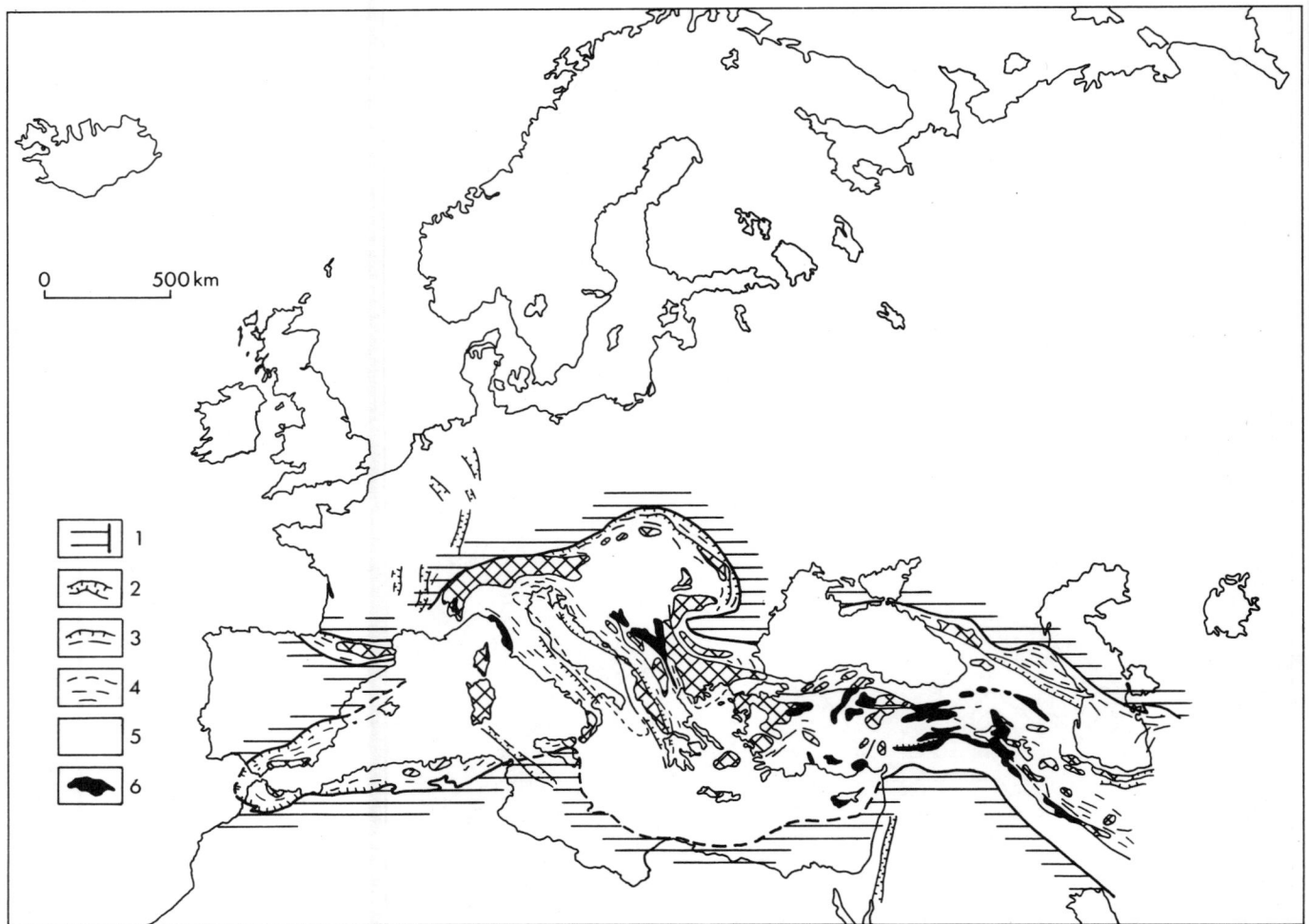

Fig. 1.9 Major Alpine structures of Europe [after Khain (1977b)]: (1) platform area; (2) rifts within platform; (3) Alpine nappe front and main faults; (4) Alpine folded complexes; (5) pre-Alpine continental massifs; (6) ophiolite belts.

(4) volcanic ranges and highlands. In the piedmont and intermontane depressions, there are areas of youthful hilly terrain associated with centres of hypabyssal magmatism (e.g., the Eugane Hills of Italy, the Beregovo upland of the eastern Transcarpathians and the area of Mineral'nyye Vody in the Caucasus). The entire list of morphostructures of the European Alpine belt is shown in the legend to the *International Geomorphological Map of Europe*.

Compared with the older orogenic belts, the Alpine zone is generally characterized by ranges and massifs exhibiting folded and block–folded structural patterns. The younger the mountains the better the relief expresses the folded structural pattern. The Alpine zone of Europe is still very active tectonically, with magmatic events, destructive earthquakes, intensive vertical movements and the formation of fault scarps and grabens. An example of active tectonics is the complex graben of Lake Balaton, part of which represents an independent, small and geologically recent graben. Generally, the Alpine zone of Europe during the neotectonic period was noted for the predominance of uplift, but subsidence in marginal and intermontane basins also took place. Absolute subsidence was accompanied by the accumulation of Cenozoic sedi-

ments of great thickness. Relative subsidence also occurred; the floors of some grabens in a tectonically rising area, for instance, lagged behind neighbouring blocks. Both the major and minor morphostructures, as well as numerous exogenic relief forms, were predetermined by tectonics.

Of particular interest are the linear grabens and tectonic valleys of the mountainous Alpine areas. These include the grabens of the Balkan Peninsula, the graben in Italy which separates the calcareous Pre-Alps from the crystalline Alps, partly used by the Adda River, and the tectonic valleys of the Great Caucasus. In this range, for example, large and small valleys in the near-summit areas, including cirques, have been shaped not only by water and/or ice but also by tectonics. Neither rockfalls nor debris flows are of purely exogenic origin. All of these relief forms are associated with faulting and represent the products of erosion of disintegrated rock blocks slumping downslope due to gravity. The Baksan river valley is confined, in its upper reaches, to the zone of the Pshekha–Tyrnyauz Fault that separates the Great Caucasus range from the side range [see Milanovskiy (1968)]. Over the first 60 km, the Baksan has failed to develop the valley.

The river is characterized by an ungraded longitudinal profile, by a width that is observed to vary in different sections where it flows either on alluvium or on bedrock, by a great variety of terraces difficult to correlate with each other and by abrupt changes of direction of the valley. Below and above these bends, a change in the height and number of the terraces and variations in thickness of the unconsolidated material composing them can be seen. All these features are typical of valleys running along active faults.

In such young mountains as the Great Caucasus, where tectonic movements have also been active in the Holocene, it is easy to recognize the pattern of tectonic events:

1) Arch–block uplift and the formation of valleys along faults
2) Relative subsidence (lagging) at the fault intersections and the formation of local grabens (linear, box-like, etc.)
3) Upheaval of massifs and peaks at the fault intersections
4) Lagging of individual blocks during the uplift of massifs and ranges
5) The formation of fault steps at different levels on their slopes
6) The breaking-up of ranges and massifs by open fissures and the separation of individual blocks.

Tectonics thus appears as both a creative and a destructive agent: the mountains are not only being created but also dissected and destroyed by tectonics. In the course of valley formation, uplift of the valley sides predominates over entrenchment of the river; the valleys bear the distinct impress of tectonics, while the rivers that started to flow in them are far from being adjusted to these valleys. Although erosion, slope processes, ice and snow are active relief-transforming agents, the essential pattern of morphostructures and minor relief forms (excepting only details) is established by tectonics; exogenic processes have only to adjust themselves to this pattern. The tectonic structure thus affects powerfully the operation of exogenic processes, speeding and facilitating their action, and determining their location and intensity. An amazing picture emerges of the vivid interaction between tectonics and exogenic processes [see Bashenina *et al.* (1967)].

According to Veyret (1980), vertical movements predominated in all phases of the orogenic history of the Alps. Such movements were responsible in the axial range of the Alps for uplifting parts of the ancient crystalline basement to great heights. Such basement blocks also outcrop in the axial ranges and mountain systems of the Balkan Peninsula, in the Pyrenees, the Caucasus and even in the Carpathians. The basement blocks here are not detached and do not float in the sedimentary cover as some investigators believe.

Horizontal movements also occurred in the Alpine zone, although not in that hypertrophic form as is visualized by supporters of the plate tectonic theory and the theory of nappe formation of Veyret and others. Tear faults and overthrusts are known from the Alps [see Bravard, Lilienberg *et al.* (1980)], from the Caucasus [Milanovskiy (1968)], the Carpathians, the Baetic moun-

tains of Spain and from other Alpine mountain ranges. However, on the *International Geomorphological Map of Europe*, it is possible readily to distinguish extended straight-line boundaries demarcating the mountains: for example, along the eastern Alps, the Appennines and the Lombardy lowland. These straight-line boundaries can be only accounted for by near-vertical faults.

The Alpine mountain systems of Europe differ in age and maximum vertical movements were not absolutely synchronous. For example, in the Holocene, they were more active in the Caucasus than in the Alps, especially the French Alps. The Caucasus and the Appennines, in particular in the southern and western parts, were subjected to active vulcanism at the end of the orogenic stage and this continued into historical time.

Divergent neotectonic history probably resulted in asynchronous glaciation, according to the participants of a Soviet/French symposium on Alps–Caucasus (1980). Correlations have only been established in historical times. No old moraines have been preserved in the Caucasus and, therefore, no data are available to judge whether there was a break between the Riss and Würm glaciations—hence there is no need to use the Alpine terminology. The much greater activity of avalanches in the Caucasus compared with the Alps at equal altitudes can be explained not only in terms of climate but also by more intensive uplift. This confirms the great tectonic activity of the Caucasus during Quaternary and recent times.

The Alpine mountain zone of Europe belongs to a single global morphostructure—a young epigeosynclinal orogenic zone. This zone is present also on the margins of other continents: for example, on the west of the American continents and in Asia. Generally, the structural, tectonic and relief features of these orogenic zones are similar except for detailed regional differences.

1.4.2 Megastructures and Morphostructures of the Ancient (Epiplatform) Orogenic Zones of Europe

The European orogenic zones are represented by several megastructures differing in age, the degree of most recent tectonic reactivation and relief. The lower age limit of the neotectonic stage in orogenic zones, like the boundaries between tectonic cycles, is time-transgressive and not simultaneous in different continents. The climax of the neotectonic movements was asynchronous not only in the epigeosynclinal zone, but also in other megastructures.

The mountain systems of Spain outside the Alpine belt have suffered a complex history. The Cantabrian mountains, the mountains of central Spain and those of Catalonia are the most tectonically active at the present and have been intensely broken and reworked by block tectonics. The Hercynides are the principal structural elements of the mountain chains of the Pyrenees. The Hercynian folded basement was broken into blocks and outcrops in the foothills and on the Galician highland, which is also deformed. Exposures of the Riphaean folded basement are less frequent. The new tectonic map of Europe (1975) shows the existence of a Mesozoic folded cover in the Iberian mountains and the almost complete absence of platform structural forms in the Iberian peninsula. Exceptions are the relatively worn-down high plateau of La Mancha, with a folded block structure, and the Portuguese plains.

Many geologists consider the Hercynides of western Europe to have been strongly affected by neotectonics. Neotectonic reactivation began with the renewal of subsidence in a system of Palaeozoic rift valleys. Rejuvenation of the Rhine graben took place in the late Cretaceous, marked by vulcanism, and then by periodic subsidence in the Eocene, Oligocene, early Miocene, Pliocene and Pleistocene [see *Collected Papers* (1975)]. The subsidence was accompanied by block movements and uplift of the Hercynides. Subsidence in the rift valley of the Ohže River extending also into the Quaternary [*Collected Papers* (1975)] was accompanied by uplift of the Erzgebirge. These structural features are expressed in the relief by the flat floors of rift valleys, by mosaics of low mountains with dominating fold–block and arch–block structures (e.g., Erzgebirge, Harz, Schwarzwald, etc.) as well as by bevelled plateaus (e.g., Ardennes) frequently strongly fractured (e.g., Moravia). Tectonically, these are active morphostructures, although the magnitude of movement is considerably less than that of the Alpine epigeosynclinal zone.

Another major structural unit whose formation was essentially completed in the Hercynian orogeny is the Ural Mountains. The tectonic development of this complex zone began as far back as the Riphaean geosyncline. On the other hand, the Urals do not represent an area greatly affected by neotectonics. The Hercynian block structures remain the most pronounced features. The eastern part of the Urals includes rift valleys (e.g., Chelyabinsk graben) which, as far back as 1887, were recognized by Karpinsky. The minor rift valleys have been more recently studied [see Bashenina (1977)]. Compared with other Hercynides (e.g., the Tien Shan), the Urals seems to be only slightly affected by tectonic reactivation. The rift valleys of the Urals do not show evidence of young magmatism such as is known in the case of the Rhine graben.

Major neotectonic reactivation affected the Caledonides of Northern Ireland, Scotland and Scandinavia, where there was considerable fracturing along existing lines of weakness, differential block movements and the formation of valleys along rejuvenated faults. Tectonic displacement took place against a background of intensive and complex dislocation along faults of different trends also involving the shield areas. This led to the formation of complex and unique morphostructures in the south-west of Scandinavia and in the northern part of the Kola Peninsula.

The ranges, massifs and plateaus of the Caledonian epiplatform areas are distinguished principally by block and arch–block structural patterns. Valleys formed along the faults were subjected to erosional modification, especially later by glaciation. After the ice melted in the lower parts of valleys overdeepened by glacial erosion, the sea entered forming fiords with their linear segments; sharp changes of direction reflect the original fault pattern and their frequently narrow widths.

Thus, the epiplatform mountains of Europe belong to three global megastructures: (1) epiplatform orogenic zones strongly affected by neotectonics, (2) epiplatform orogenic zones, slightly affected by neotectonics (e.g., the Urals) and (3) orogenic zones of shields affected by neotectonics.

1.4.3 Platform Megastructures and Morphostructures of Europe

The platform megastructures, as observed at present, have also resulted from neotectonic movements which are essentially vertical and differential in character and intensity. They are associated with the relatively stable shield and platform areas. Their spatial distribution is governed by the reactivated faults, and every platform megastructure represents a megablock. Megablocks are broken into blocks of a lower order, many of which have undergone complex development. Neotectonics has not only emphasized and refined the individual characteristics of the blocks but has altered the course of development in some of them. Some of the major morphostructures are represented by portions of the relatively stable shield, others by relatively stable platforms. Each is expressed in the relief either as depressions (i.e. graben–synclises and synclises filled with deposits of different age and genesis) or as uplifts (i.e. horst–anteclises and anteclises modelled by erosion). Many areas during the Phanerozoic experienced a change in the direction of tectonic movement: upthrown blocks were subjected to subsidence and, sometimes, to further uplift, while downthrown blocks underwent relief inversion at different times, with the formation of denudational plains on monoclinal or horizontal beds in their places. Plains whose development followed the first pattern are exemplified in the central Russian uplands. One example of plains that developed according to the second pattern is the Privolzhsk upland which subsided during almost the entire Phanerozoic and underwent inversion only in neotectonic times. Another example is the lowland of south-eastern England and north-western France which shows monoclinal bedding of the strata and cuesta relief forms, developing in place of the original single Anglo–Parisian Basin. A third pattern of development has been followed by synclises which showed a tendency for downwarping during the Phanerozoic and for partial inversion (e.g., Baltic and Fore-Caspian synclises).

Examples of platform morphostructures developed on the ancient folded basement and showing a tendency towards uplift in the Phanerozoic are the Galician highland, the Massif Central, especially its eastern half, and, in part, the Armorican massif. The first two examples are greatly uplifted and fractured and exhibit a complex relief based on horsts and grabens. Recent (Quaternary) volcanic eruptions are known to have occurred in the Massif Central, where the altitude in the north-east reaches 1700 m. To the east, the Massif Central is bounded by the Saône–Rhône rift valley. The height of the Armorican massif, in contrast, does not exceed 400 m. At the present day, it forms three separate uplands or uplifted plains (Brittany, Normandy and Cotentin) resting on the folded Hercynian basement.

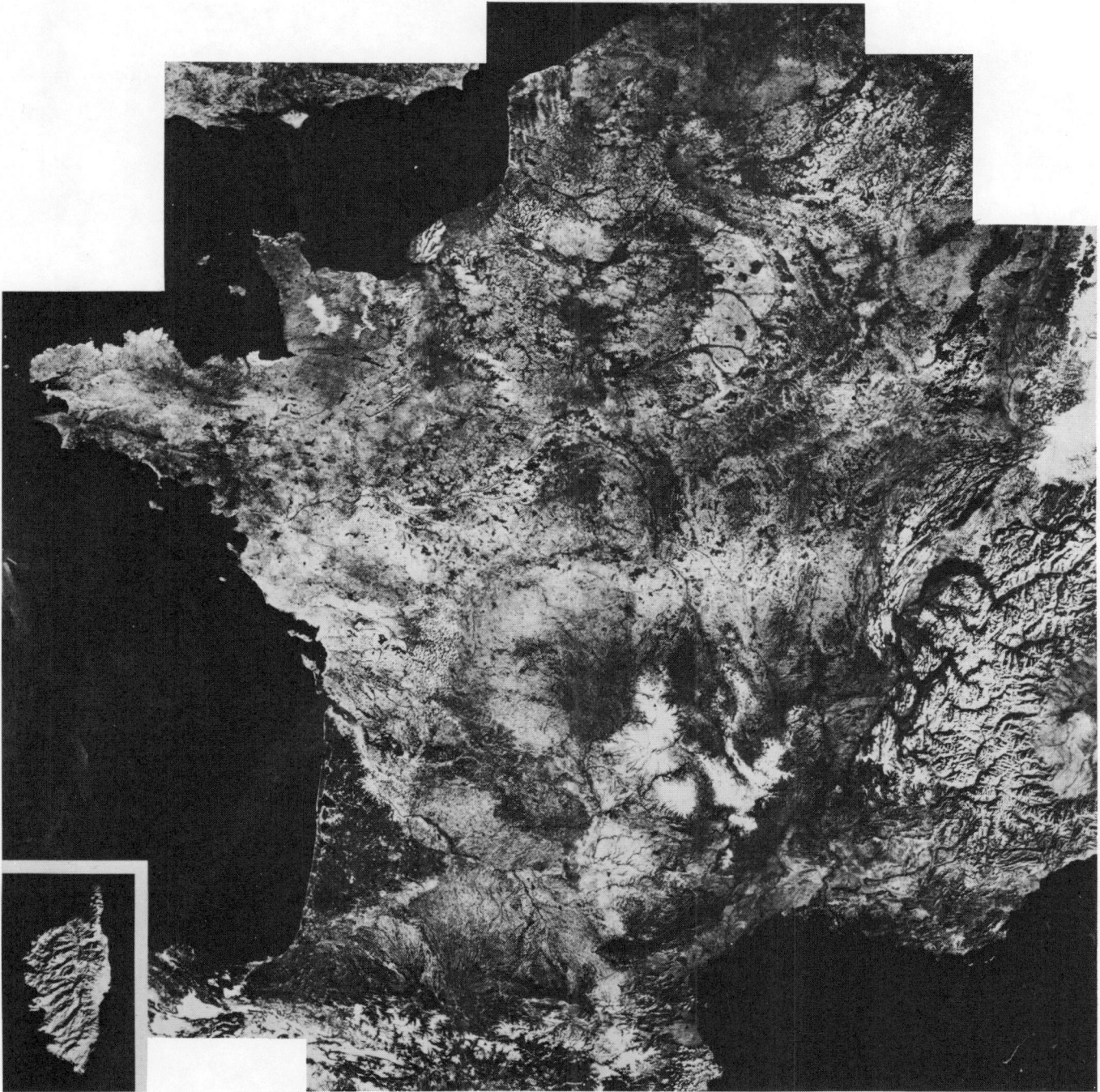

Fig. 1.10 Satellite image of France; scale approximately 1 : 5.5 million. Inset bottom left: Corsica.

The relief of the eastern European Precambrian platform differs basically from that of the western European, mainly Hercynian, platform in several respects. The former is generally plain-like. The basement of the eastern European platform is a subsided part of the Precambrian shield overlain by marine Palaeozoic, Jurassic and Cretaceous sediments. Only the horst–anteclise of the Ukrainian Shield is composed of middle Precambrian rocks, while the complex oblique horst of the Timan Ridge consists of Riphaean rocks. The Precambrian formations of the western European platform are strongly deformed, represented by fragments (blocks) within the Hercynian platform that have been uplifted to different levels.

1.4.4   Relief of Europe as Depicted on Satellite Imagery

Satellite imagery (e.g., France Fig. 1.10 and central Europe Fig. 1.11) has confirmed the important role of faulting and block tectonics in determining the relief of Europe.

Fig. 1.11 Satellite image of central Europe. Scale approximately 1 : 11 million.

On the imagery of France, it is possible to distinguish almost all the large morphostructures and major faults. In the south, the low plain of the Pyrenean piedmont (*see* section 8.5.2) can be identified readily as a gigantic alluvial fan. This fan represents a system of numerous small fans whose configuration becomes more distinct towards the Pyrenees. North-westwards there lies the triangular lowland alluvial plain of the Landes. A fault may be discerned through the alluvial cover bounding the plain on the south. Two near-longitudinal strike faults bound the Massif Central on the west. The major one, which is traceable through the unconsolidated deposits of the Landes and Pyrenean piedmont, is shifted somewhat to the east by the intersecting fault that bounds the Massif Central to the north. A second sublongitudinal fault separates the slope of the Massif Central from the plain located westwards. It is possible to discern clearly the various sublatitudinal and diagonal faults that separate the massifs into blocks of a lower order (*see* Fig. 1.10). The north-westerly striking fault separates the Cevennes. Faults are also seen to limit the folded Jura; the folds are expressed distinctly in plan by the pattern of longitudinal ridges and valleys. In Fig. 1.10, the almost ring-like morphostructure of the Paris Basin, distinguished by several rows of cuestas forming an arcuate pattern, is extremely clear. The fractured blocks of the Armorican massif are isolated and separated by plains. On the satellite image, it can be seen how they have been crushed along faults. Also the latitudinal trend of strongly eroded ancient folded structures of Brittany (i.e. Appalachian relief) appears clearly. In the north-east, the Rhine graben is clearly visible. Less distinct is the Saône–Rhône graben which narrows abruptly south of Lyon but widens again south of the Isère confluence. In the Massif Central, further examples of grabens (Allier rift, etc.) can be seen, and many lakes are concentrated on the fault that bounds the Massif on the north.

The general view of Europe on the satellite image (*see* Fig. 1.11) is of great interest. It can be seen that both large and small morphostructures are delimited by faults of different strikes and orders. The Bohemian massif is surrounded by a double ring of faults. The irregular faults of the inner ring separate the massif from mountains—the Bohemian Forest or Sumava to the south-west, the Erzgebirge to the north-west, the Sudetan Highlands to the north-east and the Moravian Heights to the south-east. Another ring of faults bounds these mountains on the outer side. A general grid-like pattern is chiefly the result of diagonal faults running north-west–south-east and north-east–south-west and is well displayed in the Bohemian massif on Sheet 10 of the *International Geomorphological Map of Europe*. The massif is dissected by valley–grabens filled with Cenozoic deposits, which may in some places be hundreds of metres thick. The edges of the massif are uneven due to numerous grabens. In plan, the massif resembles a diagonal rhomboid, the same pattern being shown by the mountains, the high and low parts of the bevelled uplands, the valley–grabens, and other major valleys thus forming an almost rectangular grid.

Large faults bound the Alps and Carpathians, separat-ing the southern arc of the Carpathians from the lower Danube lowland and from the mountains of the Balkan Peninsula. Likewise faults divide the karst plateau of Yugoslavia from the Dinaric Alps. A fault running west-north-west separates the Sudetan Highlands from the Moravian uplands and extends along the western Carpathians into the Russian plain. Still farther south, there runs a parallel fault, extending along the Balkan Mountains as far as the Adriatic Sea. A block of the High Tatra can be seen to extend south at the fault intersection. Within the Carpathian arc are the fault-bounded grabens of Alföld and others. Dissection of the Carpathians, Alps and Appennines into blocks by minor faults is also easily traceable. The karst plateau of Yugoslavia is bounded on all sides by faults; the fault that limits the bevelled upland on the north continues west-north-west where it is masked by a graben and Lake Balaton.

All the major lakes of the Italian Pre-Alps (i.e. Garda, Como and Maggiore) are confined to faults of almost longitudinal strike (*see* Fig. 1.11). As they intersect with others in the Alps, an intricate mosaic of blocks is delimited in the relief.

1.5 A Note to the Legend of the International Geomorphological Map of Europe

Compilation of the legend is based on a systematic framework that recognizes the following principal elements: (1) relief amplitude, (2) relief types classified on a morphostructural basis, (3) relief forms (mainly exogenic) and (4) relief age.

Five classes of relief amplitude are used: (1) faint relief, (2) low relief, (3) moderate relief, (4) mountainous relief and (5) high mountain relief. It also seems advisable to include for every relief class not only the depth of dissection but also the absolute elevation of the relief, which, using a colour intensity scale, contributes greatly to an effective cartographic representation of the relief in three dimensions.

Relief types are schematically demonstrated in Fig. 1.12, including (1) predominantly erosional relief (ranging from mountains to plains), (2) volcanic relief (recent and mainly unaffected by other processes), (3) depositional relief developed on thick sediments overlying a deep-seated base, (4) depositional relief developed on thinner sediments overlying a shallow-seated base, involving uplift of parts of platform, marginal or intermontane depressions (the latter are frequently seen to be overlain by slope and alluvial fan deposits), and (5) erosional–depositional relief on a shallow-seated base, especially the elevated parts of depressions and basins, as well as the youthful cuestas of platform lowlands. These types of relief have been affected to a varying extent by Cenozoic, or locally by Neogene–Quaternary, tectonic movements that help to determine the trend of exogenic processes in respect of the transport and deposition of sediments. The nature of the movements and the degree of activity, as well as the depths of the fault zones, determine the character of manifestation of magmatic processes.

Within the framework of the five relief types one can

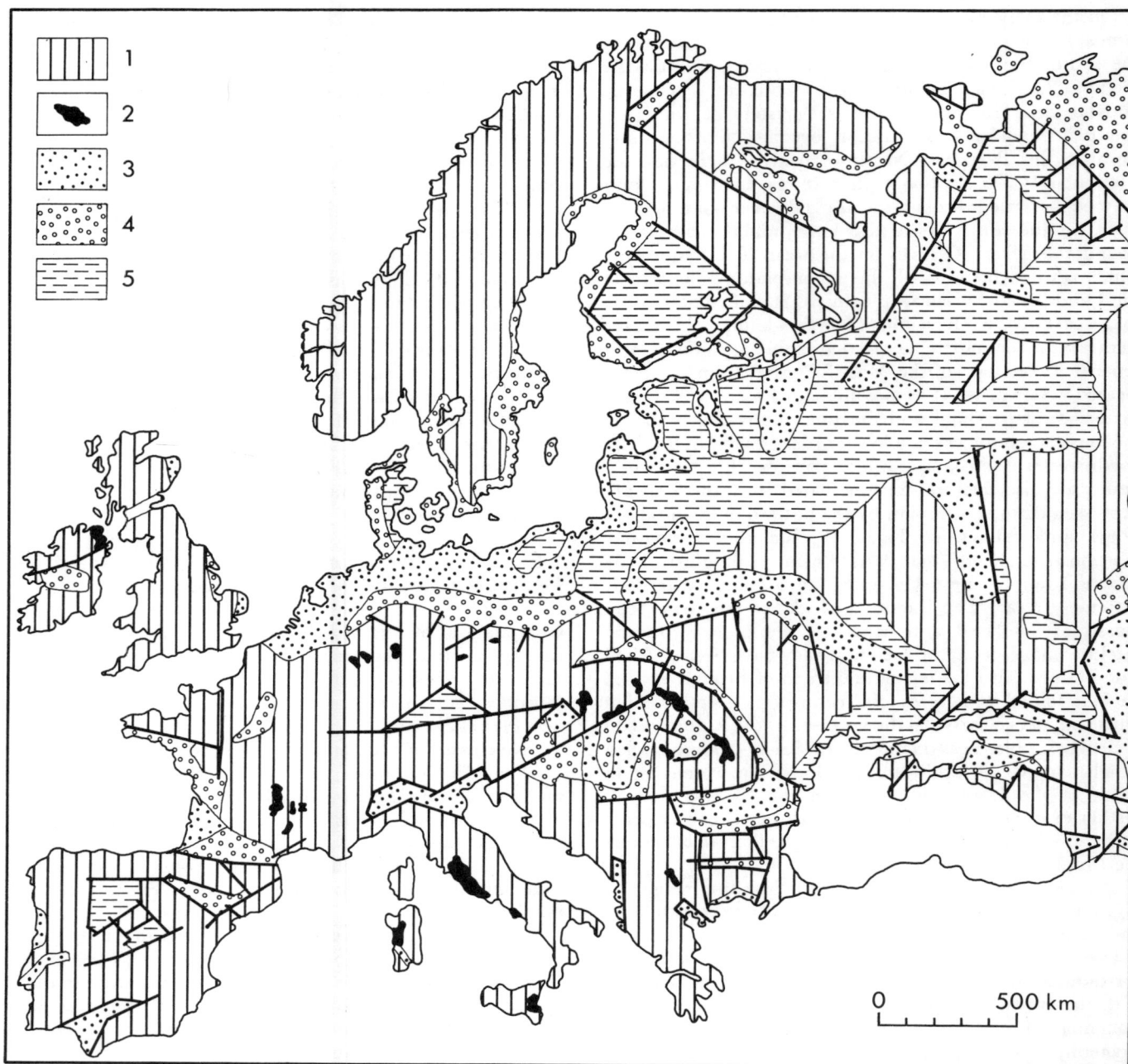

Fig. 1.12 Relief types of Europe [after Bashenina *et al.* (1979)]: (1) erosional; (2) volcanic; (3) depositional on a deep-seated basement; (4) depositional on a shallow basement; (5) erosional–depositional on a shallow basement.

identify corresponding morphostructures as shown in Fig. 1.12. There are altogether only 13 megamorphostructures. Erosional relief types, for example, include six megamorphostructures: (1) relatively stable shields, (2) relatively stable platforms, (3) epiplatform zones slightly affected by repeated tectonics, (4) shield zones affected by neotectonics, (5) epiplatform zones strongly affected by neotectonics and (6) young epigeosynclinal orogenic zones. There may be an impression that such categories of megamorphostructure are tectonic rather then geomorphological but this is not so. One is dealing with the erosional relief of these megastructures, and the full

descriptions should read, for example, erosional relief of young epigeosynclinal orogenic zones or erosional relief of relatively stable shields. Volcanic relief includes only a single morphostructure represented by (7) Cenozoic vulcanism (i.e. magmatism) that gives rise to a variety of volcanic landforms. Depositional relief on a deep-seated base is confined to the following two megamorphostructures: (8) marginal and intermontane depressions, and (9) platform depressions. Depositional relief on a shallow-seated base is restricted also to similar depressions and basins (10, 11), but only to those of them, or their parts, in which subsidence has been replaced by uplift leading to considerable removal of the sediments filling the depressions. The basement is then traceable through a thin cover of unconsolidated rocks. Depositional–erosional relief on a shallow-seated base is formed not only in marginal and intermontane depressions (12) under conditions of still more active neotectonic uplift, but also on the uplifted parts of the crystalline basement beneath plains, and that beneath platforms and shields (13). In the latter case, a kind of superimposed erosional–depositional relief is formed in which a thin cover of sediments is not always due to subsidence or subsequent uplift accompanied by partial removal of previously accumulated sediments. It is the deposits themselves that have have superimposed on the uplifted areas (blocks, domes and updomed portions of shields, and platforms). Such relief may be created essentially either by glacial deposition (e.g., Valdai upland, Smolensk–Moscow upland, etc.) or aeolian deposition (e.g., the desert plains of Africa and Australia). Hence, erosional–depositional relief is represented by two megamorphostructures, comprising erosional–depositional relief on uplifted areas of shields and platforms.

In six erosional megamorphostructures and in one volcanic megamorphostructure, some 80 subsidiary morphostructures have been recognized. This differentiation is based on differences in relief, structure and lithology. This can be illustrated from several morphostructures that are listed in the legend. For example, the third megamorphostructure (i.e. erosional relief of epiplatform zones slightly affected by recent tectonics) includes planation surfaces on the floors of grabens and old rift valleys in crystalline and metamorphic rocks (3.1), dissected plateaus on gently folded strata and in sedimentary consolidated deposits (3.5), etc. The sixth megamorphostructure (i.e. erosional relief of young epigeosynclinal orogenic zones) includes folded-block ranges and ridges in crystalline and metamorphic rocks (6.2), gently folded strata (Jura-type) on calcareous rocks (6.19), etc. The seventh volcanic megamorphostructure consists of eight morphostructures that differ in morphology, lithology (lava, ash, tuffs) and in neotectonic reworking. Examples include plains and plateaus of the floors of volcano-tectonic depressions composed of lava (7.1), folded and faulted lava flows (7.3), and plateaus of ash and tuff (7.6).

Seven megamorphostructures in depositional and erosional relief types are represented only by plains (i.e. lowlands, low plains and elevated plains). The plains in each of these megamorphostructures differ in type of sedimentary accumulation. For example, the ninth megamorphostructure includes fluvial depositional plains of platform depressions on deep-seated bases (9.1), lacustrine (9.2) and glaciomarine (9.6). Other examples are lacustrine–fluvial depositional plains of intermontane depressions on a shallow-seated base (10.3). Altogether, 54

types of plains are included in the seven depositional megamorphostructures. However, not all of these plains represent a morphostructure, but only those that fully coincide with some structural feature (e.g. platform depression, dome, elevated block). The fact that plains with different types of deposit are developed with respect to a single structural depression does not mean that each of them represents a morphostructure. For example, although the north European plain stretching from the Low Countries through Germany into Poland is associated with a single large platform depression, the depositional relief is due to deposits of very different origins—marine, glacial, glaciofluvial. Thus, these are plains composed of sediments of different origin; they are not separate morphostructures since each of them corresponds to only part of the structural form—a large platform depression. However, such seemingly different genetic indications, which serve to discriminate between morphostructural plains of erosional and volcanic relief, on the one hand, and plains within the categories of depositional relief, on the other hand, are quite justified. The erosional relief of plains and mountains differs depending on the megamorphostructure to which such relief is related since the differences between megamorphostructures are responsible for the variety of relief developed under different structural and lithological conditions. But it is not such factors that explain the differences between the depositional and erosional–depositional relief of plains in megamorphostructures 8–13. The major genetic factor responsible for the differences is sedimentation of different genesis. As regards the erosional relief of mountains and plains, on the other hand, the type of deposition is not a determining factor for classification purposes. In other words, every megamorphostructure is classified on the basis of the principal genetic factor. Such an approach reflects the principles on which the legend is based. The main purpose is to show, as objectively as possible, the relief as it exists today, and to classify it according to the principal genetic factors and within the limits of the Map's scale.

The legend also includes a great number of individual relief features and their assemblages, both of subaerial and submarine types. They are shown on the map by colour symbols according to which genetic group they belong. It is also important to show on the Map those deep-seated fault zones that are clearly expressed in the relief and determine many of the most important features of the latter. Some faults known from the published literature and confirmed by satellite images (*see* Figs. 1.10 and 1.11) are shown in Fig. 1.8. The faults determine both the boundaries of blocks and the pattern of block morphostructure. The entire framework of the Earth's relief, including Europe, is governed by a complex multistage fault network [see Bashenina *et al.* (1979)].

Chapter 2
Principal Structural and Tectonic Features of the Ocean Floors around Europe

2.1 Bathymetric Characteristics

The oceans are conventionally divided into four main bathymetric zones.

1) The littoral zone, whose depth varies from a few metres to several tens of metres.
2) The neritic zone, which is up to about 200 m deep.
3) The abyssal zone with depths between 200 and 6000 m.
4) The hypabyssal zone with depths exceeding 6000 m.

It is advisable also to distinguish between subbathyal and subabyssal zones. The former are observed in the over-deepened basins on the continental shelf and the latter represent basins within the bathyal zone. An example of a sub-bathyal basin is the Derbent depression (central Caspian Sea) and of a subabyssal basin is the Black Sea depression.

The littoral zone needs only brief mention. It may extend along any sea coast and, depending on the off-shore gradient, its width may either reach a few metres or descend quickly to tens of metres.

The neritic continental shelf includes the Barents and White Seas, offshore areas of the Norwegian Sea and the entire North Sea, with the English Channel and Baltic Sea. The distinctive Kandalaksha basin in the White Sea may serve as an example of a subbathyal basin. Another example is the Norwegian trench which runs between the North and Norwegian Seas. The bathyal zone in the Norwegian Sea is represented by a continental slope and the Vöring plateau, as well as by a system of submarine rises between the North Sea and Iceland, comprising the Iceland–Faröe rise in the north-west and the Rockall bank located to the south-west. Iceland is surrounded by a shelf that is terminated, in turn, by the continental slope (i.e. the bathyal zone). The Norwegian Sea is separated in the west by the Kolbeinsey and Mohn rises from the Greenland rise, which is outside the limits of Europe and lies in the abyssal zone. On both sides of the Rockall bank lie the Irish and Icelandic depressions of the Atlantic Ocean. These depressions have maximal depths of 3537 and 3733 m, respectively. In the north-west, the Irish depression is limited by the mid-oceanic Reykjanes ridge, which joins Iceland at the cape of the same name. The minimum depth above the ridge is 400–600 m.

The shallow waters west of the British Isles and France are bounded offshore by a steep continental slope, with depths of 200–4000 m. The deeper parts of the ocean floor represent an abyssal plain of the western European basin, which is flat in the east and of varying depths in the west. The maximum depth of 5669 m is in the southern part of the depression. A number of individual seamounts are observed there, including one in the Bay of Biscay with a minimum depth of 2410 m.

The Atlantic shore of the Iberian Peninsula is bounded by a narrow shelf (up to 200 m deep). The edge of the shelf rises from the bathyal zone in an area where the relief appears to be extremely rugged. The major relief element here is the Galician marginal plateau. Similar features are characteristic of the topography of the Atlantic Ocean floor in the south-western corner of Europe. Westwards there is the Iberian basin with a maximum depth of 5925 m observed in a narrow valley in its southern part. There are also the oceanic ridges of Horshue, Azores–Biscay, Goringe and a number of individual seamounts in this part of the Atlantic Ocean. One of their peaks, the Amper seamount, rises to within 40 m of the surface.

The neritic zone in the Mediterranean Sea is extensive only in the Gulf of Lions, east of Tunis and in the Adriatic Sea. In the remainder of the Mediterranean Sea, this zone is only a few tens of kilometres wide. For the most part, the Mediterranean Sea lies within the bathyal and abyssal zones. The following abyssal or subabyssal basins are apparent: Alboran (1501 m), Balearic or Algerian–Provençal (2887 m), Tyrrhenian (3719 m), Central (4117 m), Levantine (3174 m) and the Hellenic deep-sea trench (maximum depth: 5124 m). There are also a number of seamounts: the Alboran and Balearic mountain ranges, the Tunisian–Sicilian rise with numerous troughs and uplifts characterized by a depth range of up to 1500 m, and the central Mediterranean ridge extending along the southern edge of the Hellenic deep-sea trench with a minimum depth of 1290 m.

Sharp depth variations are typical of the Aegean Sea. Apart from deep basins with depths of up to 1300 m, there are numerous banks, seamounts and islands. The southern part of the Aegean Sea is often considered as an independent unit, as is the sea around Crete. Here there is a subabyssal trough up to 2590 m deep, bounded on the south by the island arc of Crete. In spite of the fact that it is small, the Sea of Marmara reaches a depth of 1261 m. The Black Sea has a wide, shallow offshore area in the north-west. Also the Sea of Azov is very shallow (only 14 m). For the most part, the Black Sea is occupied by a subabyssal plain with the depth varying from 2000 to 2200 m. A bathyal zone includes the continental slope that delimits a deep trough of the Black Sea. The Caspian Sea is very shallow in the north (less than 15 m); the central and southern parts are represented by the central Caspian (Derbent) depression (788 m), and the southern Caspian basin (1024 m).

2.2 Major Oceanic Morphostructures

Present-day geomorphologists divide the ocean floor into global-scale morphostructures, which represent the primary features of the Earth's relief corresponding to essential differences in crustal structure [see Leont'ev (1965, 1968a, 1971), Udintsev (1959, 1972), Krasniy (1977)]. These megamorphostructures include: (1) the submarine continental margins, (2) transition zones, (3) deep ocean floors and (4) mid-oceanic ridges. This classification is based on differences in crustal structure according to composition, thickness and mobility. Since the 1960s, much new information about the suboceanic crustal structure has come from deep-sea drilling [see Laughton (1977)].

2.2.1 Submarine Continental Margins

The continental crust of the Earth is on average 35–40 km thick (attaining, in rare cases, 85 km) and consists of three layers: (1) sedimentary (of variable thickness), (2) granitic (composed of mainly acid igneous rocks) and (3) basaltic (composed of basic rocks). The thicknesses of the granitic and basaltic layers are approximately equal, being 15–16 km [Worzel and Shurbet (1955), Raitt (1963), Bott (1971), Galpezin and Cosminskaia (1964)]. This continental type of crust is less mobile than the oceanic one—except in zones of tectonic activity—and is typical of the continental platforms. The submarine extensions of these platforms were formed as submarine continental margins [see Heezen *et al.* (1959), Udintsev (1959)] and consist of shelf, continental slope and continental rise.

The continental shelf represents structurally and geomorphologically a direct extension of the continent. Its surface possesses mainly faint relief, but occasionally basins and hills are present. Evidence in places of subaerial sculpture points to recent submergence. At the outer edge of the shelf, Dibner (1978) stated that the gradient increases sharply towards the abyssal depths. This relatively steeply sloping zone is called the continental slope. The continental slope is seen frequently to exhibit a step-like pattern in profile; these steps, especially those of significant width, are termed marginal plateaus. Elsewhere, the continental slope exists as a gently sloping plain. It is commonly incised by submarine canyons that may also cross the continental shelf. The continental slope merges downward into the continental rise. Usually, this is a gently sloping plain, differing from the continental slope in its smoother relief. Occasionally, hilly or ridge-and-hollow types of submarine relief may be observed.

The boundaries between these various elements are not rigidly confined to particular depths. The outer limit of the shelf lies frequently at a depth of 100–200 m [see Shepard (1973)], the average depth at the edge being 132 m. The base of the continental slope usually lies at a depth of 3000–4000 m, but in the Black Sea, for example, it lies at only 2000 m. The depth at the outer edge of the continental rise varies from 3500 to 4500 m.

The thickness of the continental crust decreases from the shelf towards the continental rise. Although the thickness of unconsolidated sediments is insignificant, the consolidated sedimentary cover may reach several kilometres. On the continental slope, however, the unconsolidated sediments appear to have considerably greater thicknesses on gentle slopes or in hollows. On the occasional steep ledges of the continental slope, outcrops of the granitic layer and sedimentary cover are observed. The continental rise is notable for the considerable thicknesses (3–5 km) of the sediments, which according to seismic and drilling evidence infill the depression in the granite shell that pinches out at the outer boundary of the continental rise [see Drake *et al.* (1959), Sheridan (1974)]. Data from continuous seismoacoustic profiling show that a sedimentary body constituting the continental rise and slope advances far inside the limits of the ocean floor and, because of this, the geomorphological and geophysical boundaries of the continental slope may not coincide.

The shelf areas, except those subjected to tectonic activity, and the continental rise are free of natural earthquakes (i.e. aseismic areas). In contrast, the continental slopes, whose formation is associated largely with tectonics, are characterized in some instances by earthquakes with shallow epicentres.

2.2.2 The Ocean Floor

The ocean floor is characterized by oceanic crust. Its thickness is five to six times less than that of the continental crust. Moreover, there is no granitic layer present. Here the sedimentary layer overlies the so-called second seismic layer or 'acoustic basement', each layer being about 1 km thick. The second layer is underlain by the basaltic (oceanic) layer about 5 km thick [see Raitt (1963), Worzel (1974)]. The oceanic crust is largely aseismic; only in places are occasional earthquakes recorded. Submarine vulcanism until recent times was widespread on the ocean floor. The acoustic basement, according to deep-sea drilling, consists of tholeitic basaltic lavas. Apart from this, the basement probably includes compact and metamorphosed sedimentary rocks.

The ocean floor relief is notable for its large, cellular structural pattern. Giant basins occupying four-fifths of the floor area are separated from each other by oceanic ridges and rises. The crust under these positive relief forms is usually thicker than under the depressions. With respect to its ruggedness, the ocean floor may be classified as mountainous, hilly, undulating or low relief. The first two classifications represent the initial types which were subsequently converted into undulating and low-relief abyssal plains, as their depth of burial by sediments increased. The ocean floor relief is also characterized by numerous seamounts, as well as by conjugate narrow ridges and depressions reflecting the trends of fracture zones.

2.2.3 Transition Zones

Within the area shown by the *International Geomorphological Map of Europe*, the transition zone is represented only by the Mediterranean Sea, and the zone here is in the latest stage of development [see Leont'ev (1968a, 1975)]. This type of relief is characterized by a predominance of continental structures and by abundant relicts of oceanic or, more exactly, suboceanic elements. From the standpoint of geosynclinal theory, the Mediterranean Sea and the adjoining Alpine orogenic zone with its relicts of suboceanic depressions (where the crust is notable for the great thickness of the sedimentary layer) represent a remnant of the former Tethys Ocean which has now largely become an area of continental structures. The elements of island arcs and relict deep-sea trenches have only been preserved in the area of the Ionian islands and Crete.

The characteristic feature of the transition zones is that they are composed of fragments of continental or subcontinental crust, as well as of oceanic and suboceanic crust. Such patterns make it possible to delineate a former geosynclinal area.

2.2.4 Mid-Oceanic Ridges

One of the greatest landmarks in geological and geographical progress this century has been the discovery of a global system of mid-oceanic ridges [see Heezen *et al.* (1959), Menard (1964), Hill (1963), Leont'ev (1965, 1968b), Udintsev (1959, 1972)]. These ridges occur as long, wide, linear mountain systems, characterized by the development of rift structures along their axial zones. They constitute a single mountainous system extending for 100 000 km. These areas are tectonically active, as shown by their high seismicity, vulcanism and fault tectonics. Within the limits of the *International Geomorphological Map of Europe*, only small parts of the mid-oceanic ridge system are shown.

The main structural features of the mid-oceanic ridges are their essentially basaltic composition, the thinness and irregular distribution of the sedimentary layer, the mélange structure of the second layer [as described by Dmitriev (1975), Aumiento *et al.* (1977)], the high density of the crust beneath the second layer (due to the intrusion of ultrabasic rocks) and the low density of the underlying mantle. This fourth type of crust can be termed riftogenic.

2.3 Structural and Tectonic Features of the Sea Floor of Europe

2.3.1. North-Western Sector

The eastern European platform, which slopes in the north under the waters of the northern Arctic and Atlantic Oceans described by Dibner (1978), is subdivided into the submarine margins of the Baltic Shield and the northern part of the Russian, Pechora and Barents Sea platforms.

The larger part of the Barents Sea represents structurally a platform incorporating the ancient, probably Archaean basement [see Dibner (1978) p.13)]. The shapes of the positive relief forms and a general ring pattern of depressions surrounding these forms suggest that they reflect mainly a structure of 'gneiss ovals' as described by Salop (1978) and dated 3000 Ma. According to Dibner (1978), and Eldholm and Ewing (1971), the sedimentary cover of the Barents Sea platform consists of Pleistocene–Neogene (550 m), Palaeogene–Upper Cretaceous (620 m), Lower Cretaceous, Upper and Middle Jurassic (1.25 km), Lower Jurassic and Upper Triassic (700 m), Permo-Triassic (900 m) and Lower Permian–Carboniferous (1.2 km) rocks. The northern portion which lies outside the *International Geomorphological Map of Europe*, is largely composed of Upper Mesozoic plateau basalts. The southern part of the Barents Sea shelf belongs mainly to the Pechora platform. The submarine structural fold system is very distinct here—the extensions of the Timan Ridge (or Murmansk uplift) and the Pre-Timan downwarp, as well as the intricately built Pechora syneclise, notable for its development of narrow ramparts and troughs in the sedimentary cover, are also apparent.

The limit of the Fennoscandian Shield in the north and north-east coincides almost exactly with the coast. Therefore its geological structure affects only the coasts and a narrow offshore strip of the Barents Sea floor and part of that of the White Sea. The White Sea depression represents a graben–syneclise of the north-western margin of the Russian platform which lies partly on Baikalian and partly on Archaean basement rocks. The basement of the south-eastern margin of the Barents Sea is composed of folded Hercynian and Caledonian structures belonging to the Ural–Novaya Zemlya koilogene system [see Dibner (1978)]. The axial zone of the uplift includes only a narrow offshore strip of the floor, whereas the shelf depressions occurring to the west and east correspond to marginal troughs. The Kara Sea, which lies almost completely beyond the limit of the Map, belongs to the epi-Hercynian western Siberian platform.

The extreme western part of the Barents Sea margin consists of Caledonian folded structures. These features (Caledonides) are also associated with the submarine continental margin in the Norwegian Sea. The continental slope south of Medvezh'i Island is confined, according to Talwani and Eldholm (1974), to the north–south-striking fault zone. A system of subparallel faults and differential uplift of the crustal blocks has imparted a step-like pattern to the continental slope profile. One of the large steps of this type is the oblique horst of the Vöring plateau. According to seismoacoustic surveys, this horst was uplifted in place of the former downwarp of the Caledonian basement infilled with terrigenous sediments up to 6–9 km thick.

A belt of Caledonides running south-west is traceable on the floor of the northern part of the North Sea, the Irish Sea and part of the oceanic shelf west of Ireland. The Porcupine Bank is a deeply submerged link in this belt in the extreme south-west.

The geological structure of the North Sea floor has become well known through the exploration and production of oil and gas (*see* Fig. 2.1). It represents basically a platform consisting of the Caledonian consolidated basement, overlain by a sequence of Upper Palaeozoic rocks up to 12 km thick, including also Carboniferous coal-bearing deposits and Upper Permian evaporites with which salt domes are associated [see Ziegler (1975a, b) Kalinko (1977)]. Most oil and gas-bearing structures are confined to a zone that crosses the North Sea from north to south, containing the Central and Viking grabens. Apart from the north–south-trending fault zones, there are also numerous south-west–north-eastern-trending faults that

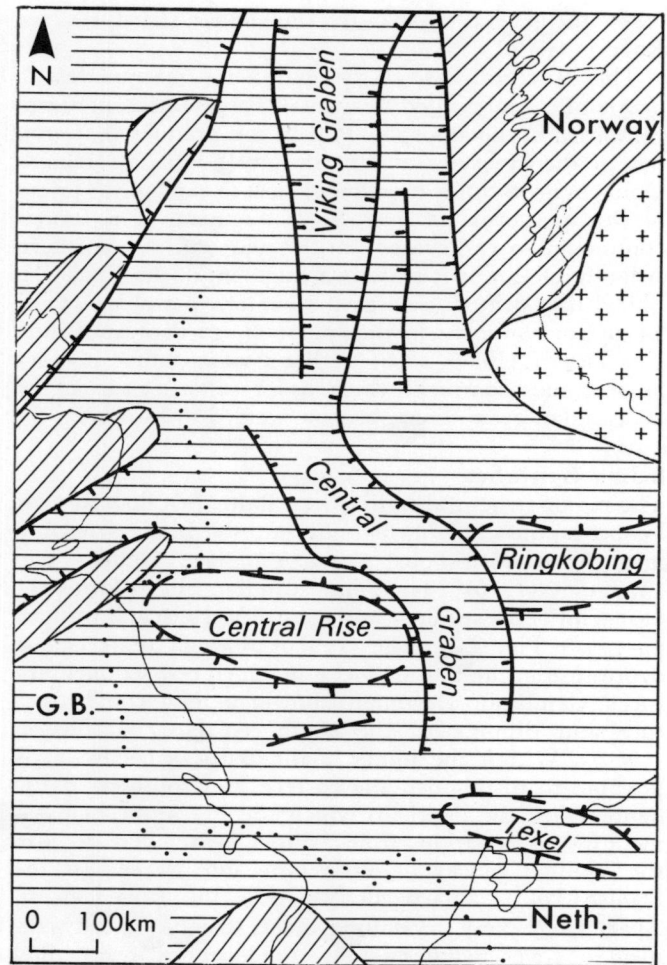

Fig. 2.1 Structure–tectonic map of the North Sea floor [after Kalinko (1977)]. (1) Exposed Precambrian basement, (2) exposed Caledonian basement, (3) subsided Caledonian basement, (4) regional faults, (5) boundaries of local structural uplifts, (6) Zechstein limit.

Fig. 2.2 Geology of the Baltic Sea floor [after Gudelis and Emilianov (1976)]. (1) Palaeogene, (2) Cretaceous, (3) Jurassic, (4) Permo-Triassic, (5) Carboniferous, (6) Devonian, (7) Silurian, (8) Ordovician, (9) Cambrian, (10) Precambrian, (11) faults.

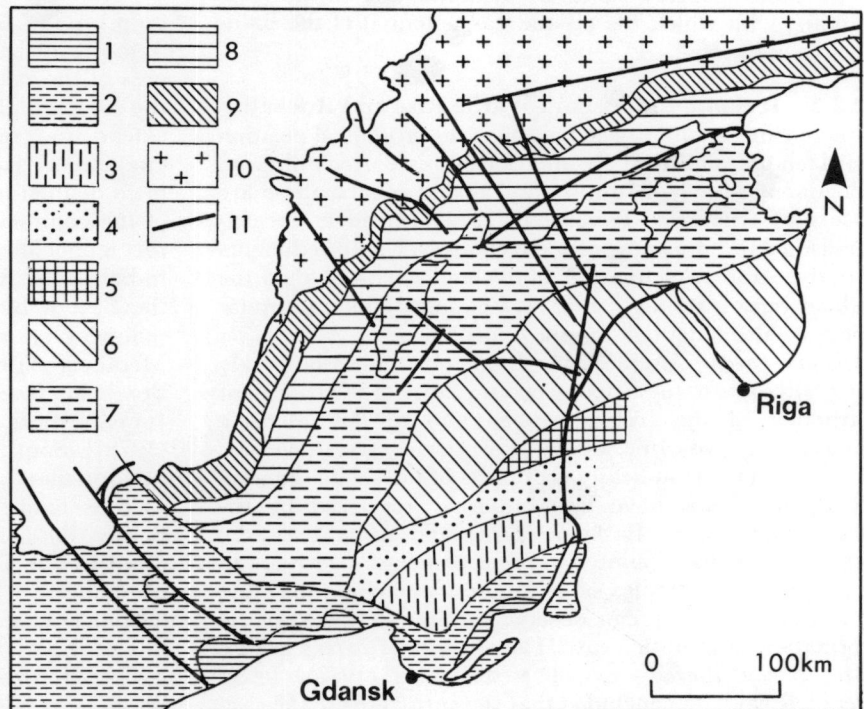

follow the general strike of the Caledonian structures [see Roberts (1974)].

The structural base of the continental margin west of Scotland is represented by a small Precambrian platform composed of metamorphic schists and granites that crop out in the Hebrides and in the north-west of Ireland. Different geotectonic hypotheses exist to explain the origin of this platform. It may be a fragment of a former continental area located in the part of the northern Atlantic Ocean or it may be a fragment of the Canadian Shield that drifted towards the western periphery of the European continent in the course of sea floor spreading [see Dewey (1974)]. The author's view is that it is an ancient microcontinent which, as a result of closure of the Caledonian geosyncline, became welded to Europe.

2.3.2   Baltic Sea
The major structural units of the Baltic Sea floor are the Fennoscandian Shield, merging into the Baltic syneclise, and the Danish–Polish aulacogen [see Gudelis and Emilianov (1976)]. The Fennoscandian Shield forms the basement of the sea floor in the north-western and western parts (*see* Fig. 2.2). The Gulf of Bothnia represents a depression in the Baltic Shield on which rest sedimentary rocks (Cambrian–Devonian) showing a monoclinal structure and dipping to the south-east [see Martinsson (1960)]. Within the limits of the Baltic syneclise, it is possible to distinguish the eastern and southern Baltic depressions; the depth to which the Precambrian basement descends varying from 2.5 to 3.5 km. There are also the uplifted blocks of Bornholm and Neksë. In the Danish–Polish aulacogene, there is the Rügen horst. Apart from these crystalline block structures, the sedimentary cover includes a number of local brachyanticlinal and dome-shaped structures, including the Liyepayasea uplifts, with which the oil and gas potential of the Baltic Sea is associated.

2.3.3   Iceland, the Iceland–Faröe Rise and Rockall
There is no unanimity so far about the structural position of Iceland and the Iceland–Faröe rise. According to Bogdanov (1972), Muratov (1975) and others, these are oceanic structures. Palmason (1974) considers the increase in the thickness of terrigenous/volcanic sediments on the submarine margin of Iceland as proof that the island represents a gigantic lava cupola. This interpretation is also adopted on the morphological *Map of the Atlantic Ocean* compiled by Ilyin (1976). Milanovskiy (1976) points to the heterogeneous nature of the basement structure of the Icelandic shelf and to the intensive present-day volcanic activity of the Vestmannaeyjär. Palmason(1974) indicates that parts of the shelf adjoining Reykjanes Cape have been intensely reworked by riftogenic processes. Baskina (1973) and some Icelandic investigators have pointed to the presence of acid rocks in the magmatic complex of Iceland and conclude that it is an area where one can observe the initial stages in the formation of granitic crust. The authors of *Iceland and the Mid-Oceanic Ridge* [see Collected Papers (1977)] using geophysical data concluded that the entire area of the rise,

including Iceland, represents a single continental massif characterized by normal seismic velocities of the mantle. They point to its block-like structure and the differences in geophysical properties between the blocks. Artiushkov *et al.* (1980), taking into consideration the absence of granitic xenoliths in the deep-seated magmatic rocks of Iceland, believe that Iceland has been formed from oceanic crust. No indications exist concerning the age of the continental basement under Iceland and under the Iceland–Faröe rise. However, if it is assumed that the Rockall plateau is a submerged step of a single Rockall–Iceland–Faröe massif (an idea that is supported by bathymetric data and sea floor topography), then the presence of rocks of Grenvill age on Rockall and on the northern side of the Iceland–Faröe–Shetland trough is also indicative of the Precambrian age of the basement of the Iceland–Faröe rise. The continental nature of the Rockall plateau is also supported by geophysical data. The crystalline basement is buried there under a 2-km thick sedimentary layer. Boreholes No. 403–406 sunk into the continental rise of Rockall to a depth of 800 m are still within Eocene rocks [see Montadert *et al.* (1978)].

2.3.4   South-Western Submarine Margins of Europe
The English Channel floor, the shelf area west of it (i.e. the Western Approaches), the shelf of the Bay of Biscay and an area west of Galicia and Portugal all belong to the Hercynian western European platform. According to Montadert *et al.* (1974) and Roberts (1974), the Hercynian folded basement also extends into the continental slope and, probably, continental rise. Only a small area along the northern shore of the Iberian peninsula is attributed to a younger age of consolidation—the Pyrenean phase of orogeny (Eocene).

Palaeozoic rocks subjected to Hercynian folding outcrop on the sea floor within the English Channel, the Armorican shelf and the Western Approaches, the thickness of the sedimentary cover increasing rapidly towards the continental slope. Lower Cretaceous rocks of the cover extend to greater depth and are overlain by Upper Cretaceous, Palaeogene and Neogene sediments composing a distinct monocline that forms the submarine margin of the continent. The continental slope is a reflection of this monocline, complicated by numerous latitudinal and radial faults. It should be noted that the continental rise in the Bay of Biscay shows a relatively small thickness of sedimentary rock. Borehole No. 400 sited south of the Meradzik plateau struck Aptian rocks at a depth of 700 m. Borehole No. 401 sunk on the plateau itself intersected Jurassic rocks at a depth of 250 m [see Montadert *et al.* (1978)]. Southwards the thickness of sediments increases rapidly and on the marginal Landes plateau already exceeds 1 km. In the south-eastern corner of the Bay of Biscay, the basement and sedimentary cover are complicated by the Cap Ferret graben, which is interpreted by Montadert *et al.* (1978), from the standpoint of plate tectonics, as an ancient rift along which the opening of the Bay of Biscay took place.

The Hercynian structural pattern of the Galician–Portuguese portion of the continental submarine margin

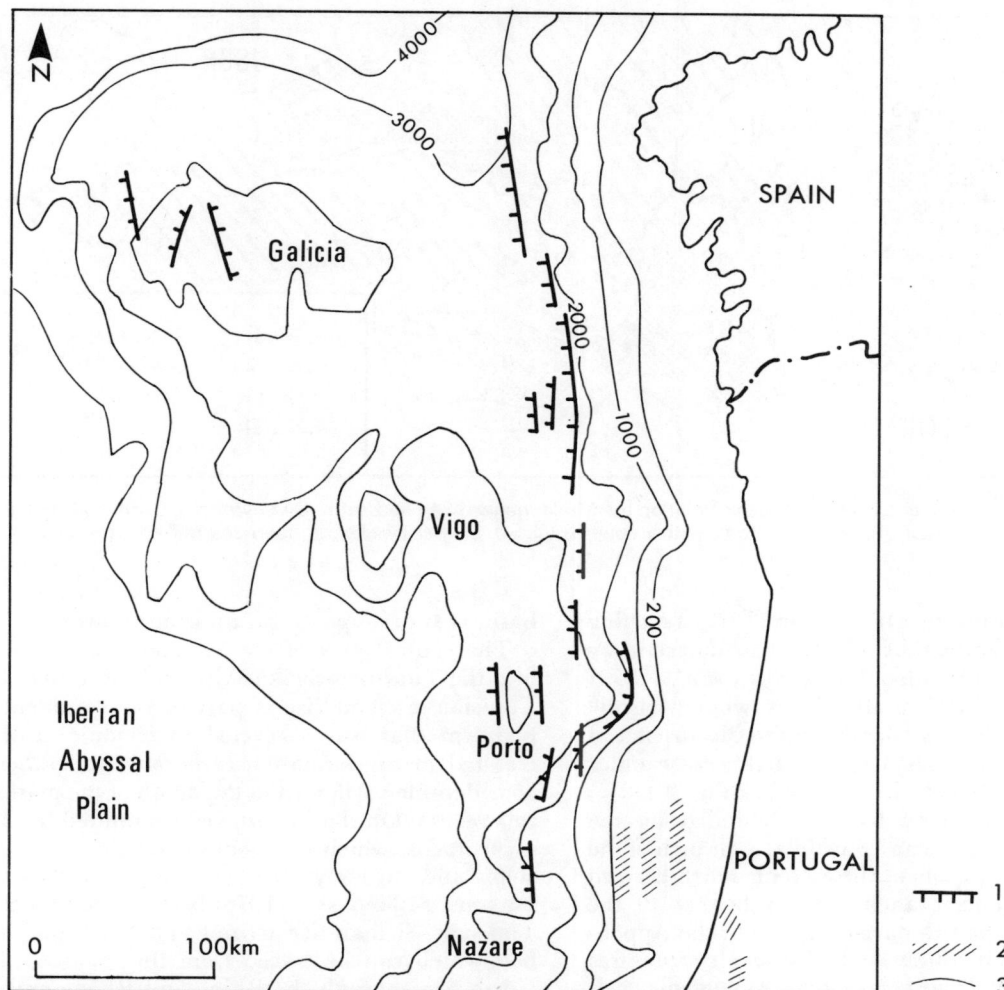

Fig. 2.3 Some relief features of the north-western submarine margin off Galicia and northern Portugal [after Montadert *et al.* (1974)]. (1) Main faults of the continental margin, (2) diapiric structures, (3) contours in metres.

(*see* Fig. 2.3) is evidenced by Upper Palaeozoic deformed and metamorphosed rocks outcropping on the Farilhões and Berlengas Islands (north of Lisbon) and on sea-mounts found off Galicia. The continental slope can be structurally attributed to the continental structure. The Oporto, Vigo and Galician seamounts represent separate blocks that have undergone considerable uplift compared to the submerged parts of the submarine continental margin. The sedimentary cover in the area west of the Portuguese lowlands includes evaporites of Triassic and Lower Jurassic age, with numerous signs of salt tectonics.

2.4 Principal Structural and Tectonic Features of the Mediterranean, Black and Caspian Sea Floors

2.4.1 Mediterranean Sea

As has been already noted, the Mediterranean Sea can be considered as a late stage in the development of a geosynclinal (transition) region. This stage is characterized by the occlusion of deep-sea trenches, the gradual disappearance of vulcanism, the conversion of the ma-jority of island arcs into large broad mountain structures and the apparently relict nature of suboceanic basins occurring as 'windows' in the greatly enlarged continental crust. Present-day knowledge of the genetic structure of the Mediterranean Sea floor is summarized by Ryan *et al.* (1970), Dewey *et al.* (1973), Biju-Duval *et al.* (1974), Auzende and Olivet (1974), and Malovitskiy (1976).

The submarine continental margin of the Mediterra-nean Sea is essentially heterogeneous. Its western and north-western parts are represented mainly by a sub-merged Hercynian platform. In the south-west, it consists of the slope of the Atlas mountain zone, the Alboran basin and range, as well as the Balearic range and the Alpine structures. The continental rise that bounds the Balearic or Algerian–Provençal basin is composed of a sedi-mentary layer, which is 4–6 km thick in the area south of Languedoc. Upper Miocene rocks represented by evapo-rites are very typical of this morphostructural zone [see Chumakov (1975), Malovitskiy (1976), Hsu *et al.* (1978)]. These rocks are associated with numerous manifestations of salt tectonics. The Algerian–Provençal basin is charac-terized by subvolcanic crust, the thickness of the sedi-

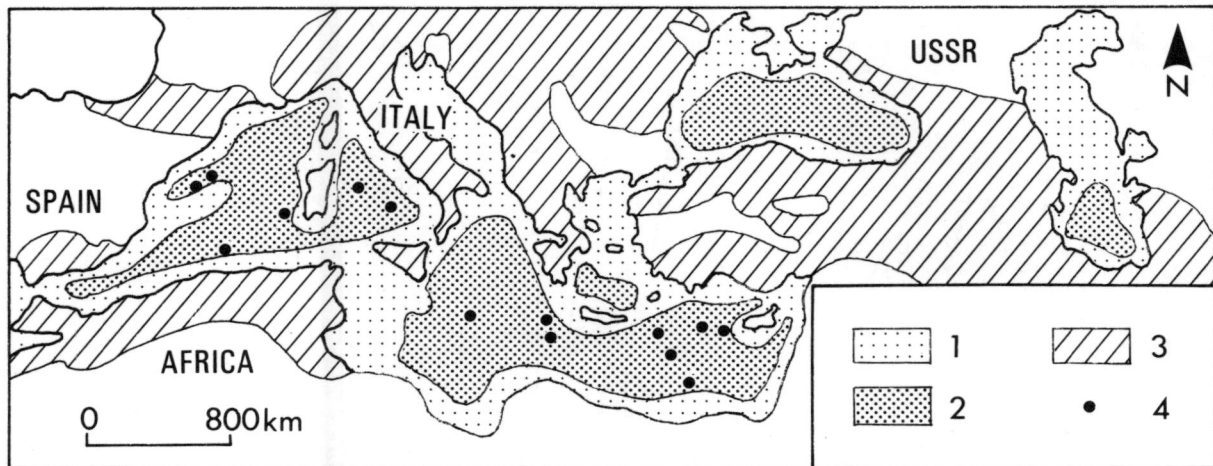

Fig. 2.4 The distribution of continental and suboceanic crust on the Mediterranean Sea floor [after Dewey *et al.* (1973)]. (1) Continental crust on the sea floor, (2) suboceanic crust, (3) Alpine system (within continents), (4) *Glomar Challenger* deep sea drilling sites at which Messinian sediments were recovered.

mentary cover according to Malovitskiy (1976) reaching 5 km. Malovitskiy believes that this cover is underlain by continental crust. On the other hand, Ryan *et al.* (1970) have identified a distinct basaltic layer whose seismic velocities vary (6.5–6.8 km s^{-1}) and point to the absence of a granitic layer. This suggests that the Earth's crust under the Balearic basin is suboceanic in type (*see* Fig. 2.4).

On the east, the Balearic basin is bounded by the Hercynian blocks of Corsica and Sardinia. A branch of the Alpine folding joins the basin at the extreme north-eastern corner of Corsica. This branch extends farther to the Ligurian Sea and to the submarine margin of the Appennine peninsula and is distinguished by volcanic structures which form, in the Tyrrhenian Sea, a volcanic belt traceable as the Lipari island arc. Farther on, this structural zone merges into the Atlas submarine ridge, which is a submarine continuation of the African Atlas. According to many investigators, the floor of the Tyrrhenian basin is characterized by suboceanic crust. In the central part (i.e. the Vavilov seamount) and in the eastern margins, several submarine volcanoes are known [see Mikhailov (1965)]. A seismic survey conducted in the western margin of the Tyrrhenian Sea has detected the presence of salt dome structures. Their distribution is probably limited by the continental rise, since Boreholes No. 332 and 333 sited in the central part of the basin have pointed to the absence of the Upper Miocene evaporites.

The Adriatic Sea is a typical shelf sea. The thickness of the continental crust reaches 40 km, but is reduced in the central subbathyal basin to 30 km. The sedimentary cover is 20–30 km thick, 2.5 km of this being attributed to Pliocene–Quaternary deposits. Numerous authors consider that the Adriatic Sea floor is a central massif which, on the west, north and east, was subjected to Alpine folding. Its north-western periphery structurally represents the submerged margin of the folded Dinaric Alps and their piedmont downwarp. Geophysical studies of the ancient basement have helped to outline the Godina and central Adriatic structural ramparts. The subbathyal

basin is superimposed on these structures.

The central part of the Mediterranean Sea is divided into the Tunisian–Sicilian rise and the Ionian basin. The Tunisian–Sicilian rise is part of the continental African platform that was subjected to tectonic crushing. This resulted in the formation of horsts and grabens, and in considerable variations in the depth of the platform within this region. On the east, the rise is limited by the Maltese escarpment, which is a continental slope complicated by steps and strongly dissected by numerous submarine canyons. Milanovskiy (1976) believes that the faults of the Tunisian–Sicilian rise are linked to the Rhine–Libyan rift belt which can be traced from the major graben of the North Sea through the Rhine and Rhône grabens to the Campidano rift in Sardinia and then via the Tunisian–Sicilian rise to Libya. In the author's opinion, the relationship between the Sardinian rift and the faults of the Tunisian–Sicilian rise, as well as between the latter and the Libyan faults, is questionable and is not confirmed either by bathymetric or geophysical data.

The Ionian basin, like the Balearic and Tyrrhenian basins, is a suboceanic depression. The thickness of the crust, according to Malovitskiy (1976), reaches 20–30 km. Borehole No. 374 of the SS *Glomar Challenger* sunk in the deepest part of the trough (i.e. 4078 m) struck, at a depth of 400 m, a sequence of the Upper Miocene evaporites, indicating considerable recent subsidence of the basin floor.

The eastern part of the Mediterranean Sea floor consists of the following macrostructural elements: the central Mediterranean rise, the Hellenic deep-sea trench and horst zone, the island arc of Crete, the Sea of Crete basin, the Aegean Sea borderland and the Levantine basin.

The central Mediterranean rise represents an elongated arched uplift of uncertain origin. Emery *et al.* (1966) believe that it is a mid-oceanic ridge but, as shown by Mikhailov and Goncharov (1962), more reliable evidence exists to relate this structural feature to the geodynamic zone of the Hellenic deep-sea trench. In this case, the rise can be thought of as being similar to marginal oceanic rises accompanying deep-sea trenches on the oceanic side. Using magnetic data, Malovitsky has indicated that the sedimentary cover is over 10 km thick here. This is also confirmed by seismic data. Boreholes No. 125, 375 and 376 sited on the central Mediterranean rise have encountered Upper Miocene evaporites [see Hsu *et al.* (1978)].

A zone of trenches and horsts located to the north is believed to be the relict Hellenic deep-sea trench [see Mikhailov and Goncharov (1962), Emery *et al.* (1966)]. Seismic evidence and the relief features both favour this view and suggest that the present-day relief of this belt is due to the fact that separate parts of a once-existing single trench have been subjected to tectonic inversion. This fits in with the concept of the Cretan island arc being the last arc of the Tethys Ocean, and of the Sea of Crete basin being a deep-sea trough of the marginal sea.

The Aegean Sea floor shows chiefly the Palaeozoic folding of its basement and represents a median massif bounded by Alpine folds (Hellenides and Balkanides). During the Alpine period, this massif underwent intensive crushing; the blocks thus created were subjected to differential vertical movements. These events produced a complex structural pattern and floor topography, and developed its borderland features (*see* Fig. 16.9).

The vast Levantine basin is in the opinion of some authors a crushed and submerged part of the African platform, while other authors regard it as a relict suboceanic basin. According to Malovitskiy (1976), the basement occurs under 15 km of sediments, while Ryan *et al.* (1970) think that the thickness of the sediments varies between 5 and 10 km. An essential role in the structure of the Earth's crust in the Levantine basin is played by faults, which divide the crust into upthrown and downthrown blocks, such as Eratosphen, Anaximandra, the Cyprus uplifts and the Rhodes depression. The narrow shelf and steep continental slope over the larger part of the submarine margin of Africa in the eastern Mediterranean is undoubtedly a submerged part of the African platform.

2.4.2 Black Sea
The floor of the Black Sea is subdivided into the submarine margin of the Eurasian continent and a deep-sea basin. In the north and north-west, the shelf is part of the Scythian Hercynian platform, whereas, in the extreme north-west, it consists of the Precambrian basement of the Russian platform. Narrow marginal zones occurring along southern Bulgarian, Anatolian and Caucasian coasts show the folded basement of Alpine age. The Scythian platform includes a number of structures, the most prominent of which are the Tarkhankut uplift and the Kerch'–Taman' downwarp. The complex fault tectonics of the continental slope are expressed in a dense network of submarine canyons, in the step-like pattern of the slope and in the relatively high seismicity of the area [see Degens and Ross (1974)]. According to Goncharov *et al.* (1972), the continental rise belongs to a zone of pre-Alpine folding and consolidation. On the margins of Anatolia, the rise is complicated by a system of block–fold ranges. Their distinct expression in the relief and their strike support the

view that they resulted from the most recent (i.e. late Alpine and present-day) tectonic movements.

The floor of the subabyssal basin of the Black Sea is underlain by the suboceanic type of crust. The granitic layer is absent, the basaltic basement occurs at a depth of up to 15 km, and the total thickness of the crust is 18–22 km. This structure agrees well with the gravity data. The great thickness of the sedimentary cover does not support the conclusions of some scientists that the Black Sea depression is youthful and superimposed. If it is assumed that the average rate of sedimentation in the Black Sea subabyssal zone is 40 mm per 100 years [see Lisizyn (1978)], the age of the Black Sea would be about 550 Ma. Even the deepest of the three boreholes drilled from SS *Glomar Challenger* in the Black Sea (i.e. Borehole No. 380), which went to a depth of 1.07 km, failed to penetrate the full thickness of the Miocene rocks.

The great similarity of the Black Sea and Mediterranean Sea basins, both in geomorphology and physical properties, is additional proof favouring the concept of the suboceanic nature of the Mediterranean Sea basins.

2.4.3 Caspian Sea
The Earth's crust under the Caspian Sea is nonhomogeneous, being represented over the larger portion of the sea floor by consolidated crust of continental type, showing different ages of consolidation in different areas. The extreme north-east falls within the Caspian syneclise of the Russian platform (Precambrian), while almost the entire northern part of the central Caspian Sea lies within the Scythian platform. The Tersk–Caspian marginal downwarp adjoins the Caspian Sea in the south-west. The deeply submerged block of the Scythian platform, together with the north-eastern portion of the marginal downwarp, form the central Caspian (Derbent) subbathyal basin. The shelf of the Dagestan and Azerbaijan coasts has been involved in the latest folded structures of the western side of the marginal downwarp and in the south-eastern pericline of the Caucasus fold-mountain building. The late Alpine age of the consolidated crust here is linked to the structures of the Apsheron rise, a chain of uplifts separating the central Caspian from the southern Caspian [see Malovitskiy (1964), Leont'ev (1964), Lebedev *et al.* (1976)].

Faults and flexures play an essential role in the structure of the crust under the Caspian Sea. Deep-seated faults separate the Scythian platform from the Russian platform and from the Tersk–Caspian downwarp. Lebedev attaches importance to the large longitudinal eastern Caspian flexure which outlines the eastern side of the central Caspian depression. This flexure is identified both in the sea floor and by gravity measurements. The thickness of the sedimentary cover in the central Caspian reaches 10–12 km, decreasing to 4–6 km in the deepest part of the basin.

The relatively shallow depth of the crystalline basement in this part of the basin accounts for the positive gravity anomalies. Positive gravity anomalies are also caused by the Peschanyimys and Kara-Bogaz uplifts at the eastern margin of the central Caspian and by the buried Karpinsk uplift, which is the western submarine continuation of Mangyshlak.

The southern Caspian is clearly subdivided into the submarine continental margins and the suboceanic relict basin. The western shelf of the southern Caspian is involved in the latest fold structures of the south-eastern pericline of the Caucasus. The latter occur as submarine mountain ranges (anticlines on the continental slope) and project into the suboceanic basin [see Soloviev *et al.* (1952), Malovitskiy (1964), Leont'ev (1964)]. There are many mud volcanoes and numerous local domes appear to be oil and gas-bearing structures. The eastern shelf of the southern Caspian is quite wide. Deep seismic sounding under the thick sedimentary cover has enabled a system of buried anticlinal folds of the Kopet Dagh strike to be traced. They are also traceable as submarine ranges of the continental slope and in the north-eastern part of the basin. Maev (1961) showed that the fold movements here occurred during sedimentation and continue at present.

The Iranian submarine margin of the southern Caspian Sea is poorly known. The deep-sea basin, as evidenced by seismic surveys, is underlain by suboceanic crust. The sedimentary cover is strikingly thick—attaining 25 km— and the total crustal thickness is up to 40 km. As a whole, the southern Caspian can be considered as an example of an evolving marine geosynclinal basin in a late stage of development, with folded structures appearing [see Leont'ev (1964, 1968a)]. Apart from the submarine uplifts following the trends of the Caucasus and Kopet Dagh structures, there are also features of the latitudinal trend that are probably older than the present-day structures.

2.5 Principal Structural and Tectonic Features of the Ocean Floor and Mid-Atlantic Oceanic Ridge

2.5.1 Norwegian–Greenland Basin
The Norwegian–Greenland basin forms part of the northern Arctic Ocean and consists of the Greenland, Norwegian and Lofoten basins, the Irish plateau, and the Kolbeinsey, Mohn and Knipovich mid-oceanic ridges. The deep-sea basin of the Norwegian and Greenland Seas is underlain by a relatively thin crust (only 10 km thick) which does not include a granitic layer [see Talwani and Eldholm (1974)]. The thickness of the crust appears to increase abruptly beneath the Iceland plateau at the expense of the plateau basalts. The thickness of the sedimentary layer in the Norwegian basin is 2 km in the east and 1 km in the west; the thickness of the second layer is also variable (2–2.7 km) [see Korsakov (1979)]. Talwani and Eldholm (1974) described the linear magnetic anomalies recorded in the Norwegian Sea but, due to faulting, these anomalies are difficult to correlate. The Earth's crust under the Greenland Sea has a similar pattern. In the Denmark Strait, it is likely also to be oceanic but is completely overlain by sediments of the continental rise.

2.5.2 Mid-Oceanic Ridge
The Kolbeinsey rise is a continuation of the rift zone of

Iceland. In its southern part, the ridge does not possess a median rift valley but, after intersection with the Spar fault zone, it acquires morphology typical of a mid-oceanic ridge. A great deal of geomorphological data is available for the Kolbeinsey rise and other mid-oceanic ridges, and detailed descriptions of these features are given in the regional review (*see* Chapter 20). Here, it is only necessary to note that linear magnetic anomalies are most distinct over the Reykjanes ridge [see Talwani *et al.* (1971)] and that earthquake epicentres are confined to the axial zones of mid-oceanic ridges. Deep-sea drilling has been carried out on the arch of the Reykjanes ridge where Borehole No. 409, for example, intersected under a sequence of calcareous oozes of Pliocene–Pleistocene age, olivine and olivine–plagioclase, porphyritic and aphyritic basalts interbedded with ash and volcanic sediments (see Initial Core Description leg. 49). Boreholes No. 407 and 408 on the north-western flank of the ridge also struck the same rocks which correspond probably to the second seismic layer of the Earth's crust. The sequence of calcareous sediments resting on that layer is, however, much thicker—up to 300 m compared with 80 m in the case of Borehole No. 409—and also includes marls and chalk of Cretaceous–Miocene age.

A similar section for the upper part of the crust can also be expected on the eastern flank of the northern part of the

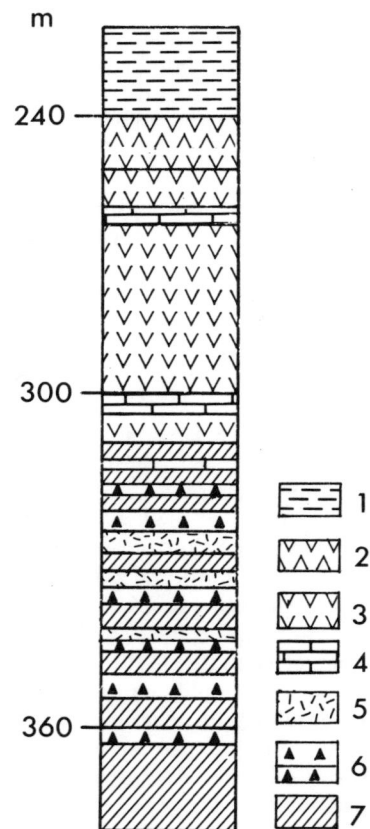

Fig. 2.5 General sequence at the *Glomar Challenger* site 334 [*Initial Reports, Deep Sea Drilling Project*, (1977)]. (1) Calcareous biogenic ooze, (2) basalt with phenocrysts of plagioclase and olivine, (3) aphyric basalt, (4) dense chalk-like sediments, (5) breccia, (6) olivine gabbronorite, (7) serpentinized peridotite.

mid-Atlantic ridge. As a whole, on the basis of drilling data obtained by Borehole No. 334 (south of the Azores) (*see* Fig. 2.5), it is possible to conclude that the second seismic crustal layer of the mid-Atlantic ridge consists predominantly of plagioclase–olivine basalts interbedded with hyaloclastites and sedimentary rocks [see Dmitriev (1975)]. The third crustal layer is represented by a mélange of blocks of serpentinized peridotites, olivine gabbronorites and their breccias cemented by hyaloclastites and sedimentary rocks [see Aumiento *et al.* (1977)]. Frequent and irregular magnetic reversals are observed along the section, which do not readily agree with the generally accepted ideas of plate tectonics, sea floor spreading and the associated palaeomagnetic reversals.

2.5.3 The Ocean Floor

The *International Geomorphological Map of Europe* includes the following elements of the north-eastern part of the Atlantic Ocean floor: the relatively shallow-water basins off Iceland and Ireland [see Klenova and Lavrov (1975)], the Iberian and western European basins, and a series of submarine ridges that separate these basins and the southern mountainous margin of the Iberian basin. Also included are the Teta basin with bordering seamounts and the north-eastern termination of the Canar basin. The crust here belongs to the oceanic type. The thickness of the sedimentary rocks is 800–900 m. The rocks underlying the sediments and the second layer are characterized by seismic velocities of 6.3–6.5 km s^{-1} (i.e. these rocks com-

prise a basaltic layer). The velocity of seismic waves in the upper mantle is 7.7–7.8 km s^{-1}. In the Bay of Biscay, the thickness of the sedimentary layer varies over short distances from 0.87 to 1.7 km, and that of the basaltic layer from 3.8 to 6.5 km [see Klenova and Lavrov (1975)]. Montadert believes that the crust there shows a rift structure.

The Iceland and Ireland basins are shallow and distinguished by a great thickness of sediments. Using geophysical data, many investigators have assigned the Teta basins, the Horshue seamounts and the Morocco basin of the Canar depression to the continental rise. Deep-sea drilling (e.g., Borehole No. 416) provides a fairly complete section of the sedimentary layer represented by turbidites and planktonic oozes of Neogene, Palaeogene, Lower Cretaceous and Upper Jurassic ages. The Upper Cretaceous is notable for a great gap in sedimentation. The Goringe, Amper and Josephine seamounts, and others, are unified into a single volcanic massif that also includes Madeira. Boreholes drilled on the Goringe seamount struck ophiolites which underlie Barremian abyssal sediments.

These characteristics and the principal structural and tectonic features of the sea floor of the north-eastern Atlantic and neighbouring areas support, in the author's view, the hypothesis of a primary ocean, and actual data on the structural geology and tectonics of the Mediterranean Sea attest strongly to the fact that the continental crust has been derived from oceanic crust.

Chapter 3
Exogenic Landforms of Europe

3.1 Fossil and Contemporary Morphoclimatic Zonation

At the end of the nineteenth and at the beginning of the twentieth century, European scientists (e.g., Dokhuchayev, Penck and Passarge) drew attention to the relationships between relief and climatic changes during the geological past. Geomorphologists looked particularly at exogenic processes, especially glacial ones, and their role in landform development in Europe. In 1913, the French geographer de Martonne for the first time used the term climatic geomorphology [see Stoddart (1969)]. After World War II, climatic geomorphology, in which the spatial distribution of the exogenic relief forms and geomorphological processes was controlled by climatic/physiographic zonation, became a focus of interest for some European schools of geomorphology [see Cholley (1950), Tricart and Cailleux (1965)]. The same idea was proposed in the USSR by Shukhin (1954), who studied geomorphological complexes in different types of geographical environment. Büdel (1969) divided climatic geomorphology into (1) climatic/dynamic aspects of the relationships between exogenic processes and climate and (2) climatic/genetic aspects of the relationships between relief and contemporary or former climatic conditions.

The genetic classification of exogenic landforms is based on morphoclimatic zonation [see Tricart and Cailleux (1965), Gornung and Timofeev (1958), Vedenskaya (1963, 1969), Gerasimov and Metcherikov (1967), Büdel (1977), Dedkov (1970, 1976)]. Some other geomorphologists, however, interpret this relief zonation as a part of a wider sphere in which morphodynamic processes play an essential part. The idea of morphosculpture—the complex development of relief controlled by exogenic and especially endogenic factors—was advocated by Gerasimov (1946, 1959). It is necessary to underline that, from the beginning, the idea of close relationships between structural and erosional relief forms was pointed out by Gerasimov (1946) and by Cholley (1950).

Aseev *et al.* [see Gerasimov and Metcherikov (1967)] analyzed the landforms of different continents. They distinguished typical assemblages of exogenic forms for different areas based on one or more leading geomorphological processes. This analysis shows that the zonal differences in exogenic landforms (i.e. morphosculptural zonation) cannot be explained in terms of different natural (bioclimatic) zones or of contemporary geomorphological processes. The differences in the spatial distribution of morphoclimatic and bioclimatic zones are due to the fact that the zonation of exogenic landforms reflects not only present-day but also past bioclimatic features. As well as the direct relationships between climate and geomorphological processes, there are also indirect relationships acting through the state of the subsurface materials: for instance, climate plays a part in determining the existence of permafrost in arctic areas or of salt deposits in arid areas, and these characteristics in turn affect the surface landforms [see Vedenskaya (1963)].

In the humid, temperate zone, which includes most of Europe, the rate of operation of slope processes is very slow due to the dense vegetation. Only in the northern parts of Europe, characterized by frost action and/or permafrost development, are cryogenic processes and features present. In the subtropical Mediterranean area, which possesses semiarid characteristics, fluvial and slope processes produce some features typical of the arid morphoclimatic zone.

On this relatively uniform zonal base, more detailed geomorphological provinces can be distinguished on the basis of morphostructural and lithological differences. Western Europe shows greater differentiation due to a greater range of relief. In the mountains, a vertical zonation of exogenic landforms and morphoclimatic processes is present.

The important feature of the exogenic morphology of Europe is the presence of fossil forms that originated during the Quaternary cold periods, glacial forms existing in complex relationships with younger superimposed erosion forms. Only in the areas of the most recent glaciation can more or less unmodified glacial forms be found, while in areas of older glaciation these forms were subjected to cryogenic processes and altered to varying degrees. Cryogenic forms can also be found in areas of the former Pleistocene periglacial zone, far from the glaciated areas.

Therefore, the relief of Europe shows features caused by (1) changes in morphoclimatic conditions during the past, (2) neotectonic movements (especially in the mountain areas) and (3) eustatic movements of sea level. Because of these different factors, the exogenic differentiation in Europe's relief is complex, resulting in a mosaic of contrasting landforms (*see* Fig. 3.1). This polygenetic character is very important in the denudation chronology of the continent. Geomorphological analysis of the relief of Europe starts with the reconstruction of the initial relief. The end of the Pliocene Epoch, that is the time before the commencement of the major climatic changes of the Quaternary, is proposed for such a reconstruction.

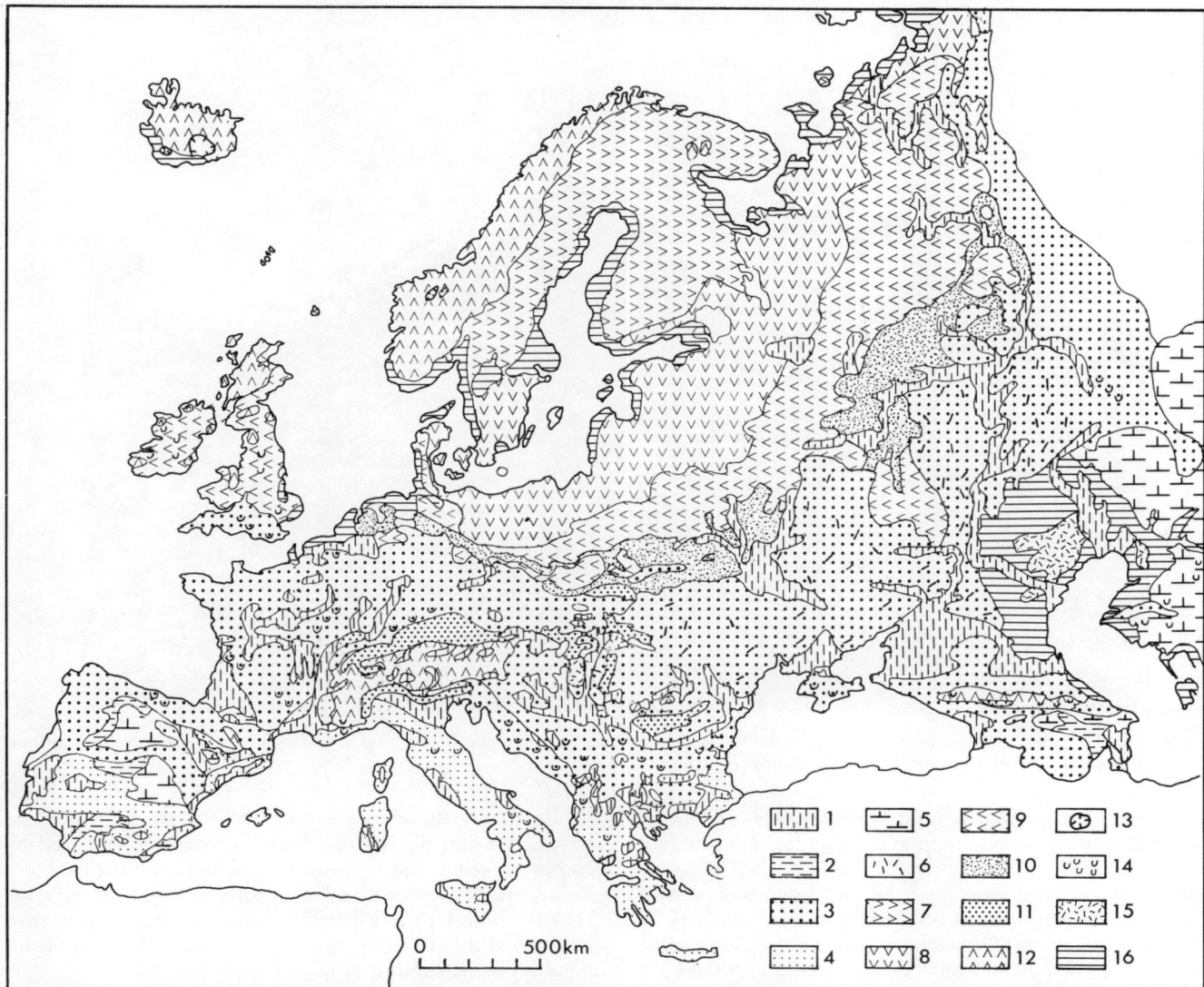

Fig. 3.1 Exogenic relief forms of Europe [after Serebryanniy (1978)]. (A) Fluvial forms: (1) alluvial and lacustrine deposition; (2) alluvial–proluvial deposition; (3) erosional and erosional–depositional in the humid temperate zone; (4) erosional and erosional–depositional in the Mediterranean zone; (5) erosional and erosional–depositional in the dry subtropical zone; (6) badlands. (B) Glacial forms: (7) erosional; (8) glacial and glaciofluvial depositional; (9) glacial depositional, remodelled by other processes (including periglacial processes); (10) glaciofluvial; (11) glaciofluvial mountain piedmont zones; (12) erosional and depositional mountain glaciation, remodelled by other processes; (13) contemporary glaciers and ice-caps (C) Other forms: (14) karst relief; (15) aeolian; (16) fossil marine deposits.

3.2 Planation Surfaces: the Main Feature of Watershed Areas in Pre-Quaternary Periods

The Upper Pliocene Epoch was three times longer than the whole of the Quaternary period. From the point of view of tectonics, it was a period of decreasing movement following the Alpine orogenesis and of subsidence of extensive areas and relief planation. As a result of this regime of neotectonic movements, planation surfaces were formed, river valleys were widened, with fluvial deposits accumulating on their floors, and extensive coastal plains were formed [see Aseev (1978)]. The Upper Pliocene Epoch was the time for the formation of stepped denudational plains on the Mesozoic and Tertiary rocks in extensive areas of the western European lowland and the

Russian platform. It was also a time of pedimentation; such processes were most effective in the semiarid environment of south-eastern Europe. The drier climate was the reason for the destruction of the weathered Mesozoic and Tertiary mantles and the transport of denudation products into river valleys.

The destruction of the thick weathered mantles was the principal reason for the origin of inselbergs and isolated massifs. Tors, as found on Dartmoor (*see* Fig. 3.2), may be the result of such processes in Britain [see Linton (1955)]. They have been formed because of differential weathering influenced by rock properties and joints. Skyline tors also mark old planation surfaces. The basal surface of weathering was exposed and a type of planation surface called an etchplain formed over extensive platform areas in western

Fig. 3.2 Hound Tor, Dartmoor, south-west England, showing typical castellated form with both angular and rounded core-stones, and an 'avenue' with fans and clitter leading from the exits. (Brunsden)

and central Europe [Bakker and Levelt (1964)]. Tors and ruwares are common on etchplains in central Europe [see Demek (1964a)].

Remnants of weathered mantles can be found even in areas of the Fennoscandian Shield that were affected by glaciation during the Quaternary [see Sidorenko (1958), Gjessing (1967)]. On the Fennoscandian Shield, a Mesozoic planation surface can be found at elevations of 100–150 m in coastal areas and 250–300 m in the interior. This so-called basic planation surface comprises parts of older planation surfaces, including the exhumed Precambrian planation surface [see Rudberg (1970a)]. The basic planation surface has been deformed by neotectonic block movements. Therefore, it can be found at elevations up to 600–700 m in northern Lapland and the Kola Peninsula, and at up to 350–400 m in northern Finland and Karelia.

Probably the Upper Pliocene planation surfaces were also developed across the Fennoscandian Shield, but were considerably modified by glacial erosion. In Lapland and in the peripheral parts of the southern Swedish highlands, Upper Pliocene pediments can be found [see Kaitanen (1969)].

The Mesozoic basic planation surface occurs in the central Russian uplands—areas that were slightly and regularly uplifted during the Tertiary Era. Also in eastern Europe and the Russian platform, this basic planation surface includes fragments of older (Palaeozoic) planation surfaces [see Borisevitch (1973)]. The denudational lands of the southern part of the Russian plain and the Volga–

Ural area, whose relief was influenced by younger, irregular uplift during the Tertiary, are characterized by stepped relief with several planation surfaces of Palaeogene, Miocene and Pliocene age [see Borisevitch (1973)]. The Upper Pliocene planation surface comprises valley pediments along the main rivers and is important for the reconstruction of recent river patterns.

In the borderlands between denudational plains and depositional lowlands in marginal and platform depressions, planation surfaces pass into marine depositional surfaces covered by Neogene and Quaternary deposits, mainly in the Caspian, Black Sea and Pechora lowlands, the Dnepr–Donets depression, etc. Metcherikov (1965) called such surfaces polygenetic.

On the western European platform, several planation surfaces are developed. In the epiplatform block mountains of central Europe, Scandinavia, Scotland and south-western Europe, these planation surfaces are deformed and have a step-like arrangement. In the Caledonian mountains on the boundary between Norway and Sweden, comparatively large fragments of planation surfaces are preserved—the so-called fjell (*see* Fig. 3.3). Glacial modification of such surfaces was small. Recent investigations show that parts of this surface date back to the early Cambrian [see Strøm (1949), Gjessing (1967)].

The relief of the central European mountains was not glaciated during the late Pleistocene but was affected by cryogenic processes of the periglacial zone. Cryogenic forms, such as cryoplanation terraces, cryopediments,

Fig. 3.3 Transition from the Palaeic surface of the young fiord valleys. In the middle distance: Fortunsdalen, western Norway. (Rudberg)

boulder fields, etc., were developed in the cold periods of the Pleistocene. Despite this remodelling, remnants of tropical weathered Mesozoic and Palaeogene mantles are preserved in some tectonic depressions.

In the peripheral part of the Paris Basin—in Saarland—remnants of an old post-Hercynian planation surface occur. Exhumed surfaces are often developed in the highlands of the recent west European platform [see Tricart and Cailleux (1965)]. Many geomorphologists [e.g., Bakker and Levelt (1964), Demek (1964a, b), Gellert (1970)] are of the opinion that the basic planation surface in central and western Europe should be called the Mesozoic and Palaeogene peneplain. This cuts the Hercynian structures and has been deformed by neotectonic movements. But this basic planation surface, with its thick kaolinite weathered mantle, is now only preserved in tectonic depressions. In most areas, it was stripped of the weathered material and a basal surface of weathering was exposed, forming an etchplain. Ruwares, inselbergs and tors are typical features of this etchplain.

Younger than the basic planation surface and the etchplain are the planation surfaces developed on the west European platform during the Neogene. Such planation surfaces can be found in uplands or in the epiplatform highland areas. In the Paris Basin, they developed during the Oligocene–Miocene on quarzitic limestones (meulières), and have many features in common with planation surfaces in the Kalahari Desert. Differences in number, age and extent of planation surfaces in areas of the western European platform can be explained by

different ages of neotectonic movements and subsequent rejuvenation of the epiplatform mountains. Demek (1964b) has tried to reconstruct the palaeoclimatic conditions during periods of planation based on an analysis of weathered mantles in the epiplatform mountain areas.

The number of Neogene and early Quaternary planation surfaces varies in different epiplatform mountain areas of central and western Europe: for example in the Thüringerwald there are four [Krähahn (1964)], in the Harz three [Machatchek (1955)], in the Bohemian highlands three [Demek et al. (1965)], in the Hunsrück and Eifel two [Stäblein (1968), Mosler (1966)], in the Ardennes, Vosges and Massif Central two [King (1962)], in central Spain three [Gladfelter (1971)], etc. These planation surfaces are dated differently, but appear mainly to belong to the Oligocene and Miocene.

The Upper Pliocene planation surfaces in eastern and central Europe are represented by valley pediments. In Poland, Radlowska (1970) described wide valleys with pediments in Góry Świętokrzyskie, the Lublin plateau and the Sandomierz highlands. Narrow Upper Pliocene pediments around the central Hungarian mountains were described by Pécsi (1965). Two valley Pliocene pediments are developed in the epiplatform mountain areas of Bohemia [see Demek (1976b)]. Miocene volcanics in the western part of Czechoslovakia were bevelled by the so-called postbasaltic peneplain, probably during the Pliocene–early Pleistocene Periods [see Moscheles (1920), Richter (1963)].

Planation of the epigeosynclinal mountains of Europe

was not so perfect as in the platform areas. Fragments of three or four planation surfaces can be found. They are an important tool for the investigation of neotectonic movements, but less so for studies of exogenic relief. Extensive planation surfaces were, however, developed in the Carpathians. In the eastern Carpathians, these surfaces still form the uppermost level of the mountains and are called *poloniny*. The oldest planation surfaces in the western and eastern Carpathians are dated Upper Miocene (Sarmatian) [see Klimaszewski (1965), Starkel (1969), Czudek *et al.* (1965), Cys (1966)]. In Romania, the oldest planation surface is Eocene [see Mihailescu and Niculescu (1967)]. Two younger planation surfaces have been mapped in the southern Carpathians.

In the Balkan Mountains of Bulgaria, the oldest planation surface is Miocene [see Vaptsarov and Mishev (1978)]. The Upper Pliocene planation surfaces in the Carpathians and the Balkan Mountains consist of pediments, mainly valley pediments. In southern Europe, the development of these valley pediments can be correlated with the Levantine transgression of the Mediterranean Sea.

The denudation chronology of the Crimea shows certain differences. The highest part of these epigeosynclinal mountains—the main ridge of the Crimean mountains—shows a summit surface. An extensive planation surface was deformed by neotectonic movements and its fragments can be found at elevations from 250–300 to 900–1100 m. This planation surface is early Cretaceous and inselbergs rising above it are Upper Jurassic. Younger planation surfaces are also extensive. Well-preserved planation surfaces are typical features of the Crimea.

In other epigeosynclinal mountains, such as the Alps and Caucasus, the periods of planation were shorter and extensive planation surfaces only developed locally.

3.3 Spatial Distribution of Exogenic Forms and Zonation of Weathering Processes

In Europe, many types of exogenic relief can be distinguished, although many are local. For instance, cryogenic forms are mainly developed in the northern and north-eastern areas, while arid forms are mostly restricted to the south-east.

Extensive areas of Europe were glaciated during the Pleistocene, with large areas belonging to the periglacial zone in which cryogenic processes were active. Traces of fossil periglacial features can be found over most of Europe, although mostly in the mountains of southern areas. Fossil glacial and periglacial forms were more or less modified by younger fluvial and weathering processes.

A complex combination of fossil and contemporary forms is typical of the present-day relief of Europe. For instance, in regions of late Pleistocene glaciation, the glacial forms have been modified by the cryogenic and slope processes of the late Pleistocene and early Holocene, as well as by the fluvial and slope processes of the middle and late Holocene. In areas of older Pleistocene glaciation, a complex denudation chronology results from the sequence of climatic changes and the accompanying changes in the denudational systems. Oscillations between colder and warmer conditions also characterized the Pleistocene periglacial zone. For instance, in the Paris Basin, three main morphostructural complexes and two transitional complexes have been postulated.

It can be seen, therefore, that the relief of Europe is polygenetic in character. Often it is difficult to define the dominant geomorphological process in the development of the relief. Nevertheless, many authors argue that the present-day relief of Europe is of predominantly fluvial origin [see Gerasimov and Metcherikov (1967), Tricart and Cailleux (1965), Büdel (1977)]. Fluvial features, especially river valleys, are an important part of the relief of Europe. Particularly significant are the valleys of such great rivers as the Volga, Don, Dnepr, Danube and Rhine, and their extensive alluvial lowlands, such as the lower Danube, Hungarian and Lombardy plains. Fluvial forms are also highly developed in the mountains and piedmont areas.

Weathering processes in Europe are controlled, as in other parts of the world, by climate on the one hand and by the various rock properties on the other. Processes of lateritic weathering, involving release of iron compounds and reduction of the quartz content, contrast with processes of podzolization where iron compounds are leached. Drainage conditions in the rock or soil are important, for they intimately affect the processes of hydration and hyrolysis. During the weathering of quartz, feldspar, mica and amphibolite, there is a tendency to form more stable minerals, such as kaolinite, montmorillonite, vermiculite, gibbsite and chlorite. Under favourable conditions, weathering mantles composed of such minerals can reach thicknesses of 4–5 m, occasionally even 10 m [see Dejou *et al.* (1977)]. The formation of such residual mantles and the weathering of the bedrock are fundamental to preparing the way for other geomorphological processes of denudation. An important secondary mineral in these weathered mantles is kaolinite, which is closely associated with eluviation and the formation of clay minerals. Deposits of kaolinite and/or gibbsite are typical of sites showing deep weathering in the humid, temperate regions of western Europe.

The hydroclimatic conditions of eastern Europe are different. Because the ground is frozen for longer periods, chemical weathering is less intensive than in western Europe, and mechanical weathering, especially that due to frost, assumes greater importance. The lowlands of western Russia are characterized by hydrous mica and montmorillonite. Dobrovolskiy (1969) explained this association as an influence of Pleistocene glaciation. In terms of chemical composition, enrichment by silica and aluminium is typical. At the present-day, weathering processes in the north and south of the Russian lowlands are quite different. To the north of the forest/steppe boundary, a more humid type of weathering prevails, while to the south a more arid type is found.

In the Mediterranean area, chemical weathering is less intensive during the dry summer period and soil-forming processes are slow. Soil erosion and deflation are the main reasons for the development of only thin weathered

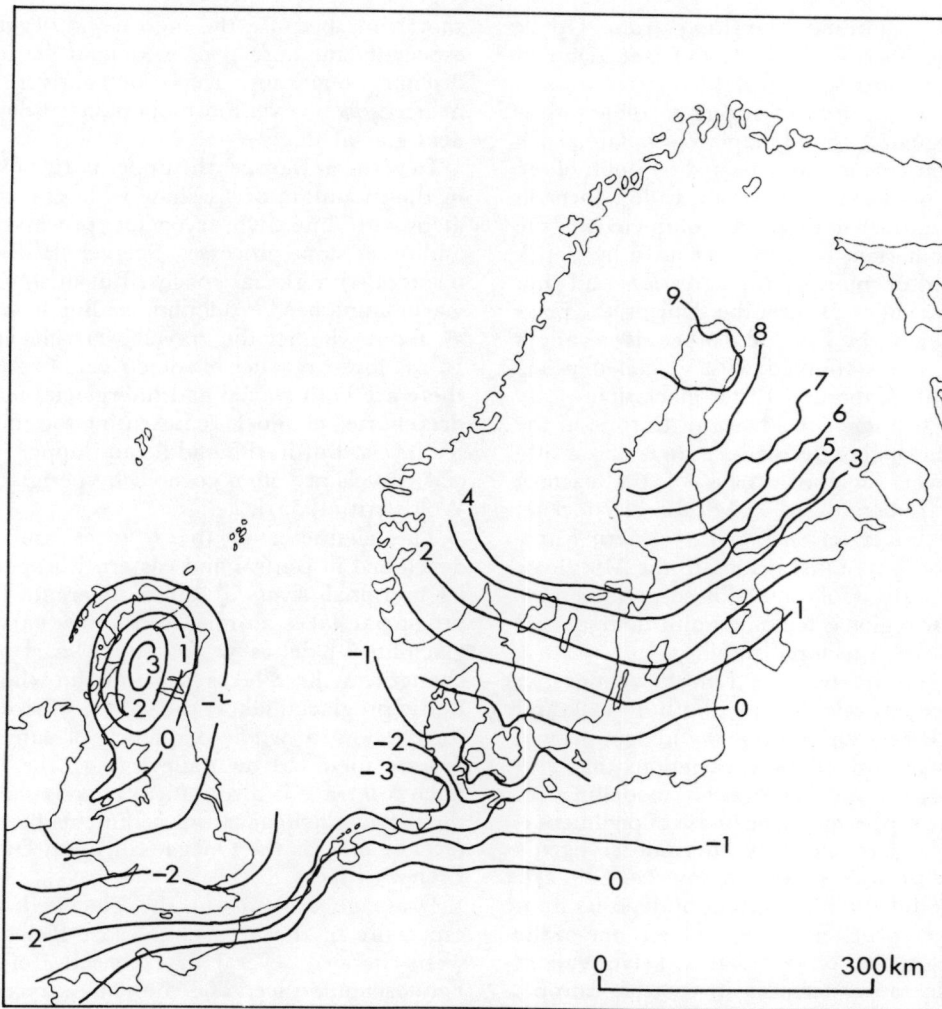

Fig. 3.4 Present-day vertical movements, based on tide gauge data, for north-west Europe. The isolines show the pattern of uplift or subsidence; the rates are given in millimetres per year [after West (1968)].

mantles. Red and yellow soils are typical, with sierozems in some places such as the Meseta. On calcareous rocks, terra rossa is developed. In the mountains, intensive accumulation of rock debris can be observed, partly cemented into breccias. In most dry areas experiencing occasional heavy thunderstorms during summer, caliche is formed.

3.4 Fluvial Landforms

3.4.1 River Valley Formation

Large river valleys in different parts of Europe are related to the morphostructural development over comparatively long periods (i.e. from the Mesozoic to the Holocene). However, river patterns changed substantially during this time and most valleys in the nonglaciated part of Europe can be regarded as Pliocene. In the Upper Pliocene, broad-valley pediments now bounded by the highest river terraces developed. The deep-valley incisions of pre-Pliocene age are only partly correlated with contemporary river valleys [see Aseev (1978)].

In the early Pleistocene, valley incision in both the nonglaciated and glaciated parts of Europe began. Deeply incised river valleys of preglacial age can be traced across the shelves of the Atlantic and Arctic Oceans. Their preservation is explained by repeated incision during eustatic movements of sea level in the Quaternary. On land, the old river channels were filled by early and middle Pleistocene fluvial deposits, by deposits of the Holstein and Eemian marine transgressions in nonglaciated areas, and by glacial and glaciofluvial deposits intercalated with marine interglacial deposits in the glaciated areas.

The infilling of early Pleistocene incised valleys by deposits of different age and genesis was aided by the general tectonic subsidence of Europe, which continued with some breaks up to the Eemian. Tectonic subsidence began at different times in different regions. Continuous subsidence beginning in the Upper Pliocene is typical of the marginal depressions of the North Sea, the Baltic Sea and some others (*see* Fig. 3.4). In such regions (e.g., the Rhine rift valley, the western Baltic and the upper Kama river depression) a continual transition from Upper Pliocene to early Pleistocene can be found [see Zagwijn (1960), Szafer (1954), Goreckiy (1964)].

The North Sea was an area of strong tectonic subsidence from the Pretiglian to the Menapian [see Pannekoek (1956)]. This subsidence was compensated by uplift of the western European platform relative to valley incision there. The incision of the Ur-Elbe was completed by the beginning of the Cromerian, because, as shown by Cepek (1967), the Cromerian marine transgression did not penetrate far into the valley of the Elbe. During the post-Cromerian, the valley of the Elbe, like other river valleys in the glaciated area, was infilled with glacial deposits beginning with the till of the first Elster glaciation.

In relation to the tectonic subsidence of Europe in the early Pleistocene, the higher terraces, that is those 30–50 m above the river, in the valleys of the eastern European platform are now dated as Middle Pleistocene. Therefore, all the younger terraces in the eastern European valleys (three or four) must belong to the Middle or Upper Pleistocene or the Holocene. These periods were characterized by new regional tectonic uplift interspersed with some short phases of tectonic stability.

The composition of the terraces reflects changes in climate. The terrace gravels include both fluvial and periglacial deposits; the various erosional and depositional phases were conditioned by climatic variations that also affected the erosional and cryogenic modelling of watershed areas. The Upper Pliocene phase of pedimentation was succeeded and replaced by alternate cryogenic and humid erosional processes [see Dedkov *et al.* (1977)].

River valleys were filled with a surfeit of deposits from the surrounding watersheds and slopes. This is one of the reasons for the development of asymmetrical river valleys with extremely wide sandy terraces in eastern Europe. The cryogenic processes under periglacial conditions supplied the rivers with large amounts of material from the slopes [see Makkaveyev (1955), Goreckiy (1958), Aseev (1963)]. Deposition diminished downstream because of river mouth incision due to glacioeustatic fall of sea level or because of rising lake or sea levels: for instance, the incursion of the Caspian Sea into the Volga River delta. In this way, the lower reaches of the eastern European rivers developed independently. A similar situation has been noted by Tricart (1949) [see Tricart (1965)] in the Paris Basin.

More active erosion in the upper reaches of the rivers was caused by irregular distribution of precipitation during the year and by the retreat of the forest cover at the beginning of each glacial epoch. Later, however, the amount of material supplied to the river beds diminished, due to cryogenic hardening of slopes and slower rates of downslope transport into the valleys. At the same time, discharge diminished due to a more arid climate, although the rivers emerging from the glaciers were locally overloaded with sediment and were unable to incise themselves.

At the beginning of the interglacials, the eastern European rivers started to incise their beds due to more humid conditions. This incision was interrupted after a short time because the increase in vegetation cover and especially the spread of woodland caused a decrease in discharge once more. Deposition of river sediments recommenced, to be overlain by periglacial deposits during the next glacial stage.

In western Europe, the upper parts of valleys beginning in the mountain areas show wide gravel terraces associated with intensive cryogenic processes, cryoplanation and other slope processes. Soergel (1919) correlated these terraces with glacial epochs. But in some terraces only a warm fauna has been found, leading to lengthy controversy about whether the gravels were glacial or interglacial. In the lower reaches of some rivers (e.g., River Thames), there are both glacial and interglacial terraces, as well as deep buried channels representing the effects of low glacial sea levels. But in the middle and upper reaches, interglacial gravels are often covered by periglacial deposits [see Woldstedt (1955)].

The character of the terraces and fluvial deposits developed in central and eastern Europe by proglacial or ice-marginal rivers is quite different. In many places, proglacial lakes formed, in which varves, gravels and glaciofluvial deltas were laid down. Due to water-level changes in these lakes (e.g., in the whole system of the Baltic proglacial lakes) the valleys of rivers emptying into them show a whole sequence of late glacial terraces despite their narrow valleys (e.g., the Daugava River). Such a terrace is also typical of river valleys incised into the recent glaciolacustrine sediments bordering the Arctic Ocean, such as the Onega, Northern Dvina, Mezen' and Pechora Rivers.

Drainage systems in the glaciated part of Europe, especially in the area of the last Pleistocene glaciation, were changed several times because of blocking by the Fennoscandian ice. The contemporary river patterns are quite different in comparison with the buried preglacial valleys. The most complicated changes took place in the Elbe, Oder and Vistula (Wisła) systems. Along the ice margin, there developed wide, valley-like depressions (known as *pradoliny*, or *Urstromtäler*) due to discharge of glacial meltwaters: for example, the meltwater channels of Lausitz, Berlin, Eberswalde, etc. These meltwater channels largely used old river valleys running west–east. Cryogenic processes, including the formation of massive icings, helped to form these wide channels, which are now used by contemporary rivers.

In eastern Europe, the upper reaches of the Bug River in the uplands of Pripyat Marshes (Belorusskoye Poles'ye) were captured by the Pripyat River, and the upper reaches of rivers originally flowing northward in the upper Volga lowland were captured by the Volga River. These examples of capture are, however, older than the last glaciation.

3.4.2 Regional Features of Fluvial and Slope Processes

There are few data about regional differentiation of fluvial landforms with respect to the climate in Europe. Some authors [e.g., Stoddart (1969)] do not accept that such a differentiation exists, but an attempt will be made here to

Fig. 3.5 Gravel-filled valley and braided river, Italian Alps north of Gemona; photograph taken in summer low-flow season. (Embleton)

show that there are regional morphoclimatic differences.

In areas near the Atlantic coast possessing a humid climate, river discharge is regular and wide, open river valleys with graded slopes are typical. On slopes, processes of creep, debris flow and earth slides are most important, giving long, graded slopes. Specific colluvial (slope) deposits called doublets consist of loamy deposits without apparent stratification and with little variation in thickness. Doublets originated in humid periglacial conditions in noncalcareous loessic loams by migration of clay and iron particles. The same process of formation probably applies to the so-called cover loams in the central and northern parts of the Russian plain. In the composition of these apparently non-stratified loams of western Europe, a higher silica content is apparent in places (e.g., Boulogne, Artois, etc.) forming concretions in the loams. Long-continued river erosion has emphasized the structural differences producing, for example, the escarpments and vales of south-east England, the Paris Basin, the Schwäbische Alb and Fränkische Alb.

The hydroclimatic conditions of eastern Europe are very different because of soil freezing, decreased river discharge in winter and floods in spring. In comparison with western Europe, mass wasting plays a less important role than linear erosion. The greater intensity of mechanical weathering, the widespread presence of loams and sands, agricultural cultivation and deforestation all encourage linear erosion and the formation of gullies. Many eastern European geomorphologists hold the view that high-density drainage systems developed during a prolonged phase of relief development, but that badlands and gullies formed rapidly during the short historical period under the influence of man [see Gerasimov and Metcherikov (1967)].

Depending on lithological conditions and tectonic regime, the character of erosional dissection differs considerably across Europe. Hilly lands are typical of the

northern regions of the central Russian uplands and long, rounded ridges of the southern regions. Narrow watersheds are common in the northern regions of the Volga Heights and broad watersheds with monadnocks in the southern regions. Structural scarps and mesas are typical features of the Vysokoye–Zavolzhsk and Obshchiy Syrt plateaus. Under the controlling influence of seasonal discharge, gelifluction, earth sliding and suffosion (i.e. piping) are locally present. Convex–concave slopes, foot-slope colluvial deposits and badlands are common. Generally, erosional dissection in the continental regions of Europe is much greater than in the more maritime areas of the Atlantic coast.

In the subtropical Mediterranean, the most important morphoclimatic factor is the alternating dry and wet periods, together with the nature of the precipitation. Drainage patterns are dense and deeply incised, especially in areas of sedimentary rocks. Badlands, landslides and dells are often found, while erosion has stripped bare many rock surfaces, especially in karst regions of Yugoslavia. The larger Mediterranean valleys are often wide with braided rivers, whose beds are underlain with thick gravels (*see* Fig. 3.5). Overloaded by coarse material, the rivers swing laterally and undercut the valley sides.

In the high mountains around the Mediterranean, geomorphological processes are very active but, in addition to the features of humid regions, there are some distinctive forms at the foot of these mountains. Mountain slopes in the interior of Spain are often covered by sandy and gravelly mantles coloured red or red–yellow. Bajadas at the foot of the mountains are more than 100 m thick and dissected by rivers. Machatchek (1955) described Neogene pediments that developed under climatic conditions with large variations in discharge during the year. Dissection by rivers took place in the Quaternary.

The Appennines are densely dissected by erosion with comparatively wide watershed ridges, moderately steep

Fig. 3.6 Karst landscape in the Crimean mountains. (Wagner)

slopes, wide main valleys and narrow side valleys. The floors of the main valleys are filled with river gravels, which are slowly moving downwards. Farther to the south, the river regime becomes more strongly seasonal, being reflected in the seasonally dry river beds (fiumare, torrents). In the sedimentary plains of Tertiary rocks, wide, rounded parallel ridges, badlands and landslides are typical.

Similar relief features are found in the eastern Crimea. Mudflows, which are common on the southern slopes of the Crimean mountains in shales, occur every two to three years, with the transport of large amounts of material. River valleys are comparatively short, but alluvial cones at their ends have volumes of several thousand cubic metres. Badlands are common on the slopes, while extensive glacis have developed at the foot of the mountains.

In the fold mountain zones of Europe, such as the Pyrenees, Alps and Caucasus, erosional forms developed on Pleistocene glacial and glaciofluvial deposits (e.g., in Bavaria and Austria). In the Alpine zone, erosional features of glaciation predominate, such as cirques and troughs. Most of the glacial troughs follow pre-existing river valleys and are structurally controlled.

3.5 Karst Landforms

Karst processes dominate the landscape of several regions of Europe. They are particularly active in the Mediterranean area due to the extensive areas of limestone bedrock, the lack of superficial deposits or deep soils, and the numerous faults and joints [see Sweeting (1972)].

The most highly karstified area is in the Dinaric Alps, especially in the western part. Large karst depressions (*poljes*) are controlled by faults. The subterranean hydrography is complex; the groundwater level serves as a base level for karst erosion and continuously changes in accordance with neotectonic movements. Karstification is controlled not only by lithology and tectonics, but also by climate, thus affecting the solution processes. Besides poljes, other karst features are common, including dry valleys, sinks, caves and lapiés. Bare karst surfaces form several steps inland from the coast of the Adriatic Sea up to the high mountains or karst plateaus at elevations of 800–2000 m.

In the western parts of the Dinaric Alps, four karst regions are to be found in (1) Slovenia–Croatia (typified by many sinks), (2) Western Bosnia (typified by large poljes with sinks), (3) Dalmatia (containing both karstic and non-karstic subregions) and (4) Montenegro (with dry valleys and small poljes).

The typical soils on the karst plateaus are terra rossa. In the central part of the Dinaric plateau, there is a typical alternation of limestones and shales with sandstones, giving rise to covered karst with sinks. The superficial deposits consist of brown loams as the products of weathering. In the eastern part of the Dinaric plateau, karst forms are rare.

Karst forms are also characteristic of much of Greece and the Aegean islands, including Crete, but they are not always so well developed as in the Dinaric Alps. Other areas where karst forms are common include Bulgaria, especially in the south-west and Dobrogea, as well as parts of Italy, Spain and Portugal.

Regions of karst morphosculpture with glacial forms form a rim around the central part of the Alps. Fine karst forms occur in the limestone French Alps, well known among speleologists for their deep avens (up to 658 m deep). Also famous are the caves (some being of archaeological interest), karst canyons, vauclusian springs, etc. The second largest karst region in the Alps is the northern limestone Alps with typical bare rock surfaces of karst plateaus and high mountain (Alpine) karst forms.

The Carpathian mountains are poor in karst regions, the only exception being the southern Slovakian karst on the Czechoslovakia/Hungary boundary, with its plateaus, canyons and long caves.

Extensive karst plateaus are developed in the Crimean mountains (*see* Fig. 3.6). On the Upper Jurassic limestones, all of the bare karst forms (i.e. plateaus with many sinks, lapiés and other small karst forms) can be found. Such karst massifs as Chatyr Dag and Karabiyayla with their moon-like landscape are classic examples. Karst forms, including deep caverns and high mountain (Alpine) karst, are known in the Caucasus.

In many regions of the platform part of Europe certain forms of karst have developed known as 'grassed' karst—a term introduced by Gvozdeckiy (1954)—and covered karst. The best examples in western Europe are the areas of the Grande Causse and Petite Causse in the south of the Massif Central. Karst plateaus are dotted with numerous sinks, on whose floors thick soils have accumulated. Grassed and covered karst have also developed on the Mesozoic rocks of the Schwäbische Alb and Franconian plateaus and cuestas, on the Devonian limestones of the Moravian karst around Brno and on the Mesozoic rocks of the Wyzyna Kraków–Częstochowa plateau. Covered karst has also developed in some areas of calcareous rock in the London and Paris Basins.

The karst areas of the Russian platform provide a sharp contrast. The forms are controlled by lithology, dip and the whole morphostructural plan of the area. The most common types of karst are grassed and covered karst on limestones, dolomites, gypsum, etc. [see Chikishev (1965, 1978)]. Covered karst is common in parts of the Baltic region, in the central Russian uplands, Volga Heights, Dnepr uplands, the low Crimea, etc. Numerous sinks have been formed due to water carrying sand and silt particles into solution fissures and caverns. In gypsum, sinkholes have often originated due to subsidence of cavern roofs. Flat-floored depressions (200 m diameter and 2 m deep) are a characteristic in the Podolie area; under such depressions, large caves have been found. Among them is Ozernaya, which is the largest cavern (26 400 m long) in the world developed in gypsum. Karst processes in gypsum and anhydrite are very active and sometimes dangerous. Grassed karst is also widely distributed over the Russian platform, the most notable surface features being depressions, especially sink holes of varied dimensions. Bare karst can be found in the Donbas Basin, in the uplands of Zhiguli and in the Caspian lowland. Very interesting forms of salt karst have developed in the basin of the Tissa River. Rock salt comes to the surface here, forming outcrops with lapiés, pyramids, tor-like features, etc. Sink holes are also common.

3.6 Fossil Exogenic Relief of the Pleistocene Glaciation

3.6.1 Glacial Relief of the Lowlands

An ice sheet covered large areas of north-western and central Europe during the cold periods of the Pleistocene (*see* Fig. 3.7). The ice spread out from two independent centres: Scandinavia and the Barents Sea. The landforms of the Lower Pleistocene glaciations were considerably modified by periglacial and fluvial processes during later stages. Features of the Middle Pleistocene glaciations (Warthe or Moscow stage) have been preserved, and the youngest Upper Pleistocene glacial forms have only been slightly altered by later erosion [see Flint (1971), Aseev (1974)].

Glacial erosion forms—roches moutonnées, drumlins with massive rock cores, fissure-controlled valleys and glacial microrelief (such as striations and grooves)—are well developed on the slopes of the Fennoscandian Shield in crystalline rocks. Forms of glaciofluvial origin include potholes, gullies, etc. A complex mosaic of glacial and glaciofluvial erosion and deposition forms developed as described by Rudberg (1973) and Charlesworth (1957). In the central part of the Fennoscandian Shield, glacial forms are less distinctive due to the cover of stony till. In the Scandinavian highlands, forms of mountain glaciation developed.

In the zone of glacial deposition on the northern European lowlands, the forms of glacial erosion include rock basins (i.e. Zungenbecken) and other depressions of glacial moulding. Such forms are common in Germany, Poland and the north-western part of the Russian platform [see Richter (1937), Gripp (1952), Reinhard (1965), Bartkowski (1968), Goreckiy (1967), Aseev (1974), Basalikas (1969)]. The valley floors and basins overdeepened by glacial erosion are separated by morainic hills and ridges, built up by the repeated advances of the ice. The pattern of morainic hills and ridges is controlled partly by the subglacial bedrock relief and partly by glaciological factors. Push moraines formed at the margins of certain glacier lobes, which contain in their cores blocks detached from bedrock or rafts of older glacial or sedimentary deposits (e.g., in southern Sweden, northern Germany and the Kirillov and Torzhchov depressions in the USSR). Less common are the long furrow lakes and eskers which mark subglacial stream action and are formed in association with stagnating or less active glacier lobes.

Along the ice margins, there are dissected zones of ice-contact deposition, including push moraines, outwash plains, etc. This zone of ice-contact deposition usually formed in places where the bedrock surface is higher (e.g., Fläming, Baltic Ridge, Valdai Hills, Klin–Dmitrov Ridge). The thickest glacial deposits and the highest glacial relief are found commonly at the junctions of two glacier lobes; their end moraines and push moraines have been studied by Gripp (1952) and Basalikas (1969). The margins of dead ice are characterized by hummocky moraines and kames, the latter sometimes being connected with outwash plains (e.g., the Mecklenburg lake plateau in Germany).

Forms of glacial erosion and deposition are also preserved on the shelves of the Norwegian, North and Barents Seas, and on the continental slope [see Hoppe (1972), Aseev and Makkaveyev (1976)].

The glacial relief of the Saale (Middle Polish, Dnepr) ice margins is preserved in the form of push moraines

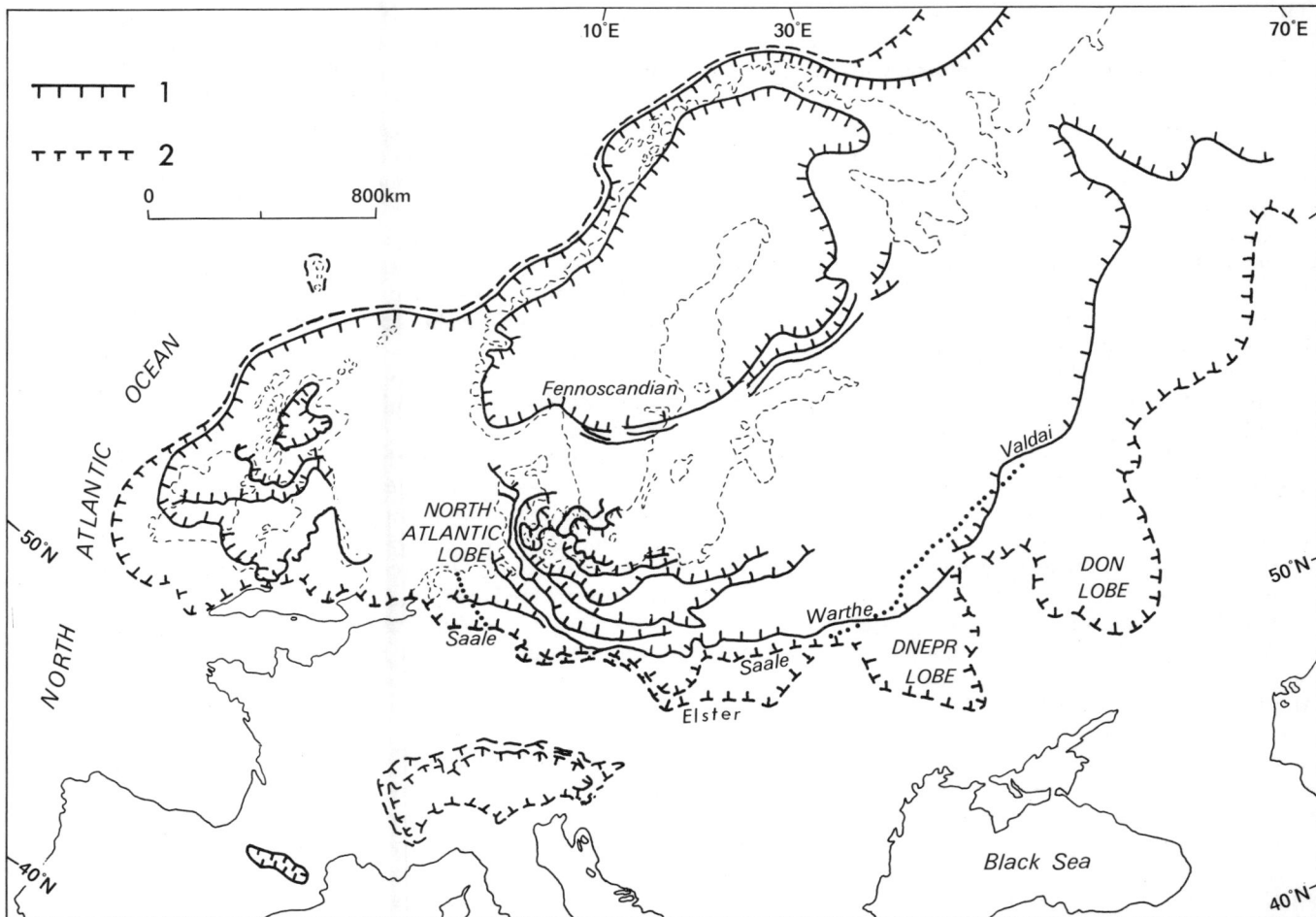

Fig. 3.7 Pleistocene glacial limits in Europe: (1) limit of the last glaciation (Würm, Weichsel, Valdai, etc.) showing retreat or readvance limits after the maximum; (2) older glacial limits.

(e.g., in the Netherlands, Germany, Poland and in the Dnepr lobe). Even better preserved are the terminal features of the Warthe glaciation in central Europe, which correlate with the Moscow glaciation in eastern Europe. The moraine systems of this age suggest several stages of deglaciation [see Woldstedt (1950), Gellert (1958), Bartkowski (1968), Aseev (1974)].

In the Upper Pleistocene, the maximum ice limits and various stages of recession (e.g., Brandenburg, Frankfurt, etc.) can be distinguished. The last is the Salpausselkä zone of terminal moraines in southern Sweden and Finland. A zone of terminal features marking a new transgressive stage of glaciation [see Aseev (1974), Chebotareva and Makaritcheva (1974)] based also on some stratigraphic, geomorphological and lithological data—an independent Upper Pleistocene glaciation described by Serebryanniy (1978)—is termed the Pomeranian, Pomorian, Vensovian or Main stage.

3.6.2   Mountain Glaciation
For the older stages of mountain glaciation in Europe, the glacial deposition forms are poorly preserved, even more so than for lowland areas. Nevertheless, it is known that the glacial features in the mountains were formed in several glacial periods beginning with the Lower Pleistocene. The forms of mountain glaciation can be divided into three groups: (1) a complex of Alpine forms that originated by valley glaciers (Alpine type), (2) a complex of forms associated with ice-cap development and (3) a complex of forms associated with piedmont glaciation. In some European mountain areas, all three types are present.

The forms of true Alpine glaciation can only be found in the Alps and Caucasus. In the Upper Pleistocene, however, Alpine-type glaciers also developed in other mountain systems. Geomorphologically, the most distinctive glacial landforms are those of Upper Pleistocene or Holocene age. Some geomorphologists have been of the opinion [e.g., Garwood (1910)] that the dissected relief of glaciated mountain areas resulted from fluvial erosion and that ice sheets or glaciers played a mainly protective role; this opinion is not accepted generally. The distribution of mountain glaciers—especially valley glaciers—was, however, controlled by the lines of the ancient (preglacial) valleys.

The most common features of mountain glaciation are glacial troughs, hanging valleys, cirques and arêtes. Some high interfluves or summits may have stood up above the general level of the ice as nunataks. Maps of the glaciation

Fig. 3.8 Aurlandsfjord, a branch of the Sognefjord in Western Norway. (Rudberg)

of the Alps (e.g., the national atlases of Italy and Switzerland) show surfaces of bare rock not covered by the Würm ice extending over comparatively large areas.

Glacial troughs (i.e. glaciated valleys) represent mainly river valleys of preglacial age that were remodelled by glaciers becoming U-shaped. Some geomorphologists attribute most of the excavation to glacial ice, often citing the Norwegian fiords and their deep rock basins (*see* Fig. 3.8). A part of the deepening, however, must have resulted from fluvial (subglacial, interglacial, etc.) action and some of the lineaments are definitely tectonic. Recent studies of the Caucasus [see Serebryanniy (1978)] have also shown the considerable extent of glacial action in the formation of the high Alpine relief. Another distinctive feature of glacial troughs is the presence of steps and rock basins in their longitudinal profiles (i.e. glacial stairways). Glacial deposition forms related to different phases of glacier activity may be found within the troughs below the present-day glacier snout. They include terminal and lateral moraines, hummocky moraine and dead-ice depressions. Such forms are best preserved on the flatter parts of the trough floors.

In the Scandinavian highlands, a similar assemblage of glacial features is present, together with cryogenically modified forms. The preglacial relief consisted of a series of dissected planation surfaces (e.g., the Hardangervidda). In the western parts, Alpine-type relief is present, with high peaks, cirques, ice caps and some valley glaciers. Large areas are still glaciated, such as Jostedalsbre, Svartisen, Folgefonni, etc.

In Iceland, the forms of mountain glaciation are developed in, and combined with, the volcanic relief. Basal tills and glaciofluvial deposits cover many of the basaltic plains and plateaus. Cryogenic processes are limited in spite of the northerly latitude of the island. Large ice caps exist, some covering active volcanoes: for example, Vatnajökull, Langjökull, Hofsjökull and Mýrdalsjökull.

Features of glaciation are well displayed in the Scottish Highlands, the Faröe Islands and other island groups of the north-eastern Atlantic Ocean. The mainland of Scotland supported several major centres of glaciation. River patterns were changed and deep troughs (glens) were eroded. Also characteristic are the broad 'straths' (lowlands) marked by glacial and glaciofluvial deposition and by raised marine deposits. Wind gaps due to stream piracy and to watershed breaching by ice are numerous.

The Alps were one of the largest centres of glaciation in Europe during the Pleistocene Epoch. Glacial sculpture is an important element of this mountain system, despite the fluvial remodelling of older glacial forms. The chronology of glaciation in the Alps was first investigated in detail by Penck and Brückner (1909), who showed that glaciation was repeated several times and that no continuous ice sheet developed. Even during the maxima of glaciation, many sharp ridges were not covered by ice, thus leading to the formation of ice-free watersheds. Various types of cirque and valley glacier dominated and produced the typical features, such as arêtes, pyramidal peaks and glacial troughs. There is no one cirque floor level in the

Alps. The highest level of cirques is formed by wide, open forms, above which only peaks rise. A lower level is formed by cirques opening on to troughs. Glaciated valleys often take the form of glacial stairways, with many rock basins separated by steep steps.

Present-day glaciation is concentrated in the western and central Alps, especially in the glacier systems of Mont Blanc, Monte Rosa, Matterhorn, Finsteraarhorn, etc. In the eastern Alps, the main glaciers are in the mountain groups of the Hohe Tauern, Stubaier Alpen, Ötztaler Alpen, etc.

The extent of Pleistocene glaciation in the Carpathians was substantially less than in the Alps. Valley glaciers developed in the highest parts of the mountain system only, producing the relief of the High Tatra in Czechoslovakia and the Făgăraş massif in Romania. In the High Tatra, a whole complex of mountain glaciation forms is preserved—especially troughs, cirques, roches moutonnées, terminal moraines, etc. The most favourable conditions for the development of glaciers were in regions with preserved remnants of Tertiary planation surfaces in the upper reaches of ancient river valleys. The planation surfaces during the Pleistocene were often above the snowline and formed areas of snow and ice accumulation.

In the central Caucasus, glacial sculpture and the centres of present-day glaciation are concentrated in the most uplifted central parts of the mountain system. Cirque glaciers, cirque stairways, glacial troughs, roches moutonnées and ridges of both young and old terminal moraines are common. Especially well-preserved are the forms dating from the Upper Pleistocene glaciation. Around the glaciers, cryogenic forms, avalanche features and thick glaciofluvial deposits are widespread.

In the Balkan Mountains, the features of mountain glaciation were first described by Cvijić (1924). According to him, there were isolated centres of glaciation during the Pleistocene, including larger centres in the Prokletije massif, the Rila Planina, and Stara Planina, where cirques and troughs can be found. Small cirques exist in the Olympus Mountains and in the mountains of Thessaly.

The Appennines lack any well-developed glacial forms. The snowline during the Pleistocene cold periods was at an elevation of about 1900 m and glaciers developed only in the highest groups, such as the Gran Sasso, Monte della Maiella, etc. In these areas, small troughs, cirques at several altitudes and roches moutonnées can be found.

In contrast, the glaciation of the Pyrenees, especially in the central parts, was on a much larger scale. The main ridge is in the central area divided by many cirques with some present-day glaciers. Especially large glaciers developed in the massifs of Pico de Aneto, Pico Posets and Monte Perdido, from which glaciers moved down the northern slopes. Glacial troughs begin in extensive amphitheatres such as the Cirque de Gavarnie in the upper reaches of the Gave de Pau. Cirques are associated with several levels and mostly contain lakes. On the northern slopes of the Pyrenees, glaciers descended to as low as 400 m but on the southern slopes the lower limit is 900 m. Their maximum extent is marked locally by terminal moraines. The longest glacier was 70 km long in the

Garonne valley.

Pleistocene glaciers developed in some other mountain groups of the Iberian Peninsula. Penck (1883) described glacial forms in the Pyrenees where short hanging glaciers occupied cirques and short troughs on the northern and eastern slopes. Similar forms occur in the Sierra de Guadarrama, Sierra de Gredos and Serra da Estrella. The position of the snowlines was strongly dependent on climate. On the northern and eastern slopes of the Cantabrian mountains, which were exposed to moist Atlantic air, the snowline dropped to as low as 1300–1500 m, but on the southern and eastern slopes it lay at 1700 m. In the Sierra de Guadarrama, the position of the snowline has been reconstructed at 2050–2100 m, in the Sierra de Estrella at 1620–1650 m, and in the Sierra Nevada (Granada) at 2400 m on northern slopes and 2600 m on southern slopes. Solé-Sabarís and Llopis (1952) and Lotze (1962) have examined the Pleistocene glaciation of the Cantabrian mountains and distinguished a special 'Pyrenean' type of glacial sculpture in which small cirques but no troughs are characteristic.

3.6.3 Cryogenic Features

Cryogenic forms are widely distributed throughout Europe. Contemporary cryogenic activity occurs in the north-eastern part of the eastern European (i.e. Russian) plain, where thaw lake (thermokarst) landscapes caused by permafrost degradation are developed. Cryogenic forms also occur in the mountainous areas of the northern Urals. On the mountain slopes, especially their lower parts, gelifluction processes are active, causing the development of sorted and nonsorted terraces. On the middle and upper segments of slopes forms, such as debris stripes composed of angular congelifractates, are found, which originate by intensive frost weathering and cover not only slopes but also the watershed areas. Angular blocks form Felsenmeere (block fields), frequently with stone polygons. Above the block fields rise castle koppies and tors often formed by more resistant rocks (*see* Fig. 3.9). Höllermann (1977) has analyzed the location of the periglacial zone in Europe and has studied the altitudinal relationships of cryogenic forms from the Iberian mountains to the Caucasus. This profile stretches from the present-day subtropical and mild, humid zones and shows the lower boundary of gelifluction forms in the western part at a height of 1950 m and in the eastern part (i.e. central Caucasus) at 2430 m. The highest position of the lower gelifluction boundary was observed on the leeward sides of mountain ridges with less precipitation and also in massive mountain groups, such as the Pyrenees and central Caucasus.

The zone of contemporary cryogenic relief, in the same way as the zone of permafrost distribution, is but a small part of the great cryogenic zone of the Pleistocene Epoch. Data about the natural components of the Pleistocene cryogenic zone—relief, vegetation and animal life—and on the age of the cryogenic phenomena led Velitchko (1968) to conclude that the whole of the present-day mild, humid zone was, during the Pleistocene Epoch, part of a great cryogenic zone. The maximum of continentality,

Fig. 3.9 Frost-riven granite cliff and (at right) cryoplanation terrace with angular debris, in the southern part of the Bohemian–Moravian highland, Czech massif, at about 655 m. (Demek).

lowest temperatures and greatest thickness of permafrost occurred at the end of Pleistocene, during the last glaciation (i.e. Würm, Valdai).

Studies of periglacial regions of the Pleistocene have shown that, during the periglacial conditions, underground ice formed in lenses and ice wedges. These areas of ground ice were located essentially in the nonglaciated zone, especially in the case of the Russian plain. In many areas, fossil periglacial forms, which developed during Pleistocene cold periods and are now remodelled by other geomorphological processes, are discernible. Cryogenic forms have been studied in detail in France, Poland and Czechoslovakia. Such forms are also known in the mountainous areas elsewhere in Europe.

Very complex fossil cryogenic features have been studied in the Russian plain by Velitchko (1968). He distinguished several zones with special combinations of cryogenic forms. The first such zone lies between the boundary of present permafrost distribution in the north and the latitude of Moscow, reaching a little farther south in the Dnepr lowlands and Tambov lowlands This zone is characterized by patterned ground, thaw lakes and oval depressions reminiscent of present-day thermokarst depressions called alases. To the south of this zone, as far as the latitude of Dnepropetrovsk and Volgograd, fossil-patterned ground and many thermokarst features, especially on the junctions of polygons, occur. Some of the thermokarst features are quite large. Even farther south,

there is a zone with minor fossil cryogenic features. The thermokarst of this zone was formed by the melting of segregated ground ice and, to a very small extent, by the melting of ice wedges. The best fossil cryogenic forms are developed in silts and loams.

Fossil cryogenic features have influenced erosional processes in the Holocene Epoch. For instance, gullies (i.e. the ovrags of the Russian plain) form along cryogenic fissures marked by long depressions at the present. Sheet floods concentrate on these shallow depressions. Processes of suffosion (piping) are connected with periglacial fissures, forming wide, flat depressions (i.e. padings, flat-bottomed steppe depressions) in the southern regions of the Russian plain.

The characteristic sediment of periglacial conditions is loess [see Velitchko (1968)]. In the area of the last glaciation, only loess-like deposits can be found. The typical loess developed to the south of the boundary of the last glaciation. Southwards, the thickness of the loessic deposits increases to several tens of metres in the northern Ukraine. Even farther south, the loess thickness decreases again. In many regions with thick loess, a distinctive relief is developed. The thick loess draped over the bedrock has the effect of smoothing the watershed areas and the steps of river terraces, thus producing a smooth undulating hilly or flat relief. The loessic surface is usually smoother than the bedrock relief. Typical forms developed in loess include suffosion forms (i.e. padings, piping sink holes,

Fig. 3.10 China clay pit, near St Austell, south-west England. The granite has been deeply altered by pneumatolysis to kaolin. (Embleton)

piping caves, etc.). Gullies are another typical form of loessic plateaus.

3.7 Anthropogenic Relief

Man is a very important geomorphological agent at the present time in Europe. Every year the intensity of human activity is increasing and now ranks equally with that of other geomorphological agents. Over large areas of Europe, the natural relief has been substantially changed by anthropogenic processes.

There are several classifications of anthropogenic relief, such as those by Bondarchuk (1949), Louis (1968), Zapletal (1969) and Molodkin (1976). They differ in the level of classification and in their approach, but in general all anthropogenic geomorphological processes may be differentiated into three groups [see Demek (1977)]: (1) those affecting endogenic processes, (2) those having a direct or indirect influence on exogenic processes (acceleration or retardation) and (3) anthropogenic processes proper (i.e. technogenic processes) among which can be distinguished intentional and unintentional processes. Demek (1977) emphasized correctly that the degree of anthropogenic transformation of the natural landscape depends on such factors as density of popula-

tion, the nature of the industry and the duration of human activity. From this point of view, Europe is probably the most affected of all the continents. The degree of transformation of the relief is particularly high in industrial regions and in areas of intensive agriculture. The resulting anthropogenic landscapes can be divided into several groups [see Bondarchuk (1949)].

(1) *Agricultural landscapes* in areas where proper measures to counteract accelerated erosion are employed. Such areas are typified by smoothed relief. However, if the measures against soil erosion are not adequate, sheet-flood erosion, linear erosion (i.e. gullying) and deflation set in. Many geomorphologists believe that present-day gully patterns are of anthropogenic origin, the result of deforestation, ploughing of steep slopes, overgrazing, etc. The intensity of deflation in some fields attains such dimensions that during a single 'black storm' lasting one or two days a humus horizon 20–30 cm thick can be completely removed.

(2) *Irrigated regions* are noted for many conservational features introduced by man (e.g., dams, irrigation canals, reservoirs, etc). Introduction of these features into the landscape changes the character of the relief and of fluvial processes.

(3) *Mining areas* are characterized by a variety of anthropogenic relief forms, such as open-cast mines for coal or iron ore, quarries for limestone or other constructional material, and areas liable to subsidence where there is underground mining. In addition, the spoil heaps from such mining can reach substantial dimensions (*see* Fig. 3.10).

(4) *Urbanized areas*, especially those with dense populations, can show greatly altered relief. The processes, too, are directly affected: for example, paved or asphalted areas can produce accelerated runoff and reduced infiltration.

It is clear that, in the future, the intensity of man's influence on geomorphological processes will grow and the extent of anthropogenically altered relief will increase. The prediction and monitoring of the changes and their effects is a major task for contemporary geomorphology.

3.8 Exogenic Processes Operating on Ocean and Sea Floors and on Coasts

3.8.1 Types of Submarine Exogenic Processes

As well as tectonic processes, exogenic processes also take part in the formation of the relief of the ocean floor. Longinov (1973) identified two groups of lithodynamic processes: hydrogenic and gravitational. The same groups can also be distinguished among exogenic relief-forming processes, although the latter show much greater variety. A group of biogenic relief-forming processes, which include processes of reef formation, may also be identified. Apart from these, an important relief-forming role in the oceans and seas, which are the regions of ultimate accumulation, is played by the deposition of sediments that settle out from the water masses and bury various relief forms as the deposition progresses. This process leads to a levelling of the floor relief. The rate of sedimentation on which this levelling depends varies in relation to the type of sedimentation process and the bathymetric conditions. Therefore, it is necessary to distinguish between terrigenous, carbonate and siliceous biogenic material, polygenic sediments and sediments of the neritic, bathyal and abyssal zones.

3.8.2 Hydrogenic Processes

This represents quite a large group of processes involving different 'water movements in the near-bottom layer caused by energy supply from the water masses of the ocean' [as described by Longinov (1973), p.7]. First of all, there are wave processes caused by the action of wind upon the surface of the water; this operates almost everywhere. However, because the wave oscillation becomes damped with depth (due to the viscosity of water), the zone in which waves can affect the bottom water is very limited, being 50 m deep at most and usually much less. Apart from wind-generated waves, the ocean also exhibits such wave processes as seiches (the so-called 'beats'). Their morphological role, however, is apparently negligible.

The group of hydrogenic processes also includes the action of different currents. In 1977, Leont'ev suggested that these processes should be termed torrentogenic (from the word *torrento*—a current). It would be better still to call them fluvial but, unfortunately, this term is already applied to similar processes occurring on land. This group includes (1) tidal currents which occupy an intermediate position between wave and torrentogenic currents, (2) various wave and wind-induced temporary currents, (3) surface quasi-stationary currents and (4) bottom quasi-stationary currents.

Tidal currents perform a major role in transporting material and creating different bottom and coastal relief forms. Temporary currents play an essential relief-forming role in those cases where they are associated with wind-induced phenomena. There are many examples of bottom erosion and near-short deposition leading to the formation of wattens or sandy bars caused by wind-induced or temporary drift currents. The relief of sand bars formed by tidal currents is widespread in the *voronka* and *gorlo* (i.e. funnel and neck, respectively) of the White Sea, in the southern part of the North Sea, in the English Channel and in the Western Approaches in the Irish Sea. The near-shore areas of these seas are noted for their many extensive wattens.

The following example will illustrate the relief-forming role of the quasi-stationary oceanic currents. In the eastern part of the equatorial zone of the Pacific Ocean, between the International Date Line and the eastern Pacific oceanic ridges, there lies the largest known bottom depositional form produced by abundant plankton that populate the equatorial waters and are distributed along the equator in the form of insoluble or poorly soluble remnants.

Bottom quasi-stationary currents at great depths only started to attract the attention of geologists and geomorphologists during the last decade [e.g., see Le Pichon *et al.* (1977), Heezen and Hollister (1971), Emery and Uchupi (1973), Leont'ev (1975)]. The formation of deep-sea bottom currents is associated with the cooling of sea waters on the Arctic and Antarctic shelves, submergence of these cold and therefore dense waters, their flowing from the continental slope and their distribution over the bottom of abyssal depressions. In the northern part of the Atlantic Ocean, these currents are due to the accumulation of cold waters on the ocean floor. Within the limits of the *International Geomorphological Map of Europe*, they are associated with such sedimentary ridges as the gigantic bottom depositional forms of Feny and Gardar. Also they probably take part in the formation of abyssal valleys, such as the Danish Channel, Mori and Imarssuak.

An important role in the formation of the bottom relief of the Gulf of Cadiz is played by a current which is due to the discharge of salt Mediterranean water into the Atlantic through the Strait of Gibraltar. In the Strait and to the west, there lies a distinct linear depression eroded by this current which passes farther into a delta-like accumulation of ridge depositional forms. Smaller depressions incising the continental slope and rising in the above area are also due to the salt water discharge from the Strait of Gibraltar [see Heezen and Johnson (1969), Leont'ev (1977)]. Numerous examples of the action of these abyssal currents are quoted by Longinov (1973).

Nonstationary currents induced by wind and waves operate over practically the entire open shelf area. In certain areas of the North Sea, Irish Sea, the Aquitaine shelf and the Western Approaches, they act together with tidal currents to give rise to a complex pattern of distribution of shelf sediments as described by Stride (1965).

3.8.3 Gravitational Processes

Gravitational processes include the action of turbidity currents and submarine landslide phenomena. These processes are confined by other currents mainly to the continental slope. Another cause of these turbidity currents is the dilution of landslide masses that slump down the slope. These currents are observed frequently to flow down submarine canyons, causing deeper entrenchment of the latter. Detrital fans from turbidity currents accumulate at the mouths of these canyons. The merging of these fans gives rise to the tilted depositional plain of the continental rise. In some cases, however, very large individual detrital fans are formed: for example, the Rhône and Ebro alluvial fans in the western Mediterranean Sea and the vast Messina alluvial fan formed by the turbidity current discharging from the Tyrrhenian Sea through the Strait of Messina into the Ionian depression. The turbidity currents frequently incise the alluvial fans and form incised radially oriented linear depressions.

Submarine landslide processes are common on the continental slope and on the sides of submarine rises and seamounts. Deep seismic profiling data often show the presence of landslide phenomena in the sedimentary strata. This phenomenon was studied in considerable detail in the case of the Black Sea by Arkhangelskiy and Strakhov (1938) and more recently by Roberts (1972) who described large landslide forms on the slopes of the Rockall rise (*see* Section 1.6).

3.8.4   Biogenic Processes

Biogenic processes, while typical of the tropical seas, are not well pronounced in the seas around Europe. Only locally in the Mediterranean Sea can bryozoan or vermetide constructions be observed that can be attributed to microrelief forms. An essential role is played by biogenic factors in the formation of tidal and wind-induced wattens and their subsequent transformation into marshes. This process has been described by Steers (1958), with particular reference to the coastlines of the British Isles.

3.8.5   Relict Exogenic Relief Forms

The sea floor is also characterized by some other erosional and depositional relief forms. It is possible to distinguish, for example, marginal morainic features on the floor of the Barents and White Seas, including also push moraines [see Matishov (1980)]. Different forms of glacial erosion consisting of numerous deep-sea glacial troughs, which occur at the present-day as submarine valleys, have been preserved on the periphery of the Fennoscandian Shield, in the near-shore part of the sea floor next to the Norwegian and British coasts. Relict river valleys are also well preserved in the North Sea, English Channel, southern part of the Barents Sea, etc. The most common are valleys eroded by glacial meltwater flowing down the continental slope. Matishov (1980) claimed that the valley network of the northern part of the Atlantic Ocean floor was eroded mainly during glacial times when the edge of the ice was located near the edge of the continental slope. At that time, meltwater turbidity currents flowing from under the glacier directly affected the continental slope, being converted into suspension currents that eroded the continental rise and the upper part of the abyssal zone.

3.8.6   Sedimentation as a Geomorphological Factor

Accumulation of sediments on the sea floor causes the relief to be reduced by burying the earlier existing uneven surfaces. As some ocean floor depressions are initially deeper than others, their partial infilling will result in the development of abyssal plains at varying depths. Another important condition is the composition and genesis of the sediments being deposited. Terrigenous hemipelagic deep-sea sediments in the area of the continental rise accumulated relatively quickly; this can be explained by proximity to the continent—a major supplier of sedimentary material—and by the fact that the accumulation of sediments occurs not only through slow precipitation from the water mass, but also through material supplied by turbidity and bottom density currents. Rapid rates of sedimentation (30–50 mm per 1000 years) are also typical of the flat abyssal plains of the ocean floor. Within depressions and over undulating plains, the sedimentation rate is appreciably lower, being of the order of 10–30 mm per 1000 years. Deep-sea red clays are deposited at the lowest rate (1–3 mm per 1000 years) whereas carbonate sediments occupy an intermediate position.

Similar relations, although characterized generally by higher rates of sedimentation, are applicable for the Mediterranean Sea depressions. The maximum rates—from 100 to 200 mm per 1000 years—have been observed on the continental rise in the Balearic and eastern depressions and on their abyssal plains, as well as in the Ionian Sea depressions. In the Black Sea and in the southern Caspian depression, the rates are 40–100 and 100–400 mm per 1000 years, respectively.

3.8.7   Types of European Coastline

As is well known, the sea coasts are formed under the action of different exogenic processes, the most important being wave erosion [see Zenkovich (1946, 1962), Leont'ev (1955, 1961), Bascom (1964), King (1959), Bird (1976)]. In many instances, the creative and destructive actions of sea waves and swash are combined with the work of tides, as well as other wind-induced phenomena. In the tropics and subtropics, biogenic processes, particularly reef builders, become of great importance.

As a result of the postglacial transgression that took place approximately 5500–6000 years ago, the sea level returned to near its present level. Thus, the age of the coastal forms on the entire periphery of the ocean is roughly similar. However, erosion of the coasts takes place under diverse conditions, the most decisive being the degree of turbidity of the near-shore waters, and climatic and lithological conditions. With respect to the extent of reworking of the coastal slopes, it is necessary to identify coasts formed by subaerial processes and those slightly affected by marine action. They can be termed initially dissected coasts (e.g., fiords and rias) or initially graded coasts, depending on whether a coastline is controlled by faults or coincides with the strike of a folded structure. The next group of coasts consists of grading coasts distinguished by the development of capes, bays and various free depositional forms, as well as by strong dissection of the coastline. According to the degree of dissection, it is possible to identify abrasion-embayed and abrasion–depositional-embayed coasts. Subsequent evolution of the coastline results in graded coasts, which are subdivided into graded abrasion, abrasion–depositional, lagoon and flat depositional coasts. There also exist other types of coasts, such as deltaic, watten, collapse–talus and thermoabrasion coasts.

The coasts of Finland, the Kola Peninsula, Scandinavia, Iceland and the north-western part of the British Isles are notable for glaciation–tectonic forms (fiord and skerry types) slightly affected by marine abrasion (*see* Fig. 3.11). Favourable lithology facilitates good preservation of the primary dissection. The north-eastern part of western Russia is characterized by active thermoabrasion processes operating on the coast composed of cryogenic rocks.

The southern part of the Baltic and North Seas is distinguished by the predominance of coasts of the abrasion–depositional or depositional grading type. They are accompanied by large lagoons, such as Kjurshu Mores, the Vistula bay, and Zuider Zee. These types of coasts are also characterized in the North Sea by vast wattens. The western coasts of Great Britain, the shores of Ireland, and the northern and southern coasts of the English Channel are notable for an alternation of abrasion and abrasion–depositional-embayed coasts and graded abrasion coasts formed mainly in the relatively easily

Fig. 3.11 The skärgård coast of western Sweden. (Rudberg)

Fig. 3.12 The broad sandy foreland north of Harlech, North Wales, built by accretion inside a northward-growing spit topped by sand dunes. (Embleton).

eroded Cretaceous rocks (*see* Fig. 3.12). Dissected ria coasts slightly affected by marine action are typical of south-western Ireland, Brittany and Galicia. A depositional coast extends for a long distance in the area south of Brittany. The well-known Landes are also distinguished by ideal graded coasts and by the development of dune landscapes.

The eastern and western ends of the Pyrenees are characterized by a great variety of coastal types. Here, alongside grading embayed coasts, graded abrasion and abrasion–depositional types are encountered. Languedoc is distinguished by a graded lagoon coast, which passes into the Rhône delta shore and then changes to the slightly dissected coasts of the French and Italian Rivieras. The deeply incised ria shores of Corsica and Sardinia show a different degree of grading of their coastlines. The opposite shores of the Tyrrhenian Sea are of a complex pattern, being represented by large bays with depositional sections changing to well-pronounced forms of embayed abrasion, volcanic, and graded depositional and abrasion coasts. The shores of the Adriatic Sea are characterized mainly by flat depositional and graded abrasion coasts.

Abrasion-embayed, abrasion-depositional-embayed and graded coasts are typical of a considerable part of the shores of Greece and Crete. The coasts of the southern peninsulas of Greece and many islands of the Aegean Sea as well as the Khalkidhiki peninsula being composed of crystalline rocks are almost untouched by marine erosion. They consist of scarps on which subaerial mass movement processes are very active.

The coasts of the Black Sea belong, in the south-west, to the abrasion–depositional-embayed types and, north of Balchik, to graded-abrasion coasts, which farther north change to deltaic forms (i.e. the Danube delta). There follow abrasion–depositional graded coasts characteristic of the north-western Black Sea shores. Collapse–talus and similar landscape forms are typical of the southern Crimea coast. The Sea of Azov is notable for a wide distribution of abrasion coasts and for extensive depositional free and closed forms (e.g., the north Azov bars and the Arabat bar). The Caucasian and Anatolian coasts belong mainly to graded and abrasion-embayed types.

The Caspian Sea shore is rather diverse. In the north, wattens associated with an active wind-induced surge predominate. In the west, depositional flat and deltaic coasts are common. A slight dissection of the coastline, which is due to specific geological structural patterns, characterizes the Apsheron Peninsula. Until the 1930s in the east, abrasion coasts were predominant, but after a considerable drop of the sea level in the 1930–40s the majority of cliffs became inaccessible to sea swash. In the south, flat depositional coasts are typical.

Iceland, with an area of 103 000 km², has one of the highest concentrations of interesting and unusual geomorphological features in all Europe, probably also in the whole world. It is a land with high rates of geomorphological activity and dramatic changes within the memory of man. This concerns volcanic activity as well as erosion, including glacial, fluvial, aeolian and marine processes. Quantitative documentation of the changes within the last 1000 years is better than in any other country, because of late colonization by man (after 872). Details of such changes are well known because of excellent historical records and also because of the elegant tephra chronology developed early in Iceland [Thorarinsson (1944)]. In spite of advanced research work of national and international origin, large areas in the unpopulated interior are not known in any detail. This poses a problem for the construction of the *International Geomorphological Map of Europe*, although one that is somewhat mitigated by the small scale which the Map uses.

4.1 Geological Development and Constructional Forms

4.1.1. General Structure and Divisions

Iceland is situated on the mid-Atlantic ridge, which passes through the country as the young volcanic zone (*see* Fig. 4.1). From the northern coast, this zone runs directly south to the middle of the island, where it changes direction, turning south-west and dividing into two. Only the western branch is directly continued by the submarine

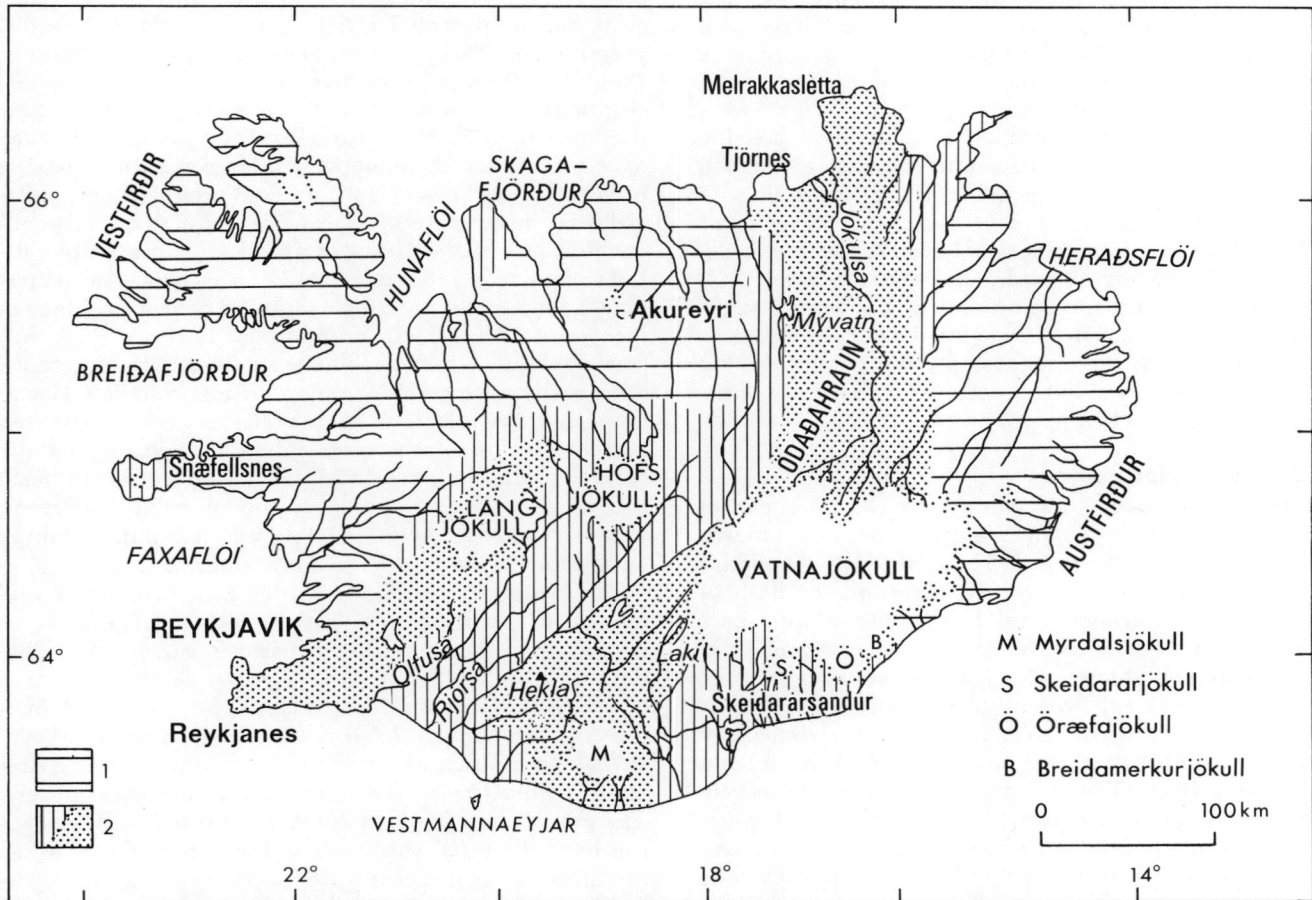

Fig. 4.1 Geology of Iceland [after Kjartansson (1956)]. (1) Older plateau basalts, (2) younger extrusives and hyaloclastite formations (Palagonite) of the median rift valley.

Reykjanes ridge, a part of the continuous mid-Atlantic ridge. In the north, the submarine Kolbeinsey rise, which continues farther to the north, is not a direct continuation of the young volcanic zone, but is situated some 50–100 km to the west. It is clear that the young volcanic zone is influenced by transcurrent faults (possibly transform faults) in a way which is well known in the case of mid-oceanic ridges. In spite of this connection with the mid-Atlantic ridge and indications of sea floor spreading (see below), Iceland is an anomaly in this particular environment, notably because of its size and complicated design. The North Atlantic is also said to be an anomaly. Discussion about the general geophysical background, including special types of upwelling currents associated with mantle convection, still continues. Iceland is situated at the crossing point between the mid-Atlantic ridge and the submarine ridge from Scotland through the Faröe Islands to Greenland (i.e. the Wyville–Thomson ridge). It should be noted that an isolated minor part with recent volcanic activity occurs in the Snaefells Peninsula in the west of the island [see Jacoby (1979)].

In addition to the southern coastal plain, Iceland is divided into three main parts by the young volcanic zone, the zone being at the centre of the island, with the western and the eastern basaltic plateaus on either side. The main pattern is in rough agreement with the concept of sea floor spreading, with the result that some older ideas have been abandoned. The basaltic plateaus were once thought to be of the same age (i.e. 50–60 Ma) as the basalt areas around the North Atlantic and the Barents Sea, as they were assumed to have once formed a continuous basaltic province. However, the Icelandic basalts are much younger, the oldest being only 16 Ma, according to potassium–argon dating. Furthermore, the oldest rocks are found on the coastal fringes of Iceland, with the age of the rocks decreasing towards the volcanic zone. Still further proof of the spreading and of the tensile stress perpendicular to the central zone are the numerous dykes, roughly parallel to this zone, steeply transecting the flat basalts. The number of dykes increases with the age of the basalt [see Jacoby (1979)].

4.1.2 Basalt Plateaus

The basalts are divided into two groups: the older, darker late Tertiary and the younger grey Quaternary basalts, the latter being formed 0.2–3 Ma BP. The boundary between the two basalts is partly drawn on the basis of palaeomagnetic considerations. In the older group, there are intercalations of lignite, which here means remnants of former bogs and forests while, in the younger basalts, there are intercalated beds of tillite. Iceland has had at least ten glaciations, to judge from a long sedimentary sequence (with basalt layers) on the Tjörnes Peninsula on the northern coast. The plateaus are not built exclusively of basalts or dolerites; there are also liparitic and rhyolitic volcanics and various intrusives, particularly gabbros. Whether the origin of the acidic lavas and intrusions is due to differentiation in magma chambers or caused by remnants of continental crust has not been established totally, but the facts are in favour of the former possibility.

The basalt beds with an average thickness of 10 m (varying from 1–2 m to 100 m, in an exceptional case) are seen as semiparallel ledges or lines on all steep basalt slopes. They are not horizontal; in most places, the layers dip inwards from the coast or, more exactly, towards the young volcanic zone. The angle of dip is usually greater in the case of the older, lower basalt beds and decreases upwards. The inward dip is not quite regular, but is disturbed by frequent steep-angle faults, so the basalt plateaus are actually built up of individual blocks, slightly diverging in terms of the slope angle of the lava beds. The general tendency of inward dipping is absent in parts of western Iceland, where the basalt beds are gently folded. The three-dimensional geometrical problem of the successively changing dip angles within the piles of basalt layers is solved partly by a supposed decrease in thickness of each layer the farther away it is from the feeding fissures of the active median zone. Another partial solution is the differential successive sinking of the basalt body. The latter part of the hypothesis is also supported by the inward dip of the dykes in the same direction.

4.1.3 Young Volcanic Zone

The young or active volcanic zone with rocks, which range in age from 0.7 Ma BP to the present-day, is one of the most active volcanic areas in the world, with 2000 eruptions in postglacial time (i.e. over the last 10000 years) from 200–300 volcanoes and volcanic fissures. From 40–50 volcanic centres or fissures, some 150–200 eruptions are known to have taken place in historical time (i.e. over the last 1100 years), and lots of eruptions may have occurred in the unpopulated interior without having been reported. The volcanic material produced in postglacial and historical times can be estimated fairly adequately as 400 and 40 km^3, respectively: for example, the Laki eruption of 1783 produced 12.5 km^3 lava—the largest known amount for any single eruption in historical time in the world. The Hekla eruption in 1947 (*see* Fig. 4.2) produced about 1.3 km^3. The volcanic activity proceeded during alternate periods of more or less complete glaciations and interglacials when glaciers were about their present size or even smaller. Thus the volcanic activity continued and still continues below the ice sheets. Such subglacial volcanic activity is fundamentally different, both in respect to the material produced and the resulting landforms, from the subaerial eruptions.

The volcanic materials of the zone are lavas and pyroclastics, both mainly, but not exclusively, of basaltic composition. Most lavas are basalts, which are highly fluid with an ability to flow over gentle slopes as thin layers and at fairly high speeds [Einarsson (1973), Holmes (1965), Schutzbach (1976) and Thorarinsson (1960a, 1975)]. In contrast, the acidic lavas of some scattered sites are of high viscosity and form broad, thick flows, if any. The pyroclastics (tephra) of the present subaerial volcanism have the usual variations in the form and size of the products. Large quantities are spread as fine-grained, airborne tephra over great areas of Iceland (and even over the rest of Europe) [Thorarinsson (1944)] to be deposited as layers whose thicknesses decrease gradually the farther ment (e.g., Laki) and sometimes the row of lava or scoria cones is the main feature. Some fissures only produce gaseous material and are now apparent simply as elongated basins (e.g., Valagjá). The Hekla volcano is formed along a 3-km fissure by numerous eruptions (20 eruptions in historical time, including five in the immediate past) or alternating layers of lava and pyroclastics. It is thus a special sort of stratovolcano.

4.1.3.2 Central eruptions
Such forms include lava cones, scoria cones and tephra rings produced by single eruptions. Where several eruptions produce fluid basaltic lava, shield volcanoes are formed. These regular domes with gentle gradients are also typical of the Icelandic volcanic landscape (as indicated by the numerous Icelandic place names ending in -dyngjá). With varying production of lava and pyroclastics during several eruptions, the normal stratovolcano is produced (e.g., Snaefellsnesjökull, Eyafjallajökull, Öraefajökull). If the eruption only produced gaseous material, the forms are maars—steep-sided, rounded basins usually now containing lakes.

4.1.3.3 Subglacial volcanic activity
Special forms, of which two must be specially mentioned, are produced by subglacial volcanic activity. The eruptions, which took place mainly below the ice or in an accidentally produced water body, formed irregular, often steep-sided hills, these in most cases being elongated in the general fissure direction (i.e., hyaloclastite ridges). If the volcanic structure protruded above the ice surface or temporary lake level, the so-called table mountains (known in Iceland as *stapi*) were formed [Kjartansson (1966)] with a steep-sided hyaloclastic base on top of which a gently rising shield volcano was formed (e.g., Herdhubreidh (*see* Fig. 4.3 also). These structures were once regarded as erosional remnants but are now, with good reason, interpreted as true construction forms. There are several examples of present-day volcanic activity in the glacier areas (some have already been mentioned when considering stratovolcanoes). Others to be added are Katla in the Mýrdalsjökull area, the Grimsvötn caldera and other not definitely localized volcanic centres in the Vatnajökull area. Subglacial eruptions produce catastrophic floods which are known as *jökulhlaups* in Iceland.

Within the rich sequence of volcanic forms, some other distinctive features must also be mentioned. These include calderas—notably those of large size (e.g., Askja)—steep domes of highly viscous acidic lava without any crater on top (e.g., Baula) and the swarms of small pseudovolcanoes produced when a hot lava flow passed over a lake, a bog or some other wet area. There are also several examples of submarine eruptions creating new islands, which are all very short-lived if they are only built of pyroclastics. The most recent example is Surtsey in the south-west (built in 1963 and subsequent years) and some other smaller islands in the vicinity that have since disappeared. As regards extent, the most important

Fig. 4.2 Hekla in eruption, 1947.

away from the source areas. These layers, which can be traced to the eruption site, form the basis of the important tephra chronology. Fundamental to this chronology is the existence of some light-coloured layers of acidic composition, derived from initial, violent phases of eruption of some volcanoes (e.g., Hekla).

In the case of the subglacial eruptions, flowing lava could usually not be formed because of its almost instantaneous solidification, with the formation of small pieces of volcanic glass in poorly consolidated accumulations or softly welded together by feeble diagenesis. For these rock materials, the term 'palagonite' has been used, with the whole suite of active vulcanism sometimes being described as the palagonite formation. The term palagonite is now often replaced by hyaloclastite.

The volcanic landforms of the active zone are extremely varied: most of the generally known volcanic forms occur within the area, as well as those that are almost exclusive to Iceland. In addition to the peculiar subglacial environment, the overall importance of long fissures in this zone of constant, but maybe somewhat erratic, tension must be emphasized. The fissures occur in groups, parallel with the zone boundaries or only slightly deviating from them. Many of these fissures are used repeatedly by volcanic eruptions. As in other areas, the volcanic landforms depend on the type of volcanic material, on the form of the eruption site and the number of eruptions [see Thorarinsson (1960a)].

4.1.3.1 Fissure eruptions
These are the most frequent, the lava flow showing no specific forms around the fissure itself (e.g., Eldgjá). Sometimes the lava flow is combined with a row of lava or scoria cones along the fissure in a late stage of develop-

Fig. 4.3a Three types of volcanic mountain as seen from Hagafell, Langjökull [Kjartansson (1956)]. In the centre: the table mountain of Hlödhufell, formed partly by subglacial and partly by subaerial eruption. On the right in the distance: Skjaldbreidhur, a shield volcano. On the left, and also in front of Skjaldbreidhur, ridges formed by subglacial eruption.

Fig. 4.3b Eiriksjökull, a table mountain surmounted by a shield volcano and thin ice cap. Postglacial basalts in the foreground. (Embleton)

volcanic forms are the flat lava flows. They often have the irregular block lava surface (i.e. aa-type), but occasionally a more even surface (i.e. ropy lava or pahoehoe-type). Associated with the volcanic processes is thermal activity, with numerous hot-water springs in the basalt plateau areas and high-temperature steam springs in the young volcanic zone. In some places, minor landforms have been developed, such as sinter deposits. The hot springs are not shown on the *International Geomorphological Map of Europe*.

4.2 Erosional History of Iceland

The periglacial, fluvial, marine and aeolian development will be dealt with here under the glacial geomorphological development because the preglacial history is short, even in the basalt plateau areas, and the glaciations (ten or more) are numerous and started early.

The young volcanic zone is essentially not a landscape of erosion. It consists of flat, plateau-like surfaces with scattered volcanic forms already described, all of them being constructional. The hyaloclastic material is often eroded easily and the forms may have been changed in detail, but erosion forms of any importance are restricted to the marginal parts of the zone. It may be mentioned that a map of relative relief, separately constructed for the *International Geomorphological Map of Europe*, gave a completely irregular picture, which is in contrast to most erosional landscapes where a certain zonation is often evident (e.g., Fennoscandia).

The basaltic plateaus, however, are definitely landscapes of erosion and certain stages in development may be seen. The upper flat surface is often seen to truncate the basaltic layers at a low angle and is thus regarded as a surface of erosion. It is certainly not the only one in the basaltic sequence to judge from the intercalated lignite layers [see Einarsson (1962)]. The three-dimensional geometrical problem involved in the upward decreasing dip of the pile of basaltic layers may be explained partly by hidden surfaces of erosion (a different explanation is offered on p.50). Other indications of landscape evolution are the two valley generations reported particularly in northern Iceland. The present valleys and fiords of the basalt landscape (notably those in the north-west and east) must have had a partial fluvial history, more or less related to the fracture systems, but the present shape is definitely mainly glacial.

The valleys incised into the basalt plateaus are often broad, open and trough-like, contrasting with the flat upper basalt surfaces, their sides being straight and somewhat monotonous with basalt ledges as the main structural form and recent or subrecent gullies as the dominating erosional details. Some valleys have steep trough ends, which are supposed to be characteristic of valleys cut by valley glaciers, fed from the surrounding plateau icefields. The fiords, most of them being fairly short and open, have the same general appearance. However, lots of inlets and bays classified as 'fiords' according to the place names are not in the geomorphological sense—as also occurs in Scandinavia. Cirques, with or without present glaciers, occur frequently on the north-

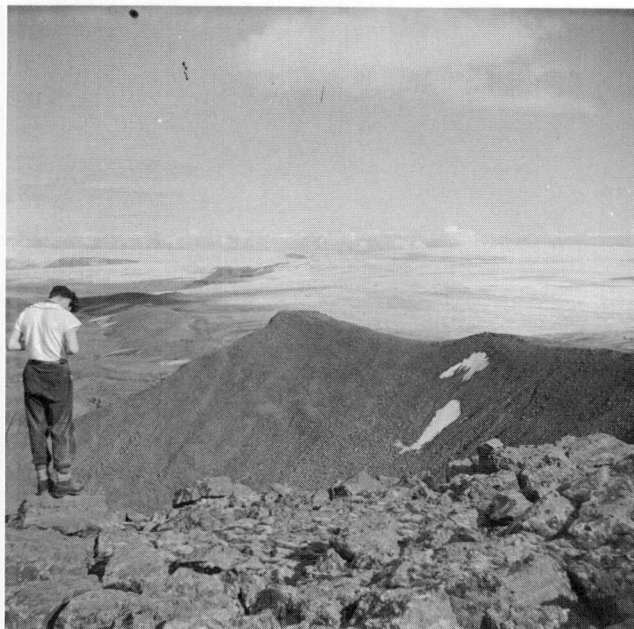

Fig. 4.4 View over part of the Langjökull. In the foreground, ridges built of hyaloclastite. (Embleton)

western peninsula and in the eastern fiord areas. They are often well-rounded, some being perfectly shaped, and remnants of flat plateau areas are preserved in between. More alpine forms with densely spaced cirques are found on the peninsula between Eyja Fjördhur and Skaga Fjördhur in the northern part of the island.

As regards the mesoscale and microscale forms of glacial erosion, glacially eroded lake basins occur, but much less frequently than in the crystalline shield areas. The roche moutonnée forms are not well developed in the basalts, the latter often showing small-scale polygonal patterns on flat surfaces connected with the normal and usually vertical columnar structure. Better roche moutonnée forms are found in gabbro areas.

A distinctive feature in the physiography of Iceland is the coastal plain, mainly developed in southern and partly in western Iceland. It is often called a strandflat, but this should not be taken to imply anything about its origin. In some places on the western coast, the strandflat, consisting of islands and skerries, is reminiscent of that in Norway. In other places, particularly in the south, the strandflat with broad, flat plains and valleys looks very different. A joint fluvial and marine origin is possible, underlined by the fossil cliffs of possible marine origin along the inner part of the strandflat on the southern side. A certain degree of glacial modification should be included in the hypothesis, at least for the 'island type' of strandflat.

The landforms of the unconsolidated Quaternary deposits have been studied somewhat less than other geomorphological features, but some points will be specially stressed and some landforms are shown on the *International Geomorphological Map of Europe*. The ice sheet was better developed in southern Iceland, with the main ice divide south of the main watershed; the ice sheet in the north was not thick enough for complete cover. Certain higher areas protruded above the ice as nunataks, as they do today (*see* Fig. 4.4), but more spectacular nunatak forms are not reported. The last ice retreat shows two characteristic stages of stagnation, sometimes indicated by moraines.

4.3   Present-Day Geomorphological Processes

Most erosion processes have been working rapidly in postglacial time. Fluvial erosion has produced many canyons in several places (e.g., in the southern fossil cliffs). The rate of fluvial erosion has been measured in some places, for instance, by means of tephrachronology. Recessive erosion in the Jökulsá-Fjöllum, with the dissection of the basalt layers, has produced a 150-m deep canyon with a length of 9 km in postglacial time [Thorarinsson (1960b)]. Glacial erosion in the Vatnajökull area is indicated by the sediment transport by the meltwater rivers. The total load from this restricted area averages 40×10^6 ton per year. The coarser material is deposited by the braided meltwater rivers to form the 'sandar' or outwash plains (*see* Fig. 4.5).

Soil erosion by wind and water has changed the landscape considerably since the arrival of man. The Icelandic soils, which lack clay, are weak and the vegetation was developing in an unstable equilibrium with erosion forces, even when no herbivorous mammals existed. With the cutting of trees and the grazing of livestock, the picture has completely changed. It is calculated that half of the once vegetated area is now devoid of vegetation [see Thorarinsson (1961)]. The large amount of sediment brought down by the glacial rivers

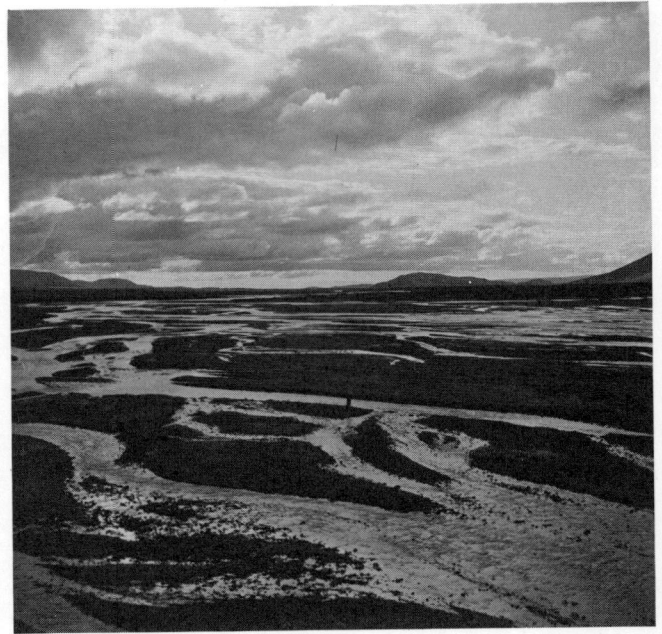

Fig. 4.5 The braided course of the upper Hvitá, a few kilometres downstream from the Langjökull. The figure standing in the centre provides a scale. (Embleton)

has greatly influenced coastal development along the southern coast, where extensive beaches of sand and gravel have developed with long offshore bars and few natural harbours. Many of the harbours that existed during the time of colonization are now rendered useless due to silting.

Chapter 5
Fennoscandian Shield

The vast areas of the Precambrian shield of Finland, Sweden and Norway belong to a still larger entity termed Fennoscandia, which also comprises the Scandes (Caledonian Highlands) (*see* Chapter 7). The geomorphology of the Shield in Finland, Sweden and Norway is considered in Section 5.1. Areas—Soviet Karelia and the Kola peninsula—outside these three countries also belong to the Shield: these regions are dealt with later in this chapter (*see* Section 5.2).

5.1 Finland, Sweden and Norway

The Precambrian and Caledonian areas of Finland, Sweden and Norway have much in common. The boundaries between these two regions may at times be sharply expressed in the geomorphology but are, on the other hand, often transitional. The same could be said about the boundaries of the subregions of the Fennoscandian Shield. At the outset, certain general characteristics may be established.

1) The Precambrian areas are dominated by crystalline rocks (*see* Fig. 5.1) with old inherited structures.
2) The main part of the Shield is supposed to have been a land area and not covered by any marine transgressions since Silurian time.
3) Arising from point 2, the Shield has had a long preglacial erosional development. One consequence of this is that correlative sediments, which might have permitted evaluation of erosional processes and dating of the more important landforms, are lacking within the Shield itself, except for some Cambro-Silurian remnants and some minor areas of younger rocks at the margins.
4) The whole area has been completely covered by ice sheets several times during the last few million years. This means that the preglacial landforms have been transformed, obscured or even obliterated beyond recognition. On the other hand, the land surface has received an overall imprint of highly varied glacial erosion forms of different orders of magnitude. The main part of Fennoscandia belonged to the predominantly erosional zone during the glaciations.
5) The relief types of Precambrian Fennoscandia—by which is meant here the relief connected with the solid rock—are very different from those of Europe outside the Shield, where sedimentary rocks, unglaciated or with signs of predominantly glacial accumulation, predominate. The landforms also differ in many respects from those described in most textbooks, and some of these landscapes even lack a good description. So far there is no

generally accepted answer to the important question of the relative contribution of preglacial (or interglacial) fluvial processes, and of glacial processes in the evolution of the relief.
6) The unconsolidated deposits belong mainly to the latest glaciation and the landforms of these deposits to the very last phases of this glaciation, particularly to the deglaciation. Consequently, the forms are fresh in appearance, only spoilt or changed by later erosion in special circumstances.
7) Postglacial uplift has been strong in most parts of Fennoscandia and is still continuing. The uplift has caused and still causes substantial effects on erosional processes in certain areas of unconsolidated sediments. This provides a contrast between the partly very old forms of the solid rock and the very young forms in areas close by.

The distinctive characteristics of the Shield areas of Sweden and Norway have raised particular problems of geomorphological mapping especially when using the legend to the *International Geomorphological Map of Europe* (1:2.5 million), which was produced mainly on the basis of experience from central Europe. Here no reference is made to the need for the new symbols, as the legend is regarded as an open system. The problem is mainly related to points 3–5 above, notably to point 5. It is also connected with the vastness of the areas, which means that even certain features whose genetic background is well known cannot be adequately mapped, simply because of the inadequacy of the existing inventory. The international legend is, of course, a compromise between different interests, but it should be mentioned here that the relief classes are not well adapted for the purposes of Precambrian Finland, Sweden and Norway. In particular, the broad relief class 75–300 m obscures some of the landscape boundaries that are traditionally regarded as being most important in the countries in question.

The small scale of the map sets very definite limits to the possibilities of reproducing features in a realistic way. Some fairly important landscape features cannot be shown. In the following chapters, where the seven points are discussed in somewhat more detail, this problem will be commented on several times.

5.1.1 Rock Types and Structural Inheritance
Among the crystalline rocks of the Precambrian shield, gneiss and granite are by far the most frequent, but there are also basic intrusives, acid and basic volcanics, and sedimentary rocks (e.g., sandstones, quartzites, shales,

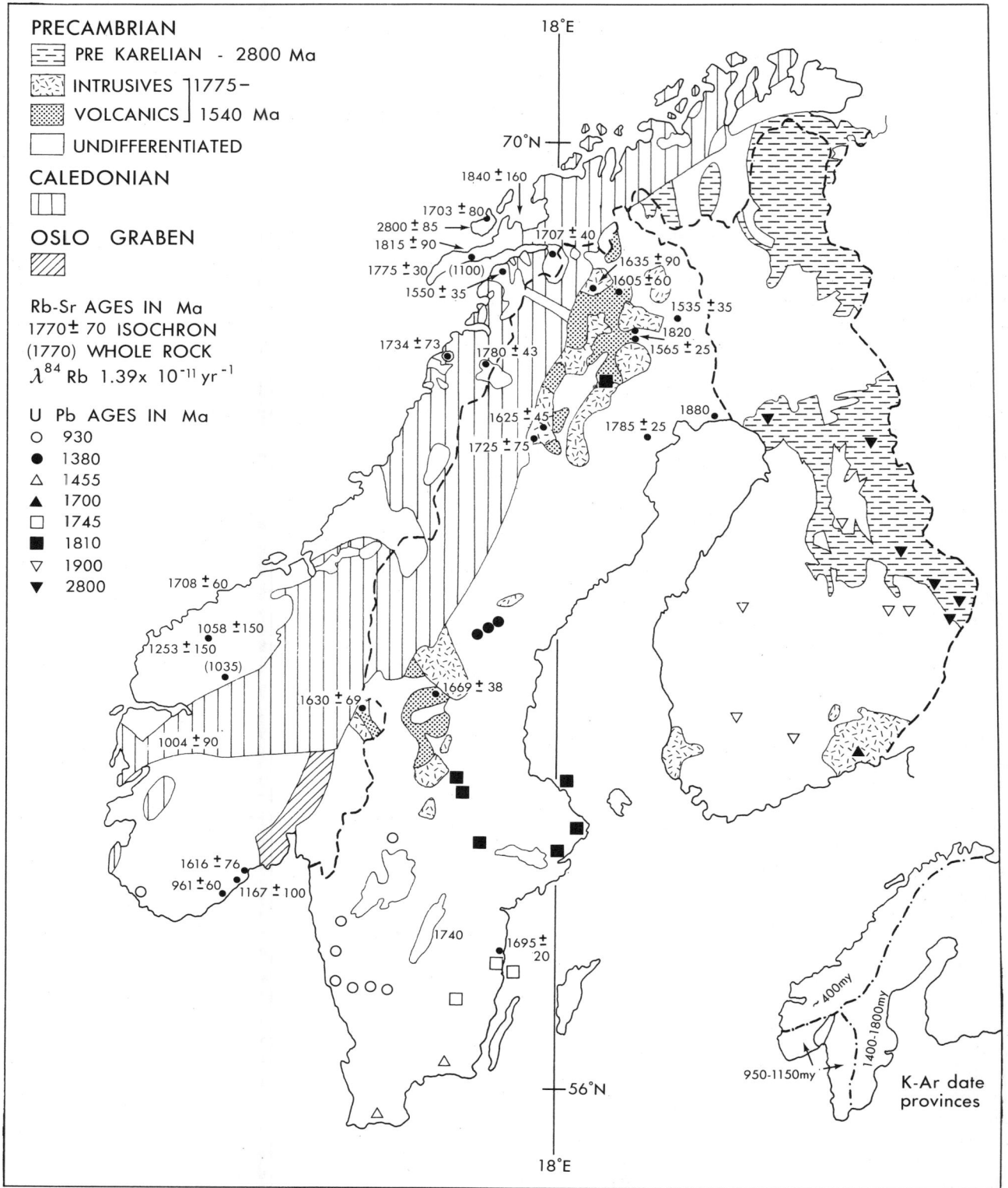

Fig. 5.1 Age differentiation of Precambrian rocks in Fennoscandia.

schists and marbles) and these may be of some importance locally [see *Atlas of Finland* (1960), O. Holtedahl (1960), Lundqvist (1979), Magnusson *et al.* (1962, 1963)]. The simple names granite and gneiss include a large variety of rocks as far as mineralogy, grain size and textures are concerned. There are gneisses with well-developed S-surfaces, with linear structures and of the migmatic type. The somewhat monotonous picture presented, at least on medium-scale geological maps without indication of variations, is thus often an oversimplification. Modern detailed mapping, for example in the well-exposed parts of the western Swedish gneiss areas, has shown large variations even on a small scale, with narrow extended bands that might be of geomorphological importance, and also with regard to their repetition. On the other hand, the monotonous nature of the map is sometimes quite realistic, as in the extensive granite areas of central northern Sweden with predominantly a coarse-grained variety of granite and few regional differences.

Within the Precambrian crystalline area, there are only a few residual outliers of Cambro-Silurian strata (*see* Fig. 5.2) with flat-lying sediments, notably sandstones, shales and limestones. These areas of softer rocks are preserved by doleritic cap rocks in south-western Sweden and partly by porphyritic cap rocks in the Oslo area, both being of Permian age, but in most areas by downfaulting. They often have normal denudational contacts on at least one side. The same type of contact follows the west side of the Cambrian sandstone outcrop on the shore in the Kalmar region of south-eastern Sweden. This area forms the outer margin of the sedimentary cover of the southern Baltic. The limit can be followed as a cuesta on the sea floor in a line curving towards Estonia [see Martinsson (1958)]. Remnants are also found in the southern part of the Gulf of Bothnia as described by Winterhalter (1972). At the front of the Caledonides below the easternmost nappes, there occur, in most places, flat-lying Eocambrian and Cambrian sediments (also Ordovician in the south) with a denudational limit to the east. Observations in Sweden all point towards a much wider extension of the Cambro-Silurian deposits in the not too remote geological past. Supposedly, only the middle part of Swedish Norrland escaped the transgression. The situation is partly the same in central southern Norway. Finland is more problematic, but the occurrence of Cambro-Silurian in the southern part of the Gulf of Bothnia and the steep front of the Cambro-Silurian in the Estonian cliff coast make it fairly probable that the Cambro-Silurian also extended across southern Finland. The significance of this former greater extension of low-resistance sediments will be discussed later.

Rock structures, such as the S-surfaces of the gneisses, fissures, fissure systems, normal faults, transcurrent faults and thrust faults, form an intricate mosaic pattern over the whole Precambrian area, which is shown in a general way by satellite imagery, but in more detail by aerial photographs and locally by field surveys. Some rock structures may be connected with the primary orogeny, contemporary with the gneiss formation—S-surfaces and certain fissures—or they may be much younger, being caused by later deformation, when the crust only reacted in a rigid state; the result was the creation of fissures, pairs of fissures and fissure zones, with the displacement of individual crustal blocks along these fissures. Such deformation has occurred several times, resulting in new fissures and also in the revival of old ones. With successive deformations, this revival of old fissures increases in relation to the formation of new fissures. Because of the complex origins and mosaic pattern, it is difficult to date the origin of the deformation responsible for the first formation of a certain fissure or fissure zone, but attempts have been made in papers published some decades ago. The majority of the fissures and fissure systems are supposed to be of Precambrian origin. Revival has occurred at least in Caledonian and Variscan time and, according to some authorities, in the Teritary and even later (e.g., in some special cases during the postglacial) [see Asklund (1923), Björnsson (1937), Larsson (1954), Ljungner (1927–30), Lundqvist and Lagerbäck, (1976), Martin (1939), Niini (1968), Rudberg (1954), Tanner (1938)].

The influence of rock type on the landforms is complicated and not yet known in all respects. It is quite clear in the small-scale features, such as roches moutonnées, where the difference between granite and gneiss areas is obvious, the more regular forms being found in granites, with their usually simpler structural pattern. As the scale of landforms increases, the difference between the two main rock types is far less obvious. None of the landform sequences can be attributed specifically to either rock type. Some of the less common types, however, have a certain geomorphological significance. The resistant rocks in the Precambrian environment, which form hills and higher ground, are often basic rocks, gabbros, dolerites and amphibolites; quartzites also come in this category. Special landscapes are connected with porphyries and the so-called Jotnian sandstone (*see* Chapter 7). The most pronounced contrast is that between the Precambrian crystalline and the softer Cambro-Silurian sediments [see Rudberg (1954), Tanner (1938)].

The Precambrian fissures and fissure systems—a widespread feature judging from most information—are of great importance in certain regions (e.g., in the fissure valley landscapes with their straight, angular patterns). The fissure mosaic is much less significant in most other landscape types. For a general explanation of these differences, the erosional history has to be taken into account [see Rudberg (1954)].

As a whole, the influence of rock type and structures cannot be shown on small-scale geomorphological maps. What may be seen, however, are the more important fissure valley landscapes and the significance of rock type and structure for the other major landscape assemblages.

5.1.2 Transgressions after the Silurian

There is little evidence to contradict the classical idea that the Shield was never covered by any major transgression after the Silurian, but some other comments ought to be made. In the Oslo area, there are some minor remnants of Permian sediments. On the flat shield areas in southernmost Sweden, remnants of Jurassic sandstone and, some-

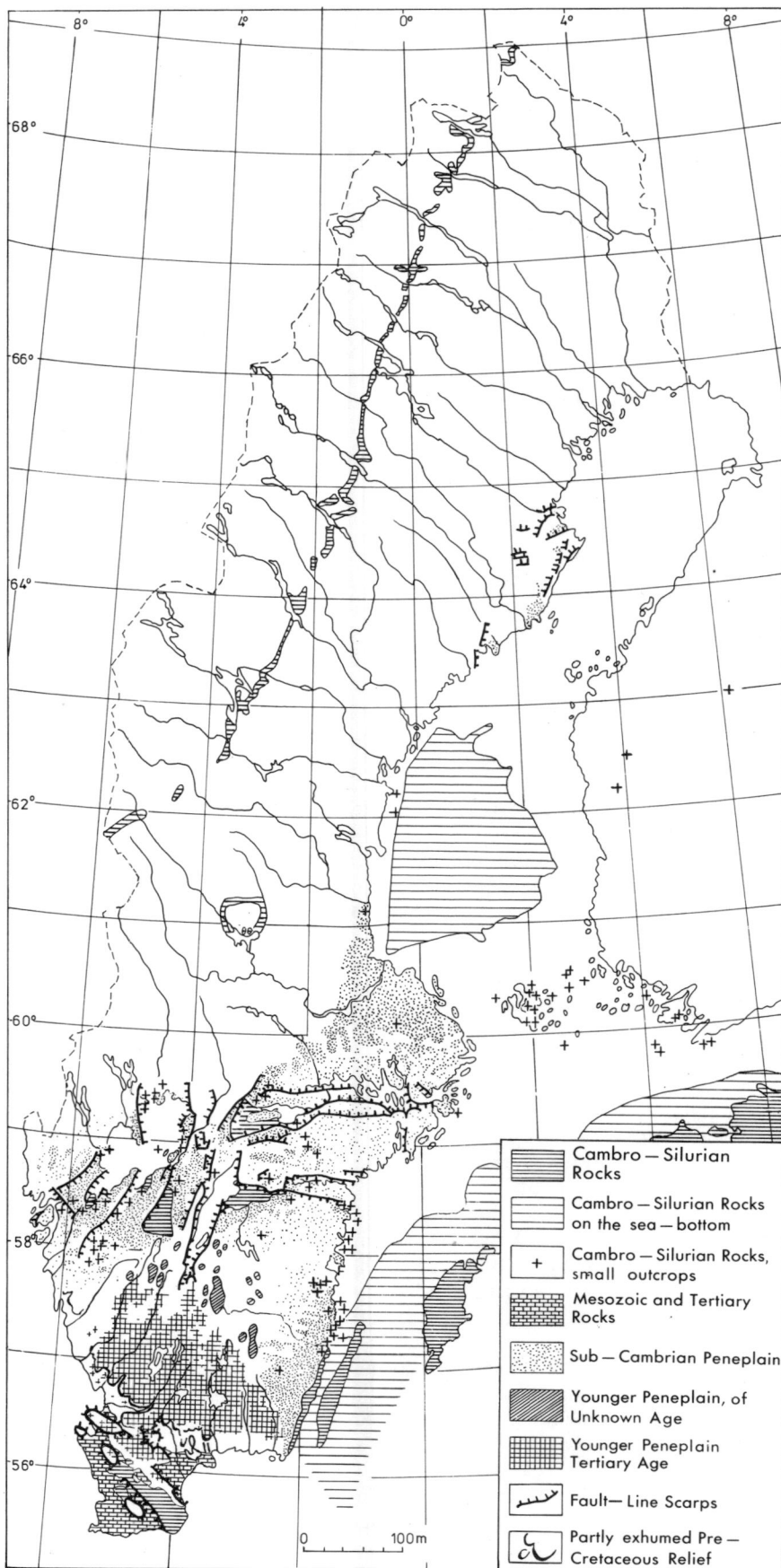

Fig. 5.2 Planation surfaces in Scandinavia and the distribution of related outcrops of sedimentary strata [Rudberg (1970c)].

Legend:

- Cambro — Silurian Rocks
- Cambro — Silurian Rocks on the sea — bottom
- Cambro — Silurian Rocks, small outcrops
- Mesozoic and Tertiary Rocks
- Sub — Cambrian Peneplain
- Younger Peneplain, of Unknown Age
- Younger Peneplain Tertiary Age
- Fault — Line Scarps
- Partly exhumed Pre — Cretaceous Relief

0   100m

Fig. 5.3 Surface of low relief bevelling the Fennoscandian (Baltic) Shield in southern Finland, viewed from the hill Tiirismaa (223 m). (Demek)

what farther north, remnants of late Cretaceous Chalk are to be found. The outer margin of this last transgression is not known. In the south-east, Lidmar-Bergström (1982) has described local remnants of a deep kaolin layer. South-western Scania, which does not really belong to the Fennoscandian entity, has deep fault-controlled sediments, with ages ranging from late Palaeozoic to oldest Tertiary. Similar sequences are found in Denmark with the present boundary—probably fault-controlled—lying in the middle of the Kattegat [see Wienberg-Rasmussen (1968)]. In the south-east, the Devonian—at present ending with a cliff-like slope on the sea-floor between Gotland and the Latvian coast—indicates an extension farther west [see Martinsson (1958)].

These ideas merely involve some possible reconstructions. What is seen at present in most parts of the Shield is old Precambrian rock covered by young Quaternary deposits. This geochronological hiatus is one of the greatest in the world. It explains the sharp delimitation in Fennoscandia of geomorphological studies dealing either with the forms of the bedrock or with the forms of the unconsolidated deposits, studies of the latter being much more frequent.

5.1.3 Preglacial Geomorphological History
It is not expected that small-scale preglacial forms will have survived. Obvious preglacial forms are on a larger scale or consist of landform assemblages. These include

major planation surfaces and valleys (*see* Fig. 5.3). Two types of planation surface occur, being easy to distinguish from a theoretical point of view but not in practice. Of these two types, the stripped or exhumed surfaces have been particularly observed, and several have been discussed. The most important is the so-called sub-Cambrian or Precambrian peneplain [see Rudberg (1954, 1970a)]. It is found around all flat-lying remnants of Cambrian sandstone as a rock surface of almost perfect flatness, usually lacking monadnocks (*see* Fig. 5.2). At some distance from the Cambrian outcrops, the surface is generally dissected but can be followed as an accordance of summit level across low plateaus or isolated hills. Sandstone-filled fissures—in a few cases of proved Cambrian age, otherwise of assumed Cambrian age—have been of some, even if restricted, help in the reconstructions, although their age has been questioned [see Mattsson (1962)]. These stripped planation surfaces are seen particularly well in southern and central Sweden, but also occur elsewhere as in parts of southern Norway. In Finland, the existence of such surfaces is probable, but more difficult to prove.

In southern Sweden, there are other planation surfaces that may partly be regarded as stripped. If so, they are of Mesozoic origin [see Lidmar-Bergström (1982), Rudberg (1954)]. The main surface is of Tertiary age and was probably never covered by sediments; this surface also possesses monadnocks. It has not proved possible to show these surfaces with their genetic relations on the *Interna-*

Fig. 5.4 Plain of the Fennoscandian Shield with monadnocks, Glommersträsk, Swedish Lapland. (Rudberg)

tional Geomorphological Map of Europe.

Planation surfaces also exist in other parts of Precambrian Finland, Sweden and Norway. There is no clear evidence of former sedimentary covers; it has been suggested by Rudberg (1954) and Tanner (1938) that these surfaces may belong to the category of unexhumed surfaces. Most of them possess monadnocks; some, notably in northern Sweden and Finland, have been described as monadnock plains (*see* Fig. 5.4). The contrast here between nearly flat surfaces and isolated, occasionally steep hills, often with a pediment-like fringe, is reminiscent of the 'inselberg plains' of the savanna regions. The slope and crest forms of the individual hills, however, may be different. Some surfaces, particularly in the higher parts of the region, occur in a step-like arrangement, and the sequences are continued outside the region (*see* Chapter 7). The monadnock plains are shown on the *International Geomorphological Map of Europe*, but it has not been possible to demonstrate directly the step-like arrangements of planation surfaces. The forms of certain hills may indicate formation under a warm climate [see Kaitanen (1969)].

The valleys of the Precambrian areas often have an irregular appearance, sections with closed sides being followed by open stretches, with distant or low valley sides. Straight and winding valley sections alternate. Minor valleys and valleys in areas with low relative relief are usually the most irregular. Characteristic fluvial forms are often obscured by structural control and glacial transformation. Details of the glacial transformation will be described in Section 5.1.4. It should be noticed that river valleys are usually lacking on the monadnock plains.

In many places, the rivers pass through water gaps. A tectonic explanation can never be excluded, but in most cases superimposition is a more probable alternative, which in particular environments means lost Cambro-

Silurian sediments [see Rudberg (1976)].

The *International Geomorphological Map of Europe* makes a distinction between fluvial valleys and glacial valleys, which in the latter case means valleys strongly transformed by glacial erosion and only locally means valleys formed exclusively by glacial erosion. A subjective choice of map symbols is necessary in a great many individual cases, but such decisions will be discussed in Section 5.1.4. V-shaped valleys are, however, fairly rare within the Precambrian. They usually seem to be connected with a strong lowering of the local base level, such as may be caused by pronounced selective glacial erosion, and thus cannot generally be included in the group of preglacial forms.

5.1.4   Glacial Erosion Forms

During the Quaternary glaciations, Fennoscandia, being situated in the centre of the ice-covered area, was the site of predominantly glacial erosion. The Shield obtained a general imprint of glacial erosion forms of characteristic appearance and great variety. Sometimes these forms show an almost textbook-like assemblage. According to size, they form a complete hierarchy from the U-shaped valleys in the western parts of the region (*see* Chapter 7), through the medium-scale forms of steep rock faces and rock basins, to the diagnostically important microrelief forms of the coastal bare-rock areas.

No Alpine forms occur within the region, with the exception of a few scattered, doubtful cirques. U-shaped valleys are few compared with the mountain regions of Fennoscandia, much smaller and usually of a less perfect form. The steep rock faces of truncated spurs alternate with gentle slopes, which are possibly reminders of the preglacial appearance. The most characteristic forms of the Precambrian crystalline areas are of medium or microscale. In the first category, there are the following four

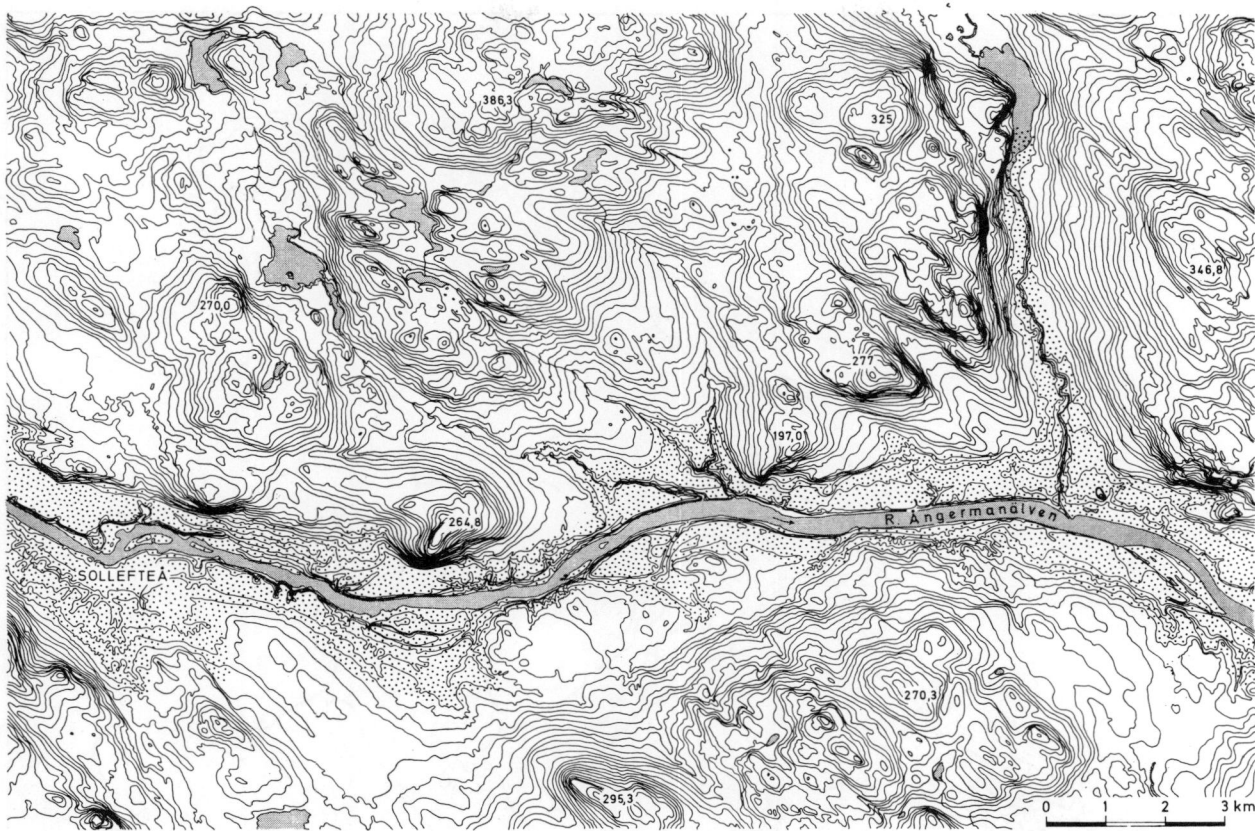

Fig. 5.5 Angermanälvern valley, central northern Sweden. Contour interval: 10 m. Stipple indicates areas of thick unconsolidated sediments. Other areas are bedrock with a till cover. North is at the top of the map, and the general direction of ice movement was from the north-west [Rudberg (1973)].

characteristic forms as described by Rudberg (1973):

1) Isolated steep rock faces
2) Asymmetrical hills and valley cross-profiles
3) More or less closed rock basins
4) Erosion forms strongly controlled by structures

Some comments on each of these points will be given.

The isolated rock faces stand parallel to the main ice movement or more often mark lee-sides, but never the stoss-sides. The ice movement is indicated by glacial striae or deduced from general consideration of topography and of the ice sheet.

The second characteristic feature is explained by the distribution of steep rock faces. A steep slope on one side of an otherwise gently rounded hill will give a special accent to the topography, forming giant roches moutonnées often called *flyggberg* in Swedish [see Rudberg (1954)]. In many valleys, the lee-side (in relation to the predominant ice movement) is often comparatively steep resulting in general valley asymmetry (*see* Fig. 5.5). The steep valley sides may be aligned at a wide angle to the main direction of ice movement. The asymmetrical valleys rank in magnitude from those of some hundred metres of relative relief to small fissure valleys only a few metres deep.

The third group of forms includes, in particular, the numerous lakes which—in most of those that have been sounded—contain one or more rock basins, but also numerous basins filled with clay or peat. Basins of varying type often follow one another in a down-valley direction like a string of pearls. Basins and basin systems also vary in magnitude from those dominating the landscape to minor forms associated with roches moutonnées.

The fourth point can be seen in fissure valleys of all sizes, fissure valley systems, lake shores, etc. The minor forms are especially found in the coastal bare-rock areas, with the major forms partly in areas better covered with unconsolidated deposits. The final clearance of fissures and fissures zones, prepared by preglacial selective weathering or erosion, is supposed to have been better performed by ice sheets than through fluvial activity along linear courses. Furthermore, the more pronounced structural landscapes belong to those marginal parts of the Shield (i.e. the Swedish west coast, southern Norway, etc.) where signs of strong glacial erosion are in any case present everywhere. The mosaic of fissure systems is found throughout the whole Precambrian crystalline area, but the erosional history—through more effective glacial erosion—reveals pronounced regional variations.

The minor forms of glacial erosion, such as roches moutonnées, are present everywhere within the Precambrian crystalline area but are often hidden by superficial deposits. The roche moutonnée forms usually occur together in major groups, forming stoss-sides and lee-sides

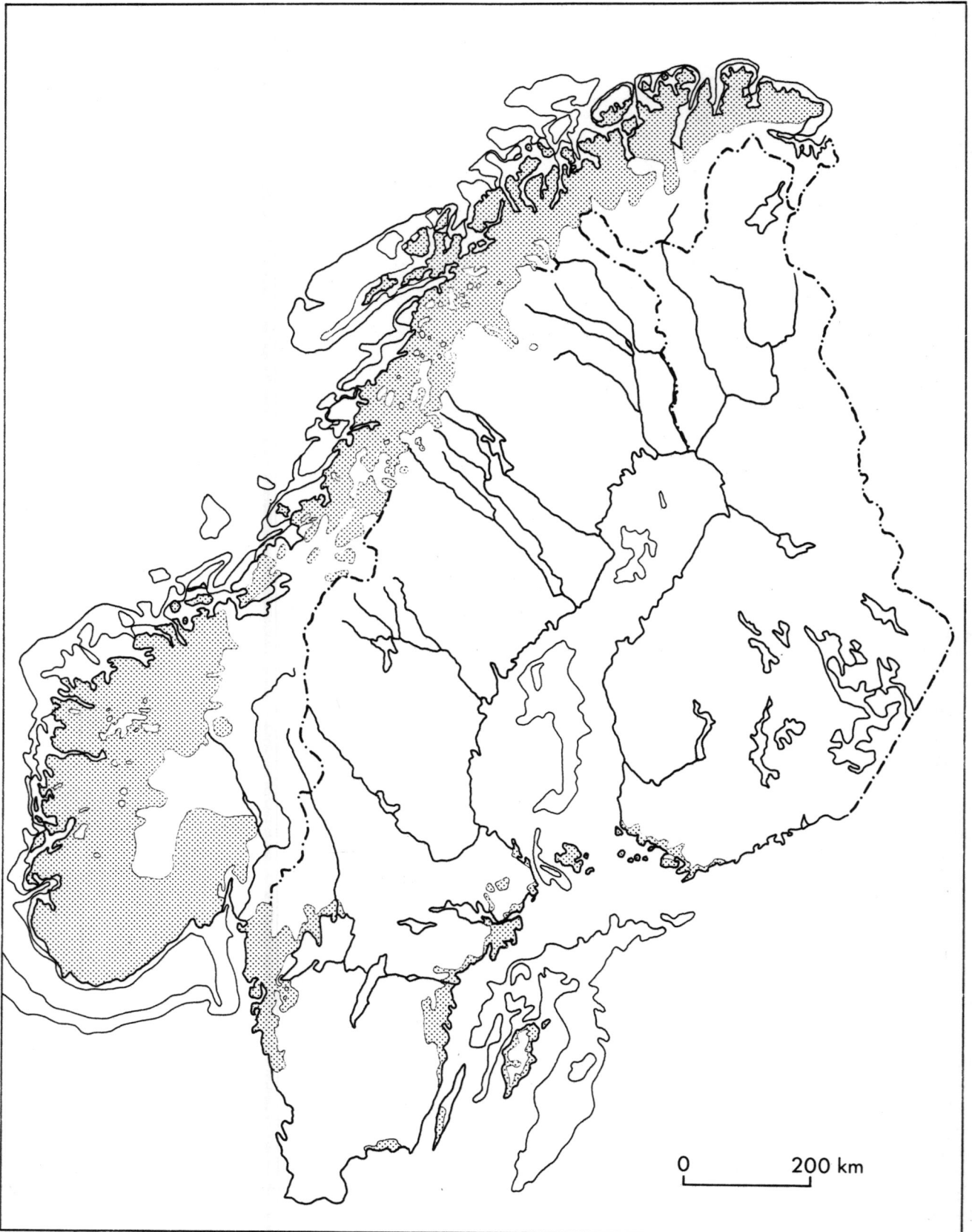

Fig. 5.6 The stippled area shows the distribution of bare rock (or rock with a thin till cover) in Fennoscandia. Submarine contours at −100 m and −200 m are shown [Rudberg (1970b)].

of a higher order. To the roche moutonnée floor also belong rock basins and micro-fissure valleys, the latter having asymmetrical transverse profiles in the manner described above. Glacial striations are found in all parts of the crystalline Precambrian area, but are locally often obscured or destroyed by weathering. They are best preserved below a thin cover of sediments. Striae are always preserved on low-lying coastal slopes, recently exposed to weathering by postglacial uplift. In the most exposed bare-rock areas (*see* Fig. 5.6), the smallest forms (e.g., crescent-shaped gouges and P-forms) can also be studied [see Dahl (1965), Gjessing (1965–66), Johnsson (1956), Ljungner (1927–30)]. Some of the P-forms are most probably the result of glaciofluvial erosion rather than glacial erosion in the true sense. The glacial erosion forms of medium and micro-scale are not yet particularly well known, not even in a purely descriptive way.

It seems quite clear that a well-developed roche moutonnée floor is connected with the crystalline rocks. It may even be said, tentatively, that old crystalline rocks form a prerequisite for the development of the most perfect forms. In Sweden, typical roche moutonnée forms are absent from areas of Cambro-Silurian sediments and are poorly developed in the Permian dolerites and some of the low-grade metamorphic rocks of the Caledonides. Similar contrasts can be seen outside Fennoscandia (e.g., in the Precambrian crystalline areas of western Greenland) in contrast to the sedimentary rock of the same region. One explanation could be the high differential resistance offered by the old crystalline rocks, between blocks delimited by fissures and fissure zones of varying age and reworked, maybe several times, finally forming a pattern that is the result of a long development.

What is more important is that the quantitative amount of glacial erosion is not known. The main reason for this lack of knowledge is that no preglacial reference surfaces are preserved, from which any calculation could confidently start. Some rough calculations have been attempted using a few favourable sites but, because they may not be representative, they are not quoted here. The most promising way to obtain the average glacial lowering of the land surface seems to be to calculate the total amount of rock material transported away from Fennoscandia and deposited in the surrounding sea and in central and eastern Europe. The method was tried as long ago as 1879 by Helland and gave a result of 25 m. This approach has not been repeated in recent times, and it is not known whether the figure is of the correct order of magnitude. The difficulties in such calculations are obvious. While the Fennoscandian origin of small rock fragments in continental till deposits may be determined and, in favourable cases, even their precise source identified, there are no methods available to detect the sources of smaller particles. These form the bulk of the material produced by glacial erosion, at least to judge from present-day experience of retreating glacier tongues in Fennoscandia; most fine-grained detritus may now be found on the North Sea floor. Just as the P-forms are interpreted as being caused by glaciofluvial erosion, there are also larger forms produced by the same processes: some of them cut in solid rock, but the majority in the superficial deposits. Some more important meltwater erosion channels in solid rock are marked on the *International Geomorphological Map of Europe* as well as some sequences of smaller marginal channels, mainly in the unconsolidated deposits. Both give some indications of the general trend of the deglaciation (*see* Chapter 7).

5.1.5 Problematic Forms

A particular problem in geomorphological mapping is that of interpreting the erosional or depositional history of a feature. Many features, as well as glaciated valleys, are good examples of this, having evolved through complex sequences of events in which many different processes may have participated. In the case of valleys in the Fennoscandian Shield, there is a clear continuum between fluvial valleys that were occupied by ice but not significantly modified and glacial valleys where either previous fluvial forms have been completely destroyed or have never even existed. The problem in geomorphological mapping is how to represent the intermediate forms, given that in many cases it is not possible to obtain the necessary proofs of a particular mode of evolution. On the Fennoscandian areas of the *International Geomorphological Map of Europe*, it has been necessary to introduce a separate symbol for these intermediate forms, where the relative contribution of fluvial and glacial processes is simply unknown.

Another even more intricate problem is presented by one of the main types of landscape in the Fennoscandian environment, found both in the Precambrian and the Caledonian parts. Using a term that is not particularly explicit, it has been called 'undulating hilly land'. To go somewhat further in description, it is characterized by rounded hills of varying size and altitude, more densely grouped than on the monadnock plain and interrupted, not by plains but by irregular, rounded, interconnected basins at slightly varying altitudes. The basins are often partly occupied by lakes. The slopes are mostly gentle, with the exception of isolated steep glacial facets. Open, broad, irregular valleys occur of the intermediate type described above. Landscapes of this type have been included by many workers as a typical constituent in the concept of the 'Palaeic surface' of Norway [see Gjessing (1967)]. They have been included here as a mainly preglacial relief type, originating possibly during a past warm climate, but whether it was humid or dry is uncertain. The lower slopes in areas of monadnock plains are occasionally fringed by pediments. The role of glacial erosion in this landscape of interconnected basins is, however, not clear. The only way of presentation on a small-scale geomorphological map at present is to use a purely descriptive symbol.

The various landscape types of the shield area and the typical glacial forms are also found in the Canadian Shield. The respective role of structural inheritance, preglacial erosional history and glacial transformation is not easy to decide. The necessary comparison, in adequate detail, with nonglaciated Precambrian shield areas has not yet been made. A pronounced structural control is seen as in the shields of Sinai and Saudi Arabia, but slopes

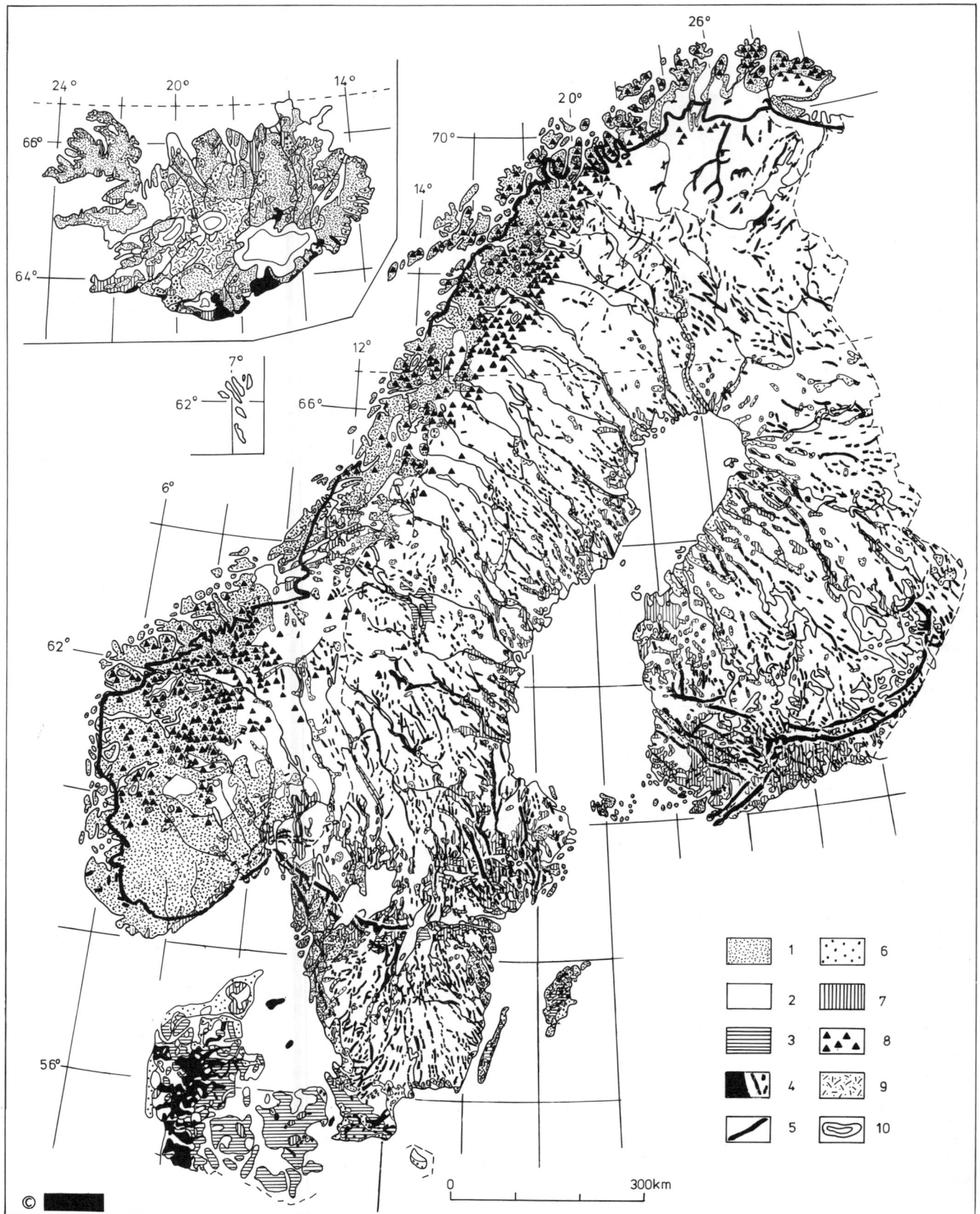

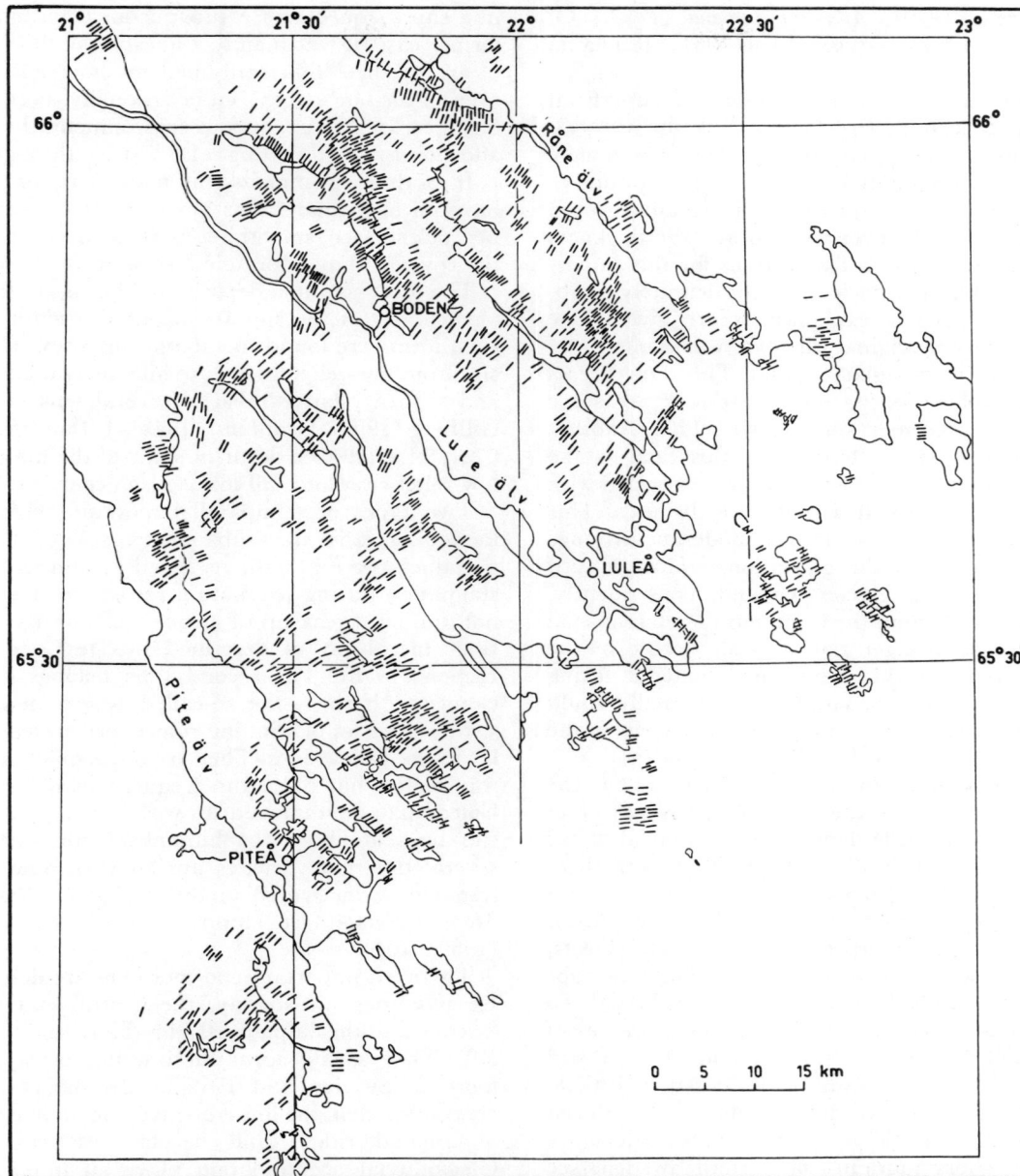

Fig. 5.8 Diagrammatic representation of the end moraines in the coastal zone of Norrbotten, Sweden. [Hoppe (1959)].

and crest forms are somewhat different. Parallels to the monadnock plains of Fennoscandia are easily found in tropical shield areas (e.g. East Africa and India) but here also slope and crest forms are different. Good parallels to the so-called undulating hilly land are still rare. It may be possible that they are old etchplains, glacially transformed. The etchplain theory has been tentatively applied quite recently for a minor area in southernmost Sweden by Lidmar-Bergström (1982). Swarms of small islands re-

miniscent of the *skärgård* areas of Finland, Norway and Sweden are found in the nonglaciated coasts of the crystalline rocks in Brittany, for instance around the Golfe du Morbihan. It would thus seem that structures are just as important as glacial erosion for the skärgård formation.

5.1.6 Quaternary Deposits and Forms

The distribution of the Quaternary deposits (*see* Fig. 5.7) and the form sequences connected with them are decided by the primary depositional environment, the type of deglaciation, the relation to older, higher sea levels, and the redistribution and transformation by various erosion processes during the postglacial uplift. Some aspects of postglacial uplift will be described in Section 5.1.7. There is a large range of literature on the Quaternary phenom-

Fig. 5.7 (*opposite*) Distribution of superficial (Quaternary and Holocene) deposits in Fennoscandia and Iceland: (1) exposed bedrock; (2) stony till; (3) clay till; (4) glaciofluvial deposits (eskers, outwash, etc.); (5) Ra moraines, Salpausselkä, (6) sand; (7) silt and clay; (8) block fields of highlands; (9) lag gravel (Iceland); (10) present glaciers and ice caps.

ena [e.g., Agrell (1979), *Atlas of Finland* (1960), O. Holtedahl (1960), G. Lundqvist (1958, 1961), Rankama (1967)].

Very characteristic of the distribution of superficial deposits in Fennoscandia (*see* Fig. 5.7) is the contrast between a western zone, where till is absent or extremely sparse—bare rock dominating all the higher ground—and an eastern zone, where the ground is usually covered by till. This means a till cover with an average thickness varying from 0.5 to 5–10 m. The reasons for this important difference in till distribution are not definitely established, but some possible explanations are given. The classical one is that an original till cover was removed by wave abrasion during postglacial uplift. This, however, is only a partial explanation as most bare-rock areas are situated above the former highest shoreline. Another explanation, which pays attention to this fact, is the following. The greatest bare-rock areas are connected with coasts that face deep sea at a moderate distance. This means a calving front at a similar moderate distance during certain phases of the glaciation, which in turn means easy removal of glacier ice and, consequently, faster ice movement behind the front and therefore glacial erosion. A result of stronger glacial erosion could be the transport of easily removable, eroded rock material during the earlier stages of the glaciations, which finally could explain the obviously low content of coarse material in the ice of the latest deglaciation [see Rudberg (1967a)].

In the bare-rock areas in western Fennoscandia, the sparse till cover is concentrated in isolated ridges or in local stoss-side and lee-side deposits. Glaciofluvial deltas and kame terraces occur in the valley bottoms, and there are often thick deposits of fine-grained glaciomarine and marine deposits. The moraines and the glaciofluvial deposits are most often interpreted as marginal deposits, connected with a standstill of the ice margin retreat for climatic or topographic reasons [see Andersen (1960)]. To the former group belong the important Ra moraines of Norway, as well as their continuation in Sweden and Finland. These have been traditionally dated to 10 000–11 000 years BP. More recent different datings have been published by Berglund (1979). There are also older lines such as the Göteborg moraines of western Sweden [see Mörner (1969)].

It is concluded that the main features of the superficial deposits in western Fennoscandia to which parts of the Shield belong (*see* Fig. 5.7) are a high percentage of bare rock, important moraines, areas of glaciofluvial deposits assembled in more or less continuous marginal lines, a few eskers, and thick deposits of silt and clay in the valleys.

Eastern Fennoscandia, to which the main part of the Shield belongs, has a different appearance in several respects. Most important are the widespread till cover, a great variety of morainic forms, long continuous eskers, numerous glaciofluvial deltas, extensive clay plains, and thick clay and silt deposits in the lower reaches of the valleys.

A significant consideration in the further distinction of deposits and forms—more important than in the western zone of Fennoscandia—is whether deglaciation occurred in a supra-aqueous or a subaqueous environment. In the former case, it was mainly a question of the downwasting of an ice sheet that continued to show some movement even in the late stages, but occasionally stagnated to form dead ice. In the subaqueous environment, the ice sheet, in addition to downwasting, also lost ice through calving.

In both environments, the main parts of the till areas have no particular forms but just show a cover over the bedrock surface, smoothing its ruggedness, notably where the cover is thin. Morainic forms, however, are frequent.

The most common depositional forms are the drumlins, particularly in the supra-aqueous environment, but also such forms are found in subaqueous sites. They range in size from low-relief flutings to hills several kilometres long and with a relative relief of several tens of metres [see Gillberg (1955), Granlund (1943), J. Lundqvist (1969a)]. Cores of solid rock occur in some of the major forms but low-relief crag and tail forms also occur.

Two types of semiparallel moraine ridges are to be found. One is the subaqueous series of minor end moraines (*see* Fig. 5.8), regarded as annual indicators of stagnation during ice margin retreat or of a regular, but not annual, break-up of an ice cliff through calving (De Geer moraines) as described by De Geer (1940) and Hoppe (1948). The second type belongs to valleys or elongated basins—the so-called Rogen moraines—consisting of series of winding ridges, orientated perpendicular to the valley sides. They are occasionally found within this region but are more frequent in the region of the Caledonian mountains dealt with in Chapter 7. The same can be said about the hummocky moraines, of which several different varieties are known; these are all represented by one symbol on the *International Geomorphological Map of Europe* [see Hoppe (1948, 1952), J. Lundqvist (1969a, b)].

The most important end moraines are described as the Ra moraines in Norway, the central end moraines in Sweden and the Salpausselkä moraines in Finland (*see* Fig. 5.9). They usually form one line in Norway, but two or more in Sweden and Finland. In the eastern part of central Sweden, the lines dissolve into an irregular pattern of isolated ridges and become less easy to follow. Glaciofluvial accumulation plays an important role in parts of these end moraine systems [see Andersen (1960), De Geer (1909), Fogelberg (1970)].

In the case of the glaciofluvial deposits, different types can be noted [see J. Lundqvist (1979)]. In the region as a whole, the most important are eskers, developed with regional patterns and running almost continuously, sometimes for hundreds of kilometres. Their general direction, which mainly varies between south–north, south-east–north-west or south-west–north-east in south-western Sweden, gives a rough impression of the ice retreat (or, in this case also, the direction of the last ice movement). The location of eskers is also influenced by the relief. They often follow the major valleys, lying along the middle of the valley or along one of the valley sides [see Rudberg (1944)]. The relative relief of the eskers ranges from a few metres to more than 50 m, the forms varying from continuous ridges to isolated hills, and the slopes from a

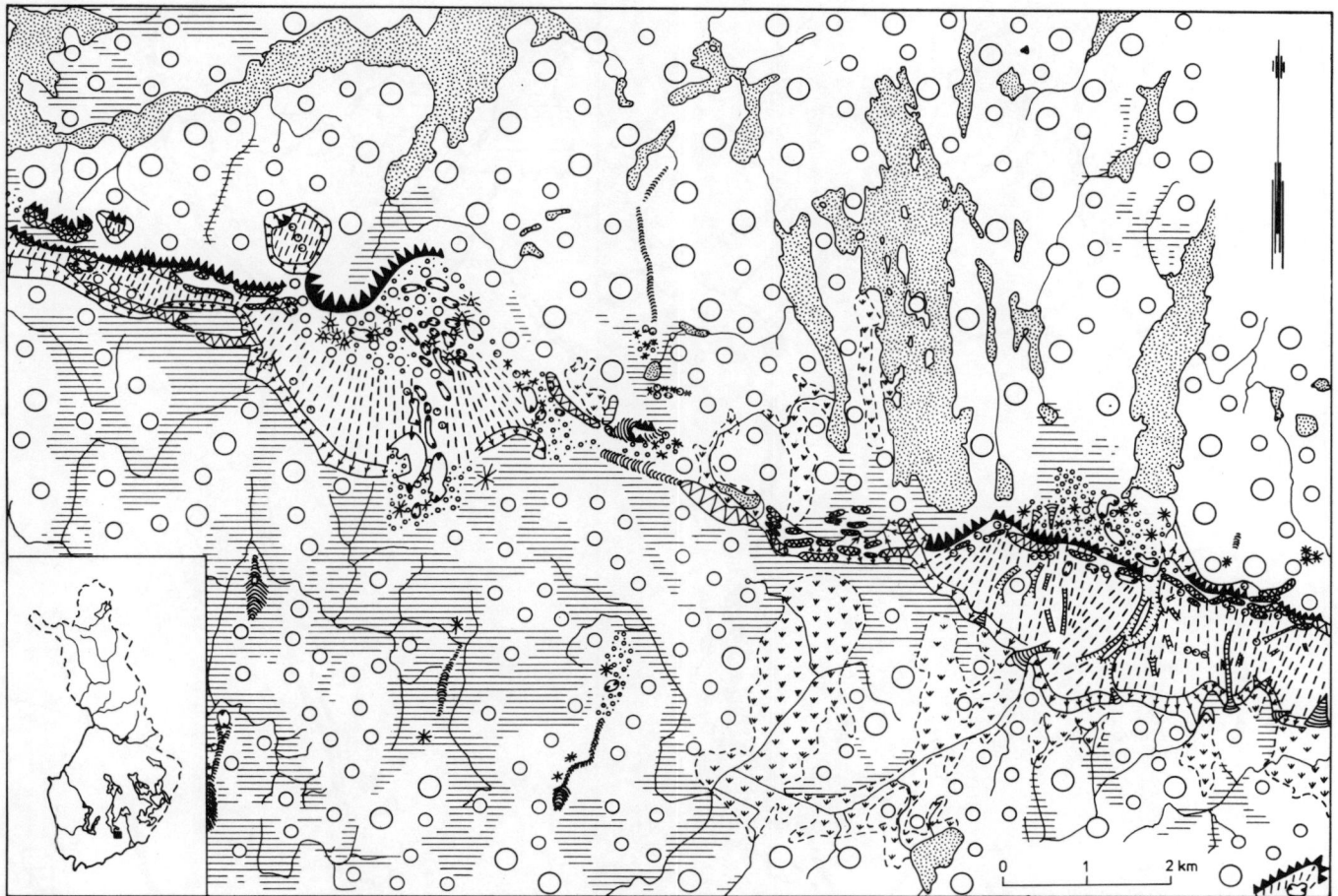

Fig. 5.9 Geomorphological map of part of the Salpausselkä system, southern Finland [Fogelberg (1970)].

few degrees to angles close to the angle of repose. An important consideration in esker geomorphology is whether the esker was built in a subaqueous or a supra-aqueous environment. In the former case, the esker may have been formed more or less as a series of successively built deltas at the retreating mouth of a subglacial river— as in De Geer's (1897) classical theory. It might also have been formed as a river bed within a subglacial tunnel in the same environment, which makes the origin more similar to that of the supra-aqueous eskers.

Of equal importance in the geomorphology of subaqueous eskers is the transformation of the original forms by

wave erosion during the postglacial uplift (*see* Fig. 5.10). The degree of transformation varies within wide limits. Eskers passing over flat surfaces where there was no protection during the uplift have often been strongly modified (e.g., by the development of surrounding sandy plains). Those that were once protected from wave attack by adjacent islands and skerries often have steep slopes and a high relative relief, consisting of isolated hills and shorter ridges with marked differences in size. They usually have no major sand accumulations around them and show a direct transition from the esker slopes to the surrounding clay plains. Closed depressions, such as kettle

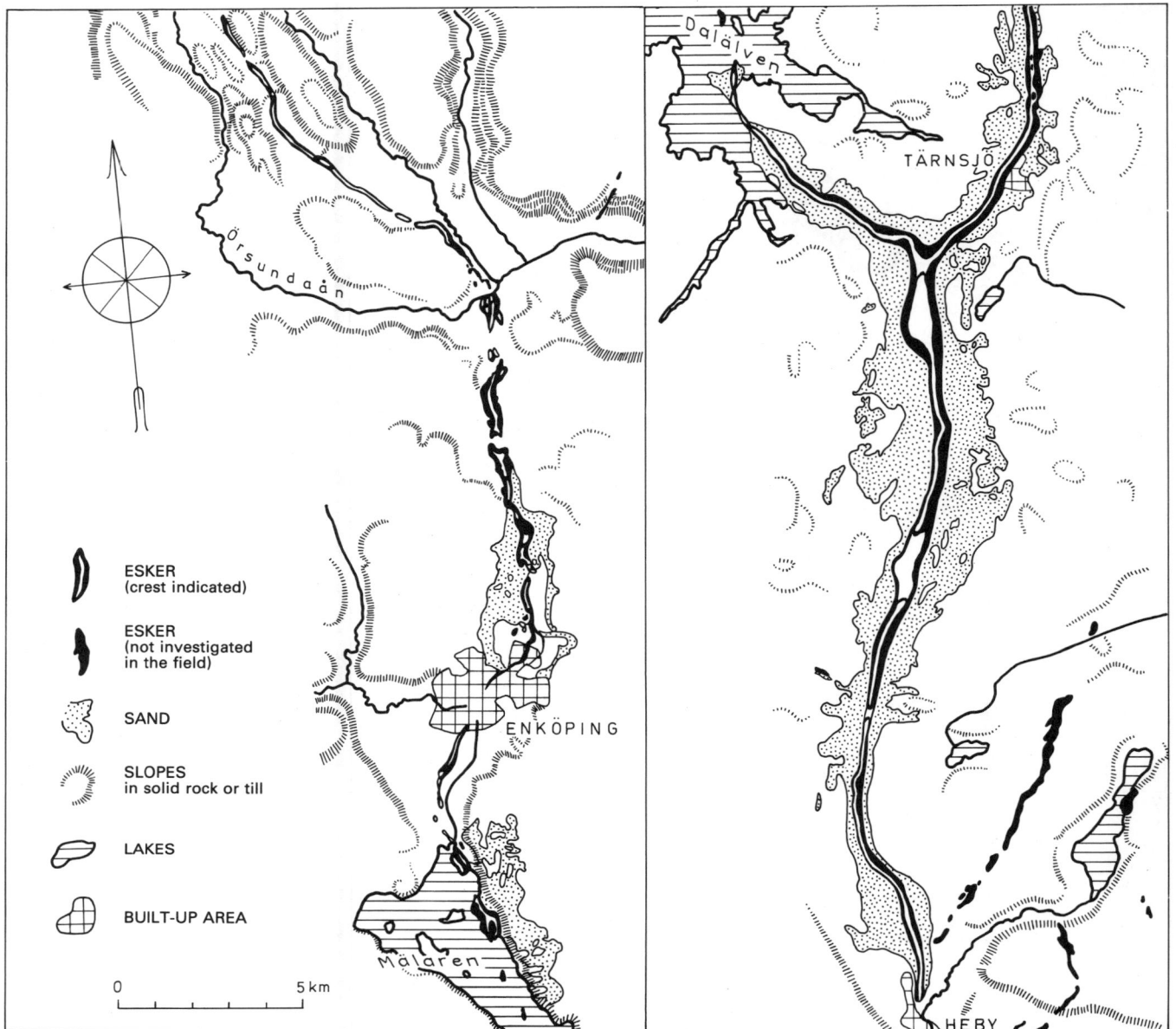

FIG. 5.10 Examples of subaqueous esker systems [Rudberg (1944)]. Left: from a hilly area north of Lake Mälaren, where the eskers were protected and only slightly affected by wave action during postglacial uplift and marine regression. Right: from a peneplain area south of the River Dalälven, where the eskers were unprotected and strongly abraded by waves during uplift.

holes or elongated depressions along the esker sides, are common. The supra-aqueous eskers share many of these characteristics, possessing steep ridges (sometimes parallel ridges or ridges forming networks) and closed depressions. Broad, sandy plains often occur around these eskers too, but these must belong to the primary stage of esker formation.

A second important type of glaciofluvial deposit is the glaciofluvial delta. The most characteristic part is the flat delta lobe, usually with steep distal slopes, and the deltas are often associated with kames and kettle holes as described by Hörner (1927). They belong to two characteristic sites. Most of them are situated in valleys, closely related to the former highest shoreline. Their general

evolution might have been as follows. At the time of deglaciation, the lower parts of valleys formed elongated bays of the sea (or ice-dammed lakes). Here, and in the valleys farther inland, the rate of ice margin retreat became slower when ice loss by calving ceased, with the ice retreat depending exclusively on downwasting. In these conditions, the mouths of the subglacial rivers remained stationary for some time, with greater accumulations as a consequence. Glaciofluvial deltas are also characteristic of the big end moraine systems of the Salpausselkä and the central Swedish terminal moraines. In both types of delta, the flat upper surfaces are supposed to coincide with the highest shore line.

A third type of glaciofluvial deposit is the sandur

(outwash) plain. Most of the reported examples are in the border zone of the Caledonian mountains (*see* Chapter 7). Fine-grained material carried by the meltwater rivers, however, was transported some distance away from the river mouths to be deposited in the sea (or in ice-dammed lakes) as clay on the floors of flat-bottomed depressions and as silt in the valleys (or in the former bays). Deposition was often sufficiently thick to cover the eskers and glaciofluvial sands that were formed some time before. In freshwater bodies, the clay and silt show varves. The great clay plains are located around the major esker courses. The varved clay is often covered by younger nonvarved clay formed during postglacial uplift, as will be discussed in Section 5.1.7.

5.1.7 Postglacial Uplift and Present-Day Geomorphological Processes

Postglacial uplift, with a present maximum rate of 0.9 m per 100 years, has resulted and still results in wave activity along receding shorelines and enhanced fluvial erosion wherever the lowering of base level and local conditions are favourably combined. The superficial deposits have been reworked by this wave activity with varying results. In the case of a till cover, this means—in increasing order of change—superficial removal of fines, visible enrichment of stones and gravel in the surface layers, formation of shingle fields and finally removal of the till cover, with the exception of some residual boulders. Most till-covered slopes have not reached the later stages of the series and the geomorphological changes are usually not so pronounced. Shore terraces have been formed on exposed slopes, but the morainic forms are usually not changed beyond identification. Sandy beaches occur within till areas, but are not particularly frequent. A more exclusive case of substantial change is the so-called *kalottberg*. The term is used for isolated bedrock hills whose summits rise above the former highest shore level. These hills preserve an original, substantially unmodified till capping (or *calotte*) whereas on the slopes below, the till deposits have been more or less washed away.

Alteration of eskers by wave activity has usually been more pronounced than the transformation of till-covered slopes and moraines, the eskers with their well-sorted layers of sand and fine gravel presenting material of high erodibility. The eskers unprotected against stronger wave activity have been gradually reduced, giving them broad crests; they are occasionally covered with shingle fields, and the eskers themselves have been welded together to form elongated, continuous ridges by longshore drifting. Closed depressions have been infilled and there are often broad secondary sand accumulations along the bases of the sides.

The postglacial clays covering the glacial clay are derived mainly through reworking (by wave activity) of older deposits during the uplift. The source has been mainly older clay and glacial till. The postglacial clay is not varved. In well-mapped clay areas, the older glacial clay is found on slightly higher ground, at valley sides or around isolated hills [see Hörner and Järnefors (1956)].

Fluvial erosion caused by the postglacial uplift has had

an easy task in the Quaternary clay, silt and sand deposits. The results of this very youthful erosion are sometimes spectacular, especially in terms of the size and variety of forms. The clay plains are often dissected by recent valleys, eroded by small rivers or rivulets. During incision, the rivers often meandered and characteristic small-scale form sequences with abandoned loops have developed (*see* Fig. 5.11). A dendritic network of small valleys and gullies is often connected with the main valley. These forms are even better developed; their sizes are greater in the thick silt layers of major valleys, notably in areas with a higher rate of uplift. Earlier stages in landform development are preserved in terraces, abandoned river beds and the meander series of abandoned loops in some rivers. The gullies are numerous in areas of fine silt, sometimes forming patterns of almost badland appearance [see Arnborg (1959), De Geer (1911), Koutaniemi (1979), G. Lundqvist (1951), J. Lundqvist (1957), Sundborg (1956)].

In most of the major valleys, the form development and the distribution of particular form sequences are complicated by the down-valley retreat of the shore and the river mouth of the postglacial river. At these temporary mouths, deltas were built. The bulk of the accumulated material was gained by erosion of the older silt and sand deposits. Later, such deltas were dissected by the river, new deltas were formed down valley and so on. The present river delta at the present river mouth is the latest step in the developmental sequence. As the uplift is still going on and the time for delta building has not been long, most of the recent deltas are rather small, which is in contrast to some lake deltas, where the development in the upstream part of the river valley has been the same as described above. The final arrangement of deposits in the river valleys is an irregular succession of sand and silt-dominated sections as described by Sundborg (1977).

In contrast to the rapid postglacial erosion in the unconsolidated deposits, erosion initiated by uplift has made no significant progress in solid rock. The exceptions are few: wave-cut cliffs occur at places where the total uplift has been small (as in Skåne) or in some less-resistant rock types such as the Cambro-Silurian sandstone or limestone [see Rudberg (1967b)] or in unconsolidated deposits [see Norrman (1964)].

The glacial sediments, particularly the glaciofluvial sands, have provided opportunities for wind activity, and dune fields are common in certain places within the region. They were obviously formed within a short time after the deglaciation, when vegetation was still poorly developed [see Hörner (1927), G. Lundqvist (1951), Seppälä (1971)]. Except for the dunes along the present shore, all are fossil but may be temporarily reactivated after forest fires or human disturbance as described by Bergqvist (1981).

Below most steep rock faces of any size talus slopes are found. Some of these may be developing only slowly; a few may be fossil features. Some steep slopes, at least in the maritime south-west of Sweden, lack any talus deposits: one reason may be that they have only been exposed to weathering for a short time because of recent uplift above

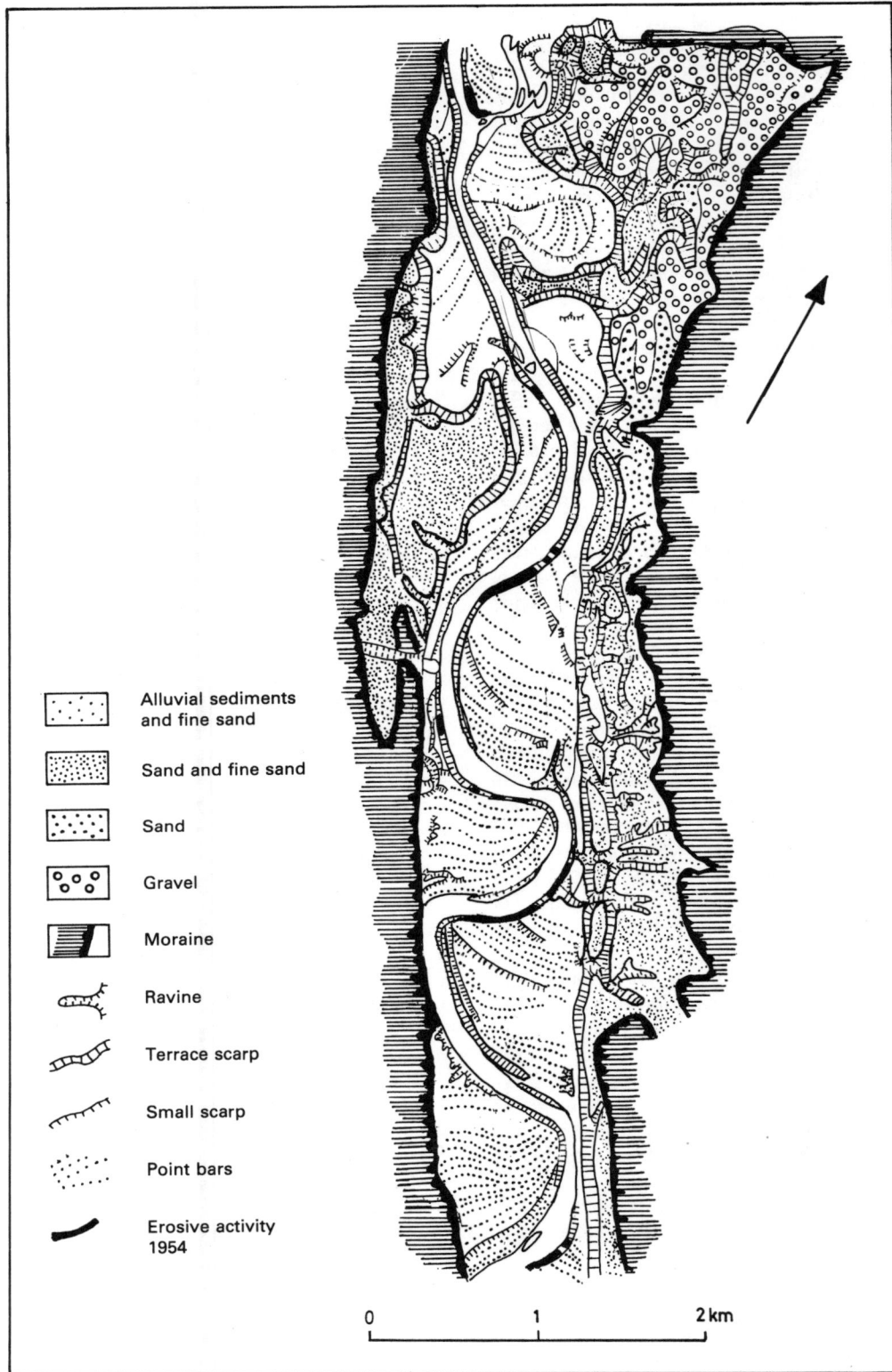

Legend:

- Alluvial sediments and fine sand
- Sand and fine sand
- Sand
- Gravel
- Moraine
- Ravine
- Terrace scarp
- Small scarp
- Point bars
- Erosive activity 1954

0   1   2 km

Fig. 5.11 Effects of river incision during postglacial uplift: the valley of the Klarälven cut into glacial and late glacial sediments [after Lundqvist (1957)].

sea level. Of local importance are the great landslides of the quick-clay type in south-western Sweden and parts of Norway [see Frödin (1919), Caldenius and Lundström (1956), Sundborg and Norrman (1963)]. The landslide scars are visible for hundreds of years after each catastrophe.

Results of the recent/subrecent periglacial processes within the region are the almost ubiquitous earth hummocks and the 'boulder depressions' (results of completed frost sorting) in most parts of the region except, in the case of boulder depressions, for some southern and western areas [see J. Lundqvist (1962)]. Sorted polygons are locally seen along lake shores in the north. Fossil tundra polygons have been reported at several places in Sweden—south of the central Swedish end moraines—both in plan and in vertical sections [see Svensson (1964b)].

5.1.8 Regional Subdivisions

Separation of the Fennoscandian Shield from the Caledonian highlands (*see* Chapter 7) is an arbitrary division in many respects. It does not conform to the Fennoscandian scientific tradition and is unfamiliar to people of Finland, Norway and Sweden. The scheme adopted in this case, however, is one in which the morphology and the geotectonic evolution are selected as the main criteria for regional differentiation, even though this presents problems in Fennoscandia and some other areas of Europe.

A first approach to the regional subdivision is to prepare a map of major relief types—slightly modified from Rudberg (1960) (*see* Fig. 5.12). The plains shown on this map are mainly planation surfaces. Important parts are formed by the exhumed sub-Cambrian peneplain, in central and southern Sweden, and most probably also in the vast plains of western Finland. Some Swedish parts are represented in more detail in Fig. 7.1 using data from Rudberg (1954) and later. The planation surface of southernmost Sweden is probably of Tertiary age [see Lidmar-Bergström (1982)] although it was regarded as Mesozoic in older papers. The peneplain of interior and southern Finland was considered by Tanner (1938) to be probably younger than the sub-Cambrian peneplain, and it may be the equivalent of the southern Swedish surface, or perhaps older.

The fissure valley landscapes of Fig. 7.1 (subregion 2) are mainly plateaus of low or moderate relief, dissected by valleys and basins, strongly controlled by structure. The plateaus often show a striking accordance of summit level and, in part, are interpreted as remnants of the sub-Cambrian peneplain.

The relief types 3, 4 and 5 of the legend to the *International Geomorphological Map of Europe* all belong to the poorly defined category of undulating hilly land (*see* Fig. 7.1). They are separated according to relative relief in order to show the regional differences. The main part of the southern Swedish highlands and interior Finland (i.e. the Lake Plateau) belongs to the lower classes of relative relief, while the northern Swedish parts belong to the highest class, except for a zone transitional to the lowlands, quite well-marked in places (equivalent to the *limes Norrlandicus* of the botanists). The subregion of undulating hilly land is broader in the south, but rather narrow in the

north. As in the case of many other subregions in northern Sweden and southern Norway, the undulating hilly land forms a zone transverse to the main rivers. The next subregion (type 5) (i.e. the monadnock plain) also has a zonal appearance in northern Sweden, being narrow in the south but widening to the north where it is continued by vast areas in Finland towards the Russian border.

Relief type 7 (i.e. the premontane subregion) mainly belongs to the Scandinavian highlands and is discussed in Chapter 7.

The small-scale forms of glacial erosion, notably the roches moutonnées, cannot be shown in Fig. 7.1. Generally they are far better developed in the Precambrian crystalline areas than in other rock provinces, but they are often hidden by Quaternary deposits. Roches moutonnées as dominating landscape features all belong to the bare-rock areas (*see* Fig. 5.6), while individual examples occur in most parts. Glacial erosion forms of meso-scale, such as rock basin lakes, giant roches moutonnées, flyggbergs, etc., cannot be shown for the whole area because of incomplete information. Rock basin lakes are particularly frequent within certain areas, for example, in the transition zone to the highlands, in central southern Sweden, on the Lake Plateau of Finland and in the fissure valley landscapes. Flyggbergs are found in most areas of hilly or mountainous relief; good examples are clustered in the middle part of Precambrian northern Sweden.

Partially submerged areas of low to medium-scale relief with scattered hills and knobs form the well-known Scandinavian *skärgård* (Swedish spelling; Norwegian: *skjergård*) or archipelago areas with swarms of rocky, glacially sculptured islands and skerries (*see* Fig. 5.13), often with strong structural control. They follow most of the Norwegian coast within the region and important parts of coastal Sweden and Finland. Consequently, cliffed coasts are restricted to minor areas, the most important being the western coast of Gotland and a few isolated points in Skåne.

A small area with a very special appearance is the 'table mountain' region of Västergötland in south-western Sweden, indicated by a separate symbol. Here, some broad mesa-like hills (rather than mountains) consist of flat-lying Cambro-Silurian sediments with cappings of Permian dolerite.

Analysis of subregions could also be based on the distribution of unconsolidated deposits; these are shown on a small-scale map for the whole Fennoscandian area (*see* Fig. 5.7). The most spectacular regional division is the contrast between the bare-rock areas of western Scandinavia and the till-covered eastern part. The boundaries of the bare-rock zone cross the limit between the highland region (*see* Chapter 7) and the shield region described in this chapter. Till is by far the most frequent glacial sediment in south-eastern Norway, in the Swedish part of the region and in Finland. The till properties vary within wide limits according to grain size, boulder frequency and content of nutrients. The variations, however, although strongly related to the bed rock and important for land use, cannot yet be mapped for the whole area. One exception to this is the fine-grained boulder clay associ-

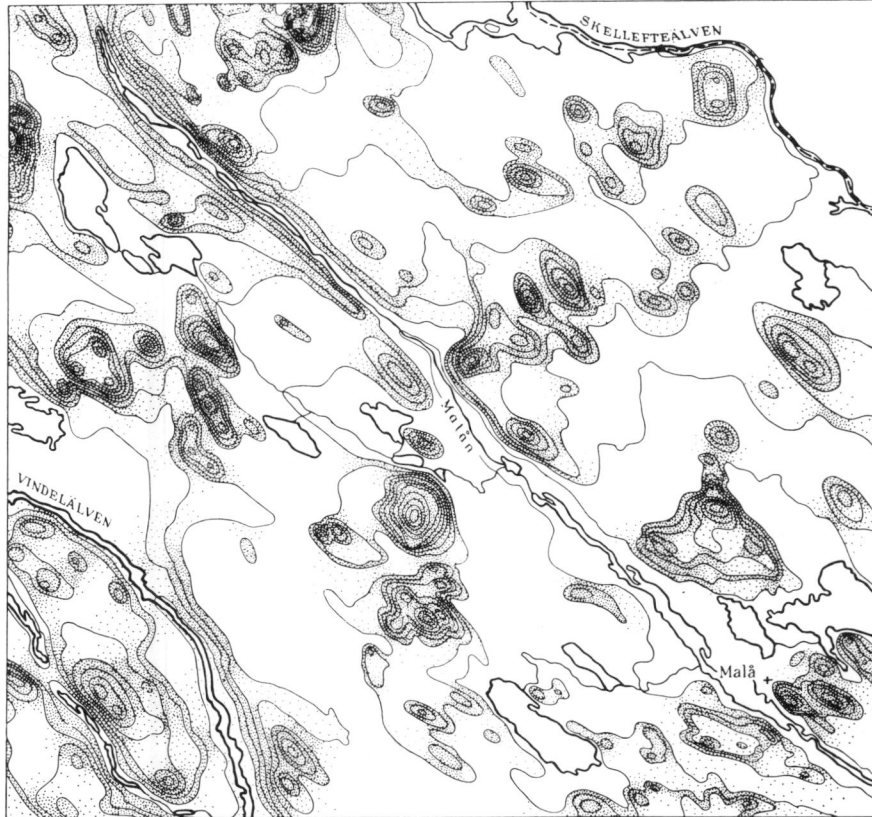

(A)

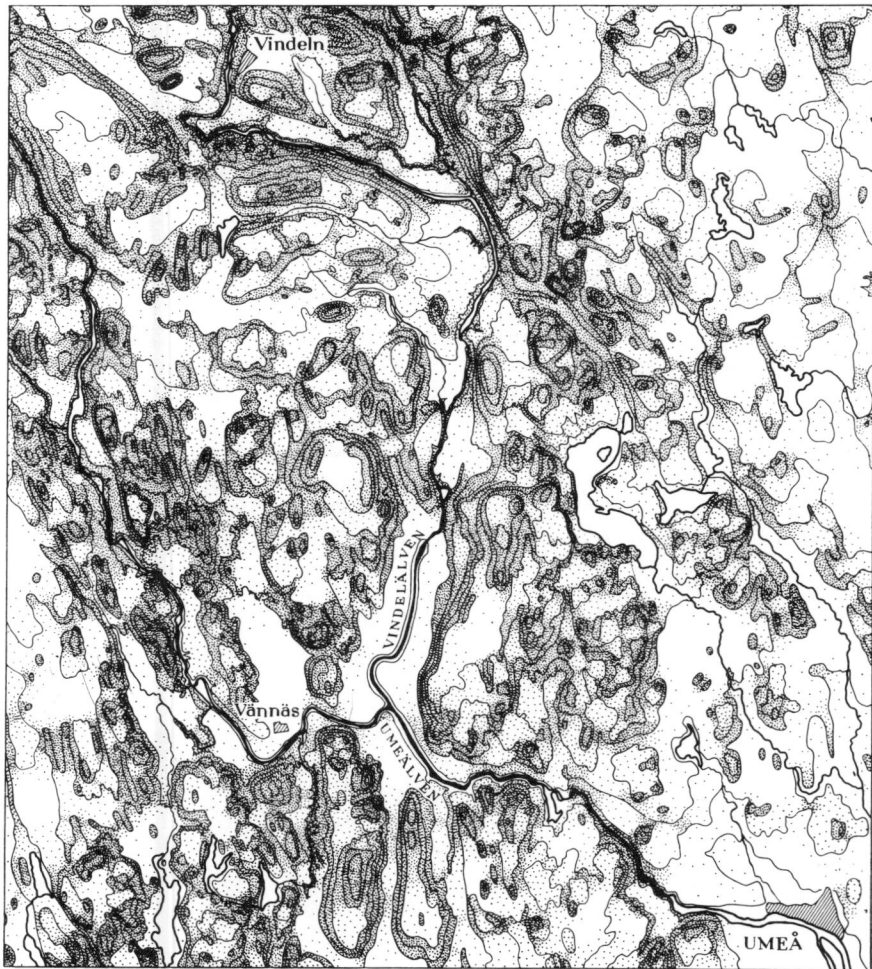

(B)

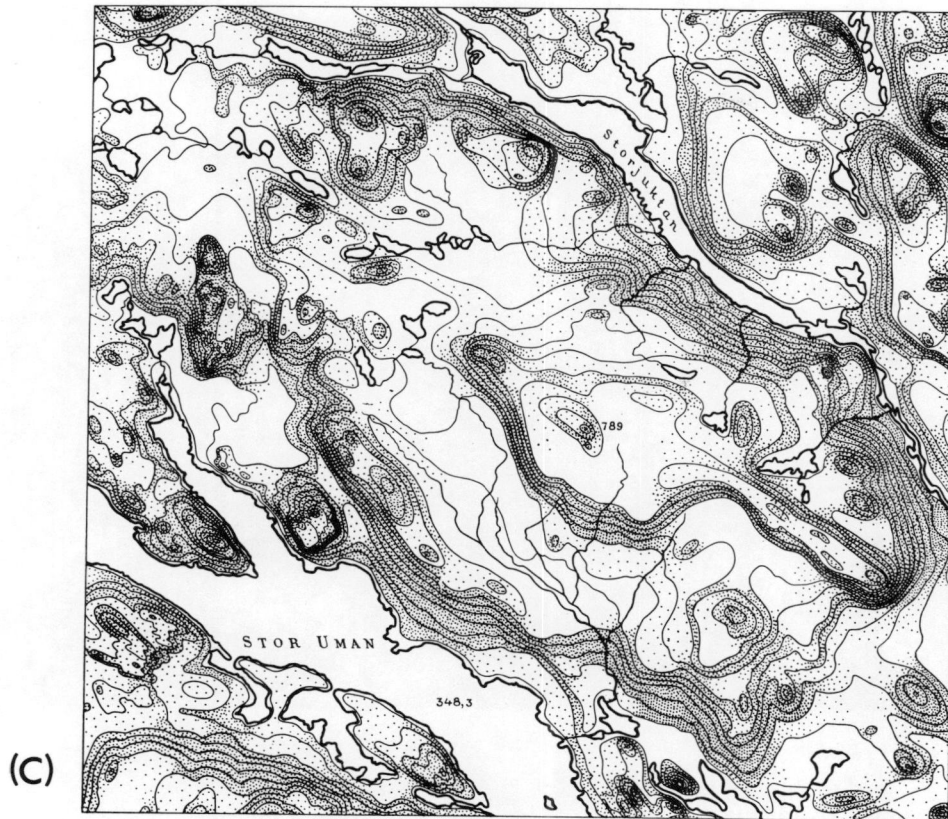

(C)

Fig. 5.12 Relief types in Väterbotten, northern Sweden [Rudberg (1954)]: (A) monadnock plain; (B) undulating hilly land; (C) premontane region. Surfaces without dots are plains with slope angles less than 1 degree. Contour interval 25 m. Scale approximately 1 : 300 000. See also Fig. 7.1.

Fig. 5.13 Glaciated lowland coast: the islands (skärgård) and fiords of western Sweden. (Rudberg)

ated with the areas of younger sedimentary rocks.

Another striking feature on the map is the difference in the distribution of eskers. In the west, these are few, glaciofluvial material being concentrated in marginal deposits. In the east, long esker courses form a dominating pattern, with the most typical areas in eastern central Sweden (i.e. the Lake Mälaren area) and in south-western Finland. The distribution of clay and silt sediments can be followed in a general way on the small-scale map. The sediments are found in local concentrations below the former highest shoreline of the sea or of former ice-dammed lakes. As most of the material was derived from meltwater rivers, the more important areas of fine-grained sediments are often found around the eskers.

A major part of the interior of southern Finland is situated below the former highest shoreline and therefore areas of fine-grained sediments are more frequent here. The high ground of Norway confines clay and silt deposits to some narrow coastal strips and the low-lying floors of the main valleys. Sweden is in an intermediate position, with the southern and central parts more reminiscent of Finland, while in the north silt and clay deposits are restricted to valleys.

The regional distribution pattern of landforms related mainly to present-day processes is less easy to present on a small-scale map. The visible results of recent/subrecent modifications are frequent in the broad zone that was under water after the melting of the ice sheet. Within this zone, the till was affected by wave action during uplift and changed in the ways already described. The minor changes are most frequent, but shingle terraces and bare-rock areas are locally important, notably in the western part of the region and also in some elevated sites in the interior which, during earlier stages of uplift, were exposed to waves with a long fetch. The fine-grained deposits of the valleys and low-lying plains are dissected by river erosion, resulting in steep-sided valleys, terraces, meander bends and gullies, notably in the major valleys in Norway and in central northern Sweden.

5.2 Kola Peninsula and Karelia

These areas form extensions of the Fennoscandian Shield into the USSR. The Kola Peninsula is clearly defined and projects eastwards into the Arctic Ocean and the White Sea. Karelia, lying between the White Sea and Finland, has no distinct boundaries except for the White Sea coast to the north and is essentially a continuation of the Lake Plateau of Finland. The eastern limit of Karelia in terms of geological structure is marked by the disappearance of the Precambrian shield rocks beneath the Palaeozoic cover rocks of the Russian platform.

5.2.1 Kola Peninsula

The main relief features reflect the north-west–south-east Precambrian structural trends, emphasized by much younger neotectonic faulting. Faulting occurred not only along the north-west–south-east lineations but also at right angles to these, producing a series of dislocated blocks. In the central parts of the Kola Peninsula, these low block mountains rise to 900–1100 m—the highest elevations in the Peninsula—corresponding structurally to the old central Kola anticlinorium. The present low mountains may bear traces of an ancient peneplain that was fragmented by differential vertical movements. Between the mountain ridges lie downfaulted lowland areas. To the north of these central uplands, two separate regions can be distinguished: the denuded lowlands of north Kola and the low mountains of Keiwy. To the south lie the south Kola lowlands extensively covered by glacial and marine deposits, and also some low block mountains [see Kratz (1963)].

The coastline of the Kola Peninsula consists mostly of fault scarps. Differential denudation of the crystalline and metamorphic rocks appears to be less important in the formation of the relief than recent fault tectonics and gives rise to only minor landforms. Much of the faulting represented the reactivation of old tectonic lines, but some new faults were also created.

During the long interval, unrepresented by any sedimentary deposits, between the Precambrian and the Quaternary, several phases of planation appear to have occurred. The oldest planation surface is claimed to date from the Precambrian, and its formation was followed by various epochs of tectonic uplift and denudation. In the Precambrian, the beginning of the Riphaean phase saw a general domal uplift of the crystalline rocks, but thereafter, from the Caledonian orogeny onwards, all tectonic movements took the form of block faulting. During the Palaeozoic era, only the peripheral parts of the Fennoscandian Shield were covered by marine transgressions, which may not have affected any part of the Kola Peninsula. The most important phase of planation occurred probably in the Mesozoic, and certainly postdated the Devonian intrusive activity dated at 370 Ma. This basic Kola planation surface was veneered with a thick weathered mantle [see Sidorenko (1958)] now largely removed by subsequent erosion (especially glaciation). In the Cenozoic, particularly during the Alpine orogeny, the surface was broken by faulting to give the present-day denudational–tectonic morphostructure as described by Metcherikov (1965), Nikolayev (1962) and G. D. Richter (1936). On most of the uplifted blocks, such as the Khibiry Mountains, only small remnants of the basic Kola planation surface have been preserved. In the eastern parts of the Kola Peninsula, fragments of the surface remain on some interfluves [see Armand and Grave (1974)].

In the Quaternary, the whole area was covered by the Fennoscandian ice sheet, probably during each glacial stage, although evidence only remains from the last glaciation. The ice moved generally eastwards and southeastwards, the weight of ice causing isostatic depression of the crust, though because the Peninsula lay some distance from the area of maximum ice thickness in northern Sweden, the subsidence was less. Estimates of present-day glacioisostatic recovery are of the order of 1–4 mm per year. Part of the uplift has taken place by renewed movement along faults; the pattern of movement has been reconstructed by Aseev and Makkaveyev (1977). The resulting changes of base level have been significant for postglacial fluvial activity.

The amount of glacial erosion over the Kola Peninsula appears to have been relatively small; estimates suggest removal of a layer of bedrock 3–6 m thick [see Makkaveyev (1975)]. In making such estimates, considerable reliance has been placed on the few surviving relics of the pre-Quaternary weathered mantle referred to above. Although the amount of overall surface lowering was small, the volume of rock transported by the ice must nevertheless have been immense, and both till and glaciofluvial deposits are widespread. In detail, too, the landscape was greatly modified, with numerous rock basins, plucked rock faces and streamlined forms including roches moutonnées being produced by the ice. Differential glacial erosion resulted in the deepening of tectonic depressions, while on the north coast locally near Murmansk, glacial troughs and fiords are found. In the central uplands some other examples of glacial troughs occur, together with cirques and lake basins.

The river patterns are strongly influenced by tectonic lineaments. Their long profiles, however, show more directly the effects of glacial erosion, containing sequences of rock basins holding lakes. Many of the deeper valleys, which are of preglacial origin, have been modified by ice and partially filled with glacial deposits.

Glacial deposition is extensive. The features, including eskers, kames and glaciofluvial deltas, all relate to the last glaciation, and most of them to the last deglaciation when the ice sheet stagnated. Meltwater channels are also frequent.

On the coasts and around estuaries, wave-cut terraces and raised beaches are common, providing evidence of the postglacial isostatic recovery. Because of differential fault movements during the recovery, the correlation of remnants of former shoreline features is difficult, so that no clear sequence has yet been formulated.

5.2.1.1 Geomorphological regions

The following relief divisions are based on the work of Armand and Grave (1974).

1) The Murmansk coastal area comprises a region with elevations up to 300 m. The landforms in this region are structurally controlled by the granitic and gneissic rocks of the Murmansk block, which is subdivided by faults into smaller units. The near-horizontal structural surfaces of the Rybachi and Sredniy peninsulas are developed in Precambrian sediments. The fiords and glacial erosion forms have already been mentioned. To the south-east is an undulating coastal plain composed of till.

2) The Pechenga–Kola uplands lie to the south of the Murmansk coastal region. The heights here range between 400 and 700 m. The upland surface shows clear evidence of glaciation; dissection increases farther southwest in the Keivy highlands, although heights here are less (about 400 m). The relief is developed on the Precambrian rocks and is controlled by numerous faults. On the higher parts, typical glacial erosion forms appear.

3) The central Kola mountains are block-faulted mountains rising to 800–1000 m. The highest parts are related to intrusions of gabbro and syenite. Some flat tops and accordant summits probably represent relics of the basic Kola planation surface.

4) The White Sea coastlands are characterized by a

surface of low relief (between sea level and 200 m) cut across gneissic rocks. There is a complex of glacial and marine depositional forms, including raised shoreline features of postglacial age.

5) The Karelian borderland is formed by the Kondorskiy massif (400–600 m) and the Kandalaksha mountains (600–800 m). These highlands are developed on gneiss and granulite bedrock, considerably affected by glacial erosion. Near to the Gulf of Kandalaksha in the south, several marine abrasion levels are present.

5.2.2 North-Eastern Finland and Karelia

The limits of this region and its relationship with the Lake Plateau of Finland have already been noted (*see* p.74). The Precambrian crystalline rocks that underlie the whole area are somewhat younger than those of the Kola Peninsula—around 1650–1750 Ma, compared with 2000–3400 Ma for Kola—although this has no geomorphological significance since the landforms date from much more recent epochs. The relief is characterized by denudational plains of low relief. In eastern Finland, the Maanselkä low mountains attain heights of a few hundred metres only, and extend into Karelia as the low ridge of Vetrennyi. Another spur forms the western Karelian upland. Numerous lakes occupy shallow basins on the Fennoscandian Shield, the work of glacial erosion selectively exploiting tectonically downfaulted zones. The two largest lakes—Ladoga and Onega—lie close to the south-eastern limits of the region where the Precambrian basement rocks disappear beneath the Palaeozoic cover of the Russian platform. The Gulf of Finland is geomorphologically and structurally similar to these lakes, the only difference being in the accident that it is occupied by an arm of the sea rather than by freshwater.

The dominant structural lines trend to the north-west or north-north-west and are expressed not only by tectonic lineaments, but also by outcrops of varying resistance to erosion. The softer formations include schists and conglomerates, which have been denuded to form lowland strips. The harder rocks, especially quartzites, stand out as rugged ridges whose details have been streamlined (rock drumlins, crag and tail features, bedrock fluting) by glacial erosion. The picture is complicated by the fact that the ice, moving mainly eastwards, crossed the structural lines at an oblique angle. Numerous lakes occupy eroded rock basins and tectonic depressions, while the river patterns are strongly controlled by structure [see Biskey (1959)]. Glacial deposition forms, especially eskers, abound on the low-angle slopes [see Armand and Grave (1974)].

5.2.2.2 *Geomorphological regions*

North-eastern Finland and Karelia comprise four main divisions.

1) The Finnish–west Karelian uplands (200–600 m) consist of low block mountains separated by grabens.
2) The eastern Karelian lowlands (150–300 m) are marked by low undulating relief, strongly modelled by glaciation and containing numerous lakes.
3) The Onega–Ladoga denudational plains (5–150 m) have a stepped relief and some low rises, such as the Olonets upland.
4) The Vetrennyi ridge rises to 200–300 m.

Chapter 6
Russian Platform

For a long time, the relationships of the main relief forms to tectonic structure over the area of the Russian platform were interpreted as a passive response to lithology and structure—an expression of differential rock resistance exploited by exogenic processes. The introduction of morphostructural concepts and methods into Soviet geomorphology [see Gerasimov (1946, 1959), Metcherikov (1968)] has provided a new interpretation of the main relief features in terms of tectonic movements and the deeper structures of the Earth's crust. The tectonic movements that formed the present-day morphostructures were not always in accordance with the tectonic movements of ancient geological periods. Morphostructural analysis of the Russian platform and European lowland supported a theory of the periodic development of morphostructures by alternating periods of uplift (and resultant dissection) and periods of stabilization (planation), reflecting the rhythmic nature of young (Mesozoic–Palaeogenic) and the youngest (Neogene–Quaternary) tectonic movements. Geomorphological evidence, such as planation surfaces and river terraces, and correlative deposits were used as a basis for these structural/geomorphological studies, and the ages of the main relief forms have been determined.

The basement of the Russian platform is formed by an irregular faulted pre-Palaeozoic floor. Only the most north-eastern and southern parts are younger (Palaeozoic). The basement is generally overlain by a cover whose thickness varies from 3–5 km to 10–15 km. The cover consists of slightly folded, faulted Palaeozoic, Mesozoic and Cenozoic deposits. Only small areas of the basement outcrop at the surface, representing miniature shields of intrusive and metamorphic rocks. The subhorizontal layering of the platform cover is the reason for the widespread development of tabular relief features in the lowland. Although the basement often lies at great depths, it has strongly influenced tectonic movements and the main structural features of the Russian lowland. The Ukrainian Shield and the central part of the central Russian uplands (the Voronezh basement block) are examples of the reflection of uplifted basement blocks in the surface relief, while the plains developed over the Moscow, Mezen', Dnepr, Caspian and other syneclises express underlying areas of subsidence.

In this way, the major geomorphological units of the Russian platform are directly connected with features of the basement. Smaller geomorphological units are associated with neotectonic structures in the platform cover, for example, with syneclises (such as the Baltic and Polesye syneclises) and with tectonic ridges (such as the Timan and Vyatka–Uval Ridges). Some smaller geomorphological units, such as the upper Volga lowland or the Klin–Dmitrov upland are not connected with structural elements of the basement but were formed by exogenic processes acting over a prolonged period. According to Metcherikov (1965), geomorphological units tend to be connected closely with the structure of the basement in the more marginal parts of the Russian platform, while units of exogenic origin dominate the central areas.

As already mentioned, there are a few places where the platform basement rises to the surface. The largest of these areas is in the central part of the Ukrainian massif, while smaller areas of the Precambrian basement outcrop in the Belorussian massif. The Donets Ridge and Timan Ridge are formed by outcrops of younger (Palaeozoic) basement rocks. The transition between the Precambrian basement and the younger (epi-Hercynian) Scythian platform in the south-eastern part of the Russian lowland (the Crimean lowland and the Fore-Caucasian depression) is not discernible in the present-day relief, but neotectonic movements are more active on the younger epi-Hercynian platform.

The margins of the lowland next to the Alpine orogenic zone are characterized by dissected relief. The Stavropol' plateau reaches heights of up to 830 m, the Volyn'–Podolsk plateau up to 470 m and the Donets Ridge up to 360 m, whereas the Caspian lowland lies up to 28 m below sea level. Neotectonic activity is also greater in these areas, influenced by the proximity of the mobile Alpine zone: the Carpathians, Black Sea depression and Caucasus (see Fig. 6.1).

A major role in the formation of the contemporary relief of the Russian lowland has been played by the residual tectonic activity of the basement, by neotectonic activation of the sediments of the platform cover and by mega-anticlinal and megasynclinal movements of the Earth's crust, which have been the cause of the pattern of submeridional and sublatitudinal zones of uplift and subsidence. In the southern part of the Russian lowland, there are four uplift zones with upland relief and three subsidence zones trending in a submeridional and south-easterly direction.

The age of the basic morphostructures in the southern part of the Russian lowland (see Fig. 6.2) is Palaeozoic and early Mesozoic. The worn-down surface of the basement had already been divided into blocks in the Precambrian era, forming grabens and horsts. The higher relief was formed by the Ukrainian Shield. Later the Dnepr–Donets aulacogenic (rift) depression was formed in the early and middle Devonian, dividing the Ukrainian Shield into two massifs. The outlines of the present-day relief were

Fig. 6.1 Major geomorphological regions of the Russian platform. The numbers correspond to the numbered sections of the text.

apparent at the end of the Permian and the beginning of the Triassic. At this time, the folded area of the Donets Ridge was created, forming later a part of the higher area of the Ukrainian massif.

New morphostructural features formed during the Oligocene and Miocene in association with tectonic movements in the Alpine geosynclinal zone. Some morphostructures became inverted: formerly low-lying areas became uplands (e.g., the Volga Heights) and higher areas became lowlands (e.g., the Tambov lowland). The tectonic phase of the early Pliocene gave rise to the inverted morphostructure known as the Volyn'–Podolsk highland, and the complex morphostructure of the Ergeni Hills. The morphostructure of the Stavropol' plateau, which after its formation at the end of the Palaeozoic–Early Mesozoic stood for a long time—from the Cretaceous to the end of the Miocene—below sea level, was reactivated. An important role in relief formation was played by the Black Sea depression. Subsidence of the Black Sea area was followed by formation of the extensive

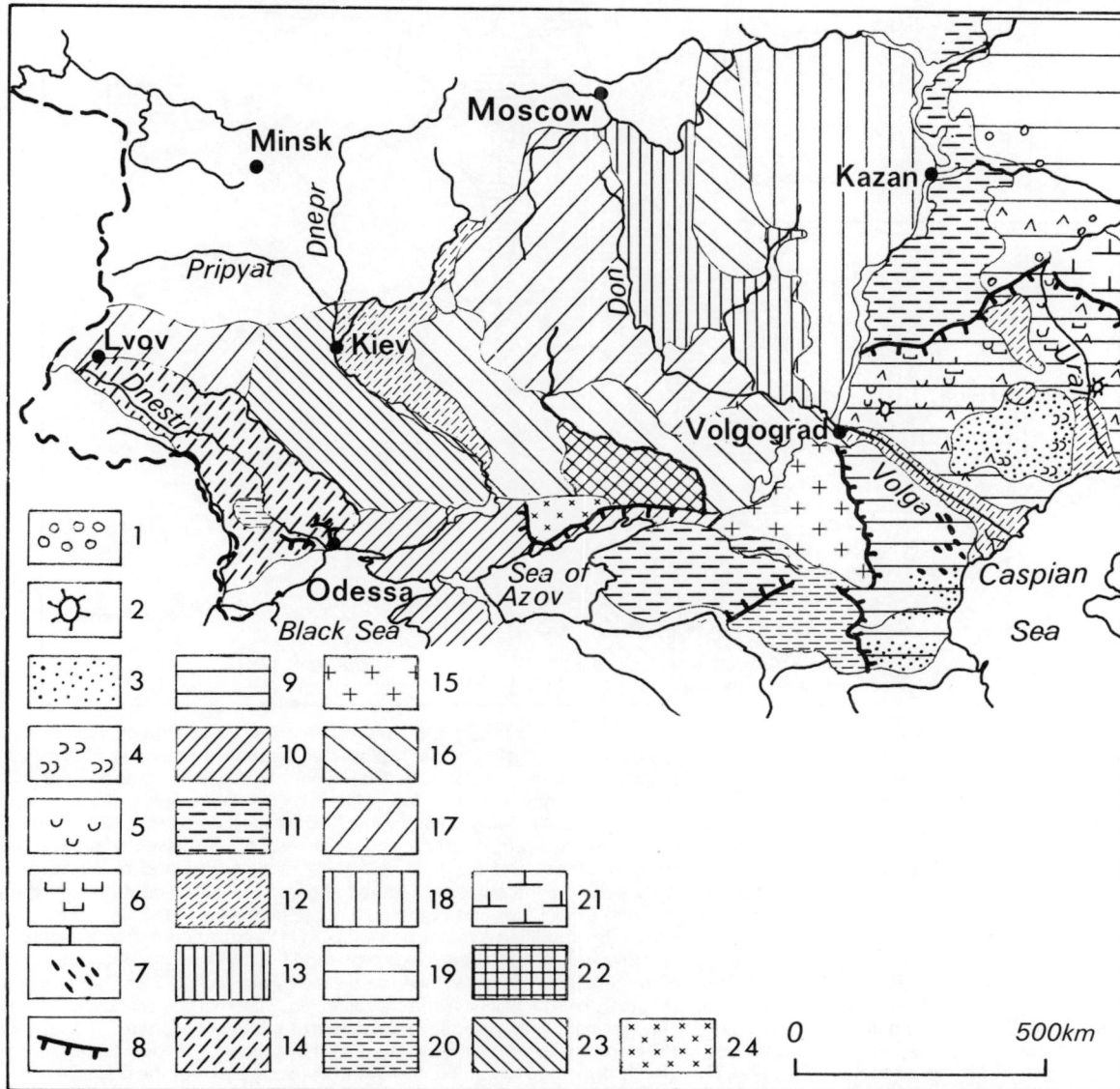

Fig. 6.2 Morphosculptural regions, southern part of the Russian platform. (A) Late Quaternary depositional plains (slightly uplifted in the Upper Pleistocene or Holocene): (9) marine lowlands; arid and undissected; (10) loess-covered coastal plains, mainly undissected; (11) fluvial–deluvial plains; (12) terraced fluvial and deltaic plains; (13) glaciofluvial and outwash plains, with loam cover; (14) arid plains, with dissected margins; (15) other depositional plains. (B) Pliocene and pre-Pliocene disected uplands: (16) undulating structure-controlled uplands; (17) undulating plateaus; (18) plateaus with planation surfaces; (19) undulating structure-controlled uplands with planation surfaces; (20) plateaus on gently dipping strata; (21) arid plateaus with structure-controlled scarps; (22) folded structures and melkosopotchnik (uplands with inselbergs) with planation surfaces; (23) erosional uplands; (24) other dissected uplands. (C) Other relief forms: (1) karst forms; (2) salt domes and depressions; (3) dunes, partly stabilized; (4) active barchans; (5) relief with padings; (6) suffosion pits; (7) Baer hills; (8) scarps.

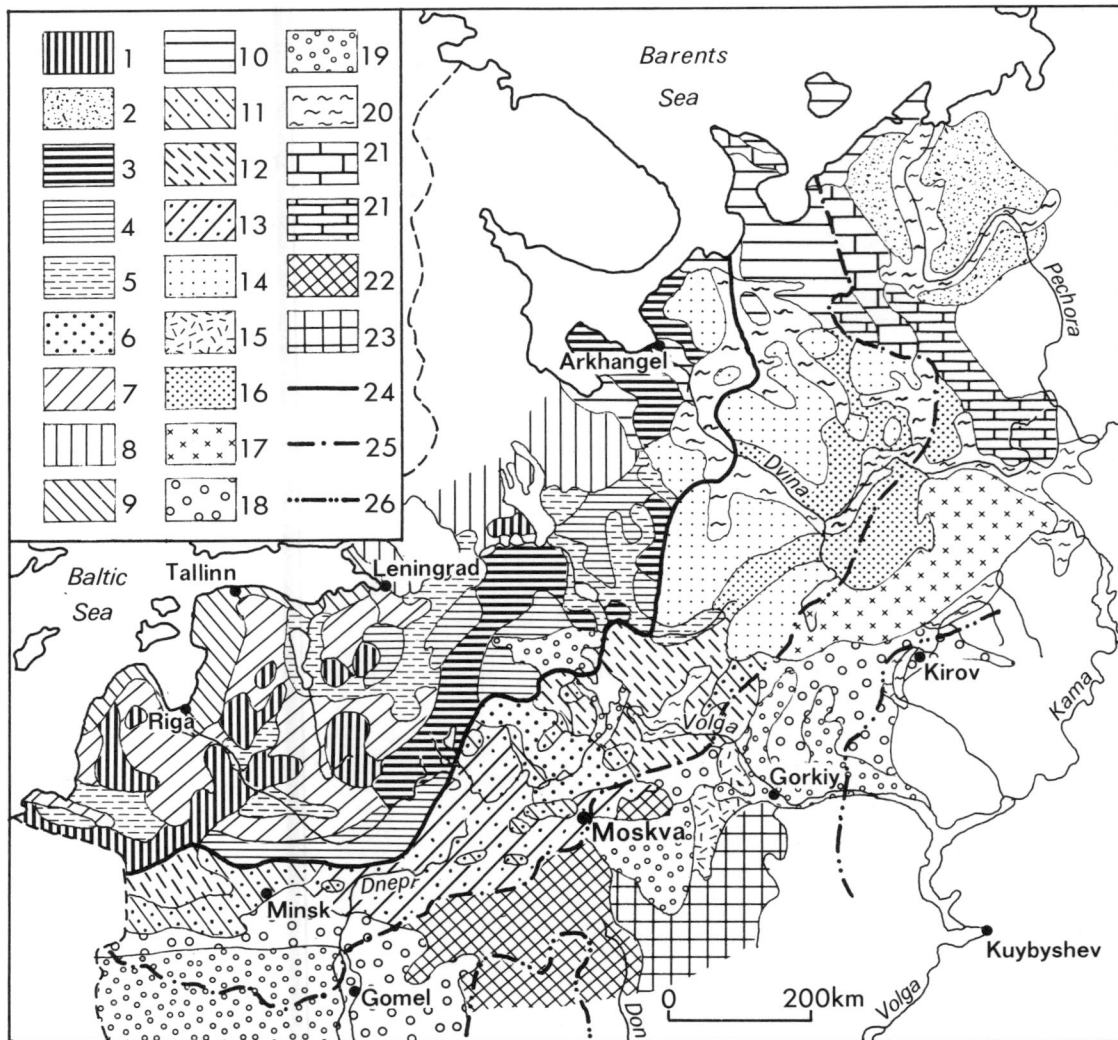

Fig. 6.3 Pleistocene glacial features in the northern part of the Russian plain [after Vedenskaya (1963)]. (A) Glacial deposition forms, slightly modified by later processes, developed on glacial deposits more than 100 m thick: (1) deeply dissected hummocky morainic high plains of Baltic type with a complex of ice contact forms; (2) dissected glacial and glaciomarine lowlands of the Pechora type with hummocky and ridge forms. (B) On glacial deposits less than 25 m thick; (3) deeply dissected glacial high plains of the Valdai type with a complex of ice contact forms; (4) degraded undulating moraine and glaciofluvial plains of the Valdai type; (5) flat glaciolacustrine lowlands of the Il'men' type; (6) flat outwash plains of the upper Volga type; (7) dissected morainic plains of the Baltic type with eskers, drumlins, end moraines, etc.; (8) dissected erosion–deposition plains of the Karelian type. (C) Glacial deposition forms of varied thickness and origin, considerably modified by later processes; (9) slightly dissected glaciolacustrine and marine lowlands of the Baltic type, with marine abrasion features and some eskers and dunes; (10) undulating marine, glacial and glaciomarine lowlands of the northern type with glacial hills and numerous cryogenic forms. (D) Glacial deposition forms developed on Quaternary deposits more than 50 m thick, considerably modified by later processes; (11) undulating morainic highlands of the Minsk type, deeply dissected by erosion; (12) degraded morainic lowlands of the upper Volga type modified by erosion. (E) Quaternary deposits of varied thickness: (13) morainic high plains with long ridges of the Smolensk–Moscow type, dissected by erosion. (F) Thin Quaternary deposits: (14) flat or undulating uplands of the northern type with some morainic hills (in places, karst plateaus with marginal scarps dissected by erosion); (15) flat or undulating morainic highlands of the Oka type; (16) undulating mainly glaciofluvial lowlands of the Vychegda type, slightly modified by erosion; (17) undulating glaciofluvial and morainic lowlands of the Northern Uval type slightly dissected by erosion; (18) undulating outwash plains of the Unshe–Vetluga type slightly dissected by erosion; (19) slightly dissected lacustrine, fluvial and outwash plains of the Polesiye type; (20) flat, mainly lacustrine and fluvial lowlands of the Northern Divina type. (G) Erosional landforms (including glacial erosion forms) on sedimentary rocks with some glacial (mainly glaciofluvial) deposits: (21) highlands with ancient planation surfaces of the Timan type with deep dissection along river valleys, karst forms and local glacial deposition: (a) dissected with deposition, (b) levelled, mainly by glacial deposition; (22) undulating uplands of the central Russian type, deeply dissected by erosion; (23) smooth undulating plains of the Oka–Don type. (H) Limits of continental glaciation; (24) Upper Pleistocene (Valdai); (25) and (26) Middle Pleistocene [(25) Moscow glaciation, (26) Dnepr glaciation].

Chernomorskaya lowland, which combined several morphostructural elements of the second order.

Contrasts of relief in the northern part of the Russian lowland (*see* Fig. 6.3) are less pronounced. Structural uplands, sometimes monoclinal, sometimes with structural scarps, and erosional uplands with planation surfaces are most widespread. Even on planation surfaces, however, slight deformations of the platform cover can be observed. For example, the development of the Moscow syneclise in the centre of the Russian lowland caused flexuring of the Palaeozoic and Mesozoic strata; this is now expressed in the relief as structural scarps. An inner elevated zone is formed by the Klin–Dmitrov uplands, bordered by a structural escarpment overlooking the upper Volga lowland. Likewise, structural scarps controlled by flexures or fault scarps mark the border of the Valdai Hills, separating them from the Il'men' lowland composed of Devonian rocks.

The Baltic syneclise, delimited by a cuesta in Devonian deposits—the so-called Baltic Ridge—fringes the Gulf of Finland. The structural scarp, termed the 'glint line', borders a plateau composed of Ordovician rocks (the Ordovikskoe plateau).

In the north, at the junction of the Moscow syneclise and the secondary Mezen' syneclise, there is a sublatitudinally trending heterogeneous zone of uplands, composed of the Gryazovets–Danilov, Galich–Chukhloma and Northern Uval uplands. The south-eastern and southern borders of the Moscow syneclise are formed by the northern slopes of the Smolensk–Moscow and central Russian uplands and the Volga Heights.

In the west, the Moscow syneclise merges into the Belorussian anteclise where the basement is nearer to the surface. This anteclise forms the uplands of the central part of Belorussian SSR. These uplands divide the Moscow syneclise from the Polish/Lithuanian syneclise on one side and from the Baltic syneclise on the other (the so-called Lithuanian pass).

In the eastern sector of the northern part of the Russian lowland, in the areas east of the Volga River and to the west of the Ural Mountains, are uplands extending in a submeridional direction (e.g., the Verkhnekama and Bugul'ma–Belevey uplands). Farther north, these uplands merge into the Timan Ridge. This also shows a submeridional trend, being composed of metamorphic and strongly dislocated rocks of the Palaeozoic basement of an old Fore-Ural depression. Farther to the north-east, in the region of the Pechora syneclise, this Caledonian basement is covered by thick sedimentary deposits.

6.1　Geomorphological Evolution

During every major tectonic epoch in the history of the Russian platform, two main phases of geomorphological evolution can be distinguished: (1) uplift of the area, resulting in dissection of the relief (*see* Fig. 6.4) and (2) planation of the relief accompanied by marine or continental types of deposition. Evidence of such a sequence can be found in all geomorphological regions of the Russian lowland.

The origins of the platform cover are connected with epiplatform invasions. Marine transgressions and regressions can be distinguished several times throughout the Palaeozoic and Mesozoic Eras. Subaerial planation surfaces originated during long periods of continental denudation between the end of the Palaeozoic Era and the beginning of the Mesozoic (Permo–Triassic) when a major regression of the epiplatform sea took place. An exhumed planation surface of Permo–Triassic age, bearing some remnants of pre-Upper Jurassic river patterns, is a constituent part of the Upper Cretaceous planation surface (the so-called basic peneplain of the northern and central regions of the Russian lowland). In the south, this basic surface was covered by marine transgressions in the Palaeogene and also partly in the Neogene, but in the north, except for the Pechora syneclise, Tertiary transgressions had only a small extent.

Areas exposed to subaerial denudation were several times uplifted, dissected and planed. Due to differences in the extent of uplift in different blocks, several planation levels developed. These levels provide good evidence of a tendency to intermittent uplift (e.g., the Volga Heights). Using weathering mantles and correlative marine deposits, Palaeogene and Neogene episodes of planation were distinguished, separated by shorter periods of relief dissection. In the Middle Pliocene, a new period of dissection was initiated and in the Upper Pliocene, a new period of planation commenced.

Quaternary tectonic activity can also be studied in the development of river valleys. Most river valleys in the Russian lowland with the exception of those crossing zones of repeated uplift (e.g., the Dnestr or Prut valleys) or subsidence (e.g., the Pechora River) show similar features as a result of a similarity in their history of development. A basic level on which many contemporary river valleys have been initiated is the youngest Upper Pliocene planation surface (valley pediment). In the area of the older glaciations, this surface is poorly preserved and covered by glacial deposits. River valleys were incised more than 10–20 m below the present-day valley floor in the pre-Cromerian period [see Aseev (1974)]. During the most recent period of planation (up to Middle Pleistocene), the valleys were filled by fluvial deposits to heights of 50 m above the present-day valley floor. Most major valleys show three systems of river terraces. These are of the aggradational type and have been correlated generally as follows: the upper terrace is Middle Pleistocene, while the middle and lower terraces are Upper Pleistocene. The composition of the fluvial deposits in the river valleys and the processes of river terrace formation were strongly influenced by Pleistocene climatic changes. The alternation of glacial and interglacial periods is reflected in the twin-layered structure of fluvial deposits comprising the aggradation river terraces: the basal layer is formed by interglacial deposits and the upper layer by periglacial deposits. This sequence is related to the greater supply of material to the river valleys by meltwater and by intensive mass movements on the valley sides in the periglacial zone during cold periods. At the beginning of each interglacial period, river incision took place. Finally, it should be noted that the alternating phases of river incision and aggradation caused by climatic changes were complicated by tectonic movements.

Thus the present-day rivers and valley systems reflect a complex history. The watersheds of all the extraglacial areas, as well as some parts of the watersheds in the glaciated areas, show an undulating relief. The lowlands are formed by fluvial and glaciofluvial (outwash) deposits. Glaciation also directly influences the formation of loessic mantles with buried soils in the south, and of glacial and glaciofluvial relief in the north. Glacial relief features are well preserved only in the area of the most recent glaciation. The areas affected by the older glaciations are characterized by periglacial relief mainly. The principal features of glacial relief include end moraine ridges, radial zones of morainic uplands and finger-like glaciolacustrine depressions [see Aseev (1974)].

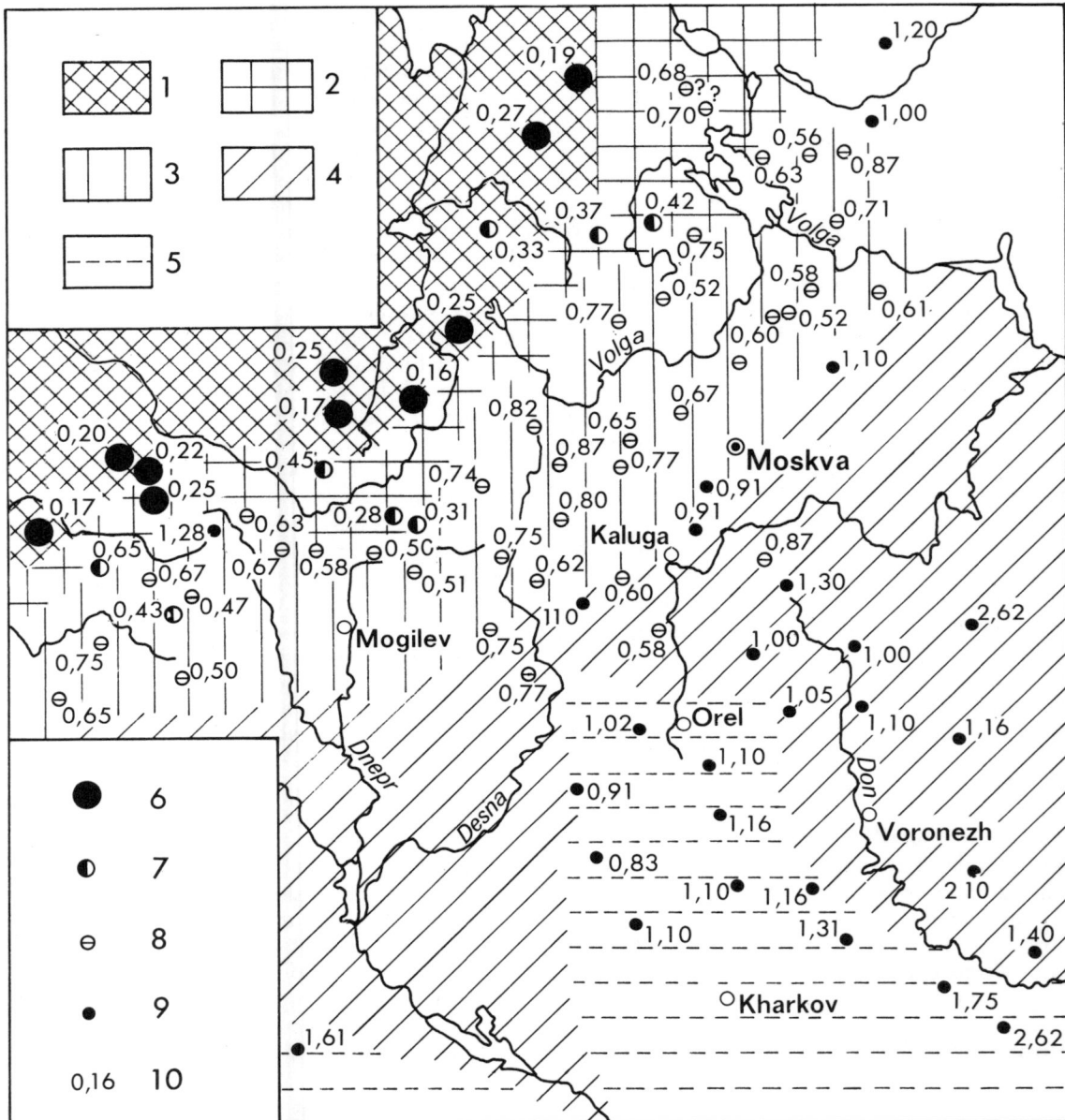

Fig. 6.4 Coefficients of relief dissection along watersheds of the Russian plain. The coefficients are calculated from the total lengths of watersheds (km) divided by the number of slope segments (separated by breaks of slope) [see Aseev 1974)]. (1) Area of Valdai glaciation, inner zone. (2) Valdai glaciation, outer zone. (3) Moscow glaciation. (4) Dnepr glaciation. (5) Pleistocene penglacial zone. Coefficients (key, lower left): (6) 0.16–0.27; (7) 0.28–0.45; (8) 0.46–0.96; (9) 0.91–2.62; (10) actual values.

6.2 Regional Geomorphological Divisions

The divisions to be considered are shown in Fig. 6.5.

6.2.1 Eastern Baltic Lowlands

Essentially these lowlands are developed on thick glacial deposits overlying the Baltic syneclise. This broad downwarp originated during the Palaeogene on the site of an old geoflexure, dividing the Fennoscandian (Baltic) Shield from the Russian platform. The eastern part of this geoflexure has preserved its original form and is now occupied by Lakes Ladoga and Onega. The western part was downwarped to form the basins of the Baltic Sea and the Gulf of Finland. The area is subdivided into two levels: (1) the Baltic lowland (where the upper surface of the basement lies between −40 and +40 m) and (2) a higher level which forms the base for the uplands near the Baltic Sea (upper surface of the basement being between +40 and +140 m, even rising to +150 m in the east). With regard to the first level, the lowest parts consist of the Nizhne–Neman and Latvian synclines (basement lower than −100 m). The greatest thicknesses of Quaternary deposits (as much as 300 m in places) are associated with morainic uplands. In contrast, thicknesses of Quaternary deposits in the glacially eroded finger-like basins do not exceed 20 m. The ground surface shows little variation in relief except for the residual Pleistocene glacial forms: morainic hills, kame complexes and glaciolacustrine deposits. The regional border follows marginal morainic ridges, such as the Baltic Ridge.

6.2.2 Bug–Pripyat Lowlands (Pripyat Marshes)

Two main structures control the underlying geology in the region: the Podlyass–Brest and the Polessk syneclises. These mark subsidence during the Tertiary Era of parts of former complex tectonic structures. The Podlyass–Brest syneclise was also reactivated during the Neogene. In the Pleistocene, this area was largely buried by glacial deposits. The surface now stands at 115–180 m and is badly drained by the Pripyat River.

The Polessk lowland consists of a gently inclined outwash plain (Predpolessk depression) which leads north into an upland of glacial deposition (Belorussian upland) and southwards into a low glaciofluvial peaty lowland with outwash plains (Pripyat lowland). The Pripyat lowland is bordered on the southern side by higher and better-drained outwash plains (Volyn'–Zhitomir lowland, also known as the Ukrainian Polesiye). The main relief features date from the maximum of the last glaciation. Mean thicknesses of glacial deposits amount to 15–20 m, providing evidence for a tendency to stability and not subsidence as was once thought.

Before dealing with the next three regional divisions of the Russian platform, the main geological and morphostructural features of the Moscow syneclise must be outlined. The Moscow syneclise is one of the major structural forms of the Russian platform. It is located between the Baltic Shield in the north-west, the Voronezh and Volga–Ural anteclises in the south and south-east, and the Belorussian anteclise in the south-west. In the north-east, the Moscow syneclise appears to merge into the Mezen' depression. The formation of the Moscow syneclise was preceded in the Riphaean (Precambrian) by the initiation of grabens, showing a pattern *en echélon* and together forming the central Russian east-north-east-trending aulacogen (rift) [see Khain (1977a)]. By the end of the Riphaean, an intricate graben–syneclise had been formed. From this time, the Moscow syneclise developed as a typical platform-type depression which is why its limbs are rather gentle and the strata are horizontal, with only minor flexures.

The syneclise is filled with various sedimentary rocks. These comprise four structural stages varying in lithological composition and separated by unconformities due to erosion and slight vertical movements. The lowermost stage is composed of the older Riphaean rocks, the second of the Upper Riphaean, Cambrian and Ordovician, the third of the Devonian, Carboniferous, Permian and some Jurassic, and the upper stage is composed of the Upper Jurassic, Cretaceous and, locally, Cenozoic rocks, mainly Palaeogene. The sediments rest on the Archaean crystalline basement with sharp unconformity. Subsidence of the Moscow syneclise continued at intervals up to the middle of the late Cretaceous, after which sedimentation ceased there, but the tendency to relative subsidence, as compared with neighbouring structures, was retained.

The Moscow syneclise is distinguished by numerous examples of inversion of relief and consequently by special morphostructures. Inversion of relief occurred at different times and was due to the reactivation of faults in the basement. This syneclise, unlike the Caspian one (*see* p.91), is expressed in the relief by a complex system of low or slightly elevated plains of depositional or erosional origin, respectively. Among the elevated plains, there is the largest example of relief inversion (the Northern Uvals) corresponding to an earlier subsided part of the syneclise. However, now it is an uplifted area with average elevations of 250–270 m. To the west, there is another system of isolated, relatively elevated (160–200 m) plains that are also examples of relief inversion [see Spiridonov (1978)]. In between, there are low-lying depositional plains of various genetic types. Fluvial and glaciofluvial denudation have been instrumental in reworking the deposits of both the elevated and low plains. Fluvial erosion developed during a continuous continental stage against a background of irregular vertical tectonic movements whose directions and rate varied.

6.2.3 Smolensk—Valdai High Plains

These occupy the eastern part of the Moscow syneclise. They are underlain by Carboniferous limestones, dipping gently towards the centre of the syneclise (*see* Fig. 6.6) and covered by continental Mesozoic strata. The southern part consists of undulating erosional plateaus with some remnants of glacial forms, while the northern part is characterized by hummocky morainic ridges and glacial lakes dating from the most recent glaciation. The mean heights over this area range from 250–350 m. On the western side, the area is bordered by an escarpment overlooking the Baltic and Belorussian plains. This is the so-called Carboniferous glint which marks a fault line in

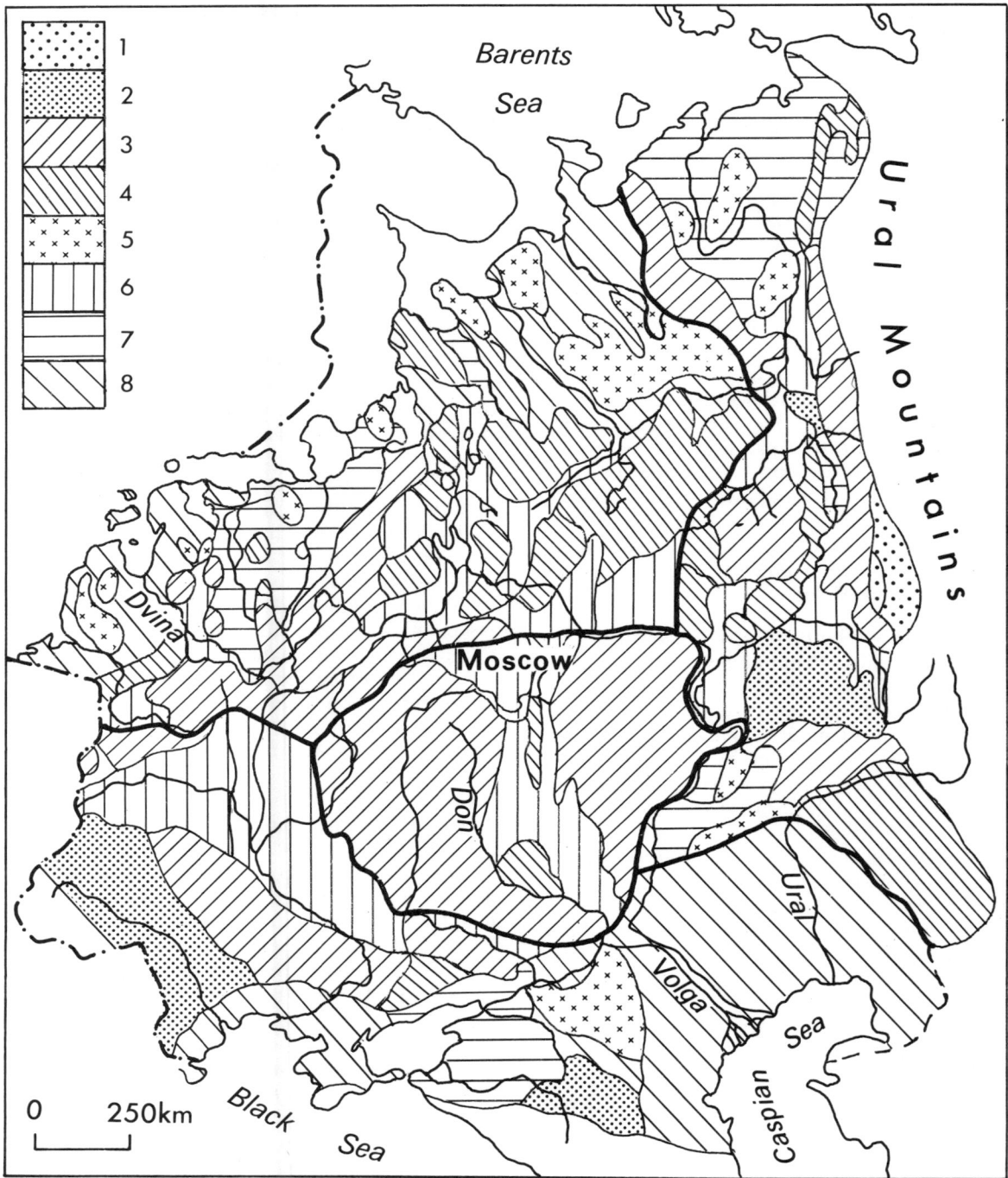

Fig. 6.5 Relief regions of the Russian platform [after Metcherikov (1965)]. The shading indicates the average height in metres. (A) Uplands: (1) over 300 m; (2) 220–300 m; (3) 180–220 m; (4) 140–180 m; (5) less than 140 m. (B) Lowlands: (6) over 100 m; (7) 60–100 m; (8) less than 60 m.

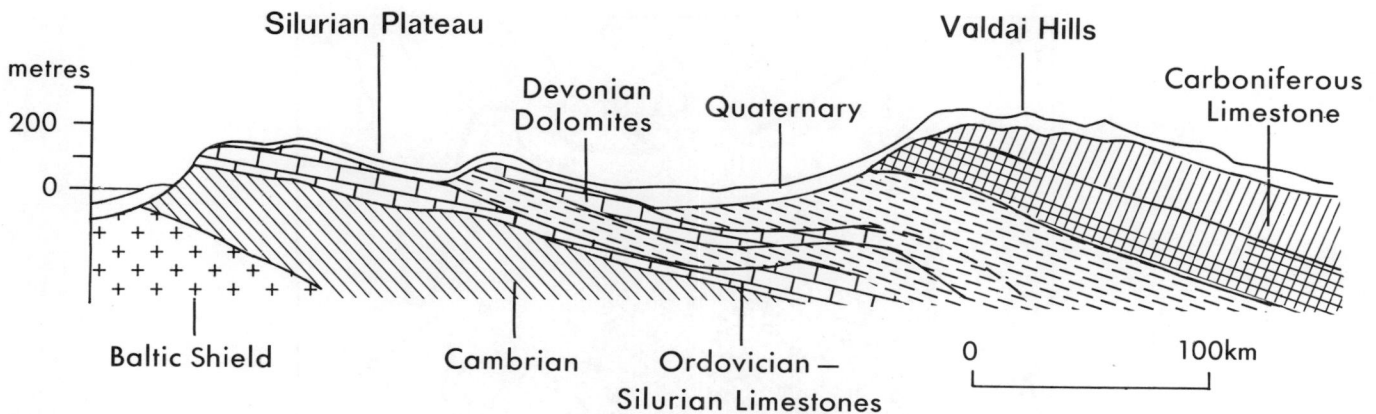

Fig. 6.6 Geological cross-section through the Silurian plateau and Valdai Hills, showing cuesta structure.

places between rocks of different resistance to denudation. The fault scarp is of preglacial age; during the glaciation, it formed an obstacle to the movement of the ice. Ice-contact features of the last glaciation occur along the fault scarp. Uplands mark the main watershed of the Russian lowland around the upper reaches of the Western Dvina, Volga and Dnepr Rivers.

6.2.4 Onega–Mezen' Plains

This area, which once formed a part of the Moscow syneclise, is underlain by Palaeozoic carbonate rocks dipping gently to the south and covered by marine Mesozoic deposits. During the Alpine orogenesis and formation of the Northern Uval upland—an example of relief inversion already noted—the northern part of the Moscow syneclise was separated and the new axial zone of subsidence moved to the Mezen' area. The orographic and morphostructural plan of the area is characterized by undulating interfluves and plateaus at about 180–300 m, together with some glaciofluvial forms and lowlands of recent deposition at around 100–180 m, drained by the Onega, Northern Dvina and Mezen'. These valleys were invaded by the sea several times and, during the ice recession, proglacial lakes formed in them. In the western parts, marginal forms of the last (Valdai) glaciation are preserved while, in the east, the forms dating from the Middle Pleistocene glaciation have been denuded and subjected to periglacial mass wasting (*see* Fig. 6.7).

6.2.5 Central Moscow Region

Beneath the superficial Quaternary and Holocene deposits, the surface combines elements of Upper Cretaceous–Palaeogene planation, on the one hand, and exhumation of a late Palaeozoic peneplain, on the other. The surface relief is characterized by an intricate arrangement of uplands and plains, partly relics of an older structural plan but partly newly formed. The primary structural-tectonic base has only a slight influence. The individualization of uplands and plains is essentially due to erosional development. One of the principal units of the region is the Moskva–Oka plain, which passes to the north into the Klin–Dmitrov upland. This is separated by a scarp (*see* Fig. 6.8) from the upper Volga lowland. The upper and middle Volga and Kama are flanked by strips of lowland. Farther north lies another zone of uplands trending in a submeridional direction (the Gryazovets–Danilov and Galich–Chukhloma uplands) which rise to 160–200 m. Here are also located the marginal zones of the second middle Pleistocene glaciation (i.e. the Moscow glaciation). The plains are formed by glaciofluvial deposits, mainly of glacial outwash.

6.2.6 Timan Ridge

Forming a divide between the Pechora Basin to the northeast and the Onega–Mezen' plains to the west (*see* Section 6.2.4), this low ridge, which rises to 300–400 m, represents an upland repeatedly worn down by erosion throughout the Mesozoic and Cenozoic Eras, and later uplifted by neotectonic movements. The bedrock is Palaeozoic including intrusives. The area was affected by early and middle Pleistocene glaciation, but was not glaciated in the later Pleistocene.

6.2.7 Pechora Lowlands

Draining into the Arctic Ocean via the Pechora River and its tributaries, this region consists structurally of the Pechora syneclise corresponding to the northern part of the Ural foredeep. The syneclise originated during the early Cenozoic and has shown continued subsidence ever since. Considerable thicknesses of marine and fluvial sediments accumulated in the Neogene and Quaternary, and were capped by Quaternary fluvial and glacial sequences. Neotectonic movements have included some differential faulting, which is continuing at the present time. One unusual result of these movements is the incorporation of locally uplifted blocks of pre-Quaternary age (the so-called *musyury*) among the later sediments. Such blocks are especially common in the southern part of the lowland.

6.2.8 Lublin–Moldavian upland

This is a belt of recent uplift that trends approximately north-west to south-east between the Carpathian mountains and the Ukrainian Shield. Because of its complexity, it is best dealt with in terms of its geomorphological subdivisions.

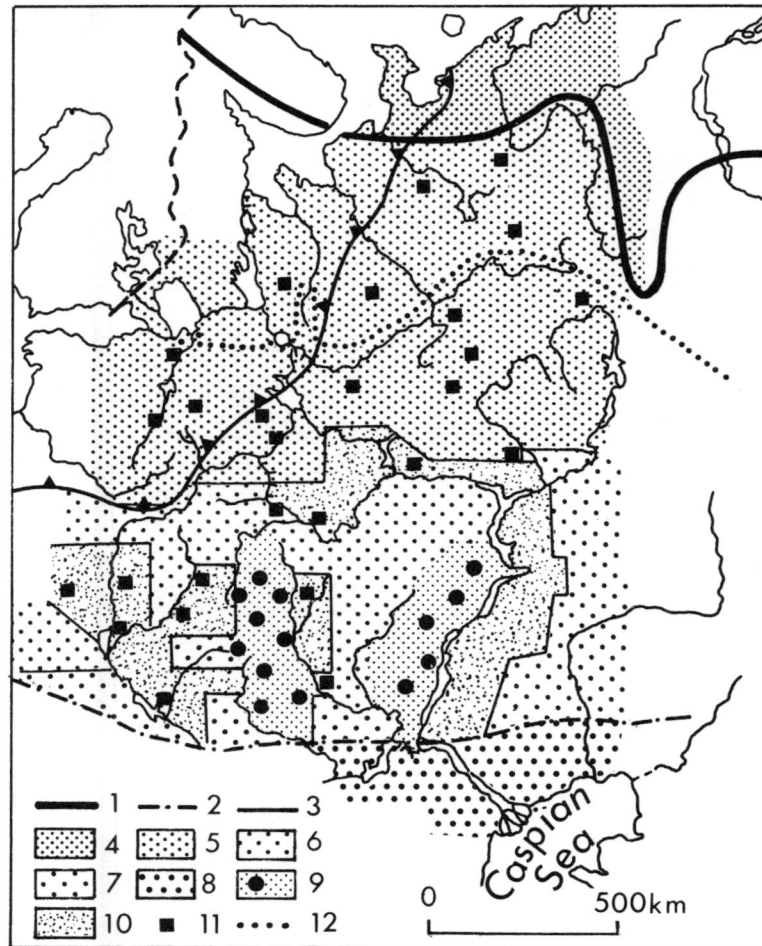

Fig. 6.7 Distribution of periglacial features on the Russian plain. (1) Southern limit of present-day permafrost. (2) Southern limit of Upper Pleistocene permafrost. (3) Limit of Valdai glaciation. (4) Present-day active cryogenic phenomena. (5) to (11) Relict (Upper Pleistocene) cryogenic and thermokarst features; (5) fresh polygonal features with numerous thermokarst depressions and lakes; (6) polygonal features with some thermokarst depressions; (7) polygonal features with shallow depressions; (8) relief with occasional shallow depressions; (9) upland relief with rills and shallow depressions; (10) baydzharakhs and depressions on gravel terraces of major river valleys; (11) alases; (12) southern limit of fossil thermokarst features according to Kachurin (1969).

6.2.8.1 *Fore–Carpathian upland*

This originated during the Neogene through a process of relief inversion. It marks the site of part of the original Carpathian foredeep and is composed of unconsolidated sediments. Monoclinal ridges and bevelled interfluves are characteristic, standing at heights of about 400–500 m, and there are deeply incised valleys.

6.2.8.2 *Volyn' cuesta upland*

This upland, which ranges from 300 to 400 m, consists of Neogene sediments resting on the Palaeozoic basement and corresponds to a marginal depression of the Russian platform. The northern part has an intricate relief, with the inclined layers of Neogene deposits containing resistant layers that have been picked out and deeply dissected by rivers (mainly the left-bank tributaries of the Dnepr). The watersheds are formed by monoclinal ridges. There is an extensive loessic cover and, where this rests on the sand

deposits, there are frequent surface depressions due to suffosion (piping).

6.2.8.3 *Dnestr upland*

This is a plateau in the southernmost part of the area. It is composed of varied deposits, the most important being fluvially deposited sandy loams. Rivers are deeply incised and form dendritic patterns. Landslides are common on slopes and, in some places, there are karst forms. Altitudes of 400–430 m are attained on the watersheds, especially on the cuesta ridges.

6.2.9 Ukrainian Shield

This is the only area of the Russian lowland where Precambrian crystalline rocks appear at the surface in a series of separate outcrops stretching from the Pripyat valley, north-west of Kiev, to the Donets area, in the south-east. The area is approximately contained between

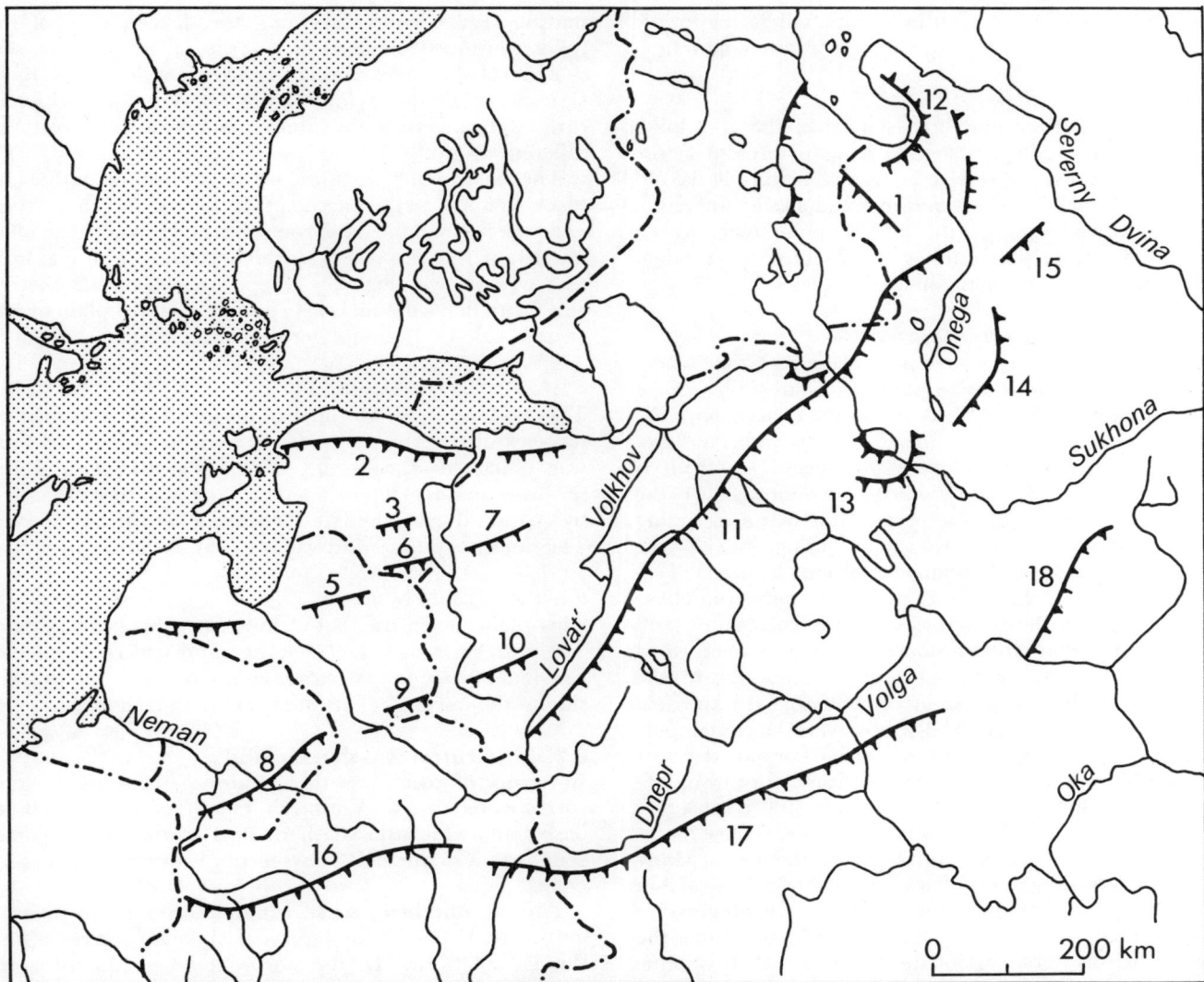

Fig. 6.8 Map of the uplands and ridges in the north-western part of the Russian platform. (1) Vetrenyi ridge; (2) Silurian plateau; (3) Otepya; (4) Kurzemskaya upland; (5) Vidzemskaya upland; (6) Khanya; (7) Luzhskaya upland; (8) Baltic ridge; (9) Latgalskye upland; (10) Bezhanickye upland; (11) Valdai Hills; (12) Onega ridge; (13) Bolozersko–Kirilovskaya ridge (14) Nyamdomskaya upland; (15) Lepshinskaya upland; (16) Belorussian ridge; (17) Smolensk–Moscow ridge, (18) Galitchko–Tchukhlomskaya upland.

the southern Bug and Dnepr Rivers, but extending on beyond the lower Dnepr towards the Sea of Azov. Three divisions of the Ukrainian Shield can be recognized.

6.2.9.1 Dnepr upland

This upland, which forms a broad watershed between the Dnepr and southern Bug Rivers, corresponds structurally to a large part of the Ukrainian Shield and, together with the Azov upland, marks the oldest core of the southern part of the Russian plain. The crystalline basement can be observed at the surface in the north-west, exposed over about one-third of the whole area of the upland. The rest of the surface is covered by Tertiary (locally Mesozoic) deposits which, in some places, are of considerable thickness. Crystalline rocks are exposed here only in the river valleys.

The main landscape type consists of undulating plana-

tion surfaces, sometimes covered with younger deposits (Palaeogene–Neogene) and by loess. Most of the region represents a Jurassic–Cretaceous planation surface, deformed and remodelled by more recent processes. A Mesozoic weathered mantle is extensive. Borings have shown the existence of buried valleys that are partly followed by the present-day river pattern. The valleys are quite deeply incised and the rivers generally adopt a dendritic network, although some valleys are controlled by faults. Average altitudes range from 200 to 300 m. In the northern part of the area, there is the Ovruch horst (with an altitude of 300 m). During the penultimate Pleistocene glaciation, the ice spread southwards down the broad Dnepr lowland, approximately as far as the Kremenchug, and rose up the gentle flanks of the Dnepr upland, leaving deposits of ground moraine and also some end moraine features. Beyond the ice limit, there are outwash spreads. In the southern part, where the loessic cover reaches its largest thickness, gullies are common.

6.2.9.2 Azov upland

This is the southern and highest part of the Ukrainian Shield. The crystalline basement is again present at the surface. The most impressive landscape feature is the old planation surface with tors and monadnocks composed of the most resistant rocks, the highest point reaching an altitude of 324 m. The surface is dissected by a radial pattern of short, but deeply incised, valleys.

6.2.9.3 Servno—Chernomorskaya lowland

In term of tectonics, this lowland represents a zone where the basement of the Russian platform is subsiding. Some parts (around the Sivash Sea) are on the contact with the Scythian platform. The surface of the Precambrian basement descends rapidly to the south; near Odessa, it is already 2 km deep and covered by deposits. At the beginning of the neotectonic period, this was a subsiding area where sediments were accumulating, but in the middle Pliocene (post-Pontian) conditions changed. Tectonic uplift caused the sea to regress and a subaerial phase commenced. On the new surface of the coastal lowland formed by the Pontian limestones, erosional relief formed with well-developed river patterns and gullies. In the Pleistocene, a loessic cover of considerable thickness in places was deposited. Generally, however, the contemporary relief represents the surface of the Pontian deposits only slightly modified by erosion. Altitudes rise from 20–30 m in the southern coastal areas to 100 m or more towards the north of the lowland. The relief is very flat in the coastal zone but more undulating in the north. Many of the rivers flowing over the lowland are overdeepened in their lower parts due to late Quaternary marine regression (the so-called late Euxinus regression). At this time, the whole north-western part of the Black Sea shelf became dry land, and the extended rivers were graded to a level 40 m below that of the present. During the next late Pleistocene–Holocene transgression, the lower parts of the valleys were drowned, with the formation of typical estuaries (e.g., those of the Dnestr, Bug and Dnepr). Some estuaries are divided from the sea by bars formed by longshore drift.

Active present-day processes include gully erosion in the north of the lowland, suffosion on the loess forming flat-floored depressions called *podi* and mass movements along the coast. The largest landslides on the Black Sea coast occur between Odessa and the mouth of the Dnepr, where Pontian limestones are sliding over Meotian clays. The bulging toes of rotational slips can often be seen about 50 m out from the shoreline.

6.2.10   Dnepr–Donets Lowland

The Dnepr–Donets lowland, which lies between the middle Dnepr and the Donets Rivers, corresponds tectonically to the Dnepr–Donets depression (aulacogen). The basement lies at depths of up to 15 km in the south-western part of the depression, which is filled by marine sediments from the Devonian to Palaeogene. After the last

marine regression (Oligocene), the development of the relief continued subaerially. In the late Oligocene and during the Miocene, the sandy sediments of the Poltava series were laid down, followed by Pliocene clays, which in turn were overlain by thick Quaternary deposits of different origin.

The basic relief features are controlled in outline by block tectonics of the basement. In the sedimentary cover, and especially in the axial zone of the depression, the most important role is played by salt tectonics. The region is divided into two parts: the Dnepr lowland on the western side (with depositional relief) and the Poltava plain on the eastern side (with mainly erosional relief).

6.2.10.1   Dnepr lowland

This lowland, with altitudes of up to 100 m, is formed by terraces of the Dnepr River covered by loess. This area was glaciated during the middle Pleistocene and the resulting mainly depositional surface was then dissected by erosion during the late Pleistocene and the Holocene. The depth of gullies and valleys is up to 25 m.

6.2.10.2   Poltava plain

This plain lies at the foot of the central Russian upland. The surface is of Palaeogene age and covered by loess. Erosional dissection is greater compared with the western side of the lowland. This area was not glaciated.

6.2.11   Central Russian Upland

In terms of tectonics, the greater part of this region corresponds to the Voronezh massif of the crystalline basement, while the southern part of the region corresponds to a depression between the Voronezh massif and Donets Ridge.

The Voronezh anteclise running from north-west to south-east is one of the largest positive structures of the Russian platform. In the north, it meets the Moscow syneclise. In the north-east, it exhibits a faulted contact with the Ryazan'–Saratov trough and with the Donets folded structure in the south. In the south-east, it shows a step-like pattern of subsidence towards the Caspian syneclise, from which it is separated by the Volgograd fault. Structurally, the anteclise consists of two units: the lower crystalline Precambrian basement and the upper, relatively undisturbed sedimentary cover composed of Carboniferous, Jurassic and Cretaceous deposits, including many limestones. The sedimentary cover rests on the basement with sharp unconformity [see Raskatov (1969)]. The Precambrian basement exhibits an apparent block structure, which is reflected in the structure of the platform cover (as throughout the entire Russian platform). The Voronezh anteclise was completed essentially in the early Permian, since when its evolution has been distinguished mainly by the reactivation of existing structures. It has been characterized by a tendency to uplift throughout almost the entire Mesozoic and Cenozoic Eras.

The basic relief features date from the Miocene when the basic planation surface of Cretaceous–Palaeogene age was formed. It was later deformed by neotectonic movements and dissected during the Pliocene and Quaternary. The northern part is a planation surface cut across monoclinally dipping strata, while the central and eastern parts are strongly dissected, the highest elevations attaining 250–280 m, with the plain sloping down towards the periphery. The eastern edge is especially sharp where structural discontinuities separate the region from the marginal parts of the Oka–Don depression underlain by thick Neogene–Quaternary deposits. The south-eastern part of the upland plain is occupied by the Donets Ridge on monoclinally dipping strata. This is distinguished by a marked asymmetry: the northern slope is short and abruptly terminated by an escarpment 70–100 m high overlooking the Don valley, but the southern, longer slope dips gently towards the Chir River. The surface of the Donets Ridge is greatly dissected by valleys, hollows and gullies.

The Kalach high plain, which is separated by the Don valley, is situated on the interfluve of the Khoper and Don Rivers. It is composed of Upper Cretaceous and Palaeogene strata dipping gently south-eastwards. The Kalach high plain is one of the most eroded areas of the central Russian upland, 59 percent of the area being affected by accelerated soil erosion.

Gullies are typical of much of the central Russian upland; rates of present-day gully erosion are quite high. The densest network is confined to the central and highest parts of the upland (i.e. the axial part of the Voronezh anticlise). The correspondence with a zone of strong neotectonic uplift strongly suggests that the latter has played a significant role in gully development.

6.2.12 Oka–Don Lowland

Comprising the Meshchera, the Oka–Don and the Tambov depositional lowlands, this depression is situated east of the central Russian upland. It seems to occupy a part of the Ryazan'–Saratov trough formed above the Precambrian Pachel'ma aulacogen (rift) and belongs to the western slope of the Voronezh massif. In the Cenozoic, the Ryazan'–Saratov trough was infilled with sediment. Most of the subsidence and the formation of the Oka–Don depression are attributed to the end of the Miocene, but subsequent relative downwarping took place in the Pliocene and Quaternary (the Meshchera fluvial plain). The northern part of the lowland is formed of Jurassic sandy clayey deposits, while the southern part consists of subaerial Ergeni sands and glacial and glaciofluvial deposits, sometimes more than 100 m thick. The base of these deposits, which is 20–40 m below the contemporary river level, corresponds to a planation surface of Palaeogene age.

The Oka–Don depression is characterized by low relief comprising wide and shallow asymmetrical valleys. Gully dissection is typical of the central and southern parts of the region, aided by glacial meltwater (see below). The narrow Oka–Tsnin' rampart, corresponding to local basement uplift that caused deformation of the sedimentary cover, rises above the surface of the plain's eastern periphery. The average elevations of the depression are about 130–140 m, but the Oka–Tsin' rampart reaches 160 m.

The ice at the time of the maximum Dnepr (Middle Pleistocene) glaciation advanced southwards along the Oka–Don depression. At present, considerable parts of the depositional plain within the depression are underlain by glaciofluvial deposits, subsequently reworked in places by fluvial processes.

6.2.13 Volga Heights and Ergeni Hills

Although these units form a single zone of relatively elevated relief to the west of the Volga River, lying to the north and to the south, respectively, of Volgograd, there are important differences between the Volga Heights and the Ergeni Hills in respect to relief and structure. They will, therefore, be treated as two separate subregions.

6.2.13.1 *The Volga Heights*

These uplands consist of elevated erosional plains cut across monoclinally dipping, gently folded Palaeozoic–Mesozoic rocks (*see* Fig. 6.2). In the south, there are Permian limestones while, in the north, Carboniferous limestone is overlain mostly by Jurassic, Cretaceous and some Palaeogene deposits. The strata were deformed by Miocene tectonics, representing the feeble reflections of the Alpine orogenic movements. Geophysical data indicate extensive faulting and block structure in the underlying basement, including grabens filled with Upper Proterozoic sediments. These are the most ancient structural forms traceable beneath the sedimentary cover. Subsequently, such structures as graben–syneclises and horst–anteclises formed in the sedimentary cover. Neogene tectonic uplift has affected the central part of the Volga plain, whose structure is also complicated by the Zhiguli swell, expressed in the relief by a chain of low hills (up to 370 m) with steep and abrupt slopes. The Volga bend near the Zhiguli Hills is associated with a fault manifested in the relief of the famous Samara meander.

As a whole, the relief of the Volga uplands is asymmetrical: the western slope is long and gentle, while the eastern slope is short and steep. Towards the south, the line of the highest elevations shifts gradually eastwards. Overall, most of the relief falls in the range 200–300 m. The low amplitude of the relief is attributable to Palaeogene planation, the resulting surface now varying in height generally between 300 and 350 m. Some of this variation in altitude may be the result of neotectonic deformation. The planation surface terminates in places in a scarp overlooking a younger (Lower Miocene–Pliocene) and lower surface—partly erosional and partly depositional—which is only slightly deformed and dissected. In the east of the uplands, nearer to the Volga valley, dissection (especially by gully erosion) is much more pronounced. This is the second most important area of serious gully erosion in Russia, the other being the central Russian upland (*see* Section 6.2.11); gullying is still very active at the present-day largely because of unsuitable land use practices.

6.2.13.2 *The Ergeni Hills*

This asymmetrical upland has a steeper eastern slope which overlooks the Caspian lowland. The northern part shows inversion of relief in relation to the structure of the basement and its sedimentary cover; the southern area—the Sal–Manych Ridge—corresponds to the intricately constructed Karlinski anticline. The Hills are covered by up to 70-m thick loessic loams, resting on alluvial Ergeni sands of Neogene age. The Ergeni Hills at the end of the Miocene already formed an upland area, evidenced by gullies filled with Ergeni sands. In the Lower Pliocene, these Hills were lowered by planation and overlain by fluvial and lacustrine/deltaic sediments.

The present-day appearance of the Ergeni Hills dates from the end of the Pliocene–early Pleistocene periods. The basic planation surface is of Miocene–Pliocene age, rising to the south from 160 to 220 m. Watershed areas are characterized by undulating relief. The long western slope is slightly dissected by rivers draining to the Sea of Azov. The eastern slope bordering the Caspian lowland corresponds to a fault scarp and is deeply dissected by parallel valleys with flat interfluves. The short, steep southern slope is formed by a geoflexure and overlooks the Manych depression.

Before examining the geomorphology of the Low Zavolzhje, the Volga–Kama and the Fore–Ural uplands individually, some general remarks about this area will be made. Elevated denudational plains extending as far as the Ural Mountains occupy the eastern sector of the Russian platform. These are the High Zavolzhje or Trans-Volga plains, which are separated from the central Moscow lowlands by the valley of the Volga and its depositional plains known as the Low Zavolzhje. North of the Volga, the north–south-trending ridge of the Vyatka–Uval forms an arbitrary boundary to the eastern plains. The Vyatka–Uval, as well as the Oka–Tsnin' ridge, is related to fault zones in the basement and represents relatively narrow troughs that underwent uplift and are now examples of inverted relief.

Several structures underlie the High Zavolzhje. East of the Volga and on both sides of the lower Kama, there is the Tatar dome. Farther east is the Perm dome. Southwards, there is the Orenburg dome, which extends from east to west parallel to the side of the Caspian syneclise, from which it is separated by a system of faults. In the north, the Orenburg dome is separated from the Tatar dome by a fault which has been picked out by the Samara meander of the Volga and which is marked by the Great Kinel' escarpment farther east.

In terms of relief, the High Zavolzhje can be divided into the Volga-Kama plain in the north, with the Vyatka–Uval upland and the Ufa and Bugul'ma–Belebey plains (*see* Section 6.2.16) separated by the Kama–Belaya depression on the north and east. Farther south, the Syrt lowland (part of the Low Zavolzhje) is delimited on the north by the Great Kinel' escarpment.

6.2.14   Volga–Kama Plains

The highest parts in the east of this area reach 200–300 m and represent a denudational plateau developed on horizontal or gently folded strata, and covered by glaciofluvial deposits. Altitudes decline to the west, where slowly flowing broad rivers with gentle banks meander across marshy valley floors in the Vyatka River basin. Farther south, the Vyatka leads into the Kama–Belaya basin, drained by the lower reaches of the Belaya and Kama Rivers. Tectonically, the basin represents a large, shallow trough initiated in the Neogene and filled with Neogene–Quaternary deposits, superimposed on complex Palaeozoic structures.

6.2.15   Low Zavolzhje

Tectonically this lowland, which lies to the east of the Volga, corresponds to a large Pliocene depression, comprising an amalgamation of many smaller structural units, erosional–tectonic depressions and ridges. The whole depression is filled with continental sandy and loamy deposits of Pliocene–Quaternary age, resting on a marine surface (the Akchagyl surface); there is also a widespread cover of loess. The lowland is divided into two by the Zhiguli ridge, crossed by the Volga at the Samara meander.

The northern part is a plain, lying at about 200 m, formed by the fluvial deposits of the Volga, Kama and Vyatka Rivers. Quaternary terraces form a sequence in these valleys: the flood plain and the first terrace of the Volga are now submerged by the Kuybyshev Reservoir, but the second terrace is clearly visible. Older and higher terraces are eroded and remain only as sandy hills.

The southern part, from the Zhiguli Hills southwards to Volgograd, is the Syrt lowland. Underlain by Upper Permian strata consisting of horizontal or gently dipping clays and sandy marls, and covered by Quaternary limon and alluvium, the plains range in altitude from 180 to 300 m; the maximum depth of dissection is 50 m. The plains are characterized by gently convex upper valley slopes, with interfluves that are flat but asymmetrical, the southern slopes being steeper with local structurally controlled steps. In the central part of the lowland, the low Pugachevo ridge represents an old uplift, where the Quaternary deposits are thinner. Farther south and southeast, the relief becomes flatter, but is also diversified by karstic and salt tectonic phenomena.

6.2.16   Fore-Ural Uplands and Plateaus

The crystalline basement in this complex area lies at great depths (1.5–6 km) and is overlain by a thick platform cover of Jurassic, Cretaceous and Palaeogene age. In the context of the broad structure of the Russian lowland, it corresponds to the large Volga–Ural anteclise which incorporates many natural gas and oil-bearing structures.

6.2.16.1   *Bugul'ma–Belebey and Ufa plains*

These stand at relatively high levels of 250–400 m. Tectonically, the former corresponds to the Tatar dome and is underlain by Permian carbonate rocks. The Ufa plain is developed on gently folded strata; morphologically it shows a degree of asymmetry in transverse profile with a steeper eastern limb and a gently stepped western slope. Several planation surface levels are recognizable, the oldest one dating back to the Jurassic. All the older levels are much dissected and fragmented by quite deeply incised valleys. In the western areas, karstic features (i.e. cones, ravines, etc.) are present giving a distinctive landscape.

6.2.16.2 Obshchiy Syrt upland

With altitudes of 200–300 m, this lies farther south, on the western side of the Ural Mountains. Tectonically it is a young uplift area, comprising a system of old domes, anticlines and depressions in the sedimentary cover of Jurassic, Cretaceous and Palaeogene rocks. Different types of relief are present—structural, tectonic and erosional—producing an undulating landscape of hills and ridges.

6.2.16.2 Sub-Ural monoclinal plateau

This stands at 200–300 m in the southern part of the region. It is characterized by structural surfaces with arid forms. Pediments coalesce to form partial pediplains in some areas.

6.2.17 Caspian Lowland

This is the lowest part of the Russian plain, with altitudes ranging from 28 m below sea level to 50 m above. Tectonically, it is the northern part of the Caspian depression or syneclise. The Caspian syneclise differs from the other syneclises of the Russian platform. It was initiated, like the others, as far back as the Riphaean, since when it has experienced a complex history, having inherited a general tendency to subsidence. This is reflected in the great depths (10 km) to which the Precambrian crystalline basement descends. Over this basement, a sedimentary cover of mainly Permian, Mesozoic and Cenozoic age was deposited. Morphologically, the western part of the syneclise is represented by the Caspian depositional lowland, locally falling below sea level, and by the northern and middle Caspian basins. The southern Caspian basin, in which, according to Leont'ev, the continental crust is missing, represents along with the Black Sea a relict portion of the Tethys Ocean. The relief of the eastern part of the syneclise underwent inversion in the second half of the Neogene and at present it is represented by the sub-Ural plateau built of consolidated Miocene rocks.

The Caspian syneclise is separated from the main part of the Russian platform by a deep-seated fault zone. According to Khain (1977a), large flexures and swells, expressed in the present-day relief, are located along the border of the zone. The western and south-eastern boundaries of the basin coincide with the Volgograd and Astrakhan faults along the valley of the Volga River. In the east and south-east, a series of faults *en echélon* separates the syneclise from the Or'–Ilet' tableland and the Aktyubinsk Fore-Urals, which are extensions of the Ural folded zone.

Some of the main features of the present-day structure and relief of the Caspian syneclise are determined by neotectonic movements. First and foremost, these divide it into the Caspian depositional lowland and the Caspian Sea basin. In the northern part of the lowland, the effects of neotectonics can be seen in uplifted or depressed structures, despite a covering of late Quaternary marine deposits. The relief of the sub-Ural plateau, a large example of relief inversion, as already noted, contrasts sharply with the depositional plains of the western part of the Caspian syneclise, the plateau being intensely dissected by a network of minor valleys and ravines.

The Caspian lowland is composed of a thick sequence of Quaternary deposits. Genetically, it is a depositional plain of marine and partly fluvial origin, gently tilted towards the Caspian Sea. Arid geomorphological processes have played an important role in its recent landscape evolution, but it has also been affected by transgressions of the Caspian Sea. After the Khvalyn and Novokaspian transgressions, there remained marine depositional plains and such major alluvial forms as the valleys of the Volga and Ural Rivers with vast terraces, and the valleys and deltas of smaller permanent and temporary watercourses. The initial marine relief of the Caspian depositional plain has changed considerably under the influence of fluvial, lacustrine (ancient lagoons) and aeolian processes. Aeolian processes were facilitated by the existence of widespread marine and deltaic sandy deposits. Thus, the interfluve of the Volga and Ural Rivers is occupied by the Ryn sands with various aeolian forms, such as dune ridges.

In the northern part of the lowland, suffosion depressions are well developed, while the eastern areas are characterized by a most striking manifestation of salt tectonics, associated with the deposition of great thicknesses of rock salt at the end of the early Permian and beginning of the late Permian. At the present-day salt forms are extremely numerous, including over a thousand stocks above which hundreds of salt domes came into existence. Zones of both scattered and concentrated salt domes, representing weak and active areas, respectively, of salt tectonics, correspond in general to the outlines of major uplifts and troughs of the salt-bearing horizons of the Caspian syneclise. Many salt domes form hills (100–150 m), the highest and most active being associated with Mesozoic and Cenozoic rocks (e.g., Bolshoe Bogdo mountain, etc.). There are compensatory depressions usually occupied by salt lakes.

Other special features of the Caspian lowlands are the Baer Hills—named after the Russian geomorphologist K. M. Baer, who studied these forms more than 100 years ago in 1856. In the southern part of the lowland, from the lower reaches of the Emba River in the east to Lake Budilo in the west, many small hills may be observed, running parallel to each other. In ground plan, they form three fans opening to the south. The Baer Hills are 2–45 m high, 100–600 m wide and 800–1000 m long. Their longer axes are oriented in a submeridional direction. Distances between the hills or their rows vary from several metres to 2 km. Sometimes the tops of the hills are rounded and sometimes sharp-crested. The composition of the hills is intricate. The basis is formed by late Pleistocene marine clays, covered by sandy clayey sediments and marine sands. The origin of the hills is far from clear. There are three main hypotheses—erosional, aeolian and water accumulation—each hypothesis having its supporters.

Chapter 7
Caledonian Highlands

7.1 Introduction

The continent of Europe possesses a highland fringe along its north-western Atlantic border, stretching through Scandinavia and what is conventionally called Highland Britain. The continuity of the region may not be immediately evident from a topographical map of the land area, but in terms of geological structure and geological history the units of Scandinavia, Scotland, northern and central Ireland and Wales belong together. Although Scandinavia is now separated from the others by the 500-km expanse of the North Sea, the latter is a shallow shelf sea, broken only by the narrow Norwegian trench, and a fall of present sea level by only 200 m would serve to emphasize that Scandinavia and Highland Britain belong to the same continental edge. At the −500-m level, for instance, the continental slope trends consistently from south-west to north-east. This Atlantic margin is now recognized as the trailing edge of the European continent which, together with the sea floor to the north-west, comprises the European plate.

Caledonia was the name given by the Romans to the northern part of Scotland beyond the Clyde and Forth estuaries. It was in these northern highlands that geological evidence of an early period of mountain building (late Silurian–early Devonian) was first described, and the term Caledonian therefore became applied to earth movements of this age. Subsequently, it was recognized that the Scottish Highlands were only a small part of a vast area of north-west Europe affected by these earth movements, and that in the Lower Palaeozoic Epoch, a great geosyncline lay along this continental border, accumulating thousands of metres of sediments. The Caledonian orogeny that followed represented the compression of these sediments between two opposing plates—the Eurasian and the American—and resulted in the formation of folded and overthrust structures striking south-west to north-east—the typical Caledonian trend, now clearly highlighted by the orientation of many relief and drainage lines that developed during subsequent differential denudation. The Caledonian orogeny created great mountain systems in north-west Europe, of which only the roots now remain; the relative elevation of some areas today is the result of later uplifts and the existence of resistant rocks. The highest elevations are far lower than those of the Alps and other younger mountain systems in Europe. Glittertind (2469 m) in the Jotunheimen of Norway represents the maximum height attained in the Caledonian Highlands, although another peak (Galdhöpiggen) only 20 km away rises to within 1 m of that height. Throughout the Scandinavian part of the Caledonian Highlands, numerous peaks exceed 2000 m, but in the British Isles, the highest point at Ben Nevis, Scotland is only 1343 m. The highest points in Caledonian Ireland and Wales are 926 and 1085 m, respectively. It should be noted that not all parts of the region are elevated today. Tectonic subsidence and denudation have been active, as well as tectonic uplift, and large areas of central Ireland and Scotland now lie below 200 m.

Because of the intensity and duration of denudation since the Caledonian period, aided by further uplifts, considerable areas of Precambrian rocks are now exposed in the Caledonian Highlands. Elsewhere, Lower Palaeozoic strata dominate, but post-Caledonian rocks are relatively rarely encountered in the region, with the exception of the Devonian–Carboniferous infills of central Scotland and the Central Lowlands of Ireland, isolated areas in western and eastern Norway, and the early Tertiary volcanics and intrusives of north-western Britain and northern Ireland. For the greater part of the region, the long interval of time from the Devonian to the Quaternary, during which the majority of the landforms evolved, is unrepresented by any deposits, so that piecing together the geomorphological evolution presents great difficulties. In the Quaternary, all parts of the Caledonian Highlands were affected by repeated glaciation. At present, only some of the higher Scandinavian mountain groups support glaciers.

7.2 Scandinavian Highlands

7.2.1 Rock Types and Structural Inheritance

The region has two main rock units: the Fennoscandian Shield and the Caledonian orogen. The latter forms a broad band between the continuous eastern Precambrian shield zone and the western Precambrian which occurs in isolated areas along the Norwegian coast (*see* Fig. 5.1). The eastern Precambrian area, or that part of it which belongs to the Caledonian Highlands, is a continuation of the Precambrian in the Fennoscandian Shield with similar granites, gneisses and other metamorphic rocks. Volcanics and sedimentary rocks—notably porphyries, sandstones and quartzites—however, play a somewhat greater role. Close to the Caledonian boundary, the Precambrian is overlain by Eocambrian–Cambrian (locally also Ordovician) sedimentary rocks, usually in a narrow strip along the Caledonian contact, which is most frequently formed by the fronts of the Caledonian nappes. The western Precambrian area is dominated by gneisses, but granites and other intrusives also occur, particularly on the

Lofoten–Vesterålen Islands. The contact with the Caledonian here is also often a tectonic one, but some of the Precambrian areas have the character of geological windows just overlain by Caledonian rocks [see Gee (1979), O. Holtedahl (1960), Magnusson *et al.* (1962, 1963), Tozer and Schenk (1978)].

The Caledonides can roughly be described as a pile of nappes, overthrust in an easterly or south-easterly direction against the Precambrian foreland. The pattern, formerly mainly observed in the eastern parts, now seems to be valid also for the western part. The metamorphic grade increases from the eastern and south-eastern parts towards the central and western parts, usually in a step-wise manner according to the nappe structure but with some exceptions. The most important exception is the so-called great nappe of the Swedish Caledonides, with predominantly micaschist–gneiss in the lower and eastern parts, and phyllites and shales in the upper and western parts.

The Caledonian rock sequence is dominated by sedimentary rocks and metasediments. In the lower nappes of the eastern margin, sandstones, arkosic sandstones (sparagmites), quartzites, shales and, in some areas, limestones dominate. In the higher central and western nappes, shales, micaschists, gneiss and locally marble are most common. Magmatic rocks play a subordinate role compared with the metasediments, but dominate certain areas. Most important are the basic rocks, gabbros, amphibolites and some smaller areas of ultramafics (peridotites, serpentinites). Granites are less common, notably in the eastern parts. Several granite areas—formerly regarded as Caledonian—are now interpreted as geological windows of the Precambrian basement. The age of the Caledonian rock sequences is gained from areas of low metamorphic grade and preserved fossils; techniques of absolute dating are also employed. The oldest members are Precambrian, such as the main rock types of some nappes [e.g., one nappe at the eastern front and one of the higher nappes in south central Norway (Jotun nappe)]. In the lower eastern nappes, Eocambrian rocks are important. The main part of the Caledonides is of Cambro-Silurian age (or Ordovician–Silurian) at least in regard to the bulk of the sediment and some of the basic rocks. The youngest members are Devonian sandstones and conglomerates, found in some isolated minor areas in the western part of southern Norway and a small isolated remnant close to the Swedish border. The Devonian has taken part in the last phases of Caledonian foldings.

Structurally, the Precambrian crystallines and the Caledonides form two different units. The former have the same main characteristics as the Precambrian of the Fennoscandian Shield, with a mosaic of fissures and fissure systems. A similar pattern may also be found in the Precambrian geological windows, at least in some of them. The Precambrian porphyries and sandstones in the southern part of the region belonging to the Shield have flat-lying layers, and the sandstone has intercalated basic sills. The Caledonides are dominated by the nappe structures and the S-surfaces of the schistose rocks—both flat-lying or slightly west-dipping notably in the eastern part. Folded structures of medium and small-scale size interfere with the nappe structures and also some major antiforms and synforms.

The rock types of the Scandinavian Highlands play a more conspicuous role in the landscape than in the case of the Fennoscandian Shield; the Precambrian rocks are described in Chapter 5. It could be added that the basic sills in the sandstone areas locally produce cuesta topography. The contrast between the Precambrian geological windows and the Caledonian surroundings is often well marked. The surface of the former is uneven in detail with steep hillocks and basins. Within the Caledonian rock sequences, the differences in resistance are more pronounced than in the Fennoscandian Shield. A ranked list of some typical rocks in order of increasing relative resistance is as follows: phyllite, micaschist, gneiss, granite = amphibolite, peridotite (serpentinite). For local reasons, quartzite is not included in the comparison but it should be expected close to the top of the list [see Rudberg (1954)].

The quartzites of the eastern front often form broad, poorly dissected plateaus. The contrast in resistance is clearly demonstrated between the eastern micaschists (and amphibolites) and the western slates and phyllites of the 'great nappe' of the Swedish Caledonides by higher and steeper relief in the former and lower relief with gentle slopes in the latter (*see* p.57). The higher mountains of the Swedish Caledonides are, with some exceptions, connected with amphibolites, the high Jotunheimen area of southern Norway with gabbros and the Lyngen Alps in northernmost Norway also with gabbros. The small bodies of peridotite (serpentinite) usually protrude as steep hills with an almost tower-like appearance above the more gently undulating surroundings. The Devonian sandstones and conglomerates form a rugged relief, with details strongly controlled by structure. The nappe fronts, often exhibiting a contrast between rocks of lower metamorphic grade overlain by rocks of higher metamorphic grade, are frequently seen at a distance in the landscape as steep rocky slopes, sometimes in continuous lines and most often east-facing (the glint line or, more correctly, the glint lines). The frequently west-dipping S-surfaces are responsible for numerous asymmetrical hill forms, the steeper slopes regularly facing east.

7.2.2 Post-Caledonian Transgressions

No younger transgressions following the Caledonian orogeny, which ended in the early Devonian, are known. A single exception is a small area of downfaulted Jurassic and Cretaceous sediments on Andöy Island, situated in a marginal position in the northernmost part of the Vesterålen Islands. The conclusion that the Scandinavian Highlands have had a long terrestrial development is in the present state of knowledge just as natural as for the eastern shield areas. A future detailed analysis of the shelf sediments of the Norwegian Sea, mainly those deposited after the opening of the North Atlantic, may provide important information about the geomorphological development and also some more precise dates of this development.

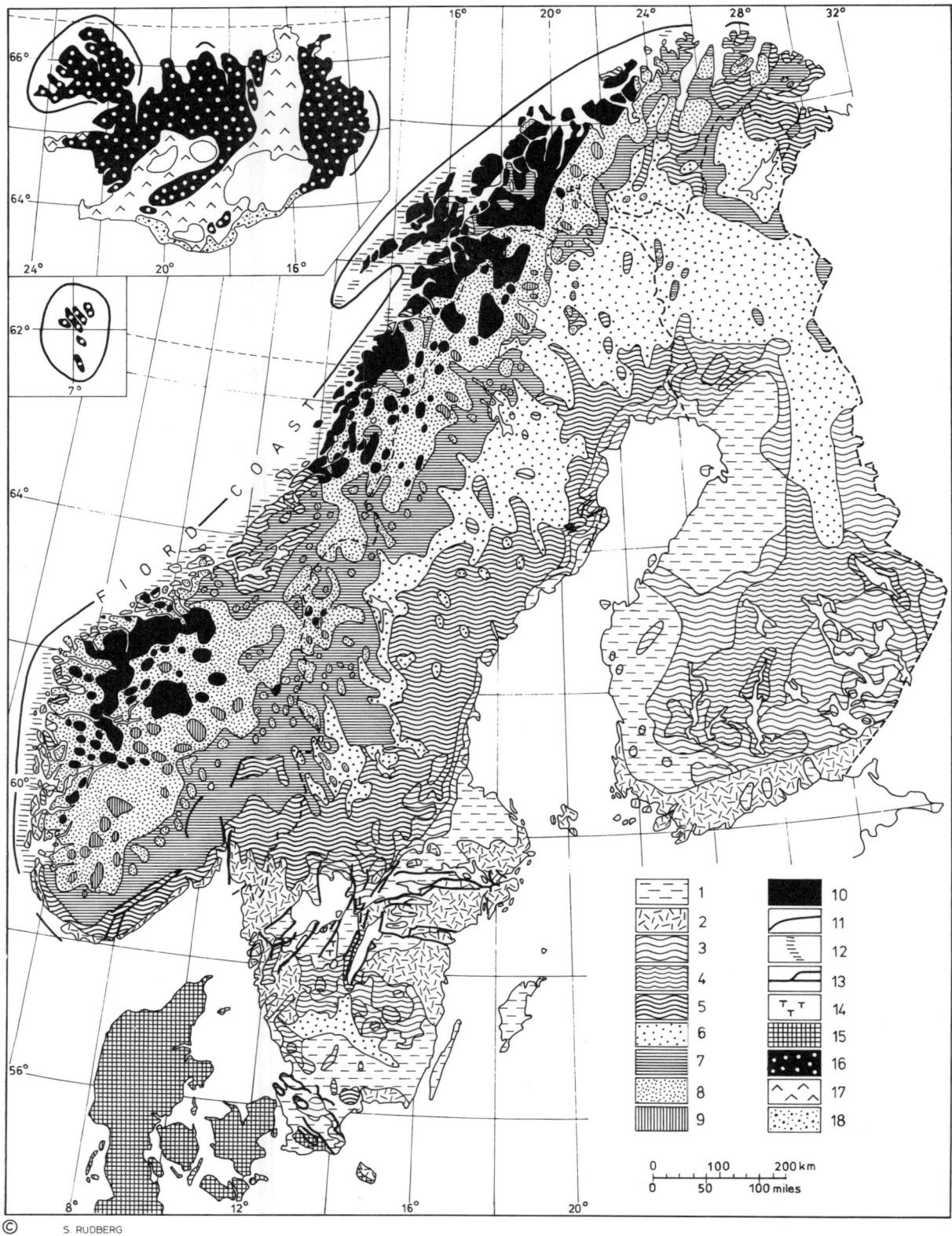

Fig. 7.1 Morphological regions [Rudberg (1960)]. (1) Plains, relative height less than 20 m; (2) fissure valley landscape (Stockholm type); (3) to (5) undulating hilly land: (3) relative height 20–50 m; (4) relative height 50–100 m; (5) relative height more than 100 m; (6) monadnock plains; (7) premontane region; (8) to (10) mountainous area: (8) undifferentiated fjell; (9) fjell with plateaus; (10) fjell with alpine relief; (11) fiord coast; (12) strandflat; (13) major fault; (14) table mountains of Västergötland; (15) relief essentially due to superficial deposits; (16) lava plateaus (Iceland); (17) young volcanic forms (Iceland); (18) sandur (outwash) plains.

Fig. 7.2 Jostedals ice-cap (skyline) and Austerdalsbreen, a glacier fed by twin ice falls from the plateau ice. (Embleton)

7.2.3 Preglacial Geomorphological Development

Particularly important has been the rejuvenation and activity of river erosion in the late pre-Quaternary. It is usually said that this development started in the Tertiary or, even more precisely, in the Miocene, which is possible but not yet proved. The rejuvenation is strongest in western Norway and decreases to the east, although gradually. In northern Sweden, there are clear signs of rejuvenation down to the Gulf of Bothnia. To draw any dividing line between rejuvenated and non-rejuvenated areas in the central and northern parts of Scandinavia is arbitrary. In any case, the young valleys dissect older relief.

There is a fundamental difference between the two flanks of the Scandinavian Highlands (*see* Fig. 7.1). The western slope is steep, strongly dissected and highly transformed by glacial erosion. It is hard to discern any preglacial stages in the development, at least any of regional importance. The eastern slope, on the contrary, is longer and less dissected. The interfluvial areas are often broad, plateau-like and frequently give the impression of planation surfaces, with or without monadnocks. On the Swedish side, a step-like arrangement of planation surfaces can often be seen and can be followed down to the Gulf of Bothnia. The existence of sequences of planation surfaces has also been suggested in Norway, but the evidence presented so far has not been convincing and the question is controversial. As to the exhumed planation surfaces discussed in Chapter 5, only small remnants of the sub-Cambrian peneplain can be traced in the Scan-dinavian Highlands, and then only with any certainty in close connection with the strips of Eocambrian–Cambrian sediments in front of the nappes. In northern Sweden, it has been suggested that this surface may be found as an accordance of summit level in the areas between the Caledonian front and the Gulf of Bothnia. This idea is unrealistic as such accordances only exist locally and at varying altitudes even in adjacent areas [see Rudberg (1954)].

The young valleys resulting from the presumed mid or late Tertiary uplift or from climatic change [see Kaitanen (1969)] are no longer preserved in their original fluvial shape, particularly on the western slope of the Scandinavian Highlands, where the glacial valley forms indicate strong glacial erosion. On the eastern slopes, in northern Sweden, where the glacial transformation has been much weaker, there are signs of a step-like succession of valley floors in a down-valley direction and sometimes in transverse section. This is reminiscent of the step-like arrangement of planation surfaces on the interfluvial areas [see Rudberg (1954)]. It is not clear whether the western slopes show a similar development, although this has sometimes been supposed. The intense rejuvenation on the western side of the Highlands, associated with the steep descent to the Atlantic coast, has caused river piracy in some places. Preglacial erosion caused numerous water gaps within the Swedish Caledonides and in the Precambrian crystalline areas of the Caledonian front [see Rudberg (1976)]. In both areas, the most reasonable explanation involves superimposition following the removal of weak cover rocks: in the first case, slates and phyllites in the

Fig. 7.3 Closely spaced cirques (no longer containing glaciers) and nivation hollows. Hinnøy Island, Lofoten group. (Rudberg)

zone of rocks of low metamorphic grade within the 'great nappe' and, in the second case, lost Cambro-Silurian sediments east of the present residuals. Altogether, though, no clear or detailed picture of the preglacial geomorphological development of the Scandinavian Highlands exists.

7.2.4   Glacial Erosion Forms
At the present-day, ice caps and glaciers are restricted to relatively small systems at the higher levels. The largest groups in Norway include the Folgefonni in the Hardangerfjord region, the Jostedalsbreen between the Sognefjord and Nordfjord (*see* Fig. 7.2), the Jotunheim systems and, in the north, Svartisen. In Sweden, the main glacierized areas are in the north, in the Sarek and Kebnekaise mountains. The total area under ice is about $5000 \, km^2$.

In the Quaternary, however, ice spread out repeatedly from western maritime mountain centres of accumulation to cover the whole region. It seems likely that each glaciation commenced with the formation of cirque glaciers, minor ice caps and valley glaciers, similar to those of the present day, but as the snowline fell, the glaciers expanded, glacier systems coalesced, and more and more of the terrain was submerged beneath an ice cover. During this development, the ice divide gradually migrated eastwards and was finally situated far to the east of the Highlands, partly above the present Gulf of Bothnia [see Ljungner (1949)]. The background of this migration may have been change in climate (e.g., in the low-pressure

tracks) or the balance within the ice sheet itself, with its western flank terminating in a calving front in open sea and an eastern or southern flank with broad ice lobes slowly moving over dry ground [see Hoppe and Liljeqvist (1956), Schytt (1974)]. The Highlands were crossed by moving ice during the stages when the ice divide lay to the east, indicating a maximum ice thickness around 3 km. There is controversy as to whether the region was ever completely covered by ice or whether nunataks existed. The botanists insist on ice-free refuges to explain the present distribution of flora, but most investigated problematic summits do have erratics [see H. Holtedahl (1949), Hoppe (1959), Rudberg (1952)]. The present compromise could be expressed in the following way. Most areas have been covered once, or several times, during the glaciations, but not simultaneously. The eastward migration of the ice divide, for instance, may have been followed by a lowering of the ice surface in some western highlands [see Ljungner (1949)].

With climatic amelioration, the ice thickness decreased and the ice front retreated at a speed partly related to topography. The deep fiords of western Norway lost their ice cover—because of calving—much earlier than the surrounding highlands as discussed by Andersen (1960). The main ice divide migrated westwards, partly once more, into the mountains. In most areas, ice movement continued into the late stages of the deglaciation, as can be seen from studies of morainic forms and detailed surveys of multisystem glacial striations, but stagnation and

Fig. 7.4 Pleistocene cirques near sea level, Senja Island, northern Norway. (Rudberg)

melting *in situ* occasionally occurred. Where the last ice remnants disappeared—within the Highlands proper or east of them—is still an open question. The relative length of the initial stage of mountain glaciation compared with the stage of more complete glaciation is not known. Apart from the major alternations of glacials and interglacials, there were also minor stadials characterized by more limited development of glacier systems at high levels. Postglacial changes in glacier extent have been thoroughly investigated by Karlén (1973, 1975, 1976) on the basis of historical records, pollen data and lichenometry.

The most exclusive forms of Alpine glaciation—the cirques—are widespread in the Scandinavian Highlands, but the distribution is fairly irregular. Some cirques have glaciers at present, most cirques have not, and only a few of these latter have end moraines of postglacial origin. The empty cirques show a depression of the glaciation limit compared with the present of 800–1200 m. These are only minimum values as lower, more plateau-like terrain in the Scandinavian Highlands would produce ice caps and not cirque glaciers, with continued sinking of the snowline. The cirques, furthermore, indicate a depression of the same limit to the north and still more to the maritime west, providing a good parallel with present conditions. Isolated, scattered cirques are often large and perfectly developed, sometimes of an almost textbook-like perfection. Densely grouped cirques (*see* Fig. 7.3) are less perfectly shaped, as the individual glaciers interfered with one another. As regards aspect, the majority of cirques are

south-east or east-facing. The tendency is most clear in inland areas and among isolated, scattered cirques. The cirques survived ice sheet erosion with moderate change, but some exceptions may be found [e.g., those described by Ljungner (1949), Rudberg (1954), Vilborg (1977)]. In the north, cirque floors approach sea level (*see* Fig. 7.4).

The glacial troughs of the Highlands have the joint background of erosion by valley glaciers during the early stages of each glaciation and by ice sheets during the maximum stages. The relative influence of the two glacier types cannot yet be decided, but it is obvious that ice sheet activity cannot be excluded. Troughs are numerous and of various types (*see* Figs. 7.5 and 7.6). There are, particularly on the western flank, perfectly shaped U-valleys with high, steep, almost symmetrical sides and with tributary valleys hanging high above the valley bottom. Furthermore, there are valleys with stepped floors and valleys with thresholds, valleys with basins and narrower sections alternating. There are also some U-valleys open at both ends and some with closed, steep ends, clearly pointing to retrogressive glacial erosion [see Gjessing (1966)]. In contrast, notably on the eastern flank, there are often uncompleted U-valleys, where the steep facet is only found on one valley side or the other, alternating down valley. In some of these valleys, the steep facet is a truncated spur, or a spur with only one side steepened, as though truncation of the spur stopped before it was finished [see Rudberg (1954, 1976)].

Vertical erosion is revealed by the overdeepened valley

Fig. 7.5 Rapadalen, a glacial trough in the Sarek area of northern Sweden. The braided river flowing across the infill of outwash gravel is shown at high-water stage. (Rudberg)

Fig. 7.6 Oldevatnet, a lake occupying a glacial trough. In the background can be seen the Jostedals ice cap (skyline) and the present-day outlet glacier. (Rudberg)

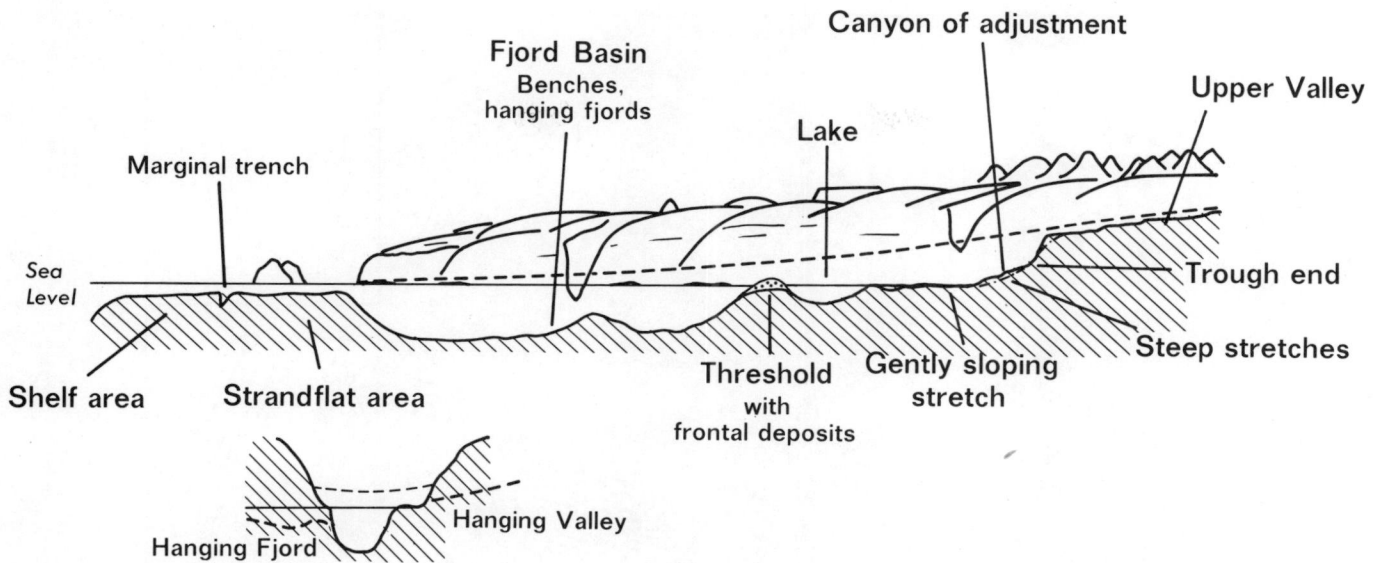

Fig. 7.7 Some form elements of Norwegian fiords and fiord valleys, longitudinal profile and cross-profile through the fiord. The strandflat, the fiord-side benches and the gently sloping valley stretch at the head of the fiord are found near the same level which is nearly in accordance with the present sea level. The canyons of adjustment in the steep stretches and at the trough end are indicated by dash-dotted lines. A possible preglacial valley floor is indicated by a dashed line [Gjessing (1966)].

and fiord basins (*see* Fig. 7.7) Counterparts to the fiords, although much less overdeepened, are the piedmont lakes on the eastern side of the Scandes. The total amount of deepening by trough or fiord formation cannot be decided, but vertical erosion exceeds 1000 m in individual cases. Remnants of older, more gentle slopes are seen in some fiords. Certain fiords show a dendritic pattern, but others are conspicuously influenced by structure.

Glaciofluvial erosion has produced many canyons in solid rock, in the more spectacular cases several kilometres long and more than 500 m deep. They are most frequent in the northern part of the region and adjacent parts of the Fennoscandian Shield where the term kursu valleys is sometimes used [see Rudberg (1949)]. Minor meltwater channels are discussed in Section 7.2.6.

On the more open lowlands or plateaus, ice sheet erosion forms prevail. Roches moutonnées are found on bare-rock areas. Generally speaking, these forms seem to be better developed in the Precambrian crystalline rocks than in most of the Caledonian outcrops. This is clearly demonstrated by some of the Precambrian geological windows in contrast to their surroundings. Also important in this region are erosion forms of medium scale, such as the isolated steep rock faces, the flyggbergs, asymmetrical valleys, rock basins and valleys of moderate relative relief with a shape reminiscent of U-valleys.

7.2.5 Problematic Forms

A discussion of some of these forms (e.g., the undulating hilly land and Palaeic surface) [see Gjessing (1967)] has already been given in the case of the Fennoscandian Shield (*see* Section 5.1.5) and need not be repeated here. But to this region also belong the strandflat areas which form a fringe of lowland along the steep western coast of Norway, a fringe often consisting of swarms of islands and

skerries, but sometimes of lowland along steep fiord sides or around high islands. Explanations are numerous: a preglacial erosion surface, a platform of glacial erosion caused by coalescing cirque floors or the bases of small piedmont glaciers, a surface of marine abrasion, a system of glacial and marine erosion surfaces formed in combination with nivation, and ice foot erosion [see Nansen (1922)]. Most evidence favours the last of these hypotheses.

7.2.6 Quaternary Deposits and Forms

In Chapter 5, the contrast between the western bare-rock and the eastern till-covered areas of Fennoscandia is discussed. The important boundary between the two divides the present region. The features of the two zones have already been described. Important in the bare-rock zone is the continuous line of the Ra moraines running through the whole of southern Norway. It has not yet been completely traced in the southern part of northern Norway, but has been mapped again in the northernmost part. Marginal moraines often including glaciofluvial deposits are frequent at the inner end of numerous fiords; they are usually located at the transition from the deep fiord basin to the valleys at the inner end, where the ice margin retreat slowed down after calving had ceased [see Andersen (1960, 1965), Aarseth and Mangerud (1974), Mangerud (1970), Sollid and Sørbel (1975)].

The eastern zone is well covered with till and morainic forms are frequent, such as flutings and drumlins, hummocky moraines and Rogen moraines, the latter notably in the eastern and northern areas [see J. Lundqvist (1969b)]. Among the glaciofluvial forms, well-developed eskers, terraces and sandur plains should be particularly noted. On the eastern slopes of the Scandes, ice-dammed lakes existed during stages in the deglaciation. Their

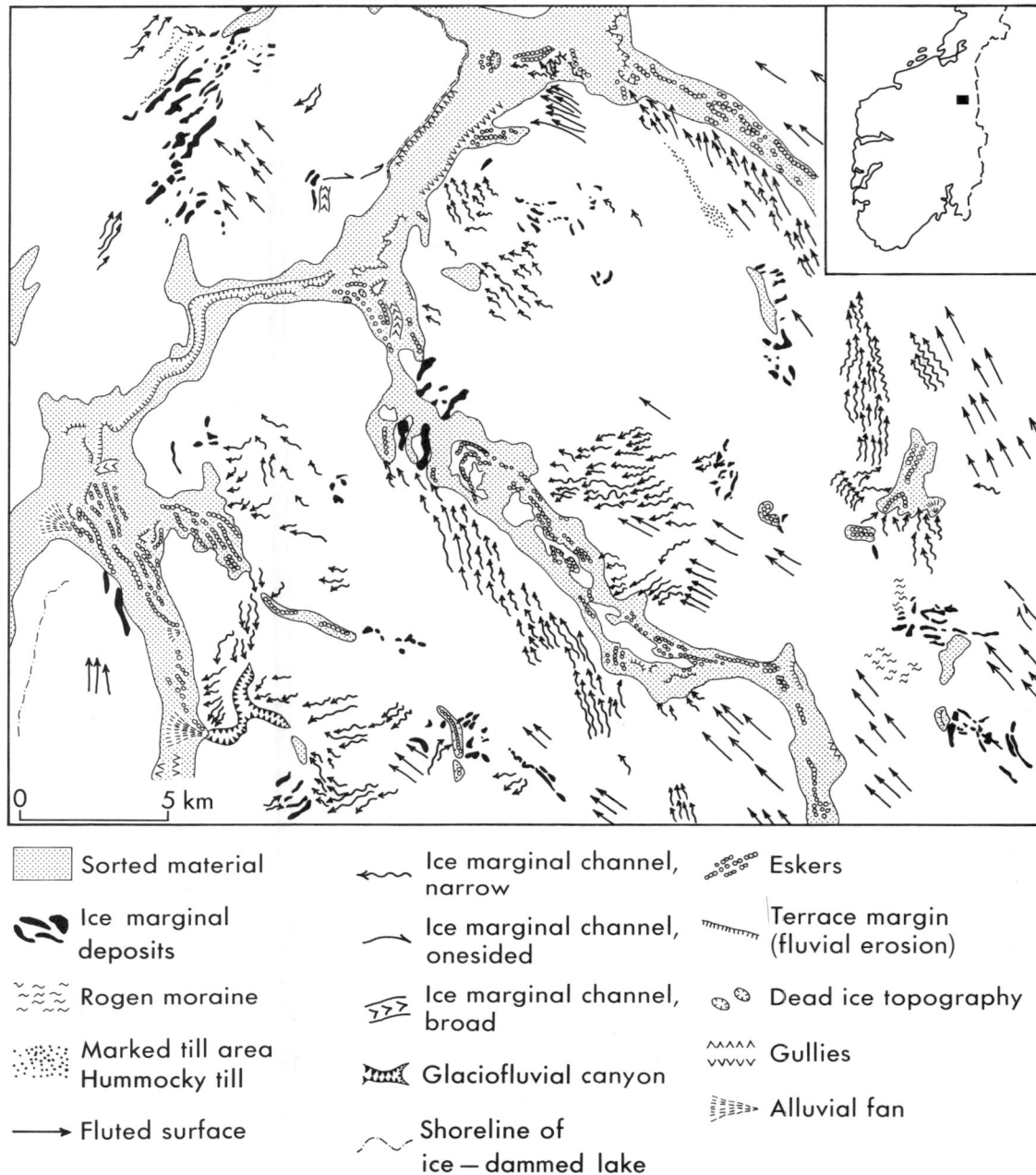

Fig. 7.8 Forms resulting from down-wasting ice during deglaciation of a highland area [after Sollid and Kristiansen (1982)].

former extension is now shown by shore terraces, deltas, varved clays and other features as described by J. Lundqvist (1972).

The highland areas of the region, with their high relative relief, often provide good opportunities for following the vertical distribution of glacial features, particularly those connected with the down-wasting ice. In the wide areas above the timber line, observations are facilitated. Following the shrinking ice from higher altitudes, one might find several features—in idealized situations—but not by any means in all places. The highest summits may have a steep, rugged outline, which is traditionally

attributed to a nunatak stage during a part of the latest glaciation (not during the whole glaciation). Such nunataks are only of importance locally. Other glacial features on the higher parts of the slopes are mainly due to the meltwater drainage (*see* Fig. 7.8). The highest forms are the col gullies, eroded by meltwater passing across the lowest parts of a ridge. On the slopes below, systems of lateral drainage channels are often developed in till-covered areas. Their longitudinal slope may indicate the slopes of the ice sheet surface at the moment of channel formation, although this is not true in the case of subglacial or submarginal channels. On the lower slopes,

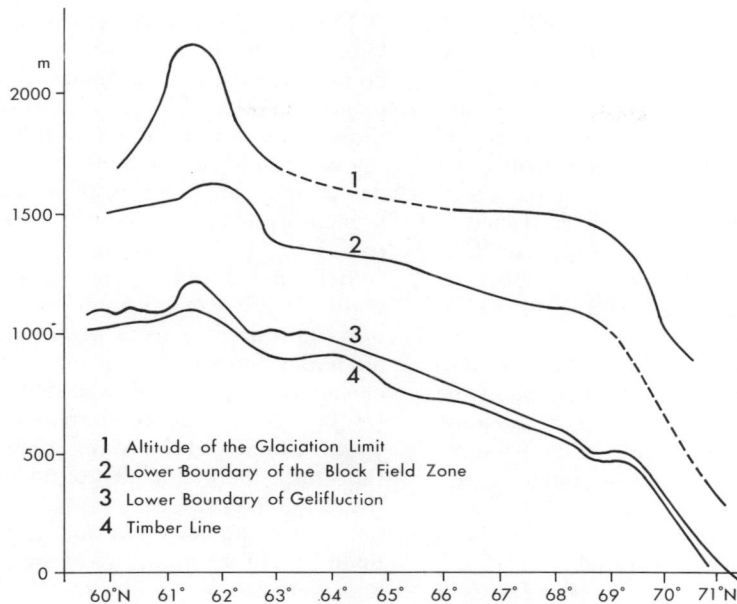

Fig. 7.9 Schematic profile through the mountain areas of Scandinavia, centred roughly around the main watershed, showing the relations between various altitudinal boundaries [Rudberg (1977)].

where meltwater courses converged, major channels have often been eroded, but more frequently glaciofluvial deposits were laid down as eskers, terraces or sandur plains. In the sequence of morainic forms, Rogen moraines occupy the lowest parts, while drumlins are found on slightly higher ground [see Gjessing (1960), Kuujansuu (1967), Mannerfelt (1945), Møller and Sollid (1972), Piirola (1967), Sollid *et al.* (1973), Sollid and Kristiansen (1982)].

7.2.7 Present-Day Geomorphological Processes
In Chapter 5, the present-day geomorphological processes and the importance of postglacial uplift as an initiating force for some spectacular results were discussed. In the Scandinavian Highlands, strong fluvial erosion has dissected the fine-grained deposits of the lower reaches of the western valleys, leaving terraces, deltas, etc. as visible geomorphological results. Recent/subrecent fluvial erosion in solid rock is less common, with one exception. In glacially transformed valleys, Gjessing (1966) has described how young 'canyons of adjustment' are often cut into valley steps, thresholds and the steps at the mouths of hanging valleys. Some major V-shaped valleys are connected with the main troughs as tributary valleys. They probably have a long history of erosion during several interglacials. The V-shape has not apparently been destroyed by intervening covering by ice sheets.

Because of the high, steep relief and the altitudinal range of the terrain, mass movements are common, notably in the western part of the region. Recent or historical examples of catastrophic magnitude have been reported, older ones being located by evidence of scars, chutes or accumulations at the base of slopes. Great rockslides are released occasionally on glacially over-steepened valley sides; there have been some well-

documented cases this century. Below steep rock faces, talus slopes are normal features, most of them, but not all, in an active state of development. The layers of loose deposits on more moderate slopes are removed by slides of earth and debris or mud flows, according to recently studied cases particularly in connection with heavy rain. Avalanches are very frequent in most mountain areas, the most important both from a geomorphological and a practical point of view being slab and wet snow avalanches. A large number of avalanches occur on east-facing slopes, because the snow accumulation responsible is the result of snow drifting due to the prevailing winds from the west or south-west. Many avalanches occur annually, with their paths shown by open lanes in the forest, by chutes on steep rock walls and by accumulations at the slope base, sometimes of the avalanche boulder tongue type [see Rapp (1961, 1965), Rapp and Strömqvist (1976), Ramsli (1951), Rudberg (1950)].

Many types of mass movement belong to, or are favoured by, a periglacial environment (*see* Fig. 7.9). Other, less dramatic, periglacial processes are also active at present, particularly in the *fjells* (i.e. the area above the timber line). The variations in processes and forms are considerable. Nivation or snow bank erosion is evident, at least in loose deposits, by the enlargement of slope depressions to give new forms with terraces and a steep back-slope. Nivation processes contribute to the formation of cryoplanation surfaces (both the small-sized variety in front of isolated snow banks and the larger ones to be mentioned below). Patterned ground phenomena are common throughout the region, with hummocks, stone pits and sorted circles being most frequent on flat ground; ploughing blocks, gelifluction lobes and sorted stripes are the most common on slopes. Frost creep and gelifluction are normally active to judge from long-term measurements at numerous sites [see Rudberg (1964)]. The greatest variety of forms is found in the tundra zone. The block field zone (frost-shatter zone) is more monotonous, but gelifluction lobes occur and wide, flat surfaces locally characterize higher altitudes within the zone. With low angles of slope and residual tor-like features, they may be classified as cryoplanation surfaces. Most of the patterned ground features described are not completely restricted to the fjells: for example, sorted circles are found on lake shores in the northern coniferous forest [see Högbom (1914), J. Lundqvist (1962), Rudberg (1962, 1974, 1977), Strömqvist (1973)].

Most of the periglacial features described do not seem to require permafrost, but other features do. Typical of the discontinous permafrost to which the fjells partly belong, together with some northern forested areas, are the palsas. Locally they are in an active state of development but are more often undergoing degradation. Their localization in areas with a generally or locally thin snow cover has been well established by Åhmann (1977), G. Lundqvist (1951), Ruuhijärvi (1960), Sollid and Sørbel (1974), Svensson (1961) and Wramner (1973). Ice-wedge polygons are observed in some places, but they may be of subrecent origin [see Rapp *et al.* (1962), Rapp and Clark (1971), Seppälä (1966), Svensson (1962, 1963)]. The same comment applies to a few observations of collapsed pingos reported by Rapp and Rudberg (1960), Seppälä (1972a) and Svensson (1964a). Rock glaciers are also known to occur.

Wind action is also favoured in the dry periglacial environment of the fjells. The efficiency of such erosion is demonstrated by numerous scars, often very fresh in appearance, on the heath and grassland of the tundra zone. This could mean that vegetation damage has been initiated by over-large herds of domesticated reindeer. The eroding winds seem to be gravity winds in areas where the wind direction has been mapped with the help of wind scars and other features [see Högbom (1914), Hörner (1927), Rudberg (1968), Seppälä (1971, 1972b), Sollid *et al.* (1973)].

7.2.8   Regional Subdivisions

Reference should first of all be made to Section 5.1.8, where some general comments about Fennoscandia as a whole are presented. The discussion of subregions will be based on the same map (*see* Fig. 7.1).

Type 7 (*see* Section 5.1.8) relief belongs mainly to the Highland region and has a more pronounced character as a zone than subregions 3 and 4. It follows the eastern front of the Scandinavian Highlands proper, from southern Norway through northern Sweden to northernmost Norway. The zone has higher altitudes, a higher relative relief and better developed river valleys than the eastern subregions. The interfluvial areas consist of plateaus or broad massifs. The western boundary of the subregion coincides with the Caledonian front in parts of northern Sweden, but not in other places, as it includes important areas of Precambrian rocks. The subregion is only locally developed to the west of the Highlands.

The next subregions comprise the Highlands proper, known as the *fjells* in Norway (Swedish: *fjälls*). It should be noted that the map boundaries are not only based on major landforms but are also in close agreement with the timber line, according to traditional Scandinavian use of the word fjell. As the fjells are the main areas of present-day periglacial processes, the boundaries are justified from a geomorphological point of view. Most fjell areas have rolling relief, reminiscent of the undulating hilly land (relief types 3 and 4), but on a different scale and with higher relative relief. Some areas are plateau-like. More important are the areas with high and steep relief, which often develop an Alpine appearance, with scattered or clustered cirques and well-developed U-shaped valleys (*see* Fig. 7.10). The main areas are in Norway, where the Jotunheimen and the Sunmøre are the most important in the south; the major part of northern Norway, except for the extreme north, has similar characteristics. In Sweden, only some smaller mountain groups are indicated on the main part of the map; two of the major areas are the Sarek and Kebnekaise mountains together with some adjacent areas.

The vast fjell areas have much in common but there are also variations that allow some subregions to be distinguished. The fjells of southern Norway form a continuous unit, stretching in a north–south direction through the whole area, forming a barrier between eastern and western Norway, and dissected in their marginal parts by radiating deep valleys or by very deep fiords in the west. This broad mountain area is followed to the north and east by scattered mountain groups, mainly of lower altitudes, around Trondheimfjord and in Jämtland province and adjacent parts of Sweden. Some lower corridors form continuous breaks in the mountains from the interior of Sweden to Trondheimfjord. Farther north, the fjells form a long, broad unit without major interruptions on both sides of the international border through northern Norway, Swedish Lappland and a minor part of Finnish Lappland. The Norwegian part is deeply dissected by fiords and valleys, including, in the southern part of the subregion, longitudinal valleys following the Caledonian strike direction. The Swedish part is transected by valleys crossing the main Caledonian units, from the main water divide to the lower land in the east, less deep than the Norwegian valleys but often occupied by long chains of lakes. Completely separate is the area of the rugged, Alpine Lofoten–Vesterålen Islands, followed to the north by other islands with a similar relief but less famous names (e.g., Senja, Ringvassøy, etc.).

The fiord coast is given a special symbol on the map (*see* Fig. 7.1). The term has been used in a somewhat restricted way, excluding smaller examples with only moderate relief (as in the case of the small fiords of western Sweden and southernmost Norway) or the open bays in the far north of Norway. In the place names, such features are often called fiords (Scandinavian spelling: *fjord*). The fiords certainly do not have the same appearance all along the extensive Norwegian coast, but so far no detailed description of the variations in type has been published. In the south, the influence of structure is striking (e.g., in interior Har-

Fig. 7.10 A glacial trough, Isterdalen, Norway. In the foreground, the 'Trollstigveien' road. (Rudberg)

dangerfjord and a group including Vindafjord–Hylsfjord). The Sognefjord and some adjacent fiords have a clear dendritic pattern. These two main types can be followed along the coast farther to the north.

A special symbol is also given to the strandflat areas, that is the fringe of lowland along the steep Norwegian coast, often with a very abrupt transition. The strandflat may consist of swarms of low islands and skerries of the skjaergård-type, or narrow bands of lowland on the mainland or some major islands. The symbol is restricted to the western and northern coast of Norway, while minor areas in south-eastern Norway and western Sweden are omitted, their relationships with the strandflat being uncertain.

Other glacial forms of micro and meso-scale (e.g., roches moutonnées, rock basins, flyggbergs, etc.) are typical of the Scandinavian Highlands as well as the Fennoscandian Shield, but type and frequency vary considerably with rock type.

The problem of regionalization in relation to the unconsolidated deposits is discussed in this chapter as in Chapter 5, and is based on the same map (*see* Fig. 7.1). Bare-rock surfaces dominate the whole of the western part of the region, but this does not mean a total lack of glacial deposits. Valley floors at lower altitudes are covered by clay and silt deposits. The Ra moraine ridges can be followed through most of the region and thick accumulations of till and glaciofluvial material are frequently found

at the inner ends of the fiords. The eastern till-covered area has much in common with the interior parts of the Fennoscandian Shield, with the same morainic forms, such as drumlins, hummocky moraine and Rogen moraine. The last type is especially typical of the eastern valleys of the region. Eskers are frequent, locally also sandur plains and glaciofluvial deltas, the latter being located where ice-dammed lakes once occurred. To the ice lake environment also belong the scattered occurrences of clay and silt. The high slopes of the mountain valleys gave excellent opportunities for development of the form sequences related to down-wasting ice and a sinking ice surface, notably the various meltwater erosion forms (e.g., systems of lateral drainage channels) and various meltwater accumulation forms along the valley sides and the valley floor (e.g., kame terraces).

It should be possible to study regional variations in present-day geomorphological processes in the Scandinavian Highlands, with their high relative relief, more easily than in the Fennoscandian Shield. The lowest zone in the west shows the same influence of glacioisostatic land uplift, with fluvial dissection of the sediments of the valley floors. Most important is the abundant development of periglacial forms in the fjell areas with patterned ground features and forms due to mass movement. The regional variations are clearly marked since most forms are related to the occurrence and type of the unconsolidated deposits: few or no forms in bare-rock areas, gelifluction lobes better developed in areas rich in fines, sorted forms in areas with a sufficient amount of stones, etc. The grain-size distribution in the till cover in turn is dependent on the bedrock. Rocks of low metamorphic grade, for instance, usually yield till with a higher content of fines; quartzites give a high amount of stones and boulders.

The most important regional difference is connected with altitude and latitude. The difference is very pronounced between the lower tundra zone and the higher block field zone, as described above. The diagram (*see* Fig. 7.9) gives a rough indication of the regional pattern. Forms related to permafrost (palsas, ice-wedge polygons, etc.) increase in number to the north. All altitudinal limits become lower to the north and from east to west. Periglacial forms (e.g., boulder depressions, thufurs, sorted circles and palsas) occur outside the fjells, but because of incomplete information no general map can be presented. Coastal cliffs dominate the open coast from Nordkapp and farther to the east. Otherwise, cliffs are only found locally on peninsulas and higher islands protruding through the protecting strandflat fringe.

7.3 Scotland

7.3.1 Scottish Highlands

The Scottish Highlands form the most extensive and elevated mountain area in Britain. Ben Nevis (1343 m) is the highest point, while numerous peaks exceeding 1000 m. The southern border is clearly defined by the Highland Boundary Fault, with its typical Caledonian trends (*see* Fig. 7.11). Another major fracture of Caledonian age, the Great Glen Fault, breaks the highlands into two roughly equal halves: the Grampian Highlands to the south and the Northern (or North-West) Highlands on the north. The greater part of both these units is built of Precambrian rocks and consists of rugged, highly dissected relief, strongly affected by glacial erosion in the Quaternary. In a broad sense, the Scottish Highlands represent the greatly eroded remnants of the Caledonian fold mountain system trending south-west to north-east across Scotland and seen again, on a grander scale, in the Scandinavian Highlands. Because of the great age of the structures and limited disturbance of the region by neotectonic movements, there is virtually no 'tectonic relief'—all is due to differential denudation—but the fault pattern has had a profound indirect influence on the relief.

Many valley alignments are guided by faults of Caledonian trend, from the Great Glen Fault itself to smaller valleys such as the upper Tay, the Spey and Strath Glass. The western coast is fragmented into a rectangular network of inlets (fiords or lochs), while the general outline of much of the Scottish coast is probably controlled by fault systems. The faulting—basically Caledonian—was strongly reactivated in post-Carboniferous (Hercynian) times, as shown by the abrupt severance of the post-Caledonian rocks in the Central Lowlands and by the strongly dislocated masses of Old Red Sandstone in many parts of the Highlands.

7.3.1.1 *Relief and structure*

Lithology is one of the chief considerations in explaining the diversity of relief in the Scottish Highlands, although there is no simple correlation between rock resistance and altitude. The Lewisian Gneiss of the north-western margin and the Outer Hebrides—a fragment of an ancient Precambrian shield or 'plate'—represents the oldest known rock outcrop in Britain and ranks high on any scale of comparative rock hardness, yet a substantial part of its surface lies below 300 m (*see* Fig. 7.12). Part of the explanation lies in the fact that the Lewisian Gneiss was reduced to a surface of moderate relief in later Precambrian time, prior to deposition of the Torridonian Sandstone (Proterozoic), and exhumation of this ancient unconformity has played an important part in the evolution of the landscape here. The area has been heavily scoured by Quaternary glacial erosion and the landscape appears irregular and barren—with numerous rocky hills and basins—together with some higher mountains. In clear contrast, the remnants of the Torridonian Sandstone often tower above in such imposing mountains as Suilven and Stac Polly. The Torridonian also forms fine coastal cliff scenery, as at Stoer Point.

The Lewisian and Torridonian areas of the north-west are separated structurally from the main Precambrian area of the Northern Highlands by the Moine Thrust, whereby the rocks of the Moine and Dalradian Series were thrust to the north-west over the Lewisian. The Moine and Dalradian Series outcrop over some three-quarters of the whole of the Scottish Highlands and are characterized by great variations in lithology and resistance. They comprise a range of Precambrian (and possibly some later) metamorphic rocks, from relatively weak metamorphosed limestones (e.g., around Loch Linnhe) to highly resistant quartzites and grits (e.g., along the Highland border in the south, in Jura, and in the peak of Schiehallion rising to 1081 m). Tilted sheets of these hard rocks give some striking cuesta-like forms, as is also the case with the Cambrian quartzites (e.g., Quinag and Canisp, in the north-west).

With the exception of the small areas of Cambrian just noted, and the Durness Limestone (Cambrian–Ordovician) with some karstic development, the Lower Palaeozoic Era is largely unrepresented in the Scottish Highlands. After the Caledonian orogeny, however, there was extensive deposition of Old Red Sandstone which now contributes a distinctive element in a number of areas. North-western Caithness, the Orkney Islands and the borders of the Moray Firth are all landscapes developed in these sandstone and mudstones, typified by smooth hills, lowlands and low plateaus. At the coast, impressive cliffs are often found; the Old Man of Hoy is an Old Red Sandstone sea-stack 150 m high off the coast of Hoy in the Orkneys. Conglomerates in the Old Red Sandstone also form some higher elevations inland. Patches of Old Red Sandstone, some very small, occur in the floors of some valleys in the Highlands, testifying to the ancient origin of these valleys and the formerly much more extensive nature of the Old Red Sandstone. Some larger fault-bounded outliers lie along the Great Glen.

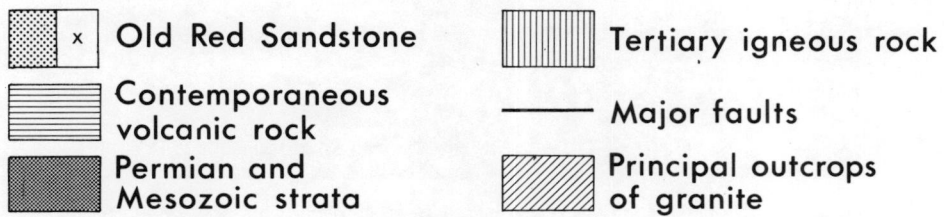

Fig. 7.11 Some geological features of the Grampian and Northern Highlands of Scotland [after Sissons (1967)].

Fig. 7.12 Lewisian Gneiss (early Precambrian) of Sutherland, north-west Scotland. In detail, the surface has suffered intense glacial erosion typical of a 'knock and lochan' landscape. The general outline, approximating to a surface of low relief, corresponds to the sub-Torridonian (late Precambrian) unconformity from which the overlying Torridonian Sandstone has been stripped away. (Embleton)

Fig. 7.13 Cuillin Hills, Skye and Loch Scavaig, Scotland. The Cuillins consist of a resistant gabbro intrusion, deeply dissected by glacial erosion in the Pleistocene. (Embleton)

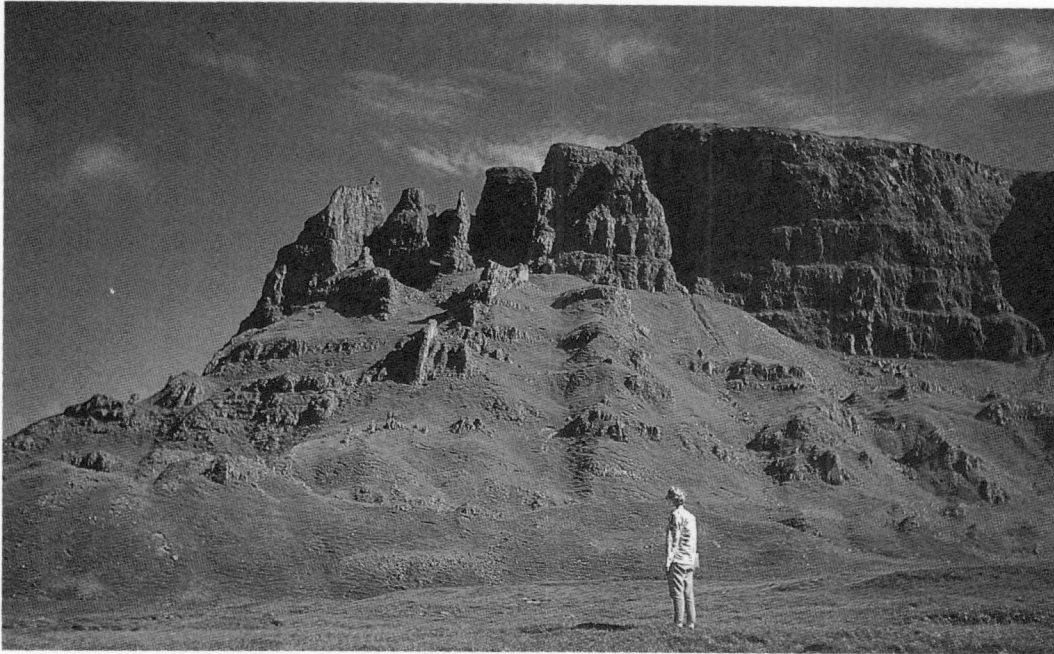

Fig. 7.14 The Quirang, an escarpment of early Tertiary basalt in north-eastern Skye, Scotland, showing recent mass movement. (Embleton)

A score of large granitic intrusions form another major element in the geomorphology of the Highlands. These include the granite masses of northern Arran, the Cairngorms, Ben Nevis (in part), the highlands around Glen Etive and the Red Hills of Skye. The response of granite to denudation is variable: as well as these mountain groups, it also gives rise to relative lowlands, such as Rannoch Moor—a basin rimmed by peaks in metamorphic rocks. Where not dissected by glacial erosion, the more resistant granites (generally medium-grained varieties) tend to produce rounded outlines. Intrusions of other types of igneous rock are less common, but the gabbro laccolith of the Cuillin Hills in Skye is worth mention. The extreme resistance of this rock to weathering and the deep dissection of the region by glaciation have combined to produce the most spectacular mountain relief anywhere in Scotland (*see* Fig. 7.13).

Rocks of Mesozoic age in Scotland are rare but of great significance stratigraphically and in terms of relief evolution. Permo-Triassic sandstones form the southern half of Arran and occur in small outliers farther north on the western coast and the adjacent islands, as well as the famous Elgin sandstones on the Moray Firth. Jurassic and Cretaceous beds are known from a variety of scattered outliers, strata ranging from Rhaetic (in the neck of a volcano on Arran) to Upper Cretaceous (Cenomanian and Senonian). In many places, the preservation of the outliers is due to burial by Tertiary lavas. The association of many Mesozoic outliers with the western coast is not fortuitous; it seems likely that the general north–south linear nature of this coast represents an ancient fault-line scarp, along which the block of the Scottish Highlands rose abruptly from the Hebridean sea floor; against it accumulated Mesozoic strata that have since been largely stripped away. They testify to the great antiquity of this feature which may have been first roughed out in early Permo-Triassic times. Other elements of the Scottish coast on the east may also date back to the early Mesozoic; the Moray Firth seems to have been developing as a basin of sedimentation since farther back in the Devonian. The original maximum extent of Mesozoic strata in Scotland cannot now be determined, but some workers, such as Linton (1951a), have claimed complete marine submergence of the Scottish Highlands by Cenomanian times and the deposition of a blanket of Upper Cretaceous rocks across the whole region, on which the present river system was initiated.

The final major lithological element in Scottish scenery is introduced by the Tertiary volcanics. From several major volcanic centres in the west, lavas were poured out and dykes intruded during the early Tertiary, potassium–argon dates being of the order of 55–60 Ma BP. The lavas, piled up in sheet after sheet, were several hundreds of metres thick in places, and the extent of their subsequent removal and dissection bears witness to the great erosion of the later Tertiary period. The Loch Lomond valley, for instance, is 900 m in depth and cuts across Tertiary dykes; it must therefore postdate the volcanic activity. The volcanics are also built into the western edge of the Scottish Highland block, the planation of whose surface (see below) must then be the result of later Tertiary denudation. The most extensive spreads of lava are found in Skye (forming most of the island south of the Cuillins) and in Mull. The associated dykes radiate from the volcanic centres in great numbers and extend across the Central Lowlands, the Southern Uplands and into northern England and North Wales. The lavas form plateaus, stepped topography, and some impressive scarps, prone to

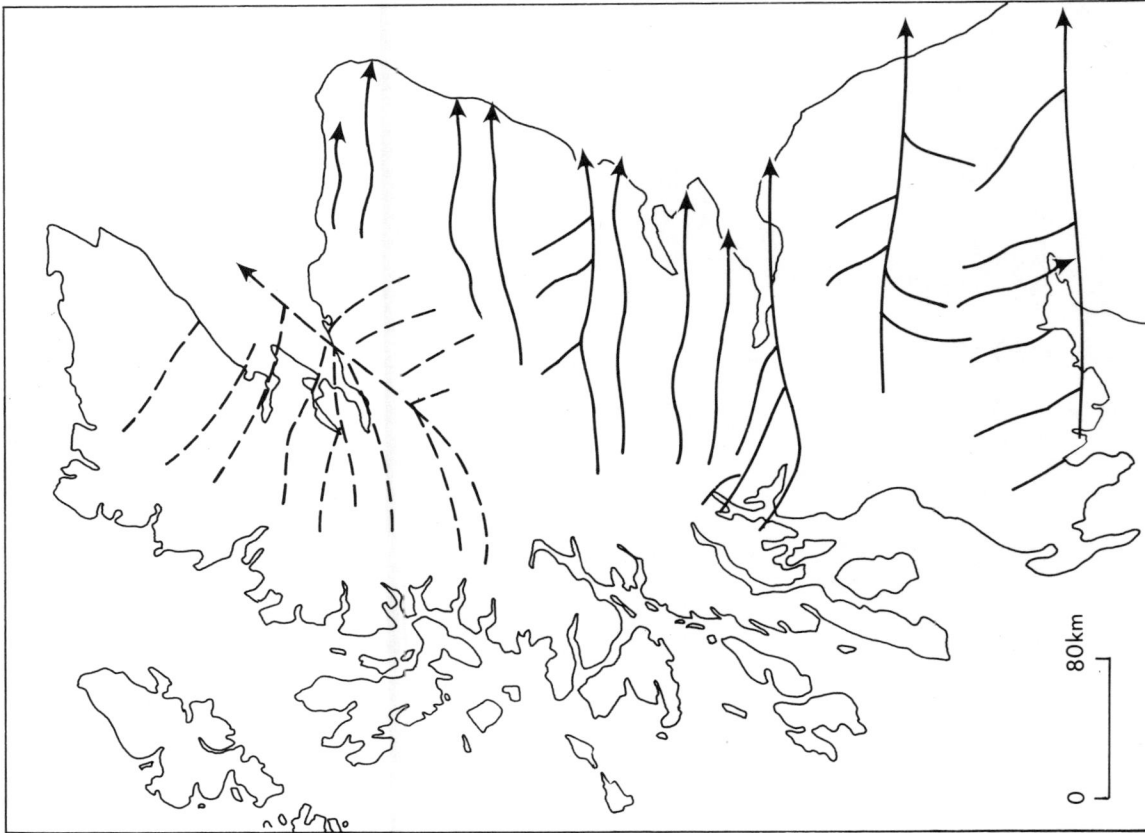

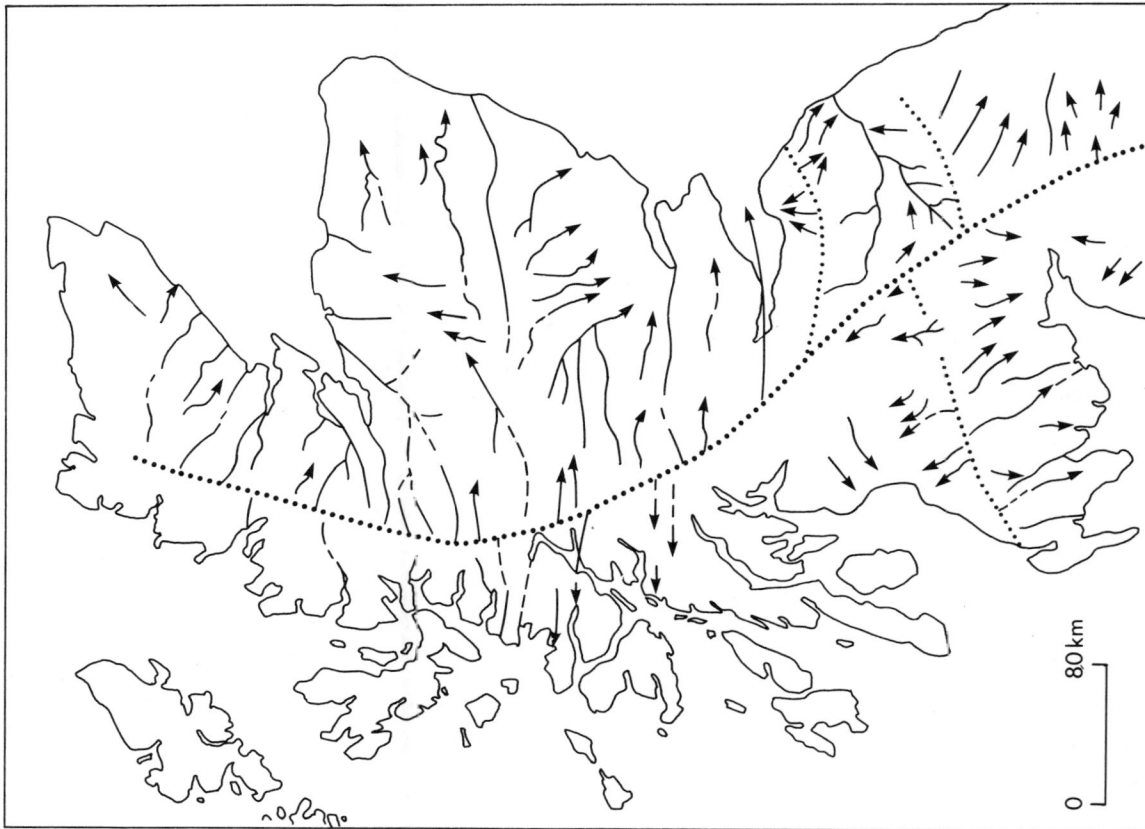

Fig. 7.15 Two models of the original Scottish drainage: left, according to Sissons (1967); right, according to Linton (1951a).

landslipping where the lavas rest on shales and clays beneath (e.g., eastern Skye: *see* Fig. 7.14). Coastal cliffs in the lavas can tower hundreds of metres above the sea (e.g., Waterstein Head). The dykes provide smaller-scale contributions to the relief, sometimes weathering to trenches, sometimes forming low walls.

7.3.1.2 *Tertiary landscape evolution*

Details of the evolution of Highland Scotland in the Tertiary are unknown, save in the volcanic areas of the north-west where portions of the early Tertiary landscape are preserved beneath the lavas and where the known ages of the volcanics permit subsequent denudation to be assessed. In general, the Tertiary was characterized by relative uplift of the highland block, accompanied by deepening of the valley system. Planation surfaces, much fragmented, and said to bear witness to stages of relative stability during the intermittent emergence of the Scottish massif, have been recognized.

About the turn of the century, Geikie (1901) and Mackinder (1902) likened the Highlands to a dissected tableland, tilted from west to east, the original surface now being seen mainly in accordant summit levels. The geologists Peach and Horne (1930) attempted a three-fold division of levels: (1) monadnocks (e.g., Ben Nevis, Cairn Gorm, etc.), (2) high plateau (900 m falling eastwards to 600 m) and (3) intermediate plateau (200–300 m).

In the Grampians, Fleet (1938) postulated three surfaces at 230–300 m (valley benches), 450–650 m (lower surface) and 730–950 m (Grampian main surface) all of which were regarded as subaerial peneplains, slightly deformed by earth movement. The most complete survey so far is that of Godard (1965) who claimed surfaces at 90–180, 400–610, 610–700 and 700–950 m, although in any one region the range of elevation of a surface is said to be much less. All surfaces are thought to be subaerial, and the higher ones may be tectonically deformed. There is no firm evidence on age, save that even the highest (and presumably oldest) level truncates the early Tertiary volcanics, while the evidence of deep-weathering deposits in a few sites, suggesting a warmer climate, might point to a mid or late Tertiary age.

An entirely different view of Scottish planation surfaces is the marine hypothesis advocated by George (1955, 1965, 1966), in which mid or late Tertiary base levels at 180, 300–335, 490–520, 700–730, 810 and 975 m were proposed, with corresponding development of wave-cut platforms. These are said to have been formed during pulsed emergence of the Scottish massif and not to have been subsequently warped or tilted tectonically. No associated marine deposits have been found, and it may be questioned whether the Scottish massif could have behaved as such a rigid block during the 1000 m or so of submergence and subsequent emergence.

Nevertheless, there seems to be fairly general agreement that relative uplift of the Scottish massif has occurred since the early Tertiary volcanic episode, that this uplift may have been episodic and that phases of relief bevelling alternated with phases of strong river incision. Thus the present-day form presents a picture of polycyclic valley forms and remnants of bevelled or graded surfaces. Reconstruction of the details of the story, however, is severely hampered by the intensity of subsequent glacial modification, which altered or destroyed much of the evidence.

The evolution of the river pattern in the Highlands has been a subject of controversy. Linton (1940, 1951a) and Bremner (1942) developed hypotheses involving superimposition from an initial Chalk cover, the early drainage pattern of Linton (1940) being one of east-flowing consequents fed by north-east and south-east-flowing tributaries (*see* Fig. 7.15). Around the Moray Firth, a centripetal system has been reconstructed, related to anomalous downwarping of the Chalk cover. In the west, downfaulting associated with a Hebridean 'rift' has been brought in to explain the occurrence of Chalk at heights near to sea level, far below the adjacent mountain summits where the Chalk blanket is said to have once existed. Ideas of a Hebridean rift have been strongly refuted by George (1966, 1974) who developed an opposing hypothesis of drainage evolution based on progressive extension of rivers across newly emergent sea-floor areas during Tertiary uplift. Sissons (1960, 1967) has combined certain elements of both Linton's and George's hypotheses, reconstructing in some detail the initial drainage pattern (*see* Fig. 7.15) which may have evolved on a Chalk cover but which was also strongly affected by the course of Tertiary uplift and the formation of both marine and subaerial surfaces.

7.3.1.3 *Glaciation*

At the maximum of the last main glaciation (Devensian) (*see* Table 8.1, p.147), ice covered virtually the whole of the Highlands. On the east, this ice sheet merged with the Scandinavian ice, while on the west, it extended well beyond the Hebrides. Its thickness was sufficient to bury most of the mountains, but there are grounds for believing that Ben Nevis was never completely overridden nor the Cairngorms above about 1000 m. The main ice shed lay somewhat to the east of the present watershed in the Northern Highlands, roughly along a line north–south through Ben Nevis, and its surface at one stage may have topped the 1300–1400 m level.

No record of pre-Devensian glaciations remains, although these must have occurred. After the Devensian maximum (about 20 000 BP), the recession of the ice was punctuated by readvances, the number and status of which are a matter of uncertainty. Only one readvance— the Loch Lomond—is firmly established as a separate glaciation, dated by ^{14}C at 10 300–10 800 BP (pollen zone III), following the Allerød interstadial. At this time, small but complex glacier systems [see Sissons (1967, 1974b)] occupied the higher parts of the Highlands, mainly in the western half of Scotland (*see* Fig. 7.16), leaving clear evidence of their former existence in the form of moraines, kame terraces, outwash, etc. At this time, too, ice-dammed lakes were produced, including the famous system in the Glen Roy area where clearly preserved shorelines (the 'Parallel Roads') mark the old water levels.

Fig. 7.16 Late Pleistocene glaciation in Scotland [Sissons (1967)]. The Loch Lomond phase (about 10 300–10 800 years BP) is well supported by evidence from moraines, ice-dammed lakes and radiocarbon-dated biogenic deposits, and is correlated with pollen zone III (post-Allerød). The Perth readvance is less securely based and may represent merely a temporary standstill during the main deglaciation about 13 000 years BP.

The influence of glaciation on the morphology of the Highlands was profound. Here the most extensive assemblage of glacial relief forms in Britain is to be found. Ice not only modified pre-existing features, but created many entirely new ones. Linton (1951c, 1963) showed how preglacial watersheds in the North-Western Highlands, in the Cairngorms and elsewhere have been deeply breached by transfluent ice, and with Moisley [see Linton and Moisley (1960)] showed that Loch Lomond itself lies in a glacial breach. Innumerable classic examples of glacial troughs can be found, some clearly aligned along major fractures, others not. Rock basins of glacial erosion are mostly concentrated in the western Highlands where erosion was most intense; about 200 exceed 40 m in depth—the deepest being Loch Morar (310 m) and Loch

Ness (230 m)—and many descend below sea-level (e.g., Loch Lomond descends to −180 m). The lower ends of many glacial troughs fall below present sea level in the west to become fiords, whose generally radial pattern reflects both ice dispersion and the network of bedrock fractures. Glacial erosion has greatly contributed to the fragmentation of land in western Scotland; Skye, Mull and many other islands were originally part of the mainland. On the sea floor, deep rock basins are known to exist, including that of the Inner Sound of Raasay with a depth of 324 m—the greatest in British coastal waters. Finally, the Highlands include splendid examples of cirques [see Linton (1959) who mapped 420]; most face between north and east reflecting climatic controls, the majority concentrated in the west and in areas that now receive more than

175 cm of precipitation a year. Near the west coast, some cirque floors lie near sea level; eastwards, their minimum elevations rise gradually to 900 m in the Cairngorms.

7.3.1.4 *Relative sea-level change*

At the present-day, northern Scotland is an area showing strong relative emergence [see Valentin (1953)] representing delayed glacioisostatic recovery. Rates of relative uplift at present are 2.5 mm per year at Dundee (although another 1 mm per year must be allowed for contemporaneous eustatic sea-level rise). It is likely that at the centre of the Highlands present uplift is of the order of 5 mm per year. Differential uplift during deglaciation, outpacing eustatic sea-level rise, was not uniform through time, the result being a complex of tilted and deformed raised beaches and shore platforms. Some are related to the late glacial, some to the postglacial period; the highest attains 49 m above present sea level. Those of the late glacial—formed when ice still existed inland—can be shown to be connected with contemporaneous moraines or outwash plains. The evidence is extensive and the details of the story complex [see Sissons (1967, 1974a, 1976)].

7.3.1.5 *Periglaciation*

During cold periods after the Devensian maximum, those areas that were ice-free were subjected to conditions favouring development of periglacial features. Much frost-shattered debris accumulated on scree slopes and as valley fills, while gelifluction helped to smooth the landscape. Fossil ice wedges (restricted to areas outside the Loch Lomond limits) are fairly common, while traces of patterned ground occur at levels over about 600 m. Some of this patterned ground may still be active today.

7.3.2 Central Lowlands of Scotland

Sharply defined in terms of geological structure by the Southern Uplands Boundary Fault on the south and by the Highland Boundary Fault on the north, this unit represents a rift valley crossing Scotland from south-west to north-east (*see* Fig. 7.17). The difference in elevation between the rift floor, most of which lies below 300 m, and the uplands to the north and south is not, however, a direct expression of downfaulting but of differential denudation, for much of the Central Lowlands is formed of less resistant rocks than the adjacent uplands. Indeed, there is evidence that, at some times, possibly in the Cretaceous and Jurassic periods, the Central Lowlands rift was completely filled with sediment. Since then, it has been re-excavated, so that the bounding sharp edges are not to be regarded as fault scarps but as fault-line scarps.

The rocks flooring the Central Lowlands are mostly Carboniferous and Old Red Sandstone (Devonian), apart from the widespread Quaternary deposits. There is an important distinction to be made, in terms of resistance to denudation, between the sedimentary formations and the volcanics and other contemporaneous igneous rocks. The Old Red Sandstone as such forms a major lowland stretching from the estuary of the Clyde in the west, broadening into the Vale of Strathmore as it is traced east. Likewise, the Carboniferous sediments—containing few,

or thin, resistant strata—form areas of low relief, as in the southern part of the Pentland Hills. Elsewhere in the Central Lowlands, the higher ground is closely related to the outcrops of igneous rocks: the Kilpatrick Hills, the Campsie Fells, the Ochil Hills (700 m) and the Sidlaw Hills north of the Firth of Tay. Inclined and faulted beds of lava produce a rugged stepped relief, as do sheets of intrusive rock in the sedimentary areas. There are numerous minor outcrops of intrusives and volcanic necks up to 1.5 km in diameter (e.g., Arthur's Seat and Castle Rock, Edinburgh).

7.3.2.1 *Glaciation*

Evidence of glacial erosion is less obvious than in the adjacent uplands but nevertheless by no means lacking. Deep basins were eroded beneath the present alluvial floors of the Forth, Clyde and other valleys, as well as beneath the great estuaries of these rivers. The solid rock floor beneath the Forth descends to 205 m below sea level at one place, while the smaller Devon valley near Stirling drops to over 100 m below sea level [see Soons (1960)]. Many of these basins are closed downstream, are not structural and must be presumed to represent glacial erosion. Ice-moulded forms have been described by Linton (1962) in soft-rock areas, by Burke (1969) in the Forth valley and by others.

Till, mainly or even wholly deposited by the last glaciation, is widespread and thick. In some areas, as near Glasgow, it is streamlined into drumlins; here and elsewhere, drift tails have formed in the lee of projecting rock bosses. Sometimes, as in the lee of Castle Rock, Edinburgh, the drift tail veneers a streamlined form moulded by ice in the underlying bedrock. Such streamlined forms bear witness to the movement of powerful ice streams, whose direction of flow is also attested by erratics (e.g., the Lennoxtown boulder train running eastwards towards the Firth of Forth). Some major moraines mark glacial limits, especially the limit of the post-Allerød glaciation: at this time, highland glaciers entered the Central Lowlands down the Loch Lomond, Menteith and other valleys, and the term Loch Lomond readvance has been applied to this episode (*see* Section 7.3.1.3). Some areas contain large numbers of meltwater channels, mostly related to the main Devensian glaciation, ranging from intricate systems of small channels to individually large ones (e.g., Glen Farg across the Ochil Hills). Many are associated with esker systems (e.g., Tinto Hills) or kame and kettle topography.

The chronology of glaciation is hardly known, apart from the main Devensian and the later Loch Lomond readvance. In the Devensian, the whole of the Central Lowlands was filled with ice, even up to the highest points of the Ochil Hills. While there is no doubt about the reality and extent of the Loch Lomond readvance (dated about 10 500 BP), there is uncertainty and disagreement about the status of the so-called Perth readvance which may have intervened between the main Devensian and Loch Lomond stages. The Perth readvance was first identified by Simpson (1933). It has been claimed that at this time (possibly around 15 000 BP) ice covered the

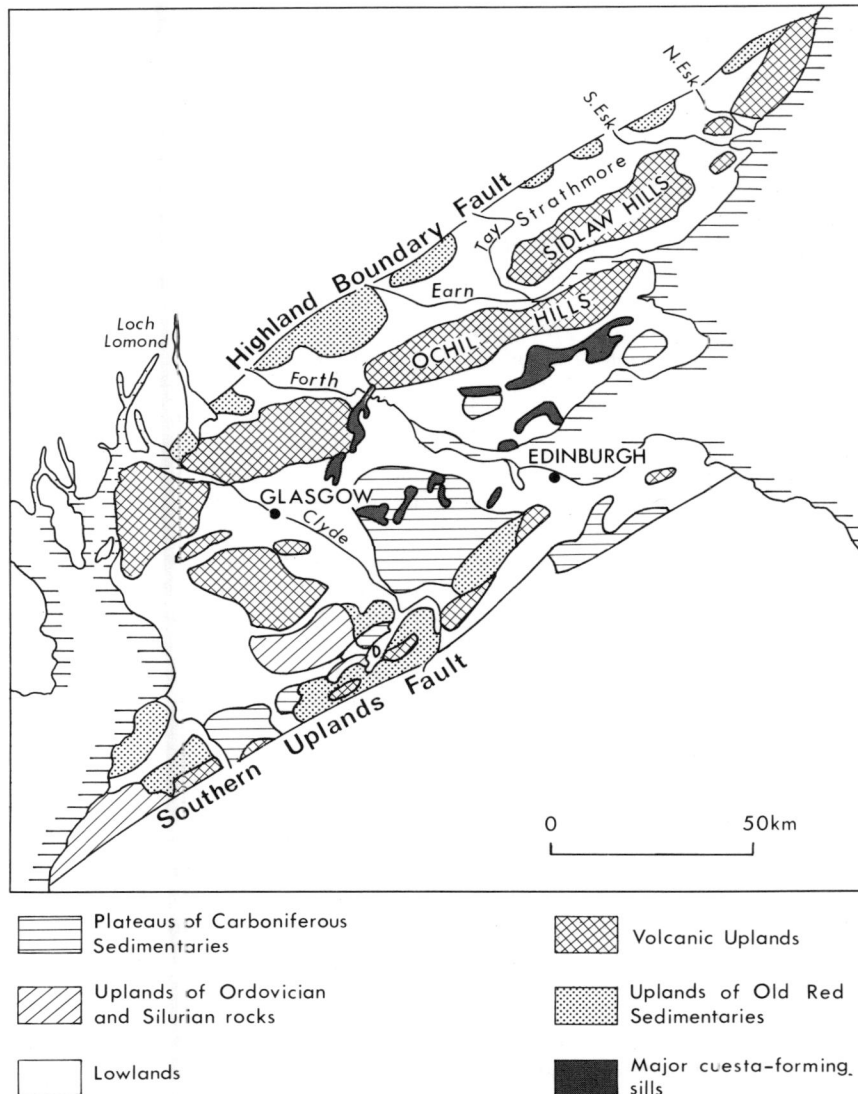

Fig. 7.17 Main geological and relief units of central Scotland [after Sissons (1976)].

western part of the Central Lowlands only, as far east as the line running from Perth to the head of the Firth of Forth and on southwards into the Southern Uplands (*see* Fig. 7.16). It may be, however, that this stage was merely a temporary halt in the general process of deglaciation at the close of the main Devensian.

7.3.2.2 Sea-level changes
A great deal of work on the late Quaternary history of relative changes of land and sea level in the area has been undertaken in the last 25 years by Sissons and others. Space does not suffice to do more than mention some salient points.

Fluctuations of level were caused by glacioeustatic changes in sea level and glacioisostatic changes in land level. These changes were not synchronous and, in the case of the latter, varied from place to place, so that the associated shore platforms—raised or buried beaches— now vary in altitude from place to place. The oldest raised

beaches traced in the area, of late glacial age, show longitudinal slopes of up to $1.27\,\mathrm{m\,km^{-1}}$, and the highest attain 38 m near Stirling. Buried beaches and buried peat deposits have been traced by subsurface surveys, showing that the deceptively simple surfaces of the carse clays (elevated mudflats) or other alluvial areas conceal an exceptionally complex stratigraphy. The history of late glacial and postglacial changes in level and the associated record of shoreline features and deposits are thus proving to be extremely complicated.

7.3.3 Southern Scotland
A belt of upland known as the Southern Uplands stetches across the southern part of Scotland in a south-west–north-east direction, rising to over 800 m in places and built mainly of Lower Palaeozoic strata (*see* Fig. 7.18). The relief is dominated by convexo-concave hillslopes and summits and is dissected by broad valleys. Many of the valleys have been etched along the strike of the sharply

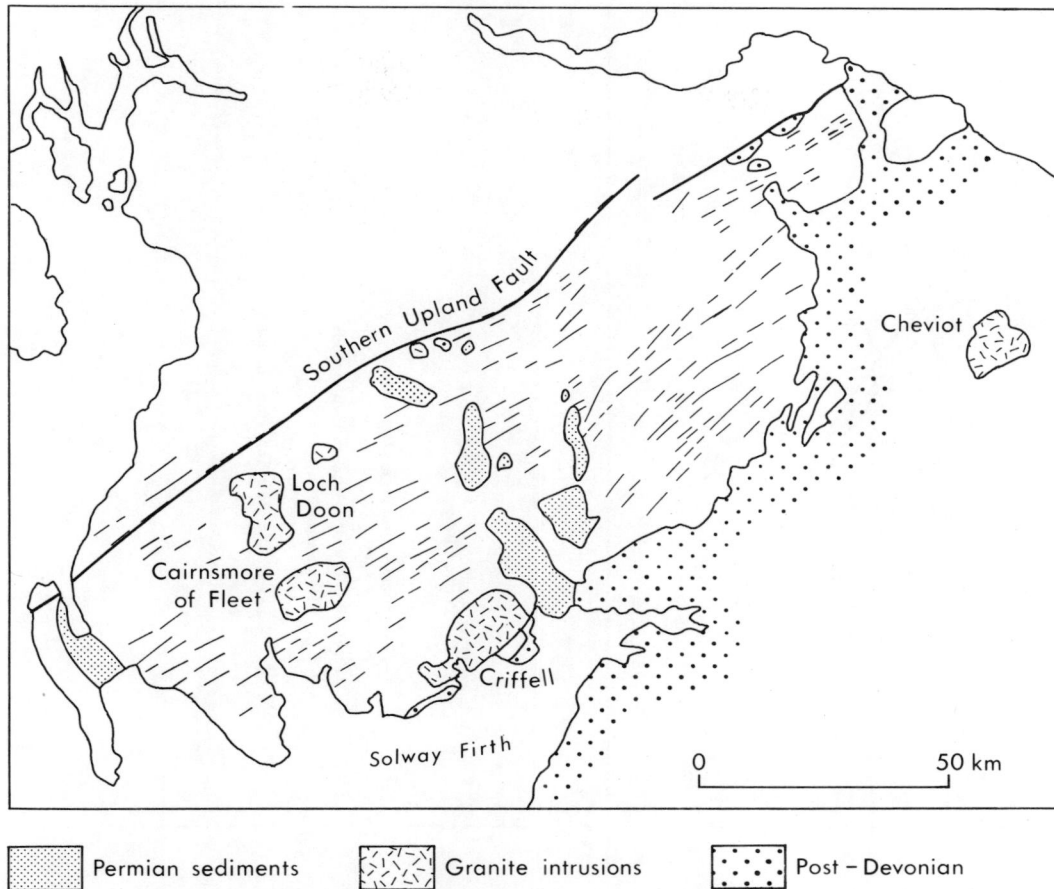

Permian sediments	Granite intrusions	Post – Devonian

Fig. 7.18 Main geological features of the Southern Uplands of Scotland [after Sissons (1967)]. The area between the Southern Uplands Boundary Fault and the outcrop of post-Devonian rocks to the south-east consists of Lower Palaeozoic sedimentary strata, sharply folded along Caledonian lines whose main trends are also shown.

folded strata, the dominant south-west–north-east trend reflecting the impress of the Caledonian earth movements. The Lower Palaeozoic strata consist of thinly bedded mudstones and shales of Ordovician and Silurian age; massive beds are lacking and the rocks weather easily to produce the characteristically rounded relief. In a few areas, such as the Tweed Basin, volcanics and intrusives give rise to minor crags. Two other elements of the structure are, however, more important: the large granite intrusions in the western half of the Southern Uplands, and a series of basins containing Permian sediments. There are three main granite masses: Criffell (569 m) a striking mountain on the edge of the Solway Firth, the Cairnsmore of Fleet (710 m) and the Loch Doon intrusion where the granite, paradoxically, forms a basin which is surrounded by hills in the metamorphic aureole (one of these—Merrick—rises to 843 m forming the highest point in the Southern Uplands). The Permian basins testify to erosion of deep hollows in the Lower Palaeozoic rocks prior to the infilling of these basins (and doubtless others) with Permian sandstones. The original thickness and extent of the Permian deposits are unknown; remnants now floor part of the Nith and Annan valleys and also the depression running from Loch Ryan to Luce Bay in the west.

7.3.3.1 *Preglacial landscape evolution*

Little is known for certain of the stages in the shaping of the landscape during the post-Permian–pre-Quaternary interval. No deposits of this age are to be found. Some workers have claimed that remnants of Tertiary planation surfaces exist at heights of between 180 and 800 m; George (1955) argued for an origin by marine planation during intermittent Tertiary uplift. Godard (1965), basing his conclusions on a survey of the whole of Scotland, preferred a subaerial origin for the surfaces (*see* Section 7.3.1.2).

The drainage pattern of the Southern Uplands has prompted a variety of hypotheses concerning its origin. Linton (1933) pointed out that there are three main elements.

1) Streams flowing mainly to the south-east, at right-angles to the Caledonian grain. These include the Cree, Ken, Nith and Annan flowing to the Solway Firth and also some north-bank tributaries of the Tweed.
2) Streams following the Caledonian trend (e.g., Yarrow, Ettrick, Teviot).
3) The west–east segment of the Tweed above Kelso.

Mackinder (1902) regarded the first category as consequents developed on a south-eastward tilted peneplain.

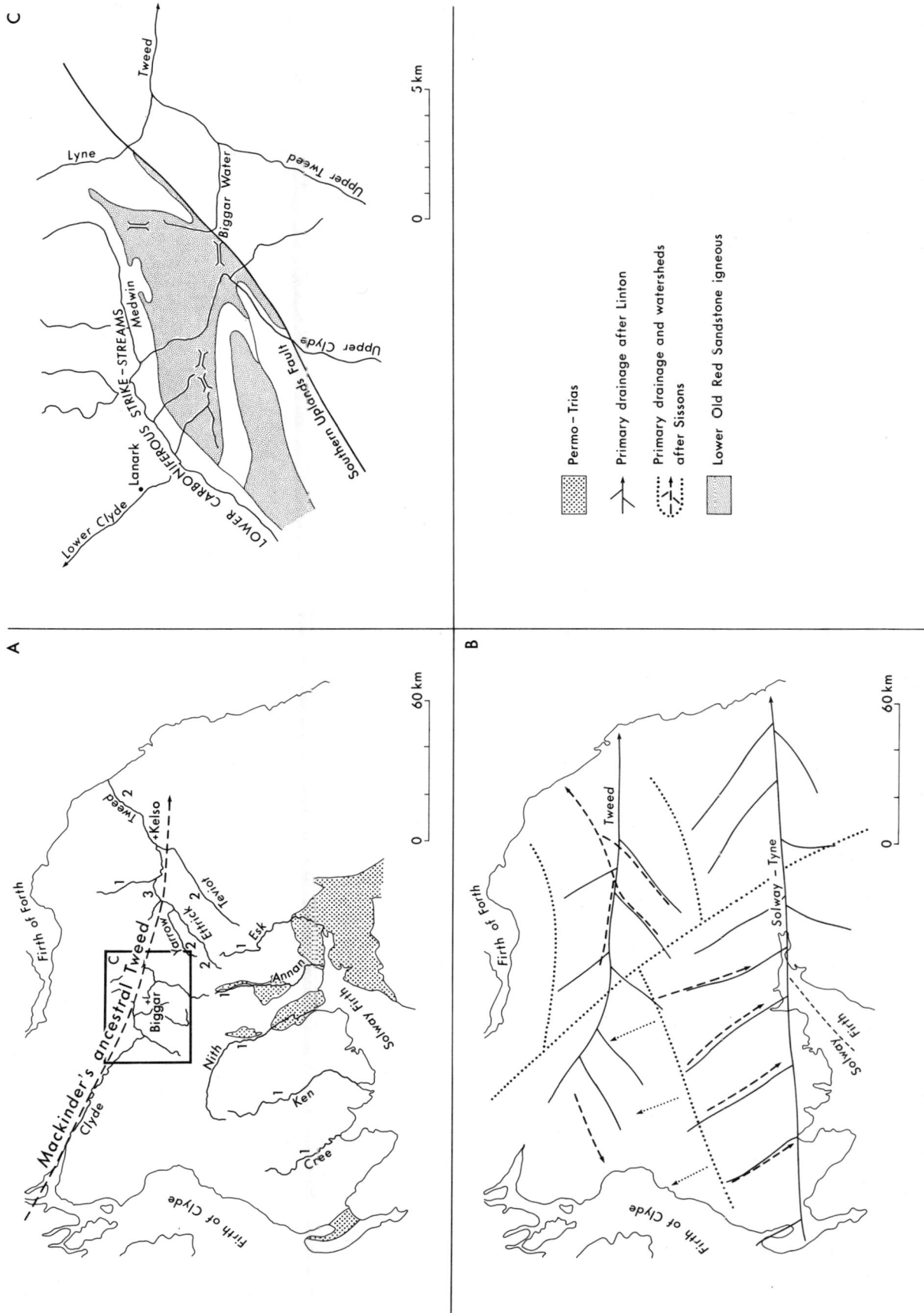

Fig. 7.19 Evolution of the southern Scottish drainage. (A) Course of the ancestral Tweed originating in the Scottish Highlands according to Mackinder (1902). (B) Two different interpretations of the original drainage according to Linton (1951a) and Sissons (1967). (C) The complex drainage pattern around the Biggar Gap (see page 115). The area of map C is located by the box on map A.

Gregory (1915) suggested that a major south-eastward-flowing consequent once comprised a reversed 'Clyde' river following a course through the Biggar Gap to join the Tweed (*see* Fig. 7.19A). Linton (1933) advocated an origin for the streams in the first category by superimposition from a high-level Chalk cover; a hypothesis which he later developed to explain the drainage of much of highland Britain. George (1955) and Sissons (1967) accepted that the first category streams are discordant with structure. The former linked their origin with episodes of marine regression during which wave-cut platforms progressively developed, the rivers being extended across them with each emergence, while the latter developed a complex hypothesis involving elements of both Linton's and George's views. All workers agreed that streams in the second category represent subsequent adjustments to structure. In regard to the third category, Linton (1933) saw it as a major superimposed trunk stream, heading far to the west of the Biggar Gap and receiving streams of the first category as tributaries. He showed that the region of the Biggar Gap bears evidence of a complex sequence of river captures (*see* Fig. 7.19C), so that the headwaters of the Tweed were gradually diverted to the Clyde.

7.3.3.2 Glaciation

The chronology of glaciation is largely unknown. Only the last main glaciation (Devensian) is recorded, in which phase the whole area was overwhelmed by ice. Erratics occur even on the summit of Merrick. Glacial erosion modified the landscape considerably in the western and central areas: for example, Moffat Water occupies a glacial trough and the Grey Mare's Tail is a well-known waterfall at the hanging junction of a tributary 200 m above Moffat Water. In the mountain areas of Merrick and the Rhinn of Kells, troughs and cirques are clearly displayed. The rim of the Loch Doon basin—itself a local centre of ice dispersal—is breached by transfluent glaciers in several places, and Linton (1934) had no doubt that the Biggar Gap was substantially modified by through-flowing ice.

The ice cover of the main Devensian was the result not only of local accumulation but of invasion, especially in the east, by ice from the Scottish Highlands farther north. Erratics, particularly the distinctive granites of Criffell, Loch Doon and Ailsa Craig, readily betray the complex patterns of ice dispersal. Other glacial deposits are widespread, mantling much of the landscape. Meltwater channels, and associated kame and kettle deposits, are numerous in some areas such as the upper Tweed as described by Price (1960).

In pollen zone III of the late glacial, following an interval when glaciers disappeared from Scotland (the Allerød), small glacier systems (*see* Fig. 7.16) were re-established for a few hundred years in the higher western mountains, leaving distinctive and fresh-looking moraines. In other parts of the Southern Uplands, at this time (about 10 500 BP) periglacial activity was widespread, producing gelifluction deposits up to 8 m thick and leaving traces of patterned ground above 600 m, but as low as 400 m on the north side of Tinto Hill [see Miller *et al.* (1954)].

7.4 Lake District and Isle of Man

Both areas represent clearly defined relief units, are built principally of Lower Palaeozoic rocks (*see* Fig. 7.20) and contain areas of glaciated highland. The Lake District is demarcated by the coast on the west and south; on its northern and eastern sides there is an abrupt transition to downfaulted lowlands of Permian, Triassic and Liassic rocks from the Vale of Eden to the Solway Firth. The Isle of Man represents a horst rising above the surrounding Mesozoic sediments that infill deep basins in the Irish Sea.

7.4.1 Lake District

Structurally, this area consists of a complex dome of Ordovician and Silurian rocks, rimmed by outcrops of Carboniferous and Permo-Triassic age. Three broad outcrops cross the area from approximately south-west to north-east, reflecting the trend imposed by the Caledonian earth movements.

1) The Skiddaw Slates (Lower Ordovician) comprise the northern part of the Lake District, giving mountains such as Skiddaw (931 m) and Blencathra (868 m) with relatively smooth profiles, less strikingly modified by glaciation than the central Lake District to the south. Part of this contrast in scenery is certainly attributable to the more easily weathered nature of the slates.
2) The central Lake District consists of the Borrowdale Volcanic Series (lavas, agglomerates and tuffs of Middle Ordovician age) carved into rugged relief by glaciation. The Scafell Pikes (978 m) represent the highest point in England.
3) The southern zone—from the Duddon estuary past Windermere—is built of entirely sedimentary Silurian rocks, with summits in the range 300–580 m.

Around the Lower Palaeozoic core, and resting on it with strong unconformity, lie the grits and conglomerates of the Old Red Sandstone and basal Carboniferous, followed by a thick series of limestones and dolomites which produce some striking escarpments in the south and north-east.

The effects of Hercynian earth movements were not so strong in the Lake District as in many other areas of Britain. The maximum dips imparted to the Carboniferous Limestone, for instance, are about 30°, and part of this may even be post-Hercynian. The dips tend to radiate from the core of the Lake District, suggesting a dome-like upheaval. Following an erosional hiatus, Permo-Triassic strata accumulated in subsiding rifts and basins around the Lake District. In the Vale of Eden, bounded to the east by the major Pennine Fault, thin and discontinuous beds of Magnesian Limestone occur, but most of the Permo-Trias is a continental facies, including brockrams (breccia) and aeolian sandstones. The floor of the Vale of Eden shows the relatively subdued west-facing scarp of the Trias north of Penrith. Traced into the Solway Firth area,

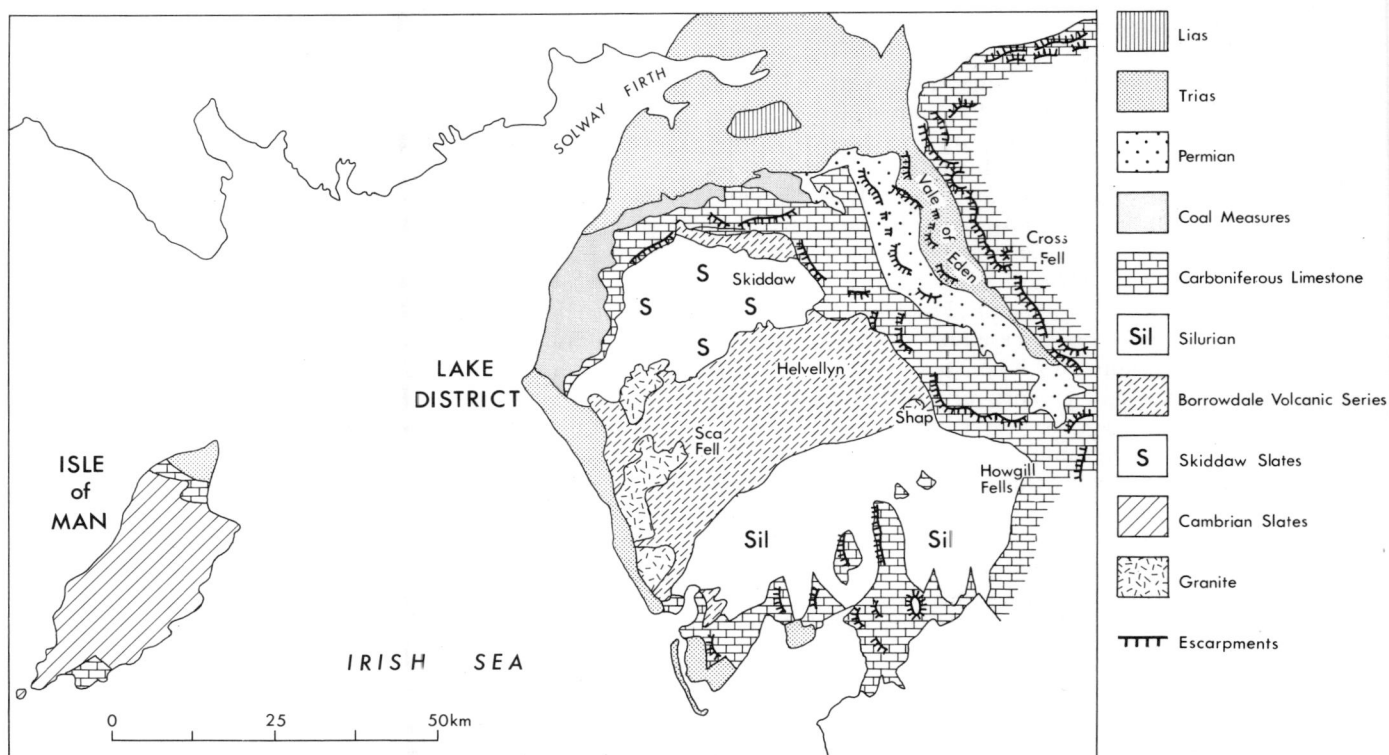

Fig. 7.20 Geology of the Lake District and Isle of Man, together with the principal escarpments in the Carboniferous Limestone, Permian and Trias.

the elevations decline steadily to approach sea level and a cover of glacial drift conceals the bedrock. A small outlier of Lias occurs near Carlisle.

There is no doubt, from the evidence of faulting, that post-Liassic movements occurred, but these cannot be dated in the absence of later sediments (other than Quaternary). A Tertiary age is considered likely, and it has been suggested that it was in this era that the Lake District, possibly buried by higher Mesozoic strata that have since been removed by erosion, was given a final dome-like structure. Goodchild (1888–89) and Marr (1906) both argued that the generally radial pattern of the drainage is the result of superimposition from such a domed sedimentary cover, but an alternative view of Linton (1957) will be considered below.

Because of the absence of Tertiary or late Mesozoic deposits, the evolution of the relief prior to glaciation can only be guessed at. It can be surmised that the immediately preglacial relief was already strongly dissected, possibly reflecting rapid uplift and valley incision in the Tertiary, on the basis of the high degree of control that relief has exercised on the patterns of ice motion. McConnell (1938) attempted to recognize the existence of remnants of Tertiary planation surfaces, but most workers now doubt that any important relics of these survived glaciation. It is noted, however, that in certain areas glacial erosion was probably less intense because of constricted ice flow. In the Howgill Fells, for instance, local ice was hampered in its escape by ice from other neighbouring, more active centres of glaciation; it is in this region that some forms of Tertiary landscape evolution

may have survived.

In the central Lake District, the Pleistocene saw the development of classic landforms of glacial erosion. Great numbers of cirques developed—from 67 to over 2000 according to the definitions adopted—most standing above 400 m. The commonest cirque aspect is between north and east [see Temple (1965)] with a mean bearing of 40°. Glacial troughs abound, from simple forms such as that containing Haweswater to more complex types resulting from transfluent and diffluent ice flow (e.g., in the Langdale–Grasmere area). Most of the big lakes occupy rock basins of glacial erosion in the valley floors, although alluviation has sometimes reduced the lakes in size, divided them (e.g., Buttermere from Crummock Water) or obliterated them entirely (e.g., upper Borrowdale). Windermere—the largest lake—has a depth of 66 m (to 27 m below sea level) and is therefore comparatively shallow; ice flow down this trough was less constricted by high ground. Hanging valleys frequently enter the main troughs discordantly (e.g., the Watendlath valley ends some 100 m above Borrowdale and Derwentwater). Divides between cirques or adjacent troughs have been sharpened by retreat of the head or side walls to give arêtes, such as Striding Edge on Helvellyn.

The radiating pattern of glacial troughs and rock basins is a striking feature of the Lake District. The suggestion that it developed out of a preglacial superimposed radial drainage system has been noted. Another view, advocated by Linton (1957), is that the radiating valley system basically reflects outward ice movement from a centre of active ice dispersal, glaciers adapting existing valleys or

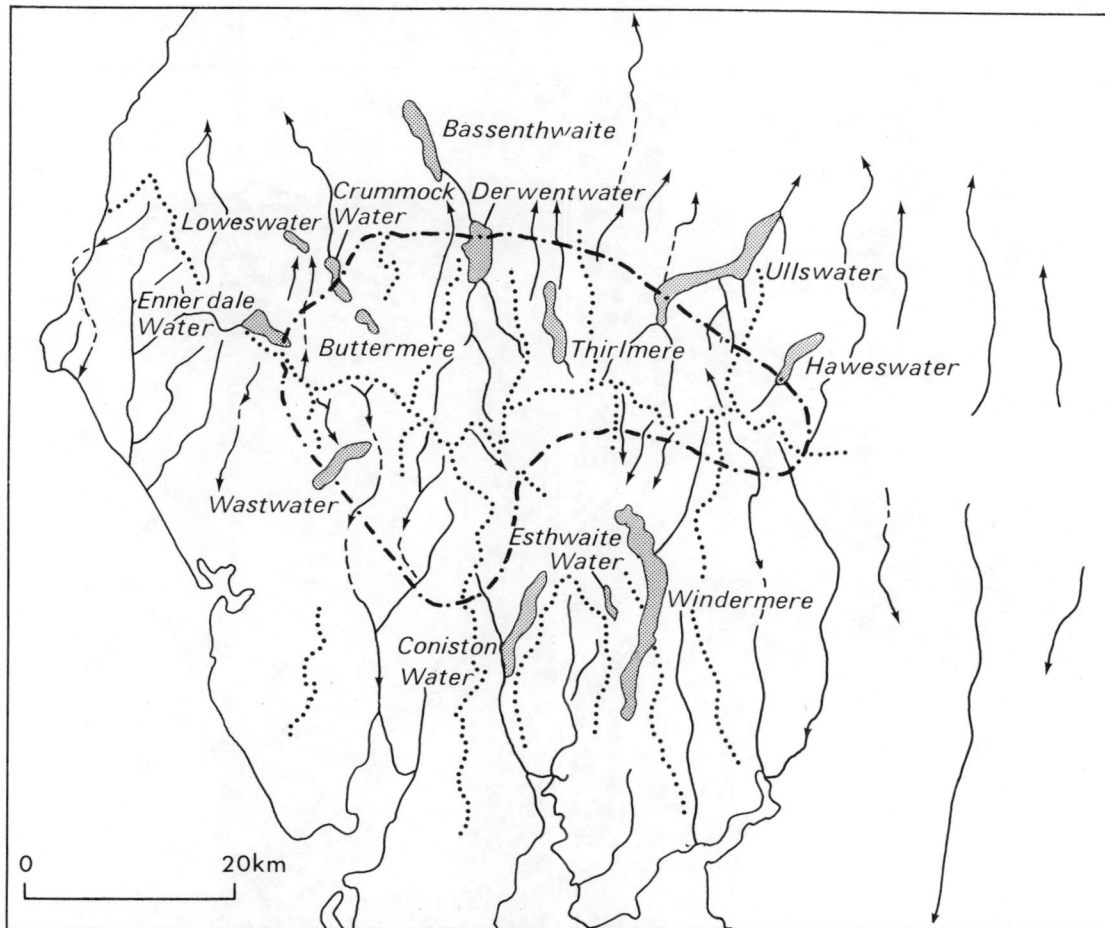

Surviving elements of early north-flowing and south-flowing drainage

--- Reconstructed elements of this system

······· Surviving watersheds related to the early north-flowing and south-flowing drainage

-·-· Area in which mountains today receive more than 2500mm precipitation a year

Fig. 7.21 Evolution of the Lake District drainage [after Linton (1957)].

creating new ones by erosion to facilitate the escape of ice (*see* Fig. 7.21). Linton pointed to the coincidence between high ground and high precipitation of the present-day (suggesting a similar situation in the Pleistocene but with more precipitation falling as snow) and the centre of ice dispersal. The major rock basins of the Lake District all lie in an annular belt (radius 10–24 km) from a centre near the Langdale Fells, a zone where glaciers would be thickest and moving most rapidly, compared with the more static ice at the actual centre of the dome. Linton also attempted to reconstruct the preglacial drainage pattern which he saw not as a radial system, but as one related to an east–west watershed through Dunmail Raise.

Within the area of most intense glacial erosion, deposits left by the ice are mostly in the form of valley or cirque moraines, representing retreat stages. These belong wholly to the Devensian or, in the case of the most recent, to zone III of the late glacial (10 300–10 800 BP). The zone III glaciation was limited to cirques and valley heads and has been studied in some detail [e.g., see Manley (1959)]. Pollen analysis has helped greatly in identification of late glacial sites; while in Windermere sediments, a record of about 400 varves [see Pennington (1973)] supports the climatic deterioration of this time. All other glacial evidence in the Lake District relates to the Devensian. This main glaciation ended before 14 000 BP when the deposits at Blelham Bog west of Windermere show that birch/willow vegetation was already established.

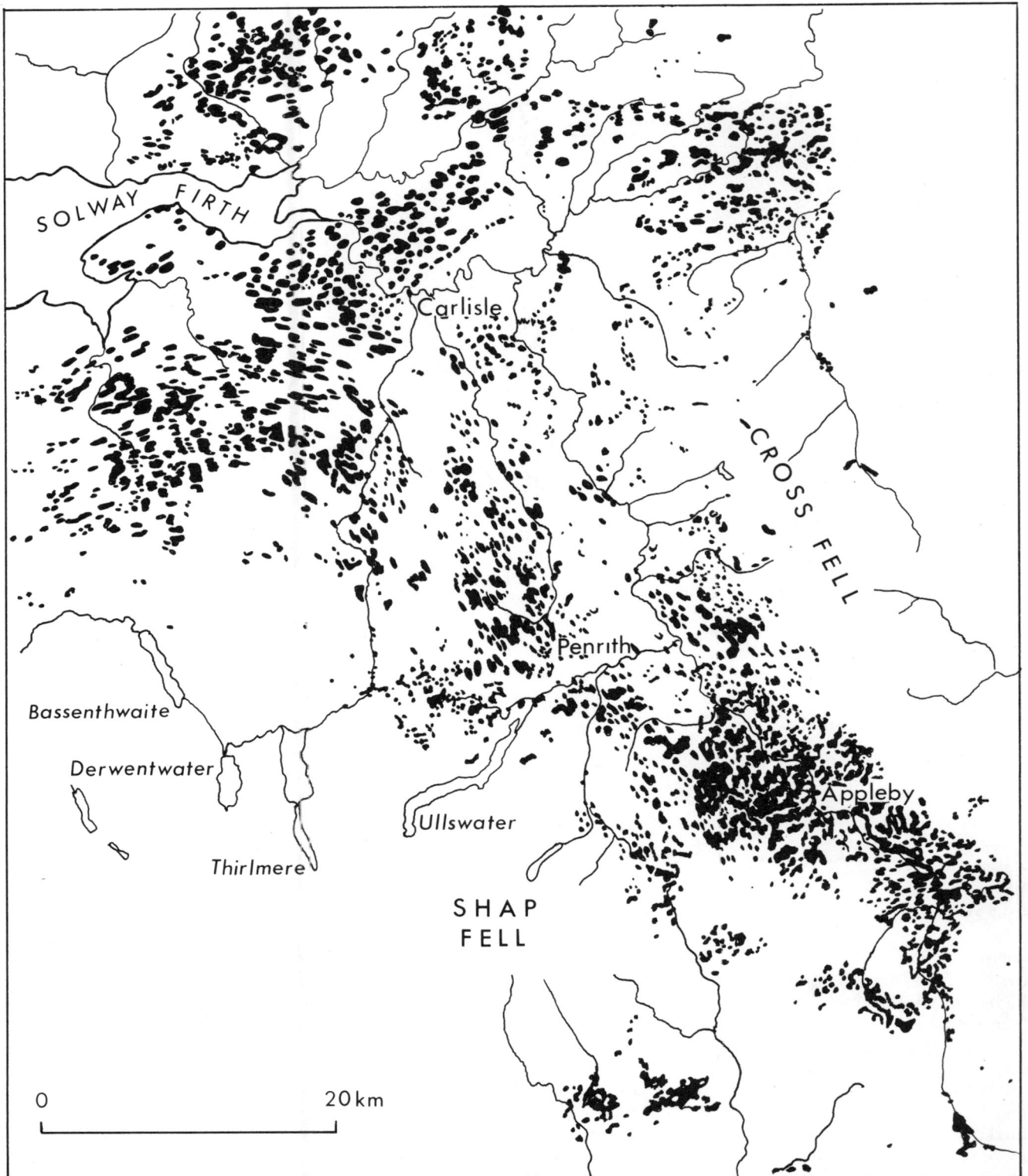

Fig. 7.22 The drumlins of the Vale of Eden and the Solway lowlands [after Hollingworth (1931)].

In the Vale of Eden and the Solway lowlands, glacial deposition features dominate the landscape. Especially noteworthy are the drumlins (*see* Fig. 7.22) some 4000 of which were mapped by Hollingworth (1931). Their pattern shows ice moving strongly down the Vale of Eden from an iceshed near Appleby, and being diverted gradually to an east–west orientation around the north of the Lake District as the ice, attempting to escape to the Irish Sea, encountered other powerful ice streams moving south and south-west from the Southern Uplands of Scotland. Another important method of reconstructing ice movements in the area outside the central Lake District is the study of distinctive erratic trains. The most famous involves the Shap granite, the source of which lay on the eastern edge of the Lake District. From here it was, at different times, distributed to the north into the Vale of Eden, to the south through the Lune Gorge and to the east across the Stainmore Pass of the Pennines. A recent detailed study of the implications of this for the chronology of the glaciation has been made by Letzer (1978).

7.4.2 Isle of Man

The island consists of two distinct relief regions: the northern drift lowland and the uplands. The lowland (lying mostly below 30 m) consists of glacial and glaciofluvial deposits up to 175 m thick, including the Bride moraine, resting on a submarine platform of Carboniferous and Triassic rocks. The uplands are built almost entirely of Cambrian slates, split into two by the transverse valley from Peel to Douglas. North of this valley rises the Snaefell upland (620 m), while to the south the hills reach a maximum of 483 m. The uplands are deeply dissected in places by valleys, originally fluvial, that have been considerably modified by ice. The Peel–Douglas valley may owe its present continuity largely to glacial erosion, later utilized by meltwater. Little is known about the preglacial evolution of the island, although tentative attempts have been made to distinguish Tertiary planation surfaces, best seen in the south-eastern part of the island.

At times in the Pleistocene, the island may have supported its own system of glaciers but in the main glacial periods such local ice merged with and was overwhelmed by Irish Sea Ice. During at least one glaciation, the island was totally submerged by ice, as testified by striae and erratics on Snaefell. The extent of the Devensian (last) glaciation is not yet agreed. Thomas (1976) contended that the island then stood out as a nunatak, surrounded by the Irish Sea Ice, and that periglaciation dominated the exposed portions, leading to the reworking of older tills as head. Boulton *et al.* (1977), on the other hand, have suggested that the ice surface rose to 1500 m, or more, in the last glaciation. A late Devensian readvance was probably responsible for the Bride Moraine which displays large-scale glaciotectonic structures. Around the margins of the island, fine sequences of meltwater channels, related to the decay of the Irish Sea Ice, are to be found. In the Flandrian, the sea level rose, drowning coastal inlets and some valleys; the maximum level attained was +2.2m, forming a cliff along the north of the Bride Moraine.

7.5 Wales

The Welsh uplands contain a diversity of landscape ranging from coastal lowlands to highly dissected mountain country (*see* Fig. 7.23). The relief is often intricate as a result of deep fluvial and glacial incision in late Tertiary and Quaternary times, and provides evidence of a long and complex history of evolution, even though there are virtually no deposits to record the interval from the Lower Jurassic to the Pleistocene. Mountains, as opposed to uplands, are restricted to a few areas. In Snowdonia, there are six peaks exceeding 1000 m, partly reflecting the resistant Ordovician igneous rocks of which the region is largely built. A second major mountain unit is the Brecon Beacons—Black Mountain area of South Wales where the Old Red Sandstone rises in an imposing north-facing escarpment to 886 m. Other high mountain groups comprise Plynlimon (752 m), Radnor Forest (660 m) and Cader Idris (892 m). Much more extensive in Wales are dissected plateaus, formed from strata ranging in age from Ordovician to Carboniferous and in height from about 250 to 600 m. As will be shown in Section 7.5.1, these represent planation surfaces of late Tertiary age.

Structurally, Wales can be divided into two principal areas. North and central Wales are dominated by the impress of the Caledonian orogeny with its north-east–south-west trend, while south of a line curving from St Bride's Bay past Brecon towards mid-Herefordshire the east–west structural trend is Armorican (post-Carboniferous). This southern part of Wales is therefore not strictly Caledonian but, for convenience, Wales will be considered as a single unit in this chapter.

The whole of the Welsh block is founded on a Precambrian basement but this frequently lies at a considerable depth (several kilometres below Mid Wales) and only appears at the surface in Anglesey, Arfon and Pembrokeshire. Elsewhere in North and Mid Wales, it is covered by a thick blanket of marine Lower Palaeozoic rocks, predominantly grits and sandstones in the Cambrian, generally fining upwards to slates, mudstones and shales in the Ordovician and Silurian, the total thickness being more than 10 km. In South Wales, the main cover is provided by Devonian (Old Red Sandstone, continental) and Carboniferous sediments totalling some 8 km. The Lower Palaeozoic series represents sedimentation in a deepening Caledonian geosyncline that lay across Mid Wales, punctuated by episodes of uplift and erosion. The grits and sandstones, especially in the Cambrian, now form rugged uplands and escarpments, while the finer-grained rocks give rise to more subdued relief with convexo-concave slope profiles. The Ordovician is distinguished by contemporaneous igneous activity, often submarine, producing lavas and intrusions which often provide the most resistant members in the rock pile. There were several distinct volcanic centres: for example, Snowdonia, the Arenigs, Cader Idris and Prescelly. The Lower Palaeozoic geosyncline was extinguished in the Caledonian orogeny, which began after Ludlow (Middle Silurian) times and continued into the Devonian. Many of the major structures of North and Mid Wales, such as the Snowdon syncline, the

Fig. 7.23 The main relief units of Wales [after Brown (1960a)].

Harlech Dome, and the Teifi and Towy anticlines, date from this episode.

The Devonian period was one in which the substantial Caledonian mountain ranges were subjected to denudation, mainly arid and semiarid in character, while tectonic unrest continued. To the south and south-east, from Pembrokeshire to Herefordshire, deep basins collected the erosion products (the series of Old Red Sandstones and Marls). In North Wales, Old Red Sandstone was also deposited, burying some of the relief to an unknown extent: only a pocket in Anglesey now survives. The onset of the Carboniferous is marked in both North and South Wales by marine transgression: the Carboniferous Limestone Series (1 km thick in the north, 1.5 km in the south). Shallowing of the sea is represented by the succeeding Millstone Grit (often a shale and of no great thickness) and the more important Coal Measures (largely terrestrial). The Carboniferous Limestone has only fairly re-

stricted outcrops in Wales, but is characterized by escarpments and notable karstic phenomena. Where the Coal Measures consist of shales, they give rise to lowlands, but in the middle Coal Measures of South Wales there is an important development of thick sandstones—the Pennant Grit—up to 1.2 km thick, giving rise to striking tabular uplands and steep escarpments.

At the close of the Carboniferous, there began the Hercynian orogeny. In South Wales, the Devonian and Carboniferous rocks were driven into folds, tightly packed in the west against the Welsh Block to the north, but more open farther east and subjected to fracturing. In North Wales, the folding of this age is relatively gentle, but the brittle foundation gave way along numerous fractures, and shatter-belts have since been etched out to form major valleys, such as the Vales of Neath and Tawe in the South Wales coalfield.

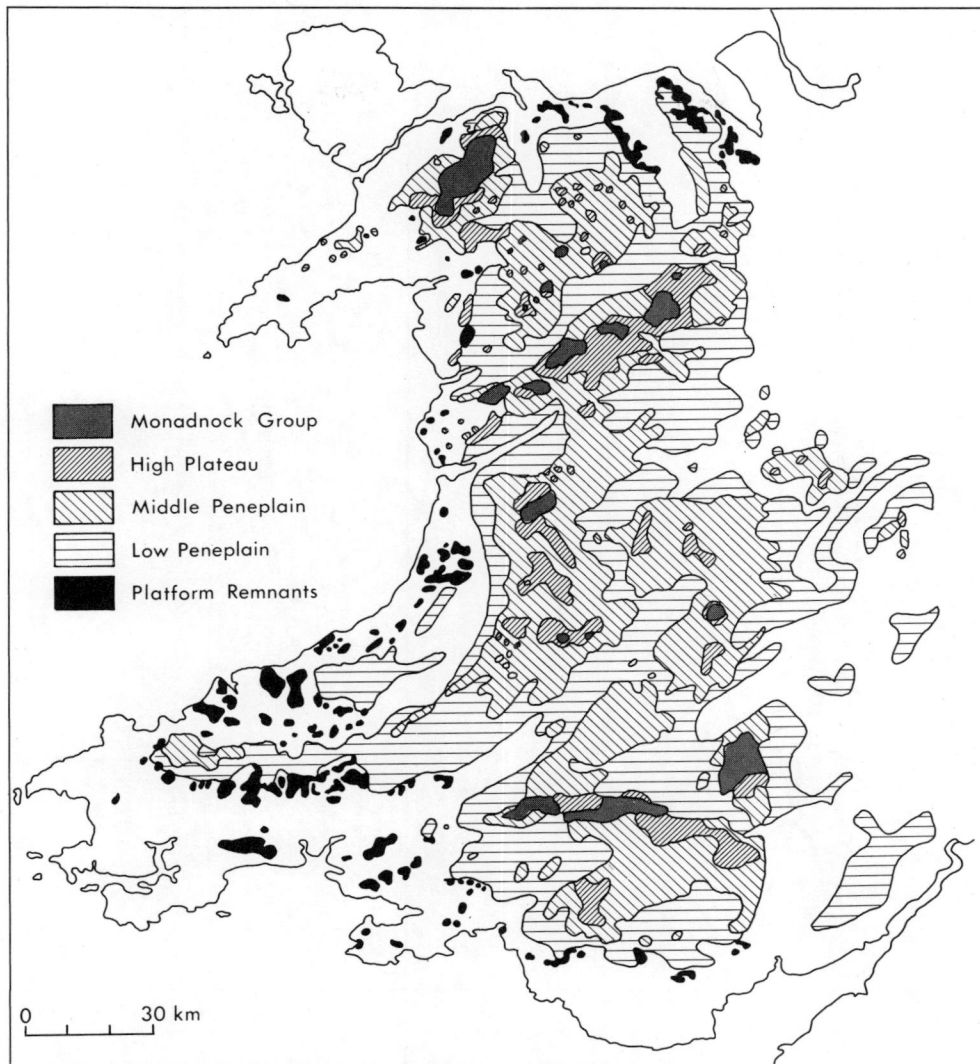

Fig. 7.24 The planation surfaces of Wales [Brown (1960a)].

7.5.1 Planation Surfaces

Enough has been said to indicate that there is great variety of rock type and structure in Wales, and the response of the landform to this variety is equally complex. But perhaps the most striking feature is the way in which, over large areas, the landform appears to lack any direct control by structure. Thus extensive plateaus bevel a range of strata and lithologies, as Fig. 7.23 suggests. The flatness of the interfluves and hill tops was commented on by Ramsay as long ago as 1866. Davis (1909) expressed the view that the Welsh 'high plateau' represented a Tertiary subaerial peneplain above which rose monadnocks, such as Snowdon or Plynlimon. In 1930, Jones (1963) proposed that the plateau was basically the result of arid peneplanation in Triassic times; upwarping in the Miocene was said to be partly responsbile for its elevation to about 600 m in North Wales compared with 100 m or so in South Wales. Yet another view, more in accord with current opinion, advocated by Greenly (1938), is that the plateau must be of mid or late Tertiary age for in

Snowdonia it cuts across some dykes, intruded in the Eocene (mean potassium–argon age 52 Ma). A systematic survey of the relief was undertaken by Brown in the 1950s [see Brown (1960a)] and a series of planation surfaces distinguished (*see* Fig. 7.24): (1) Low Peneplain (210–330 m), (2) Middle Peneplain (370–490 m) and (3) High Plateau (520–600 m). Above this rise peaks ranging up to 1085 m. Brown regarded the surfaces as Neogene subaerial forms, described as Davisian-type peneplains in 1960, but later the possibility of an etchplain origin was considered [see Walsh and Brown (1971)]. The surfaces postdate the Eocene dyke intrusion and appear not to be warped or disclocated [see George (1974)], thus suggesting a still later, post-Miocene, date. A different interpretation of the Welsh planation surfaces has been offered by George (1961, 1974) who regarded them as marine, cut during intermittent emergence of the Welsh block from beneath the sea in the Neogene.

Around the seaward margins of Wales, lower-level planation surfaces have long been recognized [see Brown

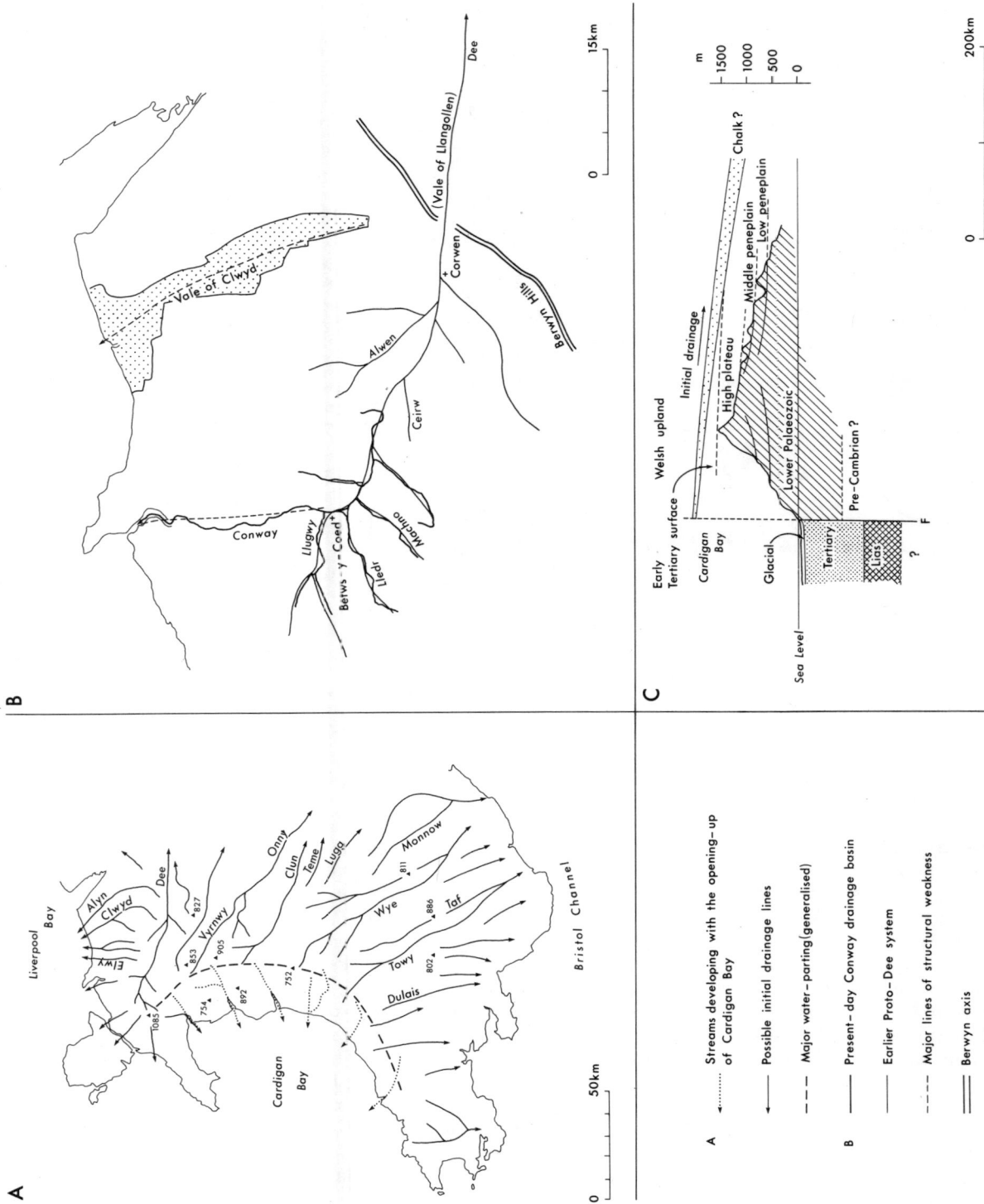

Fig. 7.25 Some aspects of the drainage evolution of Wales. (A) The original drainage of Wales [Brown (1960a)]. (B) The evolution of the Dee system [Linton (1950)]. (C) Cross-section showing downfaulted Mesozoic and Tertiary sediments in Cardigan Bay, the possible position of a high-level former Chalk cover, and the general relationships of planation surfaces according to Brown (1960b).

(1952), Driscoll (1958), Embleton (1964)]. There is now general agreement that these represent marine platforms cut by higher sea levels in the late Tertiary or early Quaternary. It is uncertain whether any are related to high interglacial sea levels; all appear to predate the oldest known glaciation in Wales but there is still a considerable part of the earlier Pleistocene in which some of the platforms may have originated. No deposits that might help in the dating of the platforms have ever been found. The highest level of about 200 m may be tentatively correlated with the Calabrian (Red Crag) marine stand recorded in south-east England. This correlation, however, based simply on altitudinal equivalence, implies widespread tectonic stability since that date, a claim which is now being questioned. Another problem is related to the great width of the platform in some areas (40 km in Anglesey and Arfon). Marine platforms of such an extent could only be created during a rising or falling sea level; even so, remarkably effective wave-cutting is implied, especially if the resistance of the Precambrian and other strata is taken into account. Many workers have thus been forced to consider the possibility that the wide platforms of Anglesey, Pembrokeshire, Gower and Glamorgan were partially prepared for Cenozoic wave action by much older geological events. The platforms of South Wales, for instance, lie close to the levels of the sub-Keuper unconformity, as shown by the collapsed masses of Triassic gash–breccia and dolomitic conglomerate in Gower and the southern part of Pembrokeshire. It is therefore possible that parts at least of the platforms may be early Triassic pediments.

7.5.2 Drainage Evolution

Although many Welsh rivers, or portions of them, now follow lines of weakness in the structure, there are many major valleys that show a striking disregard for geology. A prime example of a discordant valley segment is that of the Dee below Corwen (*see* Fig. 7.25), flowing eastwards obliquely to the Caledonian structure, passing across the Berwyn axis in a deeply entrenched course (the famous Vale of Llangollen). Explanations of regional drainage discordance have involved two types of hypothesis. The first is that of superimposition from a former cover of sedimentary rocks blanketing Wales. Lake (1934), Linton (1951a), Jones (1951) and Brown (1960a) have all put forward arguments for superimposition of the original Welsh drainage from a Chalk cover, even though no Chalk now remains within Wales. George (1961), on the other hand, disputed that the Chalk ever encroached significantly on Wales and has contended that the drainage evolved during a process of intermittent regression of the late Tertiary sea, extending outwards from an initial core in the north-west. Later stages of drainage development involved gradual structural adaptation by river capture, and glacial diversion. Fig. 7.25B shows the scheme proposed by Linton (1951b) for the evolution of the Dee–Conway system, while Fig. 7.25A demonstrates a reconstruction made by Brown (1960) of the initial Welsh drainage system. This drainage system has been said to have developed on a smoothly domed surface of Chalk whose highest point was attained somewhere to the west of the present-day Welsh upland, the site of the present Irish Sea.

In view of the possibility that the ancestral rivers, flowing mainly to the east and south, may have had their headwaters outside Wales, some interest attaches to the geological history of the Irish Sea basin. A deep borehole at Mochras on the coast of Cardigan Bay as reported by Wood and Woodland (1968) encountered the following sequence: (1) glacial drift (0–53 m below sea level), (2) Tertiary clays, lignites, etc. (53–602 m below sea level) and (3) Lias including mudstones, siltstones and limestones (602–960 m below sea level). As Fig. 7.25 suggests, a major fault must lie close to the east of the boring, and substantial Tertiary or post-Tertiary movement is involved. The absence of Chalk or other Mesozoic strata in the sequence does not disprove the hypothesis of drainage superimposition from a Chalk cover since there is a major erosional break between the Lias and the Tertiary. The existence of such a thickness of relatively unresistant sediments suggests that Cardigan Bay is the result of rapid differential denudation (compared with the resistant rocks of the Welsh upland) and that it is this that has destroyed all trace of the headwaters of the initial Welsh drainage. The short steep rivers, such as the lower Rheidol or the lower Ystwyth, that now descend westwards to the Cardigan Bay coast, are actively truncating the headstreams of the original south or south-east-flowing rivers, so that in Fig. 7.25A it will be noticed that the reconstructed drainage begins, not in the Irish Sea area, but from a line parallel to and east of the Cardigan Bay coast. Their original upper courses have been totally destroyed.

7.5.3 Glaciation

In the Pleistocene, the uplands of Wales provided the source regions for local ice which attained hundreds of metres in thickness, at times overwhelming even the highest points of the relief (*see* Fig. 7.26). This Welsh ice at times extended into the coastal areas and into the lowlands of central England, but for the most part was confined on the north and west by a thick ice sheet in the Irish Sea basin, of northern derivation (Scotland, Lake District, etc.). Irish Sea Ice at its maximum rose to heights of at least 600 m in North Wales, pushing inland to cross Pembrokeshire in South Wales and blocking the Bristol Channel, its southernmost limit being near the Scilly Isles.

There is still much uncertainty about the chronology of glaciation in Wales [see Lewis (1970)]. This arises from the paucity of interglacial or interstadial sites and the dearth of material suitable for ^{14}C dating. Only about a dozen sites have yielded usable ^{14}C dates and nearly all of these lie on or near the west coast. The earliest identified glaciation in Wales predates the last (Ipswichian) interglacial, is possibly Wolstonian in age and covered the whole of the principality. Deposits left by this 'Pencoed' glaciation are seen in the southernmost coastal areas, which were not subsequently covered by Devensian ice, and at the Devensian border can be seen to underlie Devensian drift. Moreover, Bowen (1977) showed that the Pencoed glacial deposits are overlain by Ipswichian

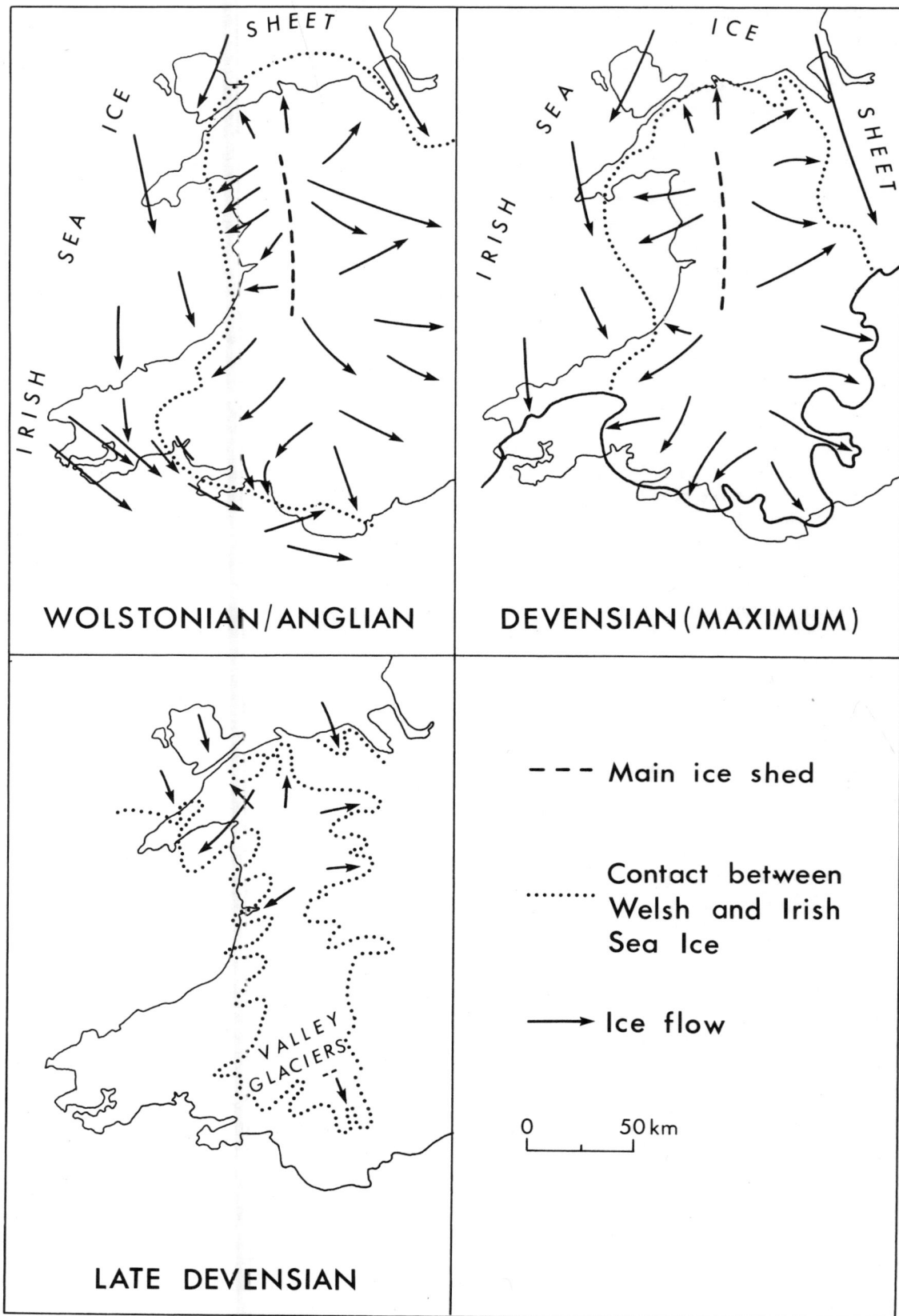

Fig. 7.26 Pleistocene ice limits and directions of ice flow in Wales [see text for explanation].

sediments in western Pembrokeshire. The Pencoed glaciation extended into the upper Severn valley as far as
Shrewsbury, merging with Irish Sea Ice pushing south
through Cheshire, and in the Wye valley reached almost
as far as Hereford. It left deposits on the north shores of
Devon (*see* Chapter 9).

The Ipswichian (last) interglacial is represented by
several coastal sites. At Minchin Hole–a cave in south
Wales—beach deposits up to 11 m testify to a higher sea
level and faunal remains point to a last interglacial age.
Other raised beach deposits in South Wales (ranging from
0.5 to 16 m) may be related to the same episode, but there
is also a possibility that the Patella beach described by
George (1932) may date from the Hoxnian interglacial,
since it seems to some workers to be overlain by head or
till of Pencoed age. Another point of interest concerns the
erratics of the Patella beach. Either these were transported
by ice rafting, or they represent the last vestiges of an even
older (possibly Anglian) till.

The Devensian (last) glaciation covered all but the
southernmost parts of Wales (*see* Fig. 7.26). There has
been much controversy about the actual limits attained.
One view places the Devensian limit of Irish Sea Ice in
Lleyn, leaving Cardigan Bay largely ice-free; the fresh
drift-like material forming cliffs in Cardigan Bay is,
accordingly, regarded as periglacial head by Watson and
Watson (1967). A different view is that the ice reached
Pembrokeshire where John (1967) has obtained ^{14}C dates
on marine molluscs in glacial gravel of about 38 000 BP.
On this evidence, the age of the last main glaciation in
Wales would be late Devensian, agreeing with the archaeological evidence from caves in North Wales. Many
workers conclude that the peak of Devensian glaciation in
Wales was attained about 20 000 BP, following the Upton
Warren interstadial (*see* Chapter 8) of the Midlands. By
15 000 BP, the ice had retreated from South Wales, for
zone I deposits have been found overlying the latest drift
near Swansea.

The last glacial ice to form in Wales dates from the post-
Allerød zone III (about 10 500 BP) when small cirque
glaciers were re-established at levels of about 300 m in
Snowdonia (*see* Fig. 7.27) and Cader Idris.

The geomorphological effects of the ice age on Wales
were profound. In the north especially, glacial troughs,
rock basins and cirques were eroded; ridges were sharpened to arêtes (*see* Fig. 7.28) and hanging valleys created.
The effects of glacial erosion become less intense towards
the south, but rock basins and a broad U-shape still
characterize some of the South Wales valleys. Depositional features include moraines, drumlins and many complex
glaciofluvial features, as well as spreads of till in the
lowlands. Coastal valleys were cut down to lower sea
levels (e.g., to at least −45 m in the Tawe valley near
Swansea) and were drowned and infilled in the Flandrian
when the sea rose to a maximum of +4.5 m in the south.

Periglacial phenomena in Wales are described from
many areas and relate to various cold phases when ice was
not locally present. Head deposits are of great thickness in
some coastal sections; fossil screes have been identified
[see Ball (1966)] as have fossil pingos [see Watson (1972)]

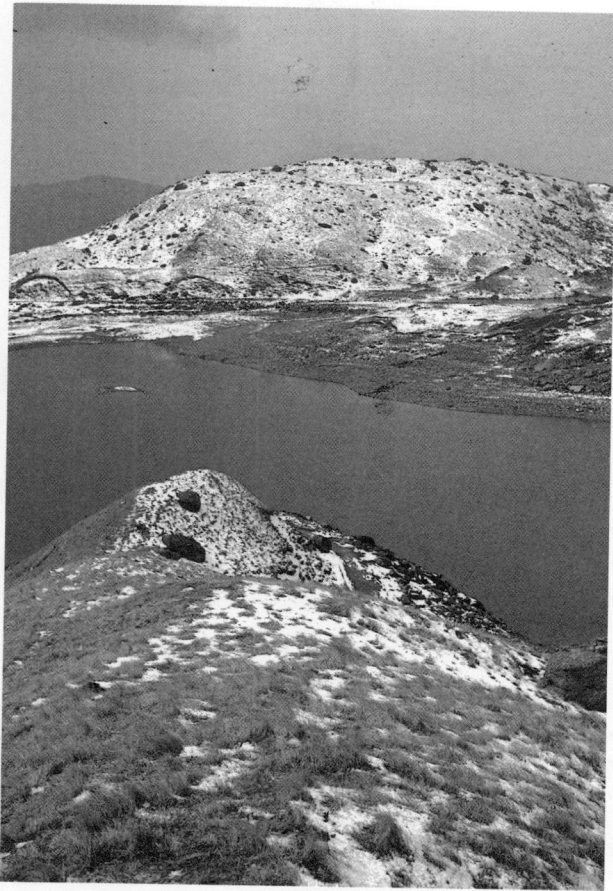

Fig. 7.27 The late glacial (zone III) moraines around Llyn Llydaw,
Snowdon, Wales, dated about 10 500 years BP. (Embleton)

and nivation cirques [see Watson (1966)]. Occasionally,
traces of patterned ground are reported, particularly ice-
wedge features, as described, for example, by Bowen
(1971), of Devensian age beyond the Devensian limit.

7.6 Northern and Central Ireland

Although Wales has been dealt with as a single major
region in this chapter, in spite of the fact that it includes
both Caledonian and Hercynian structural elements,
Ireland possesses such a distinctive Hercynian province in
the south-west that this part of the island is considered in
Chapter 9. In Wales, the Hercynian structures of the
south do not on the whole give rise to pronounced
differences of relief from other parts of the principality. In
Ireland, the Hercynian province with its striking Appalachian-type relief stands in sharp contrast with the geomorphology elsewhere. The dividing line between Caledonian
and Hercynian terrains in Ireland is not, however, sharply
defined, for the influence of the Armorican orogeny, as in
Wales, spread farther north. Gill (1962) recognized three
major structural zones in Ireland on the basis of the
diminishing impact of the Armorican movements from
south to north. In the southern zone, the structures
comprise a series of sharp folds trending approximately
east–west, including some strongly asymmetrical folds

Fig. 7.28 The aréte of Lliwedd (893 m), a ridge trending east from Snowdon and separating the two cirques of Glaslyn/Llydaw (on the left) and Cwm Llan (on the right). (Embleton).

and even overfolds, reflecting pressure from the south. The northern limit of this zone is marked by the outcrop of the major Armorican thrust, traceable from Dingle Bay in the west to Dungarvan in the east. The second zone, north of this line, is one in which the Armorican earth movements were important but less intense, giving more open folding often along older Caledonian structural lines rather than the pure Armorican trend. This zone extends roughly as far north as a line from Galway Bay to Drogheda (*see* Fig. 7.29). North of this, the Hercynian influence is quite subdued, being limited to modest disturbance of the Carboniferous strata and reactivation of some Caledonian faults. This chapter will be concerned only with the second and third of these structural zones.

7.6.1 Geology and Structure

Strong links between the patterns of Caledonian structures in Ireland and those of the rest of the British Isles are evident. The general north-east–south-west trend continues to dominate northern and central Ireland, and the Irish equivalents of several British morphostructural units can be discerned. Prolonged to the south-west, the Highland Boundary Fault and the Southern Uplands Boundary Fault enter Ireland (*see* Fig. 7.29), the former being possibly traceable from Cushendall to Maghera and Clew Bay and the latter from Belfast Lough to Armagh and Carrick-on-Shannon. There is, however, some dispute about the exact alignments and equivalents. Neither has the topographical significance that these faults possess in Scotland, and there is no Irish counterpart to the midland valley of Scotland. On the other hand, the Lower Palaeozoic belt of low hills from Down to Cavan clearly corresponds to the Southern Uplands of Scotland. Also, on the north-west, the Precambrian schists and gneisses from Mayo to Donegal match those of the Scottish

Highlands. The Lower Palaeozoic massifs of Wales and the Wexford–Wicklow uplands, on opposite sides of the Irish Sea, also have some common characteristics and lie on the same broad Caledonian axis. As well as the distinctive tectonic patterns, the Caledonian orogeny was responsible for the emplacement of some major granite bosses, of which the Leinster granite (the largest granite exposure in the British Isles) and the Newry, Donegal, Foxford and Galway intrusions are the principal examples.

Post-Caledonian rocks cover about two-thirds of Ireland, consisting of Old Red Sandstone and Carboniferous strata, a succession that was originally complete up to the Coal Measures although because of erosion the latter now only remain in a few outliers. The Carboniferous Limestone has the most extensive outcrop, underlying the greater part of central Ireland. In contrast, post-Hercynian rocks are poorly represented and the events of the long interval from Permian to Tertiary, when many features of the present-day geomorphology originated, are largely unrecorded in the stratigraphy. Opinions have varied on whether there was ever a complete succession of Mesozoic sediments that was later removed by erosion, or whether such sediments were never in fact deposited. Until recently, the only Cretaceous rocks known in Ireland were those preserved by the Tertiary basalts in the north-east but in 1960 came the unexpected discovery of a small Chalk outlier at Ballydeenlea, a few kilometres north of Killarney as described by Walsh (1966). Preserved in a pipe formed by the collapse of shales into a limestone cavity beneath, the deposit stands at only 80–120 m above present sea level. This discovery has cast doubt on hypotheses of any former thick Mesozoic cover and has also had major implications for many hypotheses of landscape evolution.

The Tertiary era in Ireland is in some ways better

Fig. 7.29 The main structural and relief units of Ireland [regions after Davies and Stephens (1978)].

represented than the Mesozoic. Even though sedimentary deposits are few and scattered (the Lough Neagh clays of late Oligocene age, the Oligocene clays and lignites at Ballymacadam in Tipperary, and a few other minor relics), the era was distinguished by important igneous activity. In Antrim, Palaeocene basalts (about 55–65 Ma) were poured out over 4000 km², up to 800 m thick, with intercalated subtropical weathering horizons. In Down, the Mourne granite and adjacent complexes were intruded (about 58 Ma); across northern Ireland, even as far west as Donegal and Mayo, there sweeps a great early Tertiary dyke swarm. These igneous phenomena were contemporaneous with those of north-west Scotland, being related to the rifting of the Atlantic between the diverging North American and Eurasian plates. There is also evidence of Tertiary tectonics, perhaps more significant than has hitherto been generally claimed. Certainly the Antrim basalts show later deformation and faulting—the Lough Neagh basin is a major downwarp, holding

350 m of Oligocene sediments—and the subsidence of the North Channel and Irish Sea is probably a Tertiary event. Faults elsewhere in Ireland may well have been reactivated and there is supporting geomorphological evidence, together with a growing weight of opinion that Ireland, along with the rest of the British Isles, was not such a stable area in the Tertiary as was formerly assumed.

7.6.2 Relief and Drainage

Outside the Armorican ridge-and-valley province (*see* Chapter 9), the surface of Ireland rarely rises to more than 800 m, and most of the uplands reach no more than 600–750 m. The uplands are broken into many separate blocks and are peripheral in distribution, forming a fragmented rim around the extensive central lowlands which, developed mainly on Carboniferous Limestone, lie mostly below 120 m. The low-lying nature of the central plains may be partly explained in terms of prolonged subaerial denudation acting on the limestone which, as Williams

(1963) has shown, is undergoing solutional lowering at an average rate of 1 m in 20 000 years. The anomalous absence of highlands, however, is not simply the result of lithology, for Ireland certainly possesses rocks that are normally regarded as relatively resistant and that often give rise to high ground elsewhere in the British Isles, such as the Precambrian metamorphic series, the Lower Palaeozoic slates and sandstones, and granite intrusions. It is true that the Old Red Sandstone of southern Ireland (*see* Section 9.4) forms a substantial number of striking ridges and peaks, and that Precambrian quartzites form eminences [e.g., Errigal (752 m) in Donegal and the Great and Little Sugar Loafs of Wicklow] but it cannot be denied that Ireland overall lacks high relief in comparison with Britain. Some have suggested a climatic explanation [see Linton (1964)], but this is totally unconvincing. More probably the answer lies in the tectonic and denudational history of the Tertiary era.

The drainage patterns of Ireland are complex, not only in terms of their relationships with geology but also because of the extensive areas of lake-studded and peat-blanketed lowland. In the central lowland, many aspects of the drainage are postglacial, the rivers flowing over Quaternary deposits in courses that can be shown in many places to be quite different from the preglacial drainage lines. Sometimes former valleys have been obliterated by drift infills or diverted by moraines or other glacial accumulations. Raised peat bogs (about 6 m thick) are extensive and affect the drainage, as well as supporting numerous small lakes in peaty hollows. Ireland's principal river—the Shannon—with a catchment that drains about one-third of the central lowland is the result of the interlinking of a series of glacially eroded basins containing such lakes as the Loughs Derg, Ree and Allen. The Killaloe gorge below Lough Derg [*see* Farrington (1968)] is a probable example of river diversion by glacial breaching of a col. Glacial interference, by ice blocking, drift damming or glacial erosion, is the cause of many other local drainage anomalies in Ireland. Dury (1959), for example, has presented evidence for extensive watershed breaching by ice in Donegal. But there still remain many rivers which strikingly ignore geological structure in the patterns of their courses and for which glaciation cannot have been responsible. One of the earliest workers to identify regional discordance of drainage was Jukes (1862) who claimed that the generally southward-flowing rivers of southern Ireland, crossing the Armorican folds and other structures, are consequent streams superimposed from a southward-tilted surface. The hypothesis of superimposition was taken up and developed by many workers subsequently. By the 1950s, many workers, as in other parts of the British Isles, thought in terms of a widespread Mesozoic (or, more specifically, Cretaceous chalk) cover on which the drainage was initiated and later superimposed on to the Palaeozoic undermass. Such hypotheses demanded the cover rocks to have formed a high-level blanket, obliterating the highest relief. More recently, especially with the discovery of the low-level Ballydeenlea chalk outlier, such ideas have been questioned. Apart from the Ballydeenlea

evidence, it is now thought that Tertiary tectonics have been overlooked for too long as a factor in the evolution of drainage. In addition, the superimposition hypothesis encounters a difficulty over much of Ireland in that, during an assumed post-Cretaceous phase of denudation and stripping of a sedimentary cover, there must have come a time when substantial tracts of Carboniferous Limestone became exposed. It is hard to see how at such a time inherited drainage patterns would have survived karstic conditions without adaptation. Although the superimposition hypothesis is no longer thought to be the only possible solution to the problems raised by the anomalous courses of many Irish rivers, satisfactory alternative models have yet to be developed. The main problem is that, although the extent of Quaternary glacial interference is becoming fairly clear, the events of the Tertiary era are far less so, and it is precisely at this time that the main lines of the Irish drainage are most likely to have developed.

7.6.3 Tertiary Landscape Evolution

Recently Mitchell (1980) has drawn together the evidence of Tertiary Ireland and many of the arguments that have concerned it. At the beginning of the Tertiary, it seems probable that parts of Ireland surfaced from beneath the Cretaceous sea, but how extensive the Cretaceous transgression was is far from agreed. The actual evidence is that Chalk is present in Antrim, at Ballydeenlea in Kerry, on the floor of the Irish Sea [see Dobson *et al.* (1973)] and as fragments of chalk and flint in some drift deposits, suggesting provenance from land-based sources, including Eogene clay-with-flints, that have no longer survived. Linton (1951a) and other workers before him suggested a Chalk cover at a high level, with later downfaulting of the Antrim chalk and subsidence of the Irish Sea as a graben. The Ballydeenlea Chalk, however, is not downfaulted; Walsh (1966) suggested a Chalk sea floor here at 120–220 m, allowing for some solution subsidence. Just as George (1961) has suggested in the case of Wales, it now seems more likely that the Cretaceous transgression in Ireland was less extensive than formerly claimed, the sea lapping around the uplands rather than submerging them.

Undoubted relics of the earliest Tertiary land surface are few. The top of the Chalk beneath the Antrim basalts shows signs of weathering and denudation under dry subtropical conditions [see Reffay (1972)] in the form of a residual deposit of clay-with-flints. At Ballymacadam in Tipperary, the early Tertiary clays lie at only 75 m and suggest that the early Tertiary surface of Ireland may not have been one at any great height above the present. On the other hand, there is some other evidence testifying to a great amount of Tertiary denudation in Ireland. (1) The Antrim basalts have lost up to 450 m in places according to Charlesworth (1963). (2) The early Tertiary dykes in Donegal and elsewhere are cut across by deep valleys, in the manner of Loch Lomond in Scotland (*see* Section 7.3.1.1). (3) The Mourne granite, intruded in the early Tertiary, has been unroofed since its emplacement, indicating a removal of some 800 m in thickness of Silurian around it. One possible answer to the apparent contradiction is that the peripheral upland areas of Ireland have suffered Tertiary uplift relative to the central lowlands, an idea that is consistent with the way in which the pre-Carboniferous strata in the uplands rise to as much as 900 m above the central lowlands where the same rocks now lie deep below the Carboniferous Limestone floor [see Davies and Stephens (1978)].

Planation surfaces of presumed mid or late Tertiary age (since they predate the glaciation and postdate the early Tertiary basalts and intrusions) have been described in several parts of Ireland, although unlike Wales there has been no regional synthesis. In part of the Wicklow Mountains, Davies (1966) identified subaerial surfaces from 170 to 325 m, and at higher levels in the northern Wicklows, an upland surface ranges from 375 to 550 m. In Antrim, the general postbasalt surface at 300–390 m may be due to later Tertiary planation, although care must be taken not to misinterpret the stepped surfaces of successive basalt flows. Reffay (1972) has claimed that the main planation surface here can be separated into two levels, the higher one already mentioned at 300–390 m and a lower one at 180–270 m. In Donegal, Dury (1964) postulated planation surfaces at a variety of levels up to 480 m. The clearest and most sharply defined planation surfaces in Ireland lie in the southern uplands of County Waterford and County Cork where they were described by the Geological Survey as early as 1861. It is convenient to mention them here rather than in Chapter 9. Miller (1955) claimed that a series of bevelled surfaces between about 180 and 245 m represented the remnants of a former 'South Ireland Peneplain', cut by marine action with a shore line at 245 m marking the limit of a transgressing, possibly Plio-Pleistocene, sea. Following retreat of the sea a new, lower surface—the coastal peneplain—was cut between 60 and 120 m. No marine deposits associated with these surfaces have been found, and the evidence for marine cutting rests on the perfectly bevelled form of many of the surface remnants. Other workers since have offered revised interpretations: for example, Orme (1964) recognized no less than nine marine benches between 64 and 215 m in the Drum Hills. Their ages remain in doubt.

Elsewhere in Ireland, there are various exhumed planation surfaces resulting from the partial or complete stripping of older unconformities, such as the sub-Carboniferous. Much of the work of exhumation may well have been accomplished in the Tertiary.

7.6.4 The Pleistocene

On the basis of such evidence as till fabric, erratic content, degree of weathering and surface morphology, together with the evidence from a few interglacial sites, it is possible to distinguish two major glaciations in Ireland. The first of these—the Munsterian—affected the whole country and probably corresponds to the Wolstonian glaciation in Britain (possibly the equivalent of the Saale). The ice cover consisted not only of local (Irish) ice, emanating from such centres of dispersal as Donegal, the Mourne Mountains, the Wicklow Mountains and the Kerry–Cork region, but also of Scottish and Irish Sea Ice which invaded north-east Ulster from the north-east,

penetrating as far as Lough Foyle and Lough Neagh, and which spread along the Irish Sea coastlands from Down to Wexford and then along the southern coastlands as far west as Cork Harbour (*see* Fig. 7.30). At the Munsterian maximum, the main ice divide occupied a broad zone running approximately from Connemara across the northern central lowlands towards the Lough Neagh area.

Following an interglacial corresponding to the Ipswichian (or Eem), there was a glaciation—the Midlandian—which was somewhat less extensive than the Munsterian and in which the Scottish/Irish Sea Ice only penetrated a short distance inland in the north-east, and barely more than a few kilometres in coastal Leinster. The main Midlandian ice sheet spread as far south as a line roughly from the Shannon estuary through Tipperary to the northern end of the Wicklow Mountains, marked by the South of Ireland End Moraine. Outside the latter, there is a belt of country that was never covered by the Midlandian ice and where Munsterian deposits may be found at the surface. But still farther south in Cork and Kerry, there was a separate glacial system in the Midlandian, covering the whole south-western corner of Ireland as far north as Dingle Bay and as far east as Cork Harbour. The deposits left by the Midlandian ice are much less weathered than the Munsterian, showing constructional features of glacial deposition such as moraines, drumlins and eskers. In contrast, wherever the Munsterian drift is exposed, it shows few such features, is more deeply weathered and also cryoturbated. The Midlandian ice appears not to have attained such high levels as the Munsterian, leaving nunataks and unglaciated enclaves, such as Bloody Foreland and Inishowen in Donegal [see Stephens and Synge (1965)] and in north-west Mayo [see Synge (1968)], often characterized by periglacial block-fields, screes, tors and corestones. Several stages of recession, some followed by readvances, are known for the Midlandian, marked by morainic limits north of the South of Ireland End Moraine: for example, the Galtrim moraine and the so-called Drumlin moraine (at Kells, Armoy, etc.).

Pleistocene interglacial sites in Ireland appear to belong mainly to the pre-Munsterian or Gortian period which, according to Watts (1970) and Mitchell (1970), correlates with the Hoxnian in Britain. Several such sites have been examined in detail [see Jessen *et al.* (1959), Watts (1967), Mitchell (1970)], including some outside the South of Ireland End Moraine where Munsterian deposits overlie peat or plant beds. Only one Ipswichian interglacial site has been identified with confidence—at Shortalstown on the Wexford coast—1.5 km inside the Midlandian limit. Here, pollen analysis suggests a correlation with part of the Bobbitshole sequence of the British Ipswichian.

There is clear evidence in Ireland of a separate glacial phase after the Allerød interstadial, corresponding to the Loch Lomond glacial phase or zone III of the pollen sequence and ^{14}C-dated at about 10 250–11 000 years BP. This latest glaciation was restricted mainly to cirques at relatively high levels: for example, on Lugnaquillia, the highest peak in Leinster [see Farrington (1966)], in the Galty Mountains [see Synge (1970)] and in Western

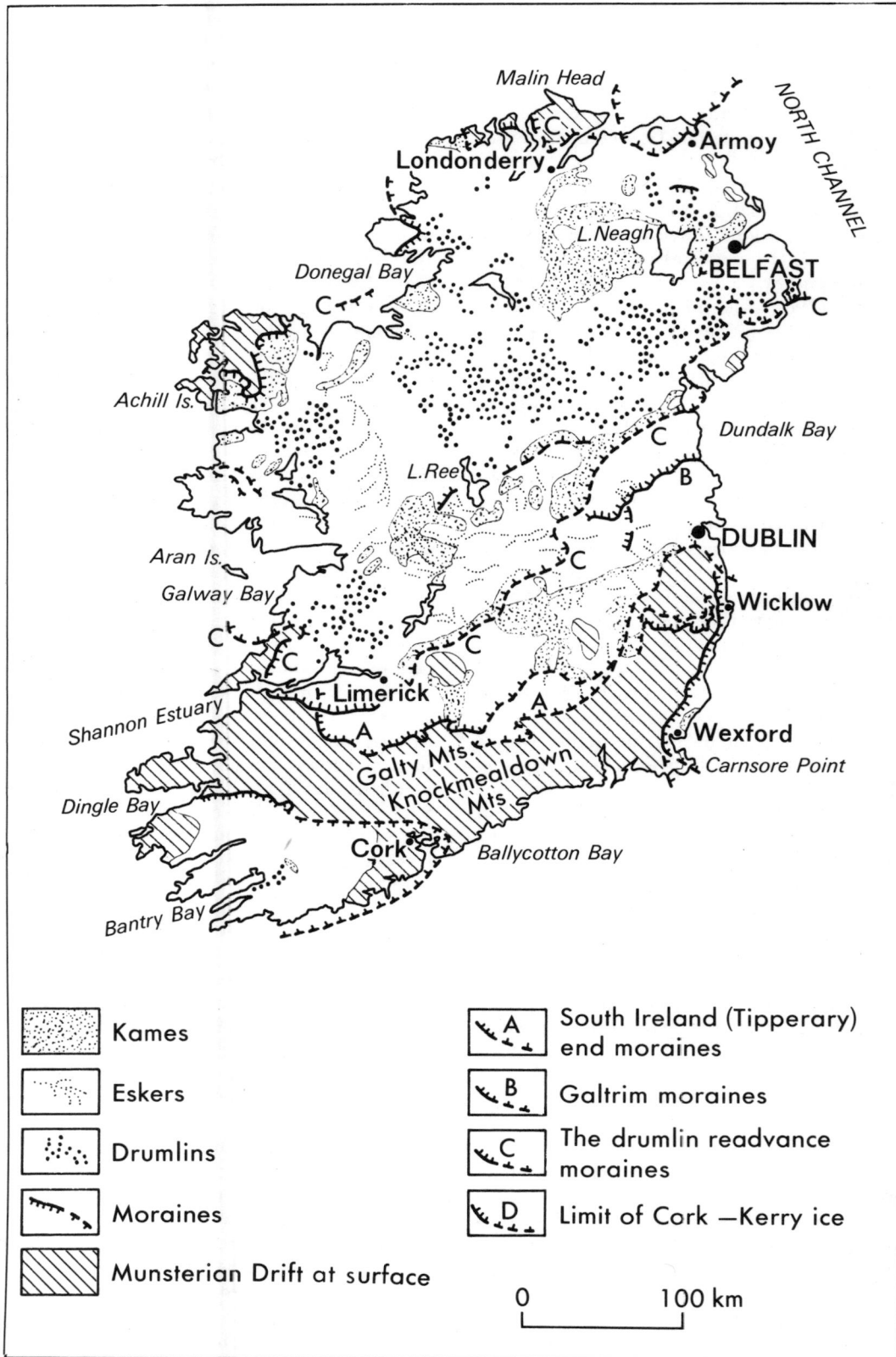

Fig. 7.30 The glaciation of Ireland [after Davies and Stephens, (1978)].

Kerry [see Lewis (1974)].

Some reference must be made to the two most distinctive features of the Irish glacial deposition forms: the drumlins and the eskers. Both words are of Celtic origin and have been used in glaciological literature since the mid nineteenth century. However, as long ago as 1744 the small hills of County Down were likened to 'wooden bowls inverted, or eggs set in salt' [see Smith (1744)]. Ireland is justifiably famed for its drumlin swarms, which extend from Down across Fermanagh and Sligo into south-west Donegal, and are also found elsewhere. The Irish eskers, too, are remarkably fine examples: for instance the Tullamore esker system and the Trim esker [see Synge (1950)].

7.6.5 Karst Phenomena

Given the extensive outcrop of Carboniferous Limestone in Ireland, it is not surprising that there are several areas possessing a wealth of karstic features. In the central lowlands, much of the limestone is concealed by peat bog or drift but in some of the more peripheral parts, where altitudes are higher, karst landforms are impressive. West of the River Shannon, there are cave systems and limestone pavements; along the Shannon itself and other water courses, many of the lakes may be partly solutional in origin (although much modified by ice action). Two other areas are of special note.

1) The Burren in County Clare, where the Carboniferous Limestone has been stripped of its former cover of Upper Carboniferous Shales. The stripping may be quite recent and partly due to glacial erosion [see Sweeting (1955), Williams (1966)] and the pavements are thought to be glaciokarstic phenomena. Dry valleys, sinks, caves and enclosed depressions are present. The Carran depression, which measures 3 km by 1 km, is claimed to be the best example of a polje in the British Isles.
2) The Cuilcagh plateau country, where horizontal or gently dipping Carboniferous strata form a series of tablelands and impressive escarpments rising to 670 m.

On the limestone areas, the range of karstic features includes sinks, caves, dry valleys and pavements; the River Shannon rises in a karst resurgence at Shannon Pot.

Around Colgagh in Sligo, Davies and Stephens (1978) have described a terrain 'reminiscent of the cockpit karst of tropical lands' which Mitchell (1980) has implied may be a relic of Tertiary karst denudation.

7.6.6 Coastline

The Irish coasts show a diversity of form with several controlling factors being responsible. Firstly, there is strong structural control, exerted both by observed land-based structures, such as the Caledonian structures of Donegal, the Armorican structures of the south-west and the Tertiary plateau basalts of Antrim. Also supposed submarine structures, such as the probable north-west–south-east-trending faults (possibly Tertiary) may control the alignment of the North Channel. Secondly, present-day marine processes vary greatly in their activity according to coastal exposure and aspect. There are considerable contrasts in the wave energy of the Atlantic Ocean and Irish Sea. On the exposed Atlantic coasts of Clare and Achill islands, cliffs in Palaeozoic rocks rise to 300 m or so, while the cliffs of Moher on the edge of the Clare plateau almost reach 200 m. In the softer glacial drifts, coastal recession can be quite rapid: 200 m in the last 120 years at Ballycotton Bay east of Cork Harbour, and even 0.3 m per year on the more protected Irish Sea coast between Ballymartin and Carlingford Lough. Thirdly, changes of relative sea level in the Pleistocene have powerfully affected coastal development and left a legacy of features, including raised beaches, buried valleys (to −60 m below Belfast Lough), rias and submerged forest or peat beds. The changes have been both eustatic and glacioisostatic, the latter showing great variation from the probably almost stable south-west of Ireland to the north and north-east which were depressed by a considerable load of ice. For example, Stephens *et al.* (1975) concluded that total isostatic recovery since 14000 years BP in County Down is not less than 80 m. Fourthly, recent marine deposition processes have added further grace and distinction to many parts, including sand spits, shingle bars, tombolos, lagoons (e.g., Ballyteige Lough in southern Wexford) and, most attractive of all, the partially drowned drumlins of Clew Bay and Strangford Lough.

Chapter 8
West and Central European Lowlands

8.1 Introduction

The plains of the western part of the Russian platform (*see* Chapter 6) merge imperceptibly into the lowlands south of the Baltic and North Seas, and continue west without any important topographical breaks into the lowlands of northern and western France. This region, occupying about 8 percent of the area of Europe west of the Urals, has often been termed the north European plain, but the title adopted in this chapter is intended to emphasize that not all of the region can be described accurately as a plain and that its position in Europe as a whole is west and central rather than northern. The region comprises a complex arrangement of lowlands, extending east–west for 2000 km and interrupted by uplands of older rocks. The English Channel and Straits of Dover sever the lowlands of southern and eastern England from those on the mainland continent. The limits of the region are clearly defined in places, either by the coasts or by the relatively abrupt topographical and geological break at the foot of the Hercynian uplands (*see* Chapter 9). Elsewhere, however, limits are arbitrary: for example, in the eastward transition to the Russian lowland. The region is crossed by many of the great rivers of Europe, such as the lower Rhine and the lower Elbe, and also includes the greater part of some complete drainage basins such as those of the Seine or the Garonne.

Practically all of the lowland lies below 300 m. Indeed, it is only the escarpments, some moraines and the Hercynian border zones that rise significantly above 200 m. The Netherlands include a substantial area that descends below mean sea level (to a minimum of −6.7 m), being the only area of Europe, other than the Caspian lowlands, to do so. Subsidence has continued to characterize many coastal areas, especially those around the southern North Sea basin where present-day rates of vertical movement (sinking) are over 3 mm per year (although 1.2 mm per year of this is attributable to eustatic sea-level rise).

Geologically, the lowland possesses both diverse and common characteristics. Essentially it is a platform area, the platform being provided by the Palaeozoic or Precambrian floor. The depth to the floor varies considerably, from many kilometres in some places to relatively shallow depths in others, such as the London Basin. The cover rocks resting on the platform are Mesozoic, Tertiary and Quaternary, and it is these that provide most of the geological control on the landforms. The rocks exposed at the surface show important regional variations: the Baltic and North Sea lowlands consist almost entirely of Quater-

nary formations at the surface; the Paris Basin and south-east England show a sequence of Jurassic, Cretaceous and Tertiary rocks, forming the distinctive scarplands of these areas; the rocks of the Aquitaine Basin range from Jurassic to Quaternary; while the lowlands of north and central England are floored mainly by Triassic rocks, with an overlying skin of Quaternary deposits.

The lowland includes both glaciated and unglaciated portions (*see* Fig. 8.1). The maximum limit of glaciation [Elster or Anglian (*see* p.147) but in places also Saale] lies to the south of all the lowland as far west as the Rhine valley, but to the east the limit continues across the Netherlands—marked by the striking ice-pushed ridges (e.g., near Hilversum)—and the North Sea into southern Essex, mid-Oxfordshire and the Avon valley. Later glaciations were less extensive, affecting only the north German–Polish lowlands, eastern Denmark and the more northerly parts of the English lowlands.

8.2 Baltic and North Sea Lowlands

Essentially this is an area dominated by glacial and glaciofluvial deposition resulting from the advances and retreats of the Fennoscandian ice in the Pleistocene. These plains extend from the Baltic and North Seas in the north to the foot of the central European mountains in the south and from the Mazurian lakes in the east, across the northern parts of Poland, GDR and FRG, together with Denmark, as far as the Netherlands in the west. In the west, the plains include the tidal coastlands of the southern North Sea, between the English Channel and Jutland, with their recent (Holocene) tidal flats, dunes and marshes.

This part of the central European lowland is built from Quaternary glacial, glaciofluvial, fluvial, aeolian and marine littoral deposits with average thicknesses of 50–100 m. Locally, however, accumulations in basins and hollows amount to over 500 m. Beneath the Quaternary lie Tertiary strata (up to 1.6 km thick in places) comprising marine and continental sequences of clays, silts and sands, together with Miocene and Eocene lignite formations in the south. These Cenozoic deposits, lying between the edge of the Baltic Shield, and the northern rim of the Hercynian block mountains and plateaus of central Europe, fill a platform depression whose rims, particularly in the east and south, they overlap in a broad belt. This extensive platform depression, sometimes referred to as the North Sea basin, is a result of tectonic downwarping; its structure emphasized by major faults. It first came into existence during the Caledonian orogeny and the struc-

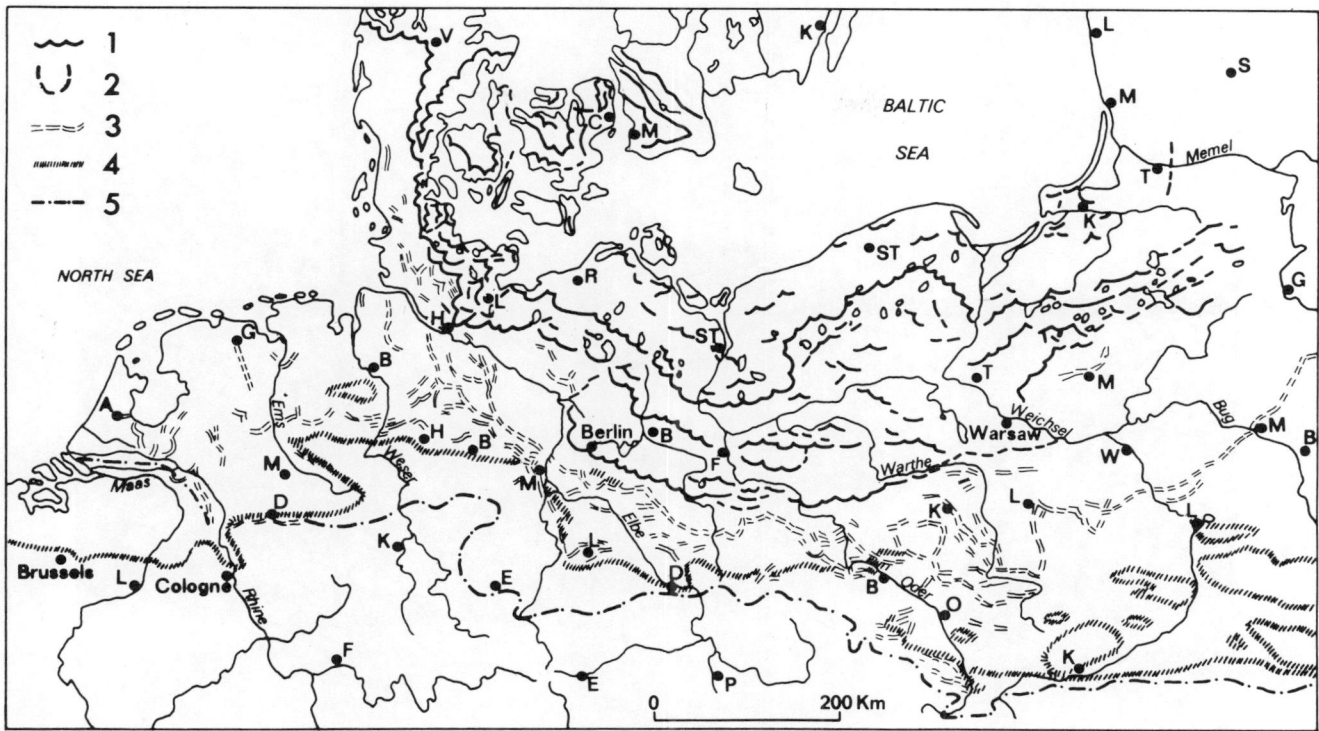

Fig. 8.1 The major Pleistocene end moraines of the north European plain [after Woldstedt (1955)]. (1) Weichselian end moraines; (2) probable Weichselian moraines; (3) end moraines of the Saale and Elster glaciations; (4) northern limit of continuous loess cover; (5) southern limit of Scandinavian erratics.

tures were reactivated by the Hercynian movements. Following these, great thicknesses of Permian–Mesozoic sediments accumulated which are themselves, in places, strongly folded and broken by faults with throws of several kilometres. Gravitational sliding, aided by the presence of massive incompetent salt beds of Upper Permian age, has helped finally to produce the highly complex structures of the pre-Cenozoic undermass. Neotectonic movements continuing into the Quaternary have affected mainly the southern border zone, in particular in the lower Rhine area of the Netherlands [see Ksiazkiewicz *et al.* (1965)].

The present landscape is mainly a product of glacial, periglacial, fluvial and aeolian processes active in the Pleistocene and associated with the Fennoscandian ice, both proglacially and subglacially. The resulting assemblage of erosional and depositional forms has been modified only to a minor extent by postglacial and contemporary processes; the landscape as a whole is dominantly a relic of the Pleistocene. In terms of glacial morphology, it is important to distinguish between the young moraine landscapes in the area affected by the last glaciation (Weichselian or Vistulian) and the older moraine landscapes modelled by the Saale (or central Polish) and Elster (or southern Polish) glaciation (*see* Fig. 9.44). It should be noted that these contrasting moraine landscapes continue eastwards into the area of the Russian platform (*see* Chapter 6, p.82). A third regional element is provided by the tidal flats, marshes and coastal dune systems (up to 30 km wide) and the Frisian Islands chain, all of which came into existence after the last

glaciation. The coast itself, from the Netherlands through Denmark to East Germany, is nearly all developed in unconsolidated glacial or postglacial deposits, often low and marshy but, in a few places, cliffed by wave action (e.g., in northern Denmark) [see Galon (1960)]. An exception is the Chalk cliff coast of Møn, Stevns Klint and Bulbjerg in eastern Denmark where the Cretaceous rocks, which are concealed beneath most of Denmark, outcrop.

In the area of the younger moraines, the surface morphology is fresh and shows clear constructional features. Surface drainage is poorly developed or even completely absent. These areas are characterized by attractive lake landscapes with their so-called internal drainage systems. In the area of older moraines, however, the forms shaped during the Saale or Elster glaciations have been widely destroyed or obscured by later glacial, periglacial or subglacial erosion and sedimentation. The glacial lakes have been filled with sediments and have thus disappeared, with only a few exceptions [see Galon and Dylik (1967), Rozycki (1965)].

The forms of the younger moraine landscape were created by both ice and meltwater (in particular, terminal moraine ridges, kames, eskers and numerous subglacial meltwater channels) [see Kozarski (1966–67, 1978), Niewiarowski (1965)]. Many shallow depressions, kettle holes and glacial lakes are well preserved and diversify the landscape. The glacial and glaciofluvial constructional features in the areas of older moraines, however, have been widely degraded due to interglacial and periglacial transformation processes and show many features of a

Fig. 8.2 Young morainic landscape in the northern part of East Germany. Erratic blocks in the foreground. (Gellert).

denudational or erosional landscape. In the border zone of glacial sedimentation, which is lower towards the south, pre-Quaternary ridges and uplands appear, in particular, groups of inselbergs formed in the Tertiary.

8.2.1   Younger Moraine Landscapes
Away from the ice margin on the distal (ice-covered) side, extensive areas of flat or undulating till plains are to be found, comprising subglacial deposits squeezed and compressed beneath the thick ice sheet. Rising abruptly above these plains at the melting and oscillating ice front are sequences of marginal forms marking the limits of the various Weichselian stadials. The principal features comprise terminal moraines, kames and other marginal forms, as well as outwash plains (*see* Fig. 8.2). During longer pauses in the retreat and melting of the glacier front, typical glacial-form complexes were developed. These are composed of closely spaced terminal moraines and broad outwash plains that become narrower farther to the south and west, and form outwash valleys that in turn run into major meltwater channels (Urstromtäler or pradolina) [see Galon (1965, 1982), Galon and Roszko (1961)]

This tripartite glacial system of forms (i.e. till plain, marginal zone and outwash plain, with associated outwash valleys and major meltwater channels) repeats itself three or four times in the area of younger moraines (*see* Fig. 8.3). The main marginal zones are clearly demarcated only in the glaciated areas of Poland and East Germany. In Schleswig Holstein and in Denmark, they are closer together, divide in places and overlap. The main marginal zones represent the advance limits of the Fennoscandian ice with the oldest of these, the Brandenburg–

Leszno main marginal zone, forming the outer border of this glaciation. In Denmark, the so-called Main Stationary Line of the Weichselian glaciation trends northwards from the German border as far as northern central Jutland, where it turns abruptly westward, reaching the North Sea coast to the south of Limfjorden (*see* Fig. 8.5).

The many small outwash plains of the main marginal zone in Germany and Poland, which has been partly destroyed by still younger outwash channels and, in the west (i.e. Mecklenburg, Schleswig Holstein and Denmark), overridden during later glacial phases, run into the Głogów–Baruth Urstromtal, which begins farther east and carried river water from the south to the west. A similar system of forms came about during the Poznan–Frankfurt stage when the respective outwash plains led into the Warsaw–Berlin Urstromtal. West of Berlin, meltwater and river water flowed towards the Rhinluch or even directly towards the lower Elbe. The distance between the two main marginal zones increases towards the west and is occupied by a retreat phase after the Leszno–Brandenburg stage.

After some morphologically well-marked retreat phases of the Poznan–Frankfurt stage, which are visible on either side of the lower Vistula, as well as in the lake zone of Mecklenburg, there was a further advance of the ice to the Pomeranian limit (*see* Fig. 8.4), when broad glacier tongues pushed deeply into the area of the older Frankfurt stage. In some places, the older terminal moraines were overridden. The Pomeranian main marginal zone is compact and forms a ridge rich in lakes, the culmination of which is the highest elevation (329 m) in the central European lowland at Wieźba, west of Gdańsk. Vast

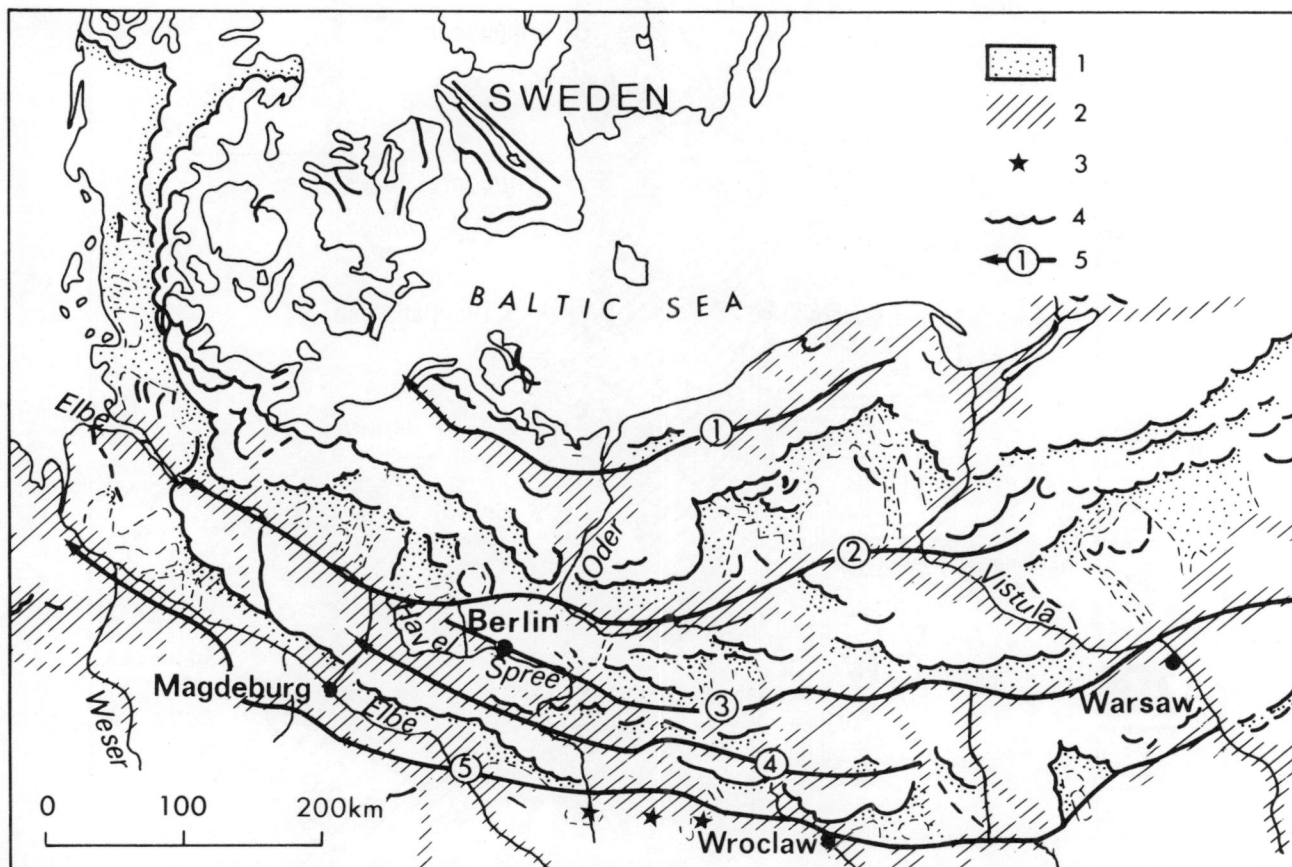

Fig. 8.3 Morainic limits and associated meltwater features in part of the north European plain [after Woldstedt (1955)]: (1) outwash gravel; (2) Urstromtäler; (3) major alluvial cones from the Mittelgebirge. (4) main ice margins; (5) principal routes of meltwater escaping along ice border to the west.

Fig. 8.4 End moraines of the Pomeranian stage, east of the lower Vistula. (Roszko).

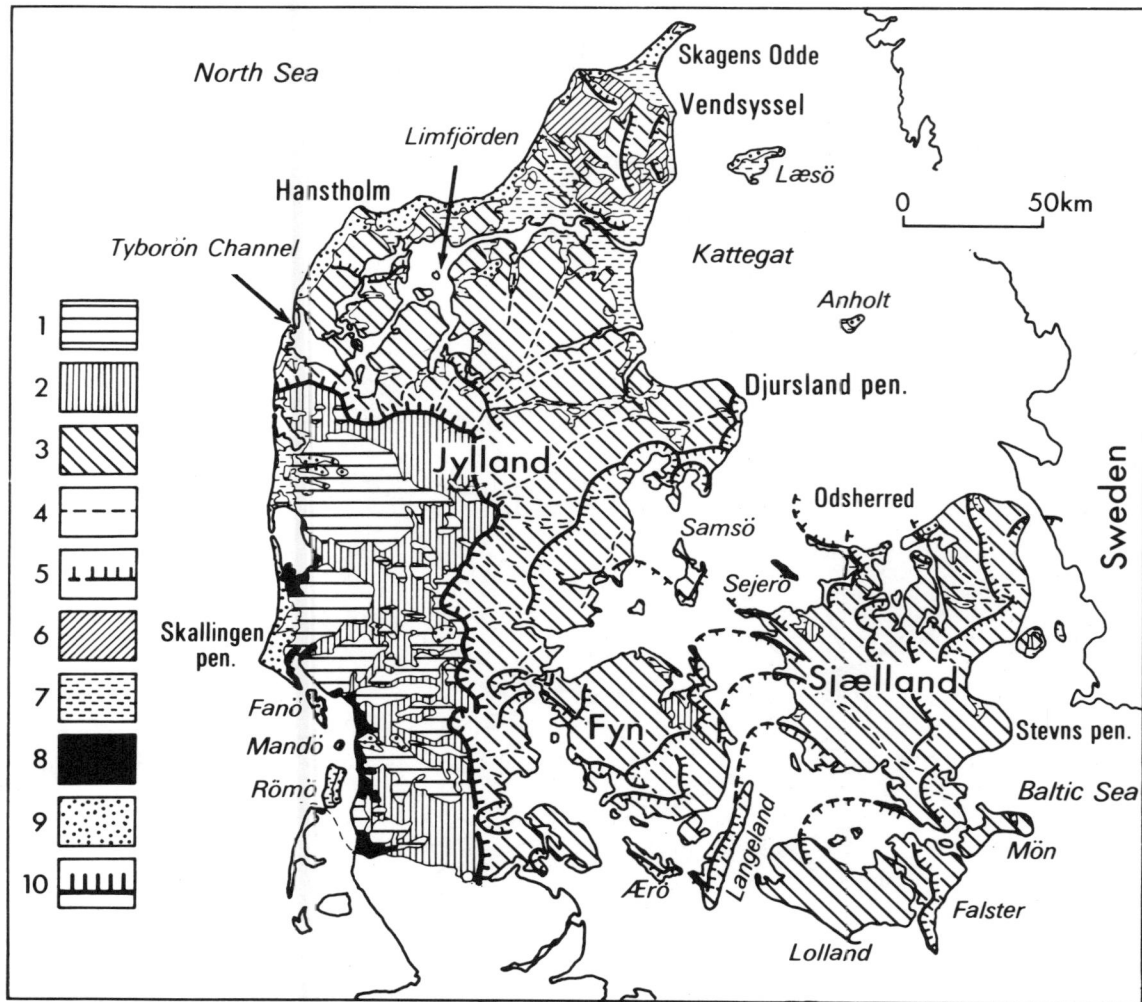

Fig. 8.5 Geomorphological map of Denmark [after Schou (1949)]: (1) old moraine landscape of the Saale glaciation; (2) Weichselian outwash plains; (3) young moraine landscape of the Weichsel glaciation; (4) valleys possibly eroded by subglacial meltwater flowing south and west; (5) retreat stages or temporary readvances after the Weichselian maximum; (6) late glacial plateau areas, raised floor of the Yoldia Sea; (7) marine foreland with beach ridges and other deposits; (8) salt marsh; (9) coastal and inland dunes; (10) the Main Stationary Line of the Weichsel glaciation.

outwash plains were built up leaving only remnants of the till sheets. These outwash plains, however, originate from the younger marginal zones within the main marginal zone, so that the farthest terminal moraines of this stage were destroyed in many places by meltwater erosion, particularly in the area of the Vistulian lobe. The outwash channels lead through outwash valleys into the Toruń–Eberswalde Urstromtal, the most important Pleistocene valley form in the southern Baltic younger moraine area, towards which all central European rivers of that time flowed. In the west, between the Oder and the Elbe, several drainage routes were followed: temporarily, also, a Netze–Noteć–Randow Urstromtal existed.

The Pomeranian stage also included various retreat stages behind the main marginal zone. As late as the beginning of the Gardna phase (known as the Velgast and North Rugen advances in East Germany, the Fehmarn advance in Schleswig Holstein and the Lower Baltic advance in Denmark and Skåne), with its associated Urstromtal-like meltwater channels, a new deglaciation phase commenced which is thought to be correlated with the oldest Dryas (Zone Ia) preceding the Bølling oscillation (about 12 000–13 000 BP). The Urstromtäler and moraines flanking them are one of the most distinctive geomorphological features of the central European lowland. Now carrying portions of great rivers such as the Weser, Elbe, Oder and Vistula, their capacity is in fact much greater; that south of the Brandenburg moraine is thought to have once carried a river 40 times the volume of the present Elbe, the water coming from both melting ice to the north and from land drainage to the south.

In contrast to the main marginal zones paralleling the ice border, with their functionally linked major meltwater channels, the younger moraines are also crossed orthogonally to the ice border by many through valleys, which in the course of the various stages of deglaciation carried water

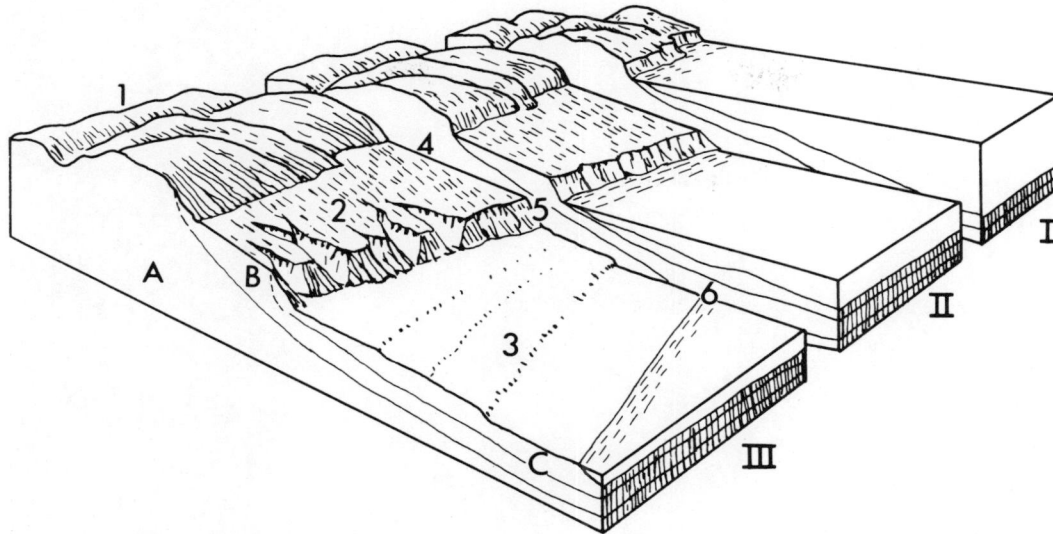

Fig. 8.6 Danish moraine landscape with coastal plains and emergent shorelines [after Schou (1949)]. (I) Late glacial stage. The Yoldia Sea is cutting into the morainic scape. (II) Neolithic stage during the Litorina transgressions. (III) Present stage. (A) Moraine. (B) Late glacial marine deposits. (C) Postglacial marine deposits. (1) Hilly morainic landscape. (2) Late glacial plateau (*see* Fig. 8.5). (3) Postglacial coastal plain, with raised beaches. (4) Raised cliff of the late glacial seas. (5) Litorina raised cliff. (6) Present shoreline.

from the older meltwater channel to the next younger one. In this way, the rectangular courses of the Vistula, Warthe, Oder, Elbe and some of their tributaries were created. The origins of these through valleys (*Durchbruchtäler*) and the other breaches in the moraines are not fully agreed [see Galon (1982)]. In Denmark (*see* Fig. 8.5), one view that has been much supported is that they represent subglacial streams carrying meltwater outward from under the ice—the so-called tunnel valleys (*Tunneldale*). These tunnel valleys are up to 100 m deep with flat floors and sharply defined sides. Their widths suggest that they may not have existed as single ice-roofed tunnels but as anastomosing systems, collapse of ice from the roof causing the streams to shift laterally. Ribbon lakes occupy parts of the floors where the subglacial streams excavated depressions as they flowed under pressure. In Germany, Woldstedt (1950) adopted a similar hypothesis for the *Rinnentäler*, the morainic breaches and through valleys. In all cases, these valleys are oriented perpendicular to the ice margins, showing how the subglacial streams were forced out to regions of lesser ice pressure and, finally, the ice margin itself [see Kozarski (1966–67), Pasierbski (1979)]. As in Denmark, lakes (*Rinnenseen*) occupy parts of these valleys. Another hypothesis is that direct glacial erosion may have played a part in excavating the larger tunnel valleys [see Woldstedt (1961), Hansen (1971)].

Other characteristic features related to the Weichselian advances and, like the moraines and meltwater channels, still freshly preserved in the landscape, are the outwash plains. These plains are built up by the meltwater rivers in the form of low-angle cones from the points where they escaped from under the ice; with the sudden reduction of hydrostatic pressure, they began depositing their heavy sediment loads over huge areas like the sandar of Iceland today. Although the outwash plains may appear flat, in reality they show an outward slope from the ice border, but sometimes the slope is as low as 1 : 1000. The inner edge of the outwash plain may be overlooked by moraines or it may be an ice-contact slope dropping away sharply where the ice no longer supports the sediments. In yet other places, there is no clear sign in the relief of the transition from outwash plain to subglacial till deposits.

During the late glacial and postglacial periods, the present coastal forms of the Baltic Sea developed with their spits and bars (*Nehrungen*), lagoons (*Haffe*) and shallow estuaries (*Förden* or *Sunds*, i.e. the drowned lower portions of the tunnel valleys) as well as some steep coastal features (*see* Fig. 8.7) and curving indentations, which led to the formation of an abrasion–accumulation coast (*see* Fig. 8.6). The melting of dead ice blocks, which lasted locally until the Boreal period, was responsible for various enclosed depressions. In the valleys, many terraces were formed on whose sandy surfaces aeolian processes produced numerous broad dune systems [see Galon (1959)]. Many of the lakes were silted up and many of the subglacial meltwater channels were transformed into normal river valleys.

8.2.2 Older Moraine Landscapes

The older moraine landscapes comprise areas of glacial and glaciofluvial forms which were shaped by the Elster (southern Polish) and the Saale (central Polish) glaciations. Both glaciations reached the central European uplands. In the area between the Harz and Carpathian Mountains, the Saale glaciation was 10–15 km less extensive than the Elster glaciation. The Elster ice even advanced over the Jitrava Pass as far as Moravia,

Fig. 8.7 Morainic cliffed coast with erratics near Gdynia, Poland. (Masicka).

Czechoslovakia. West of the Harz-Mountains, however, the situation is quite different, for in the Leine–Weser Bergland, the Lippe Bergland, the Munster and lower Rhine embayments and the northern Netherlands, the Saale ice generally advanced the farthest south and south-west.

It can be assumed that the older moraines were originally as fresh-looking in their relief as the younger moraines. The forms shaped by the older glaciations however, have been largely destroyed or degraded by erosion and sedimentation. This applies in particular to the loessic belt in the southern part of the older moraine zone, where loess often obscures the transition to the adjacent uplands. Whereas the central north German/north Polish lowland consists of an almost uninterrupted cover of unconsolidated Quaternary deposits with an average thickness of 50–100 m, there are consolidated Palaeozoic and Mesozoic deposits in many parts of the loessic zone on the southern border of the older moraine landscape. In this zone, the surface form is not so much determined by the Pleistocene Fennoscandian glaciations as by the forms of the buried pre-Quaternary surface and the deep river valleys with their terraces. Here the periglacial morphofacies dominates [see Dylik (1956)].

The glacial forms of the Elster (southern Polish) glaciation are hardly noticeable in the present surface relief of the area of older moraines. The power of the Elster ice can only be recognized in terms of the buried relief, which is characterized by wide basins of glacial erosion with infills of Elster deposits of up to 120 m and by meltwater channels up to 400 m deep. These channels are not fossil river valleys but have been proved to have originated underneath the Elster ice by meltwater erosion. In places, these subglacial channels, which are exclusively filled with glacial sediments, form extremely deep hollows with no outlet and with side slopes of up to 60°. In the Reesselner meltwater channel in the north-east of the Lüneberg Heath, the glacial sediments of the north German/north Polish lowland reach their maximum thickness of 502 m.

In contrast to the marginal forms of the Elster (southern Polish) glaciation, those of the Saale (central Polish) glaciation are relatively well-preserved, even if only locally. The Saale glacial is subdivided stratigraphically into the older Drenthe stage and the younger Warthe stage. During the Drenthe (Radomka) stage, which is again divided mostly into two, the Saale ice reached its maximum extension as noted already. Morphologically, the most striking folded terminal moraines of the Drenthe stage are the Rehberg, Damme and Fürstenau Hills in Lower Saxony, West Germany. This marginal zone of the Drenthe stage extends to the east near Braunschweig and to the west as far as the Veluwe in the Netherlands. In the neighbourhood of Leipzig and around the Vistula valley downstream from Sandomierz, the Drenthe (Radomka) glaciation has left surprisingly clear terminal moraines.

The glacial forms of the Warthe glaciation have a transitional character between the typical landscapes of the older and those of the younger moraines. They consist primarily of numerous main marginal lobes embracing glacially excavated basins. Negative glacial relief such as kettle holes has, despite younger periglacial reshaping, generally been preserved more frequently than in the area of the Drenthe glaciation.

The Warthe glaciation extended from the Typus region in central Poland, through Lower Lausitz, Fläming (201 m), Letzling Heath, the Wiernen Hills (130 m), Wilseder Berg (169 m) and the Harburg Hills near Hamburg (155 m) to western Holstein. Beyond this point, it is difficult to trace this glacial limit. According to the most recent investigations, however, this main marginal zone of the Warthe stage is not uniform. The relief of the terminal and push moraines represented by the hills in these north-western areas was apparently already shaped by the Drenthe glaciation and only slightly reshaped by the Warthe ice which came to a standstill behind. In some places (e.g., in the Harburg Hills), high outwash plains form the cores of these elevations.

Only fragments of the oldest meltwater and river channels have survived, mostly in the area of the older moraines of the Saale (central Polish) glaciation. The southernmost Urstromtal belonging to the Warthe stage leads from Wrocław via Magdeburg to Bremen and the lower Weser. The eastern equivalent of this channel in Poland runs eastwards from the upper Warthe via the Pilica and east of the Vistula via the valley system of the Wieprz and Krzna towards the Pripyat River and beyond to the Dnepr. Later on, through valleys became available leading water across the morainic belts towards the younger meltwater channels of the Vistulian stage. Under the influence of cold climatic conditions during the younger glaciation phases, widespread periglacial forms developed in the area of older moraines with cover sand and loessic deposits, dells and asymmetrical valleys, as well as large dune fields on the outwash plains, while terraces formed along the Urstromtäler in response to changing conditions of river volume and sediment load. Finally, with the ending of cold conditions, vegetation was reestablished, leading to the present-day landscape with its forested moraines and grass-covered valley floors.

8.2.3 Western North Sea Lowlands

During the Holocene, the sea level rose as a result of melting of the Fennoscandian ice. The sea invaded the sandy plain of the former dry North Sea floor, built up by fluvial, periglacial and aeolian deposits, and reached the present coastal area about 4000 years ago.

In the coastal area of north-western France, Belgium and the western Netherlands, Zagwijn (1974) identified three zones of sedimentation (*see* Fig. 8.8) during the rise of the sea level: (1) a littoral sandy zone of beach ridges and dunes, (2) a clayey zone of tidal flats and brackish lagoons and (3) at a greater distance from the sea, a zone of peat formation in a freshwater environment. These zones shifted to the east as the sea gradually flooded the dry North Sea floor. The Holocene deposits may reach a thickness of over 9 m near Ardres in northern France [see Roeleveld (1975)], over 15 m near Oostende in Belgium [see Tavernier and Ameryckx (1970)] and over 20 m in the west of the Netherlands [see Hageman (1969)]. The

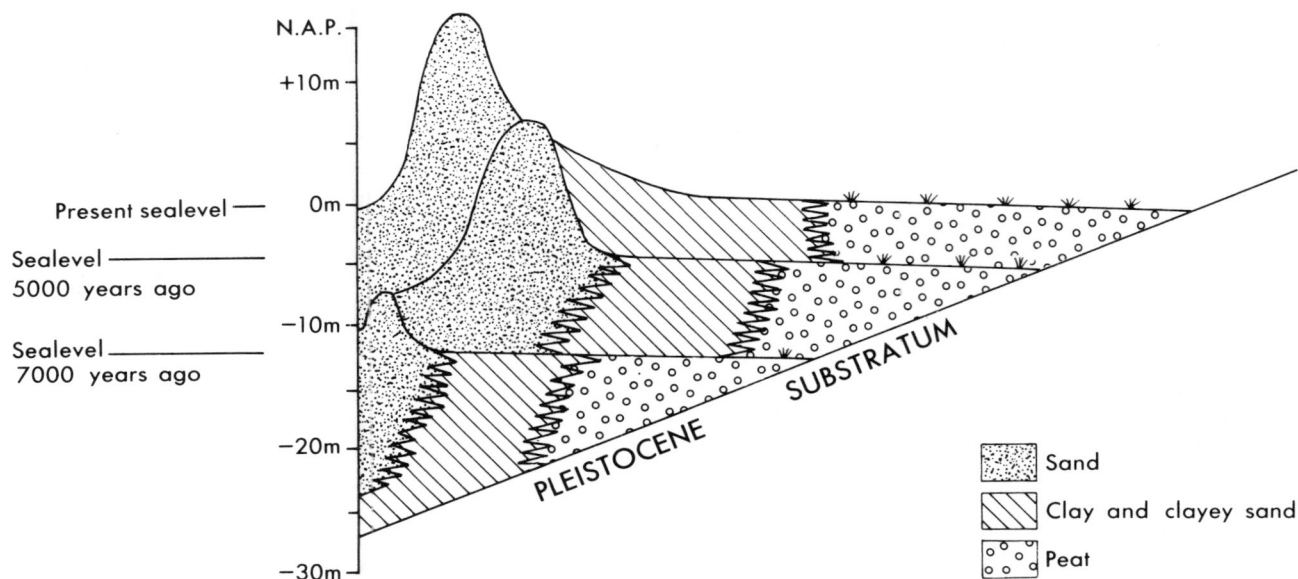

Fig. 8.8 Diagram showing the lithological succession in the Netherlands, related to the Holocene rise in sea level [after Zagwijn (1974)].

whole area, which is built up by beach ridges and beach plains, tidal accumulation plains and reclaimed peat plains, is situated at and just below sea level. The areas below sea level arise partly due to shrinkage, especially of the peat. These coastal plains are protected from the sea by 20–50-m high coastal dunes [see De Puydt (1972), Jelgersma et al. (1970)] or dykes. The Rhine and Meuse, with their natural levées and backswamps, branch into several lower courses that cross the coastal plain and flow into the North Sea. On the Zeeland coast in the south-west of the Netherlands, there are islands separated by tidal creeks and estuaries, with coastal dunes only occurring on the seaward sides of the western islands. Since medieval times, man has radically modified the landscape by the construction of sea dykes and river embankments to reclaim land and to provide protection against floods. Also man has undertaken the drainage of lakes (e.g., the Moere near the French/Belgian border and the Prins Willem Alexander polder in the Netherlands, now 1.5 and 6 m respectively, below sea level).

The coastal area of the northern Netherlands, Germany and western Denmark as far north as Esbjerg consists of the Waddenzee—the biggest tidal flat in the world, having an area of 10 000 km². It is separated from the North Sea, in places, by a system of barrier islands, but part of the tidal flat is open to the sea with various systems of sand bars which are intersected by the estuaries of the Elbe and Weser. The flats consist of mud and sand. Nearly all the water that covers the flats at high tide runs through tidal channels that act both as tributaries and distributaries. The islands consist partly of beach ridges and coastal dunes. Some of them, such as Texel and Sylt, have a core of pre-Holocene sediments, whereas the islands off the coast of Schleswig Holstein, as described by Brand et al. (1965), are remnants of Holocene marshland (Halligen)—

a large area destroyed in 1654 by an enormous storm surge. The mainland, artificially stabilized by reclamation (i.e. polders), consists of salt marshes in which marsh bars occur. These are flat, elongated bodies consisting of clayey and fine sands. There are also reclaimed peat plains in this region. Farther inland, peat bogs are widespread: for instance, the Boertange peat bog in the north-east of the Netherlands comprises a vast raised moor now almost completely cut over for fuel. Several marine cliffs occur north of the Elbe estuary: the Rote Kliff on the island of Sylt, where Tertiary and Pleistocene sediments outcrop, is well known. In the coastal area of Esbjerg as far as Skagen in Denmark, two groups of ancient raised sea floors, formed due to isostatic upheaval since the last glaciation, can be distinguished [see Jensen and Jensen (1976)]. These are a late glacial raised sea floor with a plateau-like character modified by erosion and a lower-lying postglacial (Atlantic and sub-Boreal) coastal plain with beach ridges, tidal flats and peat plains of the Litorina transgression. The highest late glacial shoreline in north-eastern Denmark rises to 56 m near Frederikshavn and to 20 m near Aalborg [see Hansen (1965)]. After the sub-Boreal period, isostatic uplift began to dominate in northern Denmark, bringing the highest shorelines of the Litorina trangression up to an altitude of 15 m; this uplift is still in progress. The zero isobase for present-day isostatic movements seems to run in a north-west–south-east direction in the region north of Esbjerg. There are several lakes cut off from the sea by regression, and lagoons cut off by spits and beach ridges. Some of the larger lagoons (e.g., Ringkøbing Fjord in western Denmark) are not yet wholly closed. Wind action has produced the continuous coastal dune landscape from Esbjerg to Skagen in most recent times.

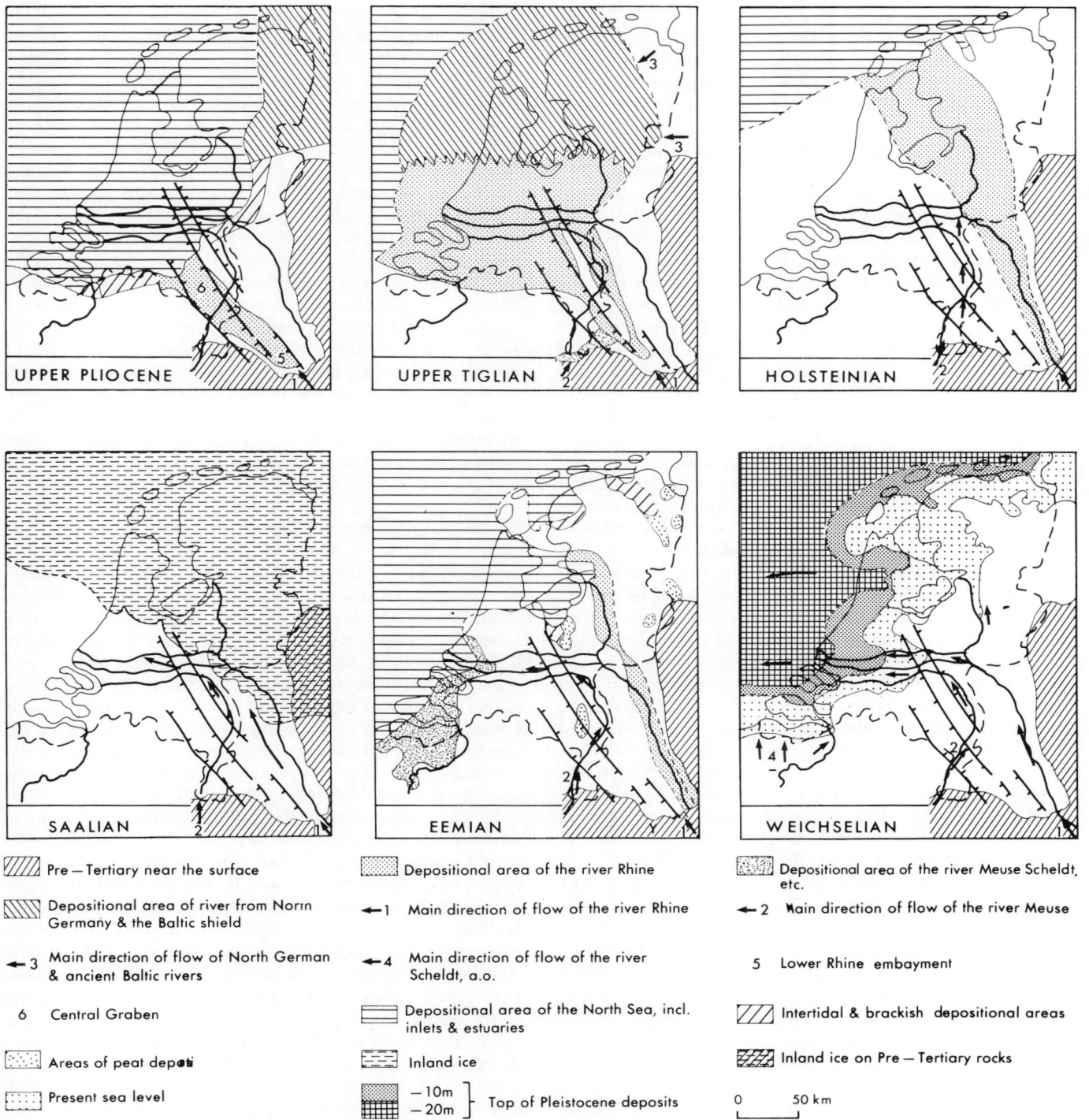

Fig. 8.9 Palaeogeographical maps of the Netherlands during the Upper Pliocene and Pleistocene [after Zagwijn (1974)].

8.3 Inner Lowlands and Low Plateaus from Flanders to Münster

At the end of the early Tertiary, there was renewed development of the North Sea basin. The Tertiary deposits in the basin are for the major part of marine origin. In particular, the older Tertiary consists of clays and clayey glauconitic sands. From the south-east, fluvial sedimentation progressively replaced marine sedimentation after the Miocene (*see* Fig. 8.9) until, during the Quaternary, the sea withdrew completely and predominantly fluvial deposits were laid down [see Staalduinen *et al.* (1979)].

8.3.1 Southern Plains

During the Palaeogene, deposits of sands and clays roughly similar to those in the London and Paris Basins were laid down with a north–south coastline along the border of the Ardennes. The Tertiary North Sea basin came into existence during the Eocene by which time crustal separation between the North American–Greenland and the European plate had been achieved. The North Sea rift had become an inactive, yet still subsiding, branch of the North Atlantic rift system [see Oele *et al.* (1979)]. At the end of the Oligocene, as a result of tectonic activity, the so-called Alpine orogeny, the ridge of Artois was elevated, the Paris Basin became dry, the North Sea withdrew to the north and England was connected with the continent. During the Neogene, Miocene–Pliocene littoral and marine deposits were laid down. At the end of the Tertiary, the southern boundary of the North Sea had shifted from the vicinity of the French/Belgian border towards the area of the Belgian/Dutch border, north of Antwerp.

The original Neogene coastal plain was uplifted and tilted. This uplifted plain was dissected by fluvial erosion during late Tertiary and Quaternary times, especially in the middle Pleistocene. In Haspengouw, the plateau-like character has remained most clearly with an elevation of about 200 m. Farther to the west in the Dijle Basin, erosion was more intense but prevented locally by coarse-grained sands in the subsurface strata. Between Zenne and Schelde, the plateau was altered to a hilly landscape in which only the Flemish Ardennes (with a resistant ferruginous sandstone cap) portion of the original plain remained. On the western side of the Schelde, lowering by erosion was still more intensive, and only a few erosional remnants (outliers), such as the Kemmelberg (156 m), indicate the original land surface in this area. Finally, loessic deposits of Saalian and Weichselian age covered nearly the whole of central Belgium. By periglacial processes, the loess filled in valleys and depressions, with further levelling of the landscape. In several places, these deposits are more than 20 m thick, which results in the plateaus becoming still flatter and the slopes more gentle and even. The area is well known as the loam region of central Belgium. North of the Demer, Rupel and Schelde, a sandy plain is the dominant feature. Only a few ridges and hills remain from the work of erosional processes, such as the Bartoon cuesta of Oedelem–Zomergem in clay and the Beersel–Hageland ridge in ferruginous sandstone. An exception is the Kempen plateau. Here, thick gravel layers belonging to an old fan of the Meuse completely prevented the erosion. The flat character of the sandy plain was intensified by aeolian deposits (cover sands) during the Saalian and Weichselian. This material was partly transported by snow meltwater into the valleys, resulting in further smoothing of the relief.

During the Neogene and early Pleistocene, a parallel river system flowing north-east developed to the north of the Sambre and Meuse, according to the maximum slope exposed by the retreating North Sea. By erosion of the Tertiary clays and sands, deep valleys were scoured out and partly filled up again with river gravels in which a series of terraces were modelled [see Tavernier and De Moor (1974)]. Together with the intensive lowering of the relief in the Middle Pleistocene, a tributary system with a north-west–south-east direction that was more adapted to the geological structure developed. The soft, sandy layers were eroded and the clay layers stood out in relief as cuestas, such an example being the Rupelian cuesta of Waas–Boom. The drainage of the Leie and Schelde during the Saalian, in the meantime strengthened by river capture from central Belgium, ran via the Flemish valley over the Gent–Eeklo–Vlissingen area to far out in the nearly dry bed of the North Sea. In the Eemian, the rising sea level flooded the present coastal plain and the Flemish valley, as a result of which the rivers were considerably shortened and estuaries formed. Renewed lowering of the sea level during the Weichselian caused the infilling of estuaries with fluvial material. Northerly winds carrying aeolian cover sands blocked the drainage of the Schelde near Eeklo at the end of the Weichselian. An easterly diversion of the river was necessary to form a new connection with the sea near Antwerp.

8.3.2 South-Eastern Plains

In the late Oligocene, tectonic movements resulted in the development of a fault system trending south-east–north-west in the south-eastern part of the Netherlands and extending to the lower Rhine embayment—the area in Germany adjacent to the south-east of the Netherlands. It is believed that the dislocations along the faults were related to those in the upper Rhine rift valley during the Oligocene [see Ziegler (1975a)]. In the lower Rhine embayment, a fan of fluvial Neogene sediments is present, overlying a thick series of continental Miocene beds, consisting of sands, clays and several lignite horizons resting on marine Oligocene deposits.

During the latest part of the Tertiary, the configuration of the coastline of the North Sea was similar to the present state boundary of the Netherlands. The Rhine followed a course towards the north-west, building out a delta in the central graben area—part of the south-east–north-west fault system already mentioned. In the east and north-east, delta formation by northern German and ancient Baltic rivers continued. According to Zagwijn and Doppert (1978), considerably greater amounts of sediment were accumulated per unit time in the south-eastern North Sea basin during the Quaternary compared with during the Neogene.

Since the end of the Tertiary, sediments have been laid down, for the greater part, in a coastal area (*see* Fig. 8.9). This means that they were deposited either in a shallow sea not more than 10 m deep or in coastal swamps, lagoons and lower river courses. At present, however, they are found considerably below sea level (sometimes as low as 400–600 m below sea level). Variations in intensity of tectonic movement, changes in the river courses and climatic changes between glacial and interglacial periods have determined the geological evolution of the subsiding basin in the Netherlands during the Quaternary [see Zagwijn (1974)].

Fig. 8.10 Push moraine of the Saale glaciation near Wageningen, Netherlands. (Embleton)

After the middle Tiglian, a relatively rapid regression of the sea took place, resulting in a shoreline located somewhat seaward of the present Dutch coastline. A large deltaic fan was formed by the Rhine/Meuse river system in the south-east and by the river system draining the Fennoscandian Shield and the northern lowland of Germany in the east and north-east. Remnants of the Rhine/Meuse fan (now terrace plains) of early Pleistocene age can be found in Brabant in the southern part of the Netherlands and the Kempen in the northern part of Belgium. River terraces of this period have been traced in the border zone of the Netherlands next to Belgium and Germany [see Staalduinen *et al.* (1979), Quitzow (1974)].

At the end of the Cromerian (about 400 000 years ago), an important shift in the pattern of fluvial sedimentation took place, as eastern and north-eastern rivers were no longer present in the Netherlands. The Rhine and Meuse formed a huge delta covering the greater part of the Netherlands. From this time onwards, the Rhine has followed a course similar to its present northerly one in the Lower Rhine district of Germany. In the Holsteinian, the coastline ran more or less east–west, parallel to the present northern shoreline of the Netherlands. The Rhine and Meuse, flowing to the north-west, built a fan of moderate dimensions. During the Saalian glacial, the Fennoscandian ice covered northern Europe again, as in several previous glacial periods in the Quaternary. Now, for the first time, parts of the fluvial landscape of the Rhine and Meuse were covered by the ice. This event had a profound influence on both the sedimentation patterns and the morphology of the landscape, with the Rhine and Meuse being forced into westerly courses. The hills in the central and eastern part of the fluvial landscape were formed by the ice sheet. This moved river sediments into ice-pushed ridges and hills up to 100 m high (e.g., the Veluwe area in the central Netherlands; *see* Fig. 8.10). In the Eemian interglacial, melting of the ice resulted in the

sea level rising, with the sea penetrating farther to the east. The Rhine resumed its northern course towards the Zuider Zee, while the Meuse had probably much the same course as it has at present. In the Weichselian, the ice sheet failed to reach the fluvial Rhine/Meuse area that remained in the periglacial zone. Terraces and terrace plains of both rivers were formed especially at the end of the Weichselian glacial. During this period, nearly all forms outside the zone of the main rivers were covered by aeolian deposits—the so-called cover sands. These well-sorted sands, intercalated by thin loamy layers and with a horizontal lamination, were probably transported by drifting snow. In the southern part of the Netherlands and adjacent parts of Belgium and Germany, a thick layer of loess was laid down over the fluvial terraces.

8.4 English Lowlands

8.4.1 Scarplands of South and East England

At the surface, this is a region of comparatively simple structure where alternating beds of sandstone, clay, chalk and other limestones of Jurassic, Cretaceous and Eogene age have been thrown into a series of mainly gentle folds (*see* Fig. 8.11). Differential erosion has carved a pattern of scarps and vales out of these sediments. The largest drainage basin, that of the Thames with an area of 15 000 km², lies wholly within the region, while farther north an area of equal size is drained by several separate rivers, including the Great Ouse and the Nene, into the Wash. The higher ground tends to correspond with two main belts of rock, the Middle Jurassic limestones and sandstones, and the Chalk. In contrast to the escarpments and uplands, there are considerable areas of land near to sea level. The Fens, which cover more than 3000 km² and which lie mostly below 15 m and in places are just below mean sea level, were formed from postglacial silts and peats. Similar but smaller areas of low-lying flat land

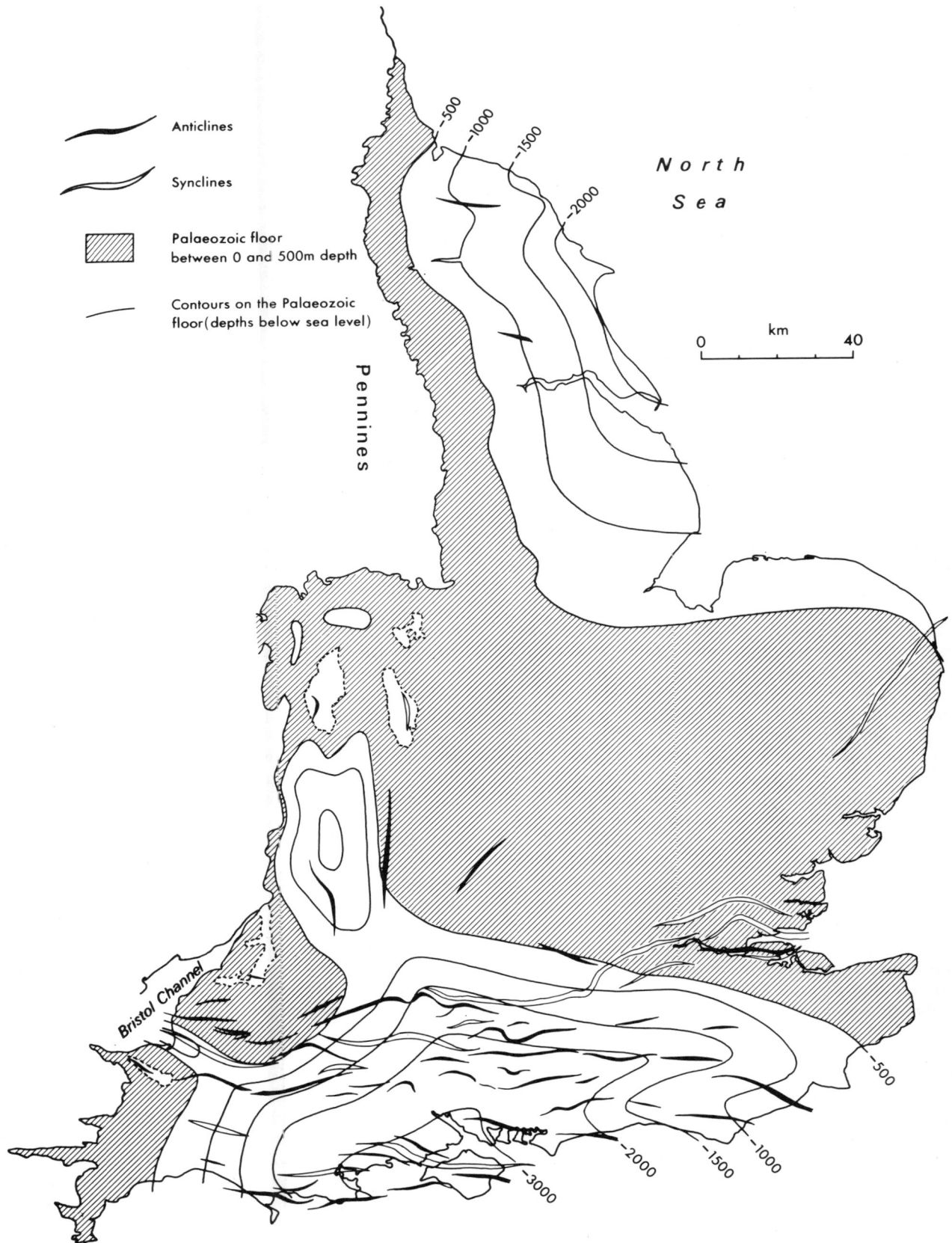

Fig. 8.11 Structural features of southern and eastern England [after Wooldridge and Linton (1955)].

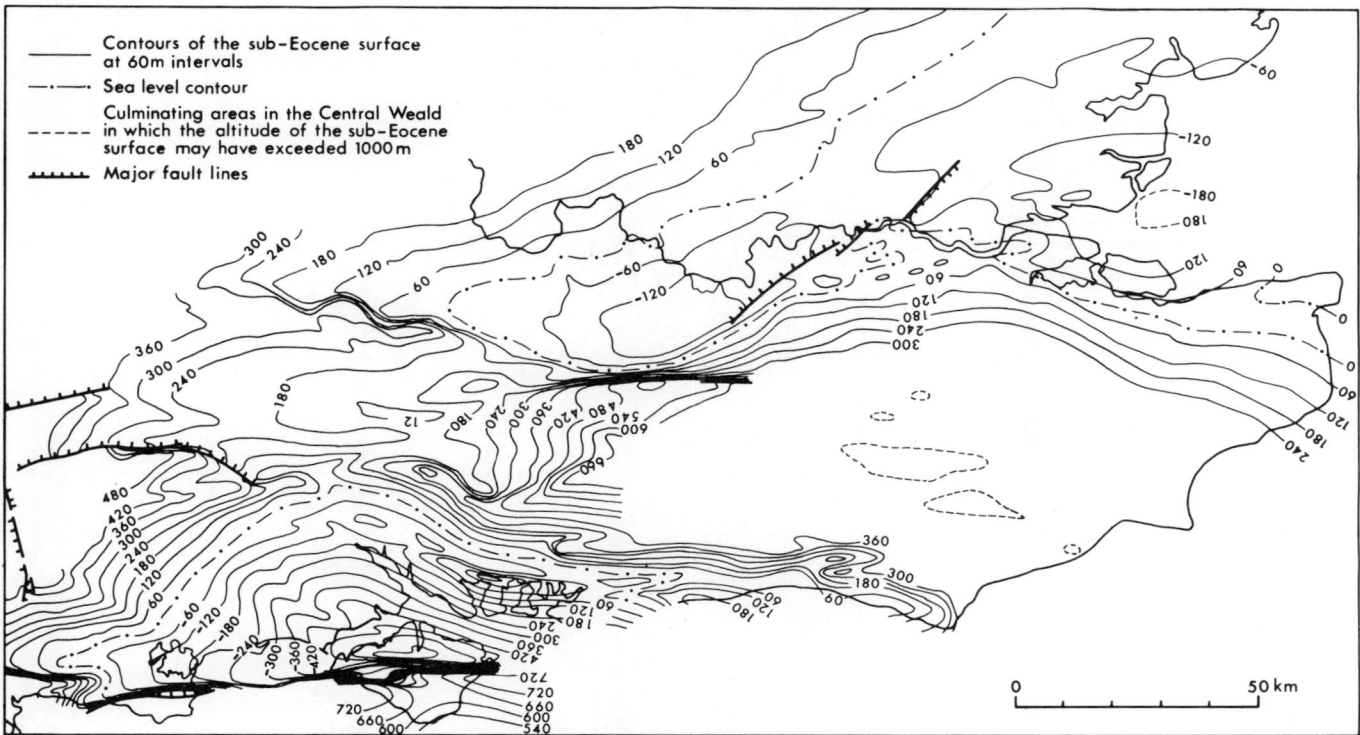

Fig. 8.12 The sub-Eocene surface in south-east England [Wooldridge and Linton (1955)].

occur in the Somerset Levels in the west of the region and in some coastal areas in south-eastern England.

The Mesozoic rocks of south-eastern England have accumulated on a platform of Palaeozoic strata. The Palaeozoic floor at its shallowest lies at a depth of only −110 m—at Cambridge—and, beneath most of the Midlands, East Anglia and the London Basin, it does not descend below −500 m (*see* Fig. 8.11). South of the Thames and in Wessex, however, the floor plunges steeply to depths of greater than −3 km in the Isle of Wight area. Likewise, in eastern England from the Wash to the Tees, the floor falls away towards the North Sea basin, being encountered near Scarborough, for instance, at a depth of about −2 km. The platform has exerted an important control on the deformation of the cover. North of the Thames where the Palaeozoic floor lies near the surface, the cover rocks were protected from large-scale deformation, displaying gentle dips and few folds, but in the Hampshire Basin and the southern Weald, the cover is thrown into quite sharp east–west folds, some exceeding 400 m in amplitude, by the Miocene orogeny.

The cover rocks provide a sequence from Jurassic to Eocene. It is impossible to identify consistent scarp-forming horizons in the Jurassic throughout the region as conditions varied so much from one basin of deposition to another. The Middle Jurassic (Great Oolite and Inferior Oolite) is responsible for major escarpments in the Cotswolds [e.g., Cleeve Cloud (330 m)] and the North York Moors (up to 454 m), the Middle Lias Marlstone in the Lower Jurassic is important from southern Lincolnshire to Somerset, and Portland stone and Purbeck limestone (Upper Jurassic) are significant scarp formers in Dorset. Clays and sands of the Lower and Upper Lias, and the Oxford and Kimmeridge Clays (Upper Jurassic) form lowlands. In the Cretaceous, there is also great variability in lithology and thickness. Major escarpments are associated with the Lower Greensand (Aptian) in the Weald [e.g., Leith Hill (294 m)] but westwards into Dorset it thins and disappears. Similarly, although it forms a marked escarpment in Bedfordshire, to the north it fades out to only a thin bed in Lincolnshire. The Chalk, which is up to 450 m thick in southern England, forms widespread uplands and escarpments [e.g., Hampshire Downs (297 m), Chiltern Hills (254 m), South Downs (271 m)] largely because of its permeability. It becomes thinner and sandier in Dorset and Devon but, in Lincolnshire and Yorkshire, it retains its massive character in the Wolds (up to 246 m). In places, notably the South Downs, there is a double escarpment associated with the Chalk. Lowlands are formed on the Weald Clay in the south, and on the Gault Clay (Albian) from the Weald as far north as Bedfordshire.

In later Senonian times, the Chalk Sea shallowed, leading finally to the emergence of the whole of south-eastern England (and of the rest of Britain too); hundreds of metres of sediments were eroded, producing the sub-Eocene planation surface (*see* Fig. 8.12). Parts of this surface have since been exposed as a result of the stripping off of later (Eogene) deposits and now form a significant element in the present relief [see Pinchemel (1954)]. Sedimentation in the Tertiary Period commenced in two shallow basins: the London and the Hampshire Basins.

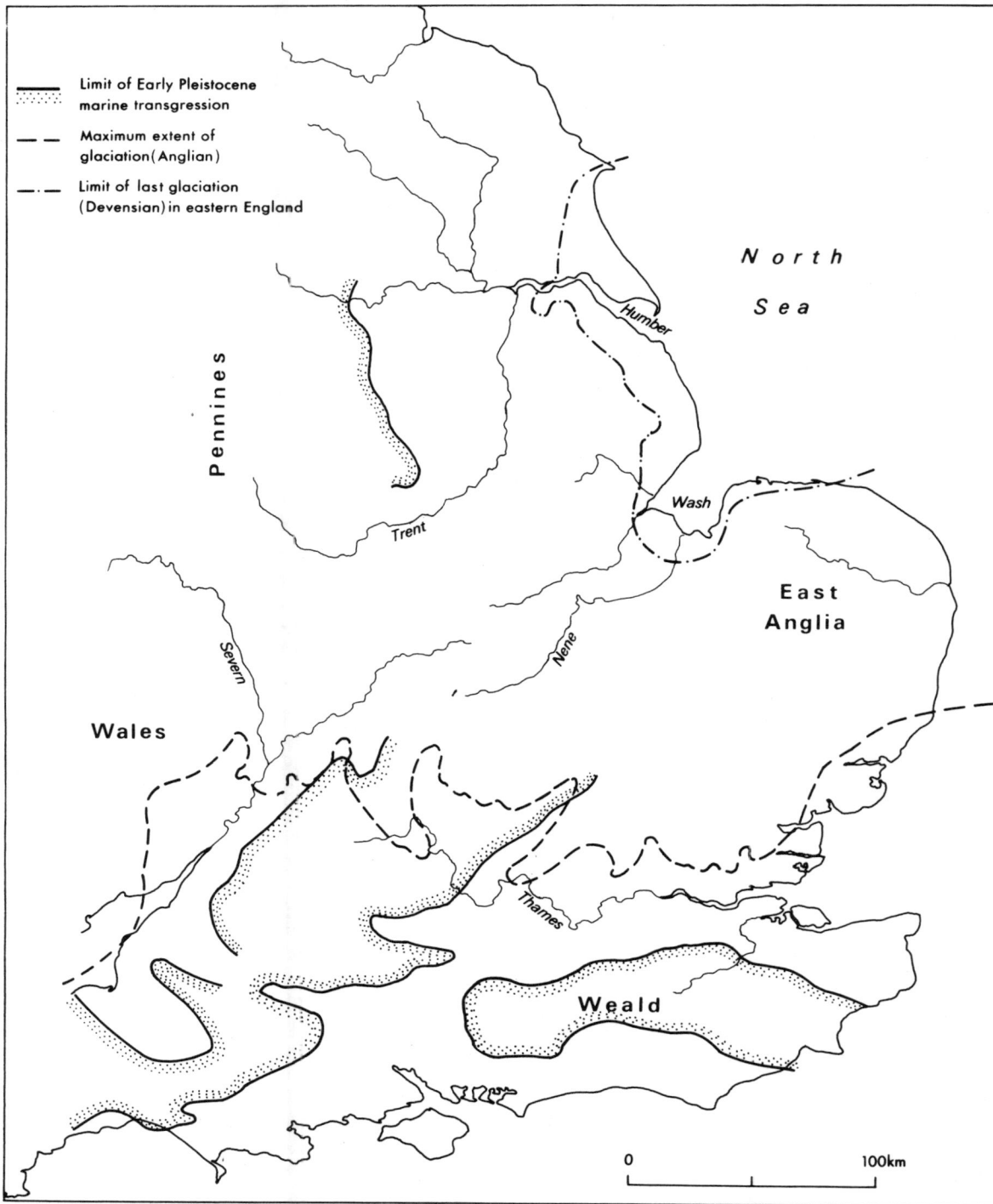

Fig. 8.13 The early Pleistocene marine transgression in England and the major glacial limits.

Sands, gravels and clays typify cycles of alternating marine and continental conditions. The coarser sediments, such as the Bagshot Beds (Upper Eocene), tend to form rising ground, while the London Clay forms a lowland. Crustal instability appears to have continued from pre-Eocene times and to have become more marked in the Oligocene, heralding the increased tectonic activity of Miocene times—the Alpine orogeny.

8.4.1.1 Neogene landscape evolution

Wooldridge and Linton (1955) identified remnants of a subaerial planation surface of early Neogene age in southern England, being recognizable over some 5000 km² and lying generally between 220 and 280 m. It appears to post-date the Alpine orogeny and not to have been significantly deformed by any later earth movements. Wide areas of this peneplain are mantled with residual deposits, especially the clay-with-flints, a large part of

TABLE 8.1 Correlation of the Pleistocene in south-eastern England.

Recommended terminology [Mitchell *et al.* (1973)]	^{14}C date (BP)	Older terminology	Possible correlations with northern Europe
Flandrian	0–10 300	Postglacial	
Devensian (glacial)	*c.* 10 300–70 000	Newer drift, etc.	Weichselian
Ipswichian (interglacial)	>70 000		Eemian
Wolstonian (glacial)		Gipping (?)	Saale
Hoxnian (interglacial)		Great interglacial, etc.	Holsteinian
Anglian (glacial)	{ Lowestoft Till { Corton Beds { Cromer Tills	Lowestoft (and Gipping?)	Elster
Cromerian (interglacial, warm temperate)		Cromer Forest Beds	
Beestonian (cool)		(Upper Baventian to Cromerian)	
Pastonian (temperate)			
Baventian (cool)		Weybourn Crag (marine)	
Antian (temperate)			
Thurnian (cool)		Norwich Crag	
Ludhamian (temperate)			
Waltonian		Red Crag (marine)	
– – – – – – – – – – – – – – – – Base of Pleistocene –			
		Coralline Crag	

which was inherited in some areas from the disintegration of a once more widespread Eogene cover [see Loveday (1962)]. Green (1974) has shown that, following uplift and possibly slight tilting of this early Neogene surface, a new planation surface developed in Wessex at a slightly lower level.

The first post-Alpine deposits in southern Britain are found in the eastern part of East Anglia, where a sequence of Pliocene and early Pleistocene rocks up to 300 m thick reflects that area's participation in the continued slow subsidence of the North Sea basin. Farther west, sedimentary outliers of Pliocene–Pleistocene age have long been known, lying at elevations of about 200 m where undisturbed by piping or mass movement. Most occur in and around the London Basin. Their age ranges from Diestian (e.g., the Lenham Beds of Kent) to Waltonian or Red Crag (e.g., Netley Heath and near Berkhamsted). Both the fossil and sedimentary evidence testifies to a marine origin. Wooldridge and Linton (1955) showed that this late Pliocene–early Pleistocene marine transgression affected large areas of southern and eastern England (*see* Fig. 8.13), cutting a bench into the uplands at about 200 m. This feature, which is traceable as far west as Wessex and into Devon, varies little in altitude, suggesting considerable tectonic stability since its formation. More recently, some other workers have claimed that this bench is of structural origin [e.g., in north Kent, see John (1977)]. There is also disagreement about the extent of modification of the drainage pattern arising from the marine transgression. Wooldridge and Linton (1955) claimed that, after the withdrawal of the sea below 200 m, large-scale superimposition of the river system occurred, thereby explaining the considerable discordance between the river pattern and structure in the areas transgressed by the sea. Jones (1974, 1981) and others have argued, however, that the discordance is more readily explained in terms of antecedence—that the present river system first

evolved in the Eogene and survived the Miocene orogeny (and the Pliocene–Pleistocene marine transgression) with relatively little derangement.

8.4.1.2 *The Pleistocene*

Regression of the Pliocene–Pleistocene sea occurred in stages, leading to further valley excavation and the formation of terraces below 200 m. The best-known sequence of Lower Pleistocene rocks has been studied in the Ludham borehole in Norfolk [see West (1961)], and evidence has been produced both of a sequence of alternating warm and cool conditions and of a fluctuating sea level. By Beestonian times (*see* Table 8.1), the whole of England had become land, together with much of the southern North Sea, as the sea level fell to about −100 m. In the Beestonian, there is no definite evidence of glacial ice in the area, although permafrost structures in Beestonian strata at West Runton, Norfolk probably indicate that glaciation was not far distant. The succeeding Cromerian was a warm, humid phase, represented: for example, by a rubified sol lessivé in south-east Suffolk [see Rose and Allen (1977)]. The upper part of the Cromer Forest Bed series on the Norfolk coast shows a return to colder conditions, which heralded the onset of the earliest glaciation recorded in the area. This is the Anglian stage, in which ice from northern England and Scandinavia penetrated as far south as the London Basin, the Vales of St Albans and Aylesbury, and the edge of the Cotswolds (*see* Fig. 8.18). Subsequent glaciations were less extensive as the limits on the map show. In the Devensian—the last glaciation in Britain—ice failed to penetrate more than a small part of East Anglia (i.e. the north Norfolk coast) where the Hunstanton Till is probably the equivalent of the Drab Till of south Yorkshire, dating from about 17 000 BP. There are also a few Devensian constructional forms of glacial origin in this area of north Norfolk (e.g., the Hunstanton esker). In Lincolnshire, late Devensian ice (about 15 000–20 000 BP) pressed inland to the chalk Wolds, blocking the mouth of the Humber. North of the Humber, it surrounded the uplands of the Wolds and the North York Moors, impounding a lake in the Vale of Pickering (see below). In the areas of Lincolnshire and Yorkshire unaffected by this late Devensian ice, there are widespread deposits of one or more older glaciations, such as the so-called Basement Till. Much of this drift is probably Wolstonian (e.g., at Sewerby it underlies the Ipswichian marine beach) but the possibility that some tills may be Anglian cannot be dismissed. As in East Anglia, these older tills are dissected and often deeply weathered.

Outside the Devensian limits, and throughout the whole of the Pleistocene in southern England, periglacial conditions were associated with the cold phases, witnessed by such features as the head and combe rock deposits, the patterened ground in Norfolk [see Sparks *et al.* (1972)] and fossil ice wedges. The chronology of the glaciations [Mitchell *et al.* (1973)] is far from being agreed finally, so that the sequence and correlations shown in Table 8.1 are to be regarded, at least for the older stages, as only tentative. The main impediments are the paucity of interglacial deposits in many areas and the difficulty of establishing absolute dates, particularly because the Pleistocene deposits of the area lie beyond the range of radiocarbon dating. East Anglia is the most important region of Britain in connection with pre-Devensian stratigraphy, but even here numerous problems remain to be solved. For instance, the precise extent of the Wolstonian ice is now very uncertain [see Bristow and Cox (1973)] and the bulk of the drift appears to be Anglian in age. There are, however, permafrost structures in the deposits overlying the interglacial beds at Hoxne (Suffolk) and a suggestion of glaciotectonic disturbance at Mildenhall (Norfolk) which may be related to the undoubtedly extensive glaciation of Wolstonian age in the Midlands (*see* p.151).

The penetration of ice into the area had several major direct effects on the landscape. Glacial erosion has undoubtedly lowered the surface of the uplands that were overwhelmed by ice. In East Anglia, the chalk escarpment was overridden and partly demolished, its line being pushed several kilometres to the south-east [see Embleton and King (1975)]. Glacial deposition, however, has had far-reaching effects. Not only are wide areas, such as the plateau of East Anglia, thickly mantled with drift, but certain escarpments have been partly or wholly buried and some valleys obliterated. The western slopes of the southern Lincolnshire Wolds show a broad ramp of boulder clay as described by Linton (1954), and the Lower Greensand cuesta of Bedfordshire traced northwards becomes a boulder clay feature. Examples of drainage diversion abound. In Yorkshire, the former drainage route of the Vale of Pickering eastwards to the North Sea was blocked by ice (mainly Scandinavian) and drift, so that meltwater was forced to find an alternative escape going instead to the Vale of York and cutting the gorge at Kirkham Priory now occupied by the River Derwent [see Gregory (1965)]. The most complex and best-known case of drainage diversion is that of the Thames in the London Basin, where the abandonment of successive courses from the preglacial route through the Vale of St Albans to the present one is attributable either to actual blockage by ice or to the vast quantities of glaciofluvial debris that poured into the system down the Colne and Brent valleys. The sequence of events first described by Wooldridge and Linton (1955) has had to be modified in the light of more recent work by, for example, Green and MacGregor (1978) and Baker and Jones (1980).

During the Devensian, as in previous glacial periods, the lower Thames cut deeply in response to glacially lowered sea levels; the subsequent rise in sea level and associated sedimentation gave rise to the buried channels, which can be traced as far as Brentford. The first buried channel was eroded in the middle Devensian to a sea level of −70 m, with the second in the late Devensian to about −30 m. The earlier Pleistocene history of the Thames is recorded in the suites of river terraces, the uppermost and oldest suggesting a base level at about 120 m. Below this, at least eight terraces have been identified, including the Boyn Hill terrace—at one time traditionally linked with a Hoxnian interglacial sea level at about 30 m, but now considered late Wolstonian by some. The lowest Taplow terrace, standing about 8 m above the flood plain, just antedates the Ipswichian interglacial according to the Trafalgar Square (London) section. The exact chronological succession is in dispute; it may be that a higher terrace is not always older than the next lower one. In the absence of datable material, other than faunal or archaeological—which may be derived—it may never be possible to unravel the succession. Terrace sequences are present in other valley systems: for instance, Everard (1954) provided evidence of base levels at 70, 55, 50, 35, 25 and 15 m in the Solent area of Hampshire.

8.4.1.3   *The Holocene*

In the postglacial (Flandrian) sea-level rise, the lower parts of several river valleys in England and Wales were drowned, following which the present alluvial infills accumulated. At this time, the development of the Fens began. Fenland shows two main groups of freshwater peat. Towards the sea, the peat is divided into upper and lower layers (about 3000–4000 and 4800 BP, respectively) by an estuarine deposit—the Fen Clay. Drying out of the peat later caused surface sinking, which is responsible for the problems of drainage in this area today; the level of the silt fen near the coast of the Wash is often higher than that of the peat fen inland. Another interesting area of peat accumulation is in the lower parts of the Yare and Bure valleys in Norfolk. Here, peat digging by man in medieval times and earlier has created the lakes known as the Broads.

8.4.1.4   *South-eastern England coast*

The coastal outline shows a close correlation with geological structure. Differential erosion of the Jurassic and Cretaceous strata has produced a series of bays and headlands along the south coast (*see* Fig. 8.14). The Chalk outcrop from south Dorset to the Isle of Wight was

Fig. 8.14 Stair Hole near Lulworth, Dorset; present marine action is breaching a barrier of Portland Stone (Upper Jurassic) and attacking less resistant Cretaceous beds on the inner side. (Embleton)

formerly continuous, providing the southern edge of a Frome/Solent river system, but the Flandrian submergence followed by wave action breached the ridge, so that the Chalk cliffs ending in the headland north of Swanage (*see* Fig. 8.15) now face the corresponding feature in the Isle of Wight (The Needles) 30 km across the sea. Coastal erosion is rapid in the soft rocks such as the Pliocene Crags of East Anglia and the glacial deposits of Holderness; there are prime examples of coastal landslipping at Folkestone Warren in the Chalk and Gault Clay, and on the Dorset coast in the Cretaceous and Jurassic. There are

also major depositional structures, notably the shingle foreland of Dungeness, the tombolo of Chesil Beach and the shingle spits of Orford Ness and Spurn Head. Drowned valleys and estuaries on the one hand and raised beaches on the other show that relative movements of land and sea level have occurred, as in the Goodwood raised beach (30 m) in Sussex (probably Hoxnian age), or the Ipswichian deposits of Selsey and of Black Rock (Brighton) at about 7.5 m [see West and Sparks (1960)]. Study of tide gauge, archaeological and historical data shows that southern and eastern England are at present sinking

Fig. 8.15 The Chalk cliffs and stacks near Old Harry Rock, near Swanage, Dorset. (Embleton)

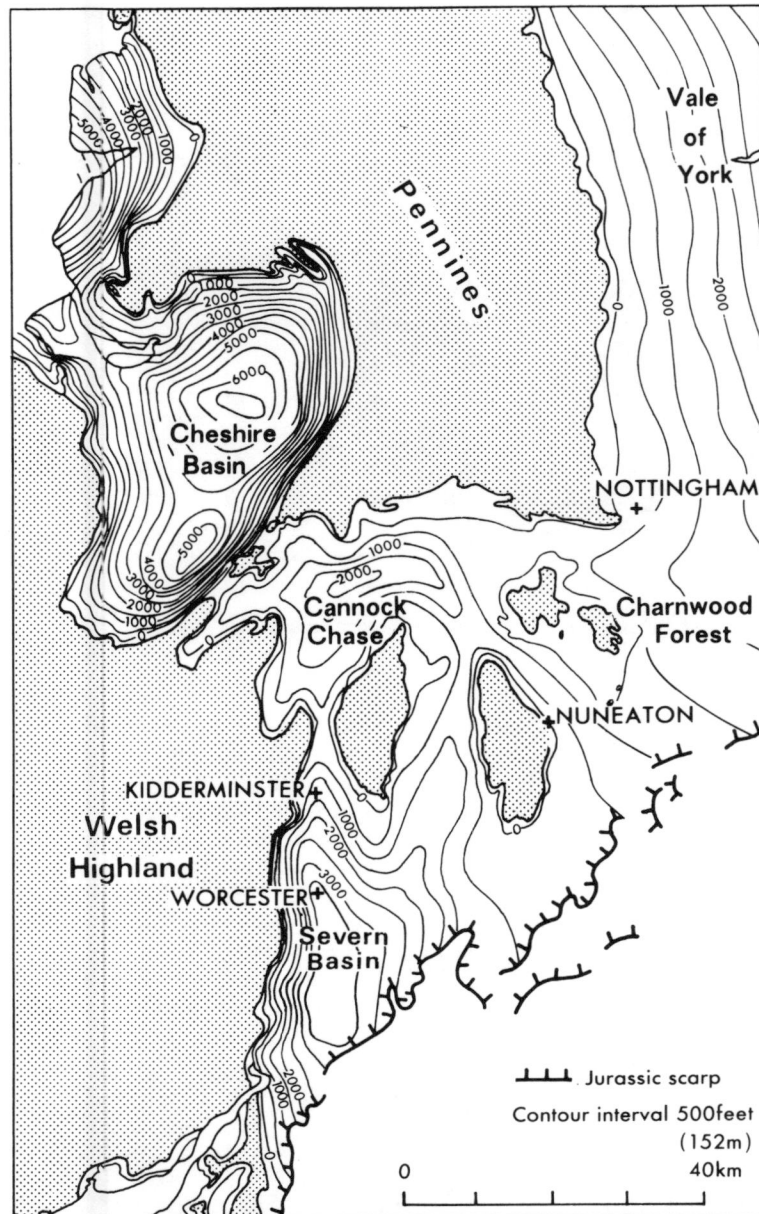

Fig. 8.16 The buried pre-Permian surface in central England north-west of the Jurassic escarpment [after Kent (1949)].

relative to sea level [e.g., Valentin (1953)], the rate for Felixstowe on the Suffolk coast being about 2 mm per year. At least two factors are involved in this: eustatic rise in sea level and continued crustal downwarping on the margins of the North Sea basin.

8.4.2 Lowlands of Central and Northern England

West of the Jurassic escarpment, a Y-shaped lowland extends as far as the Welsh border and encloses the southern part of the Pennines. Beneath a widespread cover of glacial and glaciofluvial deposits, nearly everywhere the subsurface rocks belong to the Permo-Triassic. Except for the pre-Permian horsts of the Midlands, the

surface rarely rises above 150 m, but it carries major drainage systems diverging to the North Sea, the Irish Sea and the Bristol Channel.

Fig. 8.16 shows the form of the buried pre-Permian surface. The bulk of the Permo-Triassic rocks lies in deep basins which represented tectonic depressions whose floors subsided contemporaneously with the accumulation of sediments (2 km thick in the case of the Cheshire Basin). The Permo-Triassic sequence in Britain consists of a continental facies related to a desert environment and a more limited marine series east of the Pennines where, in the Permian, deposits of limestone and evaporites were laid near the margin of the Zechstein Sea. The Magnesian

Limestone of the Permian (mostly dolomitized and up to 500 m thick) forms a bold cuesta whose scarp faces west, overlooking the rising ground of the Coal Measures on the Pennine flanks. The dip slope sinks gently east beneath the Upper Permian and Triassic sandstones that floor the Vales of York and the lower Trent. Traced southwards, the Zechstein deposits become thinner and disappear around Nottingham, being replaced by continental deposits that cannot be clearly distinguished from the Bunter.

In the Midlands and Cheshire, Permian and Trias belong to one sequence of continental deposits in which Keuper Marl rests on Keuper and Bunter sandstones; there is no equivelent of the European Muschelkalk. The bulk of the sandstones represent aeolian deposits, often displaying fine dune bedding. The marls are probably largely playa lake sediments. At the base of the Bunter, breccias represent ancient scree deposits that accumulated at the foot of steep slopes, with conglomerates reflecting the occasional torrential stream action of a typical semi-arid climate. The variations in lithology have an important effect on the detail of the present land surface. The Middle Bunter Pebble Beds (conglomerates) often form more elevated areas, such as Thurstaston Heath in the Wirral or Cannock Chase in the Midlands. The marls and softer sandstones form lower-lying ground, while the Keuper sandstones often form minor escarpments (e.g., the edge of the Helsby Hills in Cheshire or the north–south trending cuesta in Nottinghamshire). The Keuper Marl of Cheshire contains massive salt beds; underground extraction of brine for industry has caused surface subsidence and the development of numerous lakes or meres.

In the bordering uplands, erosion prevailed in Permo-Triassic times, removing great thicknesses of Palaeozoic and older strata. Grooves etched into the Precambrian rocks of Charnwood Forest may be the result of Permo-Triassic wind abrasion—although Pleistocene wind or glacial action has also been suggested, at least above the level of the Triassic [see Sparks (1971)]. Other forms of arid erosion and mass wasting were active, and many geomorphologists consider that the Permo-Triassic period was of immense importance in roughing out some of the outlines of the major relief units in present-day Britain. The present Permo-Triassic sequence is only a denuded remnant of an originally thicker, more extensive formation. It is possible that at one time the Carboniferous horsts of the Midlands and the Precambrian of Charnwood Forest were entirely concealed beneath Trias, if not beneath still younger Jurassic or Cretaceous strata, and likewise the Triassic rock must originally have climbed to greater heights on the flanks of the Pennines and Wales.

Hercynian movements continued into the Permian and throughout Triassic times, although reduced in magnitude after the early Bunter. Block faulting was widespread, forming horsts such as those of Worcester–Kidderminster (later to become a graben), west Warwickshire and Nuneaton–Charnwood. Post-Triassic faulting is also evident; many older fractures exhibit renewed post-Triassic displacement, amounting to 1 km on the west of the Lancashire coalfield.

The post-Triassic pre-Quaternary denudational history of the area is largely unrecorded. Jurassic and Cretaceous strata may once have covered most, if not all, of the region and their removal, except for the small outlier of Lias in Shropshire, must have been accomplished in the Tertiary. Although glacial erosion in the Pleistocene helped to lower the surface, it seems likely that the first ice advanced into an area whose broad outlines were not very different from those of the present. Fig. 8.17 attempts to reconstruct the main lines of preglacial, possibly late Tertiary, drainage. At an earlier stage, it is possible that the Welsh Dee continued eastwards to the middle Trent [see Linton (1951b)] and that this in turn utilized the Ancaster Gap through the Jurassic upland still farther east.

8.4.2.1 The Quaternary

No Lower Pleistocene deposits are known in the area. The oldest till appears to be of Anglian age and much of this has been removed by erosion or reincorporated in later tills. Investigation of the subdrift surface in the Midlands by Shotton (1953) has shown that, up to the time of the advance of the Wolstonian ice, a major valley drained north-east to the Trent (*see* Fig. 8.14) along the outcrop of the Keuper and Lower Lias, in a direction opposite to that of the present-day Avon, which now occupies part of this vale. The pre-Wolstonian 'proto-Soar' valley has now been mostly obliterated by a thick drift infill. The ancestral River Severn at this time rose near Birmingham—in the present Stour valley—and flowed to the Bristol Channel; Ironbridge Gorge had not yet been opened and the present-day upper Severn from Wales at this time joined the Dee, crossing the Cheshire plain to the Dee estuary.

Fig. 8.18 shows the maximum limits attained by ice. The exact location of the line is still contentious and the dating of specific parts of the line as Anglian or Wolstonian is uncertain. That there were two such glaciations reaching rather similar limits is, however, borne out by the existence of interglacial sites of Hoxnian age in the Birmingham area. Pollen analysis has been the most important technique enabling the deposits to be assigned to the Hoxnian; they clearly separate older underlying (Anglian) till from later (Wolstonian) overlying till [see Shotton and Osborne (1965)]. Wolstonian ice impounded a lake, named Lake Harrison, in the proto-Soar valley which overflowed the Jurassic escarpment to the south-east at a height of about 140 m [see Bishop (1958)], implying that there was ice at least to this height in the lower Severn valley in the Wolstonian. With melting of the ice, the lake was able to drain south-west to the Bristol Channel, thus initiating the present-day lower Avon. Terraces in this valley fall into five sets spanning some 40 m in altitude [see Tomlinson (1925)] and range from late Wolstonian to late Devensian.

In the Devensian, ice entered the Midlands from Wales on the west, while ice from the Lake District and Scotland pushed south through the Cheshire Gap, abutting against the southern Pennines. Devensian 'Irish Sea' till stained predominantly red with the incorporation of Triassic debris reached to just south of Wolverhampton (*see* Fig.

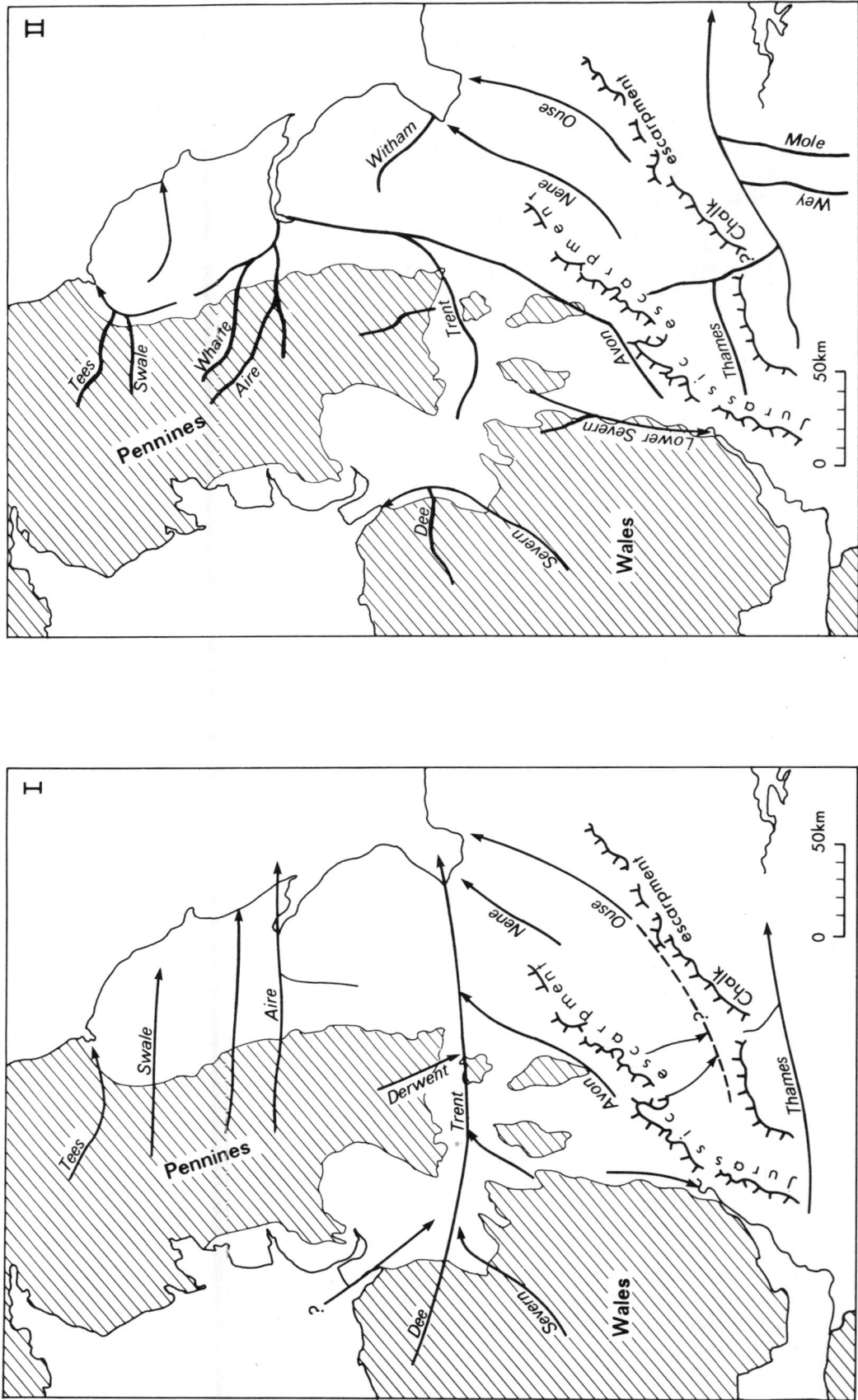

Fig. 8.17 Evolution of the drainage pattern in central England. (I) Possible initial drainage. (II) Drainage pattern prior to glaciation.

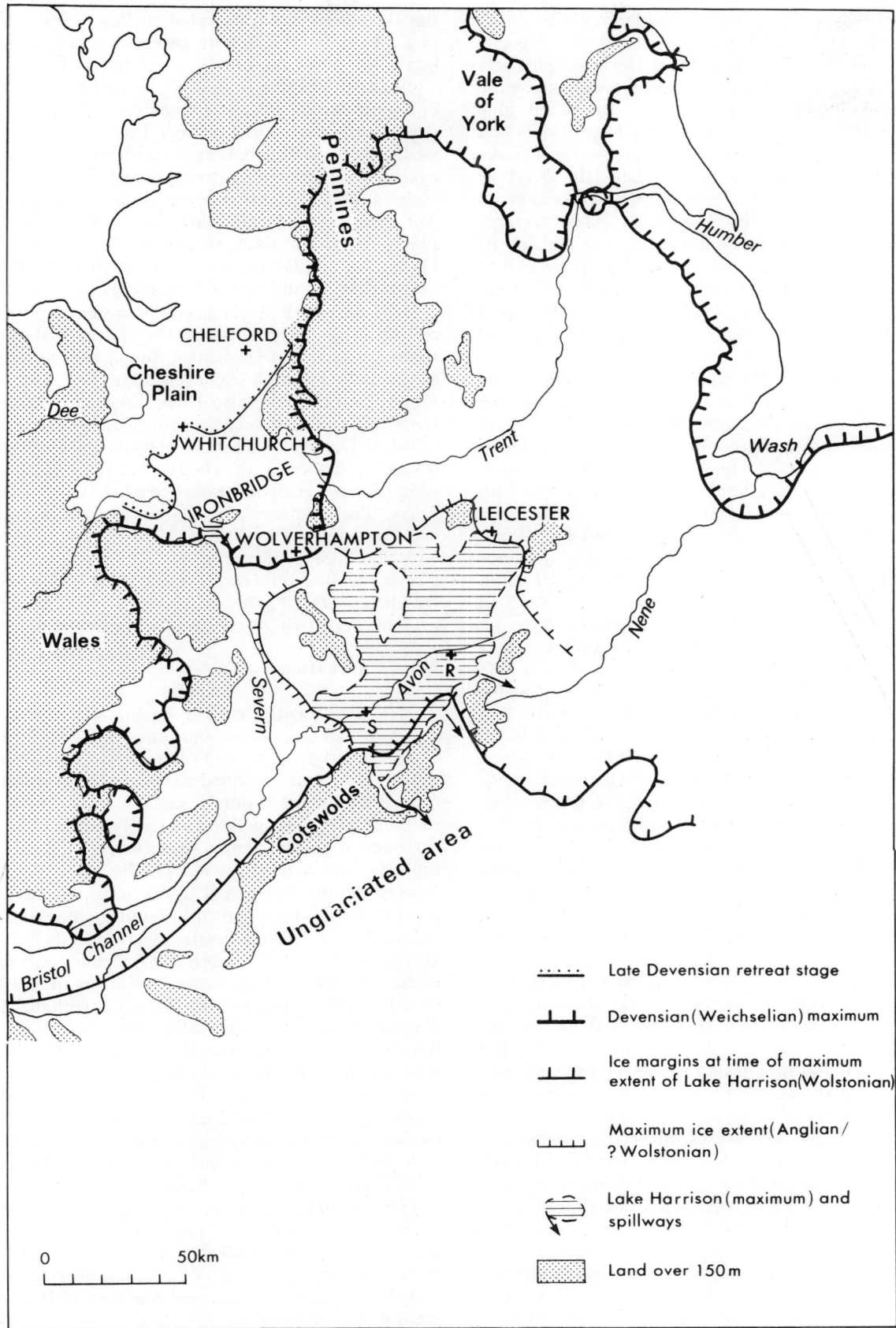

Fig. 8.18 Main glacial stages in central England.

8.18). Nearby at Four Ashes, an important site shows Devensian till overlying Ipswichian interglacial deposits [see Morgan (1973)]. It appears that Devensian ice was the cause of the diversion of the upper Severn across the Wenlock Edge–Wrekin ridge to form the Ironbridge Gorge. Wills (1924, 1937) postulated the impounding of a series of lakes, which later coalesced to form Lake Lapworth, by ice in the Cheshire plain, the overflow cutting the gorge. The terraces of the lower Severn corroborate this hypothesis. The three oldest terraces antedate the opening of the Ironbridge Gorge and can be traced only from the lower Severn up the Stour valley. After the gorge was opened, further terraces were formed, the oldest 'main' terrace [see Wills (1937)], beginning at the exit of the gorge itself and dating from about 25 000 BP.

In the lowlands of Cheshire and Lancashire, the Pleistocene deposits are almost entirely Devensian. An older, rarely seen basal till may be Wolstonian. The rest of the Pleistocene deposits represent a single complex glaciation of Devensian age, although the details of the chronology and ice limits have still to be worked out. An important interstadial site is found at Chelford where the Chelford Sands Formation has been interpreted by Worsley (1966) as an extensive alluvial deposit that accumulated in permafrost conditions. Within the Sands is an organic bed dated using radiocarbon to about 60 000 BP and providing evidence of a cool–temperate forest environment, possibly correlating with the Brørup interval of the Danish Quaternary. Above the Chelford Sands Formation lies a series of middle and late Devensian deposits. Evidence from Upton Warren in Worcestershire [see Coope *et al.* (1961)] and from Four Ashes point to an interstadial (warm, but treeless) around 40 000 BP, which was followed by the main Devensian glaciation when ice advanced to the Bridgnorth–Wolverhampton line (*see* Fig. 8.18). Farther north (in Cheshire and Shropshire) another Devensian limit has been identified, the Ellesmere–Whitchurch–Woore moraine. Most authors have argued that this represents a late Devensian readvance [e.g., see Yates and Moseley (1967)] but, in the view of Poole and Whiteman (1961), it is an older feature overridden by later ice.

In the Vale of York, ice entered from both the Pennines to the west and from the north (i.e. the Tees valley). Surveys of the subdrift surface show that the divide between the Tees and Ouse drainage once lay farther south around Thirsk, the Swale then being a tributary of the Tees. Fig. 8.18 shows the extent of penetration of Devensian ice. South of this limit, older deeply weathered tills are found. These tills are of uncertain age but are probably Wolstonian. There has been dispute about the position of the Devensian limit. It was held formerly that the York and Escrick ridges ('moraines') marked successive positions of this, but more recent work by Gaunt (1976) suggests that these ridges are not terminal or even ice-frontal, for Devensian deposits exist much farther south and the ridges provide evidence of flow till or melt-out till in their composition. They were probably a product of widespread ice stagnation. At its maximum,

the Devensian ice impounded shallow lakes to the south (e.g., 'Lake Humber') with successive levels of 52, 30 and 8 m when the Humber outlet was blocked.

Outside the Devensian limit in the Midlands and Vale of Trent, periglacial phenomena are evident. Gelifluction deposits may be up to 3 m thick. Involutions and fossil ice-wedge patterns date mainly from the Devensian, but older examples are known. After the Devensian glaciation in the Vale of York, extensive cover sands accumulated (up to 5 m thick), often forming low dunes. Radiocarbon dating places them at about 10 500 BP (i.e. mainly younger Dryas). A similar deposit in north-west England is the Shirdley Hill Sand (up to 3 m thick).

Deep buried channels mark the present or former lower courses of the large rivers as they approach the coast. For the most part, these represent fluvial incision in response to glacially lowered sea levels, but the great depths of some channels have also been taken as possible indications of direct glacial overdeepening [see Gresswell (1964)]. The bedrock surface beneath the Dee estuary, for instance, descends to about −120 m, which may be somewhat lower than the depressed sea level of Devensian times. The events of the Flandrian in Lancashire have been investigated by Tooley (1974). The rising sea level with the ending of glaciation is recorded in submerged forests and intertidal peats, dated from 12 000 to 4500 BP. By about 6000 BP, the sea level was close to that of the present, even rising to +1.8 m at one stage.

8.5   Paris Basin

This symmetrically shaped basin, with its distinctive curving rock outcrops and cuestas of Mesozoic and Cenozoic sedimentary rocks (*see* Fig. 8.19), represents one of the most clearly defined morphostructural units of the central European lowlands. On the north, it stands open to the English Channel, but on its other sides it is virtually surrounded by the gently rising flanks of the Hercynian uplands (the Armorican massif, the Massif Central, the Vosges and the Ardennes). In the north, the Paris Basin is separated from the Flemish lowland by the Artois anticlinal uplift while, in the north-east, it extends in triangular fashion between the Vosges, the Ardennes and the Rhine highlands. Between the Vosges and Morvan, to the southeast, its surface rises to 500–600 m to overlook the downfaulted rift of the Saône valley. On the south-west, the transition to the Aquitaine Basin (*see* Section 8.6) is more gradual, marked only by the low Poitou sill (100–200 m). Over the whole area of the Paris Basin, elevations are mostly less than 200 m, and no part exceeds 600 m. It consists of plains and a series of low plateaus, broken by valleys of varying depth but always shallower than 150 m [see Cholley *et al.* (1956), Sommé (1977)].

The regional structure is relatively simple, comprising a sequence of horizontal or gently dipping conformable beds sometimes slightly folded or faulted. In all directions outwards from the centre of the basin, progressively older rock outcrops are encountered together with increases in ground altitude (*see* Fig. 8.20). In the centre lie Palaeogene beds, sequences of clays, marls, sands and sandstones, and

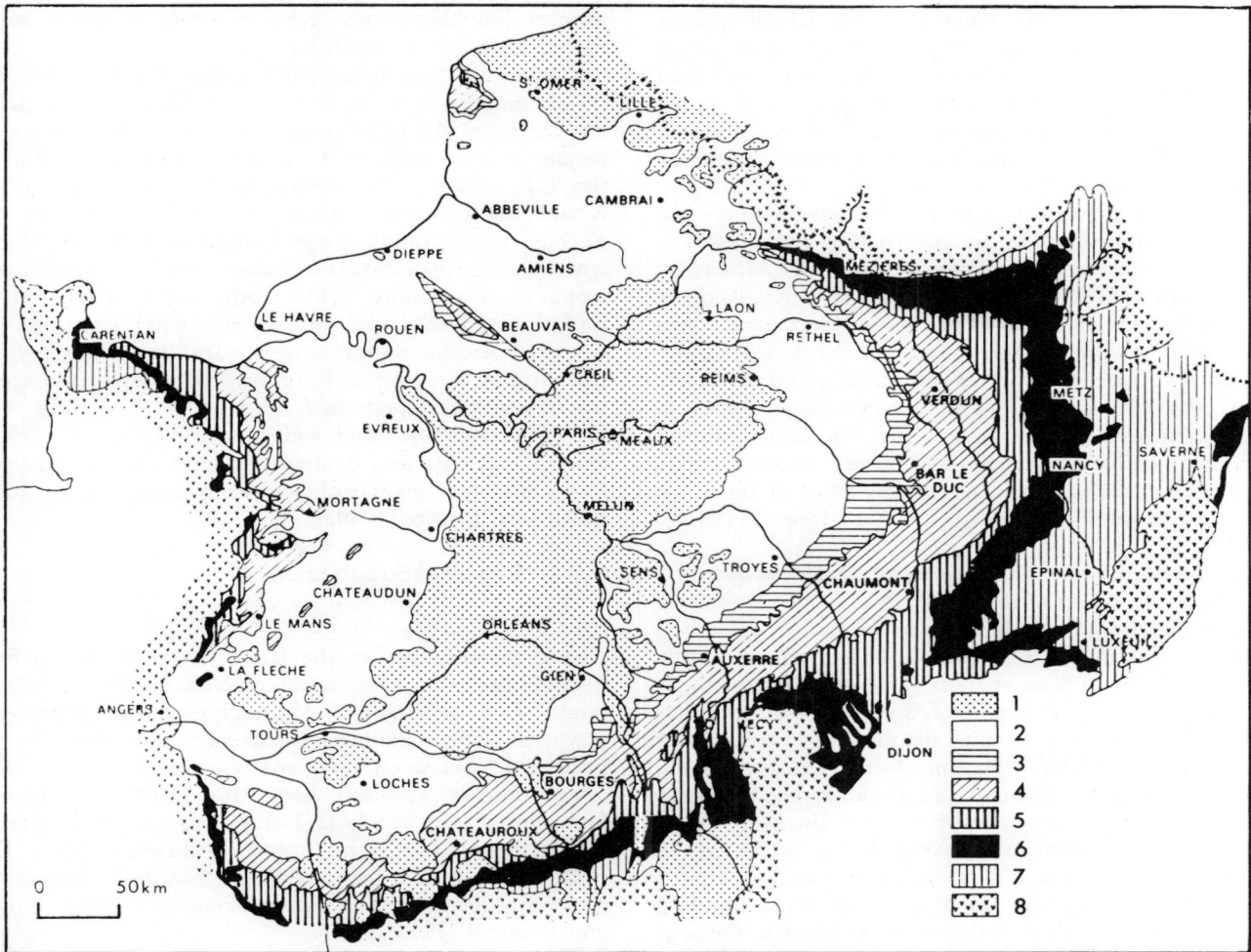

Fig. 8.19 Solid geology of the Paris Basin. (1) Tertiary. (2) Upper Cretaceous. (3) Lower Cretaceous. (4) Upper Jurassic. (5) Middle Jurassic. (6) Liassic. (7) Triassic. (8) Crystalline of Hercynian massifs.

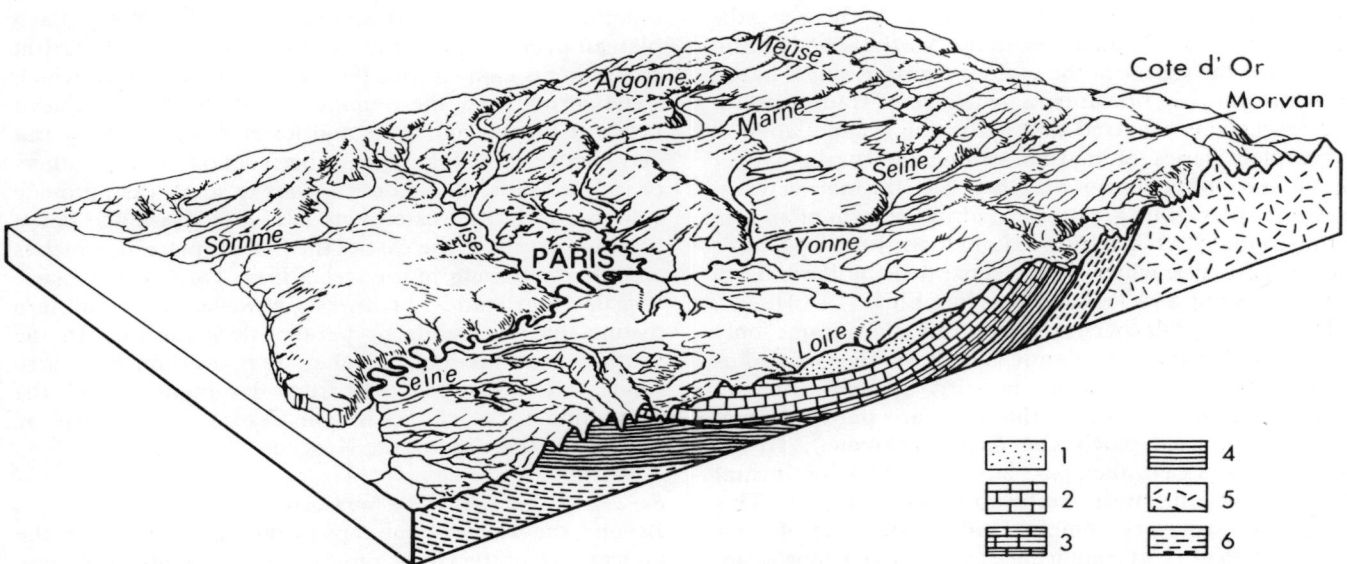

Fig. 8.20 Simplified block diagram of the Paris Basin, looking east. From Paris to the Seine estuary is about 180 km. (1) Tertiary. (2) Upper Cretaceous. (3) Lower Cretaceous. (4) Upper Jurassic. (5) Lower Jurassic and Triassic. (6) Palaeozoic.

of freshwater and marine limestones. The Chalk appears from underneath this Tertiary core, its outcrop shaped like a ring, in turn followed by another annular-shaped outcrop of Lower Cretaceous rocks. Finally, except in the north, there appear the Jurassic and Liassic limestones and marls, resting directly on the Hercynian crystalline massifs [see Sommé (1980)].

The geomorphological limits of the Paris Basin are not always so simple, for they are governed by denudation as well as by structure. At the contacts with the Armorican massif and the Massif Central, they more or less coincide with the maximum extent of the Mesozoic marine transgression but, in the north-east between the Ardennes and the Vosges, the limits are much more gradual. In the north, the Chalk dips under the Cenozoic rocks of Flanders beyond the Artois anticline. On the south-east, the Plateau de Langres terminates above the Saône trough in a striking fault scarp, the highest point of the Paris Basin as already noted [see Klein (1974), Dewolf (1982)].

8.5.1 Structure and Palaeogeographical Evolution

The deeper structure of the Paris Basin is now beginning to become better known thanks to geophysical exploration techniques and deep drilling for oil. It is characterized by a major depression of more than 3 km between the Marne and the Seine in the region of Brie. Several structural breaks are also known of which the main ones are a major curving fault traceable from Bray to the Vosges, passing south of Brie, and a faulted zone trending north-west to south-east from the lower Seine towards Beauce and the Loire between Orléans and Nevers. The latter structure, whose downthrow reaches 500 m in the southern part, divides the crystalline basement into two blocks standing at different levels: an uplifted Armorican block in the west and a depressed Parisian block to the east. It marks the uncompleted separation of the Precambrian basement into two plates [see Tricart (1949)].

The beginning of the main subsidence of the Paris Basin dates from the end of the Triassic Period. At this time, the Paris Basin was only an annex to the North German basin which was the source of the various transgressions. Later, from the Jurassic onwards, a sequence of transgressions and regressions occurred that led to the accumulation of a complete series of marine sediments. In the Upper Cretaceous, at the time of the Chalk Sea, the marine transgression attained its greatest extent in the form of an arm of the sea that stretched from Britain to the Jura. There was then a regression which finally separated the Paris Basin from the Alpine and Tethyan Seas [see Elhaï (1970)].

During the Palaeogene, marine invasions came only from the north and the Atlantic. The Basin then existed as a shallow gulf, periodically invaded by epicontinental seas. The marine facies of this time are partly detrital (sands, clays) and partly chemical (calcareous). To the south-east and east, they pass laterally into a continental facies of gypsum, marls, freshwater limestones, etc. This has produced a very complex and varied suite of sediments that is of great significance to the surface morphology. Continental conditions became dominant in the Neogene with either fluvial or geochemical types of

deposit [conglomerates, granitic sands or clays, sandstones and sarsens (meulières)].

During all these palaeogeographical changes, tectonic movements made themselves felt at different times, from the Liassic to the late Cretaceous, in the form of renewed displacements along faults, gentle warping and uplift of the Burgundy sill. Throughout the Mesozoic, they strongly affected the sedimentary history. Yet again in the Eocene, gentle uplift of the eastern margins contrasted with subsidence of the central basin. But in the Oligocene, tectonics were more active, with the rejuvenation or initiation of faults and flexures resulting in the termination of marine sedimentation in the central basin, the uplift of the margins (i.e. Poitou, Burgundy, Morvan and Vosges), some well-developed fold structures (i.e. Bray and Boulonnais–Artois) and some fault systems of varying depth. Tectonics and stratigraphy have thus both helped to diversify the geomorphological evolution of the Paris Basin [see Tricart (1949)].

8.5.2 Regional Subdivisions

8.5.2.1 *Central area*

The geomorphology of the Basin centre reflects a long history of selective erosion in horizontal or near-horizontal strata of varied lithology. Although elevations are mostly less than 200 m, Pliocene–Quaternary river incision has produced a relatively dissected relief (*see* Fig. 8.21). The alternation of resistant beds (limestones or siliceous formations) with sands and clays has greatly facilitated the etching out of structural plateaus between valleys. The overall arrangement is quite simple but in detail is much more complex [see Pomerol and Feugueur (1968), Diffre and Pomerol (1979)].

The most striking plateaus coincide with resistant beds dipping gently to the south. The Soissonnais, Valois and Multien plateaus are formed in Eocene limestone; Brie and Beauce plateaus have been developed in Oligocene limestones and siliceous sandstones (*see* Fig. 8.22). Each plateau overlooks lower ground on the north and east in escarpments capped with these harder rocks, below which softer formations (the Stampian sands of Fontainebleau between Beauce and Brie, and lower Tertiary sands and clays between the others) have been cut out. Often outliers or isolated hills, aligned north-west to south-east, provide evidence of former extensions of higher strata across the plateaus except in the case of the Beauce plateau, which is completely lacking in any relief from Paris to the Loire. On the north and north-west, however, it is cut into distinct narrow bands by parallel deep furrows. In the south (Orléanais and Sologne), it is overlain by coarse sands, alteration products from decomposition of the crystalline rocks of the Massif Central, distributed by streams at the close of the Neogene.

8.5.2.2 *Picardy and upper Normandy*

Beyond the Tertiary outcrop, to the north and west, the appearance of the Chalk produces a remarkable lithological and topographical contrast. In Picardy, tectonic movements continued throughout the Tertiary and almost

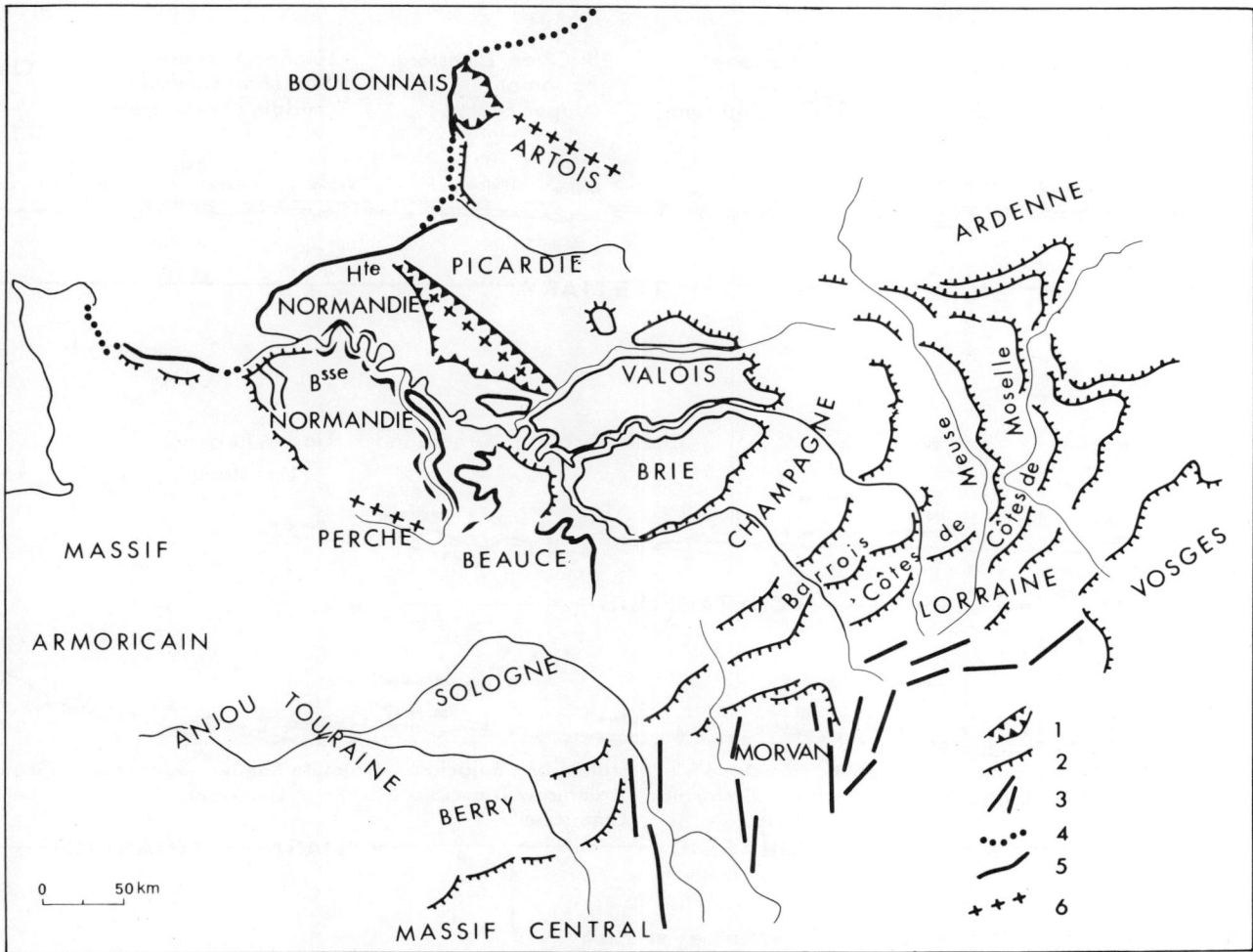

Fig. 8.21 Cuestas, coastal and fluvial features of the Paris Basin. (1) In-facing scarps of Chalk surrounding Lower Cretaceous inliers. (2) Outward-facing scarps of Tertiary and Mesozoic strata. (3) Faults. (4) Sandy and dune coasts. (5) River bluffs. (6) Anticlinal axes.

into the Quaternary giving rise to structures gently folded along north-west–south-east axes. The main relief and drainage lines are adjusted to these lineaments—the excavation of the Artois dome in the Boulonnais down to the Palaeozoic basement, the valley of the Somme and the 'buttonhole' of Bray (*see* Section 8.5.3.2).

Farther west, Normandy is characterized by dissected relief, with valley depths ranging from 30 to 150 m. The valleys of the Eure and its tributaries, as well as the lower Seine with its fine incised meanders, separate large plateaus (Thimerais, Ouche, Lieuvin, Roumois, Vexin and Caux). The flat surfaces of these plateaus are covered by clay-with-flints; a residual weathering deposit related both to disintegration of former overlying Tertiary beds and to solution of the Chalk. The solution products (i.e. carbonates and silicates) found their way during the Palaeogene into the central lake basins of the Parisian cuvette [see Elhaï (1963)].

In Boulonnais, the Artois dome ends in high cliffs at the coast. Other cliffs occur on the Normandy coast on both sides of the Seine estuary. The only significant break in this cliffed coastline is in Picardy: for example, in Mar-

quenterre and the Somme embayment, the coast is low, flanked by dunes and separated from an old cliffline inland by marshes and polders [see Delattre *et al.* (1973)].

8.5.2.3 Lower Normandy and Perche

At the contact with the Armorican massif, the border zone of the Paris Basin is more varied and less regular. Two subdivisions can be distinguished. In the north, in lower Normandy, the Auge cuesta overlooks the Dives valley. It marks the western edge of the Chalk plateau with its clay-with-flints cover. Below, the calcareous beds of the middle Jurassic, which dip gently eastward, form the Caen plain. This is itself the backslope of a cuesta with its west and south-west-facing crest looking across the Liassic clays of Bessin. The depression excavated along these clays marks the edge of the old crystalline massif. It extends towards the north-west up to the south-east of Cotentin, being interrupted only by the estuarine deposits of the Baie des Veys at the mouth of the Vire. Except around Veys and the mouths of the Orne and the Dives, the Normandy coast is lined with cliffs. The east coast of Cotentin, is low and fringed with dunes [see Doré *et al.* (1977)].

BURDIGALIAN

| Beauce Limestone Chattian / Stampian (upper Oligocene) | Brie Limestone Sannoisian (middle Oligocene) | St. Ouen Limestone Bartonian (upper Eocene) | Lutetian, Ypresian, Thanetian Limestone (middle / lower Eocene) | Chalk |

Sologne Beauce Hurepoix Seine Brie Marne Multien Valois Aisne Soissonnais Laon

TERTIARY

S   N

W   E

Palaeogene   White Chalk   Chalk Marl   Greensand   Upper Jurassic Limestone

Brie Côte de l'île de France Marne Champagne Champagne humide Argonne Barrois
Champagne pouilleuse

CRETACEOUS

Vosges

Côtes de Meuse   Côtes de Moselle

Argonne Barrois Meuse Woëvre Moselle

| Portlandian / Kimmeridgian | Oolitic Limestone Sequanian / Oxfordian | Bathonian / Bajocian Aalenian / Toarcian Limestone | Dolonite Shelly Limestone | Sandstone | Granite |

JURASSIC — **LIAS** — **TRIASSIC**

W   E

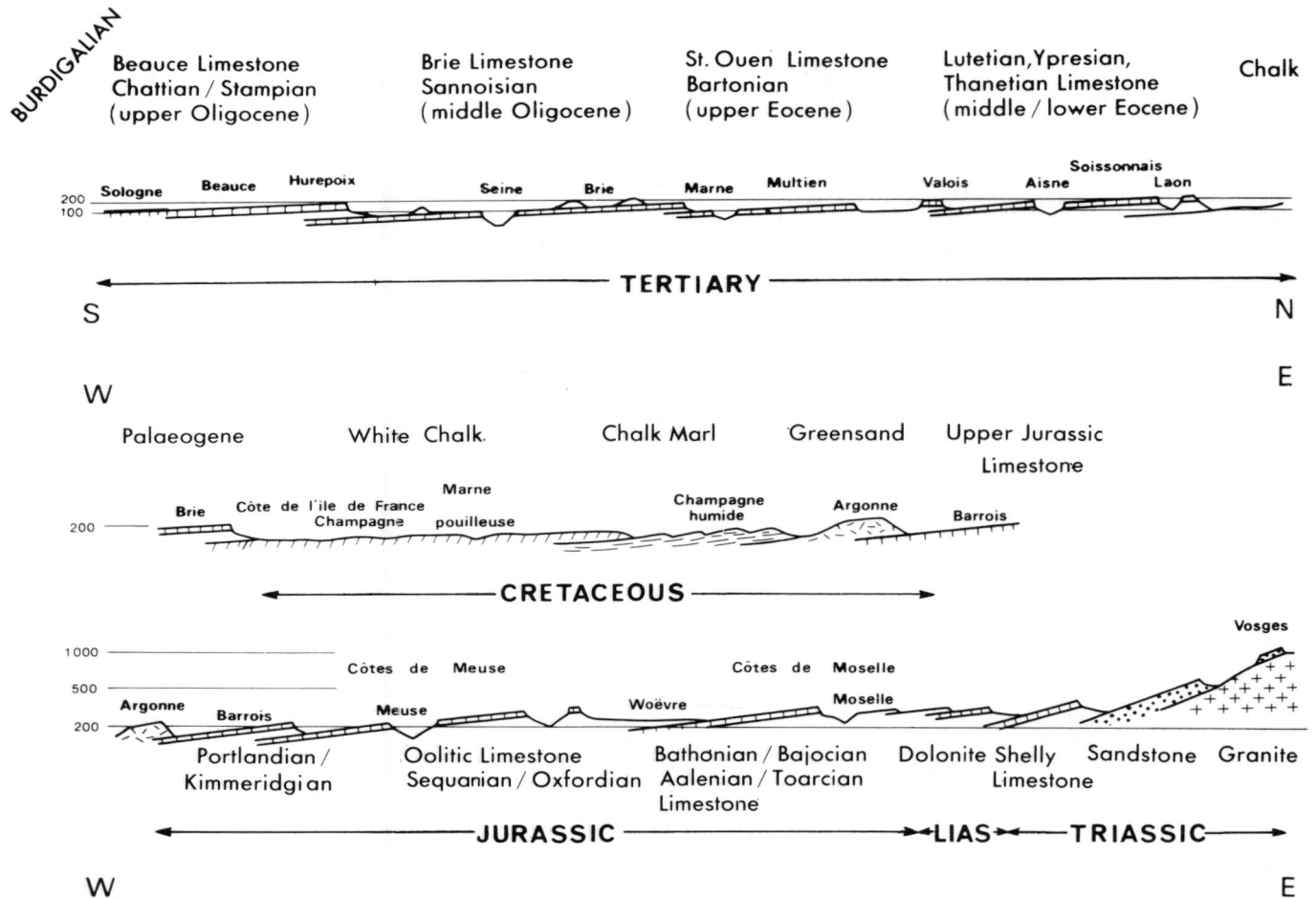

Fig. 8.22 Sections to show relations of geology and relief in the Paris Basin.

In the south, between Alençon and Angers, in Perche and Haut Maine, the edge of the crystalline massif runs from north-east to south-west, and the relief here is more complicated. One still finds segments of cuestas in Cretaceous and Jurassic strata, and remnants of Mesozoic formations are to be found far to the west (up to 200–300 m elevation) where they add another element to the residual Hercynian relief. But this overall picture is complicated by anticlines and faults of Armorican (west-north-west–east-south-east) trend, giving a landscape divided into blocks tilted to varying degrees. Small islands of the ancient massif rise above the rifts in which Lias and other Jurassic sediments are preserved. Most of the river courses have undergone superimposition, with valleys along which gorge-like stretches alternate with broader alluvial basins. The greatest diversity is to be seen around Alençon and Mortagne-au-Perche where the ancient crystalline extends farthest to the east. Farther south, in Haut Maine, the relief is simpler, with lines of cuestas that can be followed almost to Angers.

8.5.2.4 *Champagne and Lorraine*

The Brie and Soissonnais plateaus end in an escarpment on the east overlooking the Champagne plain (*see* Fig. 8.21). Between the Seine and the Marne, this escarpment is known as the edge or 'cliff' of the Ile de France. It is the first of a series of steps that follow one another towards the Vosges, the Rhine massifs and the Ardennes (*see* Fig. 8.22). Each step consists of a dip slope rising generally to the east and culminating in the crest of a scalloped escarpment overlooking the next tread of the staircase. In front of each escarpment, there are often isolated hills (outliers) left during scrap recession [see Hilly *et al.* (1979)]. These cuestas are composed of progressively older geological formations towards the east—the Chalk of the Champagne hills, the Lower Cretaceous sandstones of Argonne, the Portland limestone of the Bars hills, the Oxfordian of the Meuse hills, the Dogger formation in the Moselle hills, and the Upper Lias and Muschelkalk of the Lorraine plateau. At the foot of each escarpment is a clay or marl vale, whose width depends on the varying thickness of the beds. The escarpments themselves follow slightly sinuous courses because of tectonic factors: anticlines or faults trending east–west or south-east–north-west. They are further indented by the openings cut by consequent rivers (Meuse, Moselle, etc.) or by wind gaps (such as that once utilized by the Aisne). This structural relief can be traced as far as Luxembourg. It is continued on the western flank of the Vosges by Triassic sandstone cuestas whose last remnant is Mont Donon at 1010 m.

8.5.2.5 *Burgundy and Nivernais*

In the south-east of the Paris Basin, the regular arrangement of the cuesta landscape is disturbed by the presence of strong tectonic structures that surround the Morvan massif (*see* Fig. 8.21). In the north, one can still pick out the lines of escarpments overlooking the peripheral depression of Terre Plaine, excavated in the Lias. But in Nivernais to the west and in Burgundy to the east, faults trending north–south or north-east–south-west play a major role, causing vertical displacements of the stepped limestone plateaus and hills which overlook the Loire and Saône valleys and which reach 400–600 m.

8.5.2.6 *Berry and the Loire valley*

The morphostructural features of Lorraine—escarpments and vales—are found again, but are less well developed, in Berry between the Loire and the Creuse. The relief is lower (everywhere less than 200 m) and the differences in levels are smaller. Jurassic limestones outcrop in the Berry 'champagne' between the Cher and the Indre, while farther north the Cretaceous is often concealed beneath a cover of clay-with-flints or detritus from the Massif Central. Detrital granitic sands are especially thick in Sologne and Brenne where they give rise to a distinctive hydromorphological unit studded with lakes.

The landscapes of the middle Loire from Nevers to Angers are very different from those of the Seine. The basins were separated in the late Tertiary with the initiation of westward drainage towards the Atlantic (see p.180). The relief is low (20–150 m) and varies little. Valleys are only slightly entrenched and very broad. Only the gap of the Loire between Nivernais and Sancerrois shows more marked relief, guided by meridional faults. Everywhere else, in Orléanais, Touraine and Anjou, the Loire valley contains a wide alluvial plain, with its tributaries, such as the Cher, Indre and Vienne, flowing for a considerable distance parallel to the main stream before joining it. The Loire valley only just cuts into the soft white micaceous Chalk (the 'tuffeau') which is partly covered in the east by the Beauce limestone and the Sologne sands and in the west by the 'faluns', the shell sands left by the Miocene Sea [see Alcaydé *et al.* (1976)].

8.5.2.7 *Poitou sill*

This represents a very ancient link between the epicontinental seas of the Paris and Aquitaine Basins, a link which functioned up to the late Jurassic. Today, the sill is a structural and topographic saddle between the Armorican massif and the Massif Central. The Jurassic is relatively thin, so that the crystalline basement appears in the floors of the Vienne and Clain valleys as far as the region of Poitiers. In the south, basement faults of Armorican (north-west–south-east) trend affect the sedimentary cover, cutting the country into a succession of blocks of unequal heights as far as the Charente.

Everywhere that the limestones are not buried by the *sidérolithique* (i.e. the detrital deposits from the Massif Central), there is a low calcareous plain (*campagne*) or stepped plateau covered with a pebbly clay weathered mantle.

8.5.3 Geomorphological Evolution

After the end of the Oligocene, with the retreat of the Stampian Sea, the Paris Basin began to evolve through a long period of mainly continental conditions that have continued up to the present. Neogene transgressions only affected the Loire valley as far upstream as Orléans [the Helvetian Sea leaving the faluns (*see* Section 8.5.2.6)] and the English Channel coast west of Fécamp in the middle Pliocene. The developing Channel was receiving heavy loads of fluvially transported sediment—mainly granitic sands—coming from the Massif Central. These deposits were spread out in the late Tertiary and even the early Quaternary (i.e. the *sables de Lozère*), and there are still extensive remains stretching from Bourbonnais and Sologne to Caux. At the end of the Pliocene and the beginning of the Quaternary, epeirogenic uplift of 100–200 m affected different parts of the Paris Basin to varying degrees, causing slight deformation, marking the onset of drainage incision and emphasizing the structural forms; a phase of rejuvenation that continued through the Quaternary era [see Michel (1972)].

8.5.3.1 *Drainage development*

The drainage history reflects the tectonic events that affected the Paris Basin. Also, the drainage history provides some indication of the climatic changes, although there is much less certainty, at least as far as the earlier events are concerned. It is generally accepted that the present drainage began to evolve early in the Neogene when continental conditions became widely established.

The drainage developed on broad low-relief surfaces which suffered some deformation and uplift through the Tertiary as already noted. An early Seine system seems to have developed on such a surface. Some other neighbouring systems, including the ancestors of the middle Loire flowing to the faluns gulf developed independently. At the same time, uplift of the Vosges and the Burgundy sill initiated the ancestral Meuse/Moselle system directed towards the Rhine. Shifts in the centres of subsidence in the Basin, firstly towards the Paris region in the Palaeogene and then towards Orléanais at the end of the Stampian, prepared the way for the convergence of drainage on Paris and the localization of the Loire bend. It was late Pliocene–early Quaternary movements, however, that finally prompted the diversion of the upper Loire to the west and began the entrenchment of valleys, resulting in their positions being fixed.

This entrenchment coincided with significant climatic changes. Following a period of warm, rather dry climate, there were the glacial/interglacial oscillations which affected the relief development. Fluctuations in discharge together with variations in rock resistance, favouring certain valley lines over others, encouraged river capture; the most striking involved the beheading of the Meuse drainage to the benefit of the Seine and the Moselle. In this way, the present drainage of the Paris Basin became established by piecemeal adaptation.

8.5.3.2   *Evolution of the relief*

Entrenchment of the valleys in the late Tertiary and through the Quaternary has, nevertheless, left substantial areas of earlier land surfaces on the interfluves. It is these remnants, which are relatively extensive in many areas, that give the region its plateau or plain-like appearance. There are two sorts of surface.

1)   Erosion (planation) surfaces developed through the Tertiary. These surfaces formed the basis for the initiation of the drainage. The oldest occupy the highest positions in the landscape: 100–150 m in upper Normandy and Picardy and 200 m in Senonais, areas exposed since the end of the Cretaceous. In the east, they are developed on the backslopes of cuestas in Lorraine and in northern Morvan at between 300 and 400 m, and on the Burgundy upland plateaus at 450–500 m. Some remnants of marine deposits spared by erosion or preserved in karstic pipes or cavities (Landenian sands in the north and Stampian in the west) in some places allow the retreat of the Palaeogene seas to be monitored. On the whole, however, these surfaces have evolved under continental conditions since the early Tertiary. It is impossible, therefore, to fix their precise chronological position: they are essentially polygenetic, being successively remodelled and evolving over long periods, and consequently should be termed 'acyclic' or 'regradation' surfaces. They are covered by weathered mantles of varying thickness; the most striking of which is the clay-with-flints (the *argile à silex* in Normandy and Picardy and the *argile à chailles* in the east). In some places, these deposits are altered geochemically or have become mixed with other formations, such as ferruginous sands and gravels, sarsens and Quaternary loess [see Dewolf (1976), Elhaï (1967)].

More recent surfaces are less easy to discern, but there is some evidence. Some have been exhumed from beneath a Neogene marine cover (e.g., the faluns of the Loire valley, or the Redonian sands of Caux). Others take the form of benches cut into the older surfaces, but above the present valley floors. They represent the first signs of Pliocene–Quaternary rejuvenation.

2)   Structural or semi-structural surfaces have been etched out on what are now interfluves in the course of drainage development. They represent selective erosion in a rock sequence of varying resistance. The resulting plateaus or benches can be observed readily, especially in the central part of the Basin (e.g., Beauce, Brie, Multien, Valois and Soisonnais).

Other structural forms in the Paris Basin comprise: (1) the cuestas (*see* Section 8.5.2) which are best developed from Brie to Champagne and in Lorraine, and northern Morvan (*see* Fig. 8.18), but fragmented in the west nearer the Armorican massif, and only poorly developed in Berry, (2) eroded anticlines with inverted relief, exposing Jurassic or even Palaeozoic rocks under the Chalk, known as *boutonnières* (buttonholes) or 'brays', the latter term being taken from the most striking example, but others are found in Boulonnais and Vigny–Longuesse and (3) fault-line scarps, which are very localized and not usually well marked, in Perche, Lorraine, Burgundy and Nivernais.

8.5.3.3   *The Quaternary*

In each valley, river terraces provide evidence of successive incisions of the river in the Quaternary. Phases of excavation were separated by phases of aggradation, being related to alternating glacial and interglacial stages of varying duration and intensity. In addition, there were at least local tectonic movements and important changes in base level.

The best developed terrace systems show at least four levels but the steps are not always equally distinct, the highest and oldest often being very degraded. Inside the bends of large meanders, several terraces often merge to form a single polygenetic alluvial slope. The terrace gravels are usually coarse, are nearly always covered with a mantle of limon of varying thickness and contrast sharply with the finer, sandier deposits of the present alluvium.

Unlike the Thames Basin, the whole of the Paris Basin lay beyond the glacial limits, so there was no input of glaciofluvial material, but the cold conditions of the glacials promoted periglacial activity. The middle terraces—Riss and particularly Würm—show evidence of this with the incorporation of huge blocks in alluvial deposits, ice wedges and cryoturbation structures in the limon. Terrace slopes have been affected by gelifluction and successive gelifluctted layers are interdigitated with the terrace gravels. Slopes have been smoothed over and structural influences masked. Frost action caused rock fragmentation and helped to produce talus slopes, the deposits being known as *grèzes*. In the Chalk areas, the upper parts of slopes were attacked by frost while the lower parts became buried progressively under the accumulating debris, giving rectilinear slope profiles. The clay-with-flints, its flints broken by frost, became in places packed masses of flints that slid downslope.

Another characteristic of the cold periods was wind activity. Aided by the disappearance or reduction of vegetation, the wind was able to dry out the surface, attack the finer sands (e.g., the Stampian) and redistribute them. There are hardly any typical aeolian forms, although the plateaus of the Forest of Fontainebleau were covered by up to a metre of *sables soufflés*. Loess, which originated from glacial outwash plains to the north of the Paris Basin or from beach deposits exposed by lower sea levels, today forms a discontinous layer; the thickest accumulation mantles the east or south-east-facing slopes. On interfluves, thicknesses are irregular, varying from a few centimetres to several metres. Often, too, the loess has been decalcified or washed by streams to produce plateau limon [see Jamagne (1973)].

8.5.3.4   *Neotectonics*

Tectonic movements continued in several places during the Pliocene and Quaternary, as shown by various lines of evidence, principally in the west and north-west of the Basin. Reactivation of the Perche fault flexure is evident in the upper Eure valley, where steep and straight slopes in loose deposits have a remarkably fresh appearance, where the drainage in the headwater area shows disorganization and where there are unusual accumulations of Quaternary sediment. Study of the meanders and terraces of the lower Seine suggests that here the river may be antecedent to gentle Quaternary uplift of the region, possibly confirming geophysical data on the deep-lying Seine fault. In the extreme south-east of Bray, Quaternary reactivation of the anticline is supported by the absence of terraces in the epigenetic course of the Oise in the axial zone of the disturbance and by the gullying of the loessic cover on its flanks. Likewise, in the north of the Basin around the Artois anticline, signs of recent tectonic movement are to be found in the chalk and the overlying limon.

Thus it seems that slow uplift of the Paris Basin, continuing since the Upper Miocene, has persisted into the Quaternary, especially in Normandy. This was accentuated by the falling sea levels of the glacials, when valleys were excavated 25–30 m below their present-day levels and later infilled with alluvium.

8.5.4 Present-Day Processes

The Paris Basin entered the present era with a general periglacial inheritance in respect of soils and weathered materials. With the postglacial, there were important changes in the surface process system. Frost action is now restricted to severe winters and snowfall is short-lasting. Wind action is only effective in some coastal zones; running water has become the principal agent of surface modification. Its effects, however, have been limited considerably by the re-establishment of the vegetation cover, especially by the extension of dense forest, at least until the arrival of man. Clearance of the woodland by man in medieval times, the growth of agriculture and the alteration of surface drainage are among the factors that have fundamentally disturbed both the natural ecosystem and the erosion–sedimentation system. In fact, it is virtually impossible to find any part of the landscape that does not bear the human imprint. Soil erosion by water and wind, flooding and, in places, gullying are all examples of the effects of man's interference with the natural equilibrium. Because of the intensity of land use and the existence of large urban centres, one cannot afford to ignore anthropogenic processes in any study of present-day process and form.

8.6 Aquitaine Basin

Covering about half the area of the Paris Basin, this is the most south-westerly part of the central European lowland. It faces the Bay of Biscay and is fringed by the Hercynian uplands of the Massif Central to the east and the younger Pyrenean fold mountains to the south. The Basin has something of a dual character because of this situation. The calcareous regions of the north and north-east (i.e. Charente, Périgord and Quercy) are similar to the Paris Basin in structure and relief. However, the Pyrenean piedmont and the Landes in the south and west, which are separated from the other regions by the Garonne, possess quite different features. The northern and north-eastern areas rise to 300–400 m or more on the edge of the Massif Central and are underlain mainly by Mesozoic (Jurassic-

Cretaceous) rocks, whereas in the south-west a large area lies close to sea level and, with the southern areas that rise towards the Pyrenees, is floored with mostly unconsolidated Cenozoic deposits. Here, the Mesozoic rocks have plunged to considerable depths in the subsiding north Pyrenean trough or foredeep and are completely buried, except locally on the southernmost margin where they have been caught up in the Pyrenean folds. The Pyrenean tectonic movements, which began in the Cretaceous and which were more powerful than those that caused uplift of the Massif Central, are responsible for the general asymmetry of the Basin. Between the Massif Central and the Pyrenees, a long triangular gulf of the sea was established early in the Tertiary and, until the Miocene, was linked in the east with the Mediterranean across the Lauraguais sill. This gulf was gradually filled up with huge amounts of detritus coming from the surrounding uplands and mountains—sandy clays from the Massif Central, molasse (mainly soft calcareous sandstone with some conglomerates and clays) from the rising Pyrenees, Miocene marine shell sand, and Quaternary and recent fluvial and glaciofluvial gravels. At the same time, the sea slowly retreated westward, driven back by sedimentation in the Garonne Basin and by the accumulation of sandy fluviomarine deposits behind a coastal sand-bar that grew northwards in the Landes [see Enjalbert (1960), Gèze and Cavaillé (1977), Vigneaux (1975)].

8.6.1 Northern and North-Eastern Aquitaine

Charente, Périgord and Quercy comprise an area of limestone plateaus (250 km from north to south and 60–100 km from west to east) whose higher parts rise to more than 300 m. In the north, the landscape gradually merges with the Poitou sill (see Section 8.5.2.7) and the Paris Basin, and abuts on the extremity of the Armorican massif in the north-west. The whole area shows a strong regional slope to the south-west, with the progressive appearance of outcrops of increasingly younger rocks: Liassic, Middle and Upper Jurassic, Cretaceous and, in the south, Tertiary. The contact with the crystalline rocks of the Massif Central is mostly faulted which helps to explain the rarity or absence of cuestas here. However, it is noticeable that the landscape becomes progressively more broken towards the east with the increase in altitude and that the rivers become more deeply incised.

8.6.1.1 Charente

The Jurassic–Cretaceous limestones and marls continue from the Poitou sill (the watershed with the Loire basin standing at about 200 m), forming low plateaus or rolling landscapes only slightly interrupted by shallow valleys and the faint outlines of cuestas. The Armorican fault systems, planed across by a Tertiary erosion surface, have hardly any influence on the drainage pattern. The land falls gently to the west, with a low coastline comprising marshes and estuaries, and in places fringed with dunes. The only exceptions are in the limestone outcrops of some islands (e.g., Ile de Ré, Oléron, etc.) with their low cliffs. The whole coast shows the effects of the Flandrian (postglacial) submergence.

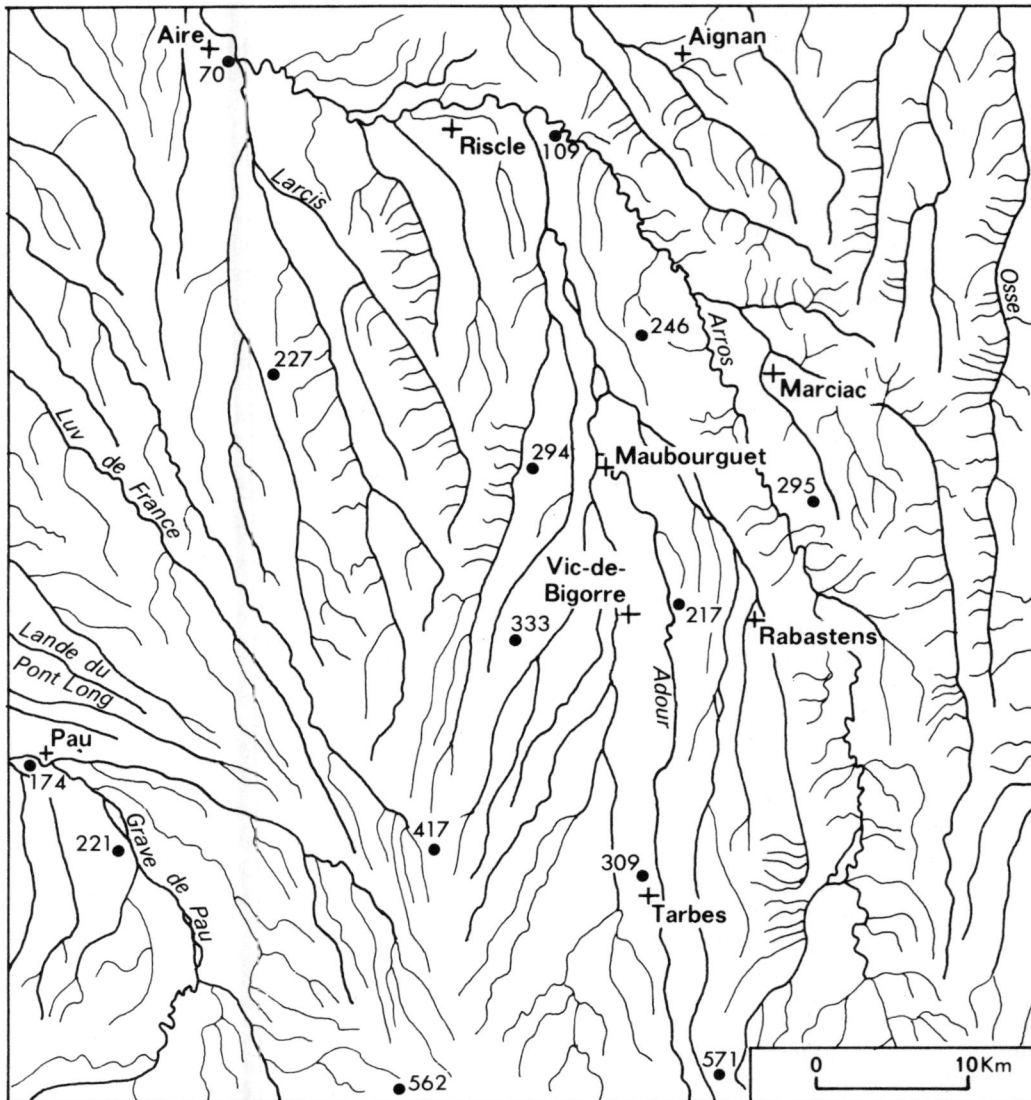

Fig. 8.23 Drainage pattern of western Lannemezan and the Pays de l'Adour (heights in metres).

8.6.1.2 *Périgord*

The Upper Cretaceous limestones of Périgord meet the Massif Central along faults, in places marked by a narrow peripheral depression. Farther south, the limestones are often masked by a thin veneer of detrital sands and some interfluves are capped with Oligocene sandstone. In spite of the thickness of the limestones, karstic surface forms are absent or poorly developed, limited mainly to shallow surface depressions. On the other hand, a dense network of dry valleys furrows the plateau and there are complex subsurface cave systems. The tabular plateau surface, sloping gently to the south-west, may not be a simple structural surface but one that has been planed across by erosion in the Tertiary. Fénelon (1951) claimed that there are several such planation surfaces including a tilted Mio-Pliocene surface at 200–400 m. The plateau is broken by the deep flat-floored valleys of the Isle, Dronne and Vézère, whose steep limestone sides with their rock shelters and caves, such as those of Lascaux, provided homes for prehistoric man.

8.6.1.3 *Quercy*

In Quercy, a much more typical 'causse' landscape is found, developed on Upper Jurassic limestones and cut by the canyon-like valleys of the Dordogne and the Lot, with their fine examples of ingrown meanders. The surfaces of the plateaus (500–600 m) have been considered to be planation surfaces—Clozier (1940) recognized three such levels—but are far from tabular, being broken by numerous enclosed depressions, karstic dry valleys and avens (deep, vertical-sided chasms). In contrast, most of the drainage, except for the major through-flowing rivers, passes underground. There are extensive cave systems including the Gouffre de Padirac with its vertical shaft (100 m deep and 30 m wide at the top). Towards the east and north-east, the gradual rise of the strata has allowed the development of true cuestas, but the karst country is clearly separated from the crystalline massif by faults. These have created a series of rifts in whose floors outcrop Liassic clays or even Palaeozoic rocks, as in the Brive rift.

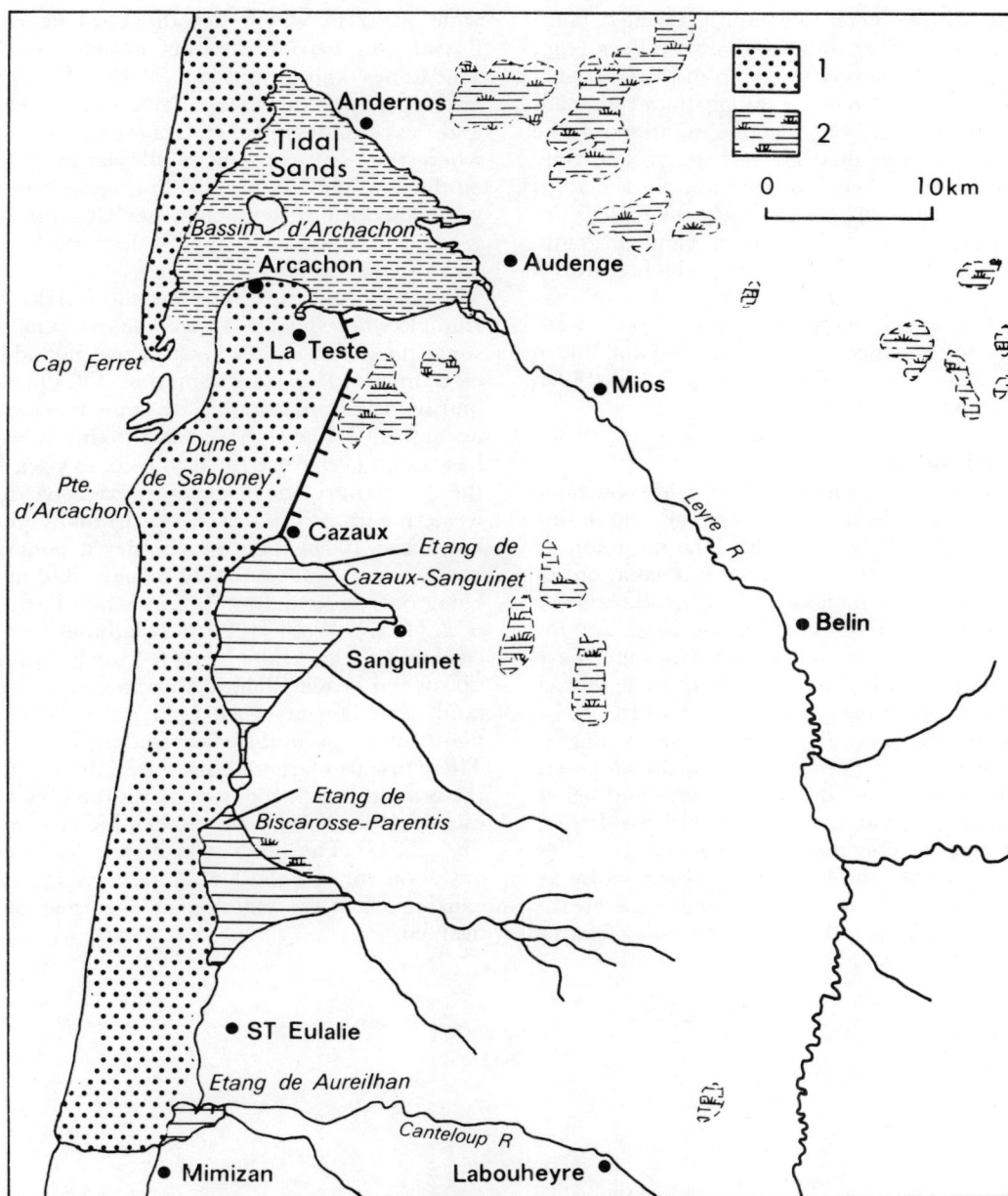

Fig. 8.24 Part of the Landes south of the Arcachon Basin. (1) Sand dunes. (2) Marshland.

8.6.2 Pyrenean Piedmont

Extending northward from the foot of the mountains at about 600 m is an immense composite alluvial cone [see Taillefer (1951)] at whose apex stands the Lannemezan plateau. It is made up from numerous separate, but overlapping cones thrown out by a whole series of rivers from the Gave de Pau in the west to the upper Garonne in the east (*see* Fig. 8.23). The margin of this cone is convex in plan to the north, reaching approximately as far as Aire and Toulouse where its surface elevations have declined to less than 200 m. The surface of the cone consists of a thick pebbly proluvial Pliocene formation, which in its upper parts fringes the most northerly Pyrenean structures while its distal parts overlie the molasse. The rivers crossing this cone show an extraordinary fan-like pattern (*see* Fig. 8.23) separating it into segments that widen downstream. The flat alluvial valley floors are entrenched between strikingly asymmetrical slopes, the asymmetry being usually explained in terms of Quaternary periglacial processes—stream flow and sheetwash dominating on the steeper west-facing slopes and gelifluction on the gentler east-facing slopes. It is also noticeable that the cone is more dissected in the west (in the Adour Basin) and also in the area of the Garonne tributaries. Apart from the fact that the former at least lies closer to the sea and to base level, a more dynamic explanation can be offered. In the west, the rivers issuing from the high Pyrenees are powerful and active, cutting down rapidly into the Tertiary formations, uncovering even pre-Pyrenean folds and creating a landscape of hills and ridges separated by large valleys (e.g., Chalosse). In the east, however, apart from the Garonne, all the rivers have their sources on the piedmont plateau, lacking the meltwater and higher precipitation of the mountains. Consequently, they are less active and only entrench themselves in their downstream portions (in Armagnac and Lomagne) where their narrow valleys are separated by low, rounded interfluves sloping gently north. On the other hand, the Garonne, which carries a greater discharge, is liable to flood regularly and has entrenched itself more actively. Its valley shows well-marked terraces, for instance, at 150–160, 180 and 200 m near Toulouse, where the floodplain stands at 130–140 m [see Cavaillé (1975)].

8.6.3   Garonne Lowland

In its middle course from Toulouse to Agen, the Garonne crosses a rather monotonous landscape of undulating molasse with a few low ridges or hills. The only note of diversity is introduced by the occasional intercalations or lenses of limestone in the molasse, which produce some sharper crests, benches, river bluffs or outliers. For the most part, however, the main valleys (i.e. Garonne, Tarn and Lot) are broad and alluvial-floored with large terrace systems, sometimes being tens of kilometres wide.

Downstream, the Garonne is joined by the Dordogne. After leaving the Périgord plateau, the Dordogne enters the low molasse country near Bergerac. Above and below Libourne, where it receives the Dronne and the Isle, it shows fine meanders. The northern edge of its valley above its confluence with the Garonne is bordered by an escarpment known as the Côtes de la Dordogne. Below the confluence with the Garonne, the combined valley forms a wide plain in which lies the Gironde estuary. Recent fluvial and marine deposits mingle, overlying the soft sandstones and limestones of the Tertiary. The latter outcrop on the north-east bank with a weathered mantle and, less prominently, in Graves on the south-west bank where they are overlain by alluvial gravels. Near Royan on the Garonne estuary, the limestones have been worn by wave action into steep cliffs [see Cavaillé (1965)].

8.6.4   The Landes

The sandy lowland known as the Landes forms a broad triangle whose longest side is the coastline. From north to south it extends for 200 km and goes inland to a maximum of a hundred or so kilometres. Of Quaternary fluvio-marine origin, the sands with some marine clays underlie an almost perfectly horizontal plain, to which the wind has added ever since the dry, cold, low sea-level phases of the Quaternary one of the most extensive dune systems in western Europe (*see* Fig. 8.24). In many places the dunes are about 100 m high and border a remarkably straight coast, which was probably at one time an offshore bar, being only indented by the Arcachon Basin [see Guilcher *et al.* (1952), Wolff (1929)]. The dunes form a continuous cordon 5–10 km wide, being fed solely by fine sand driven up off the beach. Behind this barrier, groundwater from land drainage has been held up, so that at the lowest points of the ground surface shallow lakes (*étangs*) appear. There are also larger étangs [see Filliol (1955)] which are the remnants of lagoons marking the sites of ancient river estuaries blocked by the developing coastal sand-bar (*see* Fig. 8.24). The high watertable has encouraged the development of a shallow, impermeable iron-pan or *alios*, causing extensive marshes at one time but now mostly drained.

Chapter 9
Hercynian Europe

9.1 Introduction

The Hercynian massifs of Europe (*see* Fig. 9.1) form a series of detached uplands and plateaus of varied relief and frequently complex geology that extend from southern Ireland in the west through central Europe to southern Poland. This chapter also includes the Mediterranean islands of Corsica and Sardinia for convenience, but the Iberian massif is dealt with separately in Chapter 12 to show its relationships with the other adjacent units of Iberia (*see* Chapters 11 and 13). The Ural Mountains, where the closing tectonic phase was Hercynian, are examined in Chapter 19. In central Europe, the Hercynian massifs are often referred to as the Mittelgebirge, meaning literally 'middle mountains', because they attain only moderate heights compared with the higher Alpine and Carpathian chains to the south.

The massifs represent parts of a large morphostructural unit termed the west-central European platform. The platform consists of Precambrian and Palaeozoic rocks affected in turn by Caledonian and Hercynian earth movements, and later buried, in part, by a cover of sedimentary strata, Permian to Quaternary in age. On the south and south-east, the platform is bordered by the younger epigeosynclinal zones of the Pyrenees, Alps and Carpathians. Both before and during the development of the Mesozoic–Tertiary sedimentary cover, the platform was broken by faulting into a series of major blocks. Some of these have been uplifted to form massifs, whereas others suffered relative or actual subsidence to become basins in which thicker sequences of sediments—the platform cover—accumulated. Many of the bounding faults may be deep-seated crustal fractures; it has been argued that the linear nature of the margins of many of the massifs, or their rectangular outlines, supports this view. Major bounding faults, however, are frequently not visible nor do geophysical data always support such an explanation (e.g. *see* Section 13.2.1). On the other hand, there are some well-documented major fault systems crossing Hercynian Europe, such as the Rhine rift-valley and its northward extension into Hesse (*see* Section 9.13.1) or the rifts of the upper Allier and Loire in the Massif Central.

The main phase of Hercynian orogenesis took place in late Carboniferous to early Permian times. The folding in the axial zones was intense, being well displayed, for example, in the contorted strata of the Ardennes. In some more distant areas such as the Pennines of central England, however, the folding was more gentle and fracturing dominated the tectonics, especially where there was a rigid foundation due to earlier Caledonian compres-

sion. The principal trends of the Hercynian systems are arranged in a fairly simple pattern. From southern Ireland, lineations are directed mainly west–east into south Wales; parallel trends can be picked up in southwest England and northern Brittany. In southern Brittany, the so-called Armorican trend is directed more to the south-east, curving gradually to reappear in the western Massif Central with an almost north–south alignment. Corsica and Sardinia show a north–south grain, but central Europe from the eastern Massif Central and the Ardennes through the Rhine Highlands, the Harz and into northern Bohemia, show generally west–east alignments. The Hercynian orogeny created massive mountain chains, but post-Hercynian denudation has been so intense that now only the roots of those mountains remain.

The southern margin of most of the Hercynian structural belt, and in places the whole of it, has disappeared because of the later encroachment of the Tethyan geosyncline, where the Hercynian floor subsided and became buried by deep sedimentation. Later, parts of the floor were uplifted in the Alpine orogeny and were incorporated in the Alpine chains (*see* Chapter 10). Over most of Europe, the northern margin of the Hercynian zone lies far to the north of the Alps, but in the east, the Carpathians have been thrust completely over it, or nearly so, as seen for instance in the partial overriding of the Silesian coalfield by the advancing nappes.

In terms of relief, the dominant trait of Hercynian Europe is the way in which the various massifs have been truncated by erosion, creating extensive surfaces of low relief. It is now recognized that, after the Hercynian orogeny, successive periods of denudation separated by intervals of uplift gave rise to a whole series of planation surfaces, the younger members of which are better preserved and less deformed by subsequent tectonics. The so-called post-Hercynian surface was initiated in the Permo-Triassic period; intense denudation often under arid or semiarid conditions reduced the Hercynian chains over most of the area to low residuals. Some parts of the surface were soon submerged by Triassic seas in which sandstones and limestones were laid down, and in much of central southern Germany and southern Poland sedimentation continued throughout the Jurassic and Cretaceous periods. In other parts, the post-Hercynian surface was warped, uplifted or dislocated by faulting; other, younger planation surfaces of pre-Cretaceous, Palaeogene, Pliocene and/or Quaternary age were cut into it. Some of these surfaces are thought to represent etchplains, developed under tropical climatic conditions. Pediments and glacis of probable Neogene and Quaternary age may also

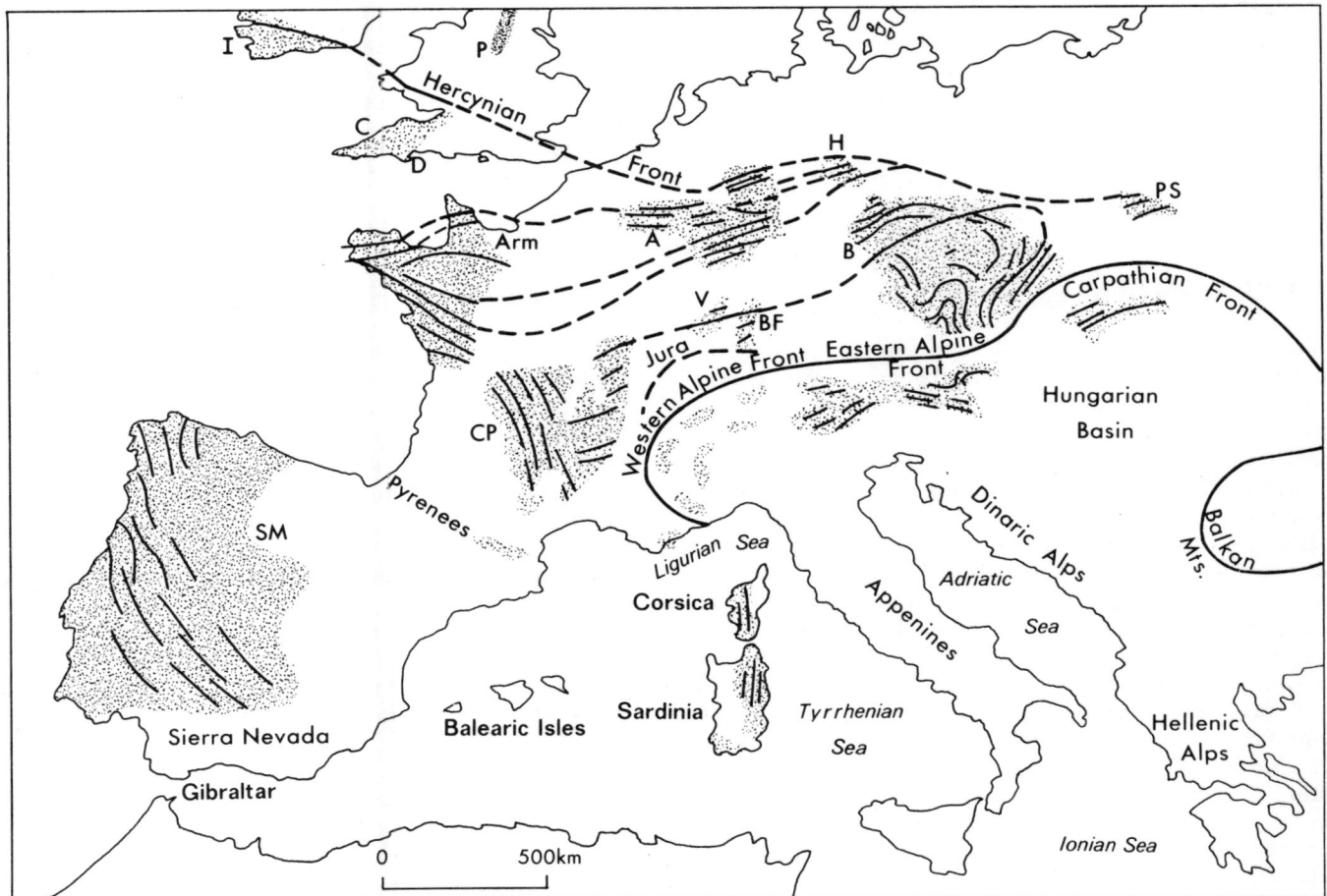

Fig. 9.1 Hercynian Europe. The major Hercynian structures are shown by stipple, and the main trendlines of late Palaeozoic folding are indicated. A: Ardennes. Arm: Armorican massif. B: Czech massif. BF: Black Forest (Schwarzwald). C, D: South-west England. CP: Massif Central. H: Harz. I: Southern Ireland. P: Pennines. PS: Little Poland, Silesia. SM: Iberian Meseta. V: Vosges.

be recognized along the major valleys. Neotectonic movements and climatic changes also gave rise to many examples of accumulation terraces.

Igneous activity has produced many distinctive features among the Hercynian massifs. Granitic intrusions, some of great size, were emplaced during the orogeny and have since been exposed by erosion. Volcanic activity has occurred in several massifs at various times, but especially in the Tertiary and Quaternary. The main neovolcanic areas are in the Massif Central, the Eifel and Westerwald, in Hesse and in parts of Bohemia. The volcanic structure of the Monts Dore in central France (rising to 1886 m) is the highest point in mainland Hercynian Europe, only exceeded in elevation by the strongly uplifted crystalline block of Corsica (highest point, 2710m). All volcanoes of the Hercynian massifs are now extinct, but some were active as recently as the late Pleistocene and evidence of higher than average geothermal heatflow in some places lingers on in the form of hot springs.

In the Pleistocene, all the Hercynian massifs—except south-eastern Ireland—lay outside the domains of the Fennoscandian and Alpine ice. The southern margin of

the Fennoscandian and British ice just touched the edge of south-west England, the edge of the Mittelgebirge in Germany and the northern flanks of the Czechoslovak massif, but penetrated a short distance into them where relief conditions were favourable, as in the Leipzig–Thuringian embayment and in southern Poland. On the other side, the Alpine ice came close to, but barely touched, the Hercynian belt; it entered the Rhône valley near Lyon and the upper Danube valley in south-west Germany. Elsewhere, small local glacier systems developed on high ground, for instance in Corsica, the Massif Central, the Vosges and the Schwarzwald, but much larger areas were governed by periglacial conditions, witnessed by the block fields, gelifluction deposits and cryoplanation features that are so characteristic of much of the Hercynian uplands.

At the present time, fluvial and mass movement processes are most important in the modelling of the relief. Sardinia, Corsica and southern France are characterized by Mediterranean-type erosional systems; the Massif Central, Brittany, southern Britain, Ireland and the Benelux uplands belong to the Atlantic maritime morphogenetic region. Central Europe is a transitional morphogenetic zone, beyond which lies the continental-type zone of southern Poland (in which, for example, an important role is played by subsurface processes such as suffosion in the loessic areas). In all parts, man has been the most important agent modifying the relief in the Holocene, although his impact varies considerably from place to place. Extractive mining for coal, salt, china clay, etc. has produced major scars on the landscape, while agriculture has had a more subdued but, on the other hand, very widespread influence, ranging from the conservational role of terracing in some areas—especially the wine-growing regions—to the destructive processes involved in soil erosion.

9.2 Pennines and North-Eastern England

In the context of Hercynian Europe, this is a completely atypical unit, but it is convenient to begin a regional treatment at one extreme margin and to move to others in geographical order. The Pennines are Hercynian in the sense that they are built of Palaeozoic strata deformed in the Hercynian orogeny, although not so severely deformed as most other parts of Hercynian Europe. Extending from the English Midlands to near the Cheviot Hills, they form a mass of high ground often referred to as the 'backbone of England', constituting the main drainage divide between river systems draining to the North Sea and those draining to the Irish Sea. They consist of a series of dissected uplands attaining heights up to about 600 m, built mainly out of Carboniferous rocks (*see* Fig. 9.2). The general structure is monoclinal, the strata being tilted gently towards the east but cut off on the west by abrupt downfolding or faulting. There is considerable diversity of structure and landform, however, and several distinctive subregions may be recognized. These will be described in turn, after which the Cenozoic evolution and glaciation of the whole area will be considered.

9.2.1 Derbyshire Pennines
This is an area sometimes known as the Peak District; because of the outstanding qualities of its scenery, it was the first national park to be designated in Britain (in 1950). The highest elevations are in the north where the flat-topped moors of Kinderscout and Bleak Low reach 636 and 610 m, respectively. Structurally, it consists of an asymmetrical dome, whose axis runs roughly north–south, flanked by other folds, such as the Ashover anticline and the Goyt syncline (*see* Fig. 9.2). Erosion of the dome has removed the Upper and Middle Carboniferous strata to expose the Carboniferous Limestone over a wide central area. Inward-facing escarpments of the Namurian (Millstone Grit Series) overlook the Limestone, being most pronounced in the north-west; in the south and east, the descent from the Limestone uplands is generally by dip slopes with less prominent escarpments. The Carboniferous Limestone (more than 700 m thick) contains some interbedded volcanics (thin lavas and tuffs) and minor intrusions, which are significant in their effect on the groundwater hydrology, but it is the surface

features that are most distinctive. The massive white and grey limestones form open plateaus at levels up to about 400 m, broken by deep valleys such as Millers' Dale and Dovedale with their steep, almost vertical sides. The deep incision of these valleys has lowered the watertable regionally so that in the intervening plateau areas dry valleys are common. Cave systems are extensive, and there are some deep swallow holes [e.g., Eldon Hole (55 m deep)], but karstification is not so pronounced as in the more northern limestone areas of the Pennines. In particular, swallow holes are not common neither are pavements widely found.

Above the Carboniferous Limestone a sequence of thick shales (the Edale shales with thin limestones, approximately 300 m thick) occurs leading to the Millstone Grit Series. The latter conceals the Carboniferous Limestone in the northern parts of the Peak District and produces very different scenery. Massive beds of gently dipping and resistant grits give rise to high moorlands and sharp 'edges' (e.g., Stanage Edge); the Kinderscout Grit (about 200 m thick) forms the highest elevations. The grits are separated by thick interbedded shales, the whole sequence resting on the Edale shales beneath to produce frequently unstable conditions in which mass movements, cambering and valley bulging develop (e.g., Mam Tor, the 'shivering mountain').

The Upper Carboniferous beds (Coal Measures) are most extensive in the east of the region, dipping eastwards off the Derbyshire dome. Shales and clays are interbedded with some moderately resistant sandstones, giving minor cuestas, but elevations diminish eastwards until the Coal Measures disappear beneath the Permian. The latter begins with the Magnesian Limestone which, east of Chesterfield, presents a bold escarpment touching 187 m but, to the north and south, is a less distinct feature.

9.2.2 Central Pennines and Lancashire Uplands
Relatively narrower and less elevated in this section, the central Pennines consist of the shales and sandstones of the Millstone Grit Series, flanked to east and west by Coal Measures (*see* Fig. 9.2). The uplands are characterized by black moorlands and peat mosses, with an average elevation of 400–420 m. The rocks dip more gently to the east than to the west, so that the overall impression is of tiered slabs of country in which the sandstone and grit outcrops produce numerous minor west-facing crags or scarps, and more gentle dip slopes descending eastwards. The uplands are broken by the broad valleys of the Aire and Wharfe, and the gorge-like valley of the Calder, while many smaller tributaries head in rocky 'cloughs' on the moors. Between the valleys of the Aire, near Skipton, and the Ribble, which drain to opposite sides of the Pennines, there is an important through valley (the Aire Gap) whose floor level does not exceed 140 m. In this area, only 13 km separate the eastern and western outcrops of Coal Measures. The latter consist dominantly of shales rather than the sandstones of the underlying Millstone Grit, but there are sandstone bands up to 60 m or so thick, forming, for example, the summits of hills around Leeds, Bradford and Halifax. Elevations in the Coal Measure country are lower

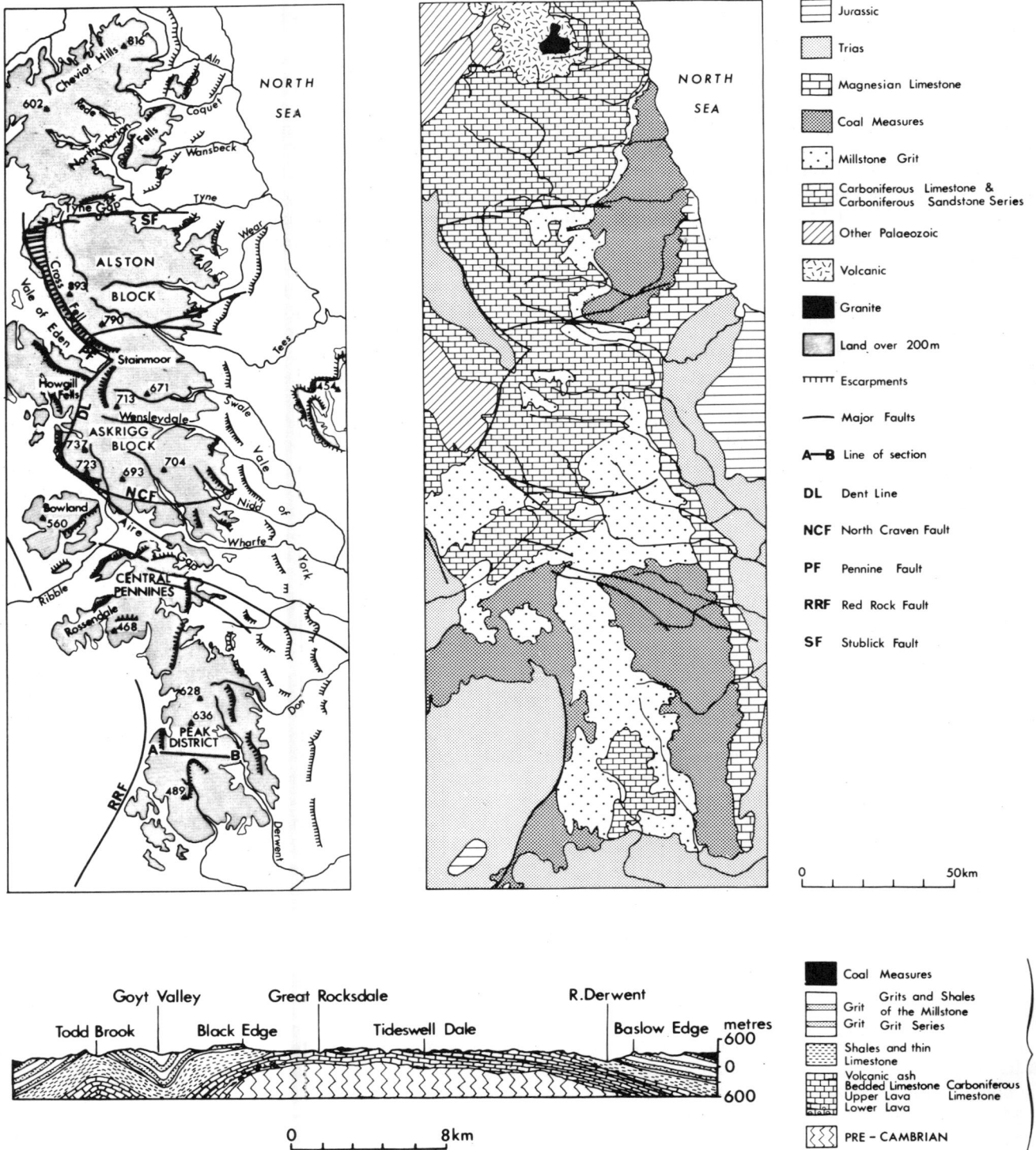

Jurassic

Trias

Magnesian Limestone

Coal Measures

Millstone Grit

Carboniferous Limestone &
Carboniferous Sandstone Series

Other Palaeozoic

Volcanic

Granite

Land over 200m

Escarpments

Major Faults

A—B Line of section

DL   Dent Line

NCF North Craven Fault

PF   Pennine Fault

RRF Red Rock Fault

SF   Stublick Fault

Coal Measures

Grits and Shales
of the Millstone
Grit Series

Shales and thin
Limestone

Volcanic ash
Bedded Limestone   Carboniferous
Upper Lava   Limestone
Lower Lava

PRE – CAMBRIAN

Fig. 9.2 The Pennines and uplands of Northern England. The left-hand map shows relief, major escarpments and faultlines. The right-hand map shows the geology. At the bottom is a cross-section of the Peak District along the line A – B marked on the left-hand map.

and the valleys more open. Farther east stands the escarpment, rather variable in form, of the Magnesian Limestone through which the Aire cuts below the confluence of the Calder. On the west of the central Pennines, the Millstone Grit in Lancashire gives rise to two smaller uplands flanking the Ribble valley: the so-called Forest of Bowland and the Rossendale Fells. The Ribble valley itself is developed on Carboniferous Limestone, although this is thickly mantled with glacial deposits. Both the Bowland and Rossendale areas present typical gritstone topography, rising to heights of about 560 m and 460 m.

9.2.3 The Askrigg Block

Lying between the Aire Gap to the south and the Stainmore depression to the north (*see* Fig. 9.2), with the Lake District and Howgill Fells to the west and the Vale of York on the east, the Askrigg Block is a well-defined geological and topographical unit [see King (1960)]. Lower Palaeozoic and older strata form the foundation of the block, which has remained relatively stable and rigid since the Caledonian (Devonian) orogeny. In the west, it is demarcated by the Dent Fault and, in the south, by the Craven Faults system. A cover of Carboniferous strata laid down on the block has been largely protected from subsequent tectogenesis, and the strata exhibit simple structures dipping at angles of usually less than 3° to the north-east. The area is characterized by extensive plateaus, tabular residuals, deeply incised valleys and a prevalence of stepped slopes related to outcrops of varying resistance (e.g., Ingleborough). Many summits reach 600 m; a few exceed 700 m.

Near the base of the Carboniferous sequence lies the Great Scar Limestone, pure and massive, up to 200 m thick. Outcropping over a broad area in the south of the block and flooring several major valleys in the north, it is responsible for probably the best examples of karst landforms in Britain. There are excellent examples of limestone pavements (e.g., above Malham Cove), stripped of overlying material by glacial erosion. Enclosed depressions and sink holes (pots) abound, the latter concentrated at zones where streams, developed on the overlying impermeable beds of the Yoredale Series, meet the limestone. The deepest known pot is Gaping Gill (111 m deep). Complex cave systems have developed in the limestone; their concentrations at certain levels may be related to former regional watertable positions [see Sweeting (1950)]. A falling watertable in late Tertiary–Quaternary times may be related to intermittent uplift of the Askrigg Block and the incision of major valleys. In turn, lower groundwater levels have caused the drying up of many smaller surface valleys on the limestone.

The Yoredale Series of alternating beds of limestone, shale and sandstone, deposited on top of the Limestone, now outcrops over about two-thirds of the Askrigg Block. The Yoredales are the stratigraphic equivalent of the upper part of the Carboniferous Limestone Series and are responsible for the typical stepped topography of much of the area. Where erosion has not removed it, the sequence passes up into the Millstone Grit, but this is relatively thin (about 300 m) compared with other parts of the Pennines.

The Millstone Grit caps the well-known hill masses of Whernside and Ingleborough which are formed basically of the Yoredales resting on the underlying massive limestone.

The Craven Fault system first developed in Yoredale times, the associated scarp being buried as Carboniferous strata accumulated. Later, differential denudation produced the south-west-facing scarp (e.g., Giggleswick Scar). Renewed fault movements occurred, up to Tertiary times at least, and relative uplift of the block is responsible for its erosion and dissection, during which the Upper and Middle Carboniferous strata were partially removed. The valleys cut into the block are of various forms. Some, like Swaledale, are relatively deep and narrow while others, such as Wensleydale, are wider and more open. Sometimes a U-shape is apparent (e.g., upper Wharfedale) suggesting glacial modification, but structure probably remains the overriding influence.

9.2.4 The Alston Block

The highest point attained in the whole of the Pennines is Cross Fell (893 m), near the western edge of the Alston Block (*see* Fig. 9.2). Like the Askrigg Block to the south, this consists of a rigid basement of pre-Carboniferous rocks which are almost completely covered by Carboniferous sediments. The Carboniferous sediments include limestone, sandstones and shales, often rhythmically repeated, but the limestone is less important in the succession than farther south, amounting to only 10–15 percent of the total thickness. The thickest 'Great' Limestone reaches a maximum of 30 m and makes a distinctive outcrop. In late Carboniferous times, a sheet of dolerite (the Whin Sill) was intruded (up to 70 m thick) that now makes an important contribution to the relief: for example, forming a line of minor, but steep, cliffs along parts of Cross Fell Edge or, broken by faulting, forming a series of escarpments in the area of the Roman Wall. Cross Fell Edge represents the major fault line escarpment marking the western boundary of the Alston Block. The northern and southern boundaries are provided by the major transverse depressions, tectonically determined, of the Tyne Gap and Stainmoor, respectively. Eastwards, the Carboniferous cover rocks decline in altitude and thicken towards the Durham coalfield and the Wear valley, finally passing eastwards below the striking escarpment of the Magnesian Limestone. The uplands of the Alston Block are bevelled by a series of high-level planation surfaces, like the rest of the Pennines farther south; cut into these are the main rivers which tend to flow radially to the Tyne, Wear and Tees from the central watershed of Cross Fell.

9.2.5 Northumbrian Fells

Around the high core of the Cheviot Hills (see Section 9.2.6) is a belt of upland country characterized by a complexity of cuestas, strike valleys, and water and wind gaps, formed in the Carboniferous rocks. The latter dip generally eastwards and southwards off the Cheviot dome; to the south they descend to the structural trough of the Tyne Gap, marked by the Stublick Faults. The Carboniferous sediments in Northumberland begin with the

Cementstone group of unresistant shales and muddy limestones, followed by the massive coarse Fell Sandstones forming a crescent-shaped girdle of uplands around the Cheviots. The succeeding Scremerston Coal Series of softer sediments has been exploited by the North Tyne and its tributary, the Rede, forming a lowland draining south. Capture of the Cheviot headstreams by the North Tyne has left their former continuations, such as the Wansbeck and the Blyth, beheaded, but, a little farther north, the Coquet and Aln have so far escaped such diversion. To the east of the depression of the Scremerston outcrop lies a further sequence of west-facing scarps and vales developed in the thin limestones, sandstones and shales of higher Carboniferous strata, leading finally to the impressive cliff scenery of the Northumberland coast.

9.2.6 Cheviot Hills
The Cheviots are built from a pile of andesitic lavas extruded on a Silurian platform in Devonian times. Later, the mass was intruded by granite and covered by a blanket of Carboniferous sediments which have since been eroded away. The relief is rounded and smooth, although the valleys radiating from the centre are deeply cut. The twin summits (815 and 716 m) are both developed on the granite core.

9.2.7 Tertiary Landscape Evolution
In the southern Pennines, there are about 60 minor outliers of sands, gravels and clays of Lower Pliocene age [see Walsh *et al.* (1972)] known as the Brassington Formation. These deposits appear to have foundered into solution cavities in the Carboniferous Limestone. Evidence suggests that they once formed part of a widespread sheet of fluvial sediments (several tens of metres thick) that accummulated on a late Miocene–early Pliocene planation surface at about 450 m above present sea level, probably cut mostly in Namurian shales covering the Carboniferous Limestone. It is uncertain whether any present-day summits or planation surface remnants in the Pennines can be shown to be part of this Neogene surface, but the evidence of the deposits supports the view that, since the late Tertiary, the southern Pennines have been uplifted at an average rate of 1 m in about 15000 years. Higher planation surfaces in the Pennines undoubtedly exist and are likely to date from earlier Tertiary times; conversely, lower surfaces are probably of younger date.

No comprehensive study has yet been made of planation surfaces throughout the area. The following examples are selected from work in the last 30 years. In the Peak District, Linton (1956) described a Summit surface at 460 m and an Upland surface at 330–440 m, presumed to be Tertiary subaerial surfaces. Johnson and Rice (1961) have made a more detailed study of the Upland surface (360–430 m) in the south-west Pennines. In the Ingleborough region, Sweeting (1950) recognized an upper surface at about 400 m, associated with a major cave system in higher limestone areas at about 380 m, and evidence of a lower planation stage associated with caverns at about 300 m. In the northern Pennines, Sissons (1960) argued that a warped and faulted planation surface

of possibly Miocene–Pliocene age was later cut into and modified by marine erosion, producing a stepped series of wave-cut platforms and associated subaerially graded coastlands. In the Cheviots, Common (1954) identified late Tertiary surfaces at 490–520, 400–460, 300–360 and 150–230 m.

Many authors postulate uplift, warping and faulting of the Pennines not only in the middle Tertiary but also in the Neogene and Quaternary. Trotter (1929) argued for Tertiary uplift of the Alston Block along the Pennine and Stublick Faults. Hudson (1933) advanced the view that the Pennine block was tilted eastwards in the early Tertiary, followed by a phase of partial planation; later uplift and warping were said to have produced a peneplain at 490–670 m. Substantial Neogene uplift is postulated by Walsh *et al.* (1972), as already noted. More and more workers are coming round to the view that the development of planation surfaces in the Pennines is a relatively recent affair, especially with the evidence of the Brassington Formulation in mind. McArthur (1977) in the upper Derwent basin has calculated that, based on rates of contemporary denudation (varying from 173 to 339 t km^{-2} yr^{-1} in the area), no more than 1 Ma was needed to carve the Upland surface [see Linton (1956)] from the higher Summit surface.

The development of the drainage in the Tertiary is less well understood. Many streams and stream segments are undoubtedly discordant with structure, even after eliminating cases of glacial interference [see Sissons (1960)]. Linton (1951a) hypothesized a history of superimposition of the drainage from a high-level (about 1000 m), eastward-tilted Chalk cover, which has since been entirely stripped away; the east-flowing Solway–Tyne and Tees–Esk rivers are examples of reconstructed original consequent rivers on this cover. Other workers [see Sissons (1960)] have appealed to Neogene marine submergence and the development of discordant streams across new marine-cut platforms during a subsequent pulsed emergence of the Pennine massif.

9.2.8 Quaternary Glaciation
There has been no comprehensive study of the glaciation of the Pennines (*see* Fig. 9.3) and, in many areas, no detailed investigation has ever been made. In the areas that have been studied, no more than two major glaciations have been deduced, making it clear that the record for the Lower and Middle Pleistocene is entirely missing, the evidence having been obliterated by later ice advances. In the Cheviot–Northumbria area, most of the ice appears to have flowed from the Southern Uplands of Scotland, moving mainly east and south-east. The highest part of the Cheviots, however, seems to have been dominated by a local ice mass, for Cheviot granite is absent from the drifts farther south, suggesting that the outcrop was never crossed by extraneous ice. The Tyne Gap carried a considerable eastward flow of ice from the congested Irish Sea basin, but the Alston and Askrigg Blocks to the south were sufficiently high to nourish major local systems of glaciers. In the former system, local glaciers moved down the South Tyne, Wear and Tees

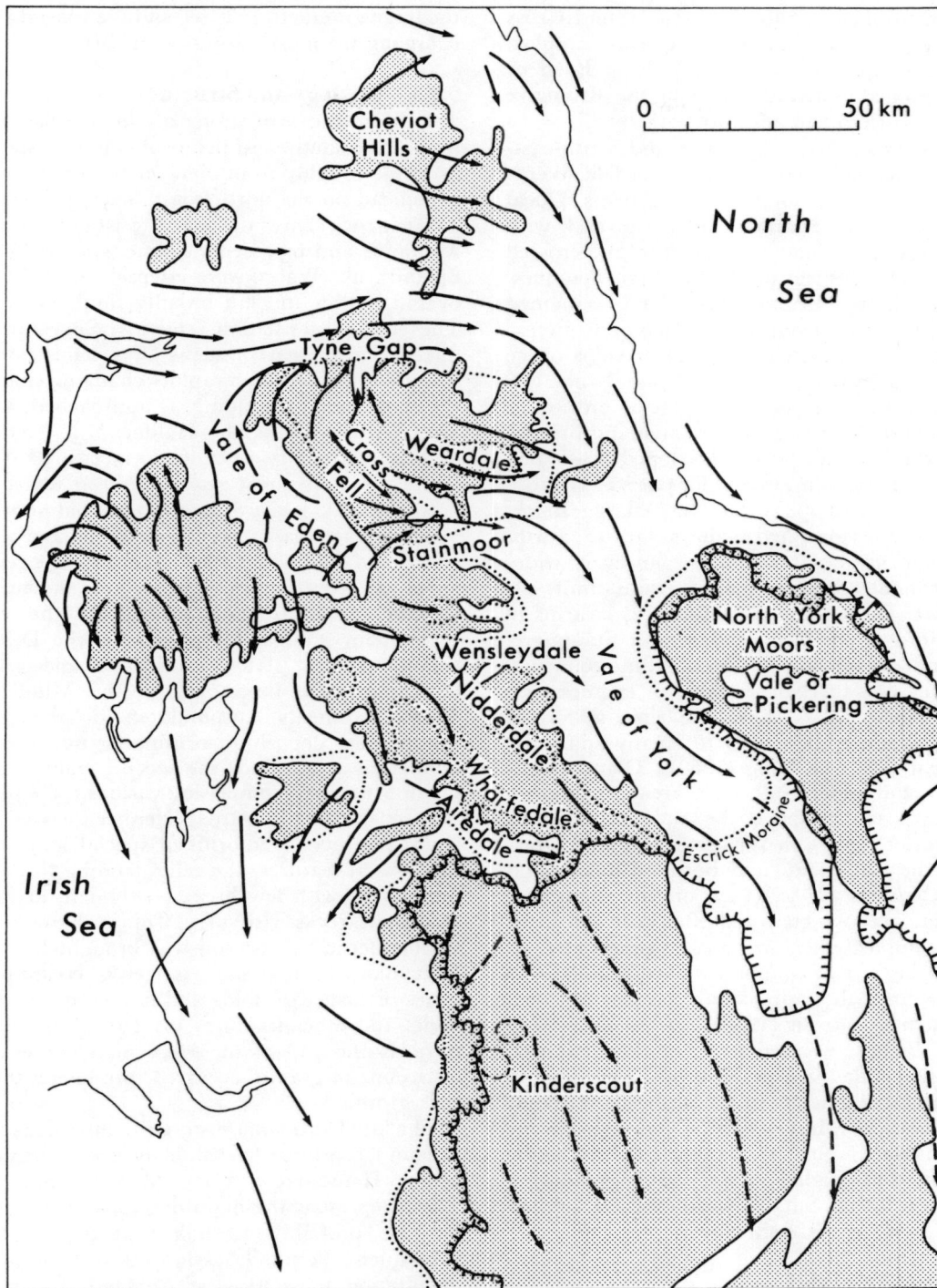

Fig. 9.3 The glaciation of northern England.

valleys from a central accumulation area around Cross Fell. The glaciation of the Stainmore Gap was complex. At times, ice from the Lake District, Vale of Eden or Howgill Fells streamed eastwards, carrying the distinctive Shap granite across the Pennines into Yorkshire.

In the Askrigg Block, the major east and south-east-trending valleys, such as Wharfedale, Nidderdale, Wensleydale and Swaledale, were dominated by glaciers of local origin. Farther south, the Pennines are lower, and, with minor exceptions, seem to have been invaded and crossed by ice from the Lake District or the northern Pennines. Parts of the southern Pennines may have remained unglaciated, even in the penultimate glaciation: for example, Kinderscout (635 m) shows no clear signs of ice having accumulated on it.

Modification of the landscape by glacial erosion is nowhere as profound as in other parts of highland Britain. Cirques are uncommon and poorly developed. U-shaped valleys are not often apparent, except for parts of Wharfedale, Littondale and Bishopdale, in the Askrigg Block, and for some of the glaciated valleys farther north. Depositional evidence of glaciation is, however, widespread. (1) Terminal moraines mark various limits attained in the last (Devensian) glaciation, (2) swarms of drumlins pass into upper Wensleydale and Stainmore, (3) erratics stand conspicuously on bare limestone surfaces (as in the case of the Norber Silurian erratics near Austwick), (4) alluvial flats mark the sites of former moraine or ice-dammed lakes, and (5) many spurs are elongated by drift tails [see King (1960)]. Glacial meltwater channels are abundant in certain areas, such as on the Pennine scarp rising above the Vale of Eden as described by Trotter (1929), in Wharfedale around Grassington, and in the Rossendale and Bowland uplands on the west. The channels are of varied origins (marginal, subglacial, direct overflow, etc.) and often turn out to be parts of highly complex glacial drainage systems [see for instance an analysis of two channels in Northumberland by Peel (1949) and the application of a subglacial hypothesis to them by Sissons (1958)].

A few of the highest western parts of the Pennines, where precipitation today is over 1800 mm, were able to support minor isolated glaciers in the post-Allerød (zone III) climatic recession. Manley (1959) has located the sites: Cross Fell, Mallerstang, the Howgill Fells, Combe Scar, Whernside and Ingleborough, all suggesting a snowline at about 700 m. Small moraines mark limited cirque glacier activity at these sites.

9.3 South-West England

In several respects, this is an area of great geomorphological interest. It stands next to lowlands (south-east England) and highlands (Wales), and is a critical region for establishing links in the palaeogeomorphological evolution of both areas. With a maximum height of over 600 m on Dartmoor, the area was nevertheless beyond the main limit of glaciation, although there were important periglacial effects in Pleistocene cold periods. Other points of interest concern the distinctive landforms developed on the large granite intrusions, and the coastal scenery which is among the most impressive in Britain.

9.3.1 Geology and Structure

The Mesozoic formations of the English lowlands approach the south-west peninsula on its eastern side, as far as a line roughly from Newton Abbot through Exeter to Minehead on the north coast (*see* Fig. 9.4). The Permo-Triassic rock shares the characteristics of its outcrop in the Midlands and represents a time when the south-western uplands, like Wales, were intensely eroded to produce an overall outline in plan broadly similar to that of today. The Jurassic formations end in bold escarpments some distance to the east, but the Cretaceous extends into the south-west, albeit with rapidly changing characteristics as the shoreline of the time was approached. Upper Greensand (Albian) caps the Haldon Hills south of Exeter, together with flinty relics of a former Chalk (Cenomanian) stratum. The nearest approach of *in situ* solid Chalk is, however, 25 km east of Exeter near Sidmouth.

The greater part of south-west England west of Exeter consists of Palaeozoic rocks. Except for the relatively small areas of Precambrian igneous and metamorphic rocks forming the Lizard and Start Point, the rocks fall into three main groups. First, there is the Devonian Series (mainly shales, slates, sandstones and grits, but also including some limestones in the Middle Devonian). These sediments accumulated in the south Devon geosyncline, deposition continuing into the Lower Carboniferous to produce the second main rock group, the Culm Measures (comprising various types of sedimentary rock, coarsening upwards). Neither the Devonian nor the Culm produces landforms of special interest apart from the coastal features, the relief tending to be smooth and undulating with few breaks, although, in the north, the Devonian rocks rise to 520 m in Exmoor. Devonian–Carboniferous deposition was brought to a close by the Hercynian (Armorican) orogeny, creating a complex series of east–west folds and numerous crush zones, tear faults and other fractures throughout the region. It was towards the end of the Hercynian orogeny that major intrusions of granite occurred, producing the third main rock group.

The late Carboniferous granite intrusions of south-west England comprise five main bosses: Dartmoor, Bodmin Moor, Hensbarrow, Carn Menellis and Land's End. These represent the main outcrops of the granite today, but are probably all linked at depth. Although the subsequent Permo-Triassic period was one of intense denudation in south-west England, the granite bosses were not unroofed at this time, because no granite detritus is known in the Permo-Triassic period. First exposure of the granite by erosion may have occurred in the Jurassic, but Smith (1961) shows from the mineralogy of the Cretaceous that, following uplift and erosion in the Albian, the granite was submerged in Middle Chalk times and possibly buried by a thin sedimentary cover, to be re-exposed in the early Tertiary (*see* Fig. 9.5).

Near the eastern edge of the Dartmoor granite lies the Bovey Basin, a small subsiding fault trough containing

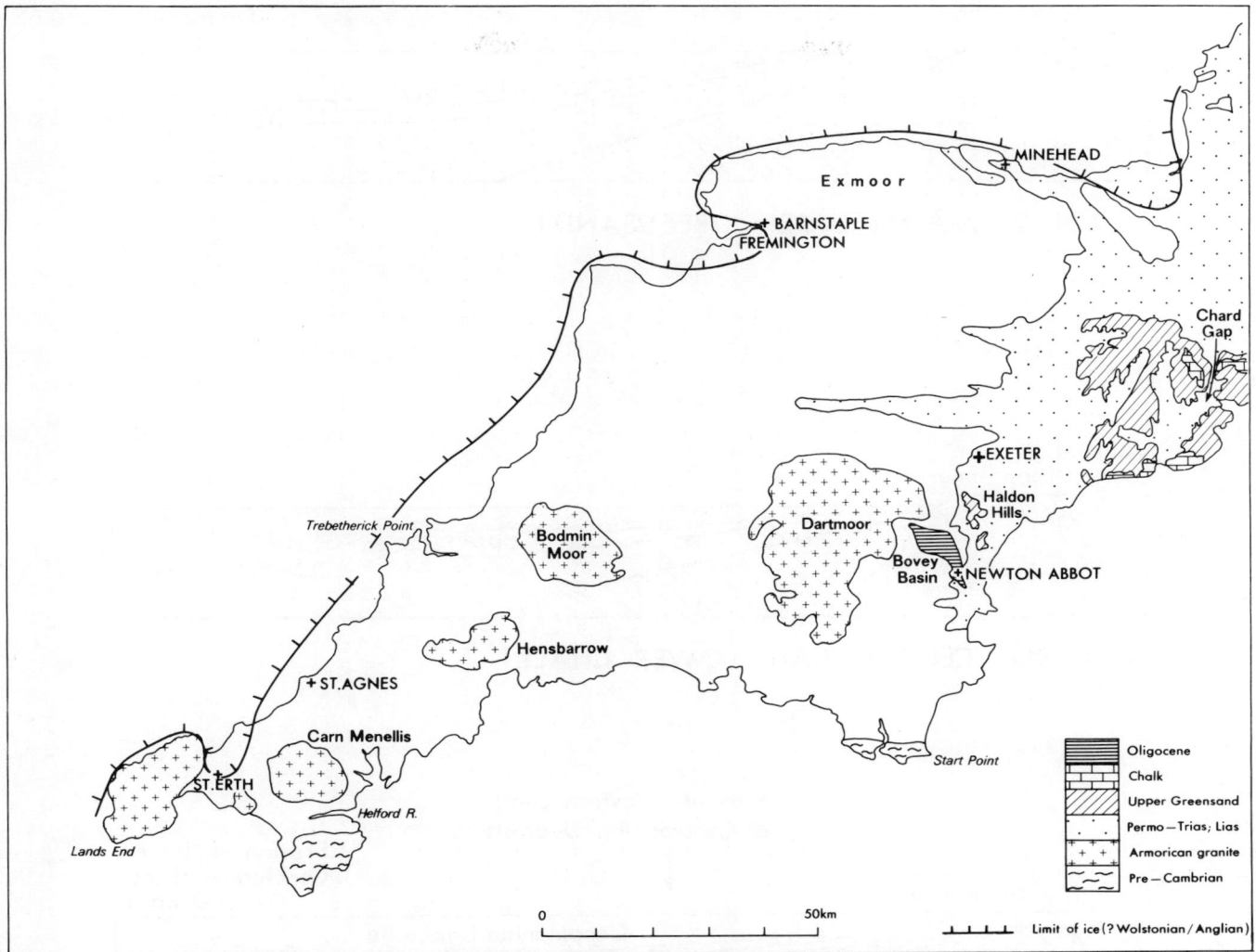

Fig. 9.4 Main geological features of south-west England. The area left unshaded represents the outcrop of the Devonian and Culm Measures (Carboniferous). The maximum limit of an older glaciation just reaching parts of the north coast is shown.

1.25 km of Eogene fluviolacustrine sediments (with granitic detritus). The faulting, which began in the early Tertiary, continued into the Miocene, when possibly many other faults in the south-west were reactivated.

9.3.2 Relief and Drainage

The highest elevations—Dartmoor (granite) and Exmoor (Devonian)—have already been noted. The response of the granite bosses to weathering and erosion is very varied: Bodmin Moor forms an upland rising to 420 m, Carn Menellis reaches only 250 m, while the Land's End granite is planed across at about 200 m. Among the hardest rocks of the region, the Precambrian rocks of the Lizard have also been truncated to a low plateau at about 100 m; it is clear that throughout much of south-west England, erosion in late Tertiary and Quaternary times has effectively subdued the landscape. Apart from coastal cliffs, the greatest amplitude of relief is found on the margins of Dartmoor, amounting to about 150 m.

Drainage patterns do not always show a simple response to structure and in part may have been inherited from a former Cretaceous cover. Some of the granite bosses show a tendency to develop radial drainage, as in the case of northern Dartmoor. The drainage density on the granite is surprisingly low ($\Sigma L/A = 4$–6 km) [see Brunsden (1968)] possibly because of the permeability afforded by joints and the water-holding capacity of both weathered granite and the thick peat cover. Some major stream segments show close adjustment to structure (e.g., the Helford River along the Meneage crush zone) but others (the lower Dart) possess strikingly discordant courses.

9.3.3 Planation Surfaces

These have been described by many workers from the highest levels of Exmoor and Dartmoor down to sea level. Early observations include those of Barrow (1908) on Bodmin Moor. More detailed investigation was pioneered by Balchin (1952) and Wooldridge (1954). A 'summit plain' was recognized by Brunsden (1963) on Dartmoor, from 580 m in the north to 500 m in the south. The

Low Relief   Foxmould

- - - - - - - - - Chert Beds - - - - - - - - -

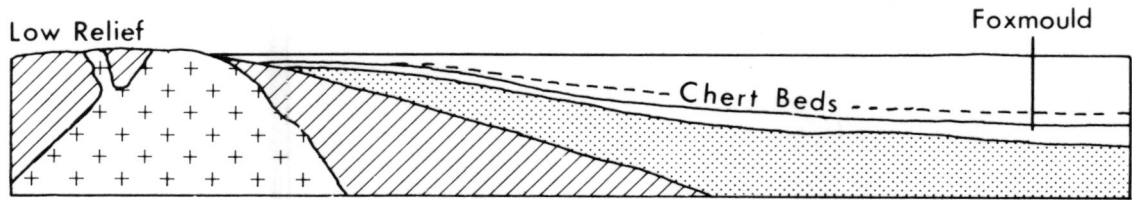

(i) UPPER ALBIAN (UPPER GREENSAND)

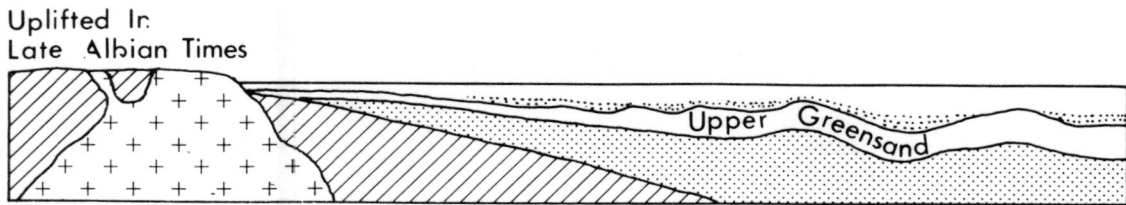

Uplifted In
Late Albian Times

Upper Greensand

(ii) EARLY CENOMANIAN (LOWER CHALK)

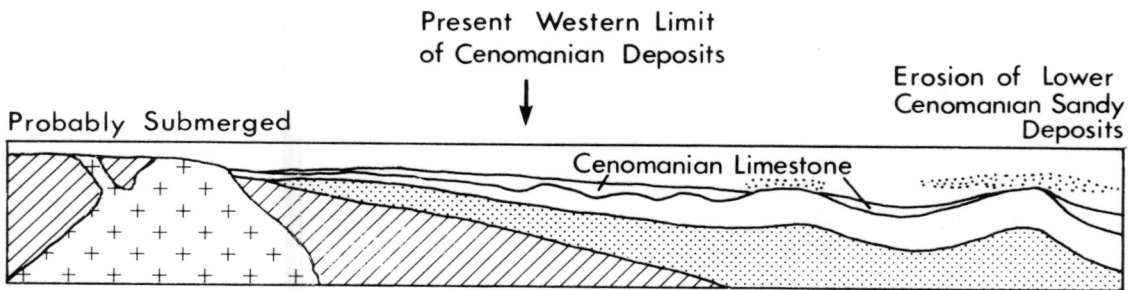

Present Western Limit
of Cenomanian Deposits

Erosion of Lower
Cenomanian Sandy
Deposits

Probably Submerged

Cenomanian Limestone

(iii) TURONIAN (MIDDLE CHALK)

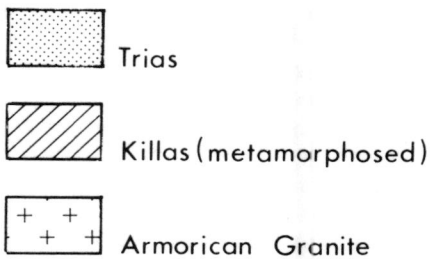

Trias

Killas (metamorphosed)

+ +
+ + Armorican Granite

Fig. 9.5 Relationships between the Dartmoor granite and the Cretaceous beds shown for three successive episodes [after Smith (1961)].

southerly tilt may be the result of earth movement, this view being supported by the apparent displacement of the surface by the Bovey rift. Its age is therefore generally thought to be early Tertiary, predating the Oligocene–Miocene movements. It would appear to represent a phase of base levelling, during which the Dartmoor granite was exposed from beneath a Cretaceous cover, and it could be the equivalent of the sub-Eocene surface of south-east England.

At a lower level of about 340–400 m, there is a feature described by Waters (1960) as a subaerial peneplain and by Brunsden (1963) as a 'partial peneplain'. Its age is unknown, but is possibly Oligocene. Below it extends a Lower Surface at 230–300 m, recognized by Green (1941) on Bodmin Moor and by Balchin (1952) on Exmoor. It penetrates the Dartmoor area as a series of valley-side benches—Brunsden (1963) distinguished four groups of these with the lowest surviving points at 230, 245, 280 and 300 m—and is preserved on the flat top of the Haldon Hills at 240–250 m. It has been correlated with the Miocene–Pliocene surface of Wooldridge and Linton in south-east England (*see* p.143) and with Brown's low peneplain in Wales (*see* p.121). It appears not to have been deformed by any later earth movements.

Below about 200 m more restricted surfaces occur that are usually thought to be late Tertiary–early Pleistocene marine, correlating with counterparts at similar levels in south-east England. Brunsden (1963) placed the marine limit at 210 m on the south-east of Dartmoor. Elsewhere in south-west England, a few occurrences of marine shingle relating to this marine transgression have been found (e.g., near St Agnes, Boscastle and on the Lizard). At St Erth, in the through valley linking St Ives Bay with Mount's Bay, Lower Pleistocene marine deposits have been preserved [see Mitchell (1965)], but fail to give evidence of a sea level higher than 60 m at the time of their formation. They may have been disturbed by an early glacial advance whose age is unknown.

9.3.4 Granite Forms

These forms deserve a special note because of their distinctiveness and the arguments that have centered on the origin of tors, often considered the most characteristic element of the scenery of Devon and Cornwall. Granite forms are closely controlled by the mineralogy and texture of the granite, by faults, by vertical joint, and by the pseudo-bedding or dilatation joints. The jointing exercises a major control on valley alignments and tor morphology.

Chemical alteration of the granite has occurred as a result of two processes: pneumatolysis (the action of high-temperature gases in the later stages of granite emplacement, leading to the formation of new minerals) and weathering. Pneumatolysis produced kaolinization of the granite on a huge scale in some areas, producing commercially valuable deposits of china clay (*see* Fig. 3.10). Weathering, not always easy to distinguish from pneumatolytic alteration, has selectively attacked certain minerals, causing the breakdown of feldspars to clay minerals and the oxidation of biotite leaving, after leaching, a residue of quartz and some mica particles known as

growan. In some sections, deep weathering of this type can be seen to have extended to depths of tens of metres and has been attributed to former tropical or subtropical climatic conditions. Proceeding first along joints, weathering attacks the edges of the joint blocks, rounding them to produce corestones (*see* Fig. 9.6). In areas of closely spaced jointing, weathering proceeds to greater depths; later erosion removing the regolith thus produces valleys or lowlands, while areas of more widely spaced jointing remain as uplands.

Linton (1955) has suggested that decomposition of the granite, penetrating more deeply in some areas than others because of variations in joint spacing, and followed by the partial stripping of the weathered debris, has been the cause of tor formation (*see* Figs. 9.6 and 3.2). He attributes the period of deep weathering to the Tertiary or even the Pleistocene interglacial periods, and the exhumation of corestones, finally producing the tors, is attributed mainly to gelifluction and washing by snowmelt in the glacial periods. On the other hand, Palmer and Neilson (1962) regarded tors as Pleistocene features, developed directly from frost action on jointed bedrock, followed by removal of the frost-shattered debris by gelifluction. They argued that the decomposed granite is mainly due to pneumatolysis, and that the corestones composing tors tend to be angular rather than rounded. The hypotheses of Linton and of Palmer and Neilson are not necessarily contradictory, however, and it seems probable that most tors have evolved through a complex history involving deep weathering, exhumation and periglacial attack, and modification. Many tors are also composite features, that is, they occur in groups on an interfluve, separated by 'avenues' or strips of flat ground. Fig. 9.6(iv) suggests how this situation may be related to a stage in the destruction of an original dome, on the evidence of pseudo-bedding, a good example being Hay Tor. Tors are frequently surrounded by 'clitter' slopes of angular debris, suggesting that, whatever the earlier history of the tor, frost action and gelifluction have certainly played a major part in their later modification and partial destruction.

9.3.5 Periglaciation

South-west England in the Pleistocene lay near the border of the ice at times of glaciation; although no actual glaciers formed on even the high ground of Dartmoor, perennial snowbeds must have been extensive, and there was probably continuous permafrost as far south as Barnstaple. In the Devensian glaciation, the ice limit lay well to the north in South Wales, but in the preceding Wolstonian, Irish Sea and Welsh Ice spread as far as the north coast of Devon and Cornwall (*see* Fig. 9.4), leaving unmistakeable glacial till at Fremington [see Maw (1864), Stephens (1966)] and Trebetherick Point [see Arkell (1943)] and even reaching as far as the Scilly Isles [see Mitchell and Orme (1967)]. Meltwater was impounded in the Severn estuary and the adjacent Somerset lowlands to form Lake Maw, which overflowed southwards through the Chard Gap. It may have been ice of this glaciation that left giant erratics (some weighing 50 t or more) on the shore platforms of Devon and Cornwall, while, for similar

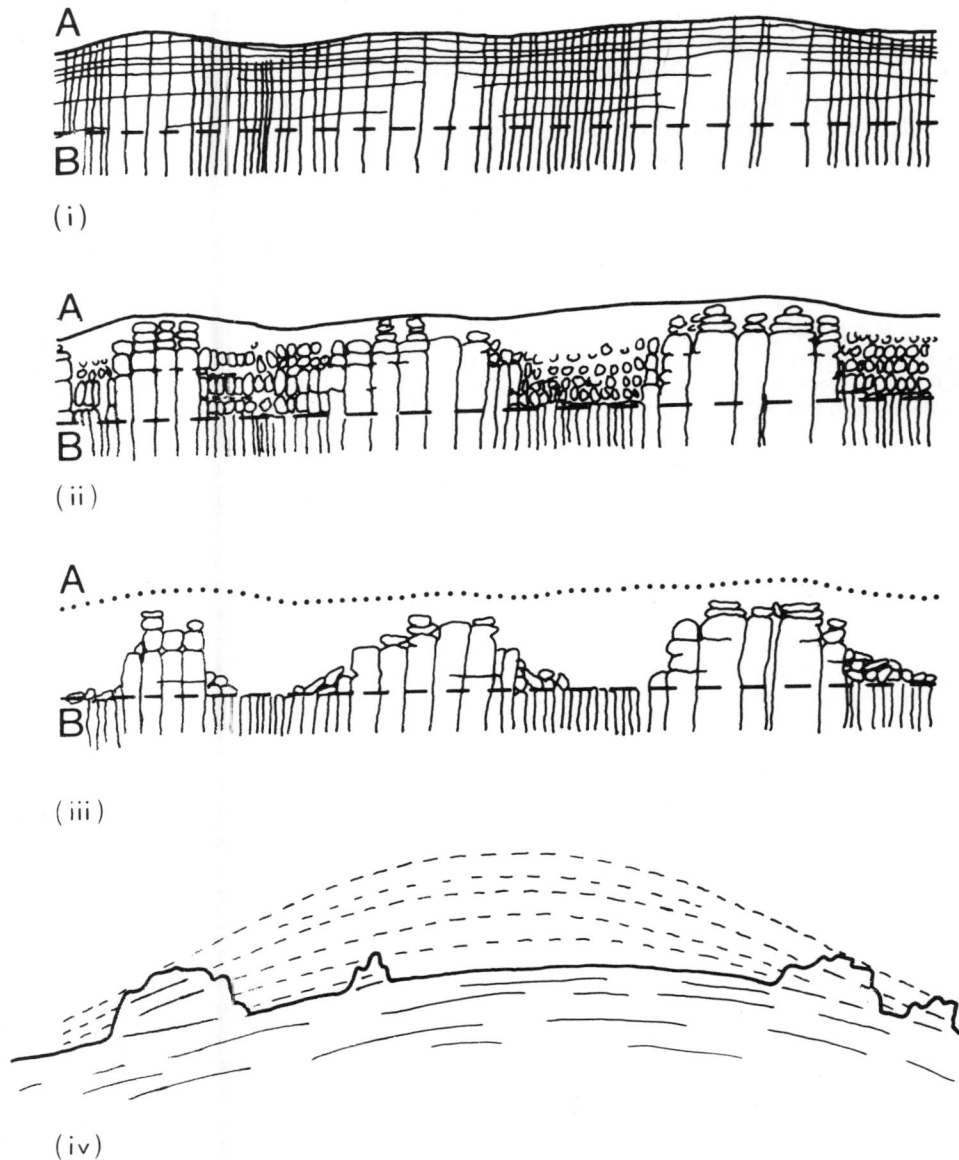

Fig. 9.6 The formation of tors. Diagrams (i) (ii) and (iii) represent stages in tor formation according to Linton (1955). A: original land surface; B: basal surface of deep weathering. Diagram (iv) shows the destruction of an original granite dome controlled by sheet structures in the granite; the tors here represent residuals of the dome whose centre has been removed by erosion.

erratics on the southern coast, floating ice has to be invoked. The bulkiest Pleistocene deposits in south-west England are the multiple layers of gelifluced head, often finely displayed in coastal cliff sections (*see* Fig. 9.7). As the head has suffered repeated movement, its chronology and stratigraphy are complex and still uncertain [see Waters (1965)]. Some is certainly Wolstonian and bears evidence of disturbance by cryoturbation and frost cracking in the Devensian. Other signs of periglaciation include the clitter slopes around tors already mentioned, involutions and fossil ice-wedge structures, thufurs and cryoplanation terraces. The latter, well developed on Cox Tor, for example, were first studied in south-west England by Guilcher (1950) and by Te Punga (1957). But apart from

these smaller-scale features, the most extensive effect of periglaciation in south-west England was undoubtedly the infilling of valleys with cryoturbation deposits and the general smoothing of the landscape, especially on the softer Devonian and Culm rocks.

9.3.6 Coastal Features

Because of its exposure to the open Atlantic in a south-westerly direction, the peninsula is subjected to intense wave attack; together with the competence of the jointed granite and the grits and sandstones of the Devonian, this has produced some of the finest coastal scenery in Britain (*see* Fig. 9.8). Three main types of cliff profile may be distinguished: flat-topped, bevelled and the hogback type.

Fig. 9.7 Porth Nanven, Cornwall. On the left is the fossil cliff in bedrock at whose foot accumulated the coarse rounded shingle and boulders of a Pleistocene raised beach. The marine deposits were then covered by 'head', the angular gelifluction debris. (Embleton).

Fig. 9.8 Polurrian Cove, a typical bay on the west coast of the Lizard peninsula, Cornwall, representing differential marine erosion in Precambrian and Lower Palaeozoic rocks. (Embleton)

The shape is a function of many factors, including the relative rates of marine and subaerial denudation, the structure and the physical history of the area, especially the relative changes in land and sea level. The hogback cliffs are best developed in north Devon where, above the rocky cliffs subject to present-day wave attack, a convex slope rises at decreasing angles to a rounded summit, which may be 200 m or more above sea level. Such a massive feature probably has little to do with marine erosion and may mark the structurally determined margin of the Bristol Channel, probably first roughed out in the Permo-Triassic period.

Changes of sea-level have characterized the Cenozoic history of south-west England, as has already been shown. The later Pleistocene changes are witnessed in a series of raised beach platforms (from 0 to 18 m) some of which may be Ipswichian and some Hoxnian. The latest change of level—the Flandrian transgression—caused drowning of river estuaries to give fine examples of rias at Falmouth, Fowey, Salcombe and elsewhere.

9.4 Southern Ireland

The area lying south of the main Armorican thrust (approximately from Dingle Bay to Dungarvan Harbour) is one of distinctive Appalachian-type folding, trending generally east–west, and giving rise to a ridge-and-valley landscape. The strata involved in the Hercynian (Armorican) folding are the Old Red Sandstone (Devonian), up to 6700 m thick in Kerry, followed by the Carboniferous. In the northern part of the region, the Carboniferous consists of a thick sequence of Carboniferous Limestone underlain by the Lower Limestone shales. In the south, the remaining Carboniferous is largely noncalcareous consisting of the Cork Beds, a thick series of shales and sandstones. The folding is quite sharp, often markedly asymmetrical with steeper northern limbs to the anticlines, and there are some overfolds all reflecting pressure of the Hercynian orogeny coming from the south. Because of regional differences in the geology and relief, the following account will treat the area in three regions: (1) the western half of the region north of the Cork Beds outcrop, (2) the eastern half also north of the Cork Beds and (3) the area underlain by the Cork Beds. The dividing line between regions (1) and (2) runs approximately south-south-west from Mallow in the Blackwater valley.

9.4.1 The West

This includes the classic ria coastline of south-west Ireland, where peninsulas finger out into the Atlantic between the bays of Dingle, Kenmare and Bantry. Anticlinal mountains of Old Red Sandstone dominate these peninsulas, running back into the land and continuing along the Armorican fold axes. A very close relationship can be demonstrated between relief and tectonics; even the heights of the ridges match the amplitudes of the folds. Few patches of Carboniferous now remain; the Old Red Sandstone has been almost totally uncovered but in terms of the broad outlines of the ridges, erosion has thereafter not proceeded much farther. The region has been severely glaciated, both in the Munsterian and in the Midlandian; in the latter phase, a separate glacial system existed in Kerry and Cork, independent of the Midlandian ice sheet over north and central Ireland (*see* Fig.7.30). Numerous cirques, arêtes, lakes basins and roches moutonnées are to be found. In the Dingle peninsula, King and Gage (1961) enumerated 34 cirques of which 31 face between north-west and east. Linton (1963) noted that the most westerly of all British cirques stands on the north-eastern slope of Mount Eagle (Dingle peninsula), containing a lake at 237 m with the cirque headwall rising a further 450 m. The drowning of the rias between the peninsulas was, in part at least, quite recent as shown by the discovery of freshwater silts at − 57 m, on the floor of Bantry Bay, ^{14}C-dated at 11 000–12 000 years BP [see Stillman (1968)].

9.4.2 The East

Long narrow synclinal valleys here preserve in them some strips of the former Limestone cover between the anticlinal Old Red Sandstone uplands. The slopes of the latter are again probably little modified by erosion from the sub-Carboniferous surface, as shown by the occasional presence of small lenses of Limestone preserved in minor flexures on the anticlinal limbs. The sandstone ridges rise to over 600 m (e.g., Knockmealdown Mountains, 795 m) and are cut across, or their flanks bevelled, by planation surfaces possibly of Plio-Pleistocene marine origin. The drainage is partly directed along the synclines, but the Blackwater and the Lee both suddenly turn south to cross the Armorican structures at right-angles to reach the sea. Whether these lower parts of their courses have been superimposed remains an open question (*see* Section 7.6.2). Jukes (1862) was one of the first to discuss such a possibility, suggesting that the discordant north–south streams of southern Ireland in general were ancient consequents superimposed from a marine surface. Superimposition from a now-vanished Chalk cover has also been postulated. Davies and Whittow (1975) have recently developed an alternative hypothesis involving denudation of a complex series of periclines whose anticlinal axes were crossed at their lowest points by drainage escaping from the synclines.

9.4.3 The South

In this area, the Carboniferous cover consisting of the more resistant Cork Beds is relatively continuous, the Old Red Sandstone appearing only in parts of the anticlinal ridges (e.g., the Mizen Head peninsula). Otherwise, tectonic control of the relief is similar to the other areas of the south-west. Remnants of planation surfaces are again seen bevelling the uplands, where elevations are lower than elsewhere, being mostly below 400 m. Recent submergence of the intricate coastline with its rias and bays, including Cork Harbour itself, is evident.

9.5 Armorican Massif

From Cotentin southwards to Vendée and from Normandy to the westernmost point of Brittany, the Armorican massif is characterized by gently undulating relief consisting of hard rock interfluves and narrower valleys. In sharp contrast, the region is bordered by a highly indented coastline. Structurally, the area is formed mainly of Precambrian and Palaeozoic rocks, together with some large granitic intrusions. The Palaeozoic series consists of slates, shales and sandstones, with some more resistant quartzites and grits, whereas the Precambrian includes metamorphic and igneous rocks. Around the massif and its borders, there are Mesozoic sedimentary rocks whose near-shore facies suggests that these formations never penetrated the massif to any great extent [see Durand *et al.* (1977), Doré *et al.* (1977), Gabilly *et al.* (1978)].

The patterns of geological outcrops (*see* Fig. 9.9) and, to a lesser extent, the trends of certain relief features, show the impress of successive periods of tectonic activity [see Meynier (1947)]. The Caledonian orogeny undoubtedly affected the Precambrian and Lower Palaeozoic rocks, but its effects were overprinted by, and therefore not always easily distinguishable from, the later Hercynian movements of later Carboniferous–early Permian age. The

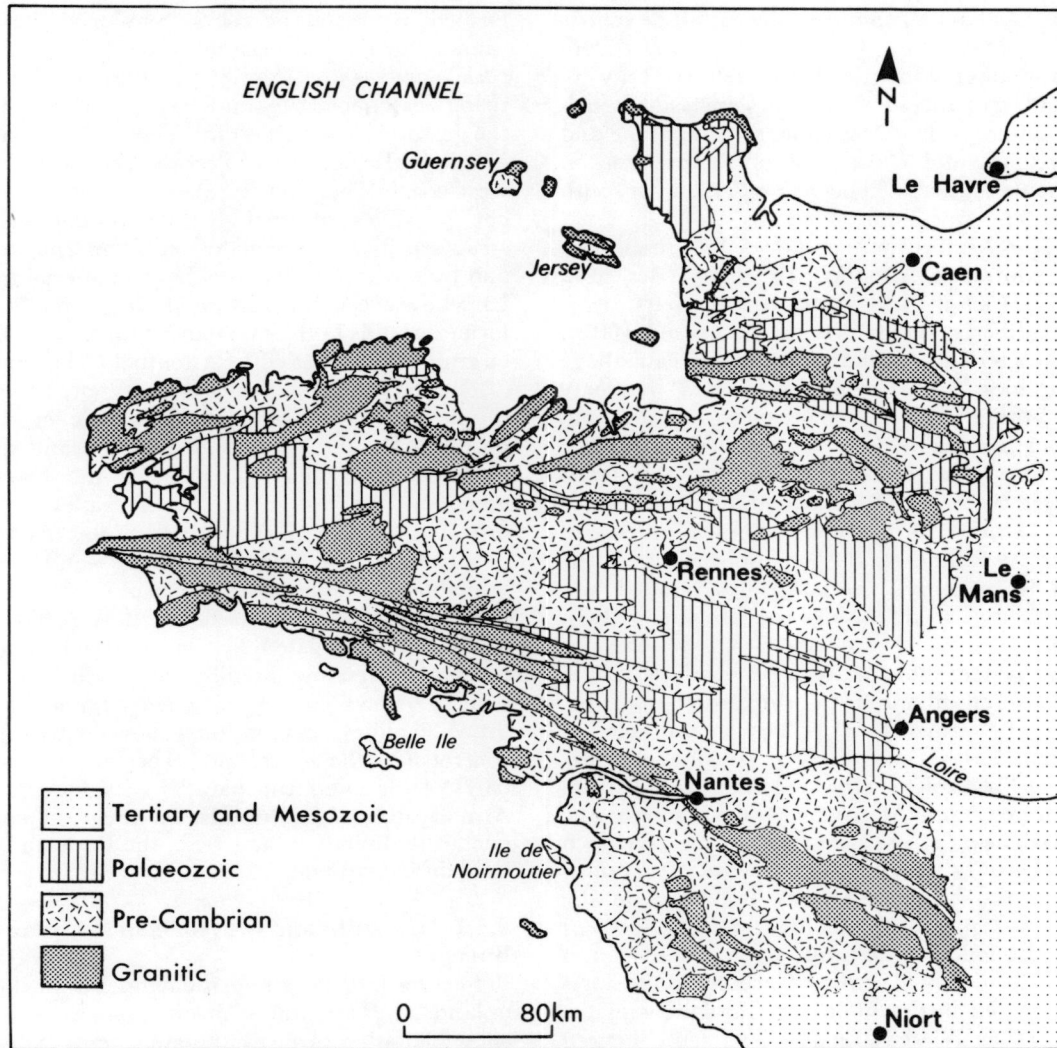

Fig. 9.9 The geology of the Armorican massif.

general structural pattern is fan-shaped, the trends diverging from west to east. Thus from Cornouaille southeastwards and roughly parallel to the southern coast runs the characteristic Armorican trend, through the Landes de Lanvaux and the Sillon de Bretagne (a narrow granite ridge 100 m high). In contrast, the main structural direction in the north of Armorica is west-south-west to east-north-east, from Léon to Cotentin. The Channel Islands represent the non-submerged parts of these old structural ridges. Between the Armorican trend of the south and the Léon trend of the north, a synclinorial axis runs west–east from the Finistère Basin through the Rennes Basin towards Laval; Carboniferous rocks are preserved along several of the downfolds. The whole area of Armorica is also greatly fractured mainly by faults of Hercynian age, giving horsts and rifts, although their present relief is the result of differential denudation rather than direct tectonics.

This is an area that has remained predominantly under the influence of subaerial denudation since the Palaeozoic, which explains the importance in the landscape of planation surfaces and hard-rock residuals. Recent marine sedimentation consists only of a few scattered relics of a Tertiary transgression that partially covered the southern margin, extending across eastern Brittany from the Loire estuary to the south of Cotentin. On the other hand, a superficial mantle of Tertiary age, consisting of sandy clayey weathering products, can be seen almost everywhere, as well as Quaternary gelifluction deposits (head) and a veneer of Quaternary loess which is particularly extensive in the east and on the north coast. The Tertiary Period was also marked by renewed slight movements along Hercynian fractures and overall uplift, with, in places, gentle warping and tilting. The general arrangement of the drainage system reflects these tectonic disturbances.

The overall uplift that the region suffered in the mid-Tertiary initiated renewed denudation of the massif, with widespread removal of the marginal Mesozoic strata and of discontinuous Tertiary sediments. The resistant granites and granulites in the cores of anticlines or in other deep-seated intrusions were exposed as elliptical dome-like masses, rising above the plateaus of crystalline schists or Palaeozoic sedimentaries. Tertiary strata, apart from the superficial weathered mantle already referred to, were reduced to small remnants in basins and depressions, such as the Eocene outcrops near Mayenne, the Oligocene and Miocene limestones and clays west of Rennes and in Vendée, and some scattered Pliocene outliers in the south between Redon and Angers.

The relief of the Armorican massif does not present any very striking contrasts. Although the land rises to a maximum of 417 m in the most eastern part, near Alençon, local differences in level rarely exceed 100 m. The most characteristic landform is the plateau above which project linear rocky ridges, separated by peaty depressions; in other places the plateau is moulded into low undulations. In addition, however, there are short, steep-sided valleys that are entrenched into this landscape, providing a contrasting element to the scenery. Otherwise, the Armorican massif is the least dissected of all the ancient crystalline massifs of France, but, at the same time, it is the only one that is surrounded and penetrated by the sea on three sides [see Guilcher (1948), Klein (1974)].

The drainage system shows many signs of adaptation to structure, but also an inheritance of discordant elements. Some discordant segments may be attributable to superimposition, but former theories involving widespread covers of Mesozoic or other post-Hercynian strata are now discounted. Some transverse valleys may be the result of antecedence, the river courses having developed prior to mid-Tertiary tectonic movements. It is now considered that the latter had more influence on the drainage than was previously thought. Many rivers or parts of rivers also follow structurally weak zones, such as the strips of Devonian and Lower Carboniferous slates and shales preserved in downfolds or rifts following the Armorican or other related trends: the deeply incised meanders of the Aulne in the Châteaulin Basin mark the lower course of a river that is, overall, structurally controlled, and from Quimper westwards to the Baie des Trépassés is another example of a valley line adapted to the Armorican structure, although occupied by portions of three different river systems. The transverse river courses, whether due to antecedence or some form of local superimposition, also frequently betray structural influence on valley form, the valleys narrowing where resistant outcrops are crossed and widening in the intervening softer bands. The Vilaine below Rennes, for example, crosses Silurian sandstone ridges in relatively constricted sections, but its valley broadens markedly in the shale or slate zones.

The coast of Armorica undoubtedly provides the most spectacular scenery. Marine erosion has selectively exploited differences of rock resistance. Inland ridges run out to rocky headlands, some with cliffs 100 m or more high, and continue out to sea in islands; for example, the granitic ridges of Cornouaille end in the Pointe du Raz but structurally continue far out to sea, for 35 km or so, in strings of islets and reefs, including the Ile de Sein. The majority of the steep headlands around the coast are formed in crystalline rocks. Similarly, the hundreds of islands beyond the coast are usually residuals of resistant rock, such as the granite pyramid of Mont St Michel, rising 78 m above the sands exposed around it at low tide, the granitic Ile d'Yeu off the Vendée coast, rising to 35 m, and Belle Ile, a mass of Precambrian schist off the southwest coast. The bays between the headlands are mostly cut in slates or shales, their outlines controlled by structure, by the extent of recent marine transgressions and by the accumulation of beach material in them. Sand bars have been built up in places, sometimes attaching former islands to the mainland: Quiberon, a former island of granite, is now tied by a tombolo 7 km long.

The evolution of the coast has not, however, been a simple matter of differential marine erosion and deposition. Recent changes of relative land and sea level have played a vital role. During the Flandrian transgression, the lower parts of valley systems were drowned—in the Rance estuary, for instance, the sea penetrated 27 km inland to Dinan—and marine erosion was able to gain access to areas of weaker rock that were previously immune. The broad Baie de Douarnenez was not simply excavated by waves attacking the slates; it represents, first and foremost, the result of the drowning of several valleys converging on a common lowland. The drowned valley systems (rias) show contrasting structural control. On the north and, especially, the south coasts, parts of the inlets run parallel with the coast, along the Armorican or Léon trends; on the west coast, the rias penetrate directly inland (e.g., the Rade de Brest and the Baie de Douarnenez).

9.5.1 Cotentin and the Normandy Uplands (the Bocage)

Projecting into the English Channel from the Normandy uplands, the Cotentin peninsula is characterized inland by a low planation surface cut mainly across Siluro-Devonian rocks (slates, shales and sandstones) with granitic intrusions. The latter appear as almond-shaped outcrops inland, marked by moorland hills rising at a maximum to about 200 m; granite also forms the coastal headlands of the Pointe de Barfleur, Cap de la Hague and Les Pieux. The relief drops to less than 100 m in central Cotentin, where the Douve drains eastwards. Here, younger rocks (Permian, Triassic and Jurassic) outcrop around lowlands covered with Pliocene deposits and recent alluvium. Some areas are marshy and sufficiently low-lying to be subject to flooding. To the south of these lowlands, the Armorican rocks gently rise again in the Normandy uplands, eventually reaching a maximum height of 417 m in the Butte des Avaloirs overlooking the Sarthe valley. This part of Armorica experienced the greatest uplift in the mid-Tertiary, and also suffered renewed fault movements. Large granite domes form distinct groups trending west–east or west-north-west–east-south-east. Those uplands form a major watershed between north-flowing rivers (Vire, Aure, Orne) and the Sarthe and Mayenne flowing south into the Loire. Their valleys narrow markedly where they breach resistant rock ridges and open out in the shales and phyllites [see Elhaï and Journaux (1969)].

Except for a few rocky headlands in the north and west already noted, most of the Cotentin coast, and especially the east coast, is low. Although there are cliffs on the east coast, these are now fossil cliffs, fronted by sandy beaches, salt marshes and dune belts [see Joly (1939)]. Some dunes reach heights of 20 m but are now mostly stabilized. In places, the cliffs are of Quaternary head deposits which can be clearly seen to overlie raised beaches.

9.5.2 Brittany

Armorica, as a whole, possesses two principal upland areas: the interior of Normandy (see section 9.4.2) and the interior of Brittany. The latter is in turn divided into two upland regions separated by the east–west Aulne valley: the Montagnes d'Arrée in the north, built of granite and Devonian sandstone and rising to 384 m in the Signal de Toussaines, and the Montagnes Noires to the south, also largely granite, reaching 326 m. Traced eastwards, these uplands, bevelled by planation and diversified here and there by granite tors and sandstone ridges, join together in the Landes de Menez (340 m). Between the two uplands lies the Châteaulin Basin drained west by the Aulne and floored with less-resistant Devonian–Carboniferous shales. The western parts of the Châteaulin Basin have been invaded by the sea at Brest and Douarnenez.

The Montagnes d'Arrée fall westwards in elevation to the lower plateaus of Léon and Trégorrois, while the Montagnes Noires decline similarly to lower-level planation surfaces in Cornouaille. These surfaces end abruptly in the strikingly cliffed coastline. The cliffs of the north and west are high and broken by deep inlets and rias, whereas those of the southern coast are lower and less indented. On the southern coast, there are also some alluvial basins (e.g., Morbihan). This coast is fronted by a broad offshore platform from which rise some substantial island, such as the Ile de Groix and Belle Ile [see Guilcher and Jist (1969), Ters and Pinct (1969)].

9.5.3 Central Armorican Basins (Bretagne Intérieure)

Opening southwards towards the lower Loire and draining to it by the Vilaine and Mayenne rivers is an area of lowlands mostly below 100 m framed by the uplands of Normandy to the north and north-east, and those of the Brittany peninsula to the west and north-west. The Rennes Basin on which converge the Ille, upper Vilaine and other streams, is indeed the largest lowland in Brittany. It appears to be a downfaulted and subsiding area [see Meynier (1947)], with a distinct fault-line scarp on its southern edge. In the early and mid-Tertiary, the Rennes Basin was at times invaded by the sea and later occupied by a lake, testified now by the Oligocene and Miocene deposits. There are also patches of the marine 'faluns' facies occurring. But over much of the area, erosion has removed these younger deposits, uncovering the schists, shales and sandstones and even, north-east of Rennes, granite. The part played by these older rocks in the landscape is, however, relatively subdued: the harder rock outcrops are more noticeable in terms of valley narrowing than relief amplitude, except towards the margins of the Normandy uplands. Between the latter and the uplands of the Landes de Menez, there is a low saddle leading north towards the coast near St Malo, with its rocky cliffs and rias (e.g., the Rance estuary). Farther east along this coast, accumulations of marine and estuarine sediments begin to fill the bays and their shallow floors, building out the coast and threatening to change the island status of the famous Mont St Michel.

9.5.4 Lower Loire Valley and Vendée

For the last part of its course, the Loire crosses the Armorican Massif in a broad, but entrenched, alluvial-floored valley. It leaves the Cretaceous rocks of the Paris Basin a few kilometres above the confluence of the Maine and enters first an area of Lower Palaeozoic slates. Subsequently, the valley widens as an outcrop of Coal Measures is encountered; then it is once more constricted in its passage across Precambrian schists. From this point to the sea, the valley is largely unimpeded, except for the granite ridge of the Sillon de Bretagne on whose southern end stands the city of Nantes, and another lower parallel ridge at St Nazaire. Clearly, the lower Loire is discordant with the structure, crossing a range of hard and soft outcrops at an oblique angle. In part, the origins of this unusual relationship lie in some of the Armorican tectonic structures re-activated in the Tertiary, but other contributory factors include the development of a fluviomarine gulf in the Neogene, submerging the Armorican structures in the lowermost Loire, and the possible former greater extension of Mesozoic strata into this part of Armorica from which superimposition has occurred. The whole of the lower Loire valley is occupied with enormous quantities of sandy alluvium, the river channel dividing around shifting sand banks and elongated islands. Below Nantes, former bays of the sea have been infilled (Grande Brière) and numerous *étangs* mark the reed-covered marshlands. Some former islands (e.g., Le Croisic, Noirmoutier) have been completely surrounded by the sediments and are either entirely or almost totally attached to the mainland.

Separated from the rest of Armorica by the Loire lies Vendée. Bevelled by erosion, uplifted and fractured by Tertiary tectonics, it is characterized by multiple planation levels. The uppermost bevels an axis of high ground, underlain by granites and schists with the typical Armorican trend running north-west to south-east, at about 250–280 m. These are the Gâtine Heights (or Vendée Hills), where a few residuals and tors rise a little farther above the general level, to a maximum of 295 m in the case of the Puy Crapaud. To the north-east is the region of Mauges where lower surfaces cut across Precambrian schists finally descend to the Loire. On the Atlantic side of the Gâtine Heights, the bocage country of lower Vendée lies mostly between 50 and 80 m, the Precambrian rocks being masked in places by Tertiary sediments. There are also rifts in which Mesozoic strata have been preserved, such as the inlier of Jurassic (with Carboniferous also) east of La Roche-sur-Yon, its downfaulting following the Armorican trend and enclosing the Chantonnay plain. The Atlantic coast of Vendée is low and almost everywhere bordered by dunes and salt marshes. Only in a few places, such as the Ile d'Yeu and the north of Noirmoutier, can low cliffs of granite or schist be observed [see Gabilly *et al.* (1978)].

9.6   Massif Central

The Massif Central forms a prominent block of high ground between the Paris Basin to the north, the Aquitaine Basin to the south-west, Languedoc to the south-east and the Sâone–Rhône corridor to the east. Measuring overall 450 km from north to south and 300 km from east to west, it occupies about one-sixth of France. From its uplands, drainage diverges to reach four of the great river basins of France: those of the Seine, Rhône, Garonne and Loire [see Glangeaud *et al.* (1969)].

Its clearest boundary lies to the east. From Morvan to Vivarais, it first of all overlooks the Burgundian plateaus, then, from Mâcon to Valence, it dominates the Bresse lowland and the Rhône trough. On the south-east, the Cevennes, deeply dissected by Mediterranean streams, rise sharply up to 1500 m and, in places, 1700 m above the Languedoc garrigues. The Cevennes are continued to the south-west by the south-facing escarpments of the calcareous Causses and then by the Montagne Noire. Everywhere else, the borders of the Massif Central are the result of erosion cutting into the sediments of the Paris Basin, the Poitou sill and north-east Aquitaine. The only clearly defined limits in the whole of this sector are formed by the major north–south fault of Villefranche and the southern edge of the Permian Brive Basin, which separates the Massif from the Causses du Quercy (*see* Chapter 8).

Within these confines, the Massif Central stands as an upland or mountainous area, characterized by plateaus interrupted by deep valleys and tectonic depressions (*see* Fig. 9.10). The mean altitude is 715 m, but distribution of the highest points reflects either the general rise to the east (1400 m in Lyonnais, 1702 m in Mont Lozère, 1567 m in the Montagne de l'Aigoual) or the superimposed volcanic structure of the Plomb du Cantal (1855 m), the Puy de Sancy (1886 m), Monts du Velay (1754 m) and the chain of the puys [e.g., Puy de Dôme (1465 m)]. Basically, the Massif represents an uplifted or updomed portion of the old Hercynian basement of central Europe built largely of Precambrian crystalline rocks and strongly folded Palaeozoic strata, subjected also to granitization and fracturing. Later, the Massif was bevelled by successive planation surfaces in the Tertiary, at which time it also suffered further tectonic movements—mainly tilting and renewed faulting—and volcanic activity extending into the Quaternary. In the Pliocene and Quaternary, there was further uplift and the incision of the drainage. The Massif Central is thus a distinctive unit, in addition to being the largest and most elevated of all the Hercynian massifs of Europe [see Baulig (1928)].

9.6.1   Geological Structure, Lithology and Tectonics
The deep structure of the Massif is one of the best known in France owing to a wealth of gravity, magnetic and seismic data for the region. On this basis, it is possible to distinguish three broad regional types.

1)   In the west (Limousin) and south (Rouergue, Causses, Cevennes), the crust, which is uniform and thick (30 km) and with no important gravity anomalies, is typical of an old basement in isostatic equilibrium.
2)   In the east (from Morvan to the Cevennes), the crust is of the same type but thinner (i.e. 25–30 km) and much more broken by faults or intracrustal displacements.
3)   In the centre (i.e. Auvergne, Velay), the crust is even thinner (24–25 km), possessing numerous gravity anomalies which reflect the existence of rift valley structures and Cenozoic volcanic activity.

Another fundamental characteristic of the deep structure is the existence, from Toulouse to Nevers, of a major sinistral tear fault, long known from its surface expression as the 'coal furrow'. This fracture, which in Nivernais links with a marked fault zone in the Paris Basin, separates the Massif into the following large blocks (*see* Fig. 9.10).

1)   On the west, there is the Limousin block, which is relatively little fractured and essentially formed of metamorphosed granitic rocks.
2)   To the east is the Auvergne–Cevennes block, where the much-broken basement is to a considerable extent covered with non-metamorphosed sediments and by volcanic structures.

The evolution of the ancient crystalline basement of the Massif began in the Precambrian Era, when the Brioverian phase of metamorphism (650–670 Ma) produced a folded series of schistose rocks. This was followed by the first phase of granitization in the Cambrian–Ordovician (470–530 Ma), which is especially evident in Limousin and the Montagne Noire. The principal phase of metamorphism, however, according to the most recent studies, dates from the Devonian (350–400 Ma). Its effects were superimposed on those of earlier date and dominate the geology of the Cevennes, Velay and the Montagne Noire. In turn, this phase was followed by intensive magmatic activity at 345–370 Ma, during which the other granitic masses were emplaced, essentially those of Auvergne (Madeleine, Thiers) and Limousin (Guéret) [see Peterlongo *et al.* (1972)].

The Hercynian orogeny commenced in the Devonian, as evidence from the Montagne Noire in the extreme south shows, but the major disturbances date from the post-Visean. In the Westphalian, large fractures broke through the Hercynian folds and lacustrine basins with coal-bearing deposits developed along them. In turn, these deposits were folded in the Stephanian. At the same time, the late Carboniferous is distinguished, mainly in the east and south-east, by the emplacement of the last great granite masses: Morvan (320–330 Ma), Margeride (323 Ma), Velay–Forez (300–320 Ma), Mont Lozère (290–300 Ma), Aigoual (298 Ma) and Sidobre (279 Ma).

It is quite difficult to define the Hercynian tectonic trends because of later dislocation. It is possible, however, to note the existence of preferred directions adopted by folds and fractures: mainly north–south or north-west–south-east (the Armorican trend) in Limousin and mainly

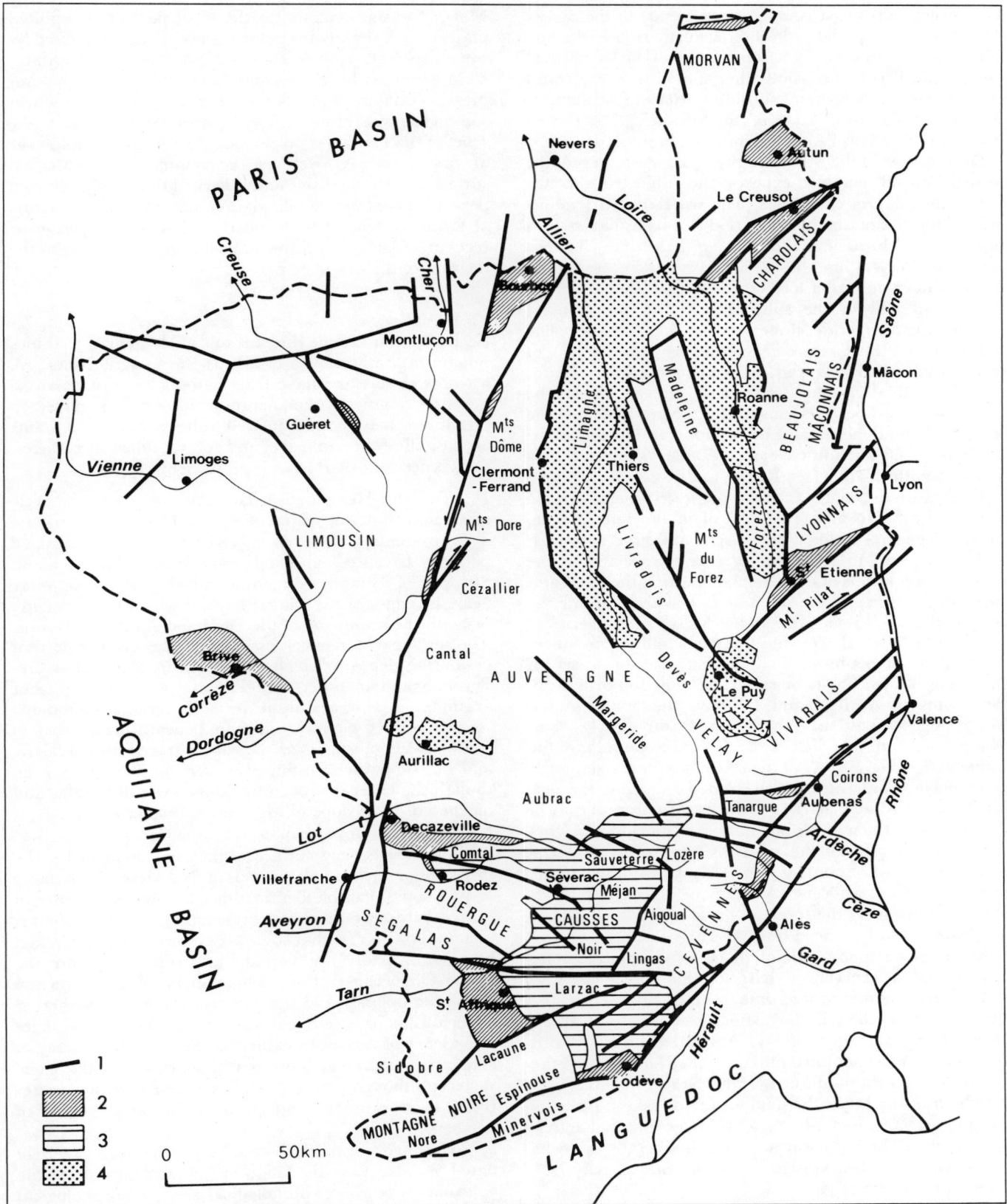

Fig. 9.10 Some geological features of the Massif Central, and a location map. (1) Major faults. (2) Basins of Carboniferous rocks. (3) The limestone areas of the Causses. (4) Basins of Tertiary sediments.

south-west–north-east (the Variscan trend) to the east of the Allier and especially the Loire. Furthermore, the end of the Hercynian orogeny was marked by relative uplift of the Massif. Permo-Triassic sedimentation was as a result concentrated in peripheral subsiding basins: for example, Autun, Le Creusot, Lodève, St Affrique, Decazeville, Brive and Bourbon l'Archambault (*see* Fig. 9.10).

During most of the Mesozoic Era, the greater part of the Massif Central remained exposed and subjected to continental denudation conditions. The sea returned temporarily at the end of the Liassic period and its transgressions through the Jurassic period penetrated some distance across the crystalline base. However, their associated deposits have only been preserved around the margins of the Massif and in the subsiding gulf of the Causses between the Margeride, Rouergue and the Cevennes. The beginning of the Cretaceous period saw the return of continental conditions and ushered in a phase of prolonged denudation and weathering which, up to the present-day, has been responsible for the production of the sandy plateau deposits and the peripheral detrital series: the ferruginous kaolinitic clays (*sidérolithique*), flinty sands and feldspathic sands [see Rouire and Rousset (1973)].

Tectonic movements related to the Pyrenean–Alpine orogeny began to affect the Massif in the late Eocene, causing a general tilting of the basement from east to west and from north to south, being clearly reflected in the present main trends of the drainage towards the Paris and Aquitaine Basins. At the same time, new fractures or the reactivation of old ones, broke the Massif into a series of smaller blocks which were in some cases uplifted, in others lowered or sometimes tilted, giving rise to a set of subsiding basins, horsts or rifts. The most important are the basins of Aurillac and Le Puy, the horsts of the Margeride, Livradois, the Monts du Forez and Madeleine, and the rifts of the Forez, Roanne and the Limagne (*see* Fig. 9.10). All the related faults are set in the assemblage of fractures which, from Languedoc through Burgundy to Alsace, marks out the western frontal zone of the Alps. Continental sedimentation of detrital or lacustrine type began in these basins and rifts from the Sannoisian (lower Stampian) onwards. It continued to the beginning of the Miocene (Aquitanian) and was terminated in Limagne by the spreads of granitic sand (Burdigalian) which can be traced into the Paris Basin.

At the same time, a major phase of volcanic activity commenced in the central parts of the Massif (*see* Fig. 9.12). Minor manifestations date from the Eocene, but the climax was reached in the Miocene–Pliocene (Cantal, Monts Dore, Limagne, Velay, Aubrac, Coirons, Cézallier). Vulcanism continued into the Villafranchian and the Quaternary in the basin of Le Puy, Devès, Vivarais and the chain of the Puys, the final eruptions taking place as recently as 3000–4000 BP. Present-day geothermal activity, evident in the hot springs on which several spa towns are based, is the last vestige of the volcanic activity [see Bout and Brousse (1969)].

9.6.2 Morphology of the Crystalline Basement

If one ignores for the moment the volcanic structures, the

Massif Central consists for the most part of an uplifted plateau with slightly irregular surface relief and incised by some deep valleys. The dominance of this upland surface, contrasting with the surrounding plains, and the impression of flatness justify the term Central Plateau which some geographers have given to this region. In fact, the relief of the Plateau is not so monotonous as might appear at first sight. As well as the remnants of planation surfaces, fairly extensive and inherited from the prolonged periods of continental denudation, there are also structural forms associated with differential rock weathering or recent tectonics, and forms of linear erosion related to the incision of the drainage net.

9.6.2.1 Planation surfaces and residual features

No forms that are the direct result of Hercynian tectonics remain in the Massif, except in the Montagne Noire. All features of this age have been extremely worn down so that only fragments of planation surfaces have survived on interfluves between the incised valleys. These remnants are not all of the same age, and several different surfaces, sometimes confused, have been recognized.

1) The post-Hercynian surface was created by the cycle of erosion that was initiated by the Hercynian orogeny. This presents, wherever it can still be identified, a perfectly bevelled—although sometimes inclined—plain. Stripped of its later sedimentary cover, it is visible today near the edges of the Massif beyond which it dips gently beneath the sediments of the Paris and Aquitaine Basins. The surface was in effect fossilized by the cover of detrital or marine deposits which show that in places it is pre-Permian and in others pre-Triassic or pre-Liassic. For example, it emerges from beneath Triassic remnants (*chams*) in the Cevennes, and from beneath Liassic rock in southern Limousin. Its extent can be traced from Triassic and Liassic outliers resting elsewhere on its surface, even though the latter is broken by faults, even in Ségalas and on the northern slope of Morvan. At higher elevations, it has been cut across by more recent surfaces.

2) The Palaeogene surface is the next youngest of the three major surfaces in the Massif. No Mesozoic surfaces (post-Liassic) can be identified because of the absence of evidence from strata or transgressions of such age. What is certain, however, is that some areas were already bevelled by erosion before the Tertiary and certainly before the major Oligocene faulting. This can be observed on the Auvergne uplands and the eastern border. In the higher parts of Limousin, in Ségalas and on the Margeride, there is evidence of very old weathering. From such an array of evidence, it can be deduced that there exists a Palaeogene surface, although probably less extensive than has sometimes been claimed, and probably in most places of polygenetic type.

3) The Neogene surface can be much more precisely dated because, from the Oligocene onwards, there are the volcanic rocks where chronological and geomorphological relationships with the landforms provide many valuable data. Oligocene sediments are also preserved beneath the early volcanics of the Cantal, Aubrac and Velay. The

Fig. 9.11 Planation surface cut across crystalline rocks, upper valley of the Dordogne. (Embleton)

Châteaugay lava flows near Clermont-Ferrand bury alluvium of Aquitanian age, while the Coirons lavas mask a post-Cretaceous surface (possibly Palaeogene) dissected by Pontian valleys. All the pre-Villafranchian lavas fossilize a Neogene surface of low relief, which is actually a polygenetic peneplain in different stages of evolution and locally affected by tectonics.

The presence of these three planation surfaces in the Massif largely explains the apparent monotony of much of the relief (*see* Fig. 9.11). This characteristic is reinforced by the widespread sandy cover overlying the crystalline base. The thickness of this mantle is very variable, depending on the petrography of the bedrock and the density of fractures (faults, joints and microfissures). The existence of kaolinite (and even gibbsite) in these sands has long been taken to indicate a tropical palaeoclimate, but it is always associated with other types of clay, especially illite and montmorillonite, which themselves become dominant in depressions. It seems more likely, therefore, that the kaolinite is evidence of prolonged weathering in an upland, well-drained environment. It provides a pointer to the distribution of the better preserved old planation surfaces.

9.6.2.2 *Structural forms*
A first category of structural forms results from the differential resistance of the various crystalline rocks to weathering and erosion. Such would already be suspected from the varying importance of the sandy mantle described in the Section 9.6.2.1. The causes of the differentiation are subtle and far from understood. It is known, however, that certain structures and the presence of certain minerals can weaken the rock (flow structures, porphyritic texture, high content of biotite or plagioclase), while other factors are conducive to rock resistance (isotropic or fine-grained structure, abundance of quartz, muscovite or potash feldspars). Thus leucogranite gneiss

and anatexite are more resistant than calc-alkaline granites with large crystals, biotite granites and those of medium grain. The plateau surfaces are mainly developed on the less resistant crystalline rocks, while the more resistant varieties provide the relief residuals or monadnocks rising above the surfaces.

The degradation of an initial surface is characterized by the development of differential erosion forms of a larger or medium scale, the main ones of which are listed below.

1) Depressions or hollows etched into the surface. Most open into valleys, which allow removal of weathered material, and many are sited at the intersection of fractures. Their shapes are often circular or elliptical, and they may reach hundreds of metres in depth, with or without intermediate flattenings. Wherever they are numerous, the ground becomes pock-marked or honeycombed, the divides between the hollows being narrow and winding but with flat summits.
2) Forms related to the disintegration of sound rock, by the removal of fines and weathered products from within. Unaltered blocks and boulders become piled on one another in the form of tors and rocky bosses which litter the plateau, the slopes or the interfluves of granite regions.
3) Inselbergs and other residual forms which dominate the landscape because of the greater resistance of the local rock to weathering. For example, there are ridges composed of upstanding quartz veins, and domes, hills and *trucs* with broadly convex slopes corresponding to poorly jointed zones or to outcrops of gneiss and leucogranite.

A second category of structural forms includes those metamorphic regions where different lithostratigraphic series outcrop. This is particularly common in Limousin, Rouergue and, to a less extent, in the Cevennes. The differences are more apparent regionally than locally. Quartzites, leptynites, gneiss and anatexites tend to form ridges or massifs, overlooking depressions developed in epimetamorphic facies, micaschists or phyllites.

The third group of structural forms depends on tectonics. The Hercynian orogeny was responsible for the emplacement of the principal granite massifs, the displaced nappes of the Montagne Noire, the Armorican–Variscan 'grain', as well as the development of subsiding rifts where coal-bearing sediments are preserved. The Pyrenean–Alpine tectonics reactivated many older tectonic features. Faults dislocated the plateau surface, framed the rifts and delimited horsts. The major fracture of the 'coal furrow' was rejuvenated in several places from Moulins to Villefranche. The same happened to the Rodez defile and the Séverac rift. The resulting fault-line scarps follow various trends: east–west in the south (southern border of Larzac, northern edge of the Montagne Noire, and the Lingas and Lozère horsts), north-east to south-west in the east and south-east (the rift valleys of Autun, Le Creusot and St Etienne, and the margin of the Cevennes from Aubenas to Alès), north–south and north-west–south-east in the centre (the basins of Forez and Roanne, the Limagne rift with its downthrow of nearly 2 km and the rift from Cher to Montluçon, extending north into Sancerrois). Almost alone in the Massif, Limousin does not appear to have been affected, although it is possible that the escarpment separating the 'mountain' from 'lowland' Limousin could be, at least partially, of tectonic origin.

9.6.2.3   Drainage pattern

Lying in the centre of France, the Massif Central represents a major watershed zone. The overall pattern follows the slopes of the deformed Palaeogene surface and thus tends to diverge from sources in the south-east (in the Cevennes and Vivarais) and flow mainly towards the Loire and Garonne Basins opening on to the Atlantic, at the expense of the Rhône and the Mediterranean.

The details of the pattern have developed as a result of captures and local adaptations to structure. The Loire is set along a series of subsiding basins (Le Puy, Forez, Roanne) separated at present by narrow, gorge-like sections. The Allier, incised in the crystalline upland of Margeride, has been instrumental in removing great thicknesses of sediment from the Limagne rift, excavating, with its tributaries, the Oligocene limestones and clays in spite of the local resistances offered by interbedded and overlying volcanics. In the south, both in the limestone Causses and the crystalline Ségalas, the Lot, Tarn and Aveyron flow in deep ravines or gorges. On the eastern margin, torrential steams have cut deeply into the Cevennes, falling steeply towards the Ardèche, the Gard and the Hérault; their steeper gradients have sometimes enabled them to capture parts of the drainage originally directed to the Atlantic. From Vivarais to Charolais, more gently flowing rivers guided along faults have been exposing the Hercynian Palaeozoic structures in rifts. A secondary centre of drainage dispersion lies in the elevated area of Haut Limousin where a radial drainage pattern is superimposed on the metamorphic and granitic bedrock. Similarly in the north-east, Morvan represents a separate local watershed area from which incised valleys diverge towards the Saône, Loire and Seine.

Everywhere, Pliocene–Quaternary rejuvenation has been responsible for the incision of valleys up to depths of several hundred metres into the crystalline basement (*see* Fig. 9.11). Since the Villafranchian period, with the general uplift of the Massif, phases of downcutting have alternated with phases of stability or even aggradation. But the incisions are still, overall, relatively modest, with certain exceptions: for example, the Allier had already cut down to its present level when its valley was invaded by lavas 2 Ma ago. Incision of river valleys and the excavation of basins attained a maximum in the Pleistocene. The various stages can be especially well traced in Auvergne, where relief inversion associated with lava flows occupying valleys has preserved the evidence, the successive lavas providing a precise chronology. Outside the basins, alluvial terraces are rare because of the narrowness of the valleys. Where present, however, the terraces yield a record of palaeoclimatic change and of alternating cold, humid periods and warmer phases, as well as, in all probability, evidence of neotectonic uplift.

9.6.3   Volcanic Relief

The most distinctive feature of the Massif Central, when comparing it with other Hercynian massifs in France, is its strikingly youthful volcanic relief, the result of vulcanism that, although changing its centres of activity, began in the Miocene and continued almost to the present-day (*see* Fig. 9.12). The volcanic structures are imposing and always readily identifiable as such, in spite of obvious differences between them related to (1) the age and magnitude of the eruption, (2) the nature of the material and type of eruption, (3) the prevolcanic relief and (4) subsequent erosion and dissection. On this basis, the following main groups of forms may be distinguished.

9.6.3.1   Major stratovolcanoes: Cantal and Monts Dore

The Cantal (*see* Fig. 9.13) is the oldest and largest (about 2700 km^2) of all the former volcanoes of the Massif Central and is also one of the highest (1858 m in the Plomb du Cantal). Eruptions began here in the early Miocene (20–22 Ma) with acid lava effusions and peripheral basalt flows. This was followed by a prolonged pyroclastic episode until the final stage when, in the Pliocene (3.7–5.9 Ma) renewed emission of acid lavas in the centre (Puy Mary, Puy Griou) took place, together with numerous basalt flows issuing from subsidiary vents and covering almost the whole structure. Since then, fluvial and glacial processes have carved out a series of valleys radiating from the centre where forms of differential erosion (necks, plugs dykes, which form the summits) predominate. Beyond the central zone, the radial valleys cut the gently sloping lava flows into vast triangular segments known as *planèzes*.

The massif of the Monts Dore is slightly younger. Eruptions here commenced later in the Miocene (about 18 Ma) and several separate centres were active. Vulcanism continued into the Villafranchian and even the Pleistocene (to 500 000 BP). As in the Cantal, the basalt flows are followed by a thick pyroclastic and volcano-sedimentary series, the whole structure being cut by acid

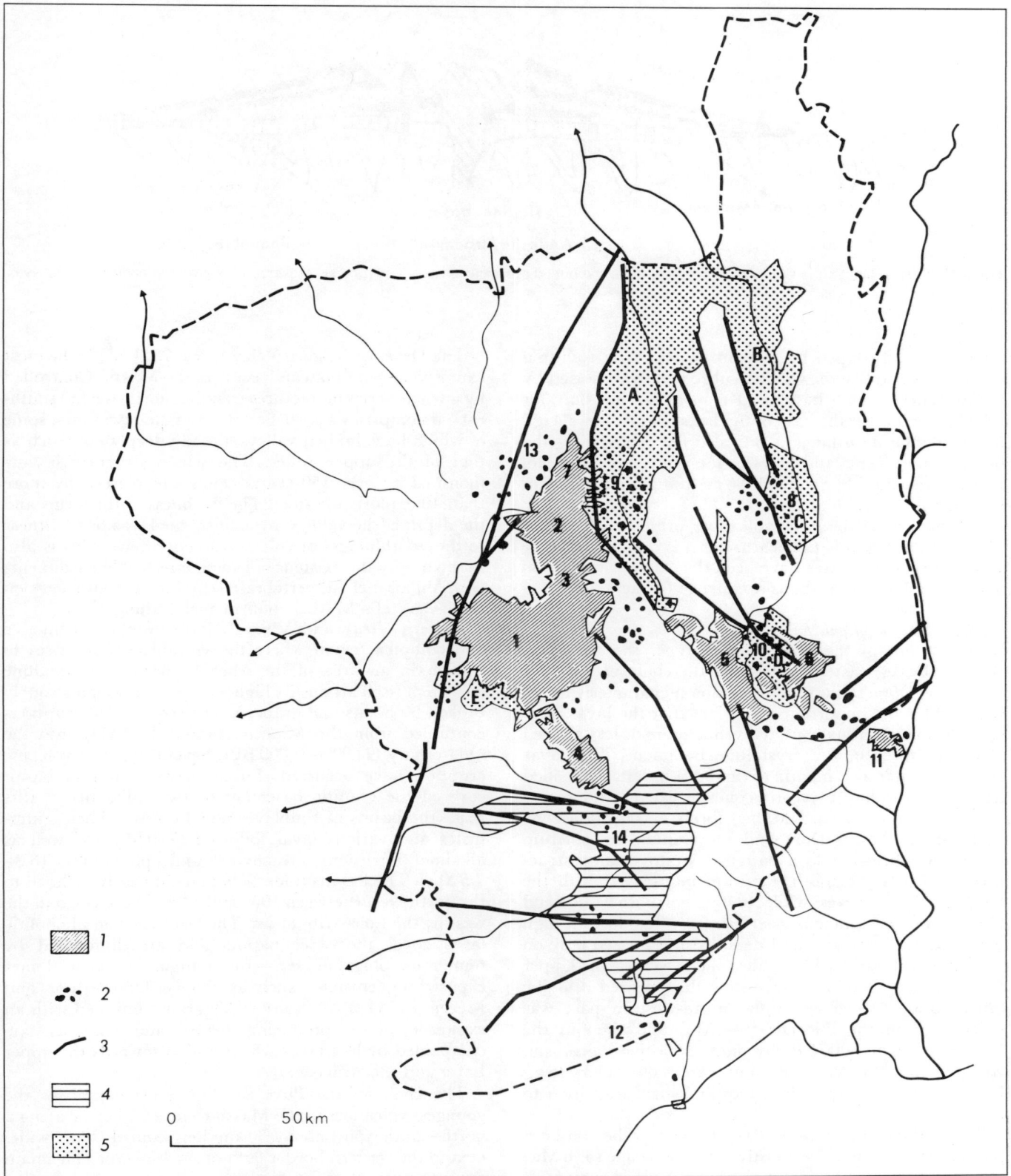

Fig. 9.12 Cenozoic vulcanism in the Massif Central. (1) Main areas of eruptive rocks. (2) Isolated centres of volcanic activity. (3) Faults. (4) Causse limestone. (5) Tertiary sedimentary basins. The numbers on the map mark the following localities and features: (1) Cantal; (2) Monts Dore; (3) Cézallier; (4) Aubrac; (5) Devès; (6) eastern Velay and Vivarais; (7) the chain of the Puys; (8) Forez; (9) Limagne; (10) the basin of Le Puy; (11) Coirons; (12) Languedoc; (13) Sioule valley; (14) Causses.

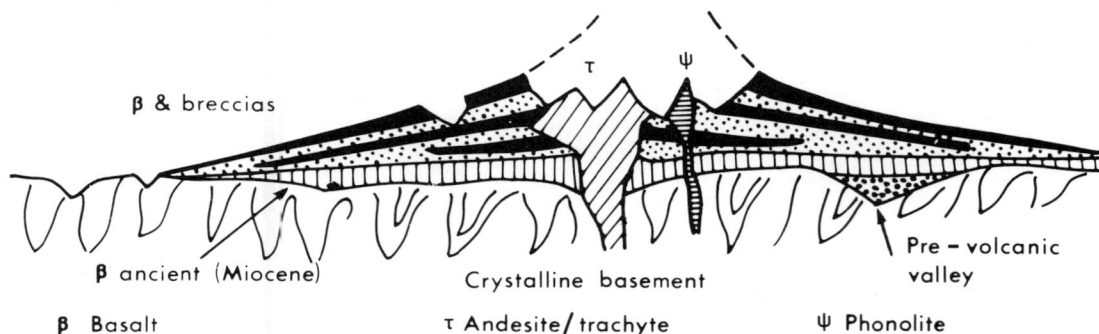

Fig. 9.13 Section across the Cantal stratovolcano. Note the buried prevolcanic valley with gravels preserved beneath the early Miocene lavas.

intrusions and topped by plateau basalts. The resulting edifice was also like the Cantal, subsequently dissected by radial valleys which have been responsible not only for relief inversion (in the case of the latest basalt flows, e.g., in the Banne d'Ordanche) but also for exposure of the original main vents and pipes in the centre: for example, forming the Puy de Sancy (1886 m) the highest point in the Massif Central.

In both cases, the stratovolcanoes preserve dislocated blocks of the basement beneath, but it is difficult to know which of these fractures have been the cause of vulcanism and which have been the consequence (calderas).

9.6.3.2 Plateau volcanic forms

Although not of the same age or type, the Cézallier, Aubrac, Devès, eastern Velay and the chain of the Puys (or Monts Dômes) share some common characteristics, of which the most important is the fact that the lavas have come from numerous scattered vents, more or less aligned along fractures in the crystalline basement. The lavas cover the latter as a mantle of (mainly) basalt, diversified from place to place by scoria cones or extrusion domes.

The Cézallier (area: 600 km²) forms a transition zone between the Monts Dore and the Cantal. Its piled-up lavas nevertheless hide a flattened stratovolcanic structure. Towards the south, they are interleaved with the lavas of the planèzes while to the north they descend almost into the Limagne near Issoire. They form a high basalt plateau, massive and denuded but incised only on its eastern border. The oldest lava vents, of Upper Miocene age (5–8 Ma), have now disappeared. On the other hand, the whole of the north-western part was reactivated in the Pliocene (3–4 Ma) and even in the Holocene (5000–6000 BP) by more explosive vulcanism, with Strombolian-type cones and explosion craters (e.g., Lac Pavin) which provide a morphological transition to the chain of the Puys.

The Aubrac (area: 450 km²) is similar to the southern part of the Cézallier. It is mostly of the same age (5–8 Ma) although activity continued until the Villafranchian. It consists of a high-level (1000–1470 m) basalt capping composed of successive flows from multiple vents along fissures trending north-west to south-east. The lavas rest directly on the basement or on Oligocene sediments.

The Devès or western Velay (area: 750 km²) is the most extensive area of plateau basalts in the Massif. Controlled by a whole series of fractures trending north-west to south-east, it comprises a sequence of very fluid lava flows, some of which have infilled valleys already deeply cut, such as that of the upper Allier. The whole structure is surmounted by over 150 scoria cones and pierced by more than 40 explosion craters. The freshness of the forms and the depth of the valleys invaded by the lavas bear witness to the youthfulness of volcanic activity here. This is also borne out by the existence of some interbedded sediments with Villafranchian vertebrate remains. Absolute ages on the lavas (0.6–2.7 Ma) confirm such dating.

Eastern Velay and Vivarais form another distinctive and complex region where the volcanic surface rises to 1400 m in an area of the Massif where the crystalline basement itself attains its highest levels, although strongly broken by horsts and grabens. The sequence of eruptions continued from the Miocene (about 11.5 Ma) into the Quaternary (11 000–35 000 BP). Several types of volcanic eruption have occurred. Firstly, outpourings of basalt covered the granitic basement or the sediments in rifts (e.g., the basins of Emblavès and Le Puy). Then ignimbrites and various lavas followed (7–9 Ma), as well as alkaline intrusions, trachytes and phonolites (6.3–7.5 Ma). The basalts underlie a plateau gently inclined to the north-west (between 1000 and 1400 m) dissected in the west by the Loire tributaries. The trachytic and phonolitic lavas cover the older basalts and are dominated by numerous plugs or *sucs*—the remains of lava domes exposed by erosion—such as the Gerbier-de-Jonc and Mézenc (1754 m). Activity in Vivarais continued with an explosive phase producing maars and calderas, accompanied by late lavas which flowed towards the upper Loire and the Ardèche.

The chain of the Puys (*see* Fig. 9.14) comprises the youngest volcanoes in the Massif Central. They lie along a north–south zone about 30 km long and 4–5 km wide, next to the western border fault of the Limagne. The main eruptive phase in this assemblage of volcanoes consisted of the extrusion of basaltic material—both as lava flows and scoria cones—and occurred between 10 000 and 40 000 BP. It was followed by a trachytic phase with domes and showers of ash between 8000 and 10 000 BP.

Fig. 9.14 View south along part of the chain of the Puys. (Embleton)

Some phenomena are still more recent, perhaps even as young as 3000–4000 BP, and must have been witnessed by early man. The chain of the Puys includes nearly 80 volcanoes of which about 70 are fairly regular Strombolian-type cones; some broken open, others complete. Eight others are acid lava domes, including the Puy de Dôme (1465 m). There are also four or five maars. The lava flows have followed the prevolcanic relief. To the west, they spread out across the plateau in large lobes known as *cheires* but to the east they have been channelled in narrow valleys as far as the Limagne where erosion of softer underlying sediments has caused subsequent inversion of relief (e.g., the Châteaugay flows, the Montagne de la Serre and the Gergovie plateau). There are numerous instances of drainage diversion and the formation of small barrage lakes [see Camus *et al.* (1975)].

9.6.3.3. *Vulcanism in the basins*
Volcanic activity was also evident in the large subsiding basins of the Massif, from the Miocene to the Pliocene (Villafranchian). In Forez, hardly any dykes or basalt plugs remain while, in the Limagne, erosion forms scallop the edges of lava plateaus, fragments of ancient lava flows now cap interfluves or hills providing fine examples of relief inversion and there are some small lava domes. In the basin of Le Puy, the sediments that accumulated since the Oligocene have been intruded in the Villafranchian by pyroclastic breccias and covered with lavas which meet those coming down from the Devès plateau. The present-day relief reflects a history of inversion, comprising denuded plugs (e.g., the upstanding rocky pinnacles in the town of Le Puy) and the tabular lava remnants (the Polignac mesa) scattered over the basin.

9.6.3.4 *Dispersed vulcanism*
Volcanic structures are also found elsewhere in the Massif, often near its margins: on the Sioule plateaus and along the 'coal furrow' to the west, on the Monts du Forez and Bourgogne in the east, in the Causses to the south and especially to the south-east in the Coirons and Languedoc.

The Coirons consists of an extensive basalt plateau standing over 500 m above the Rhône corridor. At the base, there are Pliocene lava flows (6–6.5 Ma), the earliest ones having preserved alluvial deposits belonging to a very old drainage system. The general uplift of this part of the Massif seems to have maintained the vulcanism up to the Villafranchian. Lavas continued to pile up, accompanied locally by explosive activity. The most recent flows were guided by erosional features towards the downcutting Ardèche. Finally, increasing rates of incision on the part of the Rhône and the lack of resistance offered by the subvolcanic bedrock led to complete inversion of relief, with the cutting of deep ravines overlooked by impressive lava edges.

In Languedoc, there is a north–south string of volcanoes which can be traced from the Escandorgue (the southern part of the Causse du Larzac) to the Mediterranean at Cap d'Agde. The earliest eruptions began only in the Villafranchian (1.5–2.5 Ma), pouring out basaltic flows from numerous fissures. Today, they often give rise to relief inversion. The most recent activity (0.7 Ma) has formed on the lower hills and plains of Languedoc a series of cones and flows that are quite well preserved and cut into cliffs by the sea.

9.6.4 Glacial and Periglacial Forms
In the Quaternary cold periods, the highest parts of the Massif were glaciated (*see* Fig. 9.15). Neither the chronology nor the altitudes reached by the glaciers are precisely known. In several places, one can distinguish three successive phases without being able to establish whether the intervals between them were true interglacials or

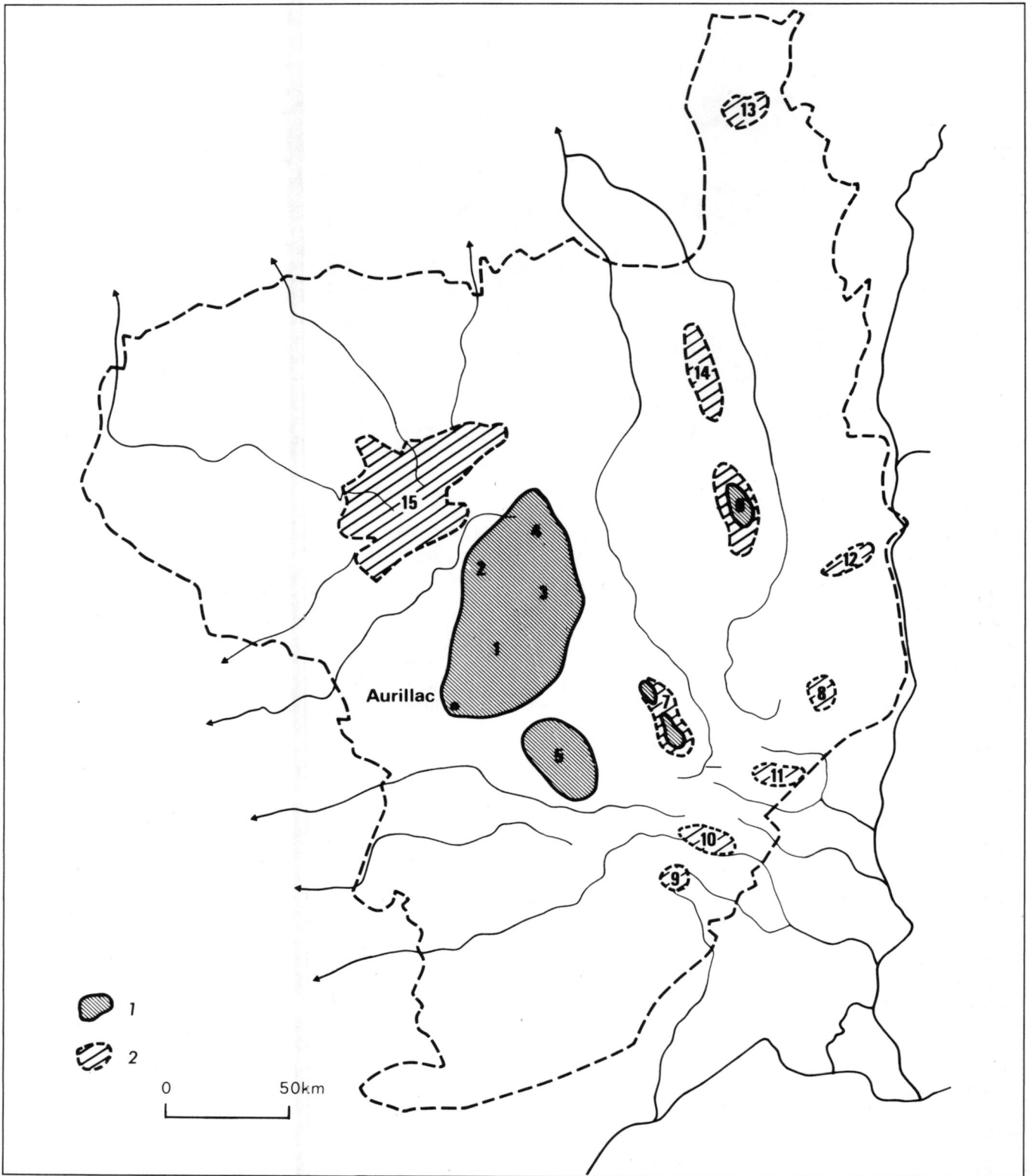

Fig. 9.15 Distribution of glacial and periglacial features in the Massif Central. Close diagonal shading (1) represents the generalized former extent of glacial ice; wider diagonal shading (2) shows area of periglacial phenomena. Localities: (1) Cantal; (2) Artense; (3) Cézallier; (4) Monts Dore; (5) Aubrac; (6) Forez; (7) Margeride; (8) Mézenc; (9) Aigoual; (10) Mont Lozère; (11) Tanargue; (12) Mont Pilat; (13) Mont Morvan; (14) Madeleine; (15) Haut Limousin.

merely interstadials of the Würm, which is the best represented. Several indications suggest that all areas lying above 1400 m were generally affected by glaciation, and that the ice descended much lower (possibly to 1200 m) on the west where accumulation was greater than on the east side. The type of glaciation has been much discussed. Until lately, it was thought that valley glaciation followed an earlier plateau glaciation, with a phase of fluvial valley cutting between. Nowadays, most workers think that, at the maximum of glaciation, both plateau ice and valley glaciers were contemporaneous, one system discharging into the second [see van Dorsser (1982)].

The surviving glacial features reflect not only the glacial history but also the preglacial relief. The Cantal glaciers have left erratics and drumlins on the planèzes and on the Artense plateau, trough forms and moraines in the valleys, and glaciolacustrine forms in the basins (such as that of Aurillac). Similar phenomena characterize the Cézallier and the Monts Dore. The Aubrac, Forez and certainly the Margeride supported small ice caps which sent down tongues into the adjacent valleys. On flat or gently convex surfaces, glacial sculpture has produced bosses and rock basins corresponding to harder outcrops or weathered zones, respectively, together with shallow lakes and remnants of moraines. On steeper slopes, especially the leeward slopes facing east, small cirques developed. Comparable features are found in the Cevennes on the Mézenc, Aigoual and the eastern slope of Mont Lozère and Tanargue.

Periglacial forms (*see* Fig. 9.15) and deposits are known from all the higher ground. Most of them are fossil features. The sandy deposits of the granite areas are often reworked or moved downhill; there are block streams, and some of the calcareous slopes support screes. On the plateaus and interfluves, rocky protrusions and piles of boulders represent tor-like forms attacked by frost and exposed by gelifluction. Phenomena similar to rock glaciers occur in the driest valleys of the Margeride and Mont Lozère. One can also see, in favourable circumstances, minor evidence of cryoturbation and wind action in a cold environment. Some of these phenomena may be still active today, notably the frost-broken scree deposits around the most frost-susceptible rocks (limestone or marly limestones) or the most fractured rocks (columnar basalts and phonolite necks).

9.6.4   Regional Subdivisions

The geomorphology of the Massif thus shows many subtle variations of style in spite of its compact appearance and common crystalline foundation. Varied combinations of relief, lithology, landscape and land use allow the following major units to be identified.

9.6.5.1   *Western Hercynian basement areas*

Basically crystalline, granitic or metamorphic, this unit comprises several subregions.

1)   Limousin to the west of the major 'coal furrow' fault is almost exclusively made up of gneiss and micaschist, folded and metamorphosed several times, and intruded by Caledonian granites. A lithostratigraphic succession,

based on increasing metamorphism, can be recognized: for example, phyllites, micaschists, gneisses and migmatites. The relief consists of plateaus rising to almost 1000 m, corresponding to polygenetic planation surfaces, cut by deep valleys (Creuse, Vienne, Corrèze). The landscape is harsh, damp and almost wholly covered with fossil sandy deposits enriched over time with former or present gelifluction deposits. With these uniform characteristics, Limousin illustrates best the typical geomorphological monotony of a massif.

2)   Ségalas, in the south-west, is reminiscent of Limousin but stands at a lower altitude and its lithology is more varied: schists, sandstones, metamorphic rocks, less granite and a few vestiges of a former sedimentary cover. The relief is also more broken by faults delimiting rifts and tilted blocks. The inclined remnants of the post-Hercynian surface are cut into from below by more recent plateau surfaces, which in turn are deeply dissected by the tributaries of the Garonne (i.e. Aveyron, Viaur, Tarn).

3)   The Montagne Noire and its surroundings, in the south, form a separate unit in the Hercynian structure of the Massif Central. Cambrian–Devonian sediments, sometimes slightly metamorphosed and formed into nappes or plates tilted to the south by the Hercynian orogeny, flank on the north (Montagnes de Lacaune) and on the south (Minervois) an axial metamorphic and granite zone (the massifs of Nore and Espinouse). This complex structure, bevelled and fractured, dips towards the south, in its western part, beneath an Eocene cover. In its eastern sector, it dips north-west and overlies the crystalline plateaus of Ségalas. The relief amplitude (more than 1200 m) fits in with a Tertiary age for the tectonic dislocation and with proximity to the Languedoc coastal lowlands, and is reflected in the contrast between the high, undulating ridge tops, cut by deep gorges, and the low glacis leading down to the Mediterranean Sea.

9.6.5.2   *Subsiding rifts and volcanic landscapes: Auvergne*

In the central part of the Hercynian massif, Auvergne is typified by large subsiding rifts which provide important contrasts in lithology and geomorphology with the surrounding crystalline and volcanic regions. Thus there are three types of Auvergne landscape:

1)   The crystalline plateaus resemble in many ways those of Limousin or Ségalas; all have evolved through a polygenetic erosional history. But in Auvergne, the fractures are more numerous and significant. Horsts or tilted blocks have been uplifted to elevations that, in the Quaternary, allowed glacial and periglacial forms to develop (e.g., Monts du Forez, Margeride, Aubrac).

2)   In the down-faulted rifts (Limagne, Forez Basin, Roanne Basin) and subsiding basins (Le Puy, Aurillac), the sedimentary infills (lacustrine or detrital) are often quite thick (from a few hundred metres, up to nearly 2 km in the Limagne). Calcareous layers form small tabular features capping clays, marls or arkoses which are less resistant to erosion. The sequences usually end in a detrital phase (cherty sands on feldspathic sands) which bear witness to the stripping of weathered deposits from the plateau by erosion following the major Oligocene–

Miocene fracturing. The Limagne is the type example and the main subsiding rift. Its excavation by differential erosion since the Aquitainian has resulted in an open landscape, today much modified by man, diversified by volcanic or calcareous hills or buttes.

3) The volcanic massifs cover a large part of the crystalline plateau of Auvergne. They form immense and unusual superimposed forms in comparison with the surrounding country and include a variety of constructional forms (stratovolcanoes, scoria cones, lava domes, explosion craters with or without lakes, lava flows spread out over plateaus or running down into valleys, forming barrage lakes or diverting the drainage). Attacked by chemical and mechanical weathering, undermined or weakened by the removal of loose material, cut into by gullies, the volcanic structures, many of which are still remarkably fresh in appearance, provide the most varied and original features of Auvergne.

9.6.5.3   *Eastern Hercynian basement areas*
In the south-east, the Cevennes form an area of strong relief (up to 1500–1700 m) overlooking the low sedimentary plateau of the Rhône and the Alès rift. Planation surfaces cutting the crystalline rocks slope gently westwards and northwards. Broken by faulting, these surfaces give a subtabular appearance to the summits (Tanargue, Mont Lozère, Aigoual, Lingas). Towards the south-east, on the other hand, the escarpment edge is notched by the deep ravines of the Mediterranean streams (Ardèche, Chassezac, Cèze, Gardons, Hérault), torrential in character in the winter, and by some impressively deep gorges cutting it into a network of flat-topped, narrow and branching ridges known as *serres*.

The Cevennes continues northwards in the mountains of Vivarais (1400–1500 m), which share similar characteristics: deep gorges, narrow ridges pointing towards the

Rhône and crystalline plateaus sloping towards the Loire. Locally, however, the horst and graben structure, partly bevelled, is overlain by volcanic phenomena: lava flows and domes, with some maars. Beyond Mont Pilat (1434 m), the eastern margin of the Massif Central, rising above the Saône–Rhône trough, takes on a distinctive appearance. The crystalline basement is now more fragmented, with north-east–south-west faults separating tilted blocks (Morvan, Charolais, Beaujolais, Lyonnais) overlooking rifts where coal-bearing strata have been preserved (and exploited in the mining areas of Autun, Le Creusot and St Etienne). The flat, wooded heights contrast with the cultivated basins, where differences in the relief reflect variations in the infilling sediments. Near to the Saône, north–south faults, especially in Mâconnais, cut the sedimentary cover and the underlying basement into parallel ridges and hollows, with gullied slopes, which form a transition to the hilly landscape of Burgundy [see Rat (1972)].

9.6.5.4   *The Causses*
The 'gulf' of the Grands Causses, in the south, is a unique sedimentary domain in the Massif Central. Here, Liassic and Jurassic seas penetrated deeply into the crystalline mass from the Mediterranean basin as a result of subsidence of the order of 1000 m. Pyrenean–Alpine tectonics controlled the actual outline of the gulf: west–east in the Rodez 'strait' (i.e. Causse du Comtal, Causse de Séverac) by which the Mesozoic Mediterranean was linked with the Atlantic and north–south between Ségalas on the west and the Aigoual–Lozère on the east [see Marres (1935), Martel (1936)].

The sedimentary series is thick and composed of various facies, although all are carbonates (dolomitic or dolomitized, calcareous marls, limestone reefs and calcareous massifs). Those of the Middle and Upper Juras-

Fig. 9.16 The Tarn gorge in the Grands Causses. (Embleton)

sic, broken by north–south faults, are responsible for the greatest development of surface and underground karstic forms in the region. Surface drainage only exists in the major through-flowing valleys, deeply cut in the form of canyons [Tarn (*see* Fig. 9.16), Jonte, Dourbie] that divide the plateau into separate units (Causses de Sauveterre, Méjan, Noir and Larzac).

These regional relief types underline at the same time the unity and diversity of the Massif Central. Prior to the Tertiary, several orogenic systems followed one another but each was in turn more or less effaced by widespread planation, helped by a crystalline lithology which, on a small scale, was relatively homogeneous. Pyrenean–Alpine tectonics provide the common link between the main events that subsequently affected the Massif: uplift of the south-eastern part, fracturing into blocks, formation of subsiding rifts and basins in the heart of a mountainous upland and the unleashing of intense eruptive activity. Glacial and periglacial modification in the Quaternary has only been of a minor, localized nature. Overall, it is the volcanic relief and the valley incisions that contribute the greatest degree of individuality to the Massif Central.

9.7 Saône–Rhône Corridor and Mediterranean Coastlands

The Saône–Rhône corridor represents a structural depression lying between the Massif Central on the west and the Jura and the Pre-Alps on the east. At the northern end of the corridor lies the Plateau de Langres, presenting a steep south-eastern edge overlooking the Saône Basin but, rising only to about 500 m, providing a broad, low-level routeway north towards the Paris Basin and Lorraine. The Belfort Gap represents another low-level col (at about 350 m) connecting the Rhône rift valley with the upper Saône. The Saône valley leads south into the Rhône, which it joins at Lyon; the great river then continues south for another 240 km to the head of a massive delta near Arles. The Mediterranean coastlands west of the delta and the lower Rhône comprise Languedoc. To the east of the Rhône delta is the area of lower Provence and the Côte d'Azur [see Journaux (1956)].

9.7.1 Saône–Rhône Corridor

The line of this depression existed as far back as the Hercynian orogeny, and its foundations are provided by a series of highly dislocated Hercynian blocks. In the Miocene, the Massif Central to the west was uplifted, tilted and fractured, while the folding of the Alpine ranges to the east reached a climax. The corridor itself suffered further subsidence, although the extent of downfaulting was much greater on the western side. Over the Hercynian foundations lie complex sequences of Mesozoic and Tertiary rocks infilling the depression. Great thicknesses of Jurassic and Cretaceous limestones, clays and other sediments were laid down under alternating marine and lacustrine conditions. In the Miocene, a marine gulf extended up the valley to Lyon and north-eastwards into central Switzerland. Later in the Miocene or early Pliocene, the gulf shrank until it was restricted to the area

south of Lyon by the Vienne sill. North of this sill and draining south across it, lay a lake, on the site of what is now the Plain of Bresse and into which the ancestral Saône and Doubs poured great quantities of Pliocene sediments, from the direction of the Belfort Gap. Subsidence of the Rhine rift in the late Pliocene diverted the Rhine to the north and the major watershed of Europe between the Rhine and the Rhône was formed. South of Lyon, uplift reduced the marine gulf until the present lower Rhône valley came into existence.

The next important phase in the evolution of the corridor was in the Quaternary when glaciers extended down the upper Rhône, Isère and other valleys from the Alps. The ice spread out as piedmont lobes, reaching at a maximum to the west of Lyon and across the Pays de Dombes almost as far as Bourg. Deposition of moraine, hummocky till and outwash was extensive and the drainage pattern was affected. With deglaciation, spreads of glaciofluvial sediment continued to be laid down, often in the form of alluvial cones beyond the valley exits, later to be entrenched as terraces.

The present-day Alpine rivers (Rhône, Isère, Drôme, Durance, etc.) are highly active, especially in the spring after snowmelt. Vast quantities of detritus thereby enter the corridor and clog the channels of the rivers. By late summer, when discharge has fallen, the rivers occupy braided channels across the pale-coloured gravel infills. The régime of the lower Rhône is thus highly variable. In contrast, the Saône carries much less sediment and its flow is much more stable.

9.7.1.1 *Saône valley*
From the Belfort Gap to Chalon-sur-Saône, the valley is floored by gently folded, faulted Jurassic rocks forming low limestone plateaus away from the main rivers. The present drainage divide in the Belfort Gap between the Doubs and the Largue is relatively flat and featureless. Downstream the alluvial floodplains of the Saône and Doubs are flanked by broad terraces, covered with Pliocene or Quaternary deposits, rising to the foot of the Côte d'Or on the west or the Jura on the east [see Journaux (1956)].

Below Chalon-sur-Saône, the Saône enters the Plain of Bresse, the site of a major lake in the late Tertiary in which clays and silts were laid down. In addition, there are areas of sand and limon, but the flat plain is dotted with many lakes in the north and marshy areas flank the rivers. Farther south lies the Pays de Dombes, roughly between Mâcon and Lyon. In the Riss and possibly earlier glacials, the Rhône glacier advanced into this part of the Saône–Rhône corridor from the east, temporarily impeding the Saône drainage and depositing sheets of glaciofluvial sediment as well as ground moraine. The surface is characterized by hollows, hummocks and low ridges, some of the last identifiable as terminal moraines.

9.7.1.2 *Rhône valley below Lyon*
From Lyon to its delta, the Rhône valley consists of an alternating series of broad basins and narrow, often gorge-like sections, related to the geological structure. In the

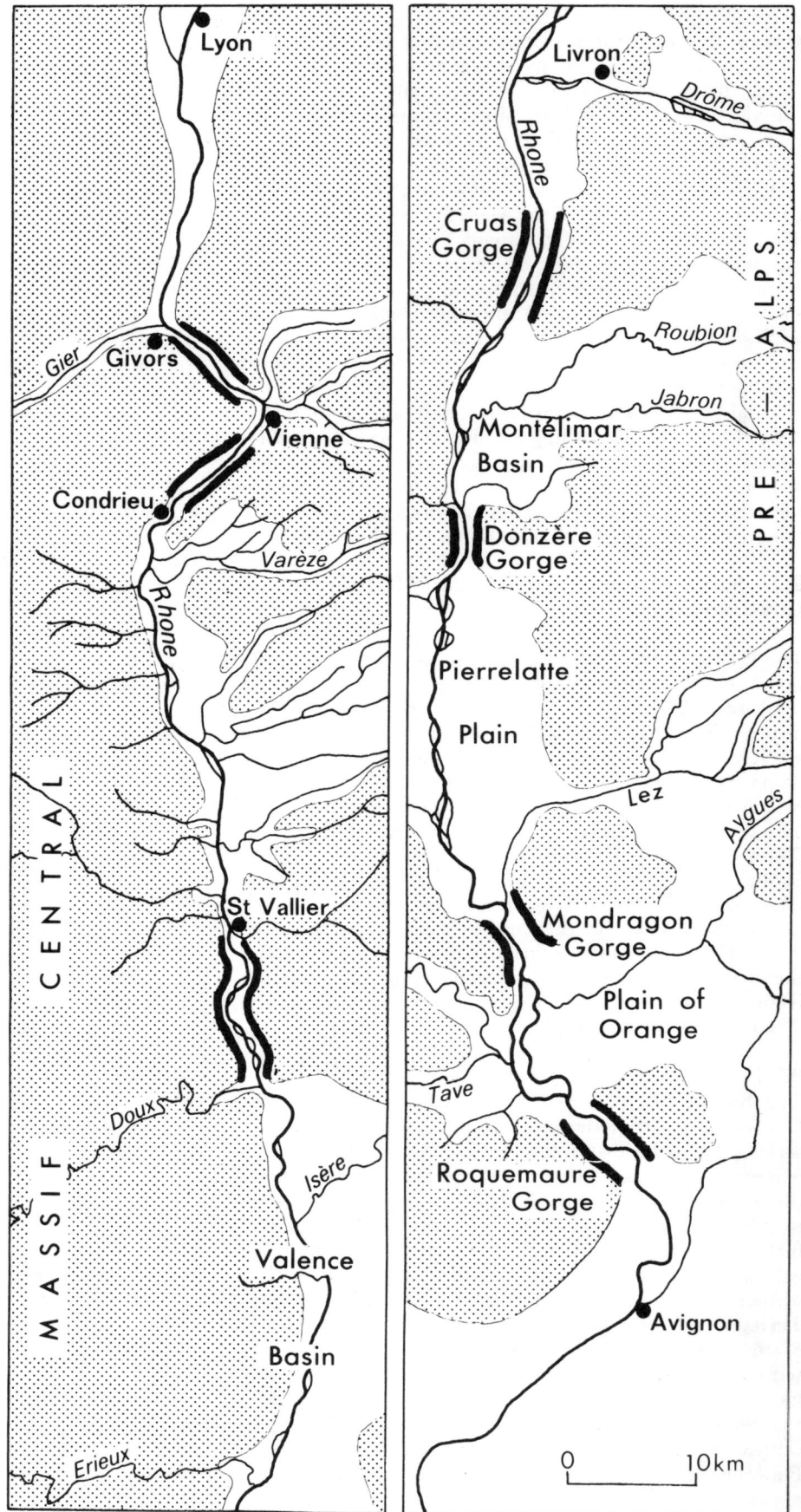

Fig. 9.17 The Rhône valley from Lyon to the delta. The right-hand map continues from the lower edge of the left-hand portion. Heavy black lines are a diagrammatic representation of the main gorge sections.

narrow stretches, the main valley is sometimes as little as 1.5 km wide with cliffed sides; most of these defiles are cut in Cretaceous or Tertiary limestones. The intervening basins vary greatly in size, from the small plain of Vienne to the substantial Valence Basin where the Isère and the Drôme enter from the Pre-Alps. The Isère brings in large volumes of summer meltwater and contributes about one-quarter of the discharge of the Rhône at the confluence. The sequence of basins and gorges along the Rhône corridor is shown in Fig. 9.17: in order, the gorges or narrow sections comprise the Vienne sill near Givors, Vienne–Condrieu, St Vallier–Tain-l'Hermitage, the Cruas gorge, the Donzère gorge, the Mondragon gorge, and the Défilé de Roquemaure near Châteauneuf du Pape. The Rhône then enters the broad Avignon plain as the Pre-Alps swing away to the east, and the last major Alpine

tributary, the Durance, joins the system. Along the whole length of the Saône–Rhône corridor, there is a striking disparity, in terms of size and number, between the eastern and western tributaries, the recent uplift and westward tilting of the Massif Central, away from the corridor, being responsible for this [see Demarcq *et al.* (1973)].

9.7.1.3 *Rhône delta*
The delta (*see* Fig. 9.18) begins just above Arles where the river is more than 150 m wide but only a few metres above sea level. Here it splits into two. The Grand Rhône continues south-south-east to the Golfe de Fos and now carries over 80 percent of the discharge, while the Petit Rhône winds south-west to Les Saintes Maries. The mouths of both have migrated eastwards in historical time as described by Russell (1942) and that of the Grand

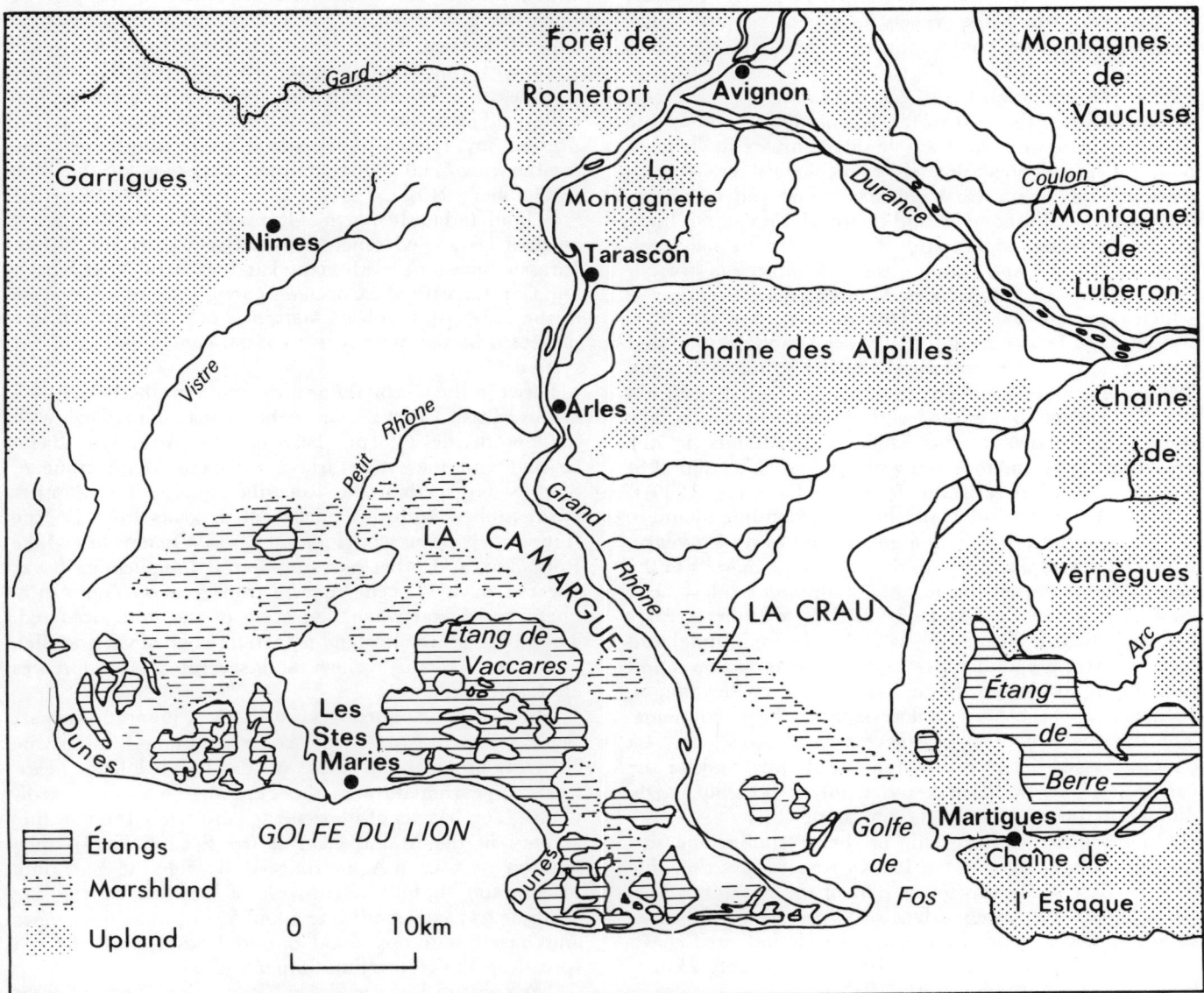

Fig. 9.18 The Rhône delta [after Monkhouse (1964)], based on the *Carte de France* 1 : 200 000. The lagoons, marshes and complex patterns of drainage channels have been simplified. Stippled areas approximately over 60 m.

Rhône, reflecting its huge load of sediment, is advancing seawards at about 40 m a year. The channels are now artificially stabilized. The area west of the Grand Rhône is known as the Camargue: a vast expanse of marshes, shallow lagoons and winding channels, divided from the sea by low dunes. East of the Grand Rhône lies the Crau, the so-called 'dry delta', consisting of two overlapping alluvial fans [see Baulig (1927)]. One of these, in the west of the Crau, was built by the Pleistocene Rhône; the other was deposited by the Durance, also in the Pleistocene, when that river used to flow south-west between the Chaîne des Alpilles and the Provençal uplands to the east. Later, because of alluviation in the gap, the Durance was diverted to Avignon and its former delta was abandoned. The Crau consists of sand and gravel, predominantly coarse (some boulders are up to 0.3 m in size) and, with its steeper surface gradient reflecting the heavy sediment load of the Durance, is now an area of extreme surface dryness, over parts of which the gravels have been cemented into hard conglomerate, contrasting with the Camargue.

9.7.1.4　Lower Provence and the Côte d'Azur

The country to the east of the Rhône delta between the Maritime Alps and the sea is highly complex in terms of both relief and structure. Every geological era is represented, the strata are highly contorted and fractured, there are varied lithologies and the relief, though not high, is greatly dissected because of proximity to sea level. Between Toulon and Cannes, there is an area of crystalline massifs—the Massif des Maures and the Estérel—which represent part of the Hercynian core of the Alps, consisting of Precambrian (gneiss, schist and granite) and Palaeozoic (sedimentary) rocks. Rising to no more than 780 and 450 m, respectively, these are not high massifs but they are rugged and gashed with numerous ravines. The steep coast shows a succession of headlands in the crystalline rocks interspersed with bays, such as that of St Tropez, cut in less resistant slates [see Bonnifay (1969)].

Between Toulon and Marseille, and extending inland to the Durance and the Verdon gorges, lies lower Provence, which contrasts sharply with the crystalline massifs to the east in that it consists principally of limestone ridges. The landscape is equally dissected but rises to rather greater heights (about 1000 m in the Massif de la Ste Baume) and has a greater preponderance of barren craggy slopes and precipitous rock walls. Some basins between the ranges, such as that of the Arc, are floored by younger sediments. The coast is noted for its succession of bays (e.g., La Ciotat, Cassis) between limestone headlands, and for the calanques: narrow marine-eroded inlets, possibly partly due to karstic collapse, in the limestone cliffs.

On the other, eastern side of the crystalline massifs, roughly from Cannes to the Italian Riviera, lies the Côte d'Azur. Essentially this forms part of the Maritime Alps (*see* Chapter 10); steep ridges of folded Mesozoic limestones approach a precipitous and highly indented coast. The peaks inland rise to 800–1100 m within only 2 km of the sea [see Campredon *et al.* (1975)].

9.7.1.5　Languedoc and Roussillon

Extending from the Rhône delta in a curve west and south towards the Pyrenees, the lowlands of Languedoc and Roussillon stand between the sea and the foothills of the Cevennes and the Corbières. Three main geomorphological elements can be distinguished: (1) the coastal system of sand bars, dune belts, marshes and lagoons (*see* Fig. 9.19), (2) the alluvial plains and terraces, interrupted here and there by low limestone ridges and (3) the limestone garrigue country [see Gèze (1979), Jaffrezo *et al.* (1977)].

The Roussillon plain is backed by the Pyrenees and the Corbières, and crossed by the rivers Tech, Tet and Agly originating in these uplands. A view inland from the coast is dominated by the striking Pyrenean peak of Canigou rising to nearly 2800 m. From these uplands, huge quantities of detritus have been carried down to the plains by the rivers. In the Pliocene, the area was a gulf of the sea in which marine clays and silts accumulated, capped by thick sheets of fluvial sands and gravels with the regression of the sea. These sequences were dissected by the streams during the falling sea level, giving a landscape composed of gravel plateaus, terraces and entrenched valleys. The rivers are liable to severe flooding at the present-day, related to occasional severe rainstorms or to the bursting of landslide-dammed lakes (e.g., on the Tech in October 1940).

Behind and to the north of Roussillon lies the Corbières, an area of tangled relief built of Cretaceous and Upper Jurassic limestone, with many karst features. Near Sigean and Cap Leucate, the Corbières garrigue approaches close to the coast, although its margin is now a fossil cliff line protected by the more recent coastal lagoon and sand bar system.

Between the Corbières and the most southerly outposts of the Massif Central stands the Carcassonne Gap, with its low divide (190 m) between the Aquitaine Basin beyond and the Aude drainage to the Mediterranean. Slightly larger than the Roussillon plain, it is another basin infilled with a sequence of deposits from Eocene (limestone, clays, conglomerates) to Quaternary. Like Roussillon and other parts of coastal Languedoc, it was affected by a Pliocene marine transgression and subsequent regression when dissection of the infill occurred. The surface of the lowland is therefore one of varied relief, with low plateaus, valleys and several sets of terraces along the Aude.

The rest of Languedoc towards the Camargue repeats these characteristics, with some additional landscape elements provided by limestone and basalt hills or headlands appearing through the Tertiary–Quaternary sediments. The basalts of Pliocene age are related to volcanic activity in the Massif Central (*see* Section 9.4.3); they outcrop at Cap d'Agde (backed by the Pic St Loup: 115 m) and in hills north-west of Beziers. During the Pliocene transgresion (up to about 50 m), these limestones and basalt outcrops stood out as islands, and the sea spread up the Hérault and other valleys.

The coast of Languedoc and Roussillon is one of great interest. Graceful sweeping sand bars (*see* Fig. 9.20) topped by dunes curve from one headland to another.

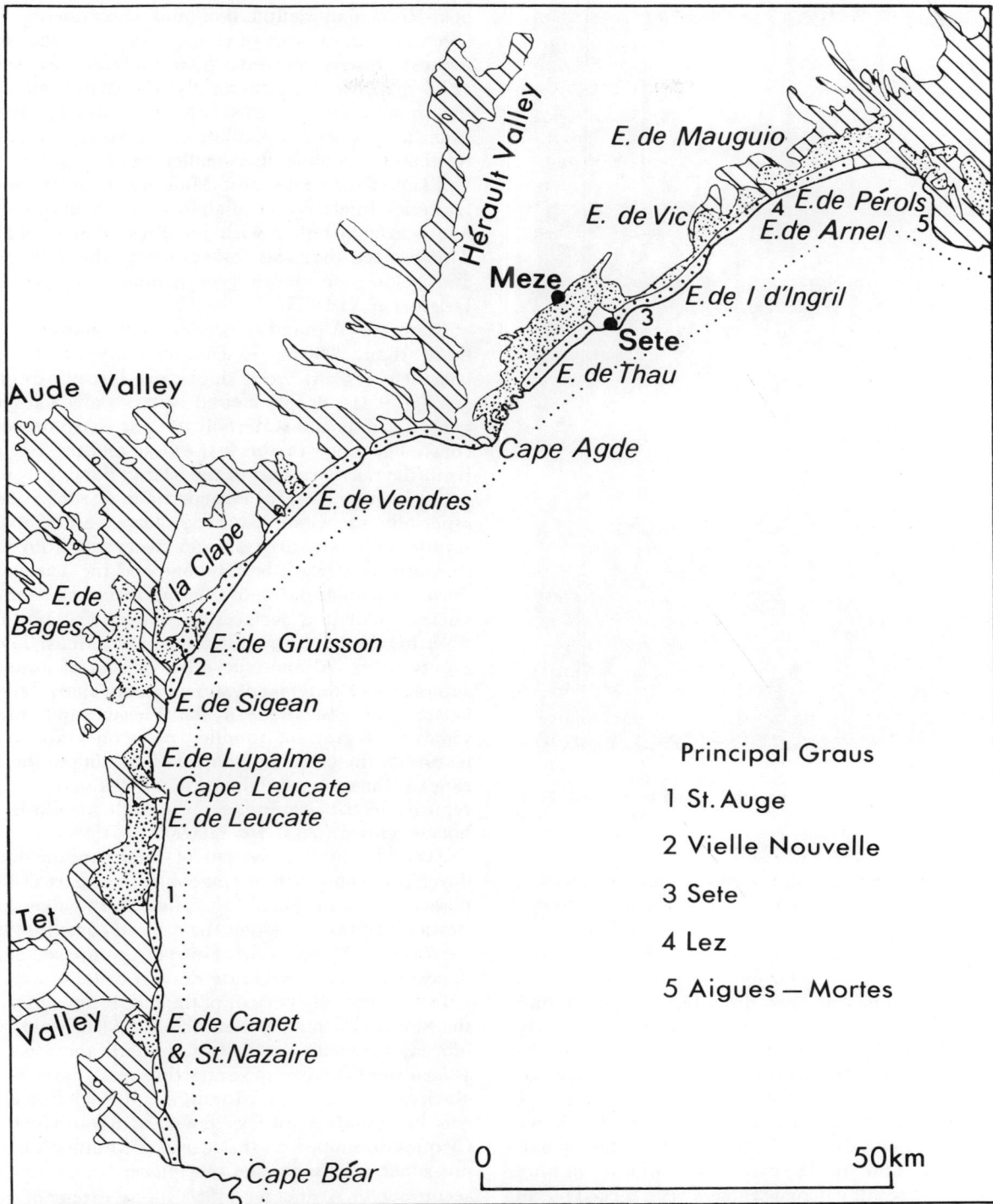

Fig. 9.19 The lagoons (fine stipple) and sand dunes (coarse stipple) of the western French Mediterranean coast. At the bottom, Cape Béar marks the termination of the Pyrenees; at the top right the Rhône delta projects. Diagonal shading shows the extent of alluvium (mainly Quaternary fluvial), the infilling of former river estuaries. The *graus* are present-day connections of the lagoons (*étangs*) with the sea [after Monkhouse (1964)].

Fig. 9.20 Franqui Plage, viewed from Cape Leucate; the curving shoreline is perfectly adjusted to present marine processes. (Embleton)

Behind them are the brackish lagoons (*étangs*) surrounded by marshland, although much of this is now reclaimed.

9.8 Corsica

The island of Corsica has been described as a miniature Massif Central, surrounded, however, not by sedimentary lowlands but by the Mediterranean Sea. Indeed, submergence of the Massif Central to about 200 m would produce a Hercynian island block not so different, except for size, from that of Corsica. Both units consist dominantly of uplifted and dissected plateaus of pre-Hercynian crystalline rock, and the Hercynian trendlines (running roughly north to south in Corsica) have been picked out in both by the main features of relief and drainage. Many other analogies may be made, even to comparisons between the Loire/Allier downfaulted troughs and the sedimentary basins northwards from Venaco in Corsica, or between the smaller limestone Causses (e.g., the Causse de Rodez) of the Massif and the karstic plateau of Bastia in northern Corsica.

The island extends for nearly 200 km from north to south, and nearly 100 km from west to east. It represents a segment of the old Hercynian platform worn down by post-Hercynian denudation and subsequently uplifted, together with an area of younger rocks, so that today the highest points exceed 2000 m [e.g., Monte Cinto (2707 m)]. Morphostructurally, the island falls into two: the greater part consists of the platform basement—granitic rocks and crystalline schists (often with a granulitic texture)—while the smaller eastern part is built of Triassic, Cretaceous and Miocene deposits, with some volcanics (melaphyres, diabases, etc.) resting on crystalline schists together with phyllites, shales and marble. This eastern part was folded during the Alpine orogeny and belongs to the epigeosynclinal zone [see Durand-Delga *et al.* (1978)].

Dissected Alpine-type relief, with summits rising to more than 2000 m, is characteristic of the platform basement region. Steep slopes rise directly from the sea (*see* Fig. 9.21); deeply incised valleys show a succession of gorges, rapids and waterfalls in their sharp descent to the coast. The longest rivers—the Golo and the Tavignano—lie in the northern part of the island, flowing from west to east. In the eastern epigeosynclinal zone, the rivers, especially the Golo, have cut deep water gaps. In the southern platform area, shorter rivers flow from north-east to south-west (e.g., the Gravone and the Taravo). In the most dissected parts of the uplands, no remnants of ancient planation surfaces have survived, but such features have been claimed in a few other areas. The highest points of the Alpine-type relief consist of granulite, the summits and watersheds showing very sharp relief, partly because of glaciation in the Pleistocene. Due to the removal of growan, rounded corestones are a common feature in the granite regions. Weathering in the Mediterranean climate has produced many distinctive microforms typical of the island (e.g., tafoni, pseudo-lapiés and honeycombed rocks) [*see* Klaer (1956)].

Around Corte, a system of intermontane basins has developed. The eastern epigeosynclinal part of the island possesses less dissected relief, but sharp ridges are sometimes also present. Along the eastern coastline, there is a narrow (10–12 km) strip of lowland composed of Miocene deposits overlain by Quaternary gravels.

During the cold periods of the Pleistocene, those parts of the mountains higher than about 2000 m were glaciated; indeed, areas above 2500 m in both the Riss and the Würm were capable of supporting small systems of valley glaciers (*see* Fig. 9.22). Moraines and glaciofluvial deposits can be found in valleys down to about 1000–1500 m. Cirques developed on the scarps of Monte d'Oro, but did not substantially change the Pliocene relief of the island according to Arnberger (1960). The extent of glaciation was also influenced by tectonic uplift of the island that continued through the Quaternary; evidence of this neotectonic uplift exists in the form of numerous abrasion terraces around the coast. Some periglacial forms such as block fields can also be observed in the highlands [see Rondeau (1961), Ottmann (1969)].

Fig. 9.21 Marine abrasion cliff at Bonifacio, southern Corsica. (Demek)

Fig. 9.22 Glacial trough, central crystalline highlands of Corsica. (Demek)

9.9 Sardinia

Sardinia, which borders the Tyrrhenian Sea on the west, making it an entirely Italian sea, is geologically and morphologically a separate body from the Italian Appennines and is more closely related to Corsica. These two islands constitute the so-called Sardinian–Corsican system, emerging from the same continental shelf. They represent fragments of a very ancient land that had already emerged in the Lower Palaeozoic Era. Its geological structure is varied, and crystalline rocks outcrop over a large part of the island, comprising granitic, metamorphic and volcanic rocks. The relief presents a mosaic of different elements, but its amplitude and height are generally moderate. Owing to the large extent of plateaus, Sardinian scenery is often dominated by horizontal lines. The island has a very different relief from that of the Italian peninsula and the Appennine chain because Sardinia was not involved, to any significant extent, in the Alpine orogenesis [see Pelletier (1960)].

The essential structure dates back to the Hercynian orogenesis, which caused the folding of all pre-Carboniferous formations and their metamorphism. In addition, the Cambrian sequence had already been affected by the Caledonian diastrophism. The dominant feature of Hercynian events was the ascent of a very large mass of granitic magma, undoubtedly produced in part by the assimilation of previous sedimentary rocks, up to full granitization. Since then, Sardinia has not undergone any other folding, but has been subject to unequal uplift, subsidence and faulting, also of recent age, which have exerted a clear control on the present morphology [see Vardabasso (1935)].

After the denudation and peneplanation of the Hercynian chains in late Palaeozoic times, several oscillations caused an alternation of marine transgressions and regressions, although never across the whole island. Deposits of various types were laid down, from Triassic to Miocene (those of the Pliocene and Quaternary are very restricted and scattered on the margins) and intense volcanism developed from the Oligocene to the Pleistocene.

Reducing the structure and relief of the island to the simplest scheme, three belts can be distinguished, elongated from north to south. The eastern zone is the most extensive, from the Strait of Bonifacio to the Gulf of Cagliari. In this wide belt, granites and, to a lesser extent, diorites prevail partially covered by more or less metamorphosed schists.

The western belt has been much reduced in land area by subsidence beneath the sea and is today divided into two unequal parts, widely separated from each other: a small one to north-west and a larger one to south-west. The fundamental structure is similar to that of the eastern zone, but in its southern portion a thick Cambrian sequence outcrops, this being the site of the most important ore deposits of Sardinia.

The broad middle zone is an ancient trough which developed in the Oligocene and was filled with volcanic products (chiefly lavas) and marine Miocene sediments. The latter have been only slightly deformed by tectonic

movements. Pleistocene (and possibly Pliocene) flood basalts were poured out here and there above the Miocene marls and sandstones. Afterwards, this middle zone sank again along a diagonal belt from the Gulf of Oristano to the Gulf of Cagliari. Pliocene and Pleistocene deposits were laid down in the new graben, which has become the major plain of Sardinia—the Campidano—(the Pliocene sediments are here buried beneath both old and recent alluvial deposits).

In the eastern belt, granite outcrops over nearly one-half of the northern part and forms a vast core of undulating plateaus with steep borders (*see* Fig. 9.23). These plateaus certainly represent ancient planation surfaces, perhaps even the Hercynian peneplain. Around this core rise other granitic uplands of moderate height but rather rugged and rocky, and in places even with sharp crests [e.g., Limbara (1362 m)], which are also found in the extreme south-east of the island.

South of the granitic plateaus rises the largest and highest massif of Sardinia, the Monti del Gennargentu (1829 m), carved out of Palaeozoic schists. From its core, several spurs branch out divided by deep valleys but, on the whole, these mountains are not rugged and recall some of the high parts of the northern Appennines, although without glacial landforms.

A vast schistose area extends farther south, deeply dissected by the Flumendosa and its tributaries running for long stretches in veritable gorges. It is in this area that one can better observe one of the morphological elements peculiar to Sardinia, the Hercynian peneplain, exhumed by erosion [see Vardabasso (1935)] which has removed the overlying Mesozoic and Eocene deposits. However, some platforms of mainly Jurassic limestone and dolomite with karst phenomena are preserved, bordered by high cliffs plunging down to the schists of the peneplain. Other patches of the same rocks north-east of the Monti del Gennargentu instead form rugged mountains, owing to their more complicated tectonics.

Plateaus of Oligocene trachyte, liparite and tuffs prevail in the north-western area of the central zone. Their flattened surfaces are probably not the primary tops of the lava flows, but rather the effect of erosion. These plateaus are fragmented into several plates and if the layers dip gently, they may produce features similar to cuestas. To the north-west, around Sassari, a low platform of Miocene limestone is introduced between the volcanic plateaus.

Farther south-east, other plateaus originated from possibly Pliocene and from Pleistocene basalt flows and extend to the border of the Campidano plain. They are especially striking where erosion has cut into the underlying marls and sandstones of Miocene age. In this case, the horizontal basaltic layers stand out as flat plateaus known as *giare*, interrupted by cliffs above the softer sediments. There are also some basaltic mountains [e.g., Monte Ferru (1050 m), which is a large complex volcano, whose core of trachytic domes and flows was later covered with basalt].

Quaternary vulcanism was also active in the area of the Oligocene lavas, especially in some local depressions where several small cones rise, so that sometimes geologists speak of a 'Sardianian Auvergne'. Finally, small

Fig. 9.23 Granitic boulders in eastern Sardinia. (Castiglioni)

more or less eroded volcanoes are aligned along the faulted eastern border of the Campidano and in the small graben of the Cixerri River.

In the western belt, the northern portion (Nurra) appears as a mosaic of low hills carved out of granitic, Palaeozoic and Mesozoic rocks (schists, limestones, etc.). The southern large area, which is divided into two parts (Iglesiente, Sulcis) by the graben of the Cixerri, is a tangled assemblage of hills and rather low mountains of granites, schists, limestones and sandstones, chiefly of Palaeozoic age. They include the Cambrian formations noted for their ore deposits. Around Inglesias, anthropogenic forms are also present (i.e. the great waste heaps of the mines).

Around Sardinia, different types of coast alternate many times. There are beaches between headlands, sand bars enclosing lagoons and indented stretches with cliffs, some of which are very high. The most characteristic part is that of Gallura in the north-east, where the recent drowning of the valleys carved out of granite has generated branching embayments like the rias of north-western Spain.

9.10 Rhineland

9.10.1 The Vosges

Between Basel and Mannheim, the Rhine rift valley separates two uplands: the Vosges on the west and the Schwarzwald (Black Forest) on the east. Prior to the Tertiary era, these uplands formed the western and eastern flanks of a mega-anticline—the broad arch whose central section later subsided forming the Rhine rift valley. This explains why the Vosges now stands as a mountain block with a steep eastern face (a major fault-line scarp) overlooking the rift (*see* Fig. 9.24), and a western slope that sinks more gradually towards the Paris Basin.

The Vosges consists of an ancient crystalline core (gneiss, schist and intrusive granite of Precambrian/Lower Palaeozoic age) which, together with some Upper Palaeozoic sediments, were strongly affected by the Hercynian orogenesis, the latter imparting a marked tectonic grain from roughly north-east to south-west (the Variscan trend). Post-Hercynian denudation reduced the area to a peneplain, the products of erosion accumulating in some basins as continental Permian deposits. The Triassic Period saw extensive marine submergence; 1 km of Bunter sandstone and other sediments were laid down, followed by the Muschelkalk (the hard shelly limestone that is now seen forming some distinct escarpments). Further sedimentation continued into the Jurassic period. Uplift of the mega-anticline raised the area above the Cretaceous or Eogene seas and initiated the development of a Palaeogene planation surface. Faulting and rifting marked the mid-Tertiary; during the later Tertiary, denudation especially in the higher central parts succeeded in removing much of the Mesozoic cover to re-expose the crystalline core. The Neogene and early Quaternary were times of remodelling of the Palaeogene surface, which is considered to show some characteristics of an etchplain.

The Vosges is divided by the Saverne Gap into the High Vosges to the south and the Low Vosges farther north.

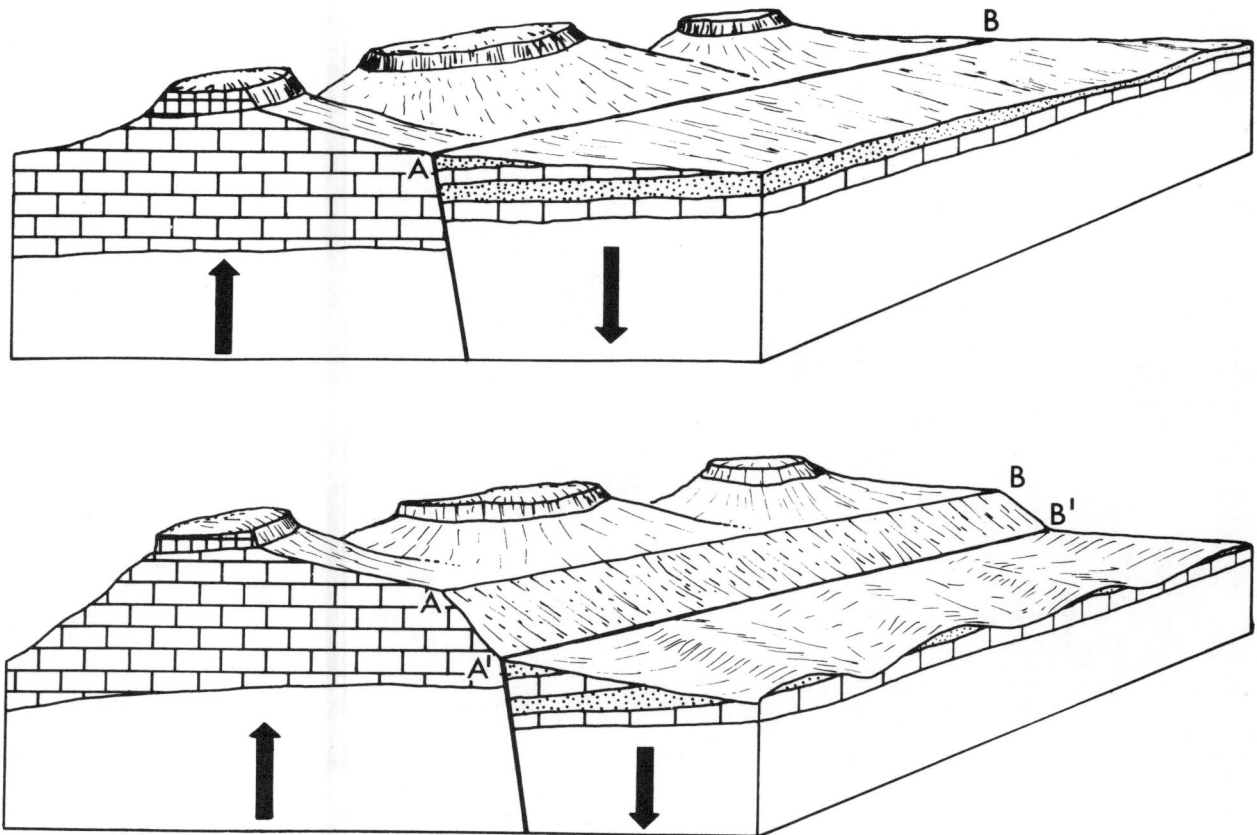

Fig. 9.24 Geomorphological effects of renewed fault movement, edge of the Vosges south of Saverne [Tricart (1974)]. Above: The Morphological stage reached in the Pliocene. The uplifted block, formed of a hard conglomerate resting on Triassic sandstones, is dissected into buttes with steep cliffed sides. The fault A–B is levelled by a concave slope that bites into the sandstone above, and lower down, bevels the marls and limestone of the Muschelkalk on the downthrown block. Quartz pebbles from the conglomerate are scattered over this slope. Below: The present relief. The fault has been active again quite recently (for Quaternary alluvia are affected). The dissected surface of the uplifted block is perched above the original fault scarp (AA–BB) as a result of the further movement. The valleys that crossed the slope are trenched and hanging; their direction is sometimes reversed so that they flow behind the buttes. The scarp has thus become a water parting, and this enables it to retain its freshness, with no notches. The dissection of the Pliocene slope at its foot is being effected by small valleys whose heads have not yet reached the scarp.

The High Vosges exceeds 1000 m over substantial areas, the highest points being the Ballon du Guebwiller (1426 m) and the Hohneck (1366 m). In the north, tabular Triassic sandstone uplands are cut by deep valleys. On the west, the Mesozoic rocks dip westwards into the Paris Basin, presenting east-facing escarpments forming a frame to the crystalline outcrop of the central High Vosges. Here, the granites and gneisses rise to massive rounded summits (*ballons*) with some tors. Because of the westerly tilt, most drainage flows to the Moselle–Meurthe system, the rivers traversing the sedimentary cuestas in water gaps. The shorter streams on the east flow into the Rhine. These are more steeply graded and deeply incised.

During the Pleistocene cold periods, all areas higher than about 800–900 m were covered by snow and ice, giving rise to small glacier systems. Cirques are quite numerous on the highest crystalline ridges—mainly on their east-facing flanks—and contain rock basin lakes such as Lac Blanc (elevation 1051 m, depth up to 60 m) and Lac Noir (at 957 m). Some valleys show signs of glacial deepening and contain moraine-dammed lakes. Elsewhere, periglacial processes produced cryoplanation terraces, block fields and sheets of angular debris.

Around the Saverne Gap, the general altitude falls to about 300 m and the Vosges become narrower. The through valley of the Zorn provides an important transverse routeway whose floor is no more than 220 m in elevation where the valley crosses the Muschelkalk outcrop. Farther north, the Bunter sandstone forms most of the Low Vosges, continuing north into the Hardt and Pfälzer Wald districts in Germany. The relief rises to 400–500 m (maximum, 581 m in the Wintersberg); the landscape consists of tabular sandstone uplands, highly dissected and often weathered into crags, pinnacles and sharp ridges.

9.10.2 Schwarzwald (Black Forest)

The counterpart of the Vosges—the Schwarzwald—stands as a prominent block, tilted to the east and northeast, on the east of the Rhine rift valley (*see* Fig. 9.25). To the south, it presents a steep face overlooking the upper Rhine between Schaffhausen and Basel; the western face represents the striking fault-line scarp of the Rhine rift. Towards the east, however, the uplands sink more gradually towards the south-western German scarplands (*see* Fig. 9.30) and, to the north, the Schwarzwald also declines slowly in general elevation to the Maulbronn Gap which offers an important transverse routeway between Karlsruhe and Stuttgart at about 200 m. Elsewhere, the Schwarzwald presents an almost continuous barrier along the east of the Rhine rift.

In the southern and central parts of the Schwarzwald, the crystalline basement (mainly granitic and metamorphic rocks) outcrops, exhibiting the clear impress of the Hercynian orogeny with its characteristic north-east–south-west trend marked, for example, by many fractures

now picked out by deep river valleys. The relief is strongly dissected and frequently exceeds 1000 m, reaching 1493 m at the Feldberg in the south.

Few remnants of planation surfaces can be recognized because of the intensity of dissection, although the block as a whole was originally bevelled by the post-Hercynian surface (*see* Section 9.10.1). In some places, however, lower surfaces of possible Palaeogene age at 600–700 m can be found. To the north and east, the surface of the crystalline rocks dips beneath a cover of Bunter sandstone,

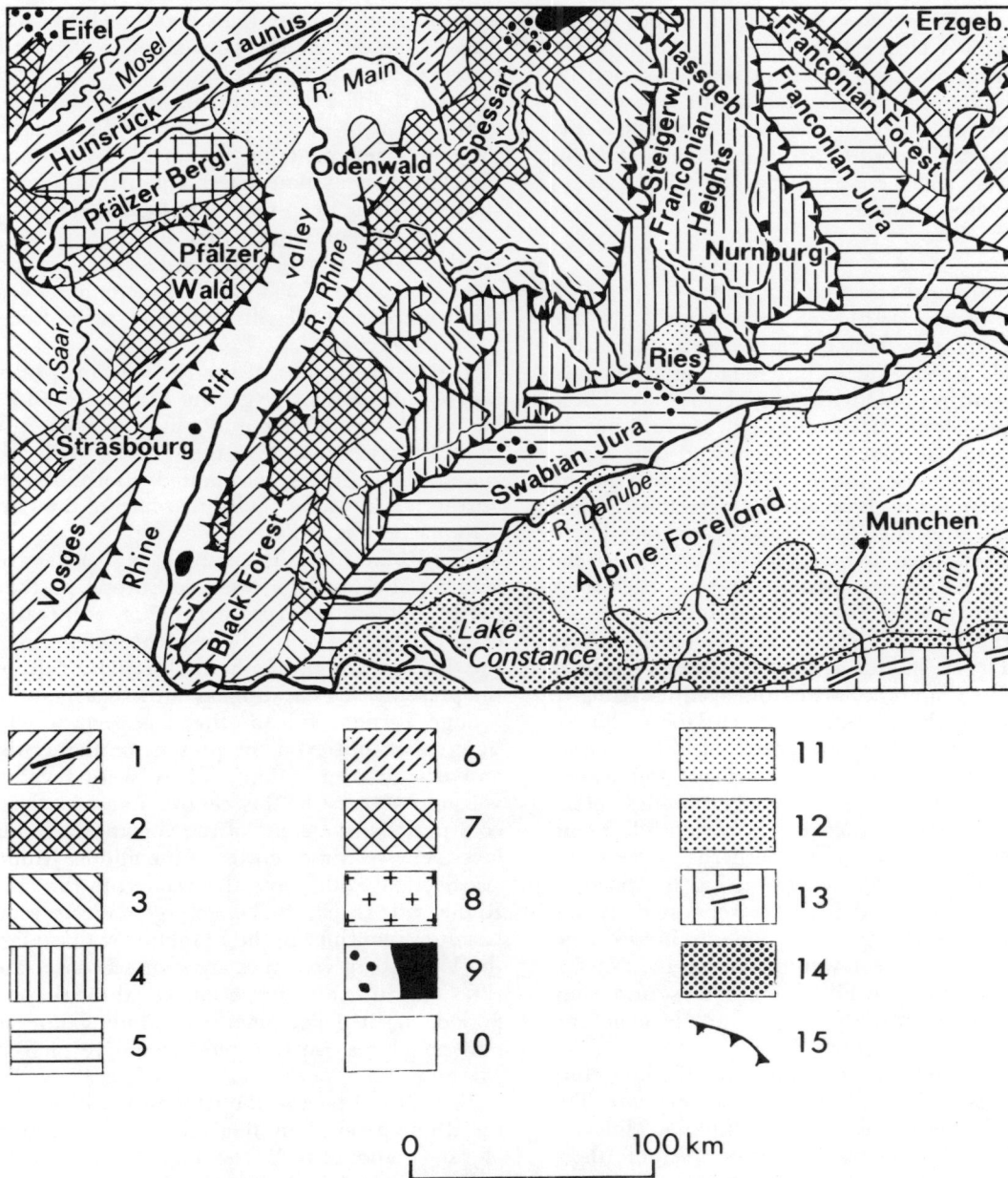

Fig. 9.25 Geological and morphological features in south-west Germany and adjacent areas [after Neef (1978)]. (1) Hercynian crystalline basement with structural ridges. (2) Bunter sandstone plateaus. (3) Muschelkalk limestone and Lower Keuper outcrops. (4) Upper Keuper sandstone plateaus and Lias. (5) Jurassic plateaus and cuestas. (6) Subsided uplands and plateaus of Mesozoic strata. (7) Pfälzer highlands (Permian and Carboniferous). (8) Basins eroded in the crystalline basement. (9) Cenozoic volcanics. (10) Rhine rift valley. (11) Alpine foreland of Tertiary and Quaternary sediments. (12) Area of Alpine foreland glaciated during the Würm. (13) Calcareous Alps with Jura-type folding. (14) Molasse and flysch zone. (15) Escarpments.

whose outcrop occupies about the same area in the Schwarzwald as that of the crystalline formations. The Bunter cover resting on the Hercynian platform is most continuous in the north-east, but even here the deeper valleys have cut through to the crystalline undermass. Farther east, the Muschelkalk forms another, narrower cuesta (10–15 km wide) with a broken west-facing scarp.

In the Pleistocene, the higher parts of the Schwarzwald were glaciated. Small glacier systems fed by cirques developed in the mountain areas. such as the Feldberg (1493 m) in the south and on the Hornisgrinde (1164 m) in the north. The Feldsee below the eastern slopes of the Feldberg is a typical cirque lake, as is the Mummelsee below the Hornisgrinde. Some valleys were modified by ice, and moraine-dammed lakes such as the Titisee and Schluchsee form attractive elements of the landscape. Periglacial forms and deposits are also widespread, including block fields, block streams, cryoplanation terraces and thick gelifluction mantles.

9.10.3   Odenwald and Spessart Uplands

From Pforzheim northwards for about 40 km, the average elevation of the uplands flanking the eastern side of the Rhine rift is only about 250 m, but then the Hercynian highlands reappear in the Odenwald and Spessart. These regions bear many similarites to the Schwarzwald but do not attain such high altitudes; the Odenwald only rarely exceeds 600 m. Otherwise, the physical elements of the landscape include most of those already described for the Schwarzwald, except that glacial forms are lacking. There are the same two contrasting major landscape types—the crystalline and the Bunter sandstone areas—the former showing the characteristic Variscan fold and fault structures with a steep fault-line scarp overlooking the Rhine (*see* Fig. 9.25). The crystalline area is termed the Vorderer Odenwald with its granites, diorites and syenites rising to rounded summits, such as the Neunkircher Höhe (605 m). The Hinterer Odenwald is composed of the Triassic sandstones, with structural plateau surfaces and watersheds bevelled by planation surfaces. The highest point here is the Katzenbuckel (625 m), a basalt hill rising above the general surface. The southern part of the Odenwald—the Kleiner Odenwald—rising to 566 m in the Königsstuhl above Heidelberg, is separated by the deep valley of the Neckar with its famous incised and abandoned meanders. The lower course of the Neckar cutting across the Odenwald is anomalous and may predate tectonic uplift of the region in the Neogene and Quaternary.

The entrenched valley of the Main forms the boundary between the uplands of the Odenwald and Spessart. The latter consists of a series of blocks uplifted to different altitudes. The morphostructural features resemble those of the Odenwald, but the crystalline outcrops occupy smaller areas and the greater part is underlain by the Bunter sandstone, forming dry plateaus at around 500 m crossed by a few deep river valleys. The highest point is the Geyersberg (605 m). Farther north, in the separate subregions of the Vogelsberg and Hohe Rhön around the Fulda depression, volcanic rocks rise to over 700 m (maximum, 950 m) (*see* Section 9.13.1).

9.10.4   Rhine Rift Valley

One of the most striking features on the map of Europe is the Rhine rift valley, which is a depression 300 km long and 30–50 km wide, bordered by the Vosges and the Hardt uplands on the west, and by the Schwarzwald and the Odenwald on the east (*see* Fig. 9.25). The floor of the rift stands at about 250 m in the south, falling imperceptibly to 85 m in the north. The rift is the central core of a tectonic zone no less than 2000 km long, which can be traced from the Mediterranean Sea to Denmark. It can be traced along the Saône–Rhône corridor, into the Rhine rift; north of Frankfurt this great tectonic lineament continues west of the Vogelsberg, disappearing beneath the newer sediments of the northern German plain. The Hercynian crystalline basement in the Rhine rift valley subsided along deep-seated faults and is today covered by 1–3.4 km of sedimentary deposits. A borehole near Frankenthal failed to reach the base of the Tertiary deposits, even at a depth of 3335 m. The tectonic movements since the beginning of the rift formation in the Eocene amount to about 4400 m.

The southern part of the rift is formed by a system of horsts and grabens (e.g., the Mulhouse and Dinkelberg blocks). The widening of the rift in the vicinity of Mainz is caused by the crossing of the rift zone with other tectonic zones. The rift is terminated morphologically at its northern end by the fault-line scarp of the Hunsrück and Taunus (*see* Fig. 9.26).

The Rhine rift valley is, then, one of the great tectonic features of central and western Europe. At the base of the bordering fault scarps, many hot, mineralized springs occur. Together with the evidence of recent vulcanism, this indicates high geothermal activity continuing up to the present. The basalt mass of the Kaiserstuhl represents a huge Tertiary volcano that has been largely destroyed by erosion, but even the present remnants rise to 350 m and extend over 90 km². There were several phases of volcanic activity in this centre. Later in the Pleistocene cold periods, the ruins of the volcano were blanketed by loess. The volcanic centres of the middle Rhine highlands (*see* Section 9.10.5) are also related to the faulting of the Rhine rift (Eifel, Siebengebirge, Westerwald and the basalt outpourings in the Frankfurt region); farther north, the Vogelsberg volcanics are similarly associated with the rift. Volcanic activity occurred throughout the whole period of rift development, continuing into the Quaternary in places, but the most intensive phase was in the Miocene.

The central part of the Rhine rift valley is occupied by the Rhine flood plain, flanked by a low terrace composed of sands and gravels (*see* Fig. 9.27). The high (10 m) terrace is developed closer to the foot of the surrounding mountains. The gravels of these terraces are extensively covered by loess; there are also areas of aeolian sand, showing fossil dune forms in places.

The flat floor of the alluvial and glaciofluvial deposits, which represent the last sediments to be deposited in the

Fig. 9.26 The Hunsruck seen from the floor of the northern part of the Rhine rift. (Embleton)

valley, meets the bounding fault scarps abruptly in some places, but in many others there are intervening foot-hills or stepped fault blocks, together with various old river terraces and alluvial fans. For example, the Hardt uplands on the north-west descend in a series of fault scarps and structural benches of Jurassic limestone to the rift floor. The Markgräfler hills south of Freiburg, on the other side of the rift, rise to 400–500 m; the Tuniberg just south of the Kaiserstuhl is an isolated fault block that attains 316 m.

9.10.5 Middle Rhine Highlands
Forming a compact block measuring about 200 km by 150 km and traversed by the Rhine gorge from Bingen to Bonn, the middle Rhine highlands comprise another distinctive unit of the Hercynian Mittelland. Geologically, the highlands consist of Precambrian and Lower Palaeozoic strata which were folded and fractured in the Hercynian orogeny along east-north-east–west-south-west (Variscan) trend lines. Post-Hercynian denudation carved a planation surface across the massifs and some Mesozoic strata, notably the Trias, were laid down as a platform cover. Further planation and uplift followed, culminating in the mid-Tertiary when vulcanism was also active in several parts. Neotectonics also caused substantial deformation of planation surfaces into mega-anticlinal

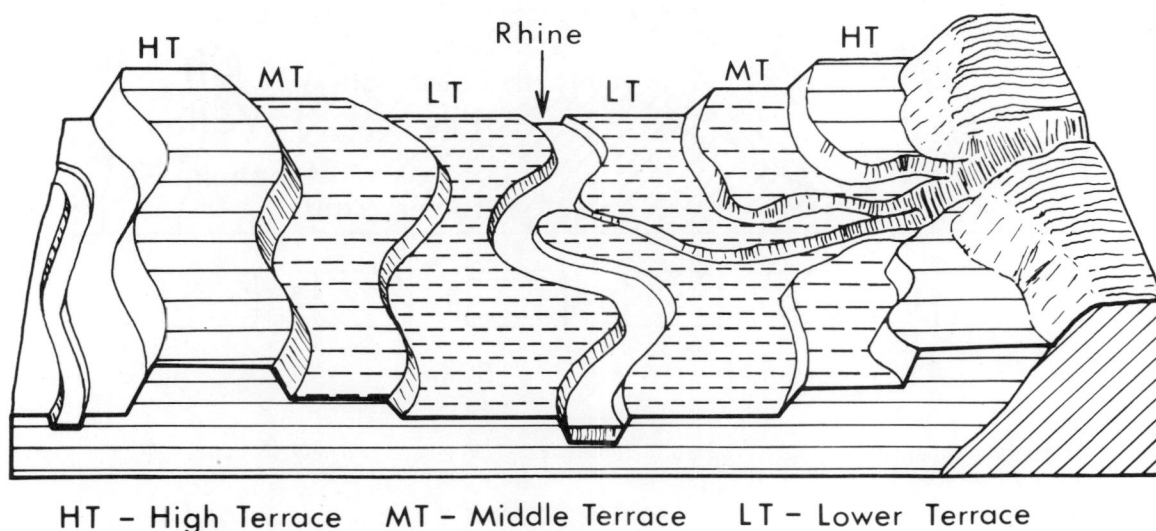

HT – High Terrace   MT – Middle Terrace   LT – Lower Terrace

Fig. 9.27 Diagrammatic representation of the Rhine terraces [after Wach (1969)].

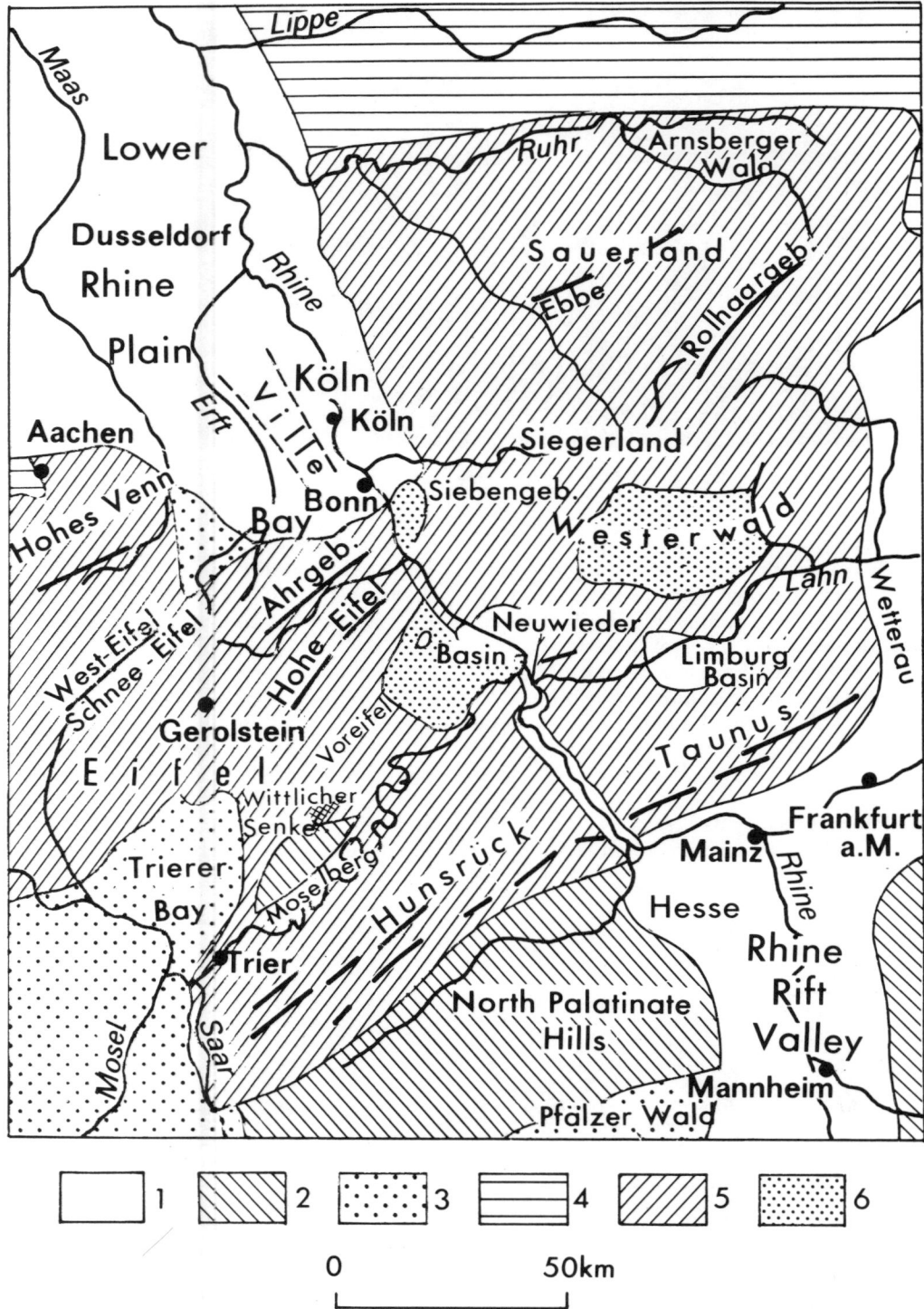

Fig. 9.28 The Rhine highlands [after Neef (1978)]. (1) Basins of Tertiary and Quarternary sediments. (2) Upper Palaeozoic rocks. (3) Triassic sandstone. (4) Cretaceous. (5) Hercynian crystalline and Palaeozoic basement. (6) Major volcanic areas.

and megasynclinal undulations, as shown by the spatial distribution of planation surface remnants. At the same time, dissection proceeded rapidly, especially near to the major valleys of the Rhine, Moselle and Lahn, and the floor of the Rhine gorge now lies hundreds of metres below the level of the surrounding mountains. The southern edge of the highlands is sharply delineated by a Variscan-trending fault-line scarp overlooking the Saar–Nähe and lower Main lowlands, marking the point where the Rhine rift valley proper (*see* Section 9.9.4) terminates.

The highlands fall into four clearly defined sectors (*see* Fig. 9.28), divided by the Rhine flowing orthogonally to the Variscan trend, and by its tributaries the Moselle and Lahn flowing roughly along this trend. The four units are the Eifel, Hunsrück, Taunus and Westerwald; an extension of the latter in Siegerland and Sauerland will be described separately (*see* Section 9.10.5.5).

Essentially, they comprise plateaus bevelled by planation and later deeply cut by valleys. The predominant rock type is the Siluro-Devonian slate, which also contains some more resistant bands of sandstone and quartzite. The latter gives rise to some prominent monadnocks or ridges rising above the planation surfaces. Some of the latter are thought to represent parts of etchplains, related to late Tertiary phases of warmer climate and deep weathering of the massifs, followed by stripping. Neotectonic uplift occurred in several phases, which can be traced in the progressive valley incisions. In the Quaternary, both uplift and rapid climatic change caused the development of important sets of river terraces along the Rhine and its tributaries. For example, the main terrace of the Rhine has an altitude of 240 m near the Lorelei Rocks, about 200 m near Koblenz and 150 m at Bonn. This relatively steep longitudinal gradient and the differences of gradient of this and other terraces suggest the influence

of Quaternary tectonics. The latter were responsible for continuing subsidence of minor grabens within and next to the Highlands, such as the Köln, Bonn, Neuwied and Limburg basins.

The defile of the Rhine is without doubt the most impressive feature of the middle Rhine highlands (*see* Fig. 9.29). Essentially, it seems to be an antecedent gorge, the river having maintained its course during neotectonic uplift [see Yates (1963)]. Its sides are often precipitous, with only narrow terraces beside the river, leaving little room for road and railway. Soon after entering the gorge, the river crosses the Bingen quartzite bar; then the river widens to 700 m by the Sooneck Castle and the outcrop of the less resistant Hunsrück shales. Further quartzites near Lorch constrict it again to only 280 m, while at the Lorelei Rocks the river is only 200 m wide and 27 m deep. Well-developed meanders occur at Boppard, together with clear river accumulation terraces. Below Braubach, the low Rhine terrace is prominent. The small Neuwieder graben is formed in young volcanics; these and other occurrences of vulcanism and geothermal springs in the highlands reflect the youthful tectonics of the region.

9.10.5.1 *Hunsrück*

Framed by the Moselle and the Rhine on the north and east, respectively, and by the Saar–Nähe rift on the south, the Hunsrück comprises a rugged highland of mainly Devonian slates. Quartzite outcrops form more exposed ridges (*see* Fig.9.28), such as the Hochwald [with the Erbeskopf (816 m)], Idarwald and Sonnwald. The dominant surface seems to be related to Palaeogene planation, later modified in the Pliocene to an etchplain. The uplands slope generally northwards towards the Moselle valley with its fine incised (and some abandoned) meanders and terraces.

Fig. 9.29 The Rhine gorge between the Hunsrück and Taunus. (Embleton).

Fig. 9.30 The Totenmaar an explosion crater in the Eifel. (Embleton)

9.10.5.2   Taunus

The highest parts of the middle Rhine highlands are to be found in the south of the Taunus, where Devonian quartzites outcrop in long structural ridges. The Grosser Feldberg reaches 880 m and the Königstein nearly 800 m. The southern boundary is the fault scarp overlooking the lower Main and has already been referred to. Slates are the predominant lithology, and the whole region is greatly fractured, the fault lines often being picked out by geothermal springs. Across the Hercynian basement an undulating surface, possibly an etchplain, has been cut, most clearly visible in the northern areas. In other parts, the plateau surface has been dislocated by horsts and grabens. Parts of the Taunus were sufficiently high in the Pleistocene to support snow fields and periglacial activity, resulting in nivation phenomena, cryoplanation terraces, block fields and block streams.

9.10.5.3   Westerwald

This upland lies between the valleys of the Sieg and the Lahn, the latter separating it from the Taunus. Although not quite so high as the Taunus, it is an area also built of highly folded and fractured slates and quartzites, in which the highest plateaus attain levels of up to 650 m. A major difference from the Taunus, however, is the presence of Tertiary volcanics (*see* Fig. 9.28). In the north-east is the Hohe Westerwald—a lava plateau cut by broad valleys—reaching 657 m in the Fuchskauten. The basalts have developed thick weathered mantles and a so-called post-basalt planation surface. Elsewhere, the basalts can be seen to overlie older Tertiary weathered deposits and clays. The margins of the lava plateau are dissected by tributaries draining to the Sieg and Lahn. In the south-west lies the Unterwesterwald, consisting of a system of horst blocks and graben basins, the latter containing Tertiary clays and lignites showing that the differential

vertical displacements are of later Tertiary age. Another volcanic area is the Siebengebirge standing next to the Rhine between Bonn and Linz. Here the volcanic forms including basalt flows and plugs of trachyte and dolerite have been partly destroyed by erosion, only the more resistant masses such as the pyramidal peak of the Drachenfels (324 m) remaining.

9.10.5.4   Eifel

This region shows the greatest variety of all four blocks comprising the middle Rhine highlands and morphostructurally is the most complex unit. Generally, the Hercynian basement of slates and other folded, fractured and partly metamorphosed rocks has been planed across by the post-Hercynian surface, parts of which were then covered by sediments—Triassic sandstones, including the bright red Bunter and limestones (the Muschelkalk)—which now remain mainly in two areas, near Mechernich in the north and in the Bitburgerland stretching north of Trier. West of Trier, the Muschelkalk continues into Luxembourg, where it eventually disappears beneath the Jurassic. Apart from the areas of Triassic rocks, four other subregions can be distinguished (*see* Fig. 9.28):

1)   The Hohe Eifel in the centre, rising to 747 m in the Hohe Acht. The Hercynian basement is here intruded by volcanic necks.
2)   The Schnee Eifel in the west, where Cambrian quartzites attain heights of 500–700 m.
3)   The volcanic Eifel in the east, a distinctive area dominated by Tertiary–Quaternary eruptions of lava and tuff. Volcanic cones [e.g., Mosenberg (509 m)] total about 200, and crater lakes (*see* Fig. 9.30) and maars are dotted over the area. There are also fields of tuff which, especially with a loess cover, are virtually waterless. The largest crater lake is the Laacher See, almost circular and covering about 3.3 km². Tuffs and lava bombs have been

incorporated in some of the Rhine and other terrace gravels; potassium–argon dating of the volcanics has therefore provided an important means of determing the absolute ages of some of the latter [see Frechen and Lippolt (1965)].

4) North of the Eifel proper and extending into Belgium lies the Hohes Venn, a plateau at about 600 m developed on Cambrian strata.

9.10.5.5 *Siegerland and Sauerland*
These form extensions to the north of the middle Rhine highlands with generally lower relief, although locally in the Rothaargebirge—a continuation of the Westerwald—the land rises to 700–800 m. The Sieg, in the south, separates the areas from the Westerwald, but otherwise the broad elements of the geological structure remain similar to those of the latter (*see* Fig. 9.28). The eastern parts, including the Rothaargebirge, have undergone greater neotectonic uplift, and the eastern boundary is delineated by a system of parallel faults involved in the subsidence of the Hesse Basin. The general plateau level of Sauerland is probably a Tertiary etchplain, above which rise monadnocks of quartzite, porphyry and dolerite.

9.11 Ardennes

Crossing the frontiers of no less than four countries, the Ardennes represent a westward continuation of the Eifel of Germany (*see* Section 9.10.5.4) across northern Luxembourg and southern Belgium into a small part of northern France. They are considered separately in this chapter from the middle Rhine highlands, even though morphostructurally they share many characteristics, because most of the area falls within the Meuse catchment rather than that of the Rhine; the Luxembourg Ardennes, exceptionally, do drain to the Rhine indirectly via the Moselle.

The uplands are derived from a Hercynian platform block, surrounded on three sides by younger sedimentary lowlands or low plateaus: Brabant and Hainaut on the north, and the Jurassic–Cretaceous scarplands of the Paris Basin on the west and south. The eastern boundary, dividing it partly from the Eifel, is formed by the depression north of the Trier Basin, containing Triassic sandstones.

The oldest rocks of the Ardennes are the Cambro-Silurian slates and quartzites outcropping in the Stavelot massif (including the Hohes Venn, *see* Section 9.10.5.4) in the north-east, and in the Rocroi massif in the south-west. Elsewhere, apart from the Carboniferous outcrops, the greater part of the Ardennes consists of Devonian strata (sandstone, quartzites, limestones and shales). The Carboniferous strata have been removed by erosion except for two belts in the north, where they have been preserved along complex downfaulted synclinoria: the Mons–Liège furrow (sometimes called the Sambre–Meuse depression) and the Dinant synclinorium.

The main trends of the folds and fractures, and of the original immense Hercynian mountain chains whose roots only now remain, run east–west in the western part of the Ardennes, curving, farther east, to adopt a more north-easterly trend. The Hercynian orogeny was followed by intense and prolonged denudation, which eventually produced the post-Hercynian surface across the Ardennes that now dips below later Mesozoic strata as noted above. In the Cretaceous, an extensive marine transgression resulted in further planation and deposition of sediments, remnants of which can still be found even at heights of 600 m or more. Elsewhere, the residual clay-with-flints may testify to a former Chalk cover. Further uplift and denudation ensued, but the early Tertiary marked another, although less extensive, marine transgressive phase. Oligocene clays and sands are now found in the north-east, for example, at over 500 m. In the Miocene, tectonic uplift and northwards tilting occurred, followed by further denudation and the formation of a Pliocene–Lower Quaternary etchplain. Thus there is a highly complex set of planation surfaces and correlative deposits, of which the younger features are best preserved. The post-Hercynian surface has been largely destroyed except on the southern margins where it has recently been exhumed.

The relief of the Ardennes upland is characterized by dissected plateaus, structural ridges in the more resistant Palaeozoic sandstones and quartzites, and gently rounded summits rising to around 600–650 m. Planation surfaces, formed at different times and variously affected by subsequent tectonics, display a complex spatial distribution. The main rivers are deeply entrenched following neotectonic uplift and are often strikingly discordant with the geology: for example, the Meuse leaves the Lias clay vale in France to flow across the western end of the High Ardennes in a gorge cut through Palaeozoic rocks between Charleville-Mézières and Givet. Farther downstream, it again enters a deep gorge section from Dinant to Namur, lying clearly transverse to the Hercynian structures. One hypothesis is that these discordant segments represent superimposition from a former Cretaceous cover; the other possibility is that such water gaps are antecedent to the more recent uplift of the ranges, deformed Pliocene river terraces supporting this view.

In the Pleistocene cold periods, the Ardennes experienced periglacial activity. Snow fields occupied the higher parts, although glaciers were absent. A dense network of dells related to the flow of snow melt over the frozen substrate was created, along with nivation hollows, cryoplanation terraces and block fields. Some areas of fossil pingos have been described [e.g., by Pissart (1963) for the Plateau des Hautes Fagnes in the north-east of the High Ardennes].

A number of subregions can be distinguished.

9.11.1 High Ardennes
Consisting mainly of Devonian rocks, the relief rises to a maximum of 694 m in the Botrange. The summits are rounded, with the plateau surfaces diversified by structural ridges and shallow depressions, possibly related to the deep weathering and stripping associated with etchplain formation.

9.11.2   Rocroi Massif

This unit is simply a continuation of the High Ardennes west of the Meuse defile and lies mostly in France. Cambrian rocks form the core of the massif, rising to plateau surfaces and subdued summits with a maximum height of 413 m. Some patches of Eogene sediments have survived.

9.11.3   Luxembourg Ardennes (the Oesling)

The plateaus at 400–500 m in the north of Luxembourg are considerably dissected by the Sure and its tributaries, a system which drains south and south-east towards the Moselle. The Devonian rocks are mainly slates and quartzites that disappear southwards below the Triassic and Jurassic strata of southern Luxembourg.

9.11.4   Famenne Depression

This is a strip of lowland lying north of the High Ardennes and following the general Hercynian trend. The relief is mostly below 200 m, even down to 100 m in places, and it represents a trough eroded in the less resistant Upper

Devonian shales and schists. Some limestones are also present, giving small hills up to 250 m high, with cave systems near Rochefort and underground drainage.

9.11.5   Condroz Massif

This broad east–west upland stands between the Famenne and the Sambre–Meuse furrow to the north. Plateaus at about 300 m are cut in Lower Palaeozoic rocks, diversified by sandstone ridges and shallow depressions following the Hercynian trend lines.

9.12   South-West German Scarplands

Although no important Hercynian structures are to be seen at the surface in this area, it is considered to be part of Hercynian Europe on account of two important facts: first, it is an area of the Hercynian platform covered and concealed by Mesozoic rocks, and secondly, it is intimately related to the adjacent exposed Hercynian massifs (Schwarzwald, Bohemia, etc.) in terms of morphostructure and geological evolution. The relief ranges from less

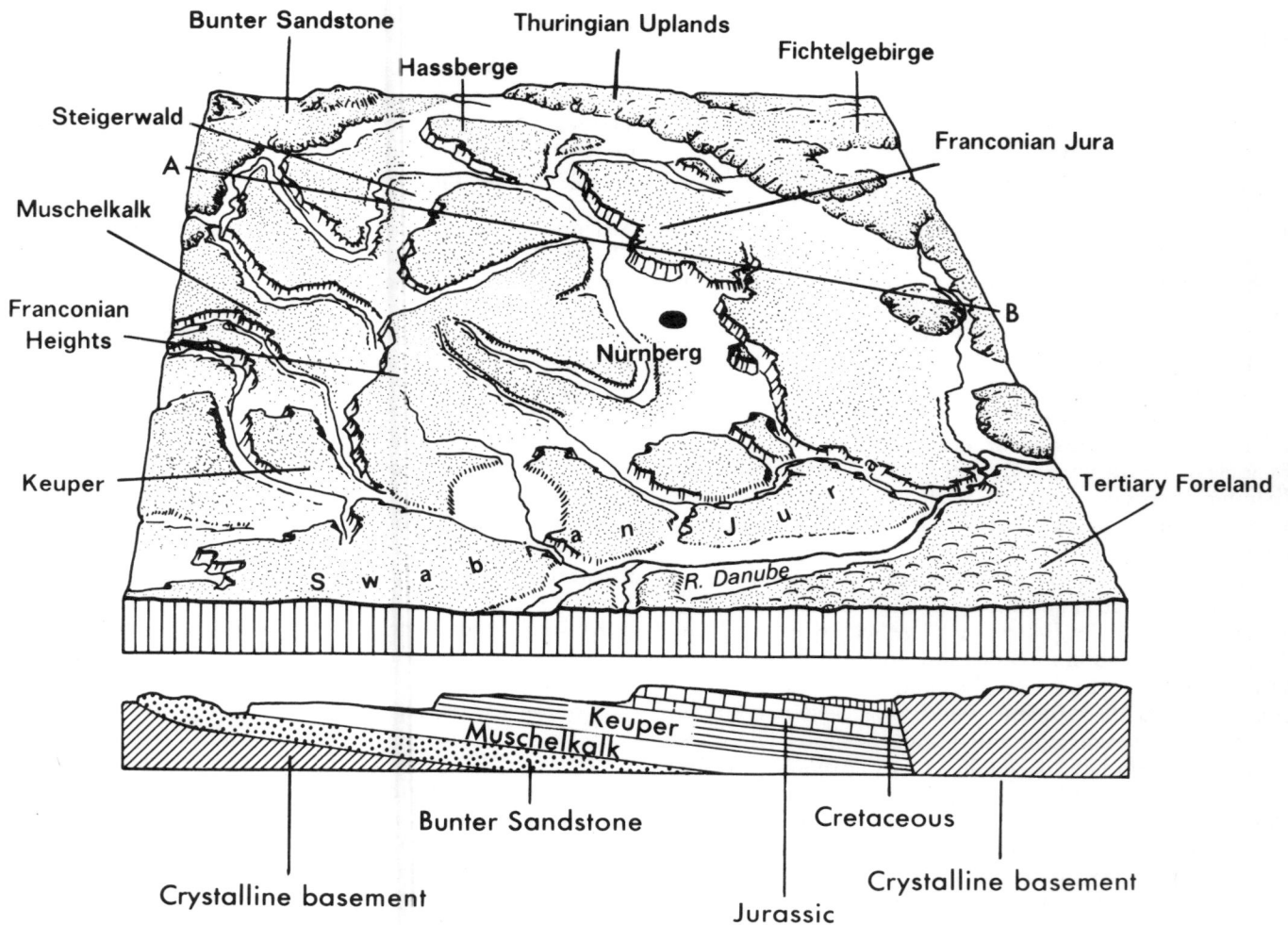

Fig. 9.31 Cuesta landscape of south-west Germany [after Wach (1969)]. The section below is along the line A – B.

than 200 m to just over 1000 m in a few places and comprises essentially a series of scarps and vales. The principal escarpments are related to three relatively resistant formations, their resistance being linked in part to their permeability—the Middle Jurassic Limestone, the Keuper sandstone and the Muschelkalk (*see* Fig. 9.31). The region is crossed by the major continental divide between the Rhine and Danube catchments which lead, respectively, to the North and Black Seas. Because of the relative proximity of the North Sea and the fact that the development of the Rhine at a lower level has been facilitated by the downfaulting of the Rhine rift, the Rhine has been steadily extending its catchment at the expense of the Danube. Undoubtedly, considerable areas of the Neckar and Main Basins used to drain south-east to the Danube, but today the Danube receives no important left-bank tributaries from the Swabian Jura (or Swabian Alb); the first break in the continuity of the latter is the Ries depression and, beyond it, the valley of the Altmühl which does join the Danube.

The scarplands can be divided morphostructurally into three main regions: the Jurassic limestone cuesta comprising the Swabian and Franconian Alb (the word *Alb* has the same root as the word alp, referring to the high summer pasture); the Neckar scarplands; and the Main scarplands (including the upper Main valley). Before dealing with these separately, some general comments are appropriate on the general scarpland evolution. Many processes have contributed to the etching out of the present landforms from a base composed of sedimentary strata of varying hardness, permeability and joint texture. Post-Jurassic tectonics, including neotectonics, have played a part—fracture lines cross the cuestas and young volcanic activity characterize a few areas. The development of subsequent (strike) streams along marl and clay outcrops has led, by capture, to drainage reorganization. Spring sapping has helped in scarp recession, together with mass movements on the steeper slopes, especially where limestones rest on clays. Fischer (1973) emphasized the role of climatic change in the late Tertiary and Quaternary, and periglacial action in the Pleistocene cold periods. Periglacial processes were active on the more resistant layers on slopes of 4° or more, and on less resistant layers on slopes even as low as 2°. In places, cryopediments developed at the scarp foot. As Büdel (1957, 1977, 1982) and others have emphasized, the geomorphology can only be properly understood in the context of Cenozoic climatic change.

9.12.1 Jurassic Cuesta

Forming the main watershed between the Rhine and the Danube over a considerable distance, as already noted, the Jurassic cuesta extends altogether for some 400 km, from the Rhine valley near Schaffhausen in the south-west (where it takes over from the Plateau Jura in Switzerland) to the Main valley above Bamberg. Its width is up to 40–50 km in places, with a few summits rising above the 1000-m contour. The escarpment faces north-west or west, overlooking Lias clay lowlands; the dip slope is represented by dissected plateaus. The cuesta is divided into

two segments—the Swabian Alb and the Franconian Alb—by the Ries depression.

9.12.1.1 *Swabian Alb*

For most of its 200 km, the scarp edge rises to over 500 m and, in the Lemberg in the south-west, touches 1015 m. A double escarpment occurs in places: the first in the Dogger formation and the second (main) cuesta in the lower Malm (in the west) and in massive dolomitic and reef limestones (in the east). In front of each scarp, erosional remnants form mesa-like outliers [e.g., Hohenzollern (855 m) and Hohenstaufen (684 m)] providing evidence of scarp retreat. The southernmost part of the Swabian Alb (*see* Fig. 9.32) is crossed by the upper Danube near Geisingen; the Danube drainage here has so far escaped capture by the upper Neckar on the north or the Wutach (flowing south-west to the Rhine) on the other hand. Also in the southern region of the Swabian Alb particularly, karst features appear, including swallow holes, resurgences and dry valleys. Water from the Danube sinks underground between Immendingen and Tuttlingen, emerging as springs (the Singener Aach) at the foot of the Hegau hills 20 km to the south. In this way, water is already being abstracted underground from the Danube to the upper Rhine.

The Swabian Alb plateau is divided by faults with which young volcanic forms are associated. These include maars, crater lakes and basalt, tuff or phonolite hills. Examples of the latter include the Romerstein (874 m), Hohenstoffeln (844 m) and Hohenkrähen (645 m); the Hegau plateau is marked by others.

9.12.1.2 *Ries depression*

At one time regarded as a down-faulted basin, this remarkable feature is now thought to be an astrobleme: a giant meteorite crater some 29 km in diameter. It breaks the continuity of the Jurassic cuesta and froms the conventional division between the Swabian and Franconian Alb.

9.12.1.3 *Franconian Alb*

Here the Jurassic cuesta does not reach such high levels as the Swabian sector; the maximum on the cuesta proper is 647 m in the Hahnen Kamm, although the Hesselberg outlier touches 689 m. Both these hills are at the southern end, near the Ries basin. Elsewhere, the average height of the crest and plateaus is about 500–600 m. The escarpment curves rapidly to adopt a more north–south alignment in the northern parts and is interrupted by several gaps and transverse valleys, such as the Altmühl and Pegnitz. A double escarpment may again be noted in some areas, while in the east, near the upper Main valley, the terrain becomes broken into a system of blocks involving both the platform cover and the crystalline basement. The main west-facing escarpment of the Franconian Alb is a 150-m high edge of massive Franconian dolomite with cliffs and other sharply cut erosional features. Karst forms include dry valleys, sinks, resurgences (e.g., the Blautopf spring near Blaubeuren) and cave systems. The plateau surface is in part thought to be a tropical karst planation

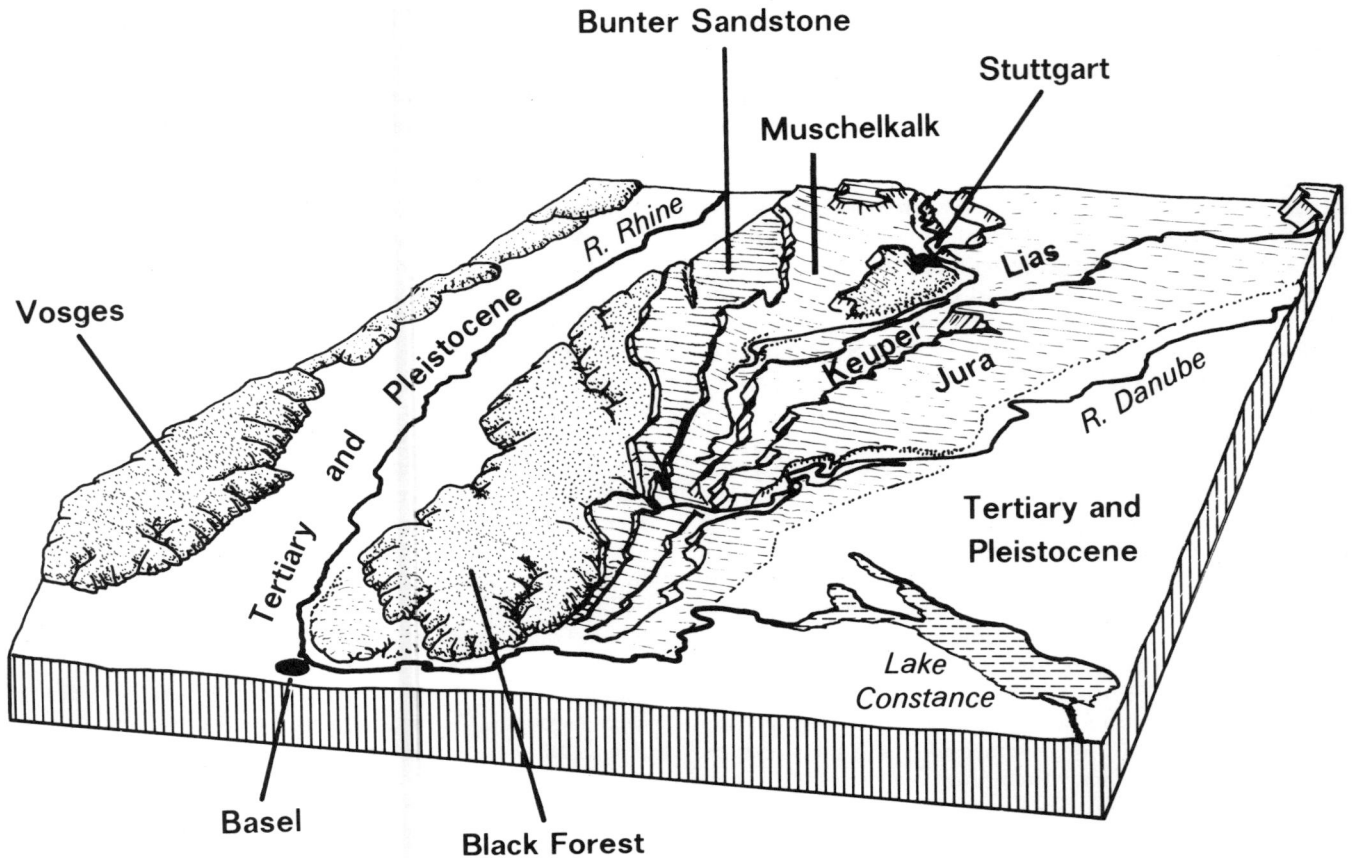

Fig. 9.32 Block diagram of the Rhine rift valley and adjacent uplands [after Wach (1969)]. The cuestas of Triassic and Jurassic strata east of the Schwarzwald are shown.

surface. Some parts are concealed by younger cover deposits.

The drainage is mostly to the Main, but partly to the Danube via the Altmühl in the south or the Naab on the east. The latter is developed on Triassic and Liassic rocks, downfaulted between the Franconian Alb and the Bohemian uplands. Volcanic necks such as the Parkstein (432 m) and the Rauher Kulm (682 m) north-west of Weiden rise above the undulating surface.

9.12.2   Neckar Scarplands

The area forms a triangle between the Schwarzwald and the Swabian Alb, and the Odenwald on the north; it consists of scarp and vale topography developed in the Lias and Trias. The main escarpment-forming strata are the Muschelkalk, giving a comparatively low cuesta, and the Keuper sandstone, the latter sometimes developing several separate cuestas. In contrast, the Lias and Keuper

clays and marls have been worn into lower-lying well-watered country. The Neckar river has dissected the cuestas and plateaus into many separate units, such as the Schurwald and Welzheimer Wald north of Göppingen, where the Keuper sandstone rises to 450–550 m; the Löwensteiner Berge and Mainhardter Wald east of Heilbronn, also in Keuper and reaching about the same level; the Limburger Berge and Ellwanger Berge between the Jägst and Kocher gaps, and so on. The westernmost exposure of the Keuper sandstone is near the Kraichgau region, where it forms the Stromberge and Heuchelberge uplands. North-east of Crailsheim and the River Jägst rise the Franconian Heights (up to 543 m) in another striking Keuper escarpment continuing into the area of the Main Basin. The Muschelkalk forms a discontinuous and less marked feature overlooking the Bunter sandstone, mostly to the west of the Neckar, except where the latter crosses it in a gorge up to 150 m deep before reaching the Odenwald.

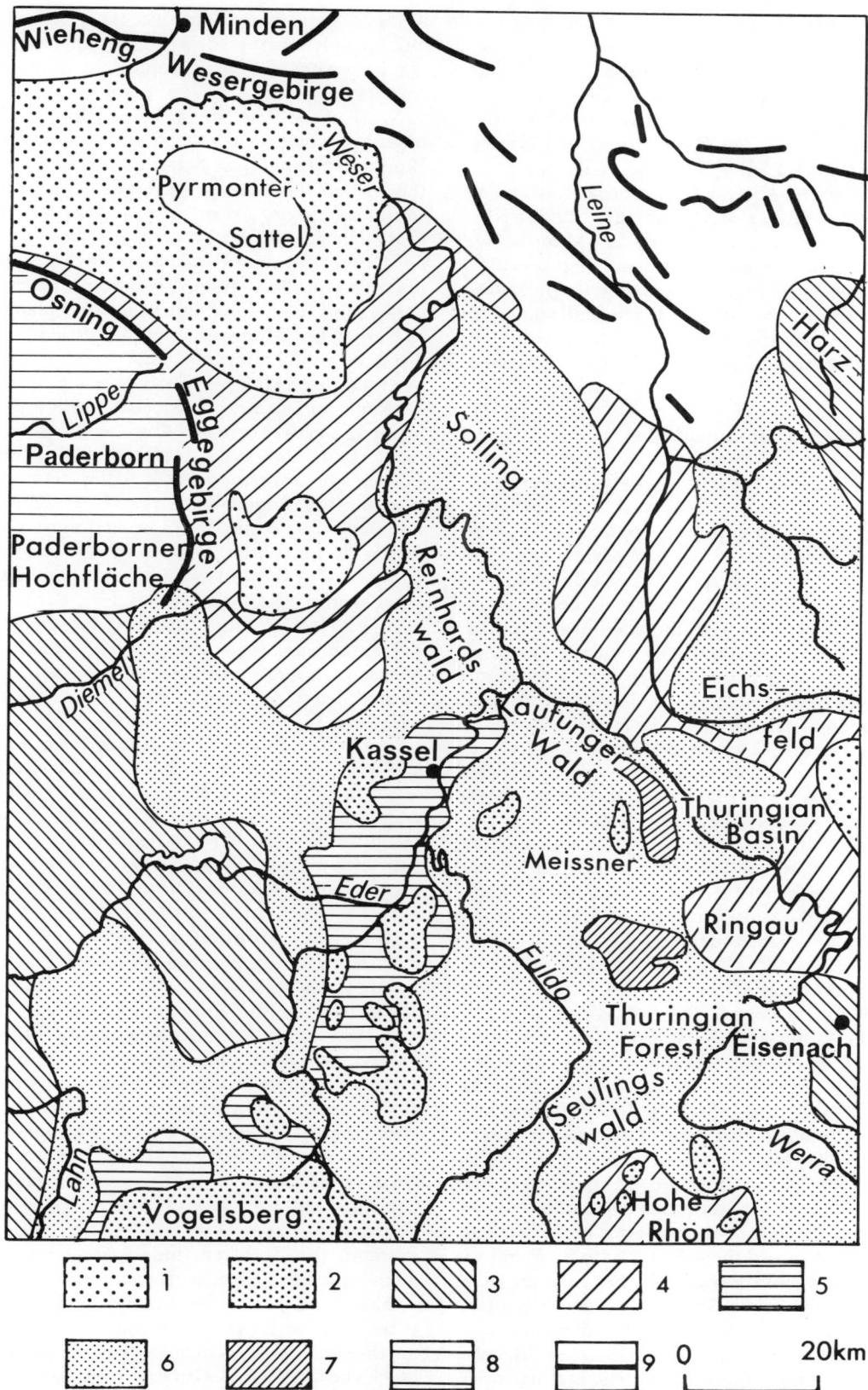

Fig. 9.33 The Weser–Saale hill country and adjacent regions [after Neef (1978)]. (1) Depressions of Keuper rocks. (2) Cenozoic volcanics. (3) Uplifted blocks of the crystalline basement. (4) Muschelkalk limestone plateaus. (5) Tertiary sediments. (6) Bunter sandstone plateaus. (7) Permian. (8) Cretaceous of the Westphalian depression. (9) Structural ridges of the Niedersächsische Bergland.

9.12.3 Main Scarplands

Like their counterpart to the south described in Section 9.12.2, this is also a landscape of Triassic cuestas and vales, bordered by the volcanic Hohe Rhön and the Bunter sandstone of Spessart on the west, and by the Franconian Alb on the east. The Keuper sandstone is once again a prominent scarp former, as in the Steigerwald (*see* Fig. 9.31), which rises to nearly 500 m. A broad lowland of Keuper marl follows farther west, and then there is the Muschelkalk cuesta which is crossed by the Main below Schweinfurt. Würzburg stands farther down this water gap, where the river flows in a valley whose flat floor lies up to 200 m below the level of the Muschelkalk plateau. About 20 km below Würzburg, the Main leaves the Muschelkalk for the Bunter sandstone of Lower Franconia and Spessart.

9.13 Weser–Saale Hill Country

It is difficult to ascribe any great degree of coherence to this area, for it includes both upland and lowland, scarplands and plateaus, rocks from Palaeozoic to Quaternary in age, and it falls across two separate drainage basins, those of the Saale (tributary to the Elbe) and the Weser (*see* Fig. 9.33). Nevertheless, the basic theme is still that of Hercynian Europe; indeed, included in the area is the Harz block, the type area of the Hercynian. The area forms a part of the so-called Mittelgebirge, a term for uplands interposed between the north German plain and the higher mountains farther south, a term not easily translatable into English. The Mittelgebirge are crossed north–south by several gaps and river valleys. These valleys do not reach any great heights, but their northern edge marks fairly closely the limit attained by the Fennoscandian ice in the Saale glaciation (for which again this is the type area). In the following regional division, some upland subregions can be clearly differentiated and, likewise, some lowland basins, but there still remain other areas that are not so neatly allocated.

9.13.1 Hesse and Weser Hills

Mostly drained to the north by the Weser and its headstreams—the Fulda and the Werra—this is a complex area structurally. In the first place, there is a great variety of rock type seen at the surface, from volcanics to sediments of Triassic, Jurassic, Cretaceous and Tertiary age. Secondly, the area lies astride the northward continuation of the Rhine rift system whose axis passes under the Wetterau depression, separating the Taunus from the Vogelsberg. The related fractures have partitioned the Hercynian platform basement into a series of horsts and grabens, and have promoted the volcanic activity manifest in the Vogelsberg, the Hohe Rhön and other smaller areas [e.g., the Knüllgebirge (636 m) and Meissner (750 m)]. The Vogelsberg (772 m) is reminiscent in many ways of Cantal in central France, with its radial valleys deeply scored in the basalt flows which form plateau segments. In the Hohe Rhön, the Wasserkuppe touches 950 m; the mass represents another volcanic ruin, but more deeply eroded [see Mensching (1957)]. Here the Tertiary basalts overlie

Triassic sandstones. The basalts of all these areas are often deeply weathered, with bauxite deposits, suggesting the influence of former tropical or subtropical climatic conditions.

North of the Vogelsberg–Hohe Rhön (*see* Fig. 9.32), the hills are composed of the Triassic Muschelkalk and Bunter sandstone [see Schunke (1968)]. Farther north in the Weser hill country, Jurassic and Cretaceous limestones dominate. An intricate system of horst ridges and graben basins, the latter filled with Tertiary deposits, is to be found. Structural plateaus include the Paderborn uplands and the Reinhardswald on the west of the Weser valley, and the Solling and Bramwald plateaus east of the Weser. The Mesozoic cuestas developed during the Neogene and Quaternary. Pliocene planation surfaces can be traced [see Hempel (1958a)] while, during the Upper Pliocene and Lower Pleistocene, valley pediments were formed, partly corresponding with aggradational river terraces in the valleys. Periglacial conditions in the Pleistocene contributed to valley cryopediments, especially in the sandstone areas [see Spönemann (1966)] but the Triassic limestones proved more resistant.

The River Weser begins near Münden through the confluence of the Fulda and Werra, and its valley shows good examples of incised meanders. Between the uplands of the Wiehengebirge and the Wesergebirge the river passes through the so-called Westphalian Gate near Minden, through which ice of the Saale glaciation penetrated from the north. The Wiehengebirge extending west from the Westphalian Gate is one of a pair of parallel ridges, its neighbour to the south being the Teutoburger Wald which continues as far as the Ems valley, a distance of 100 km. These finger-like ridges are only 10–15 km wide, the first consisting of eroded Triassic rocks, the second of Jurassic–Cretaceous limestones, separated by a vale drained partly east to the Weser and, partly westwards, past Osnabrück, to the Hase.

9.13.2 Harz Mountains

The Harz is an elliptically-shaped block between the Rivers Leine and Saale, some 90 km from east to west and 30 km wide, representing a horst in the Hercynian platform basement. It consists of highly folded and faulted Palaeozoic rocks, of which the most important at the surface are the Devonian and Carboniferous greywackes, shales and some limestones, with diabase and tuffs. A granitic core is also exposed in a few places, including the Brocken massif which rises to 1142 m at its highest point. Near to Ilfeld and Ellrich, Permian strata and volcanics outcrop, and the Grosser Auerberg (578 m) is built of Permian porphyry. The Ilfeld basin has developed in the Permian the Zechstein limestones, and gypsum form a distinctive landscape, with solution forms.

The block is bounded by high fault scarps on all sides except the south-east: the most impressive scarp faces north-west. Away from the edges, the central parts of the block show plateaus with relatively subdued relief, representing remnants of planation surfaces (possibly an etchplain) at heights from 250 to 600 m. Lower Palaeozoic quartzites form structural ridges or low monadnocks [e.g.,

Auf der Acker (860 m), Sonneberg (854 m) and Rehberg (894 m)]. Remnants of tropical weathered mantles have been located. On the margins, the river valleys such as the Bodetal are deeply incised, reflecting Miocene tectonic uplift. The higher parts suffered periglaciation in the Pleistocene, evidenced by block fields, cryoplanation terraces and other nivation forms. On the granites, tors and corestones occur.

On the northern side below the main fault scarp lies a zone of foot-hills 20–30 km broad extending as far as the Aller Urstromtal (ice-marginal valley). These hills consist of ridges formed in the Triassic Muschelkalk of the platform cover and vales eroded in softer Jurassic or Cretaceous beds. The whole of this belt and the flanks of the Harz are mantled by Pleistocene loess. The southern foot-hills of the Harz are characterized by the Goldene

Aue, a depression along the River Helme due to solution of the Zechstein salt deposits (*see* Fig. 9.33).

Extending from the south-east of the Harz is the Kyffhäuser ridge, rising to 477 m. This has a core of gneiss, schist and granite, flanked by Upper Carboniferous and Permian rocks, and is, like the Harz, an uplifted block of the Hercynian basement. Its northern fault scarp is steep, while the southern slope is formed in the Permian (Zechstein) in which salt karst forms appear (e.g., the cave known as the Barbarossahöhle near Ochsenburg).

9.13.3 Thuringia
Corresponding generally to the upper basin of the Saale, this area falls into two distinct morphostructural units: the Thuringian Forest upland and the Thuringian Basin (*see* Fig. 9.34).

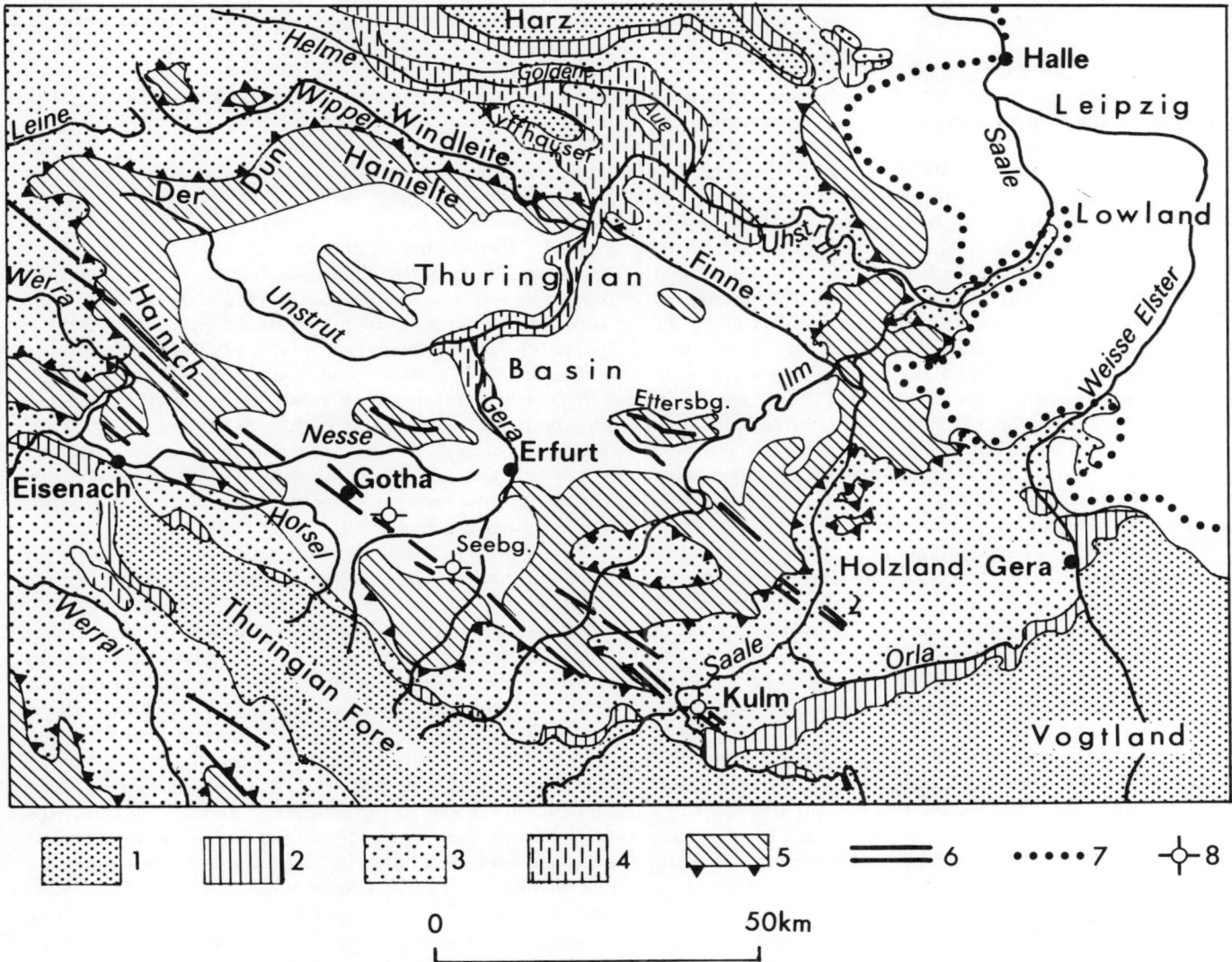

Fig. 9.34 Thuringian Basin and adjacent uplands [after Neef (1968)]. (1) Crystalline basement. (2) Permian (Zechstein) formation. (3) Bunter sandstone. (4) Depressions formed by solution of the Permian salt strata. (5) Limestone plateau and escarpments. (6) Fault scarps. (7) Boundary of thick Tertiary platform cover. (8) Important examples of relief inversion.

9.13.3.1 *Thuringian Forest*

This north-westerly projecting upland stems from the Fichtelgebirge, extending for about 100 km towards Eisenach. It is another uplifted block of the Hercynian basement, with marked fault scarps along its margins. In the north-western portion, Permian sandstones and conglomerates with some intrusions dominate; in the central part, porphyry and granite—composing the core of the block and visible in a few 'windows'—rise to the highest points [i.e. the Grosser Beerberg (982 m), the Schneekopf (978 m) and the Inselsberg (916 m)]. Farther south-east, the uplands are formed of Proterozoic and Ordovician shales and quartzites. The main mountain ridge is bevelled by a planation surface, possibly an etchplain as suggested by the kaolinitic remnants of a tropical weathered mantle found in some places. Uplift occurred in the Neogene or early Quaternary, and the fault scarps and mountain sides are now deeply dissected. Frost weathering in the Pleistocene cold periods produced a widespread cover of angular debris, mixed with gelifluction deposits.

9.13.3.1 *Thuringian Basin*

This depression lies between the Thuringian Forest and the Harz, with general morphostructural characteristics resembling those of the south-west German scarplands. The stratigraphic succession begins with the Permian (Zechstein) and continues through the Triassic sequence of Bunter sandstone, Muschelkalk and Keuper marls. Cuestas in the Bunter sandstone are well developed in the Untere Eichsfeld near Duderstadt, the Windleite near Sondershausen, in narrow cuestas near the foot of the Thuringian Forest in the Rudolstädter Heide east of the Saale valley, and farther east still in the Atenburger Holzland. The Muschelkalk limestones also form well-marked escarpments: Obere Eichsfeld near Eschwege [Goburg, (570 m), Dün (520 m) and Hainleite (460 m)]. The scarpland is dissected by the water gaps of the Werra and Wipper Rivers. Karst forms appear on the Gossel plateau south-west of Arnstadt, and there are also solution phenomena on salt and gypsum layers.

The floor of the Thuringian Basin possesses undulating relief sloping generally towards the Gera and Unstrut Rivers, developed on Keuper marls with some limestone outcrops, such as the Fahner Höhe (412 m) north-west of Erfurt (cut by the gorge of the Gera) and, a little to the east, the Ettersberg (478 m) near Weimar. The Permian (Zechstein) salt layers have developed solution forms. The valley floors, except locally where they cross limestone ridges, are broad and filled with thick gravel accumulations. During the Middle Pleistocene, the Fennoscandian ice penetrated the Thuringian Basin from the Leipzig–Halle embayment, roughly as far as Gotha, affecting the drainage pattern. Elster glacial deposits are preserved in the centre of the Basin, and the whole area is blanketed by loess.

9.14 Bohemian Massif

The Bohemian (Czech) massif represents the eastern sector of the European Hercynian platform. The horst-like block of the Massif presents itself as the most rigid and stable part of this eastern sector and is sometimes termed a *paraspis* (*para*: nearly; *aspis*: shield). Resting on this rigid block, there is only a restricted and rudimentary platform cover, which differentiates the area from several other parts of Hercynian Europe, and the surface of the massif consists predominantly of pre-Mesozoic formations, amongst which crystalline schists of various ages are most common.

The boundaries of the Bohemian massif at the surface are usually very well marked [see Demek *et al.* (1965)]. In the west, the massif is bounded by a series of faults dividing it from the Mesozoic rocks of the Main and Neckar scarplands. In the north-west, in East Germany, the crystalline rocks of the massif plunge steeply beneath the Upper Permian and Lower Triassic strata of the Thuringian uplands. In spite of a covering of younger platform sediments, the north-eastern boundary in Poland is still relatively clearly defined, largely due to the important Odra fault. Next to the latter, the north-eastern horst-like sector of the Bohemian massif has been uplifted, rising above the adjacent plateau country in the Sudeten block and the pre-Sudeten block. In the south-west, south and south-east, the geomorphological and geological separation of the Massif from the Alpine–Carpathian foredeep zones is controlled by the presence of a system of major vertical faults.

9.14.1 Relief and Structure

In terms of its general morphostructure, the Bohemian massif comprises a great basin (the Czech Basin) surrounded by uplands of moderate height. The northern part of the Basin consists of a plateau built of horizontal or subhorizontal sandstones and shales of Cretaceous age, whereas the southern part is characterized by rather monotonous uplands and highlands where the platform basement of folded Proterozoic and Palaeozoic crystalline rocks is exposed. The platform basement, divided by faults into blocks, also forms the framework of highlands surrounding the Basin.

In its present form, the Bohemian massif occupies an area that became completely consolidated by the Variscan (Hercynian) orogeny, and developed subsequently as a part of the young west European platform. Structurally, it now consists of two basic elements: the platform basement showing the strong impress of the Variscan movements, and the overlying and partial post-Variscan platform cover (*see* Fig. 9.35). The form is that of an old massif bevelled by planation and morphologically rejuvenated by Tertiary and Quaternary block tectonics. Neotectonic movements began in the Oligocene and affected the area in two ways: first, by the doming and uplift of the massif as a whole and, secondly, by faulting. Both processes are genetically connected; the bevelled surface of the massif was folded into broad anticlines and synclines whose formation was accompanied by faulting and the development of fault scarps. The blocks may be arranged in step-like fashion, as a series of horsts and grabens, or as isolated units. Altogether, Pliocene and Quaternary uplift in certain places totaled more than 1000 m [see Zoubek

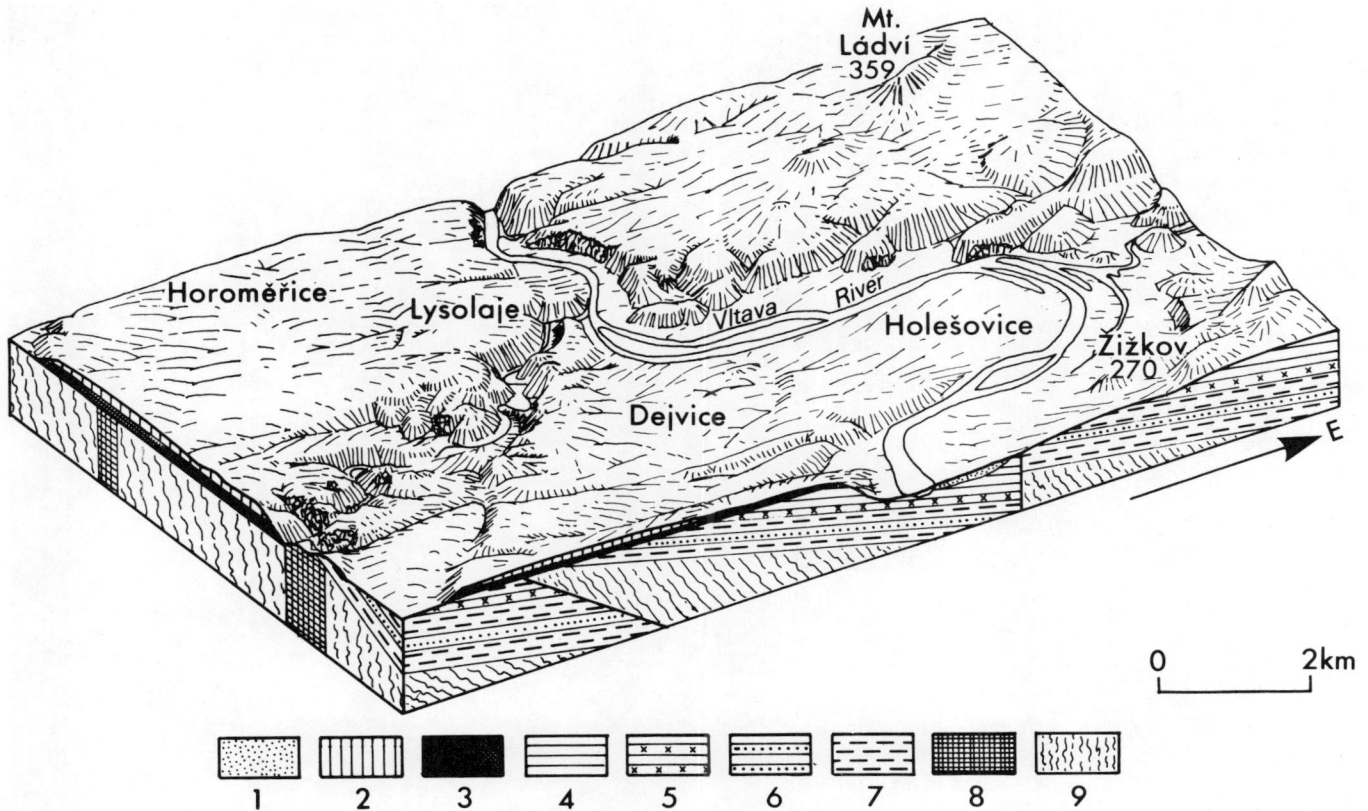

Fig. 9.35 Block diagram showing typical upland relief of the Czech massif: a portion of the country around the Vltava valley near Prague [after Raušer and Kettner, from Demek *et al.* (1965)]. (1) Quaternary fluvial deposits. (2) Lower Turonian (Cretaceous). (3) Cenomanian (Cretaceous). (4) Palaeozoic; Záhořany formation. (5) Palaeozoic quartzites: Drabov formation. (6) Palaeozoic quartzites: Skalka formation. (7) Palaeozoic: Osek and Kváň formations. (8) Intrusive. (9) Precambrian crystalline.

(1960)]. Tertiary and Quaternary faulting was also accompanied by volcanic activity, but this was predominantly superficial and limited to the subsiding areas and their immediate neighbourhood. The main volcanic centres are in the Doupovské hory and České středohoří mountains, while a smaller independent centre of activity lies in the Jeseníky mountains.

9.14.2 Denudation and the Evolution of Relief

The whole of the Bohemian massif was worn down to a surface of low relief at some time(s) during the Mesozoic, giving the surface termed the pre-Cretaceous peneplain. Tectonic movements brought about the subsidence of the northern parts of this surface and the transgression of the Cretaceous sea. After marine regression, a new phase of planation began, culminating in the development of the Palaeogene planation surface, on which a weathered mantle, mostly kaolinitic and often more than 100 m thick, was formed. This surface was remarkably even: not even its highest points exceed 200 m above the sea level of the time. Rejuvenation and dissection of the surface, due to neotectonic movements, set in during the Neogene and Quaternary, and parts of the surface were uplifted by faulting to levels of 1200–1400 m. Remnants of the former

bevelled surface were still preserved, however, on the uppermost parts of even the highest block mountains, as can be seen in the Krkonoše (Giant Mountains), but the weathered mantle was extensively destroyed. Its removal exposed the basal surface of weathering, giving rise to the so-called Czech etchplain, a new Pliocene–Quaternary planation surface (*see* Fig. 9.36). The surface morphology of this etchplain is much more controlled by the differential resistance of the underlying rocks to tropical weathering than was the case with the Palaeogene surface. Rocks of lesser resistance to Tertiary tropical and subtropical weathering developed basins and valleys [see Jahn (1980)].

Late Neogene climatic changes are thought to have had an important influence on the evolution of the landscape. For a short time, the climate was relatively dry, but later it changed to approximately the present type of temperate humid regime. In the Pliocene, pediments developed at the foot of many slopes, especially in depressions such as the Hořovická brázda, Klatovská kotlina, Plzeňská pánev, etc. In the areas of neovolcanic rocks (e.g., the České středohoří and Doupovské hory) a Pliocene surface developed across the volcanics: the so-called post-basalt planation surface.

Fig. 9.36 Planation surface, probably etchplain, with peat cover, Jizerské hory, Czechoslovakia. (Demek)

The evolution of the relief during the Quaternary was guided on the one hand by the influence of continuing tectonic movements and on the other hand by the effects of pronounced climatic fluctuations. These changes had a particularly important impact on the river systems, where streams were forced to make repeated adjustments not only to changes in level caused by tectonics but also to alterations in the amounts of water and sediment being supplied to the systems. In this way, complex sequences of terraces and valley fills were produced.

In the Pleistocene, Fennoscandian ice succeeded in penetrating, on at least two occasions, into the area of the Bohemian massif [see Demek (1976)]. The first known incursion took place in the Kraków (Elster) phase and the second during the Middle Polish (Saale) glaciation. The ice pushed in as far as the Sächsisches Mittelgebirge, to the foot of the Sudeten mountains, and into the Moravian Gate, where it crossed over the main European watershed for a short distance. The effects of the ice sheet on the morphology were felt in two ways: there was erosion in some areas, as in the piedmont zones of the Jizerské hory,

the Lužická pahorkatina and the Žulovská pahorkatina, and there was widespread glacial deposition. In some mountain areas, local glacier systems developed. In the late Pleistocene, it is known that there were cirque glaciers in the Bohemian Forest (Šumava or Böhmerwald), in the Giant Mountains (Krkonoše or Karkonosze) and in the Hrubý Jeseník mountains.

During the cold periods of the Pleistocene, most of the Bohemian massif lay in the periglacial zone—a cold desert or tundra environment. The geomorphological evidence for this is clear: fossil periglacial phenomena include ice wedges, sand wedges, patterned ground and gelifluction forms. The extent of cryoplanation varied in different parts of the massif. In the highland and mountain areas, parallel retreat of frost-riven cliffs (*see* Fig. 9.37) led to the formation of cryoplanation terraces; at the foot of slopes, cryopediments and cryoglacis developed. Because of differential exposure of slopes to insolation, the intensity of slope processes varied and this resulted in the formation of valleys with asymmetrical cross-profiles. On the lower areas, loess and other periglacial deposits accumulated.

Fig. 9.37 Glacial and periglacial features in the Giant Mountains (Krkonoše), Czechoslovakia. (Demek).

The Holocene period was marked by smaller-scale climatic fluctuations and a general return to present-day climatic characteristics, but the dominant change was in the advent of man, who has become one of the main geomorphological agents.

9.14.3 Geomorphological Regions

9.14.3.1 The Ore Mountains (Erzgebirge or Krušné-Hory)

This system occupies the north-western part of the Bohemian massif, lying on the border of East Germany and Czechoslovakia. It may be divided into four subregions [see Gellert (1973)]. The first is the Sächsisches uplands, which basically represent an undulating Pliocene planation surface extending west of the River Elbe. The surface truncates the old Hercynian structures and rises gradually to about 380 m, while river valleys, often gorge-like, are incised into it for depths of 100 m or more. Some parts are covered by Palaeogene weathering deposits, while the region between the Mulde and upper Pleisse river valleys is overlain by Tertiary sands and gravels. In some depressions floored by Permian rocks, moraine and outwash gravel belonging to the Elster glaciation can be found. The lower areas are also mantled by loess.

The second subregion consists of the Ore Mountains proper, a mountain range trending from south-west to north-east, sloping gently towards the north-west into East Germany but presenting several steep fault scarps on the other, Czech, side [see H. Richter (1963)]. This range represents essentially the result of asymmetrical uplift of the Pliocene–Quaternary etchplain. The south-east-facing fault scarps rise in places to heights of over 500 m. Geologically the range is built of schists and granitic rocks. In the more central parts, remnants of the etchplain are preserved, above whose slightly undulating surface rise some volcanic forms such as the Geising (824 m), Bärenstein (898 m), Pöhlberg (831 m) and Špičák (1115 m). The highest points are provided by monadnocks, including the Fichtelberg (1214 m) and Klínovec (1243 m). Valleys, such as those of the Mulde, Weisseritz and Müglitz Rivers, are narrow and deeply incised; those draining to the south-west are shorter, but also deeply cut. Granite areas display typical relief forms: low exfoliation domes (ruwares), tors and piles of rounded blocks (core stones). There are some cirques on the highest parts, and periglacial processes have been active, forming cryoplanation terraces.

The third subregion comprises the rift valley of the Ohře, marked out by deep-seated faults at the foot of the Ore Mountains. The south-eastern side is overlooked by the horst of the Slavkovský les mountains and the Tepelská pahorkatina, while the other is dominated by volcanic mountains.

The Sächsische Elbsandsteingebirge (East Germany) and the Děčínská vrchovina highland (Czechoslovakia), built of uplifted Cretaceous strata, form the fourth subregion. The Elbe (Labe) flows in a deep water gap across this sandstone region, where the rock has weathered into many typical forms of microrelief, such as honeycombed surfaces, as well as some larger-scale features, such as rock

gates and arches.

9.14.3.2 Sudeten Highlands

The Sudeten mountain system stands as the north-eastern rampart of the Bohemian massif [see Klimaszewski (1972)]. This is a zone of horst-like ranges marked by complex structure and varied relief. The basic geological structure consists of the crystalline basement rocks, overlain in places by Palaeozoic or Mesozoic sediments. The north-eastern edge is rectilinear in plan, reflecting its tectonic origin. The system of faults striking parallel with the Sudetan mountains runs roughly north-west to south-east but deviates from this direction in some areas; the faults are closely connected with the formation of the Cretaceous basin (*see* Fig. 9.38). The fault system includes the outer Sudetic fault separating the mountains from the Silesian lowland, the faults dividing the Kłodzko graben from the Bystrzyckie mountains and the Rychlebské hory, the Lusatian fault and others. In contrast to this fault-scarped edge, the inner parts of the Sudetes resemble an upland rising steadily from 400 to 700 m, with some ridges rising significantly higher to 1200–1400 m but still possessing flat tops. Separating these ridges are large intermontane basins.

The highest parts of the western Sudetes are the Jizerské hory. The greater part of this western area is built of granite, with numerous tors, castle koppies and core stones. Above the extensive Neogene or early Quaternary etchplain rises the monadnock of Jizera (1121 m). The Giant Mountains (Krkonoše) constitute the highest mountain range of the Bohemian massif: in the Sněžka they reach a height of 1602 m. The range is bordered by fault scarps, especially along the northern slope dominating the Jeleniogórska depression in Poland. The mountains consist of two parallel ridges. That nearest the frontier displays a plateau-type relief with some summits (e.g., Vysoké kolo) rising higher; here cryoplanation forms are visible. The Czech ridge shows greater dissection. Local glaciation has influenced the morphology, the longest glacier extending for about 6 km. The Jeleniogórska basin was invaded by the Fennoscandian ice.

The highest part of the central Sudetes is the Orlické hory, which is built of crystalline rocks of the platform basement. The mountains form a narrow compact ridge trending from north-west to south-east, characterized by a bevelled crest, representing the remnants of the former etchplain and rising to 1114 m (Velká Deštná) in the north-west. A narrow tectonic depression separates the Orlické hory from the parallel ridge of the Góry Bystrzyckie mountains in Poland.

The eastern and central Sudetes are divided from each other by the broad graben of Kłodzko whose floor lies at 350–450 m. The floor is underlain by Pleistocene glacial and glaciofluvial deposits, beneath which there is an infill of Cretaceous sediments. Pediments rise from the floor towards the bounding scarps. Maximum elevations in the eastern Sudetes are found in the Hrubý Jeseník mountains of crystalline rock. The range was formed by tectonic uplift along a system of faults, the individual blocks having been raised to various altitudes so as to give a

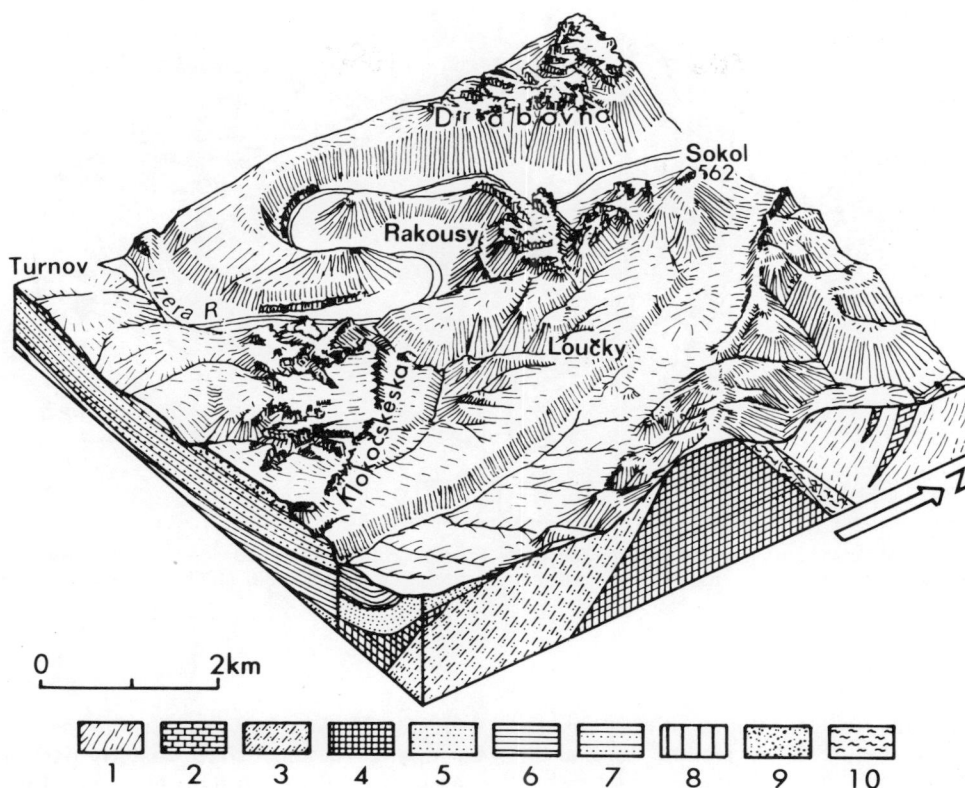

Fig. 9.38 Block diagram of part of the Jizera valley near Turnov, Czechoslovakia, showing contact zone between the Czech massif and the Cretaceous platform cover, along the Lusitanian fault zone [after Raušer and Soukop, from Demek *et al.* (1965)]. (1) Phyllites. (2) Basement limestones. (3) Folded shales and sandstones of the basement. (4) Melaphyres. (5) to (9) Cover rocks: (5) Cenomanian sandstones and conglomerates; (6) Lower Turonian marls; (7) Middle Turonian sandstones; (8) Upper Turonian mudstones and marls; (9) Upper Turonian and Coniak sandstones; (10) fault zone of mylonite. (Note the cuesta of Klokočské skály.)

series of steps, which become progressively higher towards the central part around Mount Praděd (1491 m). The uplift caused deep incision of the rivers. In the Pleistocene, cirque glaciers developed on the higher parts, such as the headwater area of the Moravice river. Extensive areas were subjected to periglacial modification, leading to the formation of cryoplanation terraces, tors, patterned ground, block fields, block streams and rock glaciers.

The northern foothills of the Sudeten mountains constitute a typical inselberg landscape (*see* Fig. 9.39), in which isolated hills, many of them resembling bornhardts, rise above the Palaeogene planation surface with its remnants of a former kaolinitic mantle. This inselberg landscape was modified by the incursion of the Pleistocene ice sheet from the north during the Kraków and Middle Polish glacial periods. Many ruwares, for instance, were reshaped to become roches moutonnées. Underneath the glacial and glaciofluvial deposits, remnants of tropical karst, with mogotes, cockpits and relics of tropical deep weathering are to be found, as at Supíkovice in Czechoslovakia.

9.14.3.3. *Bohemian Forest (Böhmerwald, Šumava system)*

The Šumava system marks the south-western rim of the Bohemian massif. Its core consists of a high and relatively unbroken mountain range lying along the frontier of Czechoslovakia with West Germany and Austria. Geologically, it is built of schists and other rocks belonging to the central Moldanubian pluton (mainly granites). The system comprises two anticlinal structures separated by the synclinal area of the upper Vltava river. The anticlinal ridge next to the frontier is widely bevelled by an etchplain in its central parts with an average height of 1150 m, but with local eminences rising above this to more than 1300 m. The central part slopes down steeply towards West Germany but more gently on the Czech side. The main summits standing above the etchplain are Plesná (1355 m), Poledník (1314 m), Gross Rachel (1492 m) and Černá hora (1315 m). Shallow depressions on the planation surface contain peat bogs, below which some remnants of Tertiary kaolinitic weathering mantles have been preserved. The other anticline is represented by the inner, Czech ridge, consisting of several mountain groups [e.g., Boubín (1361 m), Knížecí stolec (1225 m) etc.] and the Šumava mountain piedmont with its structural ridges and extensive pediments at two levels.

The Šumava mountain range was glaciated in its upper parts during the Pleistocene cold periods, and cirque

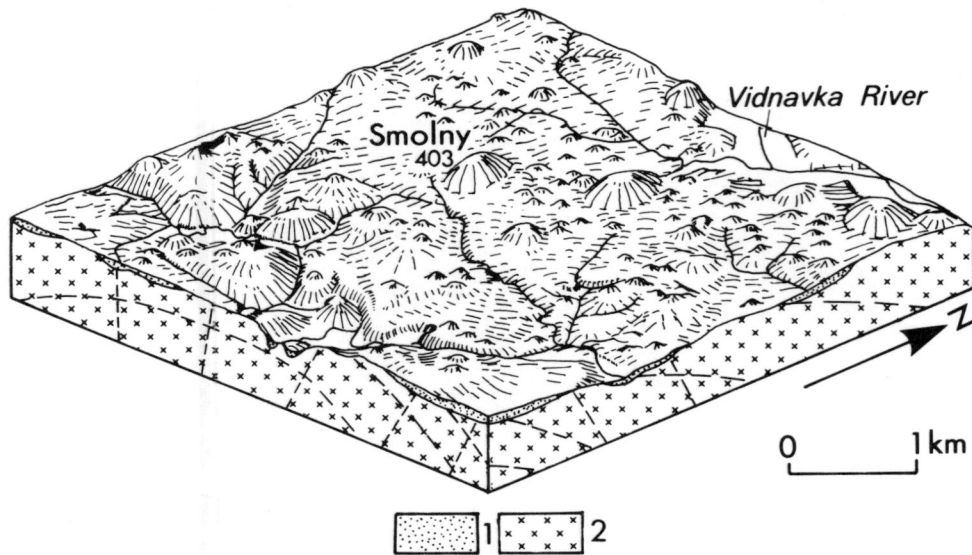

Fig. 9.39 Granite landscape with inselbergs (Smolńy: 403 m; Jahodník: 378 m) and ruwares remodelled by Pleistocene glaciation into roches moutonnées: the Žulovská upland in the Czech massif [after Raušer and Skácel, from Demek *et al.* (1965)]. (1) Glaciofluvial sands and gravels. (2) Granite batholith.

glaciers developed on the eastern slopes of the Gross Arber, Klein Arber, Rachel, Plechý, Jezerní hora and others. In the late Pleistocene, the snow line lay at a height of about 1000–1100 m.

The term Bohemian Forest is often applied to a more limited part of the Šumava system: namely its north-western continuation at a somewhat lower level, along the border between Czechoslovakia and West Germany. Here the highest point is Čerchov (1041 m). On the German side, the foot-hills are represented by the Passauer Vorwald. The highland possesses an undulating summit surface cut across the granites and schists of the Hercynian basement. The surface is partly overlain by Tertiary deposits. The southern part is dissected by the Danube valley (up to 100 m deep) within which Pliocene gravels stand 30–40 m above the river.

9.14.3.4　*The Waldviertel/ Českomoravská vrchovina system*

This marks the south-eastern flank of the Bohemian massif, comprising a system of uplands and highlands developed in the crystalline basement rocks. The southern part consists of the undulating highlands of the Waldviertel where the Neogene etchplain has been dissected by valleys. In Czechoslovakia, the system is formed by the Českomoravská vrchovina—a highland area whose central part represents an anticline complicated by faulting. Here are to be found the highest points, culminating in the granite peak of Javořice (836 m), where many tors, castle koppies and core stones are to be found. Periglacial activity in the Pleistocene also produced cryoplanation terraces. In contrast to the central parts, the margins are more in the nature of uplands whose flattened watersheds represent etchplain remnants later dissected by deep and narrow valleys. Ruwares also occur in the granite areas.

9.14.3.5　*The Brno system*

Occupying the most south-easterly part of the Bohemian massif, the Brno system may be divided into three subregions. The first is that of the Boskovická brázda—a trough 130 km long and 3–10 km wide—filled with Permian and Carboniferous deposits. Secondly, there is the Bobravská vrchovina, an upland belt lying between the Boskovická trough and the Pre-Carpathian foredeep and built of Proterozoic granitic rocks. Towards the centre, remnants of a planation surface have been preserved, but in the northern parts it comprises a series of narrow horsts separated by elongated grabens. The third subregion— the Drahanská vrchovina—is a more massive highland also situated between the Boskovická trough and the Pre-Carpathian foredeep. Its greater part is built of the Culm Measures—greywackes and shales—along which a strip of Devonian Limestone (the Moravian karst) intrudes. The area forms an anticline reaching 723 m in Mont Skály. In the central parts, extensive etchplain remnants are found, but the margins are deeply dissected by river valleys. In the limestone country of Moravia, typical karst phenomena are present, including karst canyons, blind valleys, caves and a most spectacular vertical-sided aven called Macocha (138 m deep).

9.14.3.6　*Southern part of the Czech basin*

Forming a rather monotonous hilly area in southern Bohemia, the Precambrian crystalline basement is extensively exposed but broken into various horsts and grabens. The following subregions may be recognized.

1) The Berounka region, forming the core of the Bohemian massif, consists of Algonkian and Palaeozoic rocks and extends along the upper and middle reaches of the Berounka River (*see* Fig. 9.40). For the most part, the relief

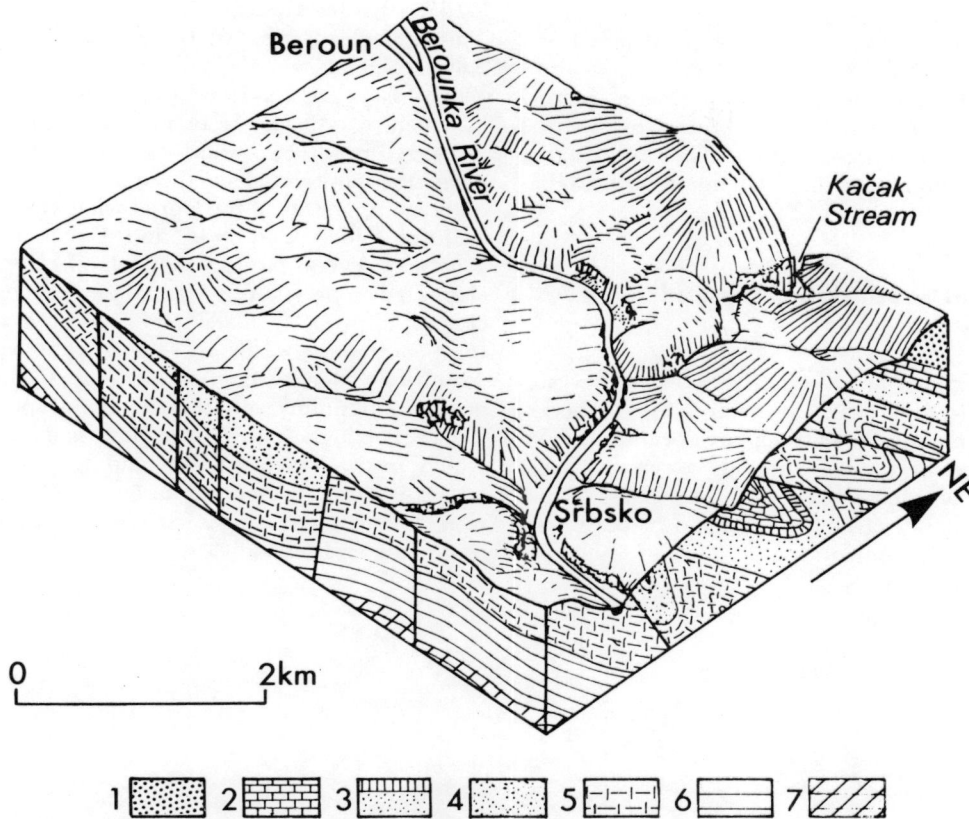

Fig. 9.40 Typical relief on the Hercynian (Palaeozoic) basement of the Czech massif: Berounka valley between Beroun and Srbsko, central Bohemia [after Raušer and Chlupáč, from Demek *et al.* (1965)]. (1) Upper Middle Devonian. (2) Lower Middle Devonian limestones. (3) Lower Middle Devonian shales. (4) Lower Devonian limestones. (5) Upper Silurian, mainly limestones. (6) Lower Silurian, mainly diabases. (7) Upper Ordovician.

is only slightly dissected, and the limited variations of altitude mainly reflect the differential resistance of the different rock sequences to former tropical and subtropical weathering. In the middle stands the upland of Plzeňská pahorkatina, whose centre is occupied by a shallow depression around the confluence of the headwaters of the Berounka, named the Plzeňská Basin. Extensive coal and china clay working has produced many modifications to the local relief. The highest part—the Brdy ridge—attains a height of 650–750 m, with rounded summits and broad ridges separated by wide valleys. Above this general summit there rise a few higher eminences [e.g., Praha (862 m) and Tok (861 m)]. In the south-west, there is the structural ridge of Hřebeny, composed of Cambrian quartzites, while the north-eastern area—the Pražská upland—shows remnants of the pre-Cretaceous peneplain. The river valleys are in places flanked by Pliocene pediments.

2) The Středočeská upland is a somewhat less elevated area lying between the Českomoravská highland to the east and the Brdy ridge on the west, reaching a maximum height of 728 m in Drkolná mountain south-east of Klatovy. Geologically it consists of granitic rocks which show many landforms typical of that lithology: ruwares,

tors and corestones. The upland drains to the Vltava River and its tributaries which flow in incised valleys.

3) The southernmost subregion comprises the basins of southern Bohemia, ranging from the relatively large and elevated basin of Třeboň (at about 400–500 m) to the smaller, lower-lying basin of České Budějovice, separated by the Lišov horst. The floors of both basins possess only low relief developed on the Cretaceous, Tertiary and Quaternary infills. The Lišov horst is of crystalline rocks and was affected by strong neotectonic movements, some of which must be of very recent age as shown by the tilted layers of Holocene peats.

9.14.3.7 Bohemian plateau

The Bohemian plateau occupies the north and north-east of the Czech basin and differs from any other part of the Bohemian massif in that it is almost completely underlain by Upper Cretaceous rocks. The strata are horizontal or subhorizontal, having been protected from tectonic warping by the rigid underlying crystalline base. There are, however, some exceptions to this, notably around the Lužice fault and in the gently folded areas of east Bohemia and north-west Moravia with their cuesta-type relief. Elsewhere, the Bohemian plateau displays a surface of low relief strongly controlled by structure. Above the extensive plateaus, some Tertiary volcanic forms [e.g., Mount Říp (455 m) and Ralsko (695 m)] project (*see* Fig. 9.41). Some uplifted blocks of Cretaceous strata form mesas [e.g., Džbán (535 m)] and some highly-jointed quartzose sandstones form castellated scenery with cliffs, towers and gorges (e.g., the area known as the 'Bohemian Paradise').

9.15   Silesian and South Polish Uplands

This area forms the most north-easterly part of the west European platform (*see* Fig. 9.42). It includes areas where the platform rocks, mainly Palaeozoic, have been uplifted and are now exposed, and areas where the basement has been overlain by a Mesozoic cover [see Kondracki (1978)]. Because of the variety of morphostructure, a regional treatment is convenient.

9.15.1   Kielce Upland

This unit lies in the east (*see* Fig. 9.43), bordered by the Niecka Nidziańska, a synclinal depression, the water gap of the Wisła (Vistula) river, and the Sandomierz basin. The basement is typically Hercynian, of Palaeozoic rocks, but is covered by Triassic and Jurassic sediments. The highest parts are represented by the uplifted and deeply eroded platform remnants of the Góry Świętokrzyskie, mountains built of folded Cambrian and Devonian strata. They consist of several parallel monoclinal ridges separated by elongated depressions whose alignment reflects the successive impress of Caledonian (west–east), Hercynian (east-south-east–west-north-west) and even Palaeogene (south-east–north-west) fold systems. In the eastern parts of the mountains, an Appalachian-type relief can be found. The highest structural ridge is the Lysogóry, along which Lysica reaches 612 m and Łysa Góra 595 m, the whole feature representing an outcrop of resistant Cambrian quartzite, its slopes covered by angular quartzitic fragments and block fields.

Fig. 9.41 Bohemian plateau, underlain by Cretaceous strata, with neovolcanic hill in distance, near Jičín, Czechoslovakia. (Demek)

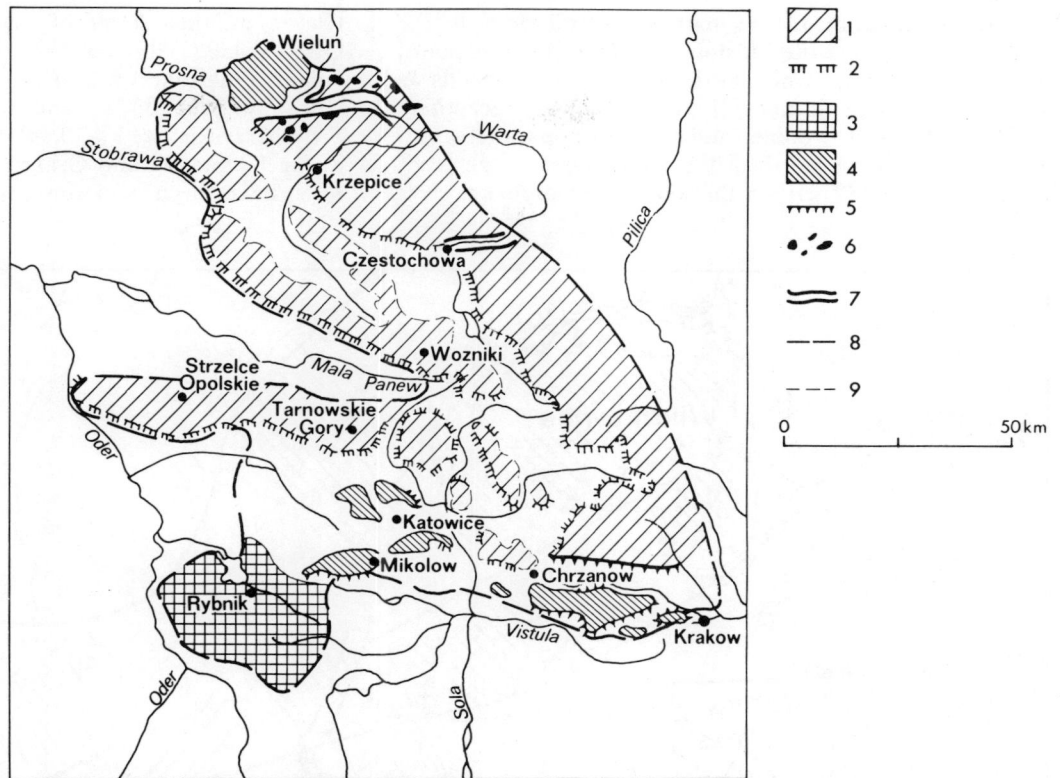

Fig. 9.42 Geomorphological map of the Silesian–Kraków highlands [after Kondracki (1978)]. (1) Structural ridges of Mesozoic rocks. (2) Escarpments. (3) Uplands of Tertiary rocks resting on Palaeozoic basement. (4) Horsts. (5) Fault scarps. (6) Pleistocene ice margin features. (7) Water gaps. (8) Generalized highland boundary. (9) Boundaries of uplifted areas.

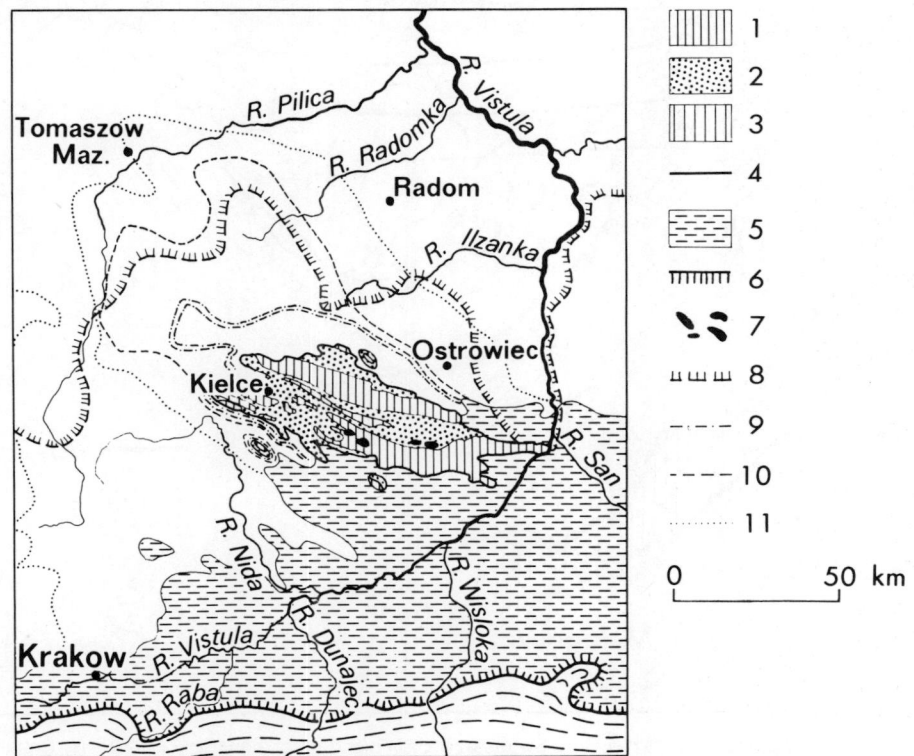

Fig. 9.43 Geological map of the central Little Polish Highlands (Wyżyna Środkowomałopolska) [after Kondracki (1978)]. (1) Lower Palaeozoic strata. (2) Devonian. (3) Carboniferous and Permian. (4) Boundary of the Palaeozoic basement. (5) Sub-Carpathian Miocene deposits. (6) Front of the Carpathian mountains. (7) Diabase. (8) Limit of the Saale (Central Polish) glaciation. (9) to (11) Subsurface boundaries of the Mesozoic platform cover: (9) Triassic; (10) Jurassic; (11) Cretaceous.

The prolongation of the main structural ridge to the north-west is called the Masłowskie range [highest point: Klonówka (473 m)] while to the south-east it extends as the Jeleniowskie range [highest point: Szczytniak (554 m)]. Lower Devonian sandstones form parallel structural ridges on the north of the Klonowskie range [e.g., Góra Bukowa (482 m)]; on the south, the Cambrian gives rise to another series of ridges [Zgórskie, Dymińskie, Orłowińskie (Dąbrowa: 451 m) and Wygiełzowskie)]. Devonian limestones rise to slightly lower elevations [e.g., Góra Zamkova (357 m) and Zelejowa (367 m). The Góry Świętokrzyskie (*see* Fig. 9.44) is broken by the water gaps of the Kamienna and Czarna Nida, rivers whose valleys may have been superimposed, but which also show

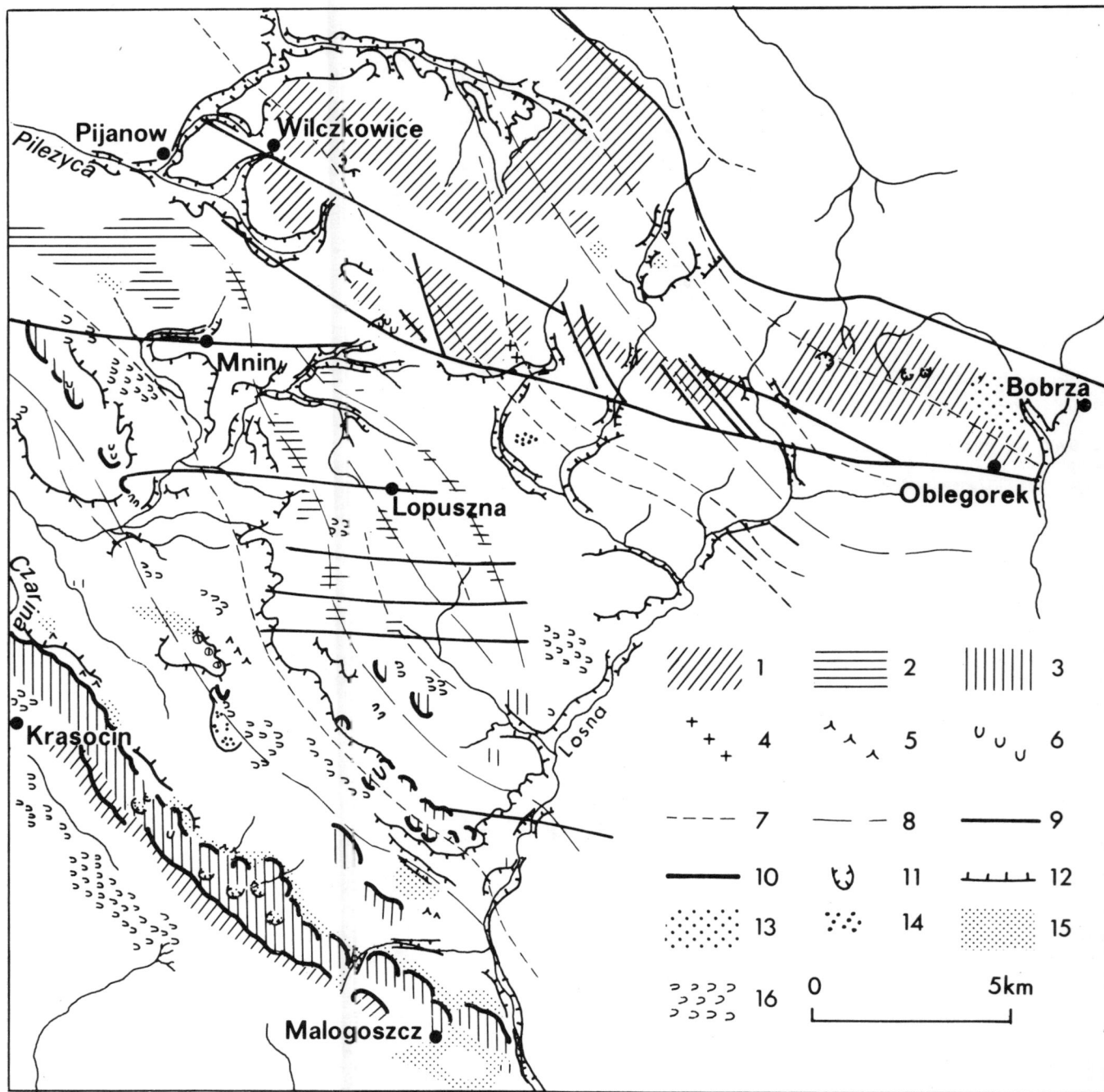

Fig. 9.44 Geomorphological map of the Góry Świetokrzyskie, Poland [after Kondracki (1978)]. (1) Cretaceous sandstone outcrops. (2) Keuper sandstone and conglomerate. (3) Limestone zone (mainly Jurassic. (4) Blockfields. (5) Tors. (6) Karstic forms. (7) Anticlinal axes. (8) Synclinal axes. (9) Faults. (10) Escarpments. (11) Spring heads. (12) Steep slopes in Quaternary sediments. (13) Loess. (14) Gravel mounds. (15) Sandy areas. (16) Dunes.

structural control by faults [see Klimaszewski (1972)]. West of a line from Miedzianka to Oblegorek, the Góry Świętokrzyskie is composed of Mesozoic sediments of the platform cover.

South-east of the Góry Świętokrzyskie, the Sandomierz uplands rise to around 260–300 m. Their southern part shows a generally anticlinal structure, comprising folded

Cambrian sandstones and shales of the platform basement which are on average less resistant than the rocks of the Góry Świętokrzyskie. The northern part of the upland, on the other hand, represents a monoclinal structure in the platform cover sediments, the surface also being blanketed with Pleistocene loess up to 35 m thick.

The northern part of the Kielce upland consists of

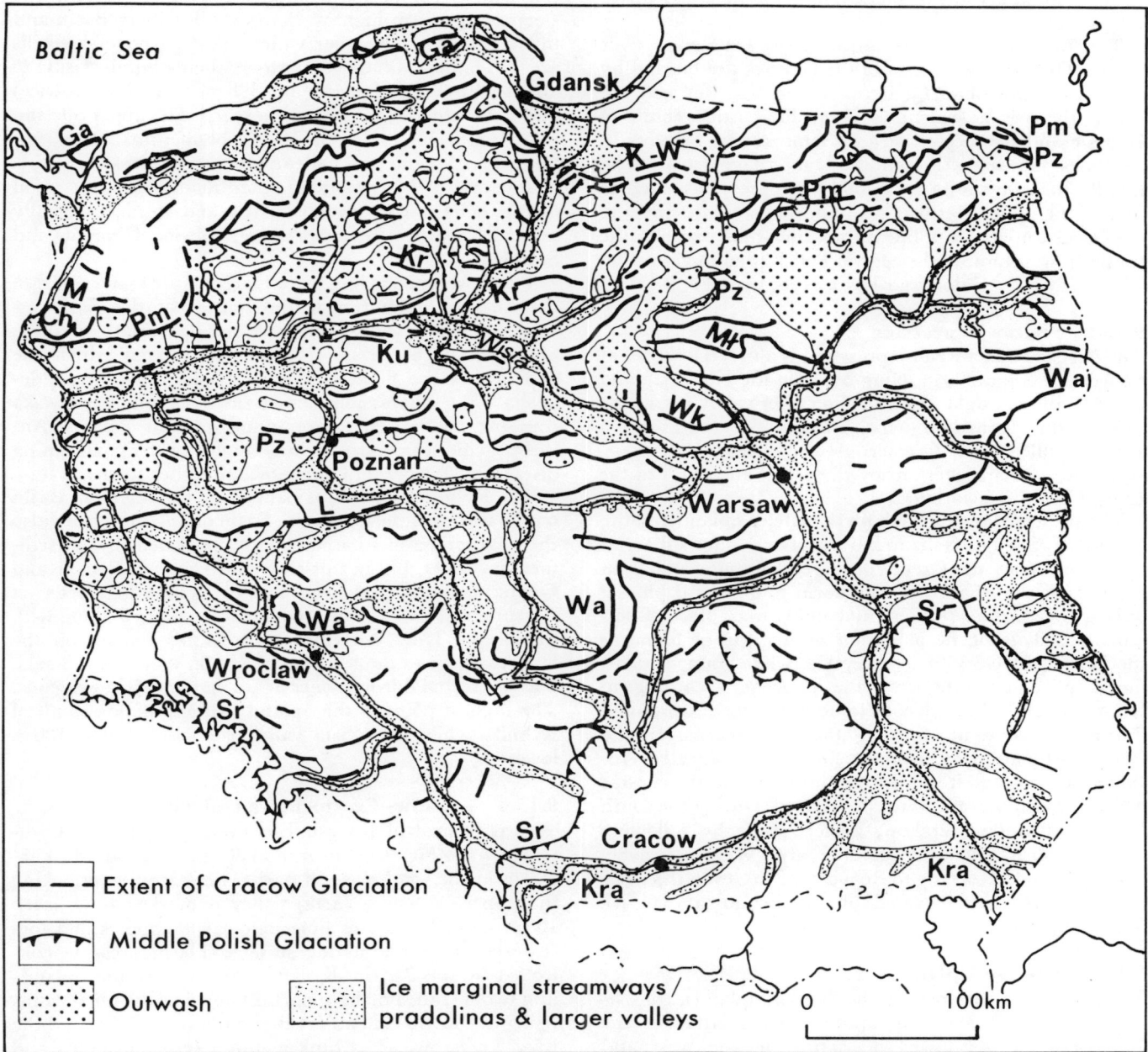

Fig. 9.45 Pleistocene glacial limits in Poland and related features. Glaciations: (I) Vistulian (Weichsel) or North Polish glaciation: L, Leszno stage; Pz, Poznań stage (Ku, Kuyavian phase; Kr, Krayna phase); Pm, Pomeranian stage (Ch, Choyna phase; M, Mielęcin phase; K–W Kashubian–Varmian phase); Gd, Gardno stage. (II) Saale or Central Polish glaciation (SR): Wa, Warta Stage; Wk, Wkra Stage; Ml, Mlawa Stage. (III) Kraków (Elster) or South Polish glaciation (Kra). (1) Limit of Kraków glaciation. (2) Limit of Central Polish glaciation. (3) Outwash gravels. (4) Ice marginal valleys (pradolinas or Urstromtäler) and other major meltwater routeways.

Mesozoic cover rocks. The northern foot-hills of the Góry Świętokrzyskie—known as the Suchedniowski hills—are formed from gently folded Triassic sandstones and clays, together with some limestones exhibiting karstic features. Altitudes range from 441 m in the south to 270 m in the north. The Gielnowski hills, whose surface generally rises in a series of steps from 280 m in the north to 408 m in the south, are eroded in Rhaetic and Liassic sandstones, the eastern boundary being marked by a fault scarp up to 130 m high.

The Iłżecka upland lies north of the Kamienna river valley. The surface features of the bedrock are blurred by a covering of aeolian Quaternary deposits, but beneath which there is a system of monoclinal ridges built of Jurassic and Cretaceous strata. Some karstic features in the limestones show through the Quaternary mantle. The middle Jurassic cuesta is divided by depressions, such as that of the upper Iłżanka. An important feature crossing the Iłżecka upland is a line of morainic hills marking the southernmost limit of the central Polish (Saale) glaciation as described by Galon and Dylik (1967) (*see* Fig.9.45).

9.15.2   Niecka Depression
The Niecka Nidziańska (depression) is a broad lowland at between 190 and 300 m, lying between the Kielce upland and the Silesian uplands. The Nida river valley marks the axis of the depression which structurally represents a syncline filled with Cretaceous and Tertiary deposits, which in turn are overlain by the sands and clays of the central Polish glaciation.

The present-day relief of both the Kielce upland and the Niecka depression seems to have evolved under subaerial conditions since the Cretaceous. Tectonic movements in the Cretaceous divided the platform into several blocks; during the Tertiary, differential uplift took place. Remnants of a Palaeogene planation surface can be found at the foot of the Góry Świętokrzyskie, while younger pediments probably date from the Pliocene according to Klimaszewski (1972). Neotectonic movements continued during the Neogene, forming the fault scarps. In the Neogene–Quaternary, climatic fluctuations affected erosion processes, and it was during this time that the cuestas in the Mesozoic strata were emphasized. During the south Polish glaciation (Kraków, Elster) and the following central Polish (Saale) stage, the area was extensively glaciated, and according to Rózycki (1965), even the Góry Świętokrzyskie was covered by the Fennoscandian ice from the north.

9.15.3   Silesian Uplands
The Hercynian basement of the Silesian uplands, consisting of Carboniferous rocks, appears at the surface in the area between the towns of Zabrze in the west and Dąbrowa Górnicza in the east, and reaching as far south as Mikołów. Elsewhere, the platform is covered by Triassic dolomites and limestones (*see* Fig. 9.46) forming the Chełm and Tarnogórskie hills and, in the east, the Jaworznickie hills. In the south-west of the Silesian upland, between the Raciborz and Oświęcim Basins, lies the Rybnicka upland where the basement is mantled by

Miocene beds containing salt layers. The highest point of the Silesian upland is in the Chełm hills, where Góra św. Anny touches 400 m. The southern margin of the Silesian upland is tectonic, marked by a fault scarp.

The Chełm hills are composed of Trias, mainly limestones as already noted, and are separated from the Tarnogórskie hills by a depression around Pyskowice. The highest point of the Tarnogórskie hills is a volcanic hill (św. Anna) composed of basalts and other volcanics. A steep scarp 170 m high overlooks the Raciborz Basin and the Wisła (Vistula) river valley. The Tarnogórskie hills are also built of Triassic limestone, their summits rising to about 340–380 m (maximum: 398 m, near Twardowice) and the southern edge forming a scarp. The upper Silesian industrial area, underlain by Carboniferous strata containing coal seams and with an overlying Triassic cover, is characterized by numerous man-made features: spoil heaps, hollows due to subsidence, and so on, especially around Katowice, Bytom, Zabrze, Gliwice, Chorzów and Sosnowiec.

The Jaworznickie hills are situated in the eastern part of the Silesian upland near the border of the Kraków–Częstochowa uplands. The undulating surface of these hills consists of Triassic limestones, among which the depressions are filled with fluvial and glaciofluvial deposits. The depression known as Pustynia Błędowska contains the largest inland dune area in Poland. Anthropogenic forms are again common in the coal-mining districts of Jaworzno, Chrzanów and other towns.

The Rybnicka upland is flanked by the Raciborz Basin on the west and the Oświęcim Basin on the east. Here also the Carboniferous basement is concealed by a sedimentary cover, but in this case consisting of Miocene and Quaternary deposits, the former containing beds of gypsum, salt and sulphur, which are exploited along with the deeper lying coal measures. Man's impact on the landscape, for example around Rybnik, Wodzislaw Śląski, and Jastrzębie Zdrój, is once again very much in evidence. The highest point of the upland (310 m) stands south of Rybnik, while the Wisła valley is incised about 100 m lower.

9.15.4   Kraków–Częstochowa Upland
This is formed of Jurassic limestones, rising near Częstochowa (*see* Fig. 9.47) to around 300 m but attaining 450–500 m in the south-eastern part. The margin overlooking the upper Warthe (Warta) valley is marked by a scarp 100 m high. Numerous limestone castle koppies and tors rise above the undulating surface. The depression controlled by the Będzin–Kraków fault divides the uplands into two: (1) the northern part is known as the Ójców hills rising to 400–460 m and (2) the southern part, beyond the Rów Krzeszowicki, is broken into a series of horsts and grabens (*see* Fig. 9.48).

Of similar general form is the Częstochowa upland, between the Warthe valley near Częstochowa in the north and the depression of the Biała Przemsza-Szreniawa River in the south. It is another limestone area, truncated by an ancient karst planation surface with mogotes (*see* Fig. 9.49). This surface probably developed in the Palaeogene,

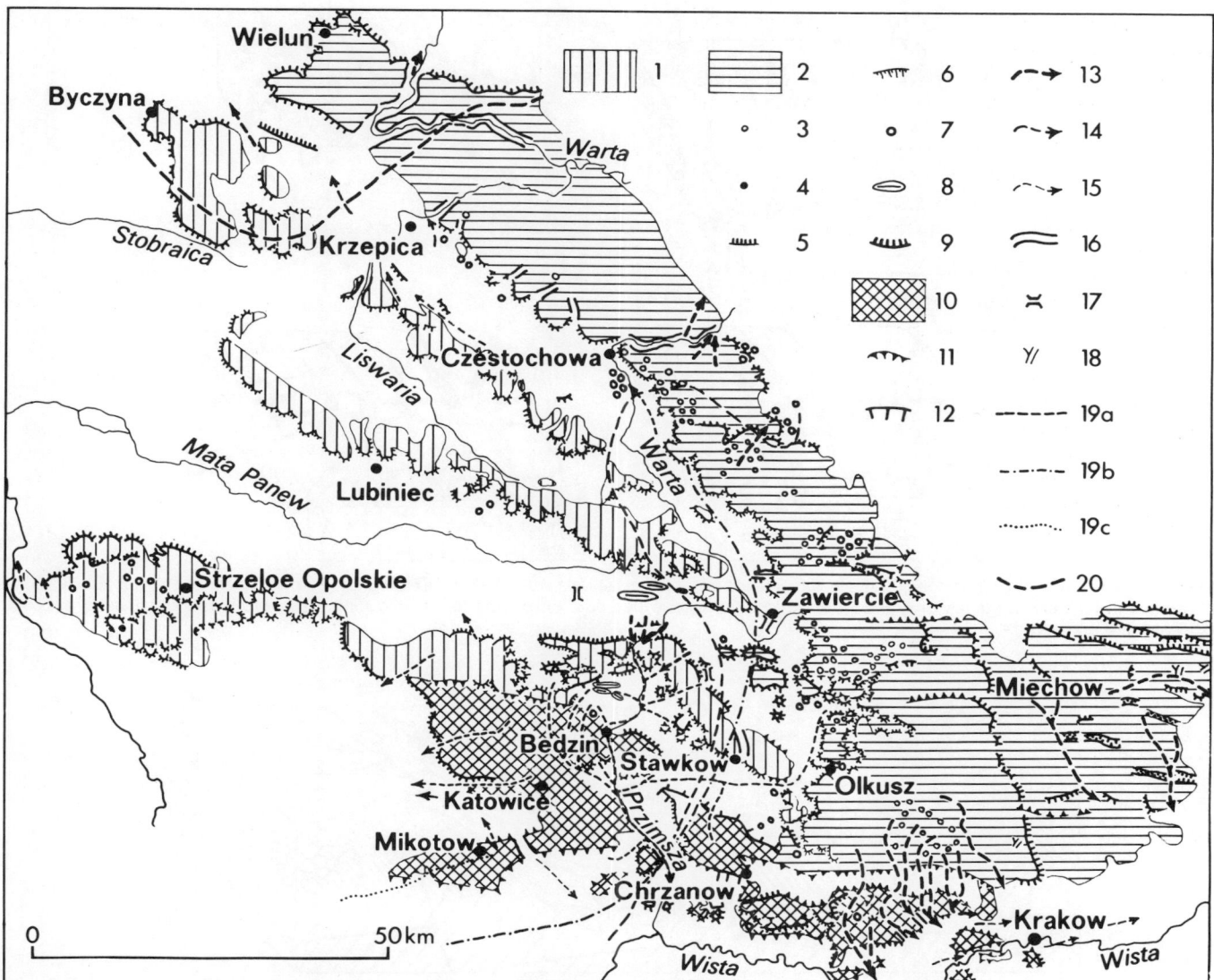

Fig. 9.47 Drainage development in the Silesian–Kraków highlands [after Klimaszewski (1972)]. (1) Low structural ridges. (2) High structural ridges and plateaus. (3) Mogotes. (4) Volcanic hills. (5) to (6) Cuestas: (5) limestone; (6) sandstone. (7) Monadnocks. (8) Anticlinal ridges in the basement. (9) Structural scarps. (10) Planation surfaces. (11) Fault scarps. (12) Resequent fault scarps. (13) Drainage directions in the Pliocene. (14) Drainage directions prior to the Central Polish (Saale) glaciation. (15) Proglacial drainage routes. (16) Water gaps (late Pleistocene). (17) Wind gaps. (18) Loess ravines. (19) Ice margins of the Odra Stage according to (a) Rózycki, (b) Klimaszewski, (c) Karaś and Starkel. (20) Ice margin of the Warta stage, according to Rózycki and Krzeminski.

but during the Quaternary the mogotes were remodelled by periglacial processes [see Klimaszewski (1972)]. Shallow sand-filled dry valleys diversify the limestone surface, carrying run-off from snow melt in the spring and occasional run-off from rain in the summer. The highest points are the castle koppies around the castle ruins in Ogrodzieniec (504 m). Karst springs feed the Warthe and Czarna Przemsza Rivers.

The Olkuska upland south of the Biała Przemsza–Szreniawa river reaches 502 m. In the Ójców National Park, the Prądnik river valley contains numerous caves in the limestones.

9.15.5 Woznicko–Wieluńska Uplands

A depression containing the Mała Panew tributary of the Oder (Odra) River divides this unit from the Silesian upland. Upper Triassic and Jurassic rocks form a series of ridges and valleys drained by tributaries of the Oder and Warthe. The whole area was glaciated during the Central Polish (Saale) period; glacial forms, such as moraines and kames, are well preserved.

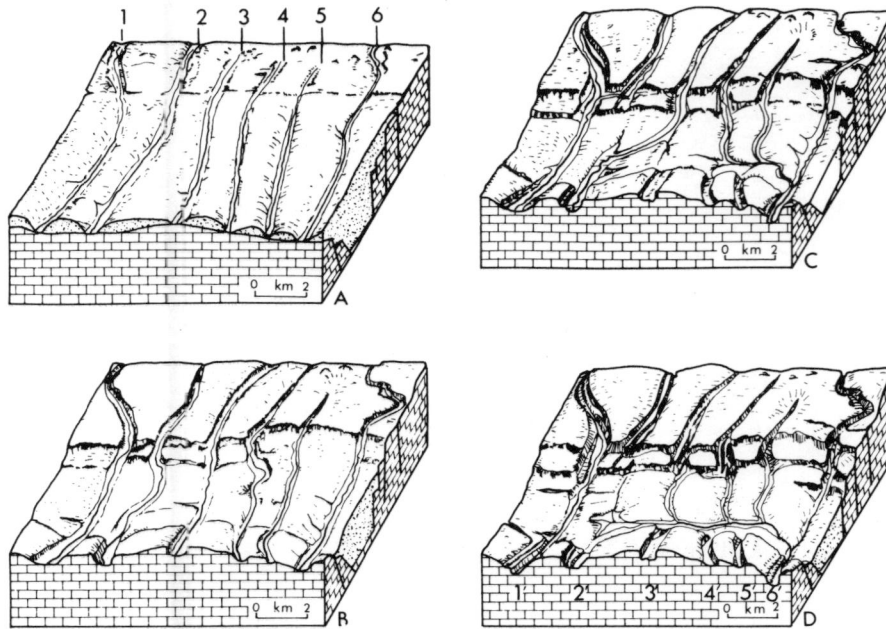

Fig. 9.46 Development of drainage in the Silesian uplands [after Klimaszewski (1972)]. (1) to (6) (Diagram A): original water courses. Blocks B. C, D represent successive stages of drainage evolution. Limestone bedrock, broken by faults, including a graben in which Miocene deposits (stippled) are preserved.

Fig. 9.48 Block diagram to show relief and geology of the Kraków–Częstochowa highland [after Klimaszewski (1972)]. Cavern and fissure systems in the limestone are shown, together with surface collapse features.

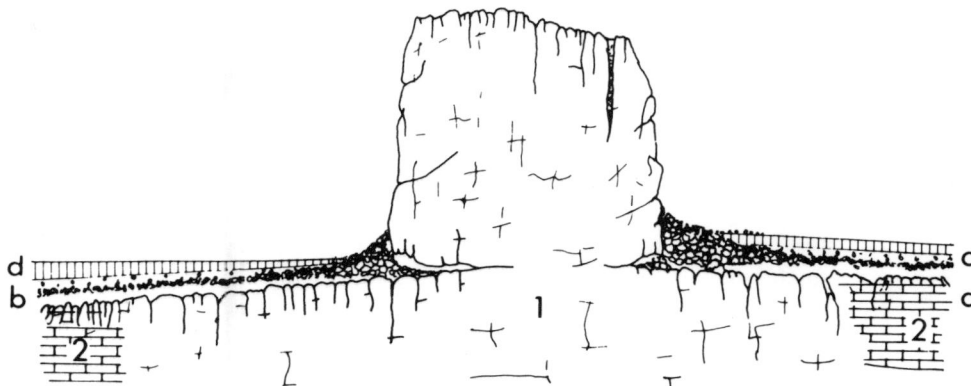

Fig. 9.49 Cross-section through a mogote in the Plaskowyz Ojcowski upland, Poland [after Klimaszewski (1972)]. (1) Massive Upper Jurassic limestone. (2) Strongly jointed Upper Jurassic limestone: (a) kaolinitic weathered mantle; (b) limestone debris; (c) Quaternary interglacial weathered mantle, (d) Quaternary loess.

Chapter 10
The Alps

10.1 Western Alps

The western Alps form part of a great arc of fold mountains (*see* Fig. 10.1) that, beginning with their southernmost part next to the Mediterranean coast, runs inland in a south-east–north-west direction before curving around to adopt a south–north trend in most of the French/Italian sector and then swinging gradually to a west-south-west–east-north-east strike in the Swiss Alps. In the western Alps, the mountains reach the highest elevations of the whole chain (Mont Blanc, 4810 m; Matterhorn, 4477 m; Monte Rosa, 4634 m). The Silvretta massif in the Rhaetian (Austrian) Alps, lying between the upper Rhine and the upper Inn, marks approximately the boundary between the eastern and the western Alps. Although there are many individual variations, the general cross profile of the Alps in terms of structure is always the same. On the outer western and northern side of the arc lies a zone of Calcareous Alps of variable breadth (e.g., the Maritime, Provençal, Vierwaldstätter and Glarner Alps). On the inner side of this arcuate calcareous zone, there is the broad belt of the central, crystalline Alps. This is not, by any means, of uniform structure, but interstratified with rocks of Triassic and Palaeozoic age. Sometimes these rocks cover extensive areas, as in the Maritime (Cottian and Graian) Alps, where they can also be replaced by Bündner schists (calc-schists, Cretaceous – Jurassic in age). This also applies to the zones between the Adula and the Rhaetian Alps. This diverse pattern is an expression of the tectogenesis and orogenesis of the Alps, but it also affects the actual relief and morphology, and the direction and intensity of the geodynamic processes.

10.1.1 Major Physical Divisions

The geomorphological division of the Alps corresponds to this major structural division. The sequence of zones comprises the Pre-Alps, the Calcareous Alps, the Schist Alps and the High Alps [see Annaheim *et al.* (1975), Barsch *et al.* (1975), Cadisch (1953), Chardonnet (1944), Gwinner (1978), Heim (1919, 1921, 1922), Kober (1955), Richter (1974), Staub (1934b)].

10.1.1.1 Pre-Alps

These are built in the flysch zone, and in the folded and overthrust molasse. The molasse itself is internally subdivided On the southern side of the Alps, the molasse is almost completely missing, which is the reason why the crystalline zone extends right up to the Piedmont (Lombardy) lowland.

10.1.1.2 Calcareous Alps

The Calcareous Alps are also a complex zone not only in the geological but also in the geomorphological sense. The more resistant strata (limestones and conglomerates) form sharp crests, ridges and rock walls, while the softer clays, marls and calcareous marls have been excavated to form valleys. Karst phenomena are widespread in the Calcareous Alps; all temperate karst forms can be found, together with the characteristic underground drainage and its associated features, such as caves. Especially characteristic of the Calcareous Alps is the overprinting of karst forms by glacial processes: the karst of the High Alps is in many places older than the youngest glaciation. There was also, however, intensive karstification of the areas of the youngest glaciation in other cases.

10.1.1.3 Schist Alps

These extend between the Maritime Alps and Grisons. The Bündner schists have only a small facies variation and are impermeable, but they weather quickly. They have been susceptible to intensive fluvial erosion and to gravitational processes (landslides) which have led to the complete destruction of Pleistocene forms in these regions.

10.1.1.4 High Alps

The High Alps form the central zone, extending, as already mentioned, in the western Alps as far as the Po plain. The crystalline rocks of the High Alps are relatively resistant, and glacial forms are well preserved here. Glacial troughs at different levels characterize the morphology, together with rectilinear slopes and impressive massive summits and ridges (*see* Fig. 10.2).

10.1.2 Structural and Tectonic Evolution

The Alpine orogeny began at the end of a long period of tectonic stability. The crystalline basement and its sedimentary cover were altered metamorphically and dislocated tectonically. The processes were complex and occurred in several phases, but the results are of great geomorphological significance, because, in this way, the boundary conditions for the development of the relief were established. The extent of the Alpine orogeny can be seen not only from the stratigraphy of the crystalline and Mesozoic sediments but also from the assemblages of relief forms that resulted. Geotectonic processes that have had significant morphological effects include foliation in various directions, systems of faults of different types and alignments, including thrusts, and fold structures ranging

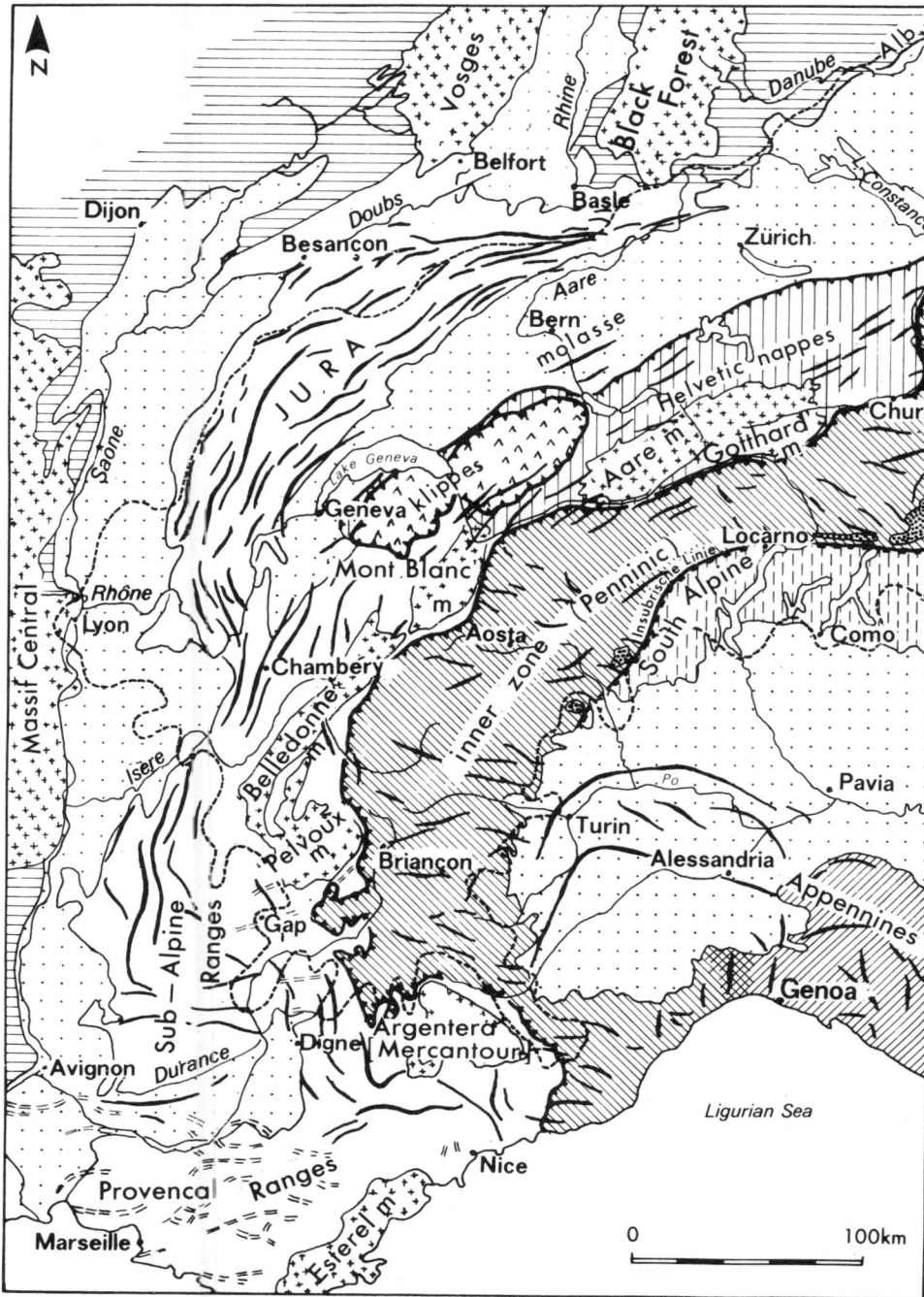

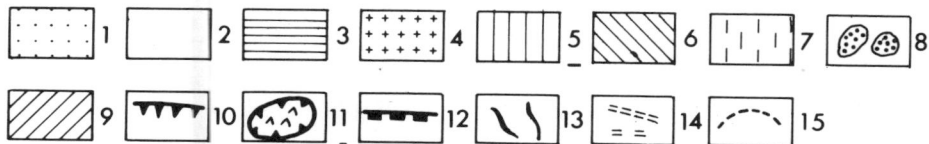

Fig. 10.1 General map of the western Alps. *Alpine foreland*: (1) Tertiary of the foreland, molasse and basins, (2) folded Jura, Dauphinois zone (sub-Alpine chains and Provençal chains), (3) structural plateaus of the foreland (Permian and Mesozoic) and (4) crystalline of the foreland and the central massifs. *Helvetic zone*: (5) Helvetic nappes, autochthonous and ultra-Helvetic. *Penninic zone*: (6) Penninic nappes. *Southern Alps*: (7) Permo-Mesozoic, (8) alpidic plutons and (9) Appennine massif. *Other features*: (10) boundaries of nappes of the large tectonic units, (11) klippes, (12) steep margins of tectonic units. (13) late Tertiary anticlines, (14) Upper Cretaceous and Eocene anticlines and (15) maximum extent of Alpine glaciation.

Fig. 10.2 The Miage Glacier, on the Italian side of the Mont Blanc group. (Cerutti)

from simple Jura-type to complex recumbent and overthrust folds (*see* Fig. 10.3). The Alpine orogeny was succeeded by the neotectonic period, in which the main feature was the uplift of the Alps in relation to the forelands. The neotectonic uplift has helped to create a high-energy environment, the resulting rapid erosion maintaining a high-amplitude relief. In some areas, such as the Rhine region, erosion rates have been at least equal to the rates of uplift; the latter rates also show variations from one crystalline massif to another.

Neotectonic uplift of the Alps is only the latest expression of prolonged geotectonic activity originating at considerable depths. Isostatic equilibrium has been disturbed in the region of the Alps, and the various vertical movements are an attempt to restore a state of equilibrium that existed before the development of the Alpine geosyncline. The subcrustal rocks, which had first subsided, rose again, lighter crustal material thereby reaching the surface. Vertical uplift of the folded rock mass is thus the latest phase of the orogeny and led to the development of the present highly dissected mountain chain. Other consequences of the vertical movements were the sedimentation of the molasse, further gravity sliding of the nappes and the uplift of various planation surfaces and benches far above the recent valleys. In general, relative tectonic stability is supposed to be required for the development of such planation forms, but their interpretation is still controversial. Some can probably be related to the Pliocene, while others, at lower elevations, are younger and have the character of fluvial terraces. In general, V-shaped valleys are supposed to be typical of the rising rock mass of the Alps during the Neogene. Neogene uplift also affected regions far beyond the geosyncline. This is evidenced by the sub-Alpine molasse, which is folded at the northern Alpine border between the Allgäu and the central Swiss plateau, in the Dauphiné zone east of the Rhône valley and also near the southern border at Como [see Annaheim (1946)].

The crystalline massifs are very important in connection with the orogenesis and relief of the Alps. They comprise complexes of pre-Triassic basement rocks. The basement was involved in the Alpine orogeny and suffered geotectonic changes and metamorphism, but there was little or no lateral transport. Hence, they represent autochthonous or para-autochthonous masses. The uplift of the massifs is geologically recent and occurred only in the Neogene as proved by the unaltered forms. The tendency of the massifs to rise is old and was present in Mesozoic times. During uplift, the Mesozoic sedimentary cover was denuded, not only by erosion but also by shearing and sliding processes. In the western Alps, this resulted in the formation of overthrust folds and sheets (the Helvetic nappes).

The massifs of the western Alps (*see* Fig. 10.1) include the Aar–Gotthard massif [see Labhart (1977)], the Aiguilles Rouges–Mont Blanc massif, the Belledonne–Pelvoux massif and the Argentera–Mercantour massif. All are arranged along the strike of the arc of the western Alps. In terms of internal structure and pre-Alpine history, the massifs correspond to the Hercynian massifs lying outside the Alps, such as the Schwarzwald and Massif Central, but whereas these other massifs show the typical Variscan trends (west-south-west to east-north-east), the Alpine massifs show different strike directions. Resting on these autochthonous crystalline masses are the Mesozoic and some Palaeozoic sedimentary cover rocks, partly imbricated with the crystallines.

In the Western Alps there are three major nappe systems (*see* Fig. 10.4).

1) At the base are the Helvetic nappes (Jurassic, Cretaceous and flysch). These appear as several branching sheets in the Swiss Alps and in the French Pre-Alps.
2) Over the Helvetic nappes lie the Penninic nappes; here the basement is involved in the nappe structure. Usually there are schists that are crystallized in different ways, sometimes becoming gneisses. The southern parts of the Swiss/Italian Alps (Graian and Cottian Alps) are formed from them.
3) Mesozoic and metamorphic rocks form the east Alpine nappes (Austrides). Their farthest limits are located in Grisons, where the Helvetic and the Penninic nappes override the Austride nappes. This happens at a big cross flexure. This geotectonic line reflects the predominance of the Austride nappes in the east, which are

Profile across the Western Alps

W V E R C O U R S   E

Kochers de Presles
Col de Balme
Moucherotte
Mte Valette
Gresse
Connex
Le Bourg-d'Oisons
La Grave
Romanche
Gr. Galibier
Les Cerces
Gran Bagna
T. Rachemolles
M. Salancia

La-Mure massif   Belledonne massif   Grandes Rousses massif   Ambin massif   Dora Maira massif

Profile across the Swiss Alps

NW   SE

Sigriswilergrat
2053m
Augstmatthorn
2140
Luegiboden
Faulhorn
2683
Eiger
Jungfrau
Mönch
4105m
Aletschfirn
Aletschhorn
Gantertal
Monte Leone
3558m
Antigoriadecke
Portal Iselle 634m

Aar massif

Lebendundecke

overthrust of the Helvetic nappe

Rhone Valley

HELVETIC ZONE

autochthonous

Ma	Mesozoic
TH	Tertiary
CH	crystalline

allochthonous

| M | Mesozoic |

T Tertiary

BRIANCONNAIS ZONE

| PM | Upper Palaeozoic & Mesozoic |

PENNINIC ZONE

CP	crystalline of inner zone
C	other old crystalline
Mt	Mesozoic & Tertiary

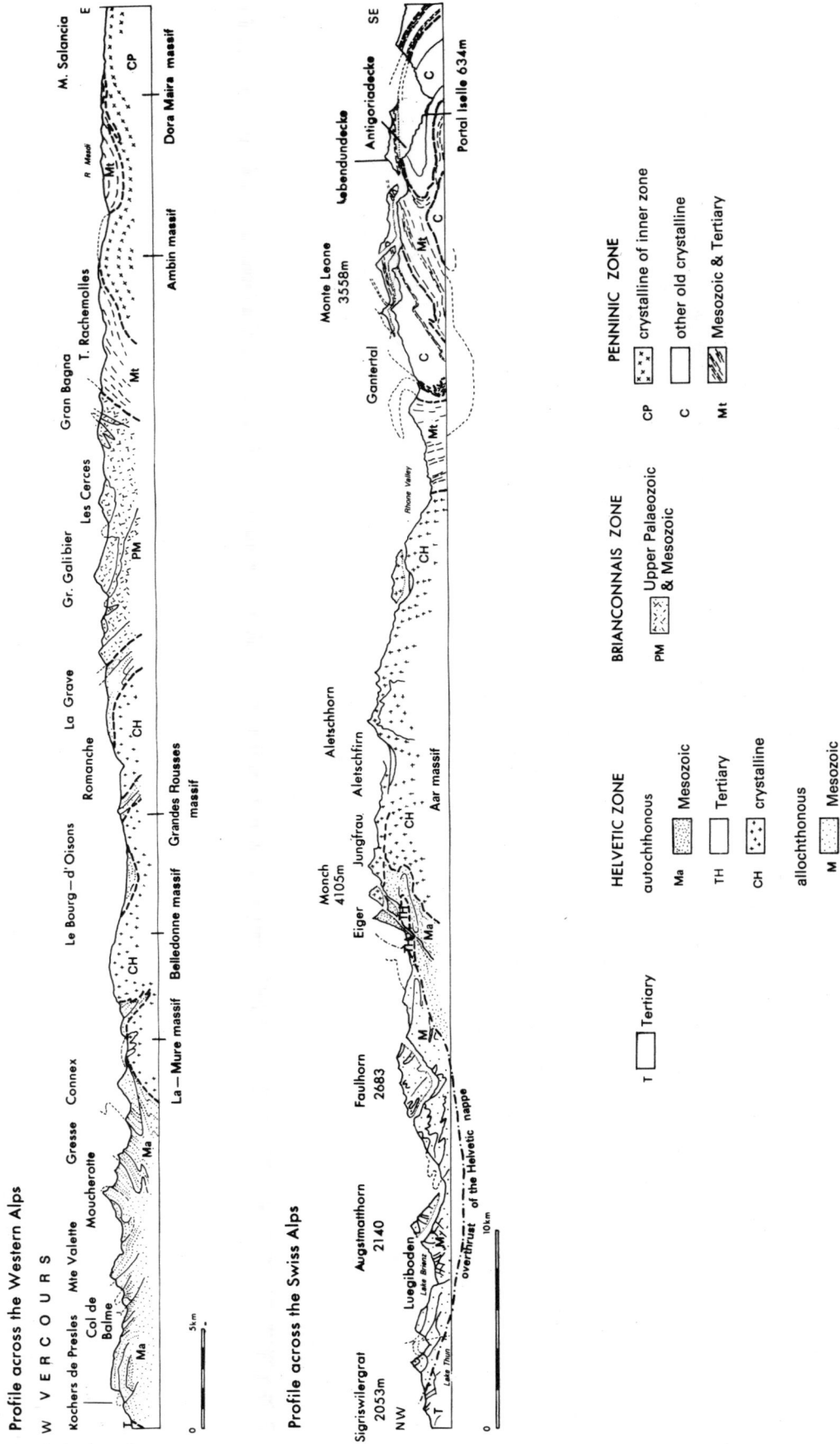

Fig. 10.3 Geological sections through the western Alps.

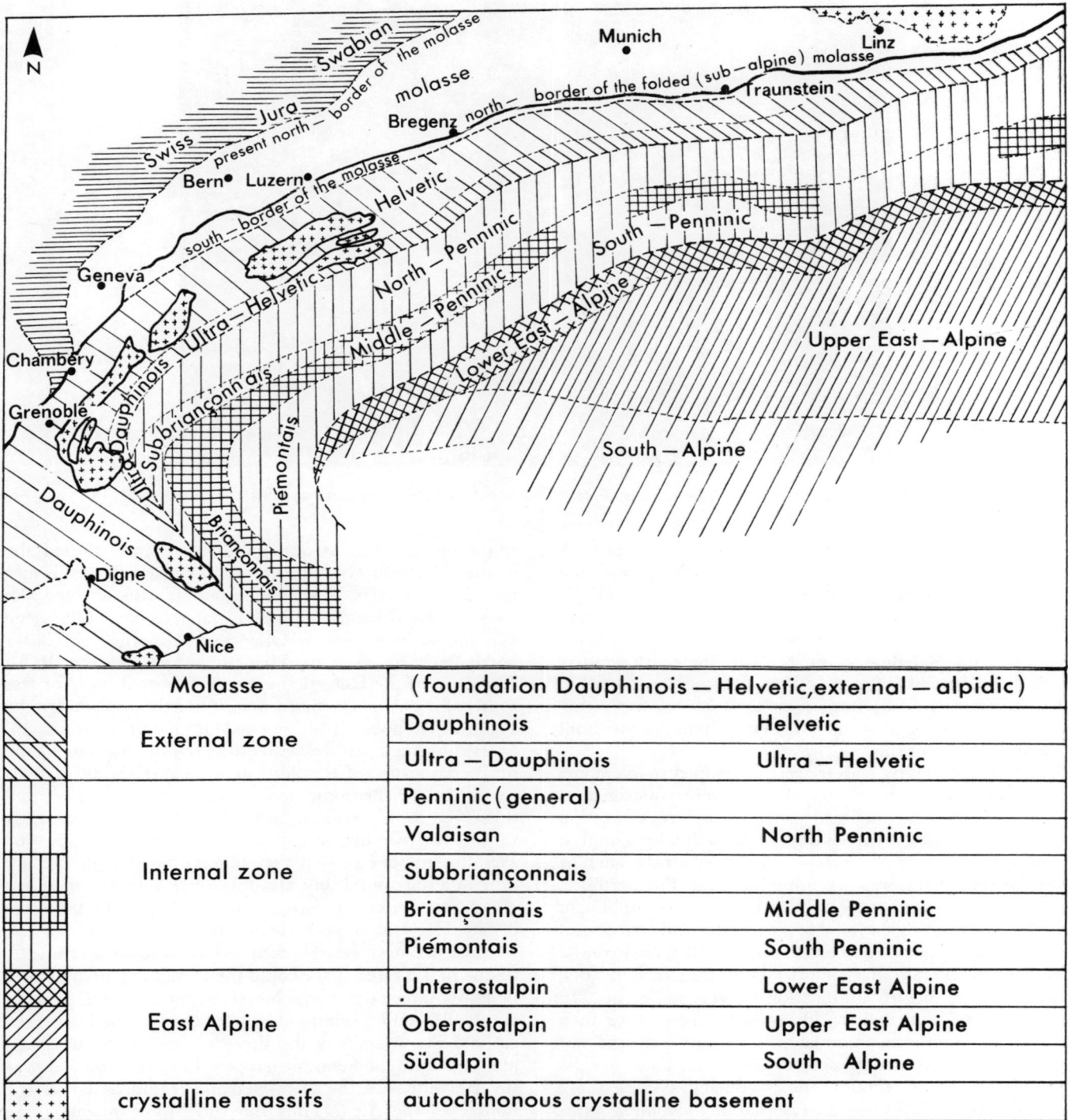

Molasse	(foundation Dauphinois — Helvetic, external — alpidic)	
External zone	Dauphinois	Helvetic
	Ultra — Dauphinois	Ultra — Helvetic
Internal zone	Penninic (general)	
	Valaisan	North Penninic
	Subbriançonnais	
	Briançonnais	Middle Penninic
	Piémontais	South Penninic
East Alpine	Unterostalpin	Lower East Alpine
	Oberostalpin	Upper East Alpine
	Südalpin	South Alpine
crystalline massifs	autochthonous crystalline basement	

Fig. 10.4 Supposed original arrangement of the alpidic area of sedimentation, prior to the Cretaceous orogeny.

subdivided several times within themselves. In the western Alps, the Austride nappes only appear at a few localities.

The classic folds and apparent overthrusts of the Alps were first noted by geologists in the early nineteenth century. Early theories involved the crumpling of strata by lateral pressure, culminating in the giant recumbent folds and displacement of nappes, the latter having travelled for many kilometres towards the foreland and, in places, far across it. As early as 1841, van der Linth described the evidence that the Verrucano beds, of Permian age, seemed to have been thrust over younger—Jurassic and Eocene—

Fig. 10.5 View from Crap Sogn Gion, showing Helvetic nappe (Permian Verrucano formation over Jurassic limestone). (Embleton)

strata for about 15 km, forming the Glarus nappe. A revised interpretation by Bertrand in 1884 involved a displacement of no less than 35 km [see Heim (1921, 1922)]. Despite this, the nappe theory continued to gain adherents and was widely applied in other orogenic belts, including the Caledonides in the late nineteenth century with displacements of over 100 km. It is now appreciated that it is mechanically impossible for such relatively thin sheets of rock to be pushed forward in this manner without complete disintegration.

Theories of gravity sliding were developed to overcome these objections—the sheets of rock were conceived as sliding down vast inclined planes, but rock friction required the latter to be sloping at impossibly high angles, even if the presence of certain 'lubricating' strata, such as clays or evaporites, was taken into account. Furthermore, some of the nappes appeared to have moved uphill for parts of their journeys. More recent and successful theories have taken into consideration the hydrostatic pressure of interstitial groundwater during tectonic disturbance and before lithification of the sediments. In conditions of high residual pore-water pressure or in a submarine environment, sediments become buoyant and very mobile, even on slopes of a degree or so. These are the conditions it is now thought under which gravity or slight tectonic disturbances could cause nappe movement without the rocks disintegrating; the instances of apparent uphill movement may be due to subsequent reversal of the slope by vertical uplift of quite a minor order. Finally, high pore-fluid pressures may arise not only due to residual water from geosynclinal conditions but also because of the metamorphism of hydrous minerals and the ascent of juvenile fluids and gases from deep levels undergoing metamorphism or granitization.

Several different but interrelated phases of nappe displacement can be recognized. The Eo–Alpine phase spans the Cretaceous and Palaeocene, and affected the whole of the western Alps between the Ligurian Alps and the Simplon Pass. To some extent it also affected the Grisons. The Lepontine phase is dated as early Oligocene and the youngest one as Oligocene–Miocene. The latter particularly involved the Helvetic nappes (*see* Fig. 10.5). The roots of the Helvetic nappes are located more or less south of the Aar–Gotthard massif, partly overlain by the Penninic nappes. They moved from south to north, overriding the central massifs to reach the molasse at the northern border of the Alps. The ultra-Helvetic nappes resemble the Penninic facies, and are especially well preserved in the western part of Switzerland. In contrast to the relative uniformity of the Helvetic nappes, the Penninic nappes show a very differentiated structure not only in a horizontal, but also in a vertical direction. They cover large areas that formerly were occupied by the rocks of the Helvetic nappes; hence the relative ages of the movements may be deduced. Only when the Helvetic nappe had set free space could the Penninic pile of nappes reach its later position (achieved in the Miocene period). But folding and formation of a pile of nappes had occurred earlier. The massifs of the Pre-Alps and of the Chablais also belong to the Penninic nappes. They were moved over the Helvetic and the ultra-Helvetic region as far as the border of the Alps. Thus the sub-Alpine molasse was involved in the overriding and the Miocene molasse of the foreland was tectonically raised.

In the residual sea of the geosyncline, molasse sediments accumulated from the Middle Oligocene until the Lower Pliocene. This happened not only in the foreland but also in the hinterland of the rising Alps. The molasse basin is very well developed on the northern side of the Alps, beginning in the west in the region of the Lake of Geneva. The molasse indicates the uplift of the orogene, the erosion debris accumulating in the foredeep. It is possible to distinguish between the foreland molasse and the sub-Alpine molasse. The latter developed at the border of the mountains and possesses sedimentation of very coarse conglomerates. The sub-Alpine molasse was, as already noted, involved in the orogenesis, which meant that it was steeply tilted, folded and imbricated. This zone of sub-Alpine molasse, with a width of 20 km at a maximum, moved over the foreland molasse. In terms of both geomorphology and tectonics, external and internal subdivisions of the zone are distinguishable. In the internal zone, dislocations and overthrusts predominate while, in the external one, folded structures occur. The internal zone is also overthrust by the Alpine nappes.

10.1.3 Patterns of Relief and Drainage

Ignoring for the moment the smaller-scale landforms, the alignments of the main Alpine valleys are controlled by the nappe structure. The typical longitudinal valleys of the Alps are located partly along outcrops of weaker rocks and partly along the root zone of the Helvetic nappes. In addition, they may be guided by longitudinal disturbances, by boundaries between nappes, by tectonic depressions and along the axial zones of eroded anticlines. Normally, the transverse valleys are wide and show intensive remodelling by Pleistocene and postglacial erosion and deposition. These transverse valleys cut across different outcrops of both hard and soft rock. Their courses are determined in part by old tectonic flexures that were caused by an east–west lateral compression from the eastern Alps and, in part, by various transverse disturbances, such as faults, transcurrent faults, flexures, lateral margins of nappes and axial depressions.

The original drainage network, which can only be recognized in some of the principal features of the present valley system, was determined by tectonic structures. In the western part of the area, the dense concentration of geotectonic and geomorphological elements suggests a more powerful and prolonged uplift of the mountains. This also has consequences for valley development, which was more intensive in the western Alps, as shown by the exposure of the deeper geological structures. By the Miocene, there was an expanding drainage net in which watersheds were shifted by regressive erosion. The nagelfluh (conglomeratic) sediments of Lombardy and Piedmont in Italy are related to the resulting high rates of erosion. Rapid uplift of the mountains was accompanied by deep incision of the valleys and, by the end of the Miocene or the early Pliocene, the courses of the main valleys were more or less fixed. Valley incision was not continuous but intermittent; the valleys became temporarily narrower during phases of more rapid uplift and broadened during periods of less activity. Remnants of old valley floors can be seen occasionally as high-level benches, but their interpretation and correlation are disputed.

The geomorphological evolution of the mountains and the valley system can be summarized in broad outline as follows. (1) Formation of a steep mountain relief during the Oligocene and Miocene uplift, with intensive erosion and sedimentation in the molasse basins. (2) Reduction in the rate of uplift and the formation of a landscape of moderate relief (Mittelgebirgslandschaft), evidenced by high-level systems of surfaces and the summit areas. (3) Further intensive, but intermittent, uplift whereby the present complex relief (Schachtelrelief) was formed. The active pre-Quaternary drainage had already incised itself and left the old valley floors at high levels [see Galibert (1965)].

In terms of their vertical distribution, the major forms of the Alps appear to be arranged in a series of levels. The oldest and highest parts comprise various summit areas. These are arranged systematically with an increase in height towards the centre of the massif and are, to some extent, developed quite regularly. Simply from this, however, it must not be deduced that one is dealing with a former peneplain or series of peneplains. As regards the highest summits, it has often been supposed that there is a denudation level that originates from some phase in the Tertiary when the relief was reduced in amplitude to a series of broad valley floors and low interfluves. Such a surface appears to have been later tilted, presumably because of younger tectonic processes, such as the Pliocene–Pleistocene updoming of the Alps after erosion processes had begun to dissect the old relief. Such old land surfaces, including the augen–gneiss surface of the Upper Oligocene or, in the northern Calcareous Alps, the Rax surface of the Mio-Pliocene, are more widespread in the eastern Alps. The term Rax surface is used by Alpine geomorphologists to refer to the high-level (approximately 2000 m) surface in the Rax region of the Styrian Alps near the Semmering Pass (*see* p.246). In the western Alps, the old land surfaces are less extensive and restricted to high-level remnants and summit areas. It should also be noted that these relict surfaces occur in areas of flat-lying sedimentary rocks, such as overthrust sheets. Thus, parts of the surfaces may be purely structural in origin and may not have any particular chronological significance in terms of denudation levels. The lower levels, however, must surely be relics of old landscapes of younger Tertiary or early Pleistocene age, because they lie above the Pleistocene valley floors, and because they represent real erosion surfaces. In the western Alps, the poor development and preservation of the levels result from the intensive and rapid uplift of the mountains. Processes of plain formation, even if possible under prevailing climatic/geomorphological conditions, were not effective here for tectonic reasons. In the western Alps, the individual parts of the mountain chains also show different rates of uplift, which is a further reason why correlation of levels over large distances is fraught with difficulty. Moreover, individual surfaces afterwards suffered further modification or development, which varied considerably from place to place.

10.1.4 Neotectonics

Reference to neotectonic uplift of the Alps has already been made. The most important movements occurred at the end of the Pliocene and in the early Quaternary, marking the end of the Alpine orogenesis. Exact proof of age and occurrence of Quaternary tectonic movements is not available, partly because of the intensive later Pleistocene remodelling of the geomorphology. Similar difficulties prevent accurate dating of later Quaternary tectonic movements and most of them can only be dated as younger than Riss. Their relatively unaltered forms support such a recent date. In this context, the long-standing quasiequilibrium between erosion and uplift in the Alps must be mentioned as the main reason why it is thought that the Alps have never been significantly higher than they are today.

10.1.5   Quaternary Glaciations

The form of the preglacial land surface of the Alps on which the Pleistocene processes of erosion and sedimentation began to operate is not precisely known. Not even the older glaciations (i.e. Günz and Mindel) can be clearly recognized in the Alps and their vicinity. The greatest extent of ice for most regions is supposed to have been during the Riss glaciation, which covered the traces of former glaciations, although in the Dombes the boundary of the maximum glaciation is formed by sediments of Mindel age. This is the reason why preglacial land surfaces can only be found in the borderlands of the Alpine glaciations and not in the Alps themselves. Around the western Alps, preglacial land surfaces can be traced in the region of the molasse ridges north-west of the Lake Constance–upper Rhine glacier in the basins of the northern Jura, in the Ajoie, in the French Jura plateau and on the plateaus between the Isère and the Rhône glaciers. In the Alps, especially in the Calcareous Alps, some have claimed to be able to recognize remnants, although actual proof is not available. In the high Alpine watershed regions, various glaciated bedrock (roches moutonnées) surfaces are interpreted as only slightly degraded preglacial relics. Intensive fluvial dissection of

the relief, however, means that the survival of such forms is rare and disputed by many workers. Furthermore, any such preglacial forms were affected by neotectonic movements and were overridden by at least four glaciations. The highest parts of the Alps, especially those that are often indicated as regions with preglacial forms, were those that were most frequently glaciated and where glaciation was most prolonged, so that the likelihood of preglacial forms having been preserved here is small.

The most important glaciations in terms of present-day relief are those of the Riss and the Würm periods [see Zimmermann (1963), Hantke (1978)]. They played an effective role not only in the shaping of the valley forms, but also in the formation of the slopes. Although the effects of the last glaciation are most evident, it should be emphasized that the glacial formation of the Alps has to be seen as a process that lasted the whole of the Pleistocene. The ice widened the smaller valleys (*see* Fig. 10.6) and formed the great trough-shaped valleys and their accompanying forms. During the Riss, intensive glacial erosion effected large amounts of valley deepening. Parts of the rock basins, particularly along the valley floors, were infilled with sediment during the following interglacial period, the Würm and the postglacial. Besides the glaciogenic sediments, a lot of fluvial material and lacustrine deposits accumulated. Late and postglacial sedimentary processes played a major part in infilling large areas of the great longitudinal valleys of the Alps, together with the Alpine border zones and the Alpine foreland with young sediments. In this way, the extensive aggradation plains of the Alpine border and of the intermontane zones were formed.

The glacial periods differed in respect of the extent of ice cover in the Alps and their foreland. The Günz and Mindel glaciations of the Alps are difficult to identify. The

Fig. 10.6   Glacially eroded valley, near Susten Pass, Switzerland. (Embleton)

Fig. 10.7 Glaciofluvial terrace, upper Rhine valley, Rueras, Switzerland (Embleton).

Riss glaciation was the most extensive in the western Alps. It formed the landscapes of the older moraines, which show mostly degraded forms in contrast to the landscapes of the young Würm moraines. The Riss was responsible for widespread deposition of till or ground moraine, but there are few marginal or terminal moraine ridges of this age. Some parts of the Alps, such as the Provençal, the French Calcareous, and the Ligurian Alps, lay outside the main limits of the Riss ice, except for local glaciation. On the other hand, the Maritime Alps were glaciated. The inner zone of the Alpine mountain arc facing the Po plain was also glaciated, but the ice only reached the border of the mountains, touching the foreland in only a few places. Here, the limit of the Würm was largely identical with that of the Riss. During the Pleistocene and in the postglacial, the glaciers of the western Alps, because of their oceanic position, responded rapidly to climatic fluctuations, more so than the glaciers of the eastern Alps or even those of the central Alps.

The glaciation of the Alps took the form of an intricate network of glaciers and small ice caps. At times, the major valleys were occupied by immense streams of ice that extended into the foreland on the north and west of the Alps, such as the Lake Constance–upper Rhine, the Linth–Rhine, the Aare–Reuss, the Emme–Aare–Rhine, the Drac–Isère–Rhône and the Durance glaciers. The glaciers did not always correspond with the present valleys in terms of direction of present stream flow. There were many instances of both glacial diffluence and glacial transfluence at the maximum stages of glaciation, but eventually the ice streams joined to feed several major lobes at the Alpine border.

In the western Alps, glacial events were controlled by the Rhône glacier and its distributaries. The Rhône glacier extended during both the Riss and the Würm as far as the Lyonnais. During the Riss, it even succeeded in reaching the border of the Massif Central. The Saône

collected the meltwater from the Rhône glacier and thereby acted as a glacial spillway. The border of the Rhône glacier was very mobile as shown by the intricate, complex ice-marginal features. Various stages of recession are clearly discernible. There were also connections between the Rhône and the Isère glaciers. For this region, as with the Dombes, the pre-Riss glaciation is supposed to have represented the maximum ice expansion. The penetration of the Rhône ice into the Jura is discussed in Section 10.6.

In the western Alps and their foreland, the Riss–Würm interglacial period is evidenced by local soils and sediments, but well-marked relief forms of this phase are absent.

10.1.5.1 *Würm glaciation*

This is well represented by the forms and sediments associated with ice decay and retreat at the end of the glaciation both in the mountains and in the foreland. The advances of the ice following the interglacial appear to have been spasmodic, colder phases being interrupted by warmer phases with ice decay. The Würm ice was directed by existing valley forms, which were already well developed by the Riss glaciation. The assemblage of erosional and depositional forms is very comprehensive— outwash gravel areas, lateral and end moraines, drumlins, glaciated bedrock features including roches moutonnées, meltwater channels and glaciofluvial terraces (*see* Fig. 10.7). The lowering of the Würm snowline is estimated at 1200–1300 m for the western Alps and the Swiss northern Alpine border. The Riss snowline, in comparison, is thought to have been 1400 m lower than present at its maximum. Indications of the lowering of the snowline on the Mediterrannean side of the Alps are uncertain, since the forms and sediments of either Riss or Würm age do not provide sufficient data on this question.

The decay of the ice in the later Würm period occurred in very different ways in the different glacial areas and was much influenced by the local microclimate. In spite of these differences, however, some general comments can be given. Sequences of glacial and glaciofluvial forms, which relate not only to minor readvances of the ice but also to decay during retreat, are developed in the valleys. The main stadials marking occasional readvances of the Würm glaciers in the eastern Alps are also present in the western Alps [see Salathé (1961)]. During the Schlern stadial, the snowline fell by 800–900 m, in comparison with that of the present. In the Maritime Alps, this corresponds with the Molières stadial. There is only local evidence for the Allerød warm phase, which is marked by dump moraines. The Gschnitz (Chastillon) stadial is dated as older pre-Boreal. The snowline lowering was less and only amounted to 500–600 m. Late glacial moraines of this stadial are not so well marked. They can only be found in certain valleys in the western Alps depending on their aspect and position. The climatically induced intensive weathering led to the accumulation of thick debris covers, while moraines due to rock falls and protalus ramparts were produced. A characteristic of the Gschnitz stadial is commonly the double embankment form, which may be related to two separate phases of advance. The Daun stadial, which follows the Gschnitz in the eastern Alps, originates from a fall of the snowline by 300–400 m. In the western Alps, on the other hand, there is another intermediate snowline decrease of 400–500 m, which is called the Margés stadial in the Maritime Alps, but it is not very well marked. The moraines here are more like rock glaciers. The moraines of the Mounier stadial, which originates from a snowline depression of 300 m, are linked to the Daun stadial. In the most westerly and southerly parts of the western Alps, the ice melted completely at this time, or only survived in extremely favourable situations. Many of the cirques were already free of ice. The low snowline resulted in considerable production of debris. In the Swiss Alps particularly, recent geomorphological investigations based on more refined absolute age determinations have shown that the stratigraphy and morphology of the moraines is even more complex, but so far no fundamental reappraisal of the moraine sequences has been made. Not only in the valleys of the Alps but also in the high Alps, there is a series of late glacial ice-dammed lakes, which have partly in the postglacial affected the development of the river pattern.

10.1.5.2   Postglacial period

Postglacial fluctuations of climate are reflected in the different glacial stages. During the early sub-Atlantic, several glacial stages appear in the eastern as well as the western Alps. The glaciers advanced to form moraines, which have been well dated. In some cases, correlation with the postglacial stages in the USA is possible, which seems to point to a general global climatic deterioration in the early Atlantic and early sub-Atlantic phases [see Eicher and Siegenthäler (1976), Zoller (1966)].

The historical glacial fluctuations are clearly marked in the geomorphological evidence [see Portmann (1977).

During the late sub-Atlantic (approximately 1000 years ago) there were several glacial advances in the Alps. There are no end moraines dating from the period between the maximum extension of the ice in the late Pleistocene and the advances of the seventeenth century, so a general glacial retreat in this period is assumed. At the beginning of the seventeenth century, there must have been a rapid climatic deterioration, because in both the eastern and the western Alps the glaciers advanced distinctly and rapidly. Further advances occurred at about 1820 and 1850, but the behaviour of the glaciers was not identical everywhere. The larger glaciers responded more slowly than the smaller ones, so that the moraines cannot easily be correlated. The maximum expansion of the large glaciers therefore has to be placed not in the seventeenth century, but at about 1820 or 1850. All the postglacial fluctuations formed a series of small end moraines (*see* Fig. 10.8), which are easily dateable. In general, it can be said that nearly all the important glacial advances of the last 9000 years correspond with the last major glacial stand of 1850, and are characterized by high and angular morainic ramparts [see Bachmann (1978), Beck (1926), Hanns (1980), Kinzl (1932)].

10.1.5.3   Recent processes

Recent geomorphodynamics of the Alps present two aspects. On the one hand, there are the glacial and periglacial forms of the postglacial period and, on the other hand, there are the various slope processes—both erosional and depositional—of which the most important are related to gravitational mass movements. These may be linked to fluvial, glacial and periglacial events and processes. Present-day processes in the subnival and nival zones are controlled in many areas by the presence of permafrost. The main indicator of Alpine permafrost is the existence of active rock glaciers. They are closely dependent on slope aspect and appear particularly on slopes facing north-east at altitudes between 2400 and 2600 m. They are more commonly found in the more continental parts of the western Alps. In Switzerland, they tend to be concentrated in the central and southern parts of the Alps. The subnival zone, whose upper limit on north-facing slopes lies 400 m below the snowline of the glacial maxima of the nineteenth century, is the actual or potential Alpine permafrost region. The lower permafrost limit on south-facing slopes stands at 2600–2800 m. About 30 percent of such slopes are underlain by permafrost [see Barsch (1977a)].

Rock glaciers represent important glacial geomorphological phenomena in the Alps. They consist of frozen debris with different ice contents; they move downhill with velocities of up to 1 m per year. Rock glaciers which are formed below major sources of debris, such as debris slopes or glaciers, represent slow mass movements in the subnival zone, where they act as a debris transport system. They belong to the frost debris zone of the Alps, where today large quantities of debris are released from frost weathering. Until recently, frost weathering as a geomorphological process has been underestimated in the Alps. From the volume of debris, it can be shown that, in the Holocene, frost sapping of the rock walls reaches a value of 10–25 m (about 2.5 mm per year) on north-facing slopes, which favours the frost-weathering process [see Barsch (1977b)].

The lower limit of the periglacial zone of the Alps is the gelifluction boundary, while the snowline marks the upper limit. The lowest occurrences of gelifluction are not the most important criteria; more important is the general regional limit for the majority of slopes subjected to such activity. It is possible to distinguish an upper zone where gelifluction is unrestrained by vegetation and a lower zone of turf-banked phenomena. The lower limit of the restrained gelifluction in the Pennine Alps attains a maximum altitude of 2350 m (unrestrained gelifluction: 2700 m). At the Alpine border in the Belledonne massif, it reaches 2000 m (unrestrained gelifluction: about 2700 m). Isolated gelifluction forms are also common below the lower limit: in the Maritime Alps, they can be seen below 2000 m. Aspect and surface material are the most important factors for gelifluction processes in the periglacial zone of the Alps. They can locally distort considerably the distribution of periglacial forms. The vegetation factor shows a close ecofunctional relationship with these processes, but is not a primary influence. The following gelifluction forms can be discerned: undifferentiated mobile debris covers, boulder fields (with transitions to rock glaciers), structure soils, small steps (which represent the gelifluction microrelief), gelifluction terraces, involutions, boulder, earth and mud flows, debris and vegetation stripes, and striped and polygonal patterned ground.

Runoff processes on slopes are interlinked genetically with both periglacial and fluvial processes. The intensity of runoff is not only controlled by the intensity of the precipitation but also by the characteristics of the surface materials. In soft, easily removable rocks, there is rapid erosion, while this process may be quite unimportant in the crystalline rock areas. In the western Alps, which experience generally high precipitation, there is surface and linear erosion in the more easily removable and erodable rocks. Recent accumulations of debris and glacial sediments are reworked or removed. The intensity of the processes varies according to climatic position in the Alps. In the Maritime Alps, which experience a Mediterranean-type precipitation regime, runoff processes operate only periodically or episodically. In other parts of the western Alps, they operate at all times when there are high intensities of precipitation. In the crystalline rock areas, runoff erosion is only active where there are high levels of soil moisture: namely, during periods of heavy precipitation and snow melt.

Fluvial activity is also differentiated as in the case of runoff processes. It begins as soon as heavy precipitation or snow melt occurs; large volumes of sediments are then transported through the channel systems. Fluvial forms show differences between the crystalline rock areas and areas of non-crystalline rocks (with the exception of limestone: see below). In regions with easily erodable rocks, a narrow valley or even canyon-type relief is developed, in which fluvial processes are most active during floods. Mud flows may occur and huge volumes of

Fig. 10.8 View of Allalinhorn (4034 m) and the Alphubel (4207 m) from the Saas Fee, showing neoglacial moraine of the Fee glacier. (Embleton)

sediment are moved, supplied by the debris and soil covers of the slopes, which are eroded both by runoff and mass movement processes. In the upper zones, frost weathering, periglacial and glacial processes cooperate in the supply of debris.

In many areas, the natural processes are intensified anthropogenically. As a result of various land use practices, there is a strong tendency to fluvial dissection, encouraged by the nature of the runoff, the rocks and the generally steep slopes. The Alpine rivers are, however, by no means completely regulated by man, and flooding and flood discharges occur every year in some areas. Gully erosion, general fluvial erosion and deposition and mud flows frequently form interdependent parts of the geomorphodynamic system. There are often relationships between erosion and deposition, sometimes over large distances. For example, in places, the upper Rhine crosses outcrops of softer rocks where gorges have been cut, on the slopes of which landslides and mud flows are frequent. The sediments are carried downstream by the Rhine and contribute to the rapid growth of the delta into Lake Constance. The western Alps are particularly characterized by intensive morphodynamic activity at the present day. Gully erosion and the cutting of ravines are reinforced by the heavy precipitation and sudden snow melt. In the southern parts of the western Alps, where less resistant rocks outcrop, steep erosional slopes are formed on which gravitational mud flows are active.

Lateral and vertical erosion, traceable to the violent and fluctuating levels of runoff and to the heavy load of sediment originating from the torrents and ravines are very evident in the landscape. Slope deposits, such as talus and alluvial cones, are common features. There are numerous alluvial cones of varied sizes, often juxtaposed or overlapping, which suggest that these forms are not very old and are probably postglacial. Indeed, multiple changes in the courses and erosional activity of the rivers and streams are documented in the complex development of the alluvial cones, and their differentiation may be related to particular catastrophic climatic events of the postglacial epoch. There is, however, another major influence on the formation of these alluvial cones. Many undoubtedly originate from the time of deforestation of the Alps, between the late Middle Ages and the beginning of modern times.

The present-day activity and dynamics of the Alpine streams and rivers are much influenced by man. The building of reservoirs and dams, regulating weirs and bank-stabilization works hinder the natural processes of erosion and deposition. Catastrophic floods could be controlled by these preventative measures and many regions, including valley floors and low-angle alluvial fans became possible for settlement. At the same time, however, these man-made regulating structures had other, sometimes unforeseen, effects on the fluvial and mass movement dynamics. New erosional activity appeared in some river beds, requiring new measures to be undertaken. Clearly, the equilibrium between erosional forces and sediment transport can be completely changed by man in the larger Alpine valleys.

Rockfalls are of widespread geomorphological occurrence in the Alps [see Abele (1974)]. Rapid movements of rock debris from mountain flanks, which reach a volume of about $1 \times 10^6 \, m^3$ in the accumulation area or which cover an area of more than $0.1 \, km^2$, are defined as rockfalls. All smaller forms, as well as debris fans and mud-flow deposits, are excluded. There is a petrographic control on the incidence and magnitudes of the rockfalls. The main area subject to these phenomena is in the northern and southern Calcareous Alps, although the southern Alps show smaller and less numerous forms than the northern Alps. Occurrences in the crystalline massifs are much more modest and nearly all of them can be found in the Aar massif. The forms are petrographically differentiated. In the calcareous areas, bigger scars and spalling points are developed compared with the crystalline regions. The Dolomites produce more and bigger rockfalls than the areas of metamorphic schists, gneisses and granites. In some parts of the Alps, the accumulation of the rockfall debris is associated with the presence of ice, either falling on an active glacier or on dead ice. Then depositional forms similar to moraines (rockfall moraines and protalus ramparts) are formed. The geomorphological consequences of rockfalls are considerable. Temporary lakes may be impounded, followed by catastrophic breakthroughs, water courses may be diverted and new watersheds formed.

The dating of rockfalls is sometimes facilitated by the presence of glacial and fluvial sediments overlying or underlying the rockfall mass. Most of the rockfalls apparently happened at the end of the Pleistocene; the majority are of late glacial age. At this time the Würm ice had already retreated, exposing oversteepened valley sides. Intensive gelifluction processes on the slopes, as well as the thawing of permafrost, helped the already loosened debris to lose adhesion and to move. Historic rockfalls are also known, but in the context of the Alpine rockfalls overall, they are not important events.

An important group of forms is related to karst processes. In some areas, these are morphogenetically dominant, but are superimposed on the glacial, periglacial, fluvial and gravitational action. Glacial processes are supposed to have had relatively little effect in the calcareous areas. Glacial erosion was very selective and related to petrographic differentiation. The smooth limestone surfaces possibly offered less resistance to ice movement. The development of frost-shattered debris under periglacial conditions is evident in the calcareous regions, although quantitatively the effectiveness of frost action is debatable. For the ice-free areas of the Calcareous Alps, obstruction of the underground drainage by permafrost may have been important allowing surface drainage and fluvial erosion to occur. Some of the karst forms of the Alps are only intelligible in terms of the combined effects of solution processes and glacial or periglacial processes. Such composite forms include the massive karst pits, which are similar to poljes. The dolines show a limited adaptation to rock structure. The further development of karst pits and dolines took place by karst corrosion processes which were assisted by the presence of interbed-

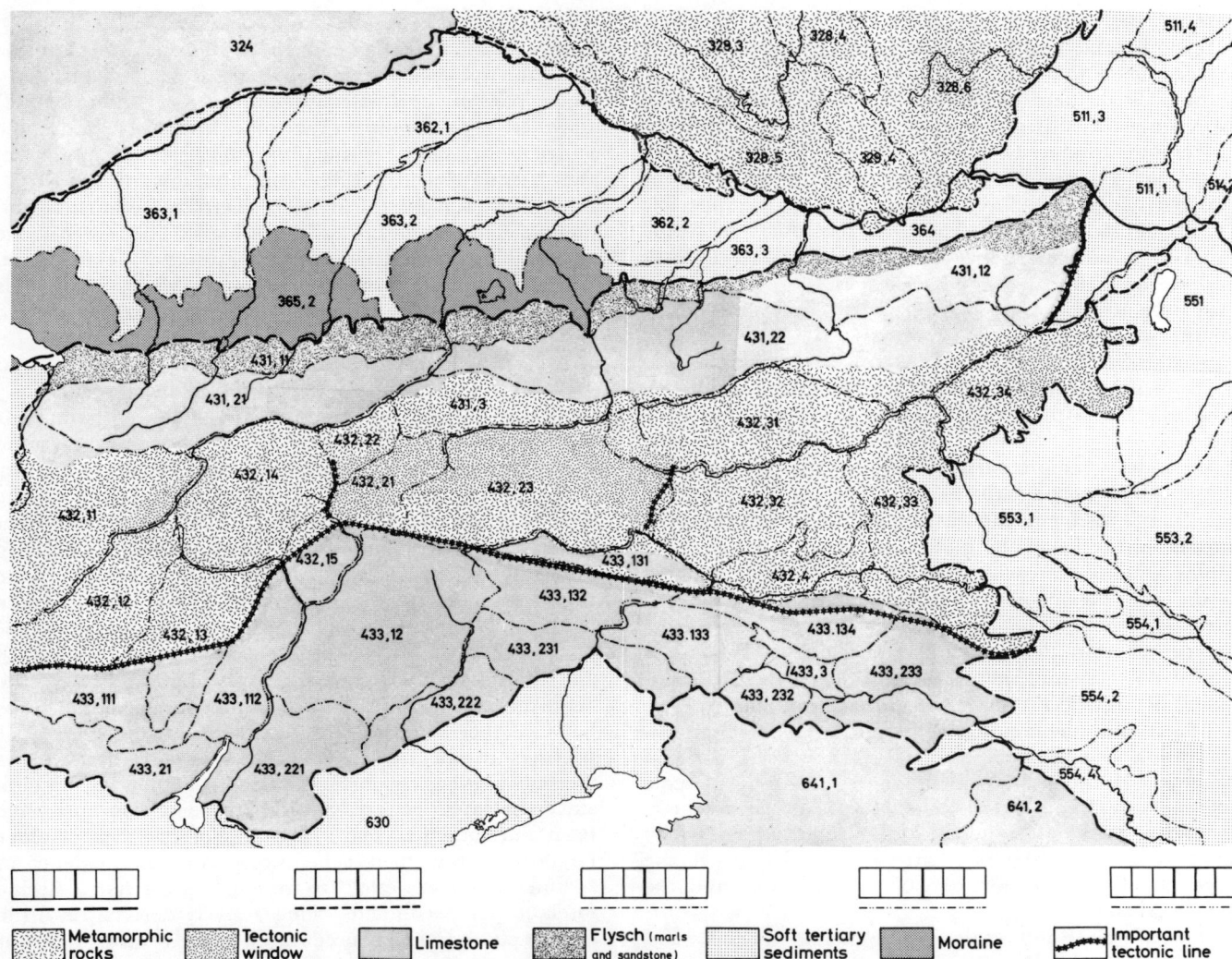

Fig. 10.9 Physiographical regionalization of Austria, based on lithological units and on the Dezimalklassifizierungssystem für Dokumentationen.

ded sediments. Other important, but underground, forms are caves, crevices and shafts. These underwent further development, especially during the warmer interglacials of the Pleistocene when the surrounding karst was hydrographically active, but it seems that they continued to develop to some extent throughout the Pleistocene. Smaller karst forms include the widespread karren, which appear at all altitudes. Most of the karren fields can be found on glacially eroded surfaces differentiated by height. The latter exerted a climatic control and, with shifts in the vertical zonation of climate, different generations of karren are discernible. Below the forest limit, the development of karst forms is greatly influenced by the vegetation cover and the soils.

10.2 Eastern Alps

The eastern Alps have their western boundary at the upper Rhine Valley–Splügen Pass–Lake Como. They trend in an easterly direction and split into two branches at the eastern limit, one trending in the Carpathian direction (north-north-east) and the other in the Dinaric direction (south-east). The Gross Glockner (3797 m), the highest mountain in Austria, marks the approximate geographical centre of the area. The Alps present an abrupt front towards the northern foothill region, as well as towards the eastern basins filled with young Tertiary sediments. The south-eastern limit, however, cannot be so easily defined since the transition to the Dinarides is obscure orographically; on the other hand, the northern limits (*see* Fig. 10.9) are well marked. The northern foothills consist of the debris eroded from the tectonically rising Alps and are part of Variscan Meso-Europe—as defined in the tectonic classification of Europe by Stille (1924)—since the crystalline rocks of the Bohemian massif form its base.

10.2.1 Structure and Tectonics
The eastern Alps are a model for the greatest nappe structures known anywhere in the world. In the central

part of the mountain range, the Pennides (Penninic nappes) become visible in at least two geological windows [according to the latest tectonic interpretation by Tollmann (1963) there are three windows]. The Pennides form the main body of the western Alps together with the autochthonous massifs. These windows (Engadin, Tauern and Wechsel) are set into the Austrides (east Alpine nappes) which have been moved as a huge nappe fold for about 165 km across the lower tectonic structures. From the morphogenetic point of view, it is of little importance if this thrust fold was the result of a tangential push—as advocated by Argand (1916) and Staub (1924) for the western Alps or by Kober (1931) for the eastern Alps—or was brought about by convection currents—as conceived by Ampferer (1906)—although the latter view is consistent with many of today's ideas about plate tectonics.

Tectonics structure partly determines the present shape of the Alps (*see* Fig. 10.10). The limits of the Tauern window are visible in the west at the Brenner Pass and in the east at the Katschberg Pass. In addition, there is the Alpine Dinaric scar, whose recent activity is documented in the subsidence of the Gail valley resulting in peat formation there. The main lithological boundaries, as between the Grauwacken zone (consisting of Palaeozoic rock sequences) and the northern Calcareous Alps (consisting of mainly Triassic limestones and dolomites) are emphasized by the creation of the north Alpine longitudinal valley (Inn–Salzach–Ennstal). The overthrust fold structure of the eastern Alps represents a large-scale folding (*Grossfaltenwurf*) [see Penck (1924)] in the morphological sense. The north Alpine longitudinal valley takes on the character of a large downfold between the large upfolds of the northern Calcareous Alps and the central Alps.

Due to the small scale of the maps in Figs. 10.9 and 10.10, only the major structural units can be shown. Tectonic influences, nevertheless, become evident in even the smallest and youngest forms: for example, in Marchfeld (east of Vienna), a uniformly deposited alluvial fan has been broken up by tectonic movements into several subsident troughs [see Fink (1969)].

Contemporary uplift of the central parts of the mountain ranges—namely, the Höhe Tauern—is of prime importance for the present relief. This uplift has been confirmed by precise levelling along the Tauern railway (between the Salzach valley and the Möll valley) [see Senftl and Exner (1973)] and amounts to 70 mm over 65 years in the area of maximum uplift. If we assume a constant uplift of the same rate (approximately 1 mm per year) this means that 4–5 km of rocks must have been removed by denudation since the Pliocene (i.e. after the end of Pontian). This is only valid, however, for parts of the Höhe Tauern. For the area east of the Katschberg, where the old planation surfaces, are situated one level lower than the comparable surfaces of much smaller dimensions in the Höhe Tauern, the uplift must have been far smaller.

10.2.2 Tertiary Landform Evolution

The most important geomorphological phenomena in the eastern Alps belong to the northern Calcareous Alps. Being part of the Austrides (i.e. the east Alpine nappe) they extend as a compact rock mass from the upper Rhine valley as far as the Thermenlinie (the western boundary of the Vienna Basin). They continue towards the east, buried by a huge thickness of Tertiary sediments and emerge again in the Lesser Carpathians. They serve as the reservoir rock for oil and natural gas in the Vienna Basin. The same is true for the autochthonous Mesozoic which is covered by Tertiary sediments and is superjacent itself to the dipping crystalline rocks of the Bohemian massif.

A classification of the northern Calcareous Alps is best given in terms of tectonics and lithology: from west to east, there are first of all mountain chains, followed by massifs and then plateaus. From the Enns eastwards, limestone high mountains and dolomitic Pre-Alps are characteristic. The plateaus, starting at the Salzach and ending at the Vienna Basin (the Rax reaching 2000 m there), represent extensive old-age planation surfaces with little relative relief. They were karstified after their formation, were raised and therefore are fossilized today. The question of the development of such uplifted peneplains became a field of geomorphological research when Simony in 1851 found small patches of gravels originating from the central Alps. Neither these gravels nor the age of the forms have yet been agreed. As regards their age, Fig. 10.11 portrays the development of the Austrian Alps and its foothills and basins, through periods of changing climate, since the late Tertiary. Several opposing concepts have been put forward. The supporters of the idea of older landforms draw their conclusions from the evidence of sedimentation in a molasse trough (i.e. the north Alpine foreland), the sediments being actively derived until the Savian phase from a source region in the central Alps, but later only from the flysch zone. Therefore, the formation of these old-age landscapes (in the Calcareous Alps), created by consequent rivers flowing north and south from the central parts of the Alps, would be of Miocene age [see Tollmann (1963)]. The supporters of the idea of a younger age for the landforms concluded from the large sediment volumes—Pannonian sediments reach a thickness of several kilometres—that the oldest landforms could only have been created in the Upper Sarmatian [see Winkler-Hermaden (1950)]. Here, no distinction is usually made between the crystalline and the limestone mountain ranges.

Rather divergent concepts exist in respect of the development of these old land surfaces. There is a concept of Lichtenecker (1925), which regards the old land surfaces as Raxlandschaften developed in one phase, tectonic dislocations being responsible for the different elevations. An opposing view is that the old surfaces situated at different elevations represent a tier of huge terraces (rock floors); a nomenclature has been devised for the various levels. Seefeldner (1926) has divided the relief of the Calcareous Salzburger Alps into the Hochkönig, Tennen and Gotzen levels. This terminology has been adopted in other areas.

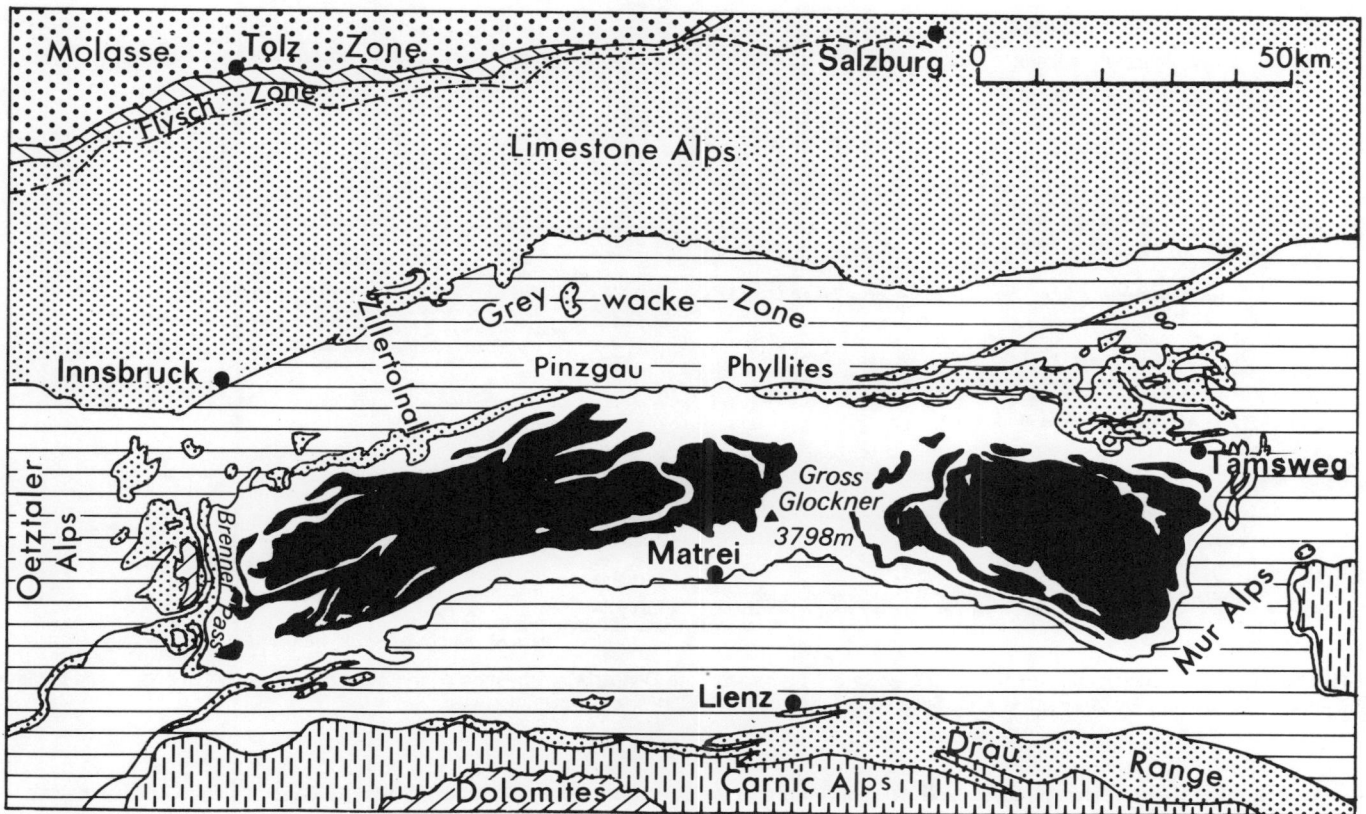

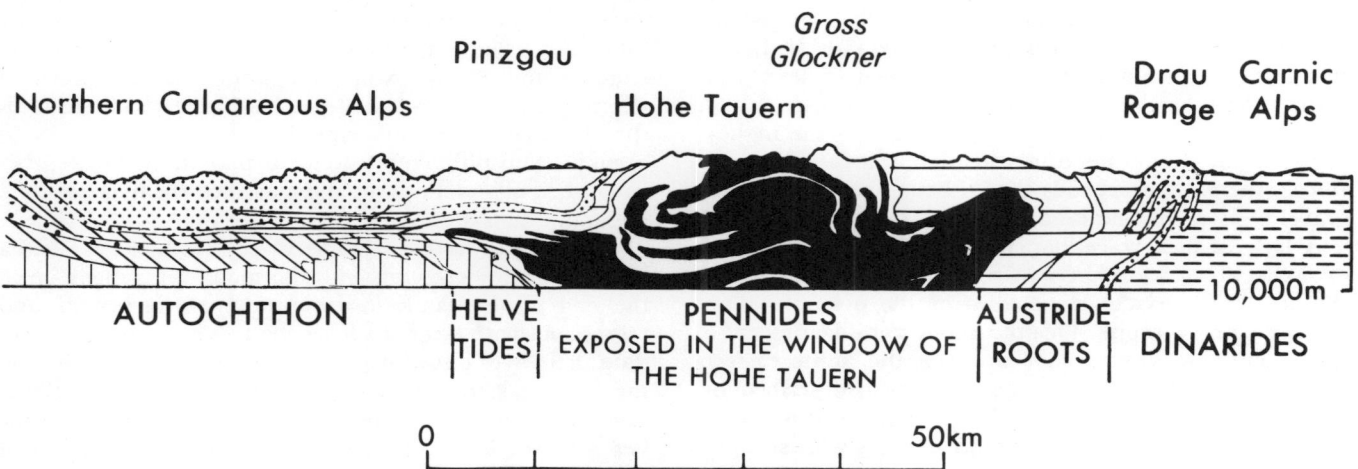

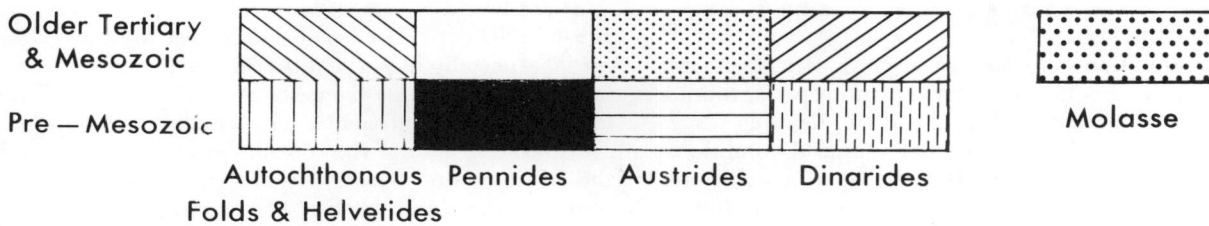

Fig. 10.10 Geological map and generalized section of the eastern Alps, showing the Tauern window.

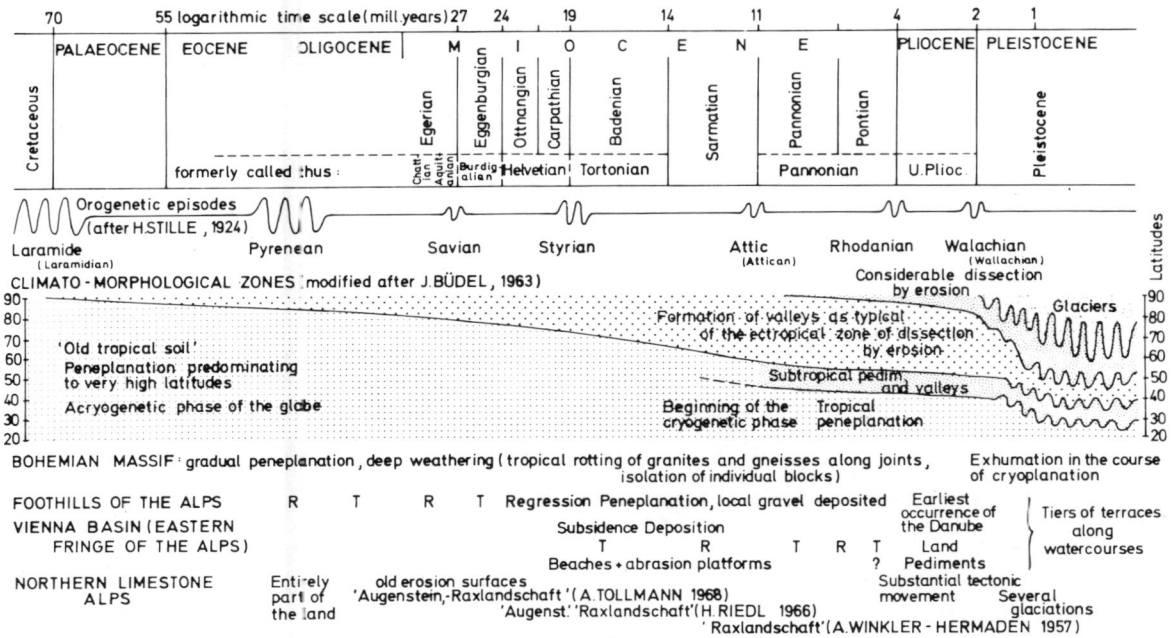

Fig. 10.11 Geomorphological evolution of the eastern Alps through the Cenozoic; correlation of the main stages of the Cenozoic and climatic change [after Büdel (1963a), Riedl (1966), Stille (1924), Tollman, (1968), Winkler-Hermaden (1957)].

In Fig. 10.11, the terms Raxlandschaft and Augenstein-landschaft should strictly only be used in the sense of Lichtenecker (1925) to imply a single-phase evolution of a hypothetical landscape, but in this summary, the author is only concerned with the different views about their age. Lichtenecker (1925) assumed that an Augenstein-landschaft was the predecessor of a Raxlandschaft, and that this Augensteinlandschaft was completely destroyed. Solar (1964), on the basis of palaeopedological studies on the Rax plateau, expresses the view that, in the highest parts of the Rax, the Augensteinlandschaft still exists since, otherwise, relict soils should have developed on the gravel beds of this old land surface.

A large-scale terrace-like staircase has been shown to exist east of the Katschberg by Spreitzer (1961a, b), where the old land surfaces slope towards the present valleys like a Piedmonttreppe. This piedmont step-like descent of the mountains also occurs towards the south-eastern forelands. Impressive examples have been described by Winkler-Hermaden (1955) for the Wechsel and Koralpe. His correlations of old Alpine land surfaces with surfaces in the foothill regions, using cols as evidence, is rather hypothetical. Therefore, the age of the landforms here, as in the case of the limestone plateaus, cannot yet be determined.

Another question still unanswered is where in the sequence of levels is the change from subtropical, humid landform development in the Miocene, leading to a peneplain, to surfaces formed under semiarid conditions in the Pliocene and oldest Pleistocene? Very simply, one could assume that this change took place above the 400-m planation surface at the mountain front, with funnel-like re-entrants penetrating into the mountain ranges. Two facts complicate an exact analysis of the landforms.

1)   The northern front of the Alps has been overridden by glaciers west of the Enns. Moraines and gelifluction have removed any correlative sediments that might have rested on the surfaces of the foothill region. Furthermore, the lithological boundary between the flysch and the lime-stone, running approximately parallel to the northern Alpine front, induces a step-like landform which obscures the climatogenetic landforms. Nevertheless, Fink (1969) attempted to differentiate several planation surfaces between Erlauf and Traisen, and classified them as part of a piedmont step sequence.

2)   At the eastern front of the Alps (including the Vienna Basin), the sea cut cliffs into the mountain ranges up to the end of the Pontian. Important transgressions and regressions with large sea-level changes make it difficult to date their age, but gravels of the early Badenian rest on the south-eastern edge of the Bohemian massif at elevations of 517 m, while brown coal formed before that time lies near sea level today. Subaerial landforms were then created on the partly preformed marine platforms.

There has, however, been some success in proving the age of Pliocene–Pleistocene landforms in some isolated areas, especially in parts of the Alpine front that have not been shaped by large rivers, because there it is possible to find correlative sediments. The best area for this is the north-eastern spur of the central Alps with basins such as the Mattersburg Basin, the Oberpullendorfer Basin in central Burgenland (*see* Fig. 10.12) and the Styrian Basin. While Riedl (1977) assumed humid conditions for the landform development, Fink (1975) believed that a semi-

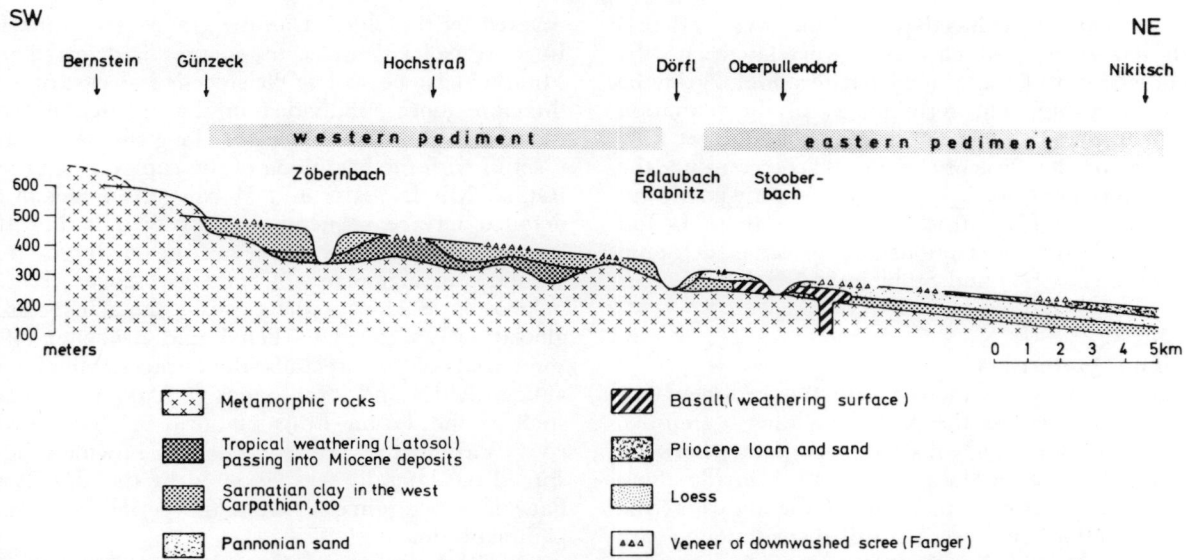

Fig. 10.12 Schematic cross-section showing development of pediment surfaces and associated deposits in the eastern Alps.

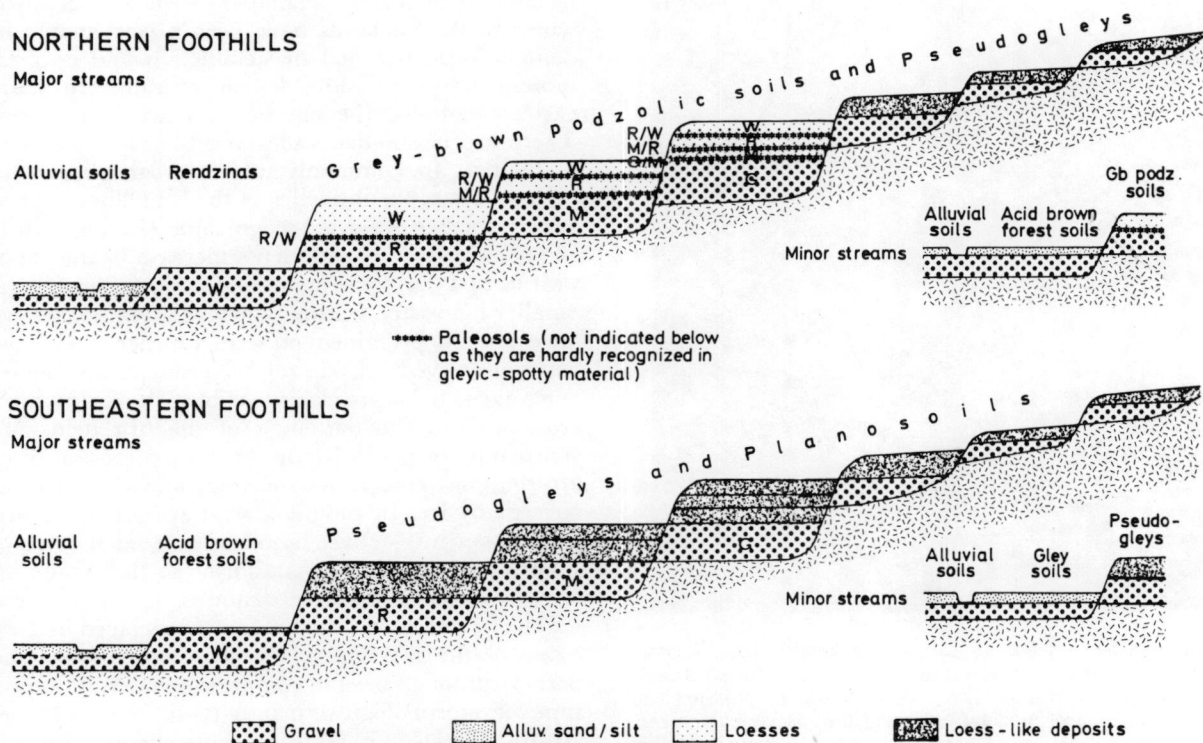

Fig. 10.13 Terrace sequence and associated soils in the east Alpine foothills.

arid climate at that time has deposited the cover of debris on nearly flat surfaces which may be correlated with the cross-bedded sand of the Hungarian terminology in the bordering Kisalföld. The terminology of the planation surfaces in front of the mountain ranges is not yet fully clarified. The term pediment is not quite correct, since the slope of the surfaces is too low and they are only partly cut in hard rock (*see* Fig. 10.12). However, there is full agreement with the descriptions of the rañas in Spain given by Fischer (1977) and Stäblein and Gehrenkämper (1977).

10.2.3   The Quaternary

Quaternary terraces are incised into the pediments or pseudopediments around the Alps, providing a step-like descent to the present valley floor. The boundary between Pliocene and Pleistocene does not exactly coincide along the large rivers as, for example, around Vienna where the Tertiary planation surfaces (so-called pediments) are

Fig. 10.14 V-shaped section of part of the lower Ötztäl, Austria. Above and below this section, the valley displays the more usual basin or U-shape of glacial erosion. The V-shaped portions are linked to increases in longitudinal gradient, affecting the glacial erosion processes, but part of the valley cutting may be attributable to subglacial meltwater erosion. (Embleton)

covered by the oldest Danube gravels [see Fink (1973)]. In more isolated areas, the change is deferred until the Middle Pleistocene. The Pleistocene staircase of terraces is therefore more subdivided in the vicinity of the larger rivers than in isolated areas or along smaller rivers where it set in with the beginning of the Alpine glacial sequence (Günz, Mindel, Riss and Würm). A model of a more detailed terrace sequence is portrayed in Fig 10.13. A stratigraphic classification can be deduced from the regular arrangement of gravels, cover strata and palaeosols. Such was the case for the scheme of Alpine glaciation developed by Penck and Brückner (1909). In some parts of the forelands, the terraces extend across the valley divides and form extensive flats or low platforms, such as the Traun–Enns platform or that between the rivers Pielach and Traisen. The way in which the gravels spread out laterally was favoured by the already existing flat surfaces originating from the period of planation or pedimentation.

The foothill region of the Austrian Alps can therefore be subdivided into three large landform units dominated by moraines, terraces and hills (*see* Fig. 10.9); the morainic areas may be further subdivided according to the ages of the terminal or ground moraines. The landforms of the older glacials (Riss, Mindel and Günz) are blurred by gelifluction, mostly bear an aeolian cover and do not exhibit lakes or bogs in the depressions, whereas the landforms of the Würm glaciations are fresh, without any aeolian cover, and depressions often contain water or bogs.

The hilly region comprises areas farther away from rivers where the flight of terraces shown in Fig 10.13 is lacking. The former planation surfaces occupying the whole of the forelands have been dissected into hills by denudation, smoothed by gelifluction and covered by a more-or-less continuous loessic stratum. In the southeastern foothills, the old surfaces are better preserved. There are cirque-like valley heads (*Tobel*) eroded headwards that are commonly subject to landslips.

Everywhere the influence of the cold climate prevailing during the Ice Ages is recognizable (*see* Figs. 10.14 and 10.15). The aeolian sediments increase in thickness from west to east (i.e. from the more humid to the drier areas). Smaller landscape features, such as asymmetrical valleys, are especially prominent on terraces where they have been better preserved. Also in the high mountains, the terraces of ice age activity are clearly visible. The distinctive valley cross-profiles, the flattening of the firn fields and the sharpening of peaks by the headward erosion of cirques are the most conspicuous examples in the area formerly covered by ice. In the periglacial area of the mountains, gelifluction was active, but only succeeded in reworking slightly the pre-existing landforms so that many of these are preserved, although denuded of their correlative sediments and soils. The latter were replaced by a regolith cover of the last glaciation. The succeeding Holocene period, although brief in comparison, does not represent a time of morphological inactivity in the Alps (*see* Fig. 10.16). The steep relief, the Alpine climate and intensive human activity have initiated mass movements such as

Fig. 10.15 The Geisberg glacier, Ötztäler Alps, showing the extent of ice recession and down-wasting since the beginning of the century. (Embleton)

Fig. 10.16 The Pasterze glacier, east of the Gross Glockner, Austria. (Embleton)

mud flows, while in the foothill region the main changes are brought about by floods. Around the north-eastern spur of the Alps, the wind is another element in shaping the landscape, at least for the small-scale forms.

10.3   Southern Alps

The southern Alps are separated from the rest of the Alpine chain by the Insubric Line, interpreted as an old persisting feature, active from the Palaeozoic to the present-day. The geodynamic model proposed by Italian geologists [e.g., Castellarin and Vai (1982)] is based on Tertiary—mostly Neogene—northward sinking of the continental lithosphere at the north-eastern border of the southern Alps, below the eastern Alpine nappes already generated and uplifted. The consequence was a significant shortening of the southern Alpine upper crust and its cover, with the formation of a general wedge-shaped structure pointing mostly towards the south. The shortening was relatively less in the central part (approximately along the meridian of Bolzano), where the southern Alps develop their maximum breadth.

In the southern Alps overall, the Permo-Mesozoic sedimentary cover is very extensive, with dominating calcareous lithologies, although there are also other important formations that have a very varied lithology. The structure is also varied and complex, characterized by folds, faults and overthrusts. A thick Permian formation of acid volcanics outcrops in the basin of the Adige River, with a faulted-block tabular structure. The southern Alps also contain an important batholith composed of Alpine granodiorites and tonalites (the Adamello–Presanella massif).

Among the main fracture lines crossing the southern Alps, the Giudicarie Line deserves special mention. Running obliquely from south-west to north-east, it extends from Lake Idro in the Pre-Alps of Lombardy to Merano.

From the point of view of the relief, the southern Alps include an internal belt with high mountain features and an external belt corresponding to the Pre-Alps. The highest peaks exceed 3000 m in the internal belt [Cima Presanella (3554 m), Monte Marmolada (3342 m)] and 2000 m in the Pre-Alpine belt. Since the whole area is intersected by a network of deep valleys, relative altitude is very great, around 1000 m. An important series of longitudinal valleys is aligned almost exactly along the Insubric Line, and segments of other valleys are also arranged along the Giudicarie Line. The depression of Lake Garda and the low valley of the Sarca River coincide with an important tectonic depression parallel to the Giudicarie Line.

Although the separation between the internal groups and the Pre-Alpine belt is not marked by clear-cut differences, the Pre-Alps are characterized by the presence of numerous limestone plateaus between 1000 and 1600 m in altitude, divided by narrow deep valleys. In the Pre-Alps, in the Veneto and Trento areas, tectonic forms are particularly evident, such as fault scarps, flexure scarps, anticlines, synclinal valleys, monoclinal ridges and slightly tilted plateaus, following the stresses imposed on the Jurassic and Cretaceous limestones and on the marly or sandstone formations of the Tertiary. The latter outcrop in some synclinal basins and along the external border of the limestone Pre-Alps, forming characteristic hogbacks.

Many mountainous groups of the internal southern Alpine belt are similar to those of the high eastern and western Alps; on the other hand, some aspects of the Dolomitic Alps, even outside the boundaries formed by the Piave and Adige–Isarco Rivers, are very distinctive. The great variety of structural forms in the Dolomites is due to the great thickness of some resistant calcareous formations (*see* Fig. 10.17), which represent repeated phases of deposition during the Triassic. Other formations with different lithological features are intercalated between them vertically; carbonatic reefs, marine basins containing terrigenous sedimentation and submarine vulcanism coexisted [see Leonardi (1967), Castellarin and Vai (1982)]. The typical Dolomite denudation forms include cuestas; the massifs are generally isolated from each other, with large walls and towers (*see* Fig. 10.18). Detailed examination shows special structural forms, some exhumed and inherited from the Triassic, others connected to the Alpine folding, and yet others connected to gravitational tectonic phenomena.

The part played by neotectonics in the morphological setting of the eastern sector of the southern Alps has recently been re-examined [see Zanferrari *et al.* (1982)]. Continuation of orogenesis during the Pleistocene is proved by the folding and uplift of Upper Pliocene marine deposits on the margin of the Pre-Alps and by other, mainly geomorphological, evidence. It is believed that uplift of the Pre-Alps during the Quaternary amounted to 0.5–0.7 mm per year on average, and may have been even greater in some internal areas. Some tectonic lines are believed to be still active, as shown by repeated earthquakes in this area, the best known being those that took place in Friuli between May and September, 1976 as reported by Martinis (1977).

Some large-scale landslides due to postglacial collapse

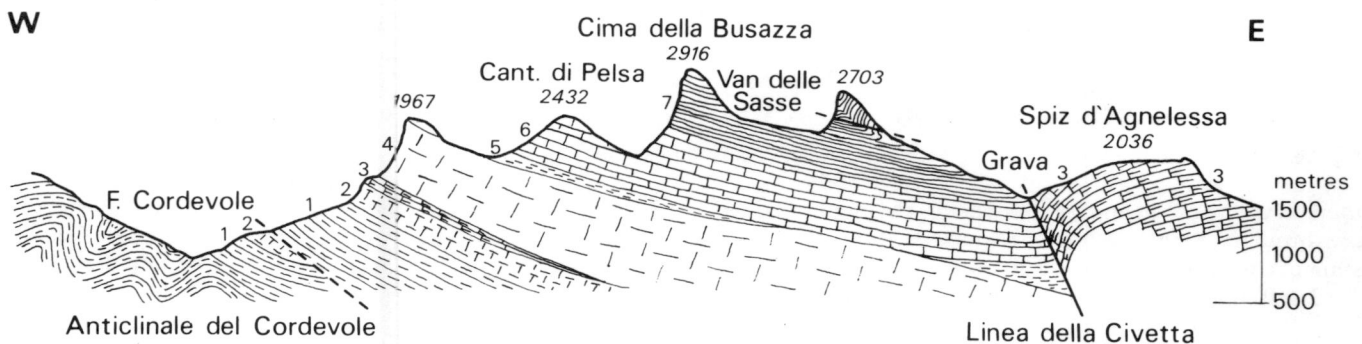

Fig. 10.17 Geological cross-section of the Civetta Group in the Dolomites [after Colacicchi in Leonardi (1967), redrawn]. This is an example of structure-controlled dolomitic morphology, with very high rock walls in dolomite and limestone formations. Note also the thrusted tectonic unit at the top, an example of *Gipfelfaltung*, not completely clear in origin. (1) Werfenian and Lower Anisian, (2) Serla Dolomite, (3) Anisian and Livinallongo strata, (4) Schlern Dolomite, (5) Raibl formation, (6) Hauptdolomit (Upper Trias) and (7) Lias limestones.

Fig. 10.18 An example of high mountain karst morphology in dolomitic rock: the plateau of the Pale di San Martino (Dolomites, southern Alps) at an altitude of approximately 2600 m. In the background, the Cima di Vezzana, 3193 m. (Castiglioni)

(*see* Fig. 10.19) are also now believed to be partly the consequence of seismotectonic phenomena: for instance, the landslides that occurred north of Lake Garda, where various methods of dating agree on an age of 2000–3000 years BP [see Sauro and Meneghel (1980)].

The presence of old land surfaces at different altitudes has been reported several times, although the various relics often appear conditioned by special structural situations. These surfaces may have been uplifted in different ways in different places following their formation. The problem of their interpretation is closely linked to the interpretation of the fluvial erosion terraces along the valleys; the fact that the valleys themselves have been intensely remodelled by glacial erosion means that any conclusions reached are necessarily hypothetical.

Progress has been made in the study of the karst plateaus, both at medium and high altitudes. Some of the plateaus are connected to rather old morphogenetic phases, probably of Lower Pleistocene or older age [see Sauro (1981)]. During the neotectonic phases, the plateaus were dismembered by faulting. Moreover, during the Pleistocene cold periods, periglacial and glacial pro-

cesses modified the karst surface features, as well as the slopes and valley sides.

Neotectonic phenomena are believed to have been partly responsible for some hydrographic changes, but the complicated valley network certainly involves the direct or indirect action of the Pleistocene glaciations. For example, the design of the valley depressions, mostly occupied by lakes, in the Pre-Alps of western Lombardy and in the canton of Ticino in Switzerland may very probably be explained by the erosive action of some glacial tongues, guided by branching river valleys and pre-existing passes.

The great depth of the largest lakes (300–400 m)—in some places their floors lie below sea level—has always been a subject for discussion. Numerous data are now available on the depth of the rock floor beneath the lakes and the lacustrine deposits filling the larger valleys. It is believed that, during the Pleistocene, the valleys were filled by sediments and then re-excavated several times, alternately by the rivers and by the large glaciers (*see* Fig. 10.20). It is probable that the maximum deepening was achieved by the glaciers during the penultimate glaciation. It is also important to note that, even today, many

Fig. 10.19 The Vaiont landslide near Belluno (Italy) and the notch connecting the Vaiont valley and the Piave valley. (Note the landslide scar strictly related to the dip of the strata, the structural control of the main forms (synclinal) and their glacial remodelling.) This photograph was taken in October 1963, a few weeks after the occurrence of the landslide. Falling into the Vaiont reservoir, it produced an enormous wave which destroyed many settlements in a few minutes. The dam, within the narrow erosional notch, has not been destroyed. (De Nardi)

Fig. 10.20 The Drau valley near Villach, showing steep bedrock sides and a flat floor underlain by a thick alluvial infill. (Embleton)

points of interpretation are still controversial, although this is a classic problem of Alpine morphology [see Nangeroni (1980)].

Another problem, that of possible glacioisostatic movements in the Alpine chain, has not attracted so much attention from research workers, although some aspects have been discussed, especially as regards the important phenomena of alluvial filling in the internal valleys, of interstadial Würmian age, followed by recutting by the rivers.

Important modifications of the landscape took place in relation to the melting of the glaciers or to their readvances during the phases of the late Würm. It should be noted that, according to Mayr and Heuberger (1968), the well-known Schlern stadial, named after a mountain near Bolzano, is no longer recognized as valid.

On the slopes, large-scale mass movements took place both during the late Würm and the Holocene, different mechanisms being controlled by local lithological and morphological situations. Study of these movements has helped greatly in the understanding of past landslides. Present-day slope processes are linked to the strong gradients and to specific geological situations, as well as to other aspects of the Alpine environment which create unstable conditions with mass movements, mass transport and avalanches.

10.4 Po Plain

As a geomorphological unit, the Po plain is clearly defined by the margins of the mountainous regions of the Alps and the Appennines. From a genetic point of view, the shallow northern part of the Adriatic Sea may be considered as the submerged continuation of the same plain. The boundary with the surrounding mountainous regions is determined by various geomorphological and structural situations, reflecting the transition between strongly uplifted areas and the tectonically lowered area corresponding to the plain.

On the Alpine side, the morphological contrast along the mountainous boundary is generally very marked, but tracing it is made difficult by the presence of deep lakes and hills composed of glacial deposits. A further complication is caused by the lengthening into the plain of the hill groups of the Berici Mountains and the Eugane Hills. This is an area dislocated by faults, in a position of a 'structural high', composed partly of old subvolcanic forms (still very prominent, although volcanic activity ceased during the Oligocene). As regards recent geomorphological history, the contact between hills and plain in this sector is dominated by the tendency of the alluvial deposits of the plain to penetrate into the valleys and bury the marginal features which often appear as isolated hills. The Berici Mountains and the Eugane Hills separate the eastern part of the plain (the Venetian plain) from the true Po plain.

On the Appennine side, there is an intricate interdigitation of ridges, erosional valleys and plain, although it is easy to see a generally linear trend to the Appennine margin.

Buried geological structures similar to those of the Appennine chain continue under the surface of the plain; the present-day margin of the relief is therefore controlled principally by neotectonic uplift on one hand and by subsidence on the other.

A special position is occupied by the Turin and Monferrato hills, which rise to 250–500 m above the Po for more than 60 km. Due to their aspect and geological composition, these hills must be considered as belonging to the Appennine system, but they are well defined on the west and east by two important southern subbasins of the Po plain (the Cuneo plain and the Alessandria plain, respectively) and continue southwards into a plateau which, before being uplifted and dissected by erosion during the Pleistocene, was part of the Po plain itself. A remnant is preserved near Poirino.

The forms dominating the plain are due to fluvial and glaciofluvial sedimentation. In the plain itself, an important hydrographic and morphological boundary dividing the higher belt from the lower is the spring line called the *zona dei fontanili* or *zona delle risorgive*. This is substantially a transition belt between the coarse sediments (mainly gravels) of the high, permeable part of the plain, and the fine sediments (mainly sands, silt and clay) of the lower part. Towards the sea, fluviolacustrine forms are associated with fluvial depositional forms; lastly come fluviomarine and coastal forms. With regard to the forms of Pleistocene glacial and glaciofluvial deposition of the Alpine border, mention must be made of the numerous morainic arcs, which sometimes protrude for tens of kilometres in front of the Alpine valleys. Glacial chronology will not be dealt with here, but it is worth mentioning that glaciomarine sediments of late Pliocene age have been found at the base of the Pleistocene glacial deposits near the edge of the morainic arc at Ivrea [see Carraro *et al.* (1975)].

The whole Po plain, with the adjacent Venetian plain, represents a predominantly sedimentational area; sedimentation was mainly marine during the Quaternary. Different phases of this infilling were produced by the interaction of the following processes: (1) lowering of the basement of the plain, a movement which was not uniform in space and time, (2) processes of tectonics, degradation and erosion in the mountains, yielding clastic sediments and (3) eustatic marine cycles. The fact that from the Pliocene to the Lower Pleistocene neritic marine deposits predominate, with total thicknesses of several kilometres, indicates that the subsidence gradually allowed the sea to penetrate into this gulf, in spite of active sedimentation. Using radiocarbon analyses on peaty samples found in drillings in the area around Venice, the rate of the subsidence has been evaluated for the period between 22 000 and 40 000 BP as 1.3 mm per year; between 18 000 and 22 000 BP, it seems to have been four to five times greater than previously [see Bortolami *et al.* (1977)].

10.4.1 Subsurface Structure
On a general scale, the great gulf of the Po plain came into being during the late Miocene and the Lower Pliocene at

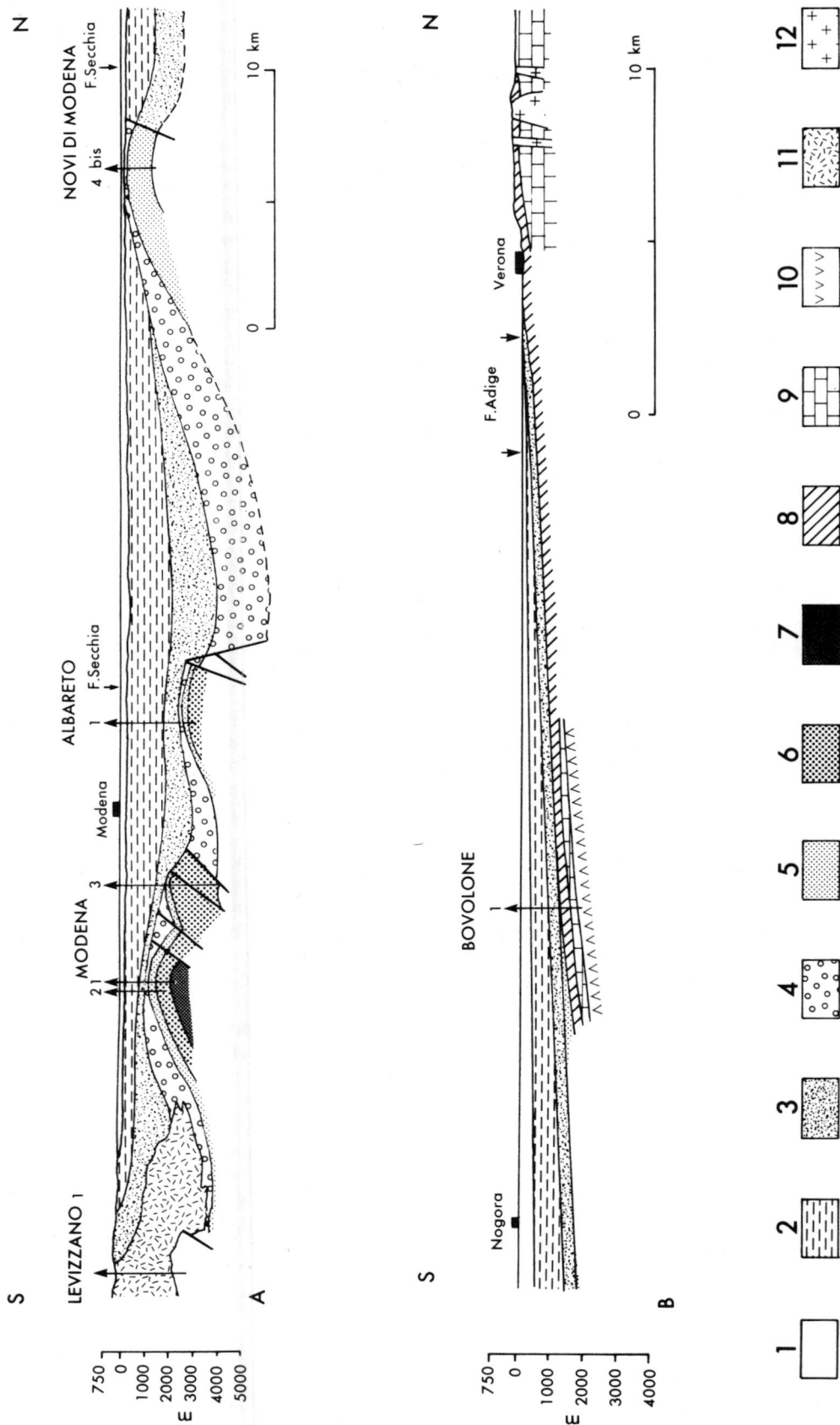

Fig. 10.21 Geological sections along the Modena–Verona line. Note the asymmetry of the structure beneath the Po plain, comparing the faulted folds in the zone near the Appennines (A) and the regularly dipping strata under the Alpine piedmont (B) [after Agip Mineraria (1959), simplified in Pellegrini (1979)]. The names in capitals and the numbers indicate drillings made by Agip Mineraria. (1) Alluvial cover, (2) marine Quaternary formations, (3) Middle–Upper Pliocene, (4) Lower Pliocene, (5) Upper Miocene, (6) Tortonian, (7) Helvetian, (8) Eocene, (9) Cretaceous, (10) Jurassic, (11) allochthonous complex and (12) crystalline rocks.

the time of the uplift of the nearby Alpine and Appennine chains. However, tectonic movements were still active during the Upper Pliocene and the Quaternary, and caused new large-scale deformation. As a reference point, the map of the base of the Pliocene prepared by Agip Mineraria (1959) can be used. The tectonic complications shown in this map correspond only in part to pre-Pliocene deformation; most are due to movements during the Pliocene and continued into the Quaternary. The Pliocene sediments are also folded and faulted, as are those of the Quaternary, although to a lesser extent.

The western part of the plain contains two large synclines, generally running east–west but with an arcuate trend: the Piedmont syncline is situated north of the Monferrato hills and the Asti syncline south of them, the latter being subdivided into two depressions (Saluzzo to the west and Alessandria to the east). The Piedmont and Saluzzo synclines are strongly asymmetrical and are associated with large faults on their southern sides.

The buried tectonic belt of the 'pede-Appenninic folds' extends east of Pavia. This is a fold belt running parallel to the northern part of the Appennines and is often faulted; it is more than 50 km wide near Piacenza and in eastern Emilia–Romagna, and about 25 km wide inside the Parma sector. The folds are normally asymmetrical. Two distinct folding phases may be recognized: one between the Miocene and the Pliocene, the other before the middle Pliocene. In reality, the fold belt is interrupted and complicated by north-east–south-west-trending faults, with a predominantly horizontal displacement.

Some of the anticlines included in this fold belt have special geomorphological significance because they have formed—or, in rare cases, still form—smooth ridges emerging from the depositional plain. One emerging hill corresponding to a large 'structural high' is the small, isolated hill of San Colombano, slightly north of the Po east of Pavia. Another minor example is the Casalpusterlengo terrace, again in Lombardy north of the Po. It is evident that some ridges, including those now buried, may have been subjected to planation and/or to pedogenesis during the periods in which they emerged. The Turin and Monferrato hills constitute another anticlinal structure, in many ways similar to the others mentioned here; apart from the size, the principal difference lies in its very pronounced tectonic uplift, only partially subdued by erosion.

North of the Piedmont syncline and the pede-Appenninic folds, lies the pede-Alpine tectonic belt, which is much simpler in structure than the preceding units. The Pliocene and Quaternary layers dip regularly from north to south with only occasional folds.

Overall, the deep structure of the plain appears to be asymmetrical. In order to show this, Fig. 10.21 shows two parts of a geological section passing south to north through Modena and Verona. The southern segment (A) begins near the Appennine boundary, and then cuts across some large fold faults of the pede-Appenninic belt. The northern segment (B) ends at the right against the Pre-Alpine relief, and shows part of the pede-Alpine tectonic belt.

10.4.2   Age of Surface Forms

Many studies have been devoted to distinguishing the different alluvial formations that compose the surface of the plain. Evidence of the terraces supplies one criterion for delineating the different formations but correlations between one surface and another, although generally clear, are not unanimously agreed.

As might have been expected, in the foothill belts, the plain is more differentiated from the altimetric, geomorphological and geological viewpoints, since its evolution was directly or indirectly influenced by the events involving the mountainous areas behind. On the Alpine side, repeated glaciation brought the fronts of the larger glaciers some tens of kilometres forwards from the exits of the valleys, and the deposition and erosion phenomena seen in the alluvial formations were largely controlled by the alternating glacial/interglacial phases, as well as by the interplay of tectonic movements. Starting from the morainic hills, the outwash plains of coarse gravels pass, towards the centre of the plain, into finer sediments in which attribution to a glaciofluvial facies rather than a fluvial one is of little significance.

Age determinations based on palaeosols and loessic covers have given good, although not definite, results. The surfaces covered by *ferretto* (a deep ferrallitic palaeosol which was widespread on the older higher surfaces, partly dissected) are believed to be of Lower to Middle Pleistocene age and are mostly older than the Mindel–Riss interglacial. The dividing lines between the surfaces belonging to the Riss or to the various intervals of the Upper Pleistocene are controversial. Most recent studies have concerned the palaeosols developed in the interstadials of the last glaciation, as well as in the previous interglacials [see Cremaschi and Orombelli (1980)]. The fact is that, in the Po plain, the cold climatic phases brought about morphogenetic phenomena that are less clear-cut and less widespread than in other parts of Europe; loessic deposition, generally not thick, is considered indicative of arid, cold phases.

In Piedmont and Lombardy, the Holocene parts of the plain are generally represented by belts 3–10 km wide along the principal rivers; along these zones, the Holocene deposits lie several metres below the main level of the plain. However, there are some cases in which the Holocene alluvium was deposited directly on top of older deposits so that no morphological break between the exposed Pleistocene and Holocene surfaces may be observed. In the case of the Tanaro and its tributary the Stura di Demonte in the Cuneo plain in southern Piedmont, the Holocene fluvial trenches are about 100 m deep with respect to the general level of the plain, as a consequence of fluvial diversion.

From Piacenza, the Holocene Po plain tends to widen eastwards and, finally, the whole of the central and eastern part of Emilia–Romagna, the extreme south-eastern part of Lombardy and southern Veneto together form a large, approximately triangular area of Holocene sedimentation, in which there are alternating true fluvial deposits and backswamp deposits once marshy, lacustrine or lagoonal; all these basins are locally called *valli*. The maximum light of marine transgression, although not defined everywhere, is recognized from the distribution of the marine or brackish sediments found buried under recent or subrecent fluvial deposits. Near Venice, it seems that the Holocene sediments covered an area that showed various irregularites in relief created during the phase of pre-Flandrian regression. Near Ravenna, the thickness of the lagoonal or beach sediments of the Holocene is approximately 20 m.

10.4.3   Drainage

In order to examine the drainage network and its relationships with the deep structure, the plain will be divided into four sections: (1) Alpine drainage, (2) Appennine drainage, (3) the course of the Po from Turin to Parma and (4) the course of the Po from Parma to the sea and the other rivers that are arranged parallel to it.

1)   The rivers coming from the Alps generally follow the slope directions created by fluvial, glacial and glaciofluvial accumulation derived from the mountains behind. This framework also includes the system of rivers converging from the south and west into the Cuneo plain (southern Piedmont) surrounded by the most westerly part of the Alpine arc terminating in the Ligurian Alps. The same may be said of the other rivers approximately perpendicular to the boundary of the Alps, from the Piedmont to the Lombardy and Veneto–Friuli sectors. This drainage net is generally in accord with the deep structure of the pede-Alpine tectonic belt. Only in the central sector, between Pavia and Cremona, do the Alpine rivers transgress some buried pede-Appenninic folds. It may also be observed that the converging trend of the Alpine rivers south of Turin is related to a subsidence area between Saluzzo and Carmagnola (*see* Fig. 10.21) where the base of the Pliocene sediments sinks to a depth of 2 km [see Agip Mineraria (1959)]. As regards the Oglio, its south-easterly trend may be explained by partial coincidence with a deep syncline, in which the base of the Pliocene lies at −4.5 km. Other features of the river network, with diverging or converging patterns, may be explained in terms of the large alluvial or glaciofluvial fans, such as those typical of the Veneto and Friuli, although they may also depend partly on recent differential tectonic movements. For example, the convergence of many tributaries towards the Livenza seems to depend on a relative sinking of that area at the boundary between Veneto and Friuli. In the case of the Piave, which now curves sharply east, a stretch of its former valley that crossed the anticline of the Colle del Montello is well preserved; it functioned as an antecedent valley for most of the Pleistocene until the Würm, during the uplift of the anticline, as shown by the deformed, arcuate terraces.

2)   The Appennine rivers will be discussed in turn, beginning in the west. First, the Tanaro, with its left-bank tributaries, is in reality an Alpine river that was diverted into a new eastward course during the last glaciation and now reaches the Alessandria plain by crossing the hills of Asti (*see* Fig. 10.22). These hills constitute an area of relative uplift, between two areas of subsidence and drainage concentration during the Pleistocene: the north-

western area (*see* p.258) of Saluzzo and Carmagnola, and the Alessandria area to the east. Study of the fluvial terraces shows that the Tanaro, which before diversion was directed towards Carmagnola and Turin, had already gradually lowered its bed and undergone a slow lateral shift to the east; at the same time, lateral erosion took place at the foot of the hills on its right. The question as to whether this tendency was due more to tectonic tilting or to sideways pushing by the left-bank tributaries is still open.

In the Emilia–Romagna plain, the rivers leaving the Appennine boundary form a gently inclined alluvial plain that masks the deep structures, arranged parallel to the boundary itself (*see* Fig. 10.21). Great attention has recently been given to the identification of plain forms controlled by neotectonic movements, both on the hypothesis that the latter represent the continuation of movements responsible for the buried Pliocene–Pleistocene structures and on the hypothesis that they followed a different trend. The deformation of the middle Pleistocene fluvial terraces near the Appennine boundary in many cases shows the present-day activity of the buried anticlinal structures [see Cremaschi and Marchesini (1978), CNR (1979)].

3)   The main river, running at the foot of the Turin and Monferrato hills, and defined by active faults since the Miocene, shows the effect of the thrust exercised on the river by the heavily loaded Alpine tributaries, and appears to have shifted south with respect to the axis of the Piedmont syncline. In two other tracts, however, the river flows almost exactly in correspondence with two zones of deep subsidence: between the confluence with the Tanaro and that with the Ticino, and then between Cremona and Parma [see Agip Mineraria (1959)].

4)   Below Parma, the plain widens to include the old branches of the Po directed towards the sea, and the Alpine or Appennine rivers which turn east like the Po. Many changes in the drainage network are evident here, not only in topographical traces, but also in historical and archaeological documentation. For example, one old course of the Adige flowed north of its present-day course in protohistorical and Roman times, through the ancient city of Atheste at the foot of the Eugane Hills. The low Modena plain has supplied many data on the existence of old river courses, directed both north and east, functioning alternately [see Pellegrini (1969)]. Signs of active faults on the surface connected with deep tectonic structures have been identified in the same area [see Pellegrini and Vezzani (1978), CNR (1979)]. Further drainage variations in the delta area of the Po will be dealt with later (*see* p.258).

10.4.4   Effects of Interference by Man

The Po plain is an outstanding example of the manifold influences of human presence and urbanization on morphogenesis. Included in this is the microrelief caused by agricultural activites, connected with such special forms of land use as rice fields and the irrigated meadows called *marcite*. More important, especially nowadays, are the effects of quarrying and extraction, and the interaction between natural morphodynamic phenomena and human

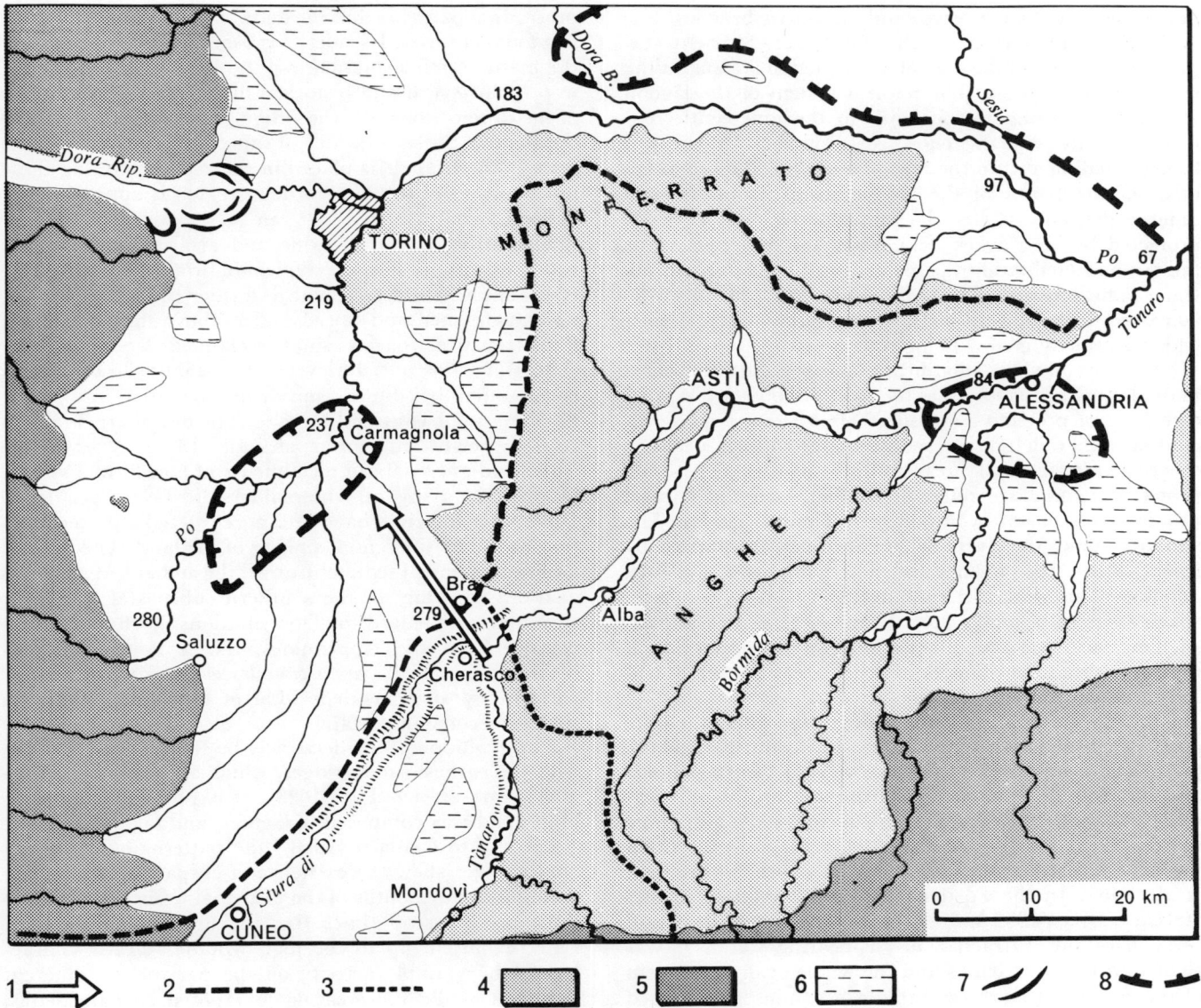

Fig. 10.22 Drainage pattern of southern Piedmont, and the diversion of the Tanaro River [Castiglioni (1979)]. (1) Old course of the Tanaro before diversion during the Würm, (2) present watershed between the Po and Tanaro Basins, (3) old watershed, (4) mountain areas of the Alps and Appennines, (5) hills and low mountains in Tertiary formations, (6) early Quaternary terraces, (7) Terminal moraines of the Dora Riparia glacier and (8) main areas of subsidence during the Pliocene and Quaternary.

activities along the waterways and the coast. On the one hand, the quarrying of stone or sand and gravel necessary for industry, building and public works has deeply affected the isolated hills, such as the Eugane Hills and the coastal dunes, while on the other, it has affected the surrounding hills of the Pre-Alps and the Appennines. Gravel, sand and clay pits are numerous and, in certain regions, have created many artificial lakes. Otherwise, they directly affect the river beds [see Pellegrini *et al.* (1979), Castiglioni and Pellegrini (1980-81)]. Excavation of river beds has become more intense just at a time when, for various reasons, the sediment load is becoming less, the main reasons being the reduced rates of erosion in many drainage basins and the deposition of sediment in

the numerous reservoirs. The result has been accelerated downcutting of the river beds over long distances—both above and below the points of removal—and the collapse of bridges and bank protection. In some places, flooding is now less severe. There has also been a change in the regime of the beaches, since normal supplies of sand by rivers now no longer arrive. Studies of coastal sedimentation show that, in some areas, the grain size of the sediments has partially changed, the percentage of silt and clay having increased with respect to sand. Some beaches have retreated, particularly around the mouths of some of the less active rivers. Above all, the tendency to retreat is manifest in beach sectors poorly supplied by the longshore drift, either because of artificial works carried out to protect ports or to conserve sand on tourist beaches.

Human intervention on the coast is an ancient tradition, especially in the area of the Venetian lagoon for the protection of its port. The natural system of the lagoon was altered during the fifteenth to the nineteenth centuries, mainly with diversion works on those rivers that brought sediment into the lagoon and with the reinforcement of the banks of the lagoon itself. From the late nineteenth century, the lagoon mouths began to be protected by long dykes, reinforcing the outlets towards the sea and facilitating the exchange of lagoonal and sea water. This has probably been an important factor in the increased frequency of high water (*acqua alta*) against which a remedy is now being sought.

The most obvious modification to the Adriatic coast made by man after the serious flood of 1966 is the great extension of protective measures almost everywhere, including the Po delta, in response to agricultural needs in the reclaimed coastal areas and to urbanization. Such protective works are often in conflict with the natural processes of coastal evolution [see Zunica (1971)]. The situation along the rivers is similar. They are flanked by levées protecting towns and low-lying areas against flooding. This, too, is a rigid system leaving little extra space at times of high water. In many cases in the past, such measures have produced a local improvement but, at the same time, the dangers in other areas farther downstream have been aggravated.

Growing attention is being paid to the phenomena of accelerated subsidence encountered in various parts of the plain. The main cause is the extraction of subsurface fluids and the resulting compaction of the unconsolidated sediments. Although these phenomena are evident in some interior areas too (e.g., Milan, Bologna), most attention has been focused on the coastal areas, where the situation is alarming. In the Po delta, sinking amounted to 50–100 cm between 1958 and 1967 in an area where there had been increasing extraction of gas-bearing water. It was clear that the river banks and the whole drainage system of the reclaimed land was rapidly becoming endangered by sinking, but the situation improved considerably with the closing of the gas-bearing wells.

In Ravenna, there has long been natural subsidence at a rate of 2–5 mm per year, but the rate dramatically accelerated in the 1950s and at present amounts to 116 mm per year. There are several reasons for this, but the main one is the pumping of water from aquifers lying at depths of between 100 and 430 m to supply water for Ravenna's industries [see Carbognin *et al.* (1978), Selli and Ciabatti (1977)]. A similar cause is believed to be responsible for accelerated subsidence in the central area of the Venetian lagoon. Here, however, the phenomenon occurred quite early and was less serious than in Ravenna, so that it could be reduced after exploitation of the groundwater was halted.

10.4.5   Po Delta: an Example of Interaction between Man and Nature in a Coastal Area

The Po delta (*see* Fig. 10.23) forms a large triangle whose base runs along the Adriatic for more than 70 km. The protruding part (i.e. the true delta) has developed over the last four centuries. Its sides, representing the outermost of the many terminal branches of the Po, which were active in protohistorical and historical times, extend from north of the present course of the Adige to the present course of the Reno in the south which at this point is artificial. If the apex of the triangle is placed in the area separating these old branches of the Po, the delta region is approximately defined, characterized by an association of fluvial, fluviomarine, fluviolacustrine and aeolian deposits. For example, around Ferrara, typical natural levées are clearly visible, which correspond to former courses of the Po, now abandoned and raised 4–5 m above the floor of old fluviolacustrine basins since reclaimed [see Bondesan (1975)]. Farther out, the levées and old sandy beach ridges, surmounted by dunes, subdivide the area into many fluviomarine basins and lagoons. The dunes are 3–5 m in relative height (maximum altitude: 13 m). Most of the water bodies have been reclaimed and now form a system of polders, many of whose floors lie below sea level. Differences in height have been accentuated after reclamation owing to differential sinking of the land. The natural addition of fluvial sediments to all the embanked areas has also ceased. Only in the southern subcoastal area near Ravenna, influenced by contributions from the rivers coming from the Appennines, did a system of land reclamation called *bonifica per colmata*—raising of selected surfaces by artificial deposition of fluvial sediments on them—become important.

The evolution of the delta may be summarized starting with the earliest sandy ridges, which are probably 3000–4000 years old (*see* Fig. 10.23). It is probable that their hinterland was composed of lagoons and lagoonal deltas. According to Ciabatti (1968), the patterns of the sandy beach ridges show the existence of cuspate deltas corresponding to the mouths of the principal branches of the Po since pre-Etruscan times (i.e. the first millenium BC). Later evolution up to the high Middle Ages is characterized by variable activity on the part of the different branches of the Po, creating juxtaposed cuspate deltas, with evident phases of partial destruction by the sea between one constructive phase and the next. A total of 11 cuspate deltas advanced the coastline a total of about 7 km to the north and perhaps three times as much in the south. At the same time, however, subsidence caused the partial submergence of the coastal formations and natural levées (the rate of sinking being approximately 3 m in 2500 years). Artificial river works have been reported since Etruscan times, principally aimed at facilitating port activities. There was also a certain amount of land reclamation in the Middle Ages carried out, for example, by the Benedictine monks at the abbey of Pomposa.

At the end of the Middle Ages and the beginning of modern times, delta construction at the northern mouths became very active, first on the Po di Goro, and then the Po delle Fornaci and its ramifications. In the sixteenth century, a lobate delta formed rapidly which, with the Po di Tramontana, harmed navigation and the agricultural interests of the Venetian aristocrats. Political reasons, therefore, dictated the artificial diversion of the Po, the so-

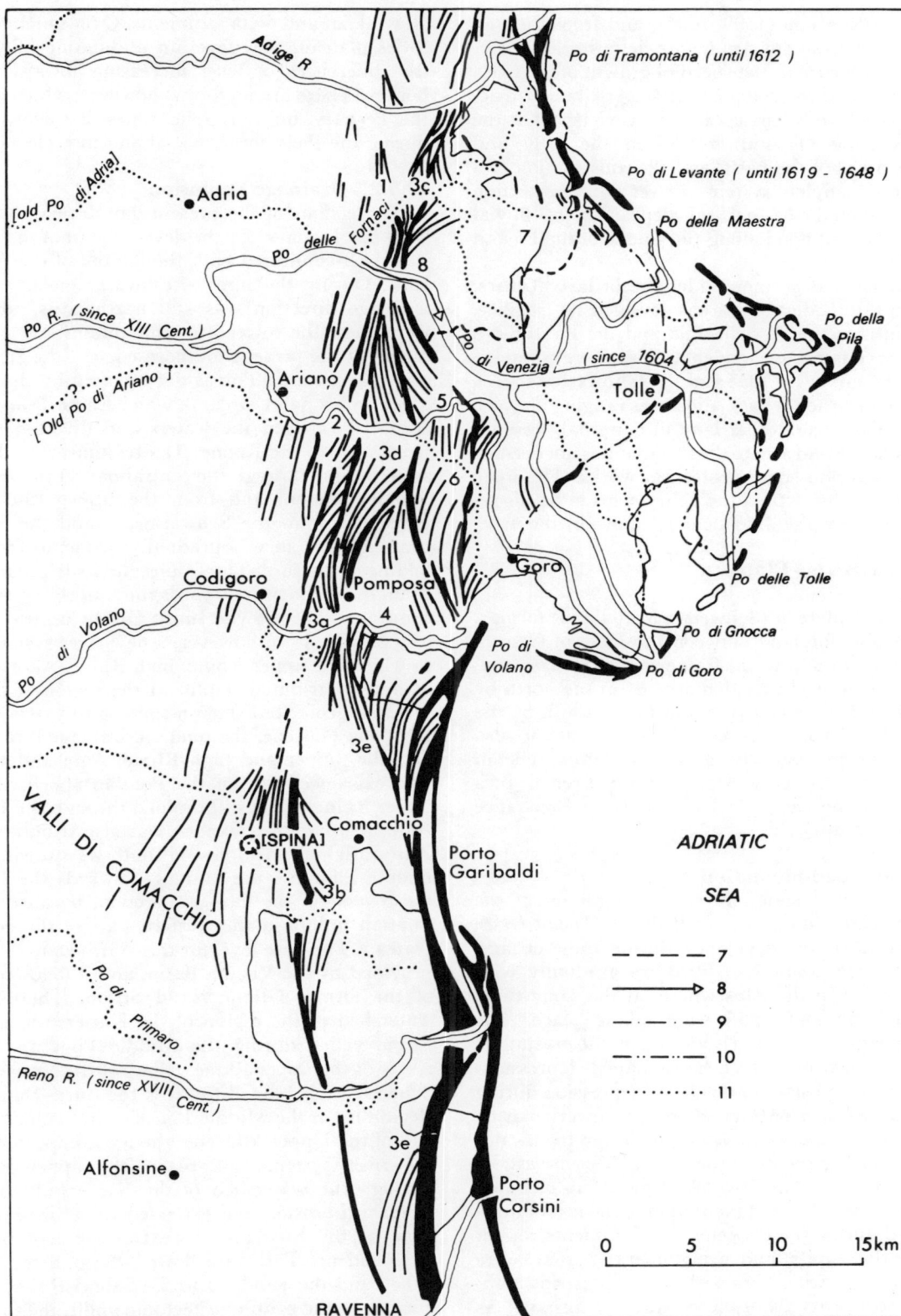

Fig. 10.23 Geomorphological features of the Po delta and phases of its development [after Ciabatti (1968)] with some additions from other sources. (1) and (2) Deltas older than Etruscan times, (3a) and (3b) deltas of Etruscan times, (3c), (3d) and (3e) deltas of Etruscan–Roman times, (4) deltas of the Middle Ages, before 1200, (5) and (6) medieval–modern deltas, (7) edge of the deltas in the sixteenth century (first lobate delta) created by the Po delle Fornaci and its branches, (8) artificial deflection of 1599–1604, (9) edge of the delta in 1750, (10) edge of the delta in 1874 and (11) old branches of the Po (with the times of major activity).

called *taglio di Porto Viro* (1599–1604), and from then on the construction of the modern lobate delta began.

The evident increase in the speed of growth of the delta in modern times and the change of style of delta construction are attributed to increased sediment transport at the mouth, caused by (1) soil erosion in the hilly and mountainous areas of the basin and (2) construction of a more and more complete system of river bank protection which has hindered the natural deposition of fluvial sediments in times of flood along the course of the Po and its tributaries.

Another change has become evident in the last 30 years, during which the modern delta has revealed its fragility; only the mouths of the Po della Pila and the Po di Goro show an active constructive tendency. Elsewhere comparison between the surveys of 1944 and 1973 shows that the sea tends to move the beaches back, distances ranging between 0 and 500 m. The loss of other land in marginal positions is due to subsidence and also to the fact that it is not deemed economic to maintain some protection works. The main cause of the lack of growth of the delta seems to be the reduction of quantities of sediment carried down by the river.

10.5   Central Swiss Plateau

The central Swiss plateau (Schweizer Mittelland) forms a part of the Alpine foreland between the Lake of Geneva and the Austrian border. In the framework of Switzerland, the Alpine foreland is clearly demarcated in the north by the escarpment of the Swiss Jura and in the south by the edge of the Pre-Alps. The central Swiss plateau also stands out geomorphologically against the Alpine relief in the west, where the Savoy Alps and the French Jura almost meet. In the east, the basin of Lake Constance forms a natural limit.

10.5.1   Molasse Sedimentation

The central Swiss plateau represents a remnant of the Alpine geosyncline; it is thus called the *Molassebecken* (molasse basin). It was developed during the Tertiary, when this molasse basin was filled up gradually with material eroded from the Alps where, at the same time, the Alpine folding and uplift were taking place. The basement beneath the molasse is formed by the crystalline bedrock in the form of a large bowl-shaped depression. The molasse was deposited in this large depression during the Oligocene (*Untere Meeresmolasse* or lower marine molasse and *Untere Susswassermolasse* or lower freshwater molasse) and during the Miocene (*Obere Meeresmolasse* or upper marine molasse and *Obere Süsswassermolasse* or upper freshwater molasse). The sedimentation of the last of these also continued during the Pliocene. The sediments consist of clays, marls, sands and conglomerates that were deposited under marine, brackish and lacustrine conditions. The sediments also suffered fluvial transport. As the sedimentation began at the border of the Alps, grain sizes decrease towards the Jura. Along with this, thickness of the sediments diminishes from 5–6 km at the Alpine border to about a hundred metres at the edge of the Jura. Considered from a genetic point of view, the sediments are

alluvial fan and delta sediments. Over much of the central Swiss plateau, they form an undulating hilly landscape, the differences of level increasing towards the Alpine border. There are no forms, however, which developed in the Tertiary, but only relief types that evolved gradually during the Pleistocene glacial and interglacial periods.

10.5.2   Drainage Evolution

The directions of the present-day drainage are inherited from the Tertiary, for the development of the river system was itself connected with the folding of the Alps and the Jura. During the upper freshwater molasse period, the drainage direction was still northwards, towards what were then the lower Jura mountains; on reaching the region of the present southern edge of the Jura, it turned south-westwards. This is documented by the *Glimmersand Strom* (mica sand stream) which came from the eastern Alps. It collected the waters and diverted them south-westwards to the Rhône. The tributaries of this drainage system came through the central part of the Aare from the Danube system and from the upper Rhine valley, in particular from the Schwarzwald and the Vosges. This drainage system was gradually dismembered during the folding of the Jura. The Doubs and its tributaries began to drain independently towards the south-west (i.e. west and south-west of the Swiss Jura). On the central plateau side of the Swiss Jura, however, a new river system developed to which the upper Rhine, high Rhine, Aare and Danube at first contributed. Uplift in the region of the southern German geotectonic region split up this river system and, since the Pliocene, the modern drainage directions of the Danube, upper and high Rhine, Aare and Rhône came into existence. During the Pontian epoch, the southern upper Rhine valley still drained through the Jura into the Aare, to which the Danube was also tributary. This Ur–Aare (original Aare) flowed south-westwards. During the middle Pliocene, the drainage towards the Danube was the reverse of the former direction, because of the post-Pontian uplifts in the western part of the central Swiss plateau. The base level for this Aare–Danube system was provided by the Vienna Basin, and it followed the course of the former Glimmersand Strom. There was still a watershed in the region of the Kaiserstuhl in the upper Rhine valley during the Middle Pliocene Epoch. The rivers of the upper Rhine valley, of the Vosges and of the Schwarzwald drained towards the Aare–Danube system. Beginning in the Middle Pliocene and culminating at the end of the Upper Pliocene, the most important change of the river system took place: the upper Rhine broke through the watershed of the Kaiserstuhl by means of headward erosion and extended its drainage area to the edge of the Sundgau. The Danube had now become independent. The Aare flowed through the high Rhine valley and the Sundgau to the Saône–Rhône depression. Thanks to the continous tectonic uplift, the southern edge of the Jura became a well-marked watershed. The final touches to the drainage pattern were only completed during the late Pleistocene period. The central Swiss plateau drained through Lake Constance and the Aare to the high Rhine. The drainage system of the Jura was partly tributary to this, unlike the Doubs system which followed the old Upper Pliocene–early Pleistocene drainage direction through the western part of the Swiss Jura.

10.5.3 The Pleistocene

From the geomorphological point of view, the Pleistocene was the most important period, because the essentials of the present relief of the central Swiss plateau were then developed. The Pleistocene period saw the last tectonic movements, the development of the river systems and watersheds, and the formation of the glacial relief. Whereas the formative processes of the Pliocene period are only deducible from the development of the river system and from local gravel deposits, distinctive and widespread sediments are available in tracing the events of the Pleistocene. There are fluvial, glaciofluvial, glacial and periglacial sediments, but those of the pre-Günz phases are rare; the main correlative sediments belong to the Mindel, Riss and Würm glaciations. It is typical of all the central Swiss plateau that, in addition to the moraines, there are also corresponding glaciofluvial gravels. The moraines and gravels of the two oldest glaciations (i.e. Günz and Mindel) are only present now as local relics. There are the so-called older plateau gravels (*Älterer Deckenschotter*) of the Günz glaciation and the younger plateau gravels (*Jüngerer Deckenschotter*) of the Mindel glaciation, lying on molasse sediments of various ages. By means of them, it has been possible to deduce the glacial advances and retreats, and the meltwater drainage systems that followed more or less the modern valleys. The same is also true of the two subsequent (Riss and Würm) glaciations.

Both the Riss and the Würm glaciations are documented over a relatively extensive area by moraines and gravels which represent the main part of the Pleistocene sediments of the central Swiss plateau. The Riss glaciation marks the maximum extent of the Alpine ice. According to the occurrences of erratic blocks and ground moraine, the Riss ice covered the whole of the central Swiss plateau, and it is likely that it came into contact with the Riss ice of the Schwarzwald. Its border lay west of the Swiss Jura and north of Lake Constance on the southern edge of the Swabian Alb.

The extension of the Würm glaciation was smaller, and the ice was associated with the major valleys of the central Swiss plateau. The principal glaciers issuing from the Alpine border were (from west to east) those of the Arve, Rhône (with the Saane and Aare), Reuss, Linth and Rhine. The Rhône glacier covered the whole of the western part of the central Swiss plateau and spread out from the Lake of Geneva to Lake Neuchâtel and to the central valley of the Aare. The Reuss and Linth glaciers, however, failed to reach the Aare valley but, on the other hand, they came into contact around Winterthur with the Rhine glacier which covered the whole basin of Lake Constance up to Schaffhausen during the Würm glaciation. All these regions are covered with ground moraine and show well-marked end moraines.

Along the Alpine border there was a series of ice-free enclaves, which were probably free of ice even during the older glaciations (i.e. Günz and Mindel). These are the higher parts of the molasse which are marked at the present day by fluvially dissected relief. The Napf Bergland is the prototype of such a landform complex.

The forms and sediments of each glacial period are generally similar, but there are differences in the state of preservation. The Würm forms and sediments are the freshest and show only a few traces of periglacial modification. The Riss forms are considerably degraded (i.e. modified, eroded or destroyed in periglacial times). The low and smooth relief of the Riss glaciation is in strong contrast to the clearly marked forms of the Würm glaciation whose ice margins are also demarcated by numerous high end moraines. Their great number indicates a step-by-step retreat of the glacier margins. The postglacial stages of the retreat took place outside the central Swiss plateau in the Alpine valleys and are described elsewhere in this chapter (*see* Section 10.1.5). The deposits of the Günz and Mindel glaciations are in an even worse state of preservation; they comprise the Deckenschotter deposits situated high above the river valleys and the Riss and Würm terraces. The Deckenschotter deposits now consist of only a few scattered patches of gravel which no longer form any distinctive relief features.

Outside the valleys, ground moraine deposits are extensive, varying in thickness from a few metres to sometimes hundreds of metres. Drumlins sometimes occur, together with the recessional moraines—both terminal and lateral—that mark temporary halts in the retreat. Another important group of accumulation forms consists of the glaciofluvial gravels that can be found in all the river valleys. They are an integral part of the morainic landscape and can sometimes be genetically correlated with the end moraines. Unlike the normally nonstratified morainic deposits, the fluvial gravels usually show good bedding and are frequently very thick (up to a few hundred metres). Those of the Riss are called the upper terraces (Hochterrassen) and the Würm gravels the lower terraces (Niederterrassen); both show differences. First the Alpine source areas of both the gravels and the moraines are definable, proved by the sedimentological characteristics of the gravels. Secondly, the sediments of the Riss glaciation are often, but not always, cemented, usually by calcareous material. This is also true for most of the sediments of Günz and Mindel age.

In addition to the depositional forms left by the Pleistocene glaciations, which determine much of the relief of the central Swiss plateau, there are also glacial erosion forms which can be found over almost the whole of the central Swiss plateau. Beneath the whole Pleistocene sequence of deposits on the Alpine foreland, there is a buried relief of basins, ridges and steps, from whose forms and trends an origin by glacial erosion with high rates of removal can be inferred. Glacial erosion was particularly intense at the border of the Alps, as shown by the depth of the bedrock, which may there be a few hundred metres below the surface. Correlation of the amounts of glacial erosion with the different glaciations is not, however, possible. To sum up, the landscape of the central Swiss plateau is not only one of youthful relief, but was also developed in and on a very thick glacial sedimentary cover.

10.5.4 Present-Day Forms and Processes

Large floodplains where the non-regulated rivers flowed as meandering systems were characteristic of the central Swiss plateau during the postglacial period. Within the last 100–150 years, there has been increasing anthropogenic influence on the relief, in particular by means of river regulation. Thus most of the rivers have now lost their geomorphodynamic efficiency, and it has become possible to settle on the valley floors and to cultivate them.

Since about 1950, there has also been extensive destruction of the surface caused by the excavation of gravel pits in the valleys and on the lower plateaus. Present-day geomorphodynamic processes fall mainly into two groups: (1) local fluvial erosion along river channels and (2) soil erosion on the cultivated slopes, in regions that have high precipitation and frequent intense precipitation events. The latter process is much more of a problem in the loess-covered areas than in the well-drained gravel and moraine regions.

10.6 The Jura

Lying roughly parallel to the Alps, but separated from them by the central Swiss plateau, lies the Jura (*see* Fig. 10.24). In the north-east near Basel, the Jura trends in a south-westerly direction; this trend gradually changes to become north-north-east to south-south-west around the Lake of Geneva, and then to north to south near Chambéry where the Jura joins the folds of the Alps. The outer zone of the western French Alps resembles the Jura geologically, tectonically and morphologically. Both have the same fold structure and the same age of folding. The arcs of the folded Jura thus continue in the western French Alps and reach the Mediterranean Sea between Nice and Menton. The fold arc of the Jura, which branches off the Alps near Chambéry, extends over 250 km. The maximal breadth is 65 km, but at both ends of the Jura, the breadth is only 10–30 km because here the Hercynian massifs of the Massif Central, on the one hand, and the Schwarzwald and Vosges, on the other, approach the Alps most closely. In the central Jura, on the inner side of the fold arc, there are the following summits: Crêt de la Neige (1723 m), Dôle (1677 m), Mont Tendre (1679 m), Chasseron (1608 m), Chasseral (1607 m) and Weissenstein (1396 m). At the same time, these points mark the highest folds and form a wall overlooking the Swiss plateau. To the west and to the north-west, the folds are lower, although the Jura has a steep western face too (Revermont, Vignoble) [see Annaheim and Barsch (1963), Barsch (1968a, 1969)].

The Jura consists of three different component landscapes, which are of morphotectonic origin: (1) the tabular Jura, (2) the plateau Jura [see Chabot (1927)] and (3) the folded Jura. These parts of the Jura resemble one another petrographically: mostly, they are built of limestone (Malm, Dogger, Trias) together with marl and clayey marl which appear between the limestone bands.

10.6.1 Tectonic Evolution

The folding of the Jura followed pre-existing tectonic structures. At one time, it was considered that the Jura folds, which are of superficial type resting on a crystalline basement, were the result of forward pushing of sediments by the advancing Alps, but this idea has been discarded because of the mechanical impossibility of pushing sheets of rock, only a few kilometres thick at most, across a more or less horizontal surface without complete disintegration. The only aid to such a process is the presence of layers of rock salt and anhydrite of Triassic Muschelkalk age, which might act as a lubricating layer or zone of detachment. However, between the Jura and the Alps lies the Swiss plateau comprising almost unfolded sediments, thus making the hypothesis untenable. A different view is that vertical movements along faults in the crystalline basement have activated the evaporite deposits by causing differential pressures on them; diapirically, the locally thickening evaporites raised the overlying strata into folds, with some faulting [see Breyer (1974), Fischer (1969), Laubscher (1961, 1962, 1965)].

Concerning the dating of the folding, there is only one chronological fixed point: the Tortonian–Sarmatian *Nagelfluh* (*see* p.237) has been overrun in the region where the folded Jura abuts against the tabular Jura. Thus the folding of the Jura must be younger than this. Nevertheless, the exact date is not known. The evidence of the Vosges gravels (*Vogesenschotter*) in the Delsberg Basin suggests that the folding is post-Pontian [see Liniger and Rothpletz (1964)]. The folding process was relatively short-lived and single-phase. It seems to have begun earlier in the south, but the northern chains must not be regarded as younger than the southern ones, for the latter suffered further folding during formation of the northern chains. The folding produced a dense system of anticlines and synclines, which partly dislocate one another. The largest folds are on the inner side of the Jura arc, forming a massive wall of Cretaceous limestone. In the Swiss Jura, many of the basins are filled with molasse (Delsberg, Moutier, Laufen).

The tabular Jura (*see* Fig. 10.25) forms the easternmost part of the Jura (Basler and Aargauer Jura) and the outermost north-west and the Ajoie (region of Puntrut). It is unfolded and the structure is only interrupted by fractures. The fracture direction is the same as that of the Rhine and the faulting can be seen to continue in different geotectonic structures of the Schwarzwald and the Dinkelberg. This region is considered to have remained rigid in a tectonic sense because it lies opposite the crystalline block of the Schwarzwald [see Herzog (1956)].

The folded Jura, beginning south of the tabular Jura, ends abruptly at this block with very well-marked folds. The folded Jura represents a morphological hinterland to the tabular Jura of the Ajoie, Basler and Aargauer regions.

The whole of the western part of the Jura is represented by the plateau Jura which is morphologically, but not geotectonically, different from the folded Jura. The strata are more gently folded and eroded, and the sharp contrasts in relief characteristic of the folded Jura are not found here.

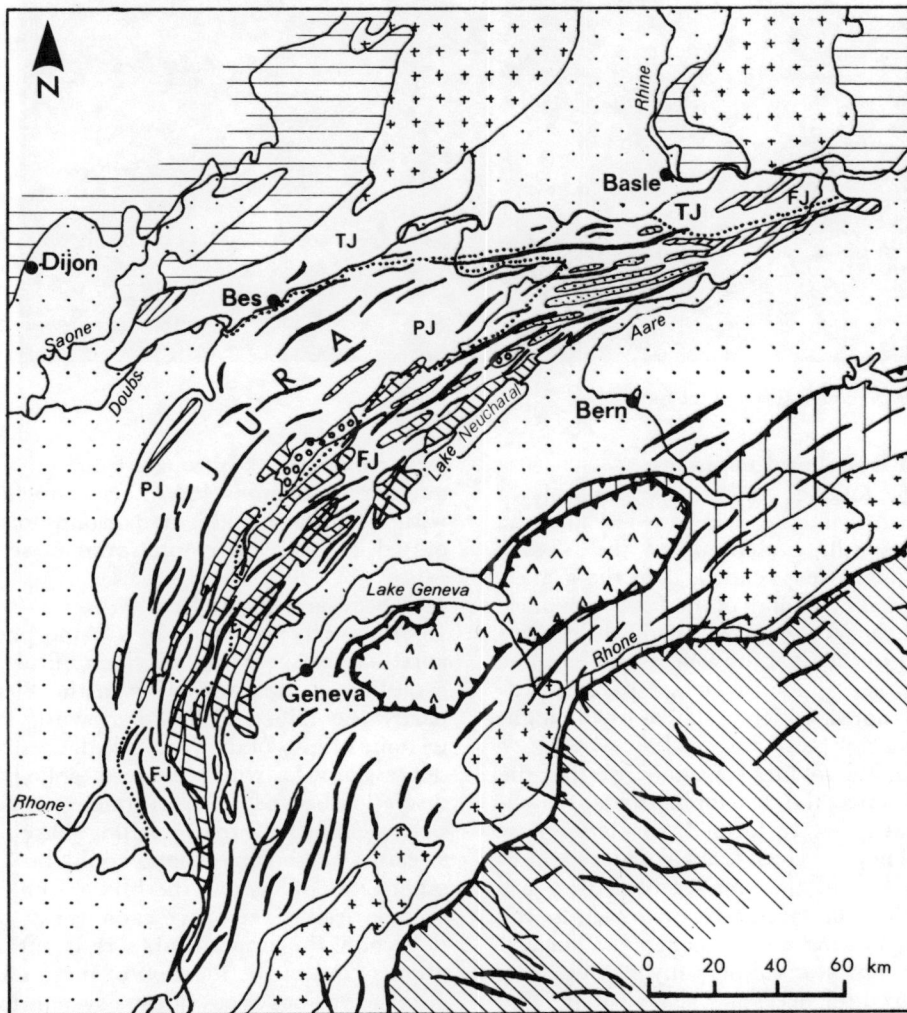

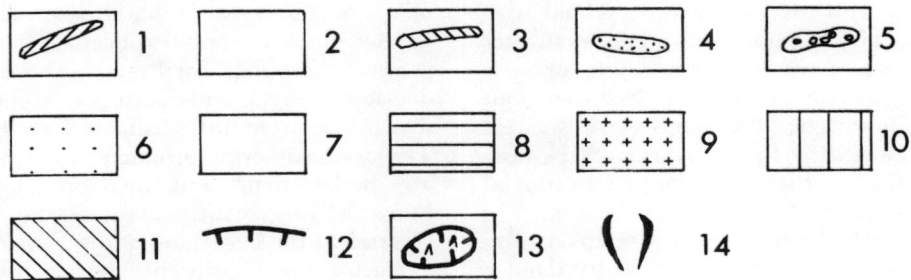

Fig. 10.24 The main petrographical and geomorphological units and tectonic structure of the Jura. *Jura*: (1) Triassic, (2) Jurassic, (3) Lower Cretaceous, (4) Lower Tertiary, (5) Quaternary; (TJ) tabular Jura; PJ: plateau Jura; FJ: folded Jura. *Alpine foreland*: (6) Tertiary of the foreland, molasse and basins, (7) Jura, Dauphinois zone, (8) structural plateaus (Permian and Mesozoic), (9) crystalline of the foreland and the central massifs. *Helvetic*: (10) Helvetic nappes, autochthonous and Ultra-Helvetic. *Penninic*: (11) Penninic nappes. *Other tectonic features*: (12) boundary lines of nappes of the large tectonic units, (13) klippes, (14) anticlines of the late Tertiary orogeny.

Fig. 10.25 View of tabular Jura near Basel. (Embleton)

10.6.2 Evolution of Relief and Drainage

Before the period of folding, there was probably a pediplain sloping down from the Schwarzwald and the Vosges towards the south, evidenced by the Vosges gravels. First the channels were refilled, then the gravels expanded over the area. The folding of the Jura took place after the deposition of the Vosges gravels; and the Vosges drainage was diverted to the west, no longer reaching the Jura. At the same time, a new drainage pattern was formed directed to the north, which is shown by the high-lying gravels (*Höhenschotter*) [see Liniger (1966)].

This happened while the folding continued, so that the drainage was incised across the developing folds. Syncli-nal and transverse valleys were formed, the latter now represented as *cluses*. They are considered as antecedent to the larger anticlines, but in the smaller ones they are explained as epigenetic. The larger streams, such as the Doubs, Birs and Ain, flowing on the peneplain, formed epigenetic valleys that almost resemble canyons in places. Karst sources guaranteed the necessary heavy runoff. The cluses partly antedate the Pliocene, because they are genetically connected with the folding process and with the fluvial dissection of the Jura. Pleistocene meltwater from ice in the Jura may have been responsible for their further development, not only in the early Pleistocene, but also in the middle Pleistocene. The combes are isoclinal valleys which were formed by the excavation of the less-resistant layers. Primarily, they were formed by fluvial processes, aided by slope wash and mass movement.

The drainage system of the Jura is, then, relatively old, connected with the uplift and the folding of the Jura, but it has always been a karst system. The tectonic movements are supposed to have happened quickly, which may help to explain the irregular drainage net in the Jura. The drainage is strongly orientated along the synclines and anticlines, but there are many sudden changes in direction as streams break through from one structural line to another. Numerous karst forms, including large basins without surface outlet and groups of dolines, are inte-grated in this hydrographic system [see Metz (1967)].

10.6.3 Karst in the Jura

In terms of its physiography and morphology, the Jura is a karst landscape with all the typical karst forms. The evolution of the karst takes in the Pleistocene, and also partly the Pliocene; in other words as far back as the tectonic origins of the Jura. The largest forms are the karst depressions, for which the term polje has to be used with discretion because surface drainage is only partly present. Because of the rapidly falling base levels of different streams, these depressions could be formed; these and other types of basin in the Jura are linked to stratigraphic boundaries between permeable rocks, liable to karstifica-tion, and the impermeable resistant strata. Such con-ditions control the location of the numerous dolines and also the various types of cave system to be found here.

The karst forms originate from the subaerial stripping of layers of unresistant marls. The various flat surfaces in the synclines are only explicable by such subaerial pro-cesses (subrosion, corrosion). These 'corrosive planes' developed by the undercutting of higher limestone slopes, which rise from the basin and valley floors sealed by Tertiary weathering products or by the marls. The forms are post-tectonic and therefore postdate the middle Pliocene. The corrosive undercutting processes probably stopped at the beginning of the Pleistocene.

During the Pleistocene glacials, karst processes were inactive. Surface drainage was impeded or interrupted, probably by permafrost that may have developed in all glaciations, even the Würm. This applies to all areas over about 1000 m. (The Würm snowline is thought to have stood at about 1000–1100 m.) The sealed depressions may have contained periglacial lakes during the Pleistocene. The ponors to which the drainage of the karst depressions flowed were then obstructed episodically or permanently because of permafrost. Another process helping to seal the floors of the depressions was the accumulation in them of Tertiary and Pleistocene weathering products (i.e. clay, silt and other fine-grained debris).

10.6.4 The Pleistocene

Pleistocene glaciation affected much of the Jura. Although indications of pre-Riss glaciations are rare and cannot be precisely dated, there is no doubt that there were such glaciations, but little is known about their exact extent or influence on the landscape. The *Wanderblock* formation in the north-western Jura belongs to the oldest Pleistocene and originated from the Schwarzwald and its borders. The Sundgau gravels—certainly of pre-Günz age—are also of uncertain stratigraphic age.

Although the glacial evidence is not totally clear, some authors suppose that the Riss glaciation covered virtually the whole of the Jura. On the other hand, local glaciation certainly occurred. On the east side of the Jura, the Rhône glacier at this time entered some of the larger basins, where erratic blocks, mammillated rock surfaces and basal moraines including Alpine materials are found [see Aubert (1965)].

The *reculées*—a term that refers to the settled valleys of the Jura—also belong to the Riss glaciation [see Zeese (1978)]. They were probably deepened during the older Pleistocene, probably in two phases. During the Riss glaciation, they were filled with glacial and glaciofluvial sediments, but not all reculées were filled to the same extent. The glacial overdeepening of the reculées is, however, pre-Würm. Further development of the forms after the Riss was the result of many geomorphological processes (fluvial, gravitational, periglacial) but they were not so intensive that the effects of the Riss glacial modifications of these old valleys were completely obliterated. The valleys lying in front of the Riss ice margin were modelled continuously by fluvial activity and terraces were developed. Some of the Jura valleys, particularly on its northern border, may be interpreted as glacial melt-water channels.

The Würm glaciation, which touched the Jura and probably overran its eastern flank, did not extend so far as the Riss ice branches of the Rhône glacier. The ice not only followed the Rhône valley as far as the southern end of the French Jura in the region east of Lyon, but also there was a major lobe pushing in from the Lake of Geneva in a north-easterly direction, following the boundary between the central Swiss plateau and the Jura. This part of the Rhône glacier was fed by the Arve, Saône and Aare glaciers, and not only occupied the basin of the Lake of Geneva and its outlet to the south, but also followed the large depression on the eastern side of the Jura to the north-east, where today the lakes of the Jura's southern margin (e.g., Lake Neuchâtel) are to be found.

It has been well established that the Rhône glacier reached the border of the Jura. More recent studies, however, suggest that it extended farther into the Jura, its ice-marginal area even including the Doubs, but sediments of the Würm glacial advance are rare. Considering the general context of this glaciation, it seem likely that ice also spread over the Jura plateau. The transfluence cols, cirques, meltwater channels and small local moraines, however, all originated from local glaciation, which was not always exactly synchronous with the activity of the Rhône glacier. For the most part, the region does not seem to have been sufficiently modified to support the hypothesis of any effective early Würm glaciation, although the sediments are easily erodable. The Jura east of the Doubs rises, in general, above the height of the presumed Würm snowline at 1000–1100 m, so that glaciation of the Jura itself might have also occurred in the Würm maximum. The individual forms seem to indicate many local glaciers which pushed forwards from the anticlines into the basins, producing the numerous small moraine embankments in front of the combes.

Continuous 'dump' moraines can be found around the lake basin of Neuchâtel. On one side, they mark the end of the Würmian Rhône glacier at the border of the Jura, corresponding to the maximum of the glaciation. On the other side, they may represent a recessional stage of a still larger Würmian Rhône glacier (*Neuenburger Stand*) which must also have penetrated into the Jura and its basins.

The limited extent of Riss sediments is explained by subaerial erosion during the Riss–Würm interglacial, destroying the Riss moraine. Streams have flushed out the fines through ponors into the underground drainage systems.

The Würmian Rhône glacier terminated in the area around the confluence of the Rhône and Saône, just in front of the foothills of the French Jura, near Lyon. The Würm end moraine is well marked and is situated not far from the limit reached by the Rissian Rhône glacier which spread out in the arc of the Rhône valley between Vienne, Lyon and the confluence of the Ain with the Rhône. Even this, however, may not represent the maximum extent of Alpine glaciation, for glacial sediments, probably of Mindel age, cover the hills north-east of Lyon where there are meltwater sediments and basal moraines. The border of the ice is also marked by dump moraines which cross the Saône near Lyon and reach the Massif Central. In general, the Rissian Rhône glacier marks the maximum extent of the ice in the Jura but, at the southernmost border, the furthest glacial limit is represented by sediments, whose age is probably Mindel or Günz.

10.6.5 Periglaciation

Periglacial landform development in the Jura is much influenced by the karst character of the landscape. During the glacial periods, karst processes were mostly inactive because water circulation was restricted by ground ice. Streams flowed above ground, lakes were formed and fluvial erosion occurred in areas where there was no glacial ice at this time. Rates of periglacial erosion cannot be estimated in the Jura region and comparisons with other periglacial areas in this respect are difficult.

As many parts of the Jura are situated near the snowline, intensive periglacial processes have eliminated the older glacial evidence and relics of Tertiary forms. As some of the rocks are susceptible to mass movements at the present day, these processes must also have been active in the glacial periods. In many Jura landscapes, this is shown by a cover of geliflucted debris on nearly all slopes up to 30°. This gelifluction cover shows fairly consistent petrographic composition, but varies considerably in thickness with some exposures showing a very thick layer. The deposits are fairly well stratified and differ from the Würmian gravel terraces which are often interwedged with the gelifluction layers. A clear stratigraphic and chronological succession of the periglacial deposits has been established. Pre-Würm gelifluction deposits whose existence is only tentatively recognized were probably reworked during the Würm. Stratigraphic evidence of older gelifluction layers is rare.

In the whole of the Jura, terraces in slope deposits are a typical and widespread accumulation form. At the foot of debris-covered slopes, there are often such terraces which show a lower slope angle than the slopes above. These terraces are of small dimensions individually, but taken collectively they are widespread. They often appear where the steeper parts of the slopes cut in limestone change into more gently sloping segments corresponding to outcrops of less-resistant rocks.

Frost-riven cliffs and *vallons de gélivation* are among the other periglacial forms. The latter are rare, probably for petrographic reasons. They represent small periglacial valleys formed by corrasion and include numerous transition forms to rills and dry valleys, as well as to recent stream and river valleys. The frost-riven cliffs are mostly formed in limestone that was attacked by frost weathering, fragmentation and detrital accumulation. The weathering process was itself retarded as soon as the rocks were covered by a small quantity of debris.

In different parts of the Jura both major and minor late glacial rockfalls, which can often be well dated, can be observed. Partly they are overlain by gelifluction debris and therefore, in such cases, must be of late glacial age or older. Less frequently, they overlie Pleistocene forms so that a Holocene age is clear. In most cases, it can be concluded from the extension of the rockfalls and from their stratigraphical position in relation to the end of the Würm or the transition to the Holocene that when the permafrost began to thaw, rock masses were loosened and became susceptible to mass movement. The timing of the rockfalls is, in general, an indication of the beginning of climatic amelioration after the maximum of the glacial periods [see Haeberli *et al.* (1976)].

10.6.6   Fluvial and Glaciofluvial Terraces

Pleistocene fluvial terraces characterize nearly all the stream valleys of the Jura. Especially in the larger basins and along rivers with high discharge, they are very well-developed, as long as the valleys are sufficiently wide. Because of the alternation between narrow valleys and basins, these terraces are difficult to follow continuously. Correlations with the well-dated terrace systems of major rivers are also difficult. The lowest Würm terrace reaches a height of 5 m and can be interrupted by smaller,

probably Holocene, terraces. The Riss high-level terrace attains a height of 10 m or more above the present water level and can be subdivided into several levels. In contrast to the lower terrace, numerous insertions are missed by the horizontal extension of the Riss terraces. This is the reason why the complex division of the Riss terraces does not provide a basis for any clear division of the Riss glaciation in the Jura. Along the larger streams, younger and older Deckenschotter appear, which stand higher above the recent valley floor than the Riss terraces and which date from the Mindel or Günz glaciations. Relics of higher levels, with or without sediments, can be locally observed.

The heights of the pre-Riss terraces vary from river basin to river basin and make classification and dating more difficult. Fossil flora and fauna are completely absent.

10.6.7   Present-Day Processes

Present-day processes in the Jura fall under three main headings: (1) karst processes, (2) soil erosion and mass movements, and (3) fluvial activity.

Karstification continues today in the solid limestone and in the limestone debris. It is evident in the high calcium carbonate concentrations in the streams, although there are major variations caused by different geo-ecological conditions and by the ratio of solid limestone to limestone debris in the drainage areas. There are no exact indications of the quantitative differences between surface and underground solution; each has to be considered separately. Estimates, which are based on single measurements, give a value of 6–9 m for the solutional lowering of the Jura since the Günz. Thus it may be concluded that pre-Pleistocene forms have been destroyed in many cases and that so-called Tertiary relict forms (e.g., planation surfaces) must be carefully examined with such lowering in mind [see Erzinger (1943)]. The limestone solution forms in the Jura can seldom be associated with recent processes, but karren surely originate from recent solution, as do some dolines. Certainly postglacial, but not absolutely contemporaneous, is the tufaceous limestone, which in places accumulated to great thicknesses on the valley floors. Probably, the postglacial climatic optimum was responsible for the tufa formation because higher temperatures than those of today are needed.

The present-day streams carry considerable loads of sediment originating from contemporary soil erosion. In the Jura, soil erosion occurs on nearly all slopes, but there are differences in rates of operation caused by variations in subsurface material, soil moisture content and land use. The forms of soil erosion are immediately eliminated by field cultivation; the rates of erosion show the importance of this group of processes. Heavy and prolonged rainfall and frequent sudden snowmelt are major causes. Some of the superficially eroded loam is transported in karst rills, while processes of suffosion are also of local importance. Part of the eroded material accumulates as alluvial loam in valley floors.

Fluvial erosion—both lateral and vertical—is active in all rivers and streams. Floods after snowmelt and heavy precipitation in the larger drainage basins represent periods of rapid activity, especially in the small, steep valleys, the combes, cluses and semicluses, of such catchments. Because of the spread of river control works, flood damage is nowadays becoming significantly less than in former times.

Present-day mass movement processes are very important spatially. Everywhere in the Jura, there are clays and marls that are susceptible to landslip, especially in parts of the tabular Jura and in the cluses and synclines of the folded and the plateau Jura. Because of the widespread presence of Pleistocene debris layers, creep and sliding processes are important, appearing on nearly all slopes and encouraged anthropogenically (e.g., by grazing of animals on steep grass-covered slopes, so that lobes, terraces and steps are formed). The intensity of creep varies from place to place. Postglacial soils, which have been overridden by still younger soil and debris, show clearly that ground creep is important. On the other hand, major landslides are not common, although there are slide niches and slump features.

Chapter 11
Pyrenees and Ebro Basin Complex

In the Iberian Peninsula, there are two cordilleras that can be considered as part of the Alpine system: the Pyrenees and the Baetic Cordillera. The latter, although it possesses peculiarities that make it different from other Alpine chains, as in the development and the position of preorogenic vulcanism and, to a certain extent, in the characteristics of its thrust nappes, compares well with the general style of these chains. Its complicated structure in terms of thrust nappes and the metamorphism affecting its internal zones is matched in the other Alpine systems. The Guadalquivir depression, which constitutes a foredeep, is related intimately to the Baetic Cordillera and contains materials entering as thrust nappes. On the other hand, the Pyrenees are a much more special type of fold mountain system, not so much because of their evolution but because of their structure, dimensions and position. The bilateral structure, with a Palaeozoic axial zone bordered by two folded systems of opposite direction, makes this cordillera one of moderate width, contrasting with the structure of other cordilleras, such as the Alps and the Carpathians. The relationship between the Pyrenees and the Ebro Basin is not as clear as in the case of the Baetic Cordillera and the Guadalquivir depression. There is, however, evidence of a much more important depth zone in the basement along the Pyrenean margin, so that at least the northern border of the Ebro Basin may be considered a foredeep of the Pyrenees. As a general reference to the tectonic structure of Iberia see Julivert *et al.* (1974) and the accompanying *Mapa Tectónico de la Península Ibérica y Baleares*, scale 1:1 000 000.

11.1   The Pyrenees

The Pyrenees constitute a massive mountain range stretching from west to east from the Bay of Biscay to the Mediterranean Sea. Their eastern limit at Cape Creus is well defined but the limit is less clear on the western side where the Pyrenees extend through the Cantabrian mountains as far as the Asturian massif. The mountains of the so-called Basque depression may be regarded as an extension of the Pyrenees towards the west; the Montes Obarenes and the Cantabrian mountains (south of Vitoria) seem, in fact, to be a continuation of the outer Pyrenean ranges. According to Viers (1962), the limit is to be found in the neighbourhood of Vitoria in the Zadorra valley, which would give the Pyrenees an overall length of 510 km.

Their structure is related to the Alpine movements, although Palaeozoic outcrops, which are of great significance in explaining the systems of folds and present relief features, appear over a large part of the range. There are strong contrasts between the northern slopes, comprising a narrow and relatively uniform zone, and the southern ones, which are much more developed and possess a wide variety of macrostructures. There is also the contrast between the western or Atlantic and the eastern or Mediterranean Pyrenees, the latter being higher and more fractured.

The greatest heights are attained in the central Pyrenees [e.g., Pico de Aneto (3404 m), Pico Posets (3371 m), Monte Perdido (3352 m)]. The western Pyrenees are only slightly less elevated [e.g., Collarada (2883 m), Bisaurin (2668 m), Pic d'Anie (2504 m)], while Pic du Canigou (2785 m) is the highest point in the east. A series of transverse valleys cross the structure towards the north and the south, forming deep canyons in the calcareous and granitic sections. During the Quaternary, many of these valleys suffered moderate glaciation, with important traces remaining, especially on the French side.

11.1.1   Principal Structural Divisions

On a lithological and structural basis the following units can be recognized.

1)   The axial zone of the Pyrenees serves as the core of the range and consists of Palaeozoic and crystalline rocks. It reaches its greatest dimensions in the central and eastern Pyrenees and is more attenuated in the west, occasionally disappearing completely. On the north and south, this axial zone is flanked by parallel zones of Mesozoic sedimentary outcrops.

2)   On the French side, the so-called Ariège zone consists essentially of Jurassic and Cretaceous rocks dominated by limestones. Within the zone there are also Palaeozoic or crystalline massifs, such as that of Pic de St Barthélemy (2349 m), but elevations are usually much lower. This zone is in turn bordered on the north by the Little Pyrenees, a chain of hills stretching from the Garonne valley towards the Corbières, showing Jura-type folding and modest heights (500–750 m), the rocks disappearing northwards beneath Tertiary sediments.

3)   On the Spanish side, the arrangement is more complicated. Immediately south of the axial zone lie the Interior Sierras, which rise up with their predominance of limestones and sandstones showing great continuity from west to east with their summits often exceeding 2500 m. Then there are the southernmost ranges of the Pyrenees— the Exterior Sierras—bordering the continental deposits of the Ebro Basin. The Exterior Sierras are narrower than

the Interior Sierras but both may be traced throughout the length of the range (*see* Fig. 11.1).

4) Between the Interior and the Exterior Sierras, there are discontinuous outcrops of younger strata. From north to south, the following can be distinguished: (a) a hill zone of Eocene flysch, (b) the interior Pyrènean depression, whose best example is the Jaca–Pamplona depression or the Canal de Berdún excavated in Eocene marls and (c) an Oligocene basin filled with continental sediments. Solé-Sabarís (1951) called all these units situated to the south of the Pyrenean axis the Pre-Pyrenees. Other interpretations have limited the term Pre-Pyrenees to the Oligocene basin and the Exterior Sierras.

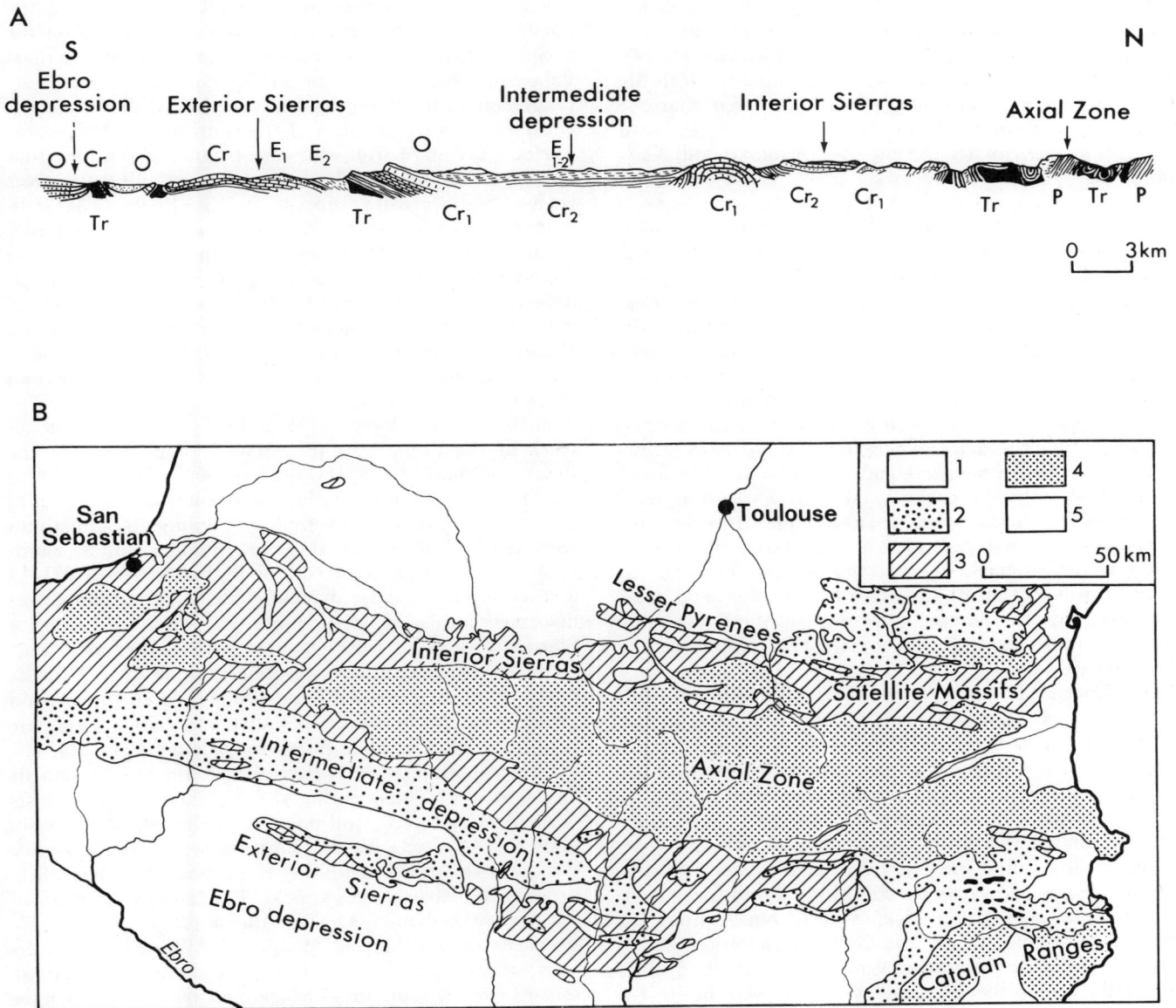

Fig. 11.1 (A) North–south section through the Pyrenees along the valley of the river Noguera Pallaresa [Solé-Sabarís and Llopis (1952)]. P: Palaeozoic and older rocks of the axial zone; Tr: Trias; J: Jurassic; Cr: Cretaceous; E: Eocene; O: Oligocene. (B) General geology and structure of the Pyreness [Solé-Sabarís and Llopis (1952)]. (1) Palaeozoic, (2) Alpine age volcanics, (3) Mesozoic (generally calcareous, except in the western sector where the upper part has a flysch facies), (4) Eocene marine and (5) Upper Tertiary and Quaternary of the Aquitaine and Ebro Basins.

11.1.2 Stratigraphic and Tectonic Evolution

The oldest rocks correspond to the Cambrian, although some Precambrian gneisses have been recognized in the eastern Pyrenees. The Silurian is characterized by schists, especially in the east, while quartzites become increasingly common towards the west. In the Devonian, with which the Palaeozoic here practically terminates (the Carboniferous only being represented by minor outcrops) limestones predominate. Hercynian tectonics appear to date from the beginning of the Carboniferous, although there are many uncertainties. The Hercynian orogeny is represented by uplifts, folding and intrusive and metamorphic activity. Large granitic massifs appear along the range (e.g., Cauterets, Panticosa, Néouvielle, Maladeta, Montlouis, La Junquera, etc.) the age of them being still in doubt. According to some, they are very late, even Mesozoic, but others regard them as having formed in two phases, the younger of which was contemporary with the Hercynian folding. Krausse (1974) considered that the Palaeozoic deformation was of Asturian age, with epeirogenic movements during the Devonian and Carboniferous.

After the Hercynian orogeny, there was intense denudation: sediments (e.g., sandstones and conglomerates with clays and shales) which are well represented in the upper reaches of the rivers Aragón–Subordán and Aragón–Somport were deposited in mainly continental conditions. Nevertheless, the Mesozoic is also characterized by sedimentation of marine facies in which the limestones and dolomites of the Jurassic and the limestones and sandstones of the Cretaceous—the latter particularly on the Spanish slopes, in which the Interior Sierras are formed—are the main lithologies. There have been many arguments about the dimensions and extent of this Mesozoic cover: whether the Palaeozoic axial nucleus was completely submerged, whether it reacted as a shallow threshold on which Mesozoic materials were also deposited or whether it was never, in part at least, submerged. There is no doubt, however, that at least one area of the axial zone was once beneath the sea, as proved by the Cretaceous remains almost at the top of the Pic Balaitous ou Marmure (3144 m).

The Eocene is well represented, especially on the Spanish slopes, although there are also outcrops in the centre and east of the French Pyrenees. Some calcareous formations may be found at the base with chalcoschists (e.g., Monte Perdido massif) passing into the flysch facies, with a thickness of 3.5–4.5 km (Cuisiense and Luteciense) and with intercalations of major calcareous and sandstone sequences, indicating the general lines of a very complex tectonic history. Towards the south, the flysch rocks pass laterally into marls with a thickness of 1.5 km. Although it would appear that during the Cretaceous Period, there were already some tectonic movements, it is really during the Tertiary that the Pyrenean structure was built. To understand the complexity and phases of the construction, it is advisable to consult Garrido (1968), Lunsen (1970), Seguret (1967, 1969), Soler and Garrido (1970), Souquet (1967) and Clin (1964).

The first great orogenic phase is post-Biarritzian and pre-Oligocene (Pyrenean phase). It was responsible for a series of north–south folds that appear along a large part of the Exterior Sierras and the Boltana anticline. These folds do not affect the Oligocene cover. Farther west, the folds are more recent and less acute. On the other hand, this phase is responsible for a series of overthrust sheets and south-dipping folds, probably the result of gravity sliding [see Soler and Puig de Fàbregas (1970)] some of which (e.g., the nappe of Cotiella) formed very early in a marine environment [see Soler and Garrido (1970)]. This series is followed by a series of recumbent folds involving over 2 km of Cenomanian–Turonian limestones, an average of 5 km of Campano–Maastrichtian and 8 km of Palaeocene limestone (Boca del Infierno, in the Hecho valley). Such recumbent folds, which dip towards the north on the French side and towards the south on the Spanish side, have permitted the role of the central Palaeozoic basement to be evaluated. The bilateral arrangement of the Pyrenean tectonic style, with a Mesozoic cover both to the north and the south of the Palaeozoic nucleus, favoured from the very beginning the idea that the cover folds were thrust nappes—an idea that was revived with greater support in the 1970s. The sedimentary cover in this case would be an area detached from the base and moved towards the south and the north. The Exterior Sierras would become the front of this great detachment. To the south, the thrust folds are seen to extend from the Peñaforca massif at least as far as Monte Perdido. To the north, the folded relief is of less importance, although the great overthrust by the Urgonian limestones is striking.

Such tectonics were responsible for the broad framework of the range. On the Spanish slope between the flysch and marl region, and the Exterior Sierras in western upper Aragón, a large continental basin was opened out in the Oligocene and filled with fluvial sediments, sometimes with deltaic facies but above all conglomerates, sandstones, clays and marls [see Puig de Fàbregas (1975)]. In a second folding phase, which is late Oligocene, this was deformed into a great synclinorium. Thus, the Interior Sierras underwent renewed deformation, the most outstanding displacement being the Gavarnie nappe. In the flysch zone, most of the folds correspond to this second phase [see Soler and Puig de Fabregas (1970)]. These are very tight folds with abundant interior detachments.

Subsequently, other phases of tectonic activity can be distinguished. Thus, during the Aquitainian, the schists were deformed at several points and the Oligocene basin folds accentuated. At the contacts with the Aquitaine Basin to the north and the Ebro Basin to the south, progressive discordances appear. The latest movements of all are neotectonic. During the Pliocene, the whole Pyrenees were subjected to uplift, while in the Quaternary, the eastern Pyrenees became an area of intense seismic activity; up to 44 volcanoes are known to have erupted around Olot.

The main orogenic phases greatly affected the Palaeozoic nucleus. In the most metamorphosed and resistant sectors (e.g., Pic du Canigou and the eastern Pyrenees, tectonic stresses created a network of fractures

Fig. 11.2 Upper Gallego valley south of Biescas. (Embleton)

with grabens and horsts (as in the Capcir and La Cerdanya); some of the grabens continued to be filled with sediment until the Pliocene. In other parts, the Palaeozoic sediments have retained a certain plasticity and have been strongly folded (e.g., Pic de St Barthélemy). Overall, the structure is enormously complex [see Sermet (1950)] and in the Hercynian massifs of the western Pyrenees (e.g., Cinco Villas, Aldudes–Quinto Real) numerous examples of inverted folds and multiple schistosities are present with varying degrees of metamorphism [see Krausse (1974)].

11.1.3 Relief and Structure

The major relief trends are related to the tectonic deformation and the relative behaviour of the different lithologies giving rise to an enormous variety of structural landscapes. The main units follow a east-south-east–west-north-west direction, although some tectonic lines—even very important ones, such as the Boltaña anticline—follow north–south trends. Nevertheless, the majority of the valleys adopt courses transverse to the structure (i.e. north–south). Thus, the principal valleys show alternating narrow and wide segments in the lithological units that they cross (*see* Fig. 11.2).

In the axial zone, there are very complicated structures whose influence on relief is difficult to establish. It is possible to distinguish the crystalline massifs from those built of Palaeozoic sediments. The former provide striking relief features and a massive, heavy appearance, although they have been much affected by Quaternary glacial action which has reduced the divides to narrow arêtes. The Aneto–Maladeta massif contains the highest points of the Pyrenees (*see* p.268); the Panticosa and the Pic de Néouvielle (3092 m) are other outstanding peaks. All of these overlook the mountains of the Mesozoic cover. There is also a surrounding metamorphic aureole that has

considerable resistance to erosion.

The Palaeozoic sedimentary formations yield very varied profiles, sometimes forming depressed areas next to the Interior Sierras, as in the case of the headwaters of the Aragón-Somport, wherever clays and slates crop out. Thus major depressions with an east–west trend appear following the general structure of the Pyrenees. The Devonian limestones, on the contrary, tend to form important escarpments. Sermet (1950) considered that the axial zone of the Pyrenees, from a morphological point of view, is essentially Hercynian due to the inclusion of the Palaeozoic block massifs and the pressence of numerous, relatively flat summit areas that revive the idea of pre-Triassic planation. The action of glaciers and a strong fluvial network have partly transformed these old massifs, some of which have evolved towards an Appalachian-type relief (e.g., the valleys of the Esera, Nogueras and West Ariège).

On the French side, the Mesozoic cover is characterized by striking escarpments that dominate long longitudinal depressions opened in marly and schist-like outcrops (sometimes also in the Keuper). Limestone crests frequently appear in a hogback arrangement or are very divided (e.g., Lourdes and Baretous) through karst processes acting on a very fractured structure. The Little Pyrenees present a folded structure of the Jura type that resulted in a para-Appalachian relief during the Pliocene–Quaternary [see Viers (1962)]. Somewhat farther west, the mountains and hills become covered by the great post-Pontian debris cone [see Birot (1937)] of the piedmont and Lannemezan platform (*see* Chapter 8, p.161).

On the Spanish side, the greater width of the Mesozoic and Tertiary outcrops allows more units to be distinguished. The so-called Interior Sierras constitute a complicated mosaic where structures ranging from simple folds to great thrust nappes and slides may be found. In the western Pyrenees, it becomes difficult to decipher the general lines of the structure, the materials being so intensely folded. In the Palaeozoic massif of Cinco Villas, the folding style is superficial, with numerous reverse faults due to the role of the underlying Keuper; when the cover rocks are strong (as within the Basque depression) a different tectonic style is evident, with Jurassic folds sometimes accompanied by fractures and occasionally interrupted by diapirs. In the central Pyrenees, a great north-facing escarpment is developed, with frequent transverse north–south faults guiding the fluvial network. Deep canyons have been formed, reaching their most spectacular proportions where two masses of limestone have accumulated, one on top of another, in successive nappes (e.g., Ordesa and Añisclo). Such strata are mainly horizontally placed, causing alternate karstified platforms [see Herranz and Carreras (1966)] and overthrust sheets forming abrupt scarps. In the Interior Sierras, it is usual for the limestones to give much higher relief than in the axial zone of Pyrenees.

To the south, the flysch zone gives much softer and less contrasting landforms. A series of ridges with rounded crests—perhaps retaining remnants of former relief—descends progressively from the contact with the Interior Sierras to the marls of the interior depression. Structure is less important here as the flysch behaves as a homogeneous material, in which erosion has removed almost all traces of tectonic influence. The presence of some large, isolated masses of thick limestone which stand out above the flysch is conspicuous. Such strata are responsible for the formation of abrupt hills and short, but very spectacular, canyons, especially if the limestones are massive (e.g., the Ansó and Fago valleys and the Santa Elena gorge in the Gállego valley).

The flysch zone has a faulted contact with the marls. At this juncture, the relief changes completely. A great longitudinal depression opens out, continuing westwards as far as Pamplona. Towards the east, after some interruptions, it may be followed as far as the Tremp Basin with some variations in width. The wide belt of Eocene marls

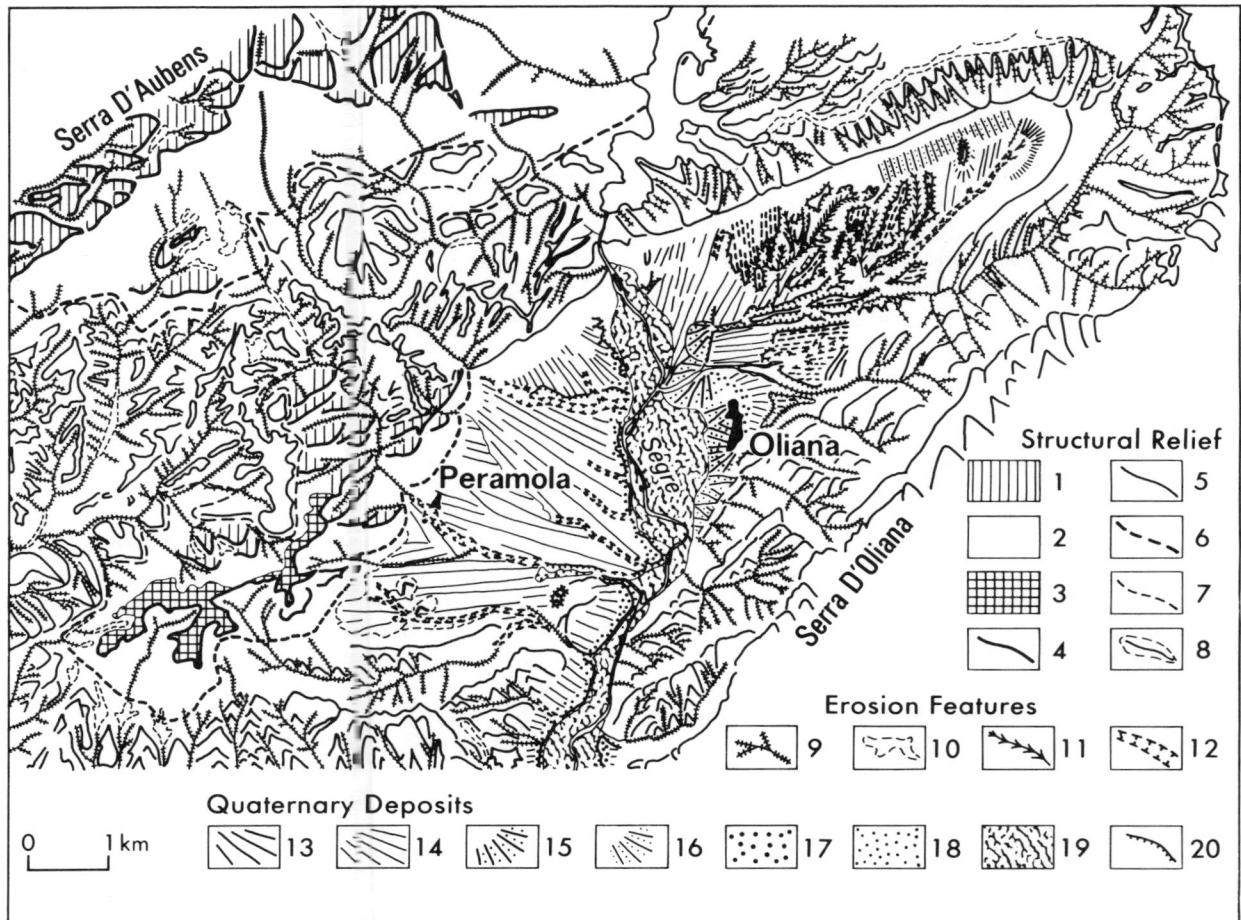

Fig. 11.3 Geomorphological map of the Exterior Sierras in the region around Oliana (Lérida) [Peña–Monné (1978)]. (1) Mesozoic–Eocene calcareous outcrops, (2) Oligocene, mainly sandstones and conglomerates, (3) intra-Oligocene planation surface, (4) scarps of resistant beds in the Mesozoic–Eocene, (5) scarps of resistant beds in the Oligocene, (6) outcrop of thrust plane (Mesozoic–Eocene over Oligocene), (7) discordant contact of Mesozoic–Eocene with Oligocene, (8) residual relief, (9) V-shaped gullies, (10) flat-floored gullies, (11) badlands in Eocene marls, (12) valleys incised in Quaternary deposits, (13) upper glacis (G4), (14) lower glacis (G2), (15) lower alluvial cones (G1), (16) block-strewn slopes, (17) upper terraces (T5), (18) low terraces (T2), (19) lowest terraces (T1) and (20) edges of glacis and terraces.

has been exploited by the rivers, which interrupt their courses transverse to the structure to follow, for a time, a longitudinal direction (e.g., Canal de Berdún). There is almost no evidence of tectonics, although the marls present a notable alternation of anticlines and synclines. Fluvial and colluvial action has masked the structure, and today a series of terraces that join up with the glacis are predominant in the lower parts of the valleys. Barrère (1966, 1979) indicated the existence of three terrace levels and an upper surface called the *coronas*—small remnants of a very old terrace and glacis level. The glacis are closely related to the marls, although their roots penetrate the flysch. Barrère (1966) considered that all the glacis are older than the maximum glaciation. The fact that occasionally the longitudinal depression has no important river flowing along it (e.g., the Val Ancha) has permitted some interesting hypotheses of river capture to be postulated [see Solé-Sabarís (1942b)]. Such hypotheses, however, are unsupported for the moment because of the lack of deposits [see Martí-Bono and Solé-Sabarís (1971)].

Farther south, the Oligocene basin represent a great synclinorium with some folds forming ranges such as the Sierra de la Peña and Oroel (1769 m) which overlook the marls of the interior depression and the sandstones and clays of the rest of the Oligocene basin. The relief consists of cuestas in which the varying thicknesses of the sandstones or the importance of particular clay strata provide some diversity. The summits of the cuestas show a mature relief.

Finally, the Exterior Sierras (*see* Fig. 11.3) are a modest replica of the Interior Sierras, although they offer greater lithological variety. The farther east they are the wider they become, along with important diapiric phenomena. Structural forms are observed frequently as the anticlinorium has been eroded into a series of cuestas and hogbacks. Outcrops of gypsum, marl and Keuper clays generate small, longitudinal depressions. The rivers cross the Exterior Sierras through gorges that may result from superimposition followed by antecedence. The edge of the Sierras is marked by an overthrust fault, which is partially concealed by the Tertiary conglomerates of the Ebro Basin which spread beyond the structural limits of the latter and covered at least part of the Exterior Sierras [see Barrère (1951)].

11.1.4 Planation Surfaces
The planation surfaces include relief forms of Tertiary or early Quaternary age, representing phases when denudation rates have kept pace with or even exceeded the rates of tectonic uplift or deformation.

In the axial zone to the east of the Garonne, numerous planation surfaces may be observed cutting across the structure. Examples of planation surfaces, which lie at between 2000 and 2900 m, include the Carlit and the Campcardós plains, the plateaus to the north of the Vall d'Aran, Pla Guillem and Campmagre. Their surfaces are very uneven and vary in height, being explained not in terms of different levels of erosion but as a result of Pliocene or inter-Quaternary tectonic deformation which reactivated old fault lines; nevertheless, the better de-

veloped levels are at 2700 and 2050 m. According to Birot (1937), there is at least one surface—the Perxa level—of Pontian or pre-Pontian age, although Birot also alluded to other older ones for which there is no precise chronology. Solé-Sabarís (1951) claimed that the highest surfaces are Aquitainian, the correlative sediments lying in the Aquitaine and Ebro Basins. The level of 2050 m is Plio-Villafranchian according to Soutadé (1980) and Pontian according to Solé-Sabarís and Llopis (1947).

At other places in the cordillera, it is more difficult to identify the existence of these pre-Quaternary levels, perhaps because of the intense fluvial activity or tectonic deformation. Nevertheless, between the Gállego and Ara Basins, ancient planation surfaces seem to be present. Barrère (1966), in his geomorphological maps of the western upper Aragón valley, marked the presence of numerous old surfaces, all of small extent and almost always found in the flysch zone. Chronologically, they may belong to the Villafranchian and even to the Pliocene, although not all of them correspond to the same phase of morphological evolution.

11.1.5 Glacial and Periglacial Features
Quaternary glaciation in the Pyrenees was of much more modest dimensions than in the Alps due to the former's more southerly position and the lower overall elevation of the mountain range. Nevertheless, there are excellent examples of glacier forms and deposits along almost the entire range, making the Pyrenees the most important glaciated area of the Iberian Peninsula. On the whole, glacial action has been far greater on the northern than on the southern side because of the differences in precipitation and insolation. Along the length of the range, from the Pic d'Orhy (2015 m) in the Basque country to Pic de Canigou (2785 m), but especially wherever the relief rises above 2500 m, there are traces of ice action (*see* Fig. 11.4).

Study of the present and former Pyrenean glaciers has attracted numerous researchers since the end of the nineteenth century. Today, many fundamental problems remain unsolved in spite of the fact that almost all the valleys have undergone more or less detailed study. Among the main problems, the number and chronology of the glaciations stand out to such a degree that the Pyrenees have become the centre of an intense controversy between monoglacialists and polyglacialists.

Only exceptionally did the Pyrenean glaciers extend to the piedmont—at Lourdes and Arudy. On the northern slopes, glaciers descended to 400–600 m, whereas in the south they melted at 800–1000 m: a difference that is reflected in the great variation in length between one side of the Pyrenees and the other, the maximum length being attained in the glaciers of the Garonne (66 km) and Ariège (52 km) valleys.

Penck (1883) noted also that the levels of the snowline and of the moraines were lower to the west because of the greater precipitation generated by the Atlantic winds. Thus, whereas in the Gállego valley the outermost moraines are found at 820 m and in the Cinca at 720 m, in the Envalira (Andorra) they rise to 1000 m, in the Querol (La Cerdanya) to 1200 m and on the Canigou massif to

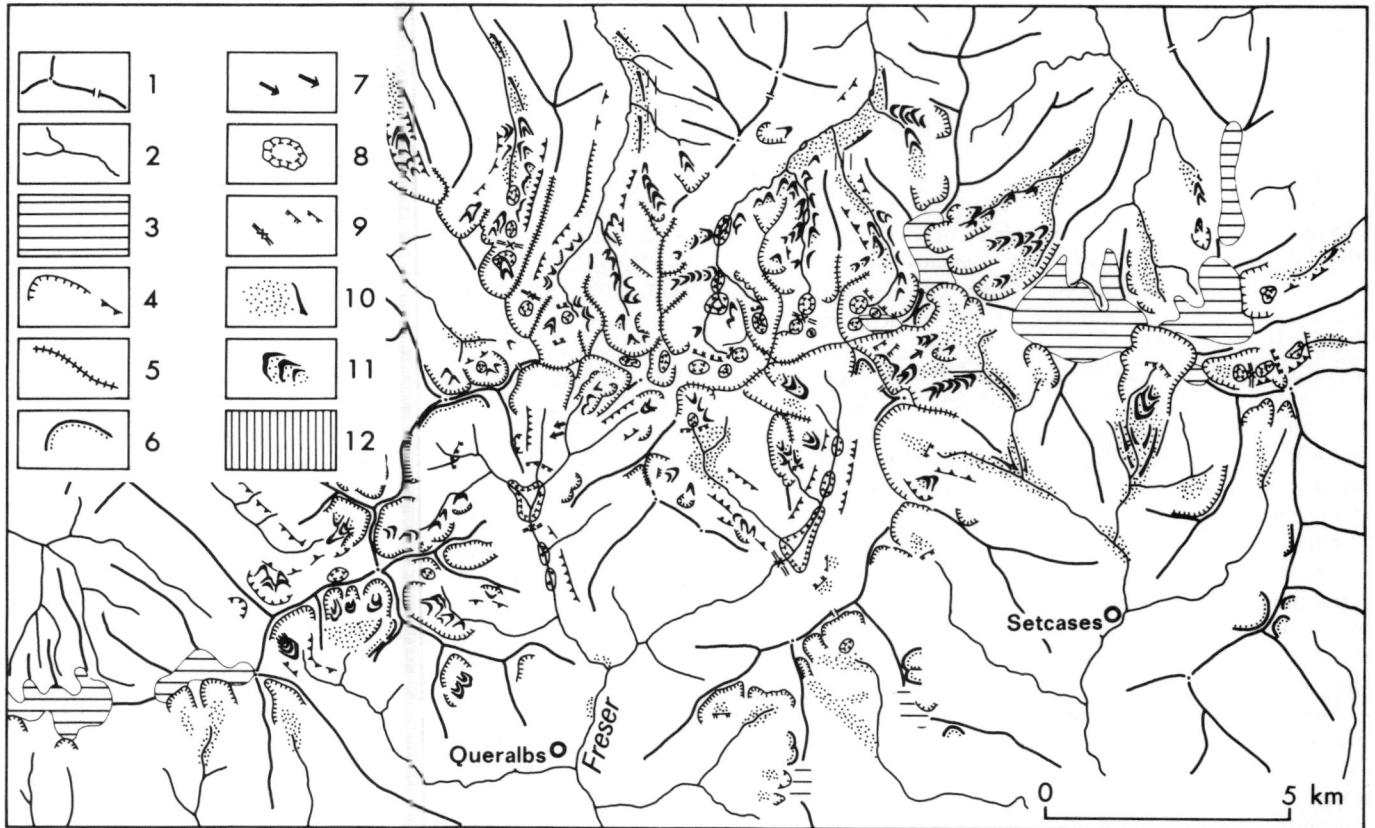

Fig. 11.4 Glacial geomorphology of the eastern Pyrenees (Puigmal–Costabona) [after Serrat (1980)]. (1) Divide (with passes and mountain peaks), (2) steams, (3) planation surface, (4) cirque wall crests and edges of glacial troughs, (5) arêtes, (6) nivation hollows, (7) transfluent and diffluent ice flow, (8) rock basins, (9) thresholds, rock bars, (10) moraines, (11) rock glaciers and (12) glacio-lacustrine deposits.

1700 m.

Ice action has left numerous traces in the landscape (*see* Fig. 11.5): cirques in the headwaters (sometimes occupied by lakes), glacier thresholds, hanging valleys, glacial troughs, microforms and, of course, frontal and lateral deposits, and some glaciolacustrine sediments. The best examples of glaciated valleys are always found in the crystalline massifs; in limestones, the glacial troughs show more characteristics of subglacial (or postglacial) erosion and the microforms have been less well preserved.

Terminal moraines are present in several Pyrenean valleys and have assisted many authors in their attempts to establish a glacier chronology. Major works include those by Alimen (1964), Barrère (1963, 1966), Panzer (1926, 1932), Solé-Sabarís (1942b, 1951), Taillefer (1957, 1969), Viers (1961, 1962), Serrat (1974) and Martí-Bono *et al.* (1978).

On the northern slopes, based on the evidence of piedmont deposits, the coloration of the soil and the different types of loess, some authors [e.g., Alimen (1964)] have deduced that six great glacier phases—from Biber to Würm—exist. Thus the formation of the Lannemezan fan, with a large part of its alterable rocks completely argilized, is attributed to the Donau or the Günz, and the state of the alteration of the rocks (especially the crystallines) has been considered an important criterion in age

determination, although this has been disputed by Taillefer (1957) and Viers (1961). According to Taillefer (1969), there is no evidence which permits the attribution of moraine-like deposits to more than one glaciation: a statement shared by other French geographers, especially Barrère and Viers. The different moraines are said to be the remains of a single glaciation, although with periods of regression and readvance. For Barrère (1963), such deposits correspond to the Riss, while Taillefer (1969) thought they were Würm, based on the freshness of the topographical forms and the radiocarbon dating of peats in the floors of the terminal basins, taking up the initial idea of Penck (1883).

On the Spanish side, the best studied moraine complexes are those of the Aragón, Gállego and Puigcerdà valleys. In the Aragón valley (at Castiello de Jaca), there are six well-developed morainic arcs (*see* Fig 11.6) and in the Gállego (at Senegüé) only one, but with moraine remnants farther downstream. Panzer (1926) considered that the moraines of the Aragón belong to two different glaciations (Riss and Würm) because they join with two terraces: one at 60 m and another at 20 m above the present level of the river. Nevertheless, Barrère (1963) claimed that all the arcs correspond to fluctuations of a single glacier phase—the Riss—leaving the Würm represented only by high-level moraines. Martí-Bono (1973,

Fig. 11.5 Pic du Canigou (2785 m) from the Lac des Cortalets (2100 m) on the north side; valley-head cirque and moraines. (Embleton)

1977) admitted, but with some doubts, the existence of three glaciations.

1) The Mindel, corresponding to the most distant moraine, associated with the 60-m terrace. He based this on the fact that this terrace has a terra rossa which, according to Alimen (1964), suggests formation in the Mindel/Riss interglacial.

2) The Riss, corresponding to the remaining arcs of the complex at Castiello de Jaca and associated with the 20-m terrace. The Senegüé arc would also be Riss, which is confirmed by the alteration of the granites.

3) The Würm, limited to the high-level moraines, at about 1500–2000 m in the western Pyrenees and confined perhaps to rock glaciers in the eastern Pyrenees.

The authors García-Ruiz and Arbella-León (1981) do not share these views and believe that the high-level moraines belong to a post-Würm cold phase, possibly of late glacial age [equivalent to the Neoglacial of Taillefer (1969)]. Some moraines, however, are related to loess deposited under cold, dry conditions . In this case, and admitting a parallelism with the French side, the morainic deposits would belong to the main Würm unless, if it were possible, they belong to the innermost arc of Castiello. There is no doubt about the existence of two glaciations because the 60-m terrace is clearly glaciofluvial as demon-

strated by Martí-Bono (1973). On the other hand, in a recent study, Serrat (1980) pointed to the existence of a moraine attributable to the Riss in the Puigcerdà complex of the eastern Pyrenees, while the rest of the morainic accumulations are, he contended, Würmian.

At times of Quaternary glaciation, there was intense periglacial activity in the moutains. This continued throughout the late glacial and, at levels above 1700 m, to the present-day. Pyrenean periglaciation, however, has only been partially studied, making it diffficult to establish general conclusions. Most workers refer to Würmian or late glacial forms, although Alimen (1951, 1964) has found deposits that are perhaps older. The greater part of the studies are related to stratified slope deposits [see Höllermann (1977), Martí-Bono and González (1979)], Serrat (1977), Solé Sugranyes (1973)], some of which are post-Würmian [see Martí-Bono (1977)]. Few studies have been dedicated to the phenomena of gelifluction or the non-sorted patterned ground, although Soutadé (1970, 1973) and Gómez-Ortiz (1981) should be mentioned (*see* Fig. 11.7). All the researchers agree on the importance of lithology, structure and aspect in the frequency and intensity of the periglacial phenomena.

Rock glaciers, the best examples of which appear in the central and eastern Pyrenees, are the product of intense periglacial action and a general decrease in the amount of snowfall [see Serrat (1979)]. Gutiérrez-Elorza and Peña-Monné (1981) placed them chronologically in the late glacial, while Soutade dated them as neo-Würmian. All the rock glaciers are situated in the floors of former cirques.

11.1.6 Coastal Features

The geological structures of the eastern and western extremities of the Pyrenees have very different characteristics which, together with the differences between the Mediterranean Sea and the Bay of Biscay, result in contrasting coastal environments at each end of the chain.

At the north-western end, the Pre-Pyrenean structures reach the coast and continue next to it, ending in the Asturian massif in the area of the Cantabrian mountains, which also mark the point of junction with the Iberian Cordillera at its northern end. From east to west, there is a succession of several units folded in a south-east–north-west direction [see Julivert *et al.* (1974)]; the littoral belt of the Zumaya flysch occupies a synclinal position extending over the northern and southern slopes of the range. Towards the south, this is followed by a coastal anticlinorial zone, by the Besair–Ebair syncline and the Bilbao anticline. West of Bilbao, there is a wide area of fractures running in a north-east–south-west direction along the coast to Santander. Farther south, the folded structures continue through a zone of gently folded strata as far as the synclinal depressions of Medina de Pomar and Miranda, and a dislocated belt forms the narrow and elongated relief features of the Cantabrian mountains and the Montes Obarenes, which seem to be prolongations of the Exterior Sierras.

Structural control is the main characteristic of this coast. Tectonic structures determine important indenta-

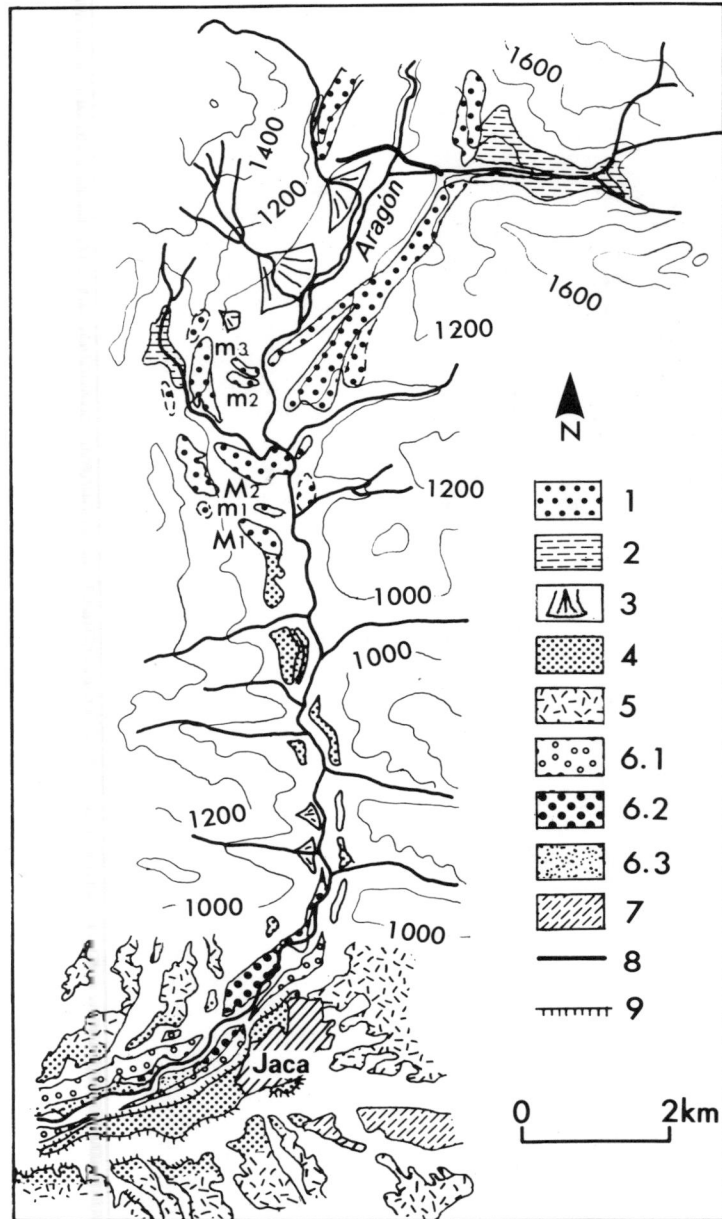

Fig. 11.6 Quaternary sediments of the Aragón valley [Martí-Bono (1977)]. (1) Moraines, (2) glacio–lacustrine deposits due to blocking of tributary valley, (3) alluvial cones, (4) 50-m terrace, (5) colluvium of previous terrace, (6) 15-m terrace, (7) 8-m terrace, (8) 1–2-m terrace, (9) structural or residual relief, (10) glacial threshold (rock bar) and (11) edge of terrace. M₁, m₁, M₂, m₂, m₃: morainic arcs of the Castiello complex.

tions, which are the only interruptions of this steep coast, consisting of a succession of cliffs and rias or bays of varying size. The area around San Sebastian is more open due to the flysch materials, being the only place with any significant beach accumulation. The other zone with low coastal features is the Santander marina.

As in the coastal zones of the Asturian and Galician massifs, the most important geomorphological features are the rias and the rasas as described by Mensching (1965). The rias are shorter and narrower than those of Galicia and are lacking in branches due to strong structural control. The largest ria is that of Bilbao, being 10 km long.

The rasa (*see* p.302) is not nearly so well developed as along the Asturian coast, although it is easily recognized in the Santander area, but it is reduced to very small hanging remnants farther east.

The structure of the eastern end of the Pyrenees is the result of rift tectonics of Neogene–Quaternary age [see Fontboté and Guitard (1958), Solé-Sabarís (1962), Martínez-Gil (1972)]. Near the Mediterranean Sea, faulting has produced grabens on each side of the cordillera, that of Rosselló to the north and that of Empordà to the south, separated by the horst of Alberes. The horst is made of materials from the axial zone and both grabens are filled

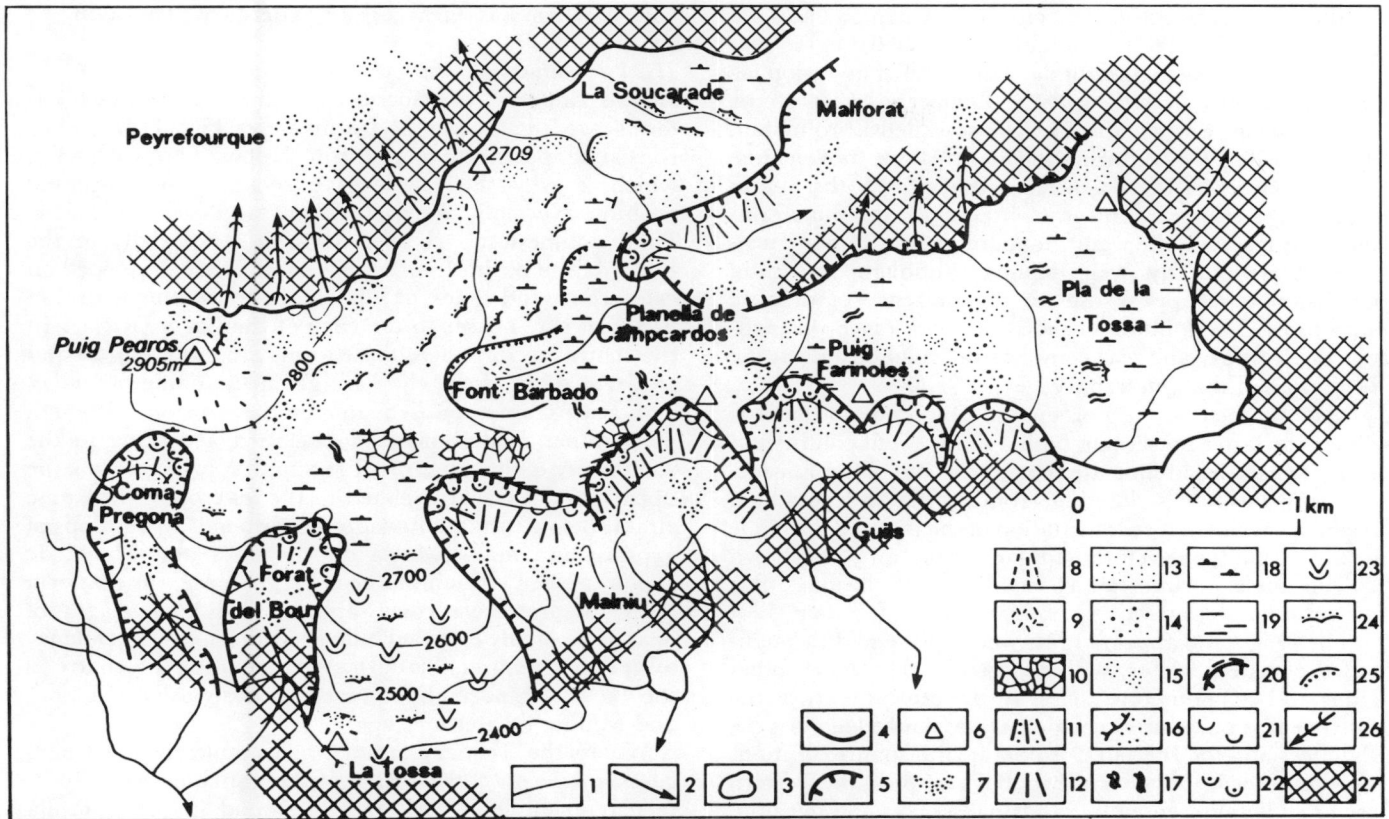

Fig. 11.7 Periglacial geomorphology and vegetation of the Campcardós area [Soutadé (1980)]. (1) Contours, (2) rivers, (3) lakes, (4) ridge crests, (5) cirques, (6) small granodiorite tors, (7) blocky moraine, (8) rock-strewn slopes, (9) cryoplanation surface, (10) stone circles, (11) active screes, (12) vegetated screes, (13) smooth grassed area, (14) stony grassed area, (15) eroded grassed area, (16) talus front, (17) block flows, (18) areas of active buttes gazonnées, (19) former buttes gazonnées being destroyed, (20) cirque border benches, (21) terracettes with *F. eskia* (22) terracettes with *F. scoparia*, (23) lobes, (24) lobes arranged in series, (25) nivation hollows, (26) avalanche chutes and (27) pine and other forests.

with detrital materials of both continental and marine facies, the latter being related to marine transgressions during the Miocene and Pliocene. Thicknesses reach 700 m in the Rosselló depression and more than 400 m in the Empordà depression; in the latter case, volcanic fragments exist within the formation. Both plains are limited by steep fault scarps at whose foot pediments provide a connecting element with the plains.

Coastal features are of two types: (1) the steep cliffs of Cape Creus with abrasion surfaces at different levels as described by Barbazà (1971) and with small narrow inlets (*calas*) often due to the differential resistance of the dykes intruded into the local rocks and (2) the wide bays of Rossello and Emporda with important lagoons (Canet and Salses in the northern depression, and Pals and Castelló d'Empuries in the southern one). The lagoons were still navigable during medieval times but have been mainly transformed into agricultural areas at present. The strong north-easterly winds blow the sand inland producing dune ridges [see Obrador *et al.* (1971)].

11.2 Ebro Basin

This is a large triangular basin opening towards the south-east, resulting from subsidence during the lower Tertiary

with synchronous infilling by continental deposits. It shows next to the Pyrenees an area of deeper subsidence which may be considered an atypical foredeep of the Pyrenees. At the beginning, it was closed from the sea by mountainous barriers which delimited it to the north (Pyrenees), to south and south-west (Iberian Cordillera) and to the east and south-east (Catalan coastal ranges), although the Ebro found a way through these coastal ranges towards the end of the Tertiary or at the beginning of the Quaternary—a decisive factor in explaining the region's present-day morphology. The Ebro Basin presents a simple relief pattern related to a structure that has been relatively little affected by tectonics. In outline, structural platforms standing out above the general relief (*muelas*) may be discerned at heights varying from 600 to 800 m, crowned by limestone strata, while the valleys and basins (*hoyas*) open out in clays, marls and gypsum at heights not greater than 500 m. Much of the Ebro Basin is characterized by a wide range of depositional landforms with glacis and terraces *en échelon*; in the driest parts, basins of internal drainage may be found.

11.2.1 Origin and Structure
The Ebro Basin was formed by slow subsidence accompanied by uplift of the marginal ranges. This may well

be the result of isostatic compensation. Today, all authors [see Riba (1955, 1964), Crusafont *et al.* (1966)] agree that the area represents an example of great relief inversion, as there are signs that, before the Oligocene, part of the present Basin was occupied by the so-called Ebro massif. This massif subsided during the meso-Alpine movements, the framework of the Basin being outlined by the mountain ranges rising around it. Contact with the mountainous margins is, on the whole, topographically, lithologically and structurally well defined, although a certain confusion may appear due to the presence of diapiric structures. The infilling of the Basin took place throughout the Tertiary and was complete in broad outline at the beginning of the Quaternary.

In some sectors, the Eocene and Oligocene formations differ, the latter presenting both a marine and continental facies which would indicate that, at the end of the Eocene, the Ebro Basin was closed off from the sea and became a basin of continental sedimentation of the internal drainage or lacustrine type even though lacking in unity, as shown by its division into many smaller lake basins, thus providing a certain variety of facies.

The Oligocene appears to be well represented throughout the eastern sector, with sandstone, clay, marl, limestone and conglomerate facies. In the central section, the Oligocene is made up of conglomerates banked against the Pyrenees and the Iberian Cordillera, with some exceptional gypsum beds in the vicinity of Barbastro. In the western sector, it is represented by sandstones, clays and gypsum. Normally, the Palaeogene materials appear slightly disturbed as their sedimentation was synchronous with Alpine tectonics. Nevertheless, it is a case of simple, regular folds that only become more complicated where the gypsum appears. The contact with the edge of the Pyrenees shows progressive discordance towards the higher part of the sedimentation sequence caused by reduced tectonic activity. In other places along this contact, the Alpine materials appear to have been thrust over the Tertiary of the Basin (post-Oligocene phases) with extremely complex fault systems.

The Miocene is wholly continental, with facies of clays, marls, gypsum and limestones, and appears to be very little tectonized, although it shows frequent large faults in a north–south or north-east–south-west direction, such as those mapped by Pérez-Lorente (1979) in the Rioja region of the Basin.

Where the gypsum beds are massive, the Miocene may appear very deformed (e.g., the anticlines of Lodosa and Arnedo–Quel) the disturbances sometimes even affecting the Quaternary fluvial deposits. It is difficult to establish a chronology for the faulting of the Basin, but many of the faults are clearly intra-Quaternary and the majority— even those on the contact with the Iberian Cordillera— have been active at times during the Quaternary [see Calvo-Palacios (1975)].

11.2.2   Relief
The relief of the Ebro Basin shows a simple pattern. On a large scale, there are structural forms basically linked to lithology. Superimposed on these is a fluvial net of open valleys which has encouraged Quaternary deposition.

11.2.2.1   Structural forms
In this category, the horizontal and subhorizontal platforms are the dominant features (*see* Fig. 11.8). They consist of posttectonic resistant Upper Tertiary series, which is why the strata have retained their original position. Normally, such platforms are developed on the Pontian limestones (Upper Miocene), especially in the centre of the Basin. In other places, they are developed on calcareous sandstones, or sandstones alternating with thin bands of clay nearer to the edge of the Basin (Rioja). In the central sector where these forms are more typical and of larger dimensions, they are given the name of *muelas*. The scarps are typical of differential erosion, although many other models may be discerned according to the complexity of the outcrops. The simplest profile shows the upper part of a scarp related to the outcrop of limestone strata. Below this, a talus slope corresponds to outcrops of gypsum and marl (Muela de Zaragoza and Muela de Jaulín). When erosion has been very intense, the platform loses its protective cover and, in the later stages of denudation, only rounded hillocks with a marked tendency to gullying remain. Platforms farther from the centre of the Basin are normally covered with deposits of Upper and Middle Quaternary age.

Where the Tertiary strata are tectonically deformed, cuestas develop after erosion of the anticlinal axis. Spectacular examples of this may be found in the Najerilla valley (Rioja), near the Sierra de la Demanda and in the sandstones of the Sierra de Santo Domingo (Exterior Sierra).

Forms developed on the conglomerates of the Basin margins are equally striking. The best examples are found both at the foot of the Pyrenees (Agüero, Riglos, Guara) and in the Iberian Cordillera (Matute, Anguiano, Viguera), or in the Catalan coastal ranges (Montserrat, San Lorenç de Munt). These present very abrupt outlines with vertical scarps facing towards the Basin, formed by both differential erosion and solution as the conglomerates are cemented by a calcareous matrix. The lateral change from conglomerate facies to sandstones is extremely rapid to the point where, once these are stripped, the conglomerates are isolated in great monolithic blocks with rounded tops.

Another form of differential erosion is related to the palaeochannels situated within the sandstones of the Híjar–Alcañiz [see Ibáñez-Marcellán (1976)] and other detrital sectors of the Basin. The palaeochannels now form, after relief inversion, ridges of sandstones several metres high and hundreds of metres long (some even attain lengths of several kilometres). These correspond to a network of poorly organized gullies that became installed during the Upper Tertiary. The channels were filled with sandy deposits and later covered by marls. Erosion of part of the cover revealed the sands—converted into sandstones—leaving them standing out above the rest of the relief due to their greater resistance.

11.2.2.2   Fluvial relief forms
The fundamental characteristic of the valleys in the Ebro

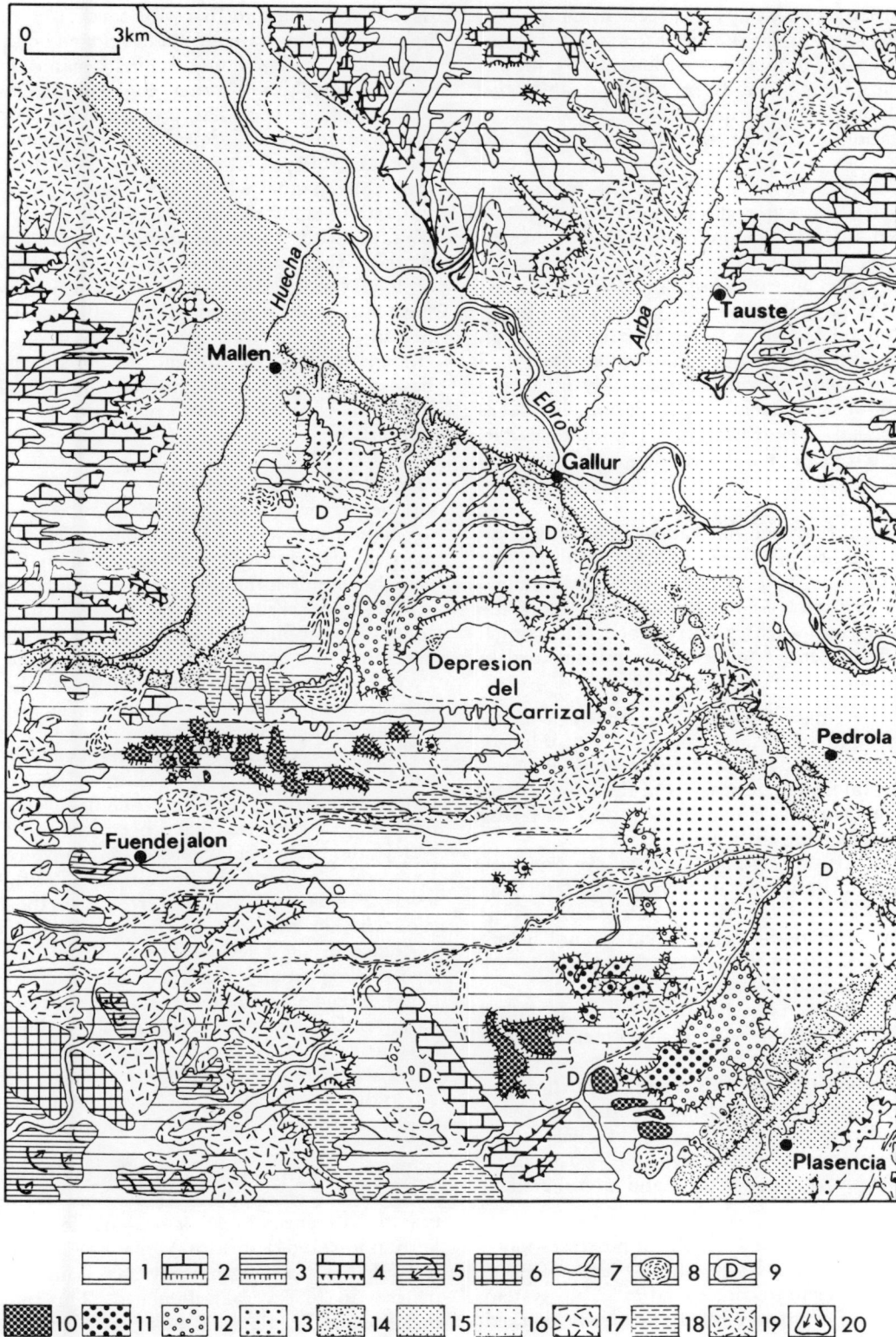

Fig. 11.8 Geomorphology of the central sector of the Ebro Basin [after Ménsua and Ibáñez (1977)]. (1) Bedrock outcrops at low levels relative to the Quaternary levels, (2) derived structural platforms, (3) bedrock outcrops in dominant positions relative to the Quaternary levels, (4) dominant structural platforms, (5) dominant exhumed Mesozoic outcrops, (6) dominant Palaeozoic outcrops, (7) small flat-floored valleys, (8) closed depressions and (9) breached depressions. *Terraces*: (10) Plio-Pleistocene; (11) sixth level; (12) fifth level; (13) fourth level; (14) third level; (15) second level, (16) lowest terrace. *Glacis*: (17) high surface; (18) middle surface; (19) low surface; (20) alluvial cones.

Basin is their broad, trench-like cross-section. Only in the valleys near the edge of the Basin (especially near the Pyrenees) do examples of marked incision appear, mainly because these are areas of sandstone, with greater relief energy and with only thin alternations of marls and clays. A similar situation occurs in the valleys cutting the Catalan ranges.

The shape of the valleys is controlled by lithology. Once the channels reach the softer clays, marls and gypsum, and reduce their downstream gradients, lateral erosion and slope retreat through the development of glacis prevail. The process of valley widening is helped by meandering and slope undercutting (e.g., the rivers Jalón, Aragón, Ega, Aguas, Matarraña and, in part, the Ebro). Along the larger rivers, numerous examples of abandoned meanders remain, some of which are already filled and cultivated, while others form small permanent lakes (*balsas de Ebro viejo*). The Ebro, in its middle course, displays spectacular examples of meanders which are very close to being cut off, such as those downstream from Zaragoza.

It is normal for other trench-like valleys to be symmetrical with the channel bed more or less equidistant from the two banks. In this case, both slopes have well-developed glacis, as well as terraces *en échelon*. But at other places, the valley is clearly asymmetrical, the river bed being much closer to one side. Several arguments are put forward to justify such asymmetry.

1) When the asymmetry is characterized by a steep, west-facing slope, climatic or aeolian causes are adduced. Winds—frequently strong and sometimes very violent— are typical of the intermediate seasons in the Ebro valley, so that aeolian deposits accumulate against the east-facing (leeward) slopes, reducing their angle.
2) If the asymmetry affects the large valleys and also follows a rectilinear trend, the presence of a fault line controlling the course of the river is considered. This seems to be the case in the Huerva valley just before Zaragoza, and especially in the case of the stretch of the Ebro between Gallur and Zaragoza.
3) The asymmetry has even been attributed to large-scale tectonic tilting, linked with readjustments of the surrounding mountain chains, resulting in the slow displacement of the river course. This would happen at the expense of one valley side which would then be over-steepened. On other occasions, the cause has been attributed to local movement in the gypsum beds, which would provoke warping on one margin and lateral erosion on the other.

Another set of valleys in the Ebro Basin possesses more subdued cross-profiles without any sharp breaks of slope at the margins. They seem to correspond to an old system of relief, with the valleys partly filled and partly marked by the presence of clays on which slope angles do not retreat parallel to themselves, but decline with time. These 'vales' appear to be closely linked to the gypsum outcrops around Zaragoza. Some are several kilometres long and more than a hundred metres wide, and are framed within a desert-like landscape with slopes almost devoid of vegetation. In a wide sense, they are fossil forms in which an old drainage net developed on the gypsum has been filled by very fine, whitish material (perhaps the result of aeolian reworking of gypsum) so that today their floors display very flat profiles transversely, but with gentle longitudinal slopes. Along the centres of smaller tributary valleys—almost always dry—there is sometimes a narrow channel which may have originated by piping.

Lastly, among the major fluvial forms are the basins in contact with the surrounding mountain ranges. In Aragón, these are termed *hoyas* (e.g., Hoya de Huesca), and, in Catalonia, *concas* (e.g., Conca d'Odena). These have probably originated by differential erosion, especially if there are any major clay outcrops that are easily eroded. Glacis-type landforms predominate in such basins, although the topography is not homogeneous, with numerous structural platforms in the interior [see Calvet (1977)].

11.2.2.3   *Quaternary deposits*

In the larger valleys and basins, forms of Quaternary deposition are very well developed. Except in the asymmetrical valleys, well-developed systems of terraces are usually found, although, because studies of them have always been regional, it is not easy to reach general conclusions applicable to the whole of the Basin. The best-developed systems of terraces are along the Ebro and its Pyrenean tributaries (*see* Fig. 11.9). In the main valley, Ménsua and Ibáñez (1977) recognized six levels in its central section. Six levels have also been found in the Gállego, Huerva and Arba de Luesia valleys. On the other hand, Van Zuidam (1980) claimed that there are four terrace levels along the Ebro, each being subdivided into a further two. Four sets have been identified in the Jalón valley, and in the Rioja valley Gonzalo-Moreno (1977–78) spoke of five levels. Pérez-Lorente (1979) mapping the levels of accumulation and erosion in the central Rioja (i.e. whether they are glacis or terraces) indicated the existence of eight levels. The truth is that many of the levels are partially fossilized and overlain by alluvial cones, linked according to Pérez-Lorente (1979), to reactivation of faults in the late Quaternary. These faults may, in some cases, have caused tilting, which would alter both the absolute and relative heights. Gonzalo-Moreno (1977–78) indicated the presence of numerous deformed remnants of terraces related to movements in the gypsum beds.

The glacis (*see* Fig. 11.10) take the form of low-angle ramps which extend from the foot of the Iberian Cordillera, the Pyrenees or structural platforms [see Ménsua-Fernandez (1964), Frutos-Mejias (1968), Ibáñez-Marcellán (1976)]. In length, they may run for hundreds of metres up to several kilometres. The thickness of the deposit on the glacis is also very variable, sometimes exceeding 5 m when the glacis have originated by the coalescence of cones. The material is angular or subangular, sometimes with a clayey sandy matrix. In the centre and east of the Basin, the glacis usually possess calcrete crusts (as in the case of many terraces) typical of arid regions.

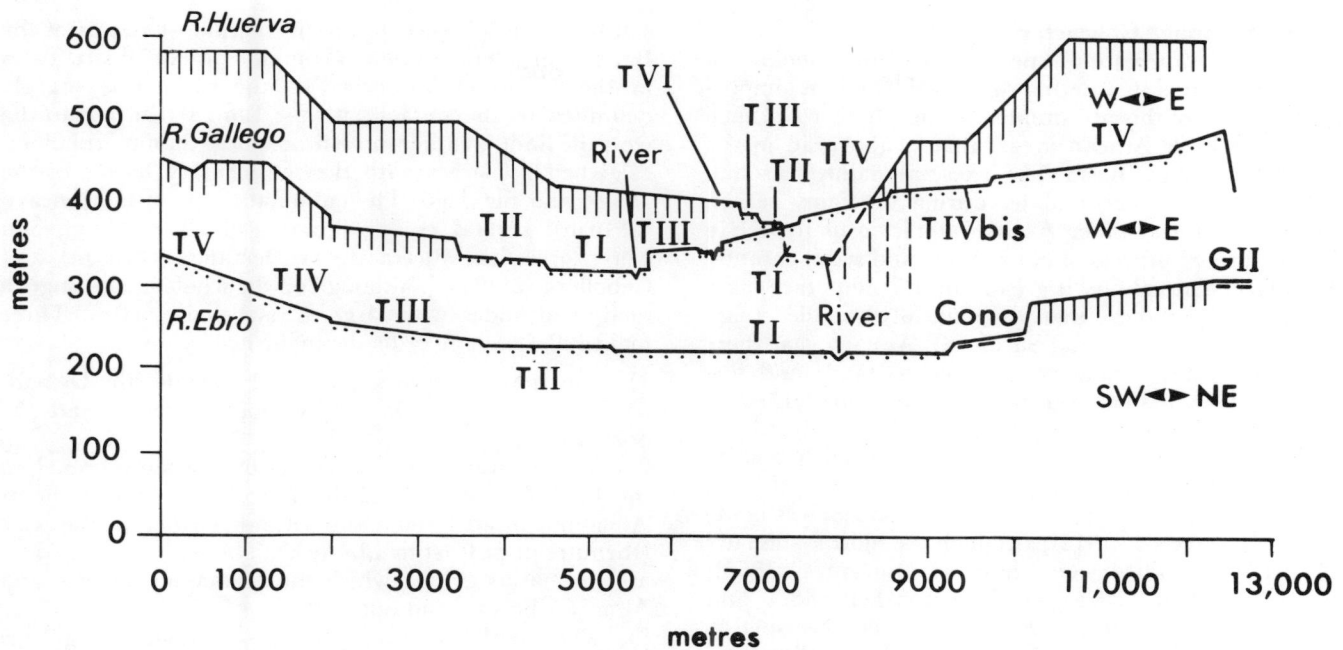

Fig. 11.9 Cross-profiles of the Huerva, Gállego and Ebro valleys, showing terrace sequences. Stipple indicates surface underlain by gravels; the shading marks bedrock [Ménsua and Ibáñez (1977)].

Fig. 11.10 The Ebro Basin near Villamayor (Zaragoza), showing glacis penetrating between and around the hill. (Sala)

The links between glacis and terraces seem to be clear, although there are some differences of opinion. The slight differences between the formation suggest that the glacis are slightly younger than the fluvial accumulations, which are themselves partly fossilized in places [see Ménsua-Fernandez and Ibáñez-Marcellán (1977–78)]. This would indicate a change in climatic conditions, causing a substitution of the longitudinal fluvial deposits by the lateral glacis deposits. As in the case of the terraces, the number of glacis varies from one sector to another: in the Rioja, Gonzalo-Moreno (1977–78) recognized up to six levels, while Ménsua-Fernandez and Ibáñez-Marcellán (1977) pointed out the existence of three clear levels in the central Aragón sector. The oldest levels, which are well developed on the edges of the Basin, have been dated as Pliocene–Quaternary [see Gonzalo-Moreno (1977–78)].

11.2.3 Drainage Characteristics

A particularly interesting aspect of the geomorphology of the Ebro Basin is its internal drainage. The most important examples of this are situated in the Monegros, Cinco Villas and lower Aragón areas, that is above all in the central sector of the Basin. The lakes are mainly seasonal, as the water that accumulates during the rainy season evaporates in the summer. As the water is normally rich in salts, an annual process of evaporation and sedimentation takes place. The salinity is greater in the temporary lakes and decreases in those that are artifically supplied (e.g., La Estanca de Alcañiz and Sariñena). As each year there are further contributions of salts, clays, muds and fine sands, the lakes tend to overflow. Some small deltaic cones contribute to this. Normally these lakes do not exceed 1 m in depth, although La Estanca de Alcañiz reaches up to 14 m and that of Sariñena up to 4–6 m.

The first attempt to explain these phenomena was made by Dantín Cereceda (1912), although the most systematic and detailed studies were those of Quirantes (1965), Ibáñez-Marcellan (1975) and Ménsua-Fernandez and Ibáñez-Marcellan (1975). Quirantes (1965), focusing on the internal drainage of the Monegros, adduced lithological and structural reasons. The lakes are contained in basins formed by subsidence of the limestones and gypsum which are strongly faulted and jointed. In fact the edges of some lakes are marked by scarps which resemble those of great dolines. Ibáñez-Marcellán (1975) added another factor, that of the marked structural horizontality which makes the concentration of surface runoff difficult. She also stressed the role of lithological contrasts. Thus, the lakes appear by a process of differential erosion wherever there are contacts between materials of unequal resistance. The carving of the basins took place during the Middle and Lower Quaternary. In later work, Ménsua-Fernandez and Ibáñez-Marcellán (1975) pointed out that, in some cases, aeolian deflation could be suffficient reason to explain some internal drainage basins. In fact, many of the basins are situated along the path of the dominant winds (north-west–south-east), although no definite conclusion has been reached on this point.

11.3 Iberian Cordillera

The Iberian Cordillera is located on the eastern margin of the Hercynian system and has a north-west–south-east trend. It consists not so much of a single, unified mountain system as of a combination of sierras in which the history of tectonic deformation and the relief forms reflect the presence of both old massifs and Alpine chains. The Palaeozoic basement outcrops in numerous places, but Mesozoic strata are much more widespread. Depressions of Alpine origin filled with posttectonic Tertiary materials are also found. Thus, a complex system, sometimes a little disconnected, of great length and also considerable width exists. To the north, it extends as far as the northernmost spurs of the Sierra de la Demanda, even connecting with the Cantabrian mountains to form a nucleus where the Hercynian system (i.e. the Asturian massif) and the Alpine system (i.e. the Pyrenees) meet; to the south and

south-east, it is joined to the northernmost sierras of the Baetic Cordillera creating a complex mosaic of structures in the region of Valencia. To the east, it is sharply delimited by its contact with the Ebro Basin, but to the west the limit is rather more lithological than morphological when it merges with the sedimentary basins of the Duero and the Tajo. The culminating point is Moncayo (2316 m); several points also exceed 2000 m [e.g., San Lorenzo, in the Sierra de la Demanda (2265 m) and Cebollera (2142 m)], although on the whole it is a range of modest altitudes with large areas below 1500 m. Three main relief units may be distinguished.

1) The north-eastern sector which includes the Demanda massif, the Urbión and Cebollera sierras and the Sorian plateau.
2) The central sector consisting of two large branches: the first in contact with the Ebro Basin (the outer or Aragonian branch) and the second, farther to the west (the interior or Castilian branch). These are separated by a set of basins among which the Calatayud corridor and Almazán Basin stand out.
3) The southern sector comprises mountain groups which are relatively isolated from each other: the Cuenca mountains, the modest, but abrupt alignments of the Valencian littoral and the sierras of Albarracín (2019 m), Javalambre (2020 m) and Gúdar (2024 m).

11.3.1 Geological Structure

For Julivert *et al.* (1974), the Iberian Cordillera constitutes an example of the 'intermediate' type of range. In spite of the great thickness of Mesozoic sediments in some parts and in spite of the existence of quite intensely deformed areas, neither the sedimentary evolution nor the tectonic style are characteristic of a truly Alpine cordillera [see Richter and Teichmüller (1933)]. To this may be added the lack of metamorphism and the almost total lack of post-Hercynian magmatic activity.

The Iberian Cordillera is a typical 'basement and cover' structure. The materials of the basement are formed by Precambrian rocks, which outcrop only to a limited extent, and by Palaeozoic rocks as described by Colchen (1974)]. The characteristics of these materials are related to the western Asturian–Leonese zone of the Iberian massif. The cover materials are Mesozoic and Palaeogenic, with a more or less complete sequence lacking sharp unconformities. This does not mean, however, that no noteworthy changes in sedimentation occurred throughout the Mesozoic, but that the changes were related to fracturing, mainly in the lower Cretaceous. In spite of these intra-Mesozoic movements, it was not until the Tertiary that major deformation occurred. The discordance, which is widely observed throughout the Cordillera, is the pre-Aquitainian uncomformity [see Riba and Rios (1960–62) that separates the deformed cover from the posttectonic materials. Some signs of Miocene compression are also observed locally [e.g., to the north of the Sierra de la Demanda as described by Colchen (1974) or in Daroca as described by Julivert (1954)].

The resulting structures display the characteristics of a basement and cover tectonic style [see Richter and

Teichmüller (1933)]. The basement was fractured into blocks and, in some cases, is clearly thrust over the cover materials. The cover partly adapts itself to the structure of the basement, thus giving some conformable structures but is also partly folded over itself, involving the formation of nappes related to plastically deforming strata. Folded and tabular areas coexist in the Cordillera. The folds show a general north-west–south-east direction in the northern part. Towards the south, some of these folds twist towards a north–south direction, retaining this until they meet the Sub-Baetic area, dominated by the Baetic Cordillera where the structures bear an east-north-east–west-south-west direction. The junction between the two trends marks the limit between the Sub-Baetic zone and the Iberian Cordillera. The groups of folds near the Ebro Basin, on the other hand, first take on a west–east direction and then turn south-west to north-east, linking with the folds of the extreme south of the Catalan coastal ranges [see Hanne (1930)].

One feature of the Iberian Cordillera that must be emphasized is the fact that, in the Neogene, this area was affected by the fracture tectonics which developed over large areas of Europe and the western Mediterranean, whether they belonged to the domain of the Alpine cordilleras or not. This tectonic style—tension type—was responsible for the long Calatayud–Teruel Basin [see Julivert (1954)] and the complicated block structure of the Valencian part of the Cordillera [Brinkmann (1931)].

11.3.2 Structural Landforms and Planation Surfaces

Because of the internal complexity of the Cordillera and the existence of more or less independent units, the area will be examined under three main divisions.

11.3.2.1 *The north-western massifs and the Moncayo*

The whole of this area consists of a large Palaeozoic block (Sierra de la Demanda) largely stripped of its Mesozoic cover. The latter becomes more important towards the south, while towards the north, along a faulted contact with the Ebro Basin, it presents a thickness sometimes less than 200 m. The Sierra de la Demanda broadly corresponds to a syncline, whose northern flank appears as a series of cuesta ridges marking quartzite outcrops. The whole of the massif shows monotonous outlines with a series of gently rounded summits at about 1900–2100 m. This may indicate the existence at this height of the remains of a former planation surface of indefinite age. At a lower level of about 1300–1500 m, another planation level may be recognized, marked by accordant interfluves far above the present stream valleys, the date of which is controversial. Nevertheless, it seems probable that some of these surfaces are the vestiges of late Tertiary planation, subsequently uplifted. The northern Mesozoic cover is intensely tectonized with numerous faults and occasional thrustfolds (Cerro Peñalba). The presence of the Trias has facilitated the development of a wide valley which runs along the edge of the Sierra de la Demanda on its northern flank. Towards the south and the east, the Mesozoic sierras that border the Palaeozoic block show simpler shapes. Thus, in the area of Soria, there are regular anticlines and synclines with low dips. A planation surface has bevelled the whole of this area, except for some monoclinal structures in limestones that stand out above it. Farther to the south-east, the Moncayo massif stands as an isolated block between the Duero and the Ebro Basins, rising above the sierras that surround it. The Moncayo, raised by Alpine movements, is made up of sedimentary and semimetamorphic materials whose age is still uncertain; the most recent hypothesis by Riba (1971) supports a Permian or Permo-Triassic age. The Palaeozoic nucleus seems to be Cambrian or Silurian. Schematically, the Moncayo is an asymmetrical anticlinorium, whose axis (from north-west to south-east) follows the normal direction of the Iberian system. Erosion has been intensely active on its north-east face (through the more active tributaries of the Ebro) and has opened a tectonic window to expose the Palaeozoic nucleus [see Pellicer (1980)]. The remnants of a deformed planation surface are preserved on the summits.

11.3.2.2 *The central sector*

This is the sector where the Cordillera acquires its greatest complexity. A distinction should be made between the Aragón branch farther east and the Castilian branch to the west. In the Aragón branch, a superimposition of tectonics (Hercynian and Alpine) is found, the latter being responsible for the series of upfaulted and downfaulted blocks. It is essentially a basin and range landscape with only a thin Mesozoic cover that has been considerably denuded. A large depression (Calatayud–Daroca–Teruel) and some smaller ones interrupt the homogeneity of the system; it is in these basins that the Mesozoic strata are now mainly preserved.

The summit areas are rounded, representing a former planation surface whose age is difficult to define. At a lower level, there is another planation surface (at about 1000 m) which topographically is linked with other levels developed on the Mesozoic and with the summits of Tertiary outcrops in the Calatayud and Almazán Basins. This is clearly a late Pontian erosion surface. The Palaeozoic horsts are flanked by Mesozoic strata that were tilted during uplift of the blocks and that have since been dissected to form typical cuesta relief.

In the Calatayud–Daroca depression, a sedimentary sequence is found normal to all closed basins, with detrital materials at the margins and chemical sediments in the centre. All the sediments are posttectonic and therefore maintain their original attitudes with slight dips; there are local deformations related to the mobility of the gypsum [see Bomer (1956)]. The relief results from the erosive activity of streams on a basically horizontal structure; thus there are wide, trough-like valleys and structural platforms capped by resistant Pontian limestones. Where the limestone cover has been removed, excellent examples of glacis have developed on gypsum. These in turn have also been dissected (*see* Fig. 11.11).

The Aragón branch is separated from the Castilian branch by a series of apparently disconnected basins, such as the Enbid rift which is simply a north-westerly extension of the Calatayud–Teruel and Almazán de-

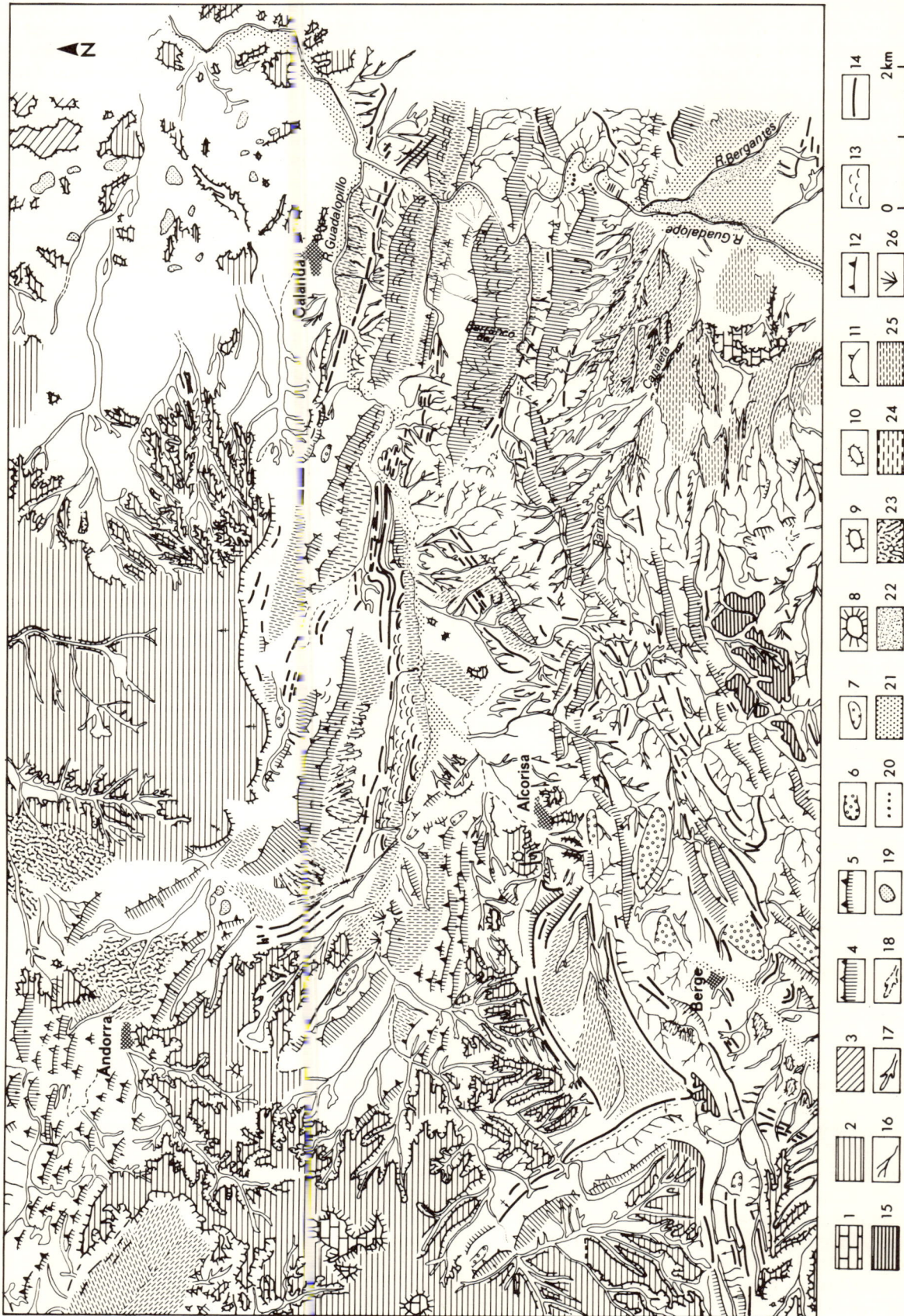

Fig. 11.11 Geomorphological map of the Iberian piedmont [after Ibáñez–Marcellán (1976)]. *Structural relief:* (1) Upper Tertiary limestone; (2) Upper Tertiary conglomerate; (3) Upper Tertiary sandstone; (4) Mesozoic limestone; (5) Lower Tertiary limestone cuestas; (5) Lower Tertiary cuestas (conglomerates, sandstones, limestones); (6) anticlinal structures; (7) synclinal structures. *Forms of differential erosion:* (8) edges of platforms; (9) edges of platforms, more than 20 m; (10) edges of platforms, less than 20 m; (11) escarpments more than 40 m (11) escarpments, more than 40 m (11) escarpments more than 20 m; (12) escarpments less than 20 m; (13) chevrons; (14) hogbacks. *Erosional relief:* (15) well-preserved planation surfaces; (16) V-shaped gullies; (17) gullies with clearly marked flat floors; (18) gullies with flat floors merging into valley sides; (19) basins of internal drainage. *Depositional forms:* (20) upper terraces, Ebro system; (21) middle terraces (middle late Quaternary); (22) present and recent river alluvium; (23) upper (early Quaternary) glacis; (24) middle glacis; (25) lower glacis (late Quaternary); (26) cones.

pressions. There is also an extension towards the east of the Tertiary basin of the Duero, with its typical relief of structural platforms capped by Miocene limestones. The tectonics are Saxonian (Permian), with large folds bevelled by a planation surface between 1050 and 1200 m. The tributaries of the Ebro have developed wide, flat-floored valleys, sometimes locally marshy and with abrupt sides, of Middle Quaternary and perhaps even Lower Quaternary age. Where erosion has reached the Keuper, the valleys become even wider. Above the planation surface, some residuals such as the Sierra Ministra stand out, but topographic horizontality is the dominant theme.

11.3.2.3 *The southern sector*

The southern section is broken into several individual units, among which the Sierra de Alberracín, the upper Jiloca depression (an extension of the Calatayud–Daroca depression), the Sierra de Gúdar and the Sierra de Javalambre and the coastal mountains are the main elements.

Albarracín is considered to be a southward extension of the Castilian branch of the Iberian Cordillera, although it stands out far above this due to its superior height. It consists of a Palaeozoic nucleus (quartzites and slates) deformed by overthrust folds. Erosion of the slates has produced quartzite crests aligned from north-west to south-east that culminate at nearly 2000 m [see Solé-Sabarís and Riba (1952), Riba (1959)]. These Palaeozoic blocks overlook some limestone plateaus situated to the north-east at between 1100 and 1300 m, modelled by a planation surface which is doubtless the same as that which appears in the Castilian branch of the Iberian Cordillera and which Gutiérrez-Elorza and Peña-Monné (1979a) considered to be of late Pontian age.

The upper Jiloca depression is a tectonic structure that constitutes the southern part of the great interior basin. This was filled during the Upper Tertiary and the Middle Pliocene [see Capote *et al.* (1981)]. These authors concluded that intense neotectonic, intra-Quaternary activity occurred. This basin possesses a branch towards the north in the Alfambra depression in which remnants of a Villafranchian glacis may be found, as well as three younger Quaternary levels of glacis and terraces [see Gutiérrez-Elorza and Peña-Monné (1976)]. The evolution of the slopes has been studied by Birot (1963) and, more recently, by Burillo *et al.* (1981).

In the Sierra de Javalambre to the south-east of Teruel, the variegated and gypsum-like marls and clays of the Keuper dominate, together with limestones and Jurassic dolomites. The folds are very gentle with dips no greater than 20° and the faulting, while very intense, does not involve great displacements. Towards 1200–1300 m, the late Tertiary planation surface is again found, which is deformed in some sectors by neotectonics. The existence of a river system draining directly into the Mediterranean Sea is responsible for the dissection of the relief and for the rejuvenation of the structural forms. The Sierra de Gúdar is an anticlinorial complex with a series of erosion levels at different altitudes (1150, 1300, 1400 and 1600 m), which are believed to correspond to the tilted late Tertiary planation surface (*see* Fig. 11.11).

11.3.3 Glacial and Periglacial Features

The existence of small Quaternary glaciers in the highest massifs of the Iberian Cordillera has been known since the beginning of the century: for example, Carandell and Gómez de Llarena (1918) described the glacier system of the Sierra de Urbión and also referred to former glaciers in the Sierra de la Demanda and the Moncayo. More recently, López-Gómez and Riba (1957) commented on the Quaternary glaciers of the Sierra de Neila, while Lotze (1962) analyzed those of the Sierra de Valnera in the Cantabrian mountains, introducing some further data on the Sierra de la Demanda. Thornes (1968) discussed the glaciers of the Sierra de Urbión, but made no reference to other massifs nearby. Martínez de Pison and Arenillas (1976) contributed a short article on the glacier morphology of the Moncayo; García-Ruiz (1979) did likewise for the Sierra de la Demanda.

Generally it can be seen that the glaciers only appeared in the northernmost sierras of the system, wherever they rise above 2000 m. All authors agree that they had a very restricted distribution. The best examples are located in the Sierra de Urbión, with numerous cirques now occupied by lakes and with some U-shaped valleys. The preglacial relief acted as a major control; in the massifs with summit planation surfaces and regular slopes, conditions for snow accumulation were not very favourable, thus the lowest height reached by glaciers was about 1750–1800 m. There is no evidence of more than one glaciation, although within the same glacier system, several morainic arcs may be seen which may correspond to different stadials. The preservation of the deposits seems to indicate that they are of Würmian age. A series of moraines found inside the cirques and almost against the back wall indicate the existence of a late stage in which gelifluction and rockfall predominated. In some cirques of the Sierra de la Demanda or the Moncayo, some rock glaciers may have existed.

During and after the glacial periods, large parts of the Iberian Cordillera were subjected to periglacial processes. Many sectors have yet to be studied, but Thornes (1968) has provided interesting data about the periglacial debris of Urbión, Pellicer (1980) has studied the periglaciation of the Moncayo, and Gutiérrez-Elorza and Peña-Moné (1977) the Sierra de Albarracín. In other studies, there are brief references to periglacial phenomena such as in Calatayud: for example, by García-Ruiz (1979) for the Sierra de la Demanda or by Gutiérrez-Elorza and Peña-Monné (1975) for the Sierra de Javalambre. Above 1300 m, inherited periglacial forms are numerous. The Moncayo and La Demanda are covered by thick mantles of debris now colonized by vegetation. They are bound with a fine matrix and, in many cases, are associated with boulder slopes and gelifluction lobes; some stratified screes are also observed. In Urbión, the screes are less widespread, although Thornes (1968) mentioned their presence, dating them as Würmian in the vicinity of Soria at an altitude of 1100 m. Some valleys show block streams, as in the Moncayo and Albarracín, and there is patterned ground in the Sierra de Javalambre (*see* Fig. 11.12), where Gutiérrez-Elorza and Peña-Monné (1975) noted stone

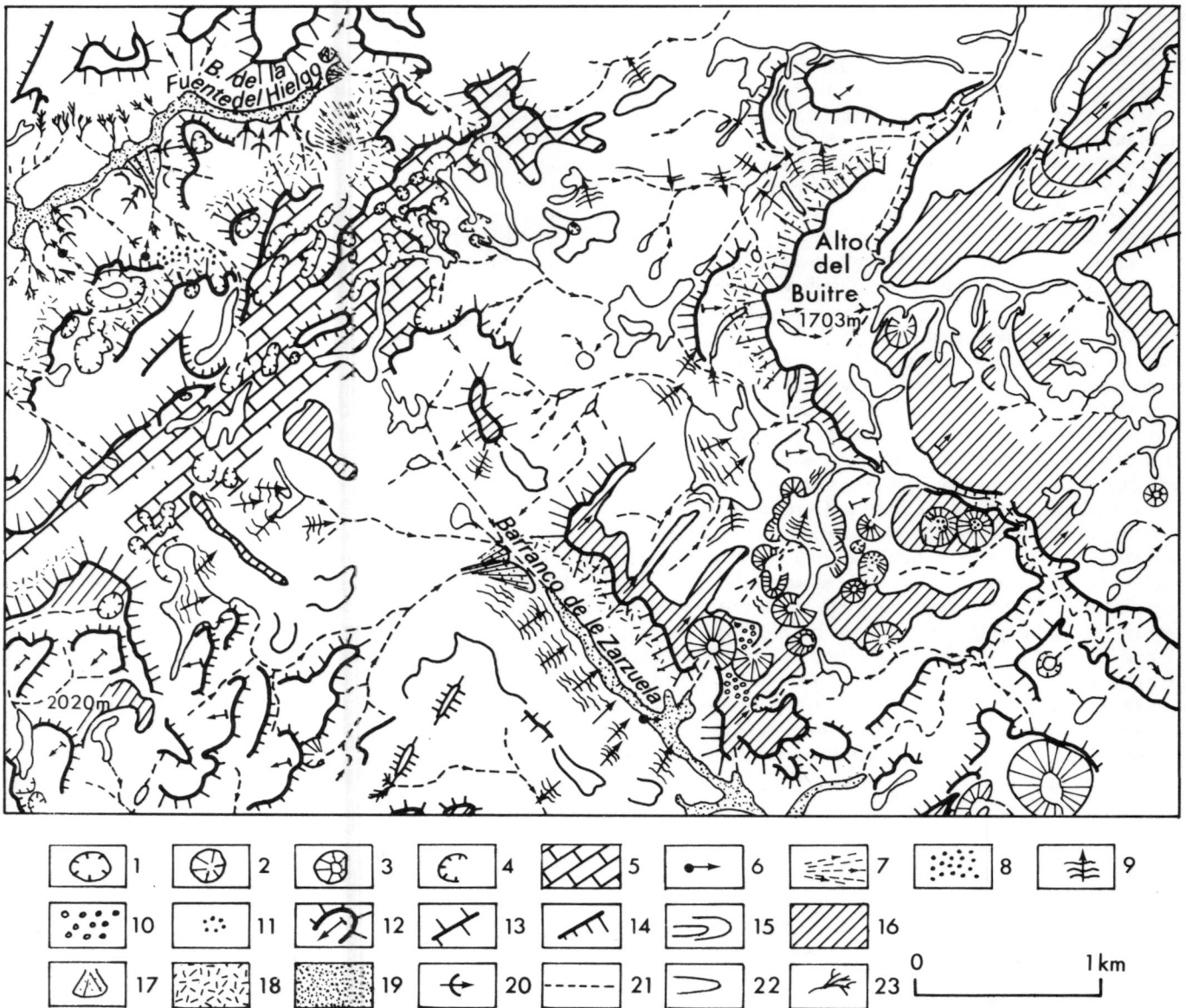

Fig. 11.12 Geomorphological map showing karst and periglacial features in the upper Javalambre region [after Gutiérrez-Elorza and Peña–Monné (1975)]. *Karstic forms*: (1) dolines; (2) funnel-shaped dolines; (3) dolines with periglacial infill; (4) captured or broken doline; (5) area of structurally controlled lapies; (6) permanent spring. *Periglacial forms*: (7) stratified screes; (8) blockfields and block streams; (9) gelifluction lobes; (10) patterned ground (polygons, stripes); (11) stone circles. *Structural forms*: (12) cuestas; (13) ridges; (14) escarpments; (15) chevrons; (16) late Pliocene planation surface (the lines mark the general direction of tilting). *Fluvial and other forms*: (17) alluvial cones; (18) screes; (19) alluvium; (20) present-day solifluction; (21) intermittent streams; (22) outlines of flat-floored and U-shaped valleys; (23) gullies.

circles, polygons and striped soils, generally at the bottom of dolines. Most authors have dated these phenomena as Würmian, although some may be more recent.

It should be noted that in many massifs of the Iberian Cordillera intense karstification has taken place. For example, the mountainous regions of Cuenca, the Maestrazgo, Javalambre, Gúdar and Albarracín show many striking forms related to the dissolution of limestones. Gutiérrez-Elorza and Peña-Monné (1979a,b) regarded the karst as late post-Pontian in age, and certainly Upper Pliocene. Nevertheless, proof of recent reactivation exists in the form of dolines that have developed since the periglacial deposits were laid down (e.g., in the glacis of the Sierra de Albarracín).

11.3.4 Coastal Landforms

The transition from the Maestrazgo, Gúdar and Javalambre sierras towards the coast is marked by a north-east–south-west tectonic fracture and a series of blocks tilted towards the littoral. The faults run parallel to the coastline and are responsible for block subsidence towards the sea. The existing volcanoes of the Columbretes and Cofrentes islands are associated with these fractures. Nevertheless, there are local exceptions to this general structural arrangement. To the east of Sierra de Javalambre, the appearance of the Trias has favoured denudation of the original structures, and the development of a strong fluvial network has contributed to the further evolution of the landforms. In the northern coastal area, the Iberian sierras almost reach the coastline, reducing the areas of lowland to small delta mouths. On the other hand, around Valencia, there is a well-developed coastal plain, whose morphological and sedimentological characteristics have been studied in detail, particularly by Rosselló (1969, 1971) and his colleagues. This plain, which extends inland for 20–30 km, consists predominantly of marine Miocene and Pliocene sediments. Over them lies a cover of Quaternary debris forming a glacis that extends from the mountain foothills to the sea. These sediments are coarse at the piedmont and decrease progressively in size, becoming silts and reddish clays near the coast. The abundance of debris, sometimes reaching 100–150 m in thickness, is related to Quaternary periglacial processes in the surrounding mountains. The sediments are strongly cemented by calcium carbonate and have a caliche duricrust that reflects the arid conditions in this area.

Towards the sea, former marine erosion has cut a 5-m cliff into the glacis; below the cliff, a more recent coastal plain of varying width has developed. This plain is continued by the present-day beaches, consisting of sands and shingle lying on top of a silt basement with some peat beds that have been dated as at least 5000–6000 BP, related to the Flandrian transgression. These marine deposits usually form bay bars with marshes and lagoons behind, temporarily or permanently connected with the sea by *graos*. From Peñíscola to Oliva, lagoons of various sizes appear in succession (e.g., Elche, Valencia, Alicante, Oropesa, etc.). Tombolos are also frequent, as in Peñíscola and Calpe. On top of the sand bars, dunes are frequent, the largest being 5–6 m high in the Valencia area.

The deltas are not very well developed, partly because they are associated with intermittently flowing torrential rivers and partly because of the longshore drift that moves sediments towards the south. The shapes of these deltas are markedly convex and their sediments comparatively coarse.

In contrast, farther south the coast lacks glacioeustatic marine levels. This absence has been explained by continental subsidence accompanied by fractures at the margins, initiated in the Neogene and continued during the Quaternary [see Gigout (1960), Solé-Sabarís (1961), Goy and Zazo (1974)]. The finding of submerged Roman ruins by Rosselló (1969) confirms this view. The abundance of sediments along the coast may be simply due to high depositional rates compensating subsidence; it may be that the sedimentation took place after the last glacioeustatic movements [see Solé-Sabarís (1961)].

11.4 Catalan Ranges

This mountain zone, sometimes called the Catalánides [see Hernández-Pacheco (1932a)] extends along the Catalan coast in a south-west–north-east direction. Its southern end merges with the Iberian Cordillera, while to the north it is connected with the eastern Pyrenees. It thus forms a barrier 250 km long and 30 km wide between the coast and the interior, isolating the Ebro Basin from the Mediterranean Sea, in spite of being only moderately high, as well as narrow and discontinuous. Its most outstanding geomorphological feature is that it is divided into two clearly defined, parallel ranges separated by a depression. The pre-littoral or interior range is the longest and highest, often being higher than 1000 m [e.g., the Sierra Montseny (1740 m)]. The pre-littoral or intermediate depression is 200 km long and has traditionally been the natural route followed from the Pyrenees to the delta of the Ebro. The littoral range, like the pre-littoral range, is 10–15 km wide, but its altitude seldom exceeds 500 m [e.g., Sierra Montnegre (727 m)]. At the foot of the littoral range lies a narrow and discontinuous coastal plain only broadening at the outlets of the main river valleys. Thus the coast, as a whole, is an abrupt one. South of Barcelona for a short distance the pre-littoral depression forms the coastline, the coast locally becoming open and flat, but this changes abruptly south of Tarragona where the pre-littoral range again borders the sea, as far as the lowlands of the Ebro delta.

11.4.1 Geological Structure

The geology of the entire Catalan ranges has been the object of several major studies [see Schriel (1929), Ashauer and Teichmüller (1935), Llopis (1947)] and short syntheses [see Solé-Sabarís (1972), Julivert *et al.* (1974)]. From a structural point of view, the ranges are not a geological unit; there are marked differences between their northern and southern parts, Barcelona being approximately at the dividing point. In the area adjacent to the Iberian Cordillera, the structure is that of an intermediate chain, with bundles of folded Mesozoic strata connected with those of the Iberian Cordillera and having similar characteristics. In the northern part, the Palaeozoic basement outcrops increasingly until the general structure becomes that of a massif. The main unifying element in the Catalan ranges is post-Alpine tectonics, which were active during the Miocene in the western Mediterranean zone, producing important fault systems. Deformation of the northern part of the ranges took place, as in the Pyrenees, during the Eocene. It began at the Monseny massif and progressively spread towards the south, where the deformation is of the same age as in the Iberian Cordillera (i.e. Oligocene). The result of these movements was a complete palaeogeographical change as the previously sedimentary basin was overturned towards the Ebro Basin by means of a reverse fault that thrust the Palaeozoic and Mesozoic strata of the Catalan Basin over the Eocene sediments of the Ebro Basin [see Fontboté (1954)]. During the Neogene, normal faulting gave rise to the present typical structure of horsts and grabens, producing the depressions of La Selva, Vallès-Penedès and Camp de Tarragona. Volcanic eruptions also took place in connection with this fracturing.

The rocks that constitute the ranges can be separated into three groups according to the historical evolution of the area.

1)   The Palaeozoic basement, which includes units from Upper Cambrian to Lower Carboniferous, together with abundant granitic intrusions that produced important metamorphic aureoles.

2)   The Mesozoic and Palaeogene sedimentary cover which dominates the southern part of the ranges and gradually disappears towards the north. The Triassic continental sediments extend to the north as far as the Montseny massif, while the Cretaceous marine transgression terminated at the Tibidabo massif. There is a small isolated basin at Garraf, where 800 m of sediments were deposited; in the lower Ebro, the deposits reach 2 km depth at Ports de Beceit. The latter sedimentary basin was located between the Ebro massif, which at that time separated this basin from that of the Pyrenees, and the Catalan–Balearic massif located in the present Mediterranean area. Materials from the margin of the Ebro Basin have been topographically incorporated in the Catalan coastal ranges due to later tectonic and erosion processes.

3)   The Neogene and Quaternary sediments are mainly located in the tectonic depressions.

11.4.2   Planation Surfaces

Well-developed planation surfaces are scarce in the ranges, due to their proximity to base level which allows regressive erosion to reach the interfluves and destroy any remnants of peneplanation. Only where the Palaeozoic rocks constitute extensive massifs has erosion been unable to destroy the vestiges of the old relief forms, especially when these were buried under Mesozoic and Cenozoic sediments. In the littoral range, erosion has progressed particularly rapidly and there are no true planation surfaces, while in the pre-littoral range, local base level is provided by the intermediate depression and erosion is less active, thus preserving some better remnants of planation. This is the case in the Serra de Prades, southwest of Barcelona, and in the Montseny and Guilleries massifs to the north-east. The lithology of the basement is also important, as can be deduced by the fact that planation surfaces have been preserved on schists and slates but rarely on granite where the only vestiges are generally the accordance of summit levels. Here, however, it has to be borne in mind that disintegration of the granite can produce apparent planation surfaces that are not of cyclical significance, as pointed out by Birot (1937).

Two well-developed fossil planation surfaces have been defined [see Birot (1937), Llopis (1947), Solé-Sabarís (1940)], one pre-Triassic and one pre-Eocene. These surfaces may correspond to two important phases of peneplanation antedating the Alpine tectonics and synchronous with those observed in central Europe. The first is found around 1000 m in the Montseny massif (Pla de La Calma) and in the Serra de Prades, and is in both cases

partially covered by Triassic sediments. The second stands at about 500–700 m and is partially covered by Eocene sediments. Both have been tilted and flexured by subsequent tectonism. In addition, Llopis (1947) described a post-Pontian surface that would correspond to that found in the Iberian massif. Solé-Sabarís (1940) distinguished five further erosion cycles in the Montseny and Guilleries massifs related to progressive, although intermittent, fluvial incision. I and II would be pre-Miocene and are quite well developed, standing at about 700–800 m (Sant Hilari) and 600 m (Pla de Fogueres); the lower levels are less distinctly cut in the Palaeozoic basement and in Miocene deposits. III stands at 500 m and is only distinguishable by the accordance of interfluves. IV is at 100 m, well developed and corresponds to the surface of La Selva which is partially covered by Pliocene deposits and is therefore, together with the previous cycle, pre-Pliocene. Level V is cut between 50 and 70 m and is the base of an alluvial terrace, hence of Quaternary age.

In the littoral range, summit and interfluve accordances seem to indicate four erosion levels in the Cadiretes massif [see Llopis *et al.* (1953)] and three in the Gavarres massif [see Marcet and Solé-Sabarís (1949)].

11.4.3   Structural Landforms

The interrelationships between tectonics, lithology and erosion have determined the different landscape types in the ranges. In the south-west, there are complex forms governed by the folding (Ports de Beceit). In the central part, where folding becomes wider and gentler or even fades out completely, the landscape consists of tabular and cuesta landforms, often with a glacis at the slope base [see Calvet (1976), Gallart (1980)]. A variant within the tabular category is the series of mountains (Montsant, Montserrat, San Llorenç de Munt) produced by differential erosion between the calcareous conglomerates and the marls and clays of the Eocene and Oligocene of the margins of the Ebro Basin, thus becoming topographically linked with the pre-littoral range. The conglomerates are discontinuous deposits corresponding to the torrential outlets of streams flowing towards the Ebro Basin. The most striking feature is that of Montserrat, which rises, with 500 m vertical walls, to 1224 m above the plain and which has a general castellated appearance to which its name refers (*see* Fig. 11.13). Its origin is closely linked to the strong joint pattern of this plateau area. The joints cross orthogonally, but vertical ones are dominant and, together with the calcareous cement of the rock, favour solution along them, resulting in conical monoliths distributed as a crown around the mountain nucleus.

Karst landforms are found in areas of thick limestones with homogeneous lithology, as in the Garraf mountain and in the less-folded areas of Ports de Beceit. The Garraf karst has been the subject of many studies which, due to the great importance of underground phenomena, have been undertaken mainly from a speleological point of view. There therefore exists a very complete inventory of its sink holes, caves and underground water circulation [see Llopis (1941), Monturiol (1950)]. The most karstified

Fig. 11.13 Montserrat mountain in the pre-littoral range, Catalan ranges, showing castellated forms in jointed calcareous conglomerates. (Sala)

area is the Begues plateau, especially the hills of La Morella. Although not very extensive, there is a wide-ranging set of surface landforms. The most important feature is the Begues polje (6.1 km long and 3.9 km wide) which seems to have evolved from an initial doline, but is most probably favoured by tectonic structure. Its evolution was arrested by regressive stream erosion on its margin, although it is once again functioning at present, although at a very low rate. In addition, there are two types of dolines: those with a low index of concavity and a flat floor infilled with decalcified clays, and those with a high index of concavity, almost totally lacking any clay infill and often with a sink hole in the centre. Some features are well-developed and some are in an incipient stage. Lapiés microforms are also present. Underground flow is very important; the main collecting area is the Begues polje. The main resurgence is that of Falconera, which is a submarine emergence and contains the outflow of a stream that circulates at 27 m depth and resurges 1 km out to sea. Seven more submarine resurgences have been observed.

Typical granite landforms are found in the northern part of the ranges, mainly in the area of Farners (Guilleríes massif) in the pre-littoral range, but also in that of Santa Cristina (Cadiretes massif) in the littoral range. In both, tor and inselberg landscapes are found in an incipient state of development [see Sala (1979a)]. They are located mainly on valley slopes below the level of the summit surface. Fluvial erosion along a fault line seems to be responsible for their exhumation, although Quaternary periglacial slope processes were also important since granite forms are much better developed in the pre-littoral range, where clear periglacial features have been found. Inselbergs are more often castellated than domed and, in the tor landscape, perched boulders are common. Nearly all the summit areas of the granitic ranges have a fringe of

boulders spreading downslope, even if rock outcrops are absent. In the Sierra de Montseny, Morou hill and Santa Fé, an outcrop of fine-grained leucogranite has produced slope and stream deposits containing extremely large boulders (2–4 m in diameter). Neotectonics have been important in the ranges, have affected the structure of the granites [see Bech (1972)] and have clearly influenced the drainage pattern. When dykes are present, differential chemical weathering has produced small ledges upslope or around the dykes; the resulting stepped landscape shows many interfluve and summit accordances (*see* Fig. 11.14).

11.4.4 Quaternary Processes and Landforms

Periglacial slope processes have been mentioned in various areas of the ranges [see Llobet (1975a)] but they have been more thoroughly studied in the Montseny massif, being the highest point (1700 m) of the pre-littoral range. Llobet (1975b) found the lower limit of these phenomena around 500 m. Nivation hollows, sliding scars and periglacial deposits are found, especially on the south-facing slopes. There seems to be evidence for only one cold period. Sala (1979a) classified the forms and deposits according to lithology and altitude, on the slopes facing the pre-littoral depression which have a general south-easterly orientation. In the phyllite area, the following succession can be observed from summit to footslopes (300 m): (1) angular outcrops fringed with angular debris, (2) rectilinear slopes with a veneer of debris, sliding scars and slope deposits, (3) irregular, convex slopes covered by a thicker mantle of debris and (4) debris accumulations clustered in concavities. The thicker and more continuous deposits are found approximately in the middle of slopes and are often related to small breaks produced by dykes. In the less metamorphosed areas, stratified screes (*grèzes litées*) have developed, while, in the schist aureole, ordi-

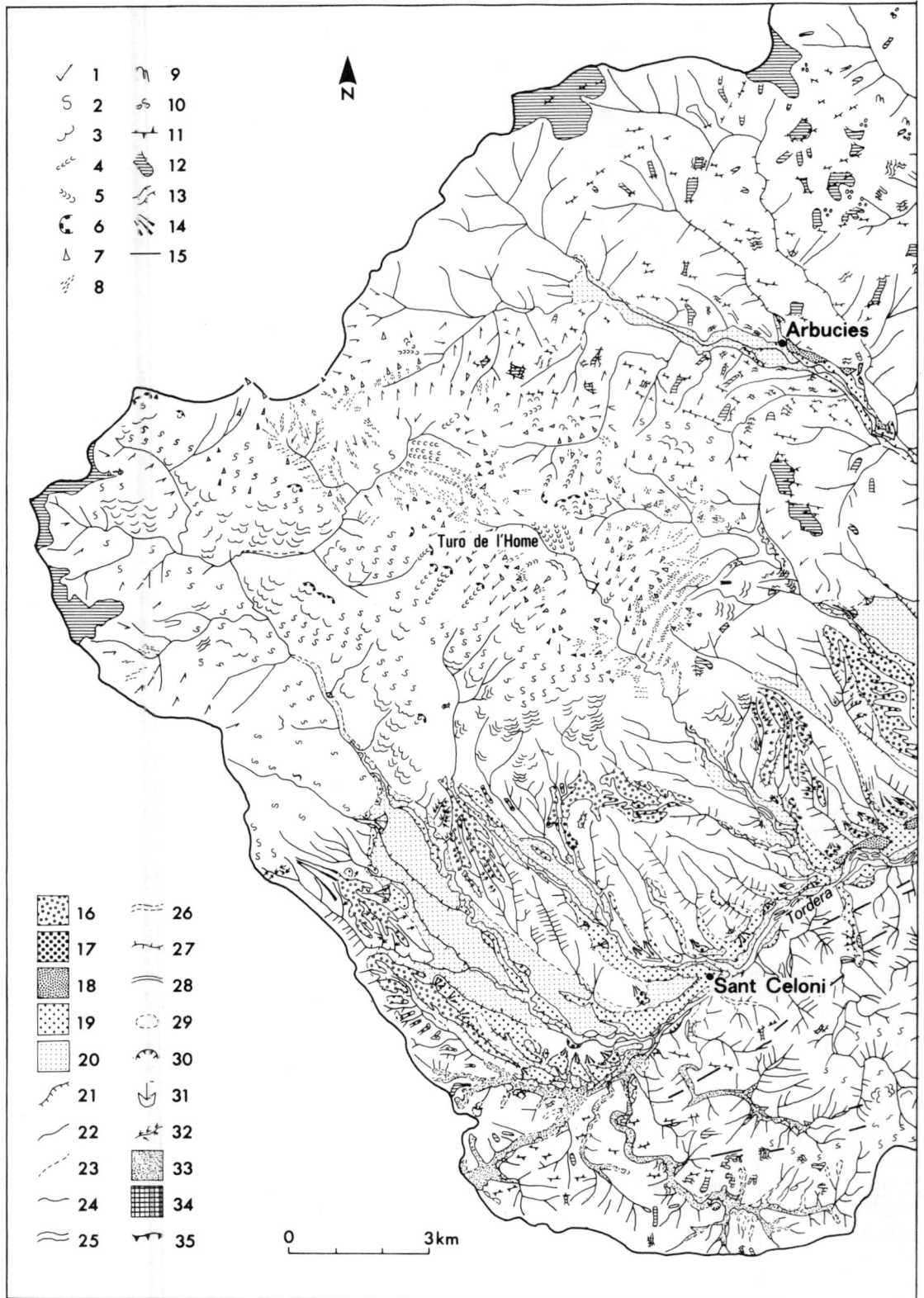

Fig. 11.14 Geomorphological map of the Tordera Basin. (Sala) (1) Rectilinear slope, (2) irregular slope, (3) gelifluction lobes, (4) hollows, (5) convexities, (6) landslip scar, (7) sharp rock outcrops, (8) rockfall debris, (9) rounded rock buttresses, (10) boulders (due to weathering and exhumation), (11) outcrops of igneous intrusions, (12) benches, (13) anthropogenic banks and terraces, (14) glacis, (15) fault lines, (16) highest terrace (17) fourth terrace, (18) third terrace, (19) second terrace, (20) lowest terrace, (21) sharp terrace edge, (22) clearly defined terrace edge,

(23) terrace with poorly defined edge, (24) narrow incised valley, (25) broad valley with clearly defined margins, (26) broad valley with poorly defined margins, (27) river channel cut in alluvial floor, (28) river channel with alluvial banks and shoals, (29) former channel, cut-offs, (30) basal undercutting of slopes, (31) alluvial cone, (32) gullies, (33) coastal beach, (34) coastal sand bar, spit and (35) coastal cliffs.

nary scree slopes prevail. In the granite areas, the altitudinal sequence of deposits is not so clear; the slopes appear less regular and steeper than those of the phyllites, with a predominance of rapid downslope movements and the removal of most of the fines from the debris.

In the littoral range, periglacial processes have also been described in detail [see Sala (1978, 1979a)], although their extension, variety and abundance are much more limited, possibly due not only to the lower altitude but also to the prevailing granite basement.

Alluvial deposits are widely distributed in the intermediate depression and along the large transverse valleys opened through the ranges [see Solé-Sabarís and Llobet (1957), Solé-Sabarís *et al.* (1957), Solé-Sabarís (1963)]. In the Tórdera Basin in the north, five alluvial levels have been distinguished near Sant Celoni by Sala (1979b) at 90, 40, 15, 5 and 2.5 m above the present channel. The upper level is found at the foot of the pre-littoral massif and lies on top of the Neogene sediments which may represent the last phase of the filling of the depression. The surface distribution of the deposits does not follow the present stream courses but appears as a series of fans rooted to the mountain front. Although these deposits are commonly described under the general term of piedmont deposits [see Solé-Sabarís and Llobet (1957), Solé-Sabarís (1972)] they compare morphologically and chronologically with the raña deposits found throughout the Iberian massif (*see* p.319), except for lithological differences due to the absence of quartzites in the ranges.

The middle levels of alluvium are distributed along the present stream courses and incised into the Neogene materials in typical terrace forms; they differ more or less strongly in their degree of weathering, colour, etc. The lower fill, which has its maximum development at the river mouth, differs from the other levels due to the looseness of the materials, its dark grey colour, its finer and well-bedded top layer, and the very neat and horizontal terrace shape (*see* Fig. 11.14).

Comparing the characteristics of these alluvial deposits with those described by Raynal and Tricart (1963) for the Quaternary pluvials in Morocco and the Mediterranean region of France, the upper fan or raña level may be correlated with the Moulouyen (Upper Villafranchian) and the successive lower levels to the Amirian (Mindel), Tensiftian (Riss), Soltanian (Würm) and Rharbian (post-Flandrian); this last fill is, according to Vita-Finzi (1969), of a less erratic character than the previous ones and the present phase, being an indication of more humid and regular climatic conditions.

11.4.5   Coastal Features

The coast of the littoral range, from Garraf to the Gavarres mountains, is an abrupt one, related to the north-east–south-west direction of faulting, although it cannot strictly be designated a fault-line coast because of the absence of deep water offshore [see Llopis *et al.* (1953)]. Broad openings occur at both ends of the range due to tectonic accidents (the Empurdà depression to the north and the pre-littoral depression to the south). Because there are no tides, sediments are not removed

offshore; materials accumulate at the river mouths and are distributed along the coast and towards the south by longshore drift. The overall result is a coast that tends to be straight due to a combination of the infilling of inlets and the erosion of headlands; probably the first process is more effective than the second, as the granite of which the coast is made is resistant and armed with dykes. The lack of islands on this coast supports this view. The process of coastal regulation is retarded in the northern part due to its subsiding tendency but is accelerated towards the south by the increased width and depth of the continental platform.

Coastal lowlands include the Ampurdan depression, where the Gulf of Rosas in the High Ampurdan is gradually being infilled by the alluvial deposits of the Muga and Fluvia, giving an area of abundant marshes and sandy beaches. South of the Montgrí outcrop, the alluvial plain—built by the Ter and Daró in the Lower Ampurdan—is much more developed and has already produced a rectilinear coastline between this mountain and the Gavarras mountain. In both bays, dune systems—both ancient and present-day—are found [see Obrador *et al.* (1971)].

South of Blanes, but especially around the Barcelona area, there is again a low coast due to the deltas built by the Llobregat and the Besós and the various streams that drain the range, producing an alluvial plain. The next low coast area is the Camp de Tarragona—the most extensive beach area of this coast—connecting as it does with the delta of the Ebro, which is the biggest of the Iberian Peninsula. This forms a wide alluvial plain covering 28 260 ha, typically triangular in shape, with two lateral wings and a single mouth set in a sharp, pointed end, which indicates its rapid progradation. Within the delta, peat areas, marshes and lagoons are widespread, together with abundant dune hillocks. Maldonado and Riba (1966) in a study of the historical evolution of the delta have found numerous abandoned meanders and branches, and the existence of two former outlets—one to the north and the other to the south of the present one—all of which indicate great channel instability. The rate of growth of the delta was 10 m per year during this century up to 1946; from then on, and due to the construction of dams, the delta has receded 1500 m. Subsurface investigations have shown that much of the delta was formed after the last glacial period.

The northern cliff area of the coast—the Costa Brava—has been studied in general works [see Marín and San Miguel (1941), Marcet and Solé-Sabarís (1949), Llopis *et al.* (1953)] and in monographs by San Miguel (1931), Llopis (1934), Butzer (1964) and Barbazà (1971). The Costa Brava extends beyond the actual limits of the Catalan ranges, from Blanes to Cape Creus in the Pyrenees. Its detailed morphology results from differential marine erosion and fluvial dissection. Most of the coastline is marked by plunging cliffs, sometimes with block accumulations at the foot, headlands and steep-sided drowned inlets. Other sections are distinguished by sand bars. Except for Cape Creus, which is made of schists, and Cape Begur, which consists of limestone—both belonging structurally to the eastern end of the Pyrenees—the prevailing rock type is a mass of medium-grained granodiorite and leucogranite with aplitic and pegmatitic subfacies and a great variety of aplitic, pegmatitic quartz and felsitic dykes. The more acidic granites and quartz or quartzitic facies are more resistant, while the porphyritic granites and the granodiorite and leucogranites are more susceptible to disintegration, thus producing noticeable differential erosion [see Butzer (1964), Llopis *et al.* (1953), Pallí (1977)]. The most characteristic relief features are the abundance of small bays called *cales* between the cliffs. Cales can be narrow with deep indentations produced by the erosion of dykes or they can be more open, concave and related to the relief of the land or variation in the granite mass. These bays sometimes contain sandy beaches but quite often they are filled with blocks coming from the strongly jointed rock walls that frame them, which sometimes present a honeycombed microrelief. Several marine erosion platforms have been mentioned [see Barbazà (1971), Butzer (1964), Llopis *et al.* (1953)]. The most conspicuous one standing at 100 m is thought to be early Sicilian [see Butzer (1964)]. Micromodelling of these platforms in the form of flat-floored pans is common [see Barbazà (1971), Pallí (1978, 1980), Sala (1979c)].

On the whole, this is a coast with few traces of older beaches, although small remnants have been found in various places [see Porta (1957), Solé-Sabarís (1961)] especially on the southern, low coast. In the northern area, there is not much agreement on this, although two beach levels have been described by Barbazà (1971). Around Barcelona [see Marcet (1956)] but particularly in the Tarragona area [see Solé-Sabarís and Porta (1955), Gigout (1959)] deposits containing *Strombus bubonius* have suggested a Tyrrhenian age, probably stage II.

Often related to beach and fluvial deposits, and widespread on the footslopes of the littoral range, a three-cycle deposit is found, consisting—from base to top—of red clays, yellow aeolian silts and an upper layer more or less heavily and continuously cemented by caliche [see Virgili and Zamarreño (1957), Solé-Sabarís (1961)]. They have been interpreted as corresponding to the Riss–Würm interglacial.

Chapter 12
The Iberian Massif

The most outstanding feature of the Iberian Peninsula is the vast remnant of the Hercynian continent, variously called the Hesperic massif by Hernández-Pacheco (1932a), the Iberian massif by Lautensach and Mayer (1961) or the Meseta block by Solé-Sabarís (1952). This last term is not to be confused with the physiographic term used for the Castilian plateau. This core of the Peninsula approximately covers its western half and contains two distinct elements. In the west, there is a series of massifs built mostly of Palaeozoic materials, which have suffered repeated folding, granitization and metamorphism, with Hercynian folds that have a general north-west–south-east alignment but have been greatly worn down by subsequent erosion. In the east, there are two wide basins with a Tertiary sedimentary cover consisting of sands, clays, gypsum beds and limestones in a horizontal or subhorizontal position. The Alpine orogeny was responsible for intense faulting and tilting of the whole peneplained block, giving rise to the two great eastern depressions and to the several uplifted mountainous blocks. Pliocene–Pleistocene movements also produced a strong warping and flexuring of the widespread Pontian peneplain, with considerable subsidence on its north-west border (Galicia) and the creation of a series of horsts and grabens in many areas. As a result of these events, the Hercynian block now shows the following units:

1) The old core, peneplained and more or less deformed, extending in the west from Galicia in the north to the Sierra Morena in the south.
2) Three massifs, aligned from west to east, comprising the Asturian–Leonese massif, the Central Cordillera and the Montes de Toledo with elevations ranging from 1400 to 2500 m.
3) Two wide depressions: the northern one drained by the Duero (Douro) and called the North Meseta or Old Castile, and the southern one, drained by the Tajo (Tagus) and Guadiana and called the South Meseta or New Castile. Their mean elevations range from 800 to 600 m, respectively.

During the Mesozoic, the entire Iberian massif was an area with a tendency to uplift, surrounded by zones of marine sedimentation. The Mesozoic seas penetrated in varying degree into the Hercynian massif, but most of it has no sedimentary cover, either because such sediments were never deposited or because of later erosion. The margins of the massif and the platform zone that surrounds it differ widely. Whereas to the north and west, the Iberian massif and its Mesozoic fringe terminate at the Atlantic Ocean, in its southern part, it is bordered by an Alpine chain and its foredeep (the Baetic Cordillera and Guadalquivir Basin). To the east, it merges into a wide platform that has been deformed, the intensity of deformation increasing in places to the point of becoming a chain of intermediate type, as in the Iberian Cordillera which stands between the Hercynian system and the Alpine chain with its foredeep (the Pyrenees and Ebro Basin).

12.1 Galician Massif

Galicia, at the north-west end of the Hercynian block, is a mountainous area with average altitudes of 500 m. The intricate relief, without definite lineations, consists of broad, rounded interfluves strongly cut by the drainage net into narrow valleys, and only in some places enlarged into broader depressions. The relief gradually decreases in altitude towards the sea where the most striking features of this landscape—the drowned valleys or rias—are found. This makes a remarkable contrast between the peneplained interior and the steep coast (see Fig. 12.1). Granite and gneiss are the two dominant rock types, with a marked influence on the landscape, but the main features of the relief are tectonically controlled.

12.1.1 Structure

Geologically, three zones may be distinguished from west to east along the northern section of the Iberian massif: the Precambrian complex of Galicia, the Asturian–Leonese zone and the Cantabrian zone. This distribution has a palaeogeographical significance in the sense that the age and metamorphism of the materials increases to the west. Variations in thickness and facies are also found [see Julivert et al. (1974), Parga-Pondal (1963), Tex and Floor (1971)].

In the Galician zone, structural differences are well marked. The main features include the great extent of highly metamorphosed rocks, the syntectonic granitization processes, the existence of important Silurian vulcanism associated with the Precambrian massifs and strong Neogene–Quaternary faulting. There are several rounded massifs of Precambrian rocks (e.g., Cabo Ortegal, Ordenas and Lalín) and an elongated zone parallel to the west coast. The massifs are made of amphibolite, metaperidotite, gneiss, eclogite and granulite, but they are encircled by less metamorphosed rocks (green schists and glaucophanic schists). Separation from the surrounding rocks is by fracture surfaces that have been interpreted as the limits of horsts of Lower Palaeozoic age [see Ribeiro (1970)]. The north–south unit has a predominance of blastomylonitic orthogneiss and is well delimited by faults or by elongated elements of granodiorite. Farther west, on the coast, migmatized Palaeozoic rocks are present.

Neogene and Quaternary tectonics have greatly affected this zone [see Teixeira (1944)]. The result is a structure with well-developed faulted blocks that shape the present relief into a succession of north–south alignments, only interrupted by several east-north-east–west-south-west faults, clearly visible on the west coast and in the south of the country. All along the west coast, there is a series of blocks delimited by normal faults, generally tilted to the north-west, and raised in respect of the marine floor. A narrow, elongated graben with north–south-limiting faults is the next unit. It stretches from Carballo in the north to Tuy in the south through Santiago de Compostela, Rodrón and Pontevedra, and is marked by several thermal springs. In the east, there is a wide horst fragmented by faults that delimit small grabens, as in Puentes de García Rodrigo. Farther to the east, several collapsed blocks give way to well-developed depressions in a north–south succession, such as those of Lugo, Sarria and Monforte de Lemos. Neogene deposits are found in many of these depressions, in some cases 100 m in thickness (e.g., Monforte de Lemos) and with a dip of 30° in the grabens that follow the Miño and Sil valleys. In Puentes de García Rodrigo, these clay deposits contain lignite.

12.1.2   Relief and Drainage

The most important factor in the modelling of the Galician landscape is the intense faulting associated with the Alpine orogeny and the following isostatic readjustments. This fracturing, however, affected a previously planed relief and was responsible for strong dislocation of the main planation surfaces [see Solé-Sabarís and Llopis (1952)], Dantin-Cereceda (1944)].

The second element in the morphology of this zone is the widespread occurrence of granitic bosses, their weathering under varied climatic conditions, and the differential erosion of the surrounding metamorphic aureole. Planation surfaces and granitic landscapes will be described separately below.

The drainage pattern has been greatly affected by the fracture pattern and shows considerable dependence on tectonic lineations. Drainage density has been calculated at several sites [see Nonn (1966)] and the values (2.29 km^{-2}) seem rather low for a granitic terrain, perhaps due to dense jointing and the high permeability of the weathered rock. Except for the Miño and Sil, rivers in Galicia are short (about 100 km) and possess steep gradients, but they also have the highest annual discharges per unit area in the Peninsula (about 0.02 m^3 km^{-2} s^{-1}). Several river captures have resulted from the steep gradients of the rivers of the north coast, at the expense of the tributaries of the Miño, draining the Lugo depression [see Vidal-Box (1941), Hernández-Pacheco (1949a)].

12.1.3   Planation Surfaces

In earlier studies by Vossler (1931), Vidal-Box (1941) and Hernández-Pacheco (1949a)—summarized by Solé-Sabarís and Llopis (1952)—four different levels of planation were differentiated, each being interpreted as a cyclic level. The highest—a summit plain at 1000–2000 m—was considered to be a remnant of an Oligocene Peneplain, the other surfaces appearing inserted in it as 'partial peneplains'. Two middle levels at 700–850 and 600–700 m, respectively, are developed on both Tertiary deposits and Palaeozoic strata.

Birot and Solé-Sabarís (1954a) reduced the number of planation surfaces to two and interpret the other levels as warped surfaces intersecting each other at low angles. The fundamental erosion surface is a well-developed plain that extends across the centre and north of Galicia at heights ranging between 200 and 500 m. It is especially well developed in Orense and Chantada; it has been dated as pre-Miocene and correlated with the fundamental surfaces in the Iberian and central Cordillera. At a higher altitude, remnants of an older surface appear to be preserved in several summits (e.g., Segundera: 2044 m) and are supposed to belong to a cycle initiated in the Triassic and continued during the Mesozoic. This surface is thought to envelope the main features of the Galician relief, rising from the Castilian surface through an east–west updomed axis and falling again towards sea level farther north as described by Terán (1968).

More recent studies by Nonn (1966), based on analysis of correlative deposits, reached the conclusion that the fundamental surface is divided into two levels, with a range of 80–120 m between them and owing more to erosion processes than to tectonics. Miocene lignite deposits are found on the two levels, but Middle or Upper Oligocene lignites are only on the lower one. A Sannoisian or Stampian age is deduced for the higher level and a late Oligocene date for the lower one.

12.1.4   Granite Forms

There is no systematic study of the granite landforms in Galicia in spite of the fact that this is the main rock type of the area, but certain facts have been established by Carle (1941), Lautensach (1949), Parga-Pondal (1958) and, especially, by Nonn (1966).

First of all, analysis of Miocene sediments and pollens in the tectonic depressions indicates the existence then of a warm, humid climate with tropical forest, in complete contrast to the evidence deduced from the continental Miocene elsewhere in Spain. But this climatic difference between the north coast and the interior is also a characteristic of the present-day landscape. In the graben of Puentes de García Rodrigo, a 100-m deposit consists mainly of kaolinite and lignite, coarse sediments being absent. Pollen indicates a forest similar to that of present-day China, Korea or the south-east of the USA. This kind of landscape persisted through the Pliocene, with the exception of a dry period during the Tortonian.

Such climatic conditions are conducive to deep weathering of granite. In spite of this, the Galician relief is not one of very marked granitic landforms. There may be several reasons for this: slow removal of the weathered materials during the Quaternary, related to the subsident tendency of this block, the widespread faulting with its strong imprint on the relief and the abundance of metamorphic aureoles.

Several differences in the granitic landscapes were

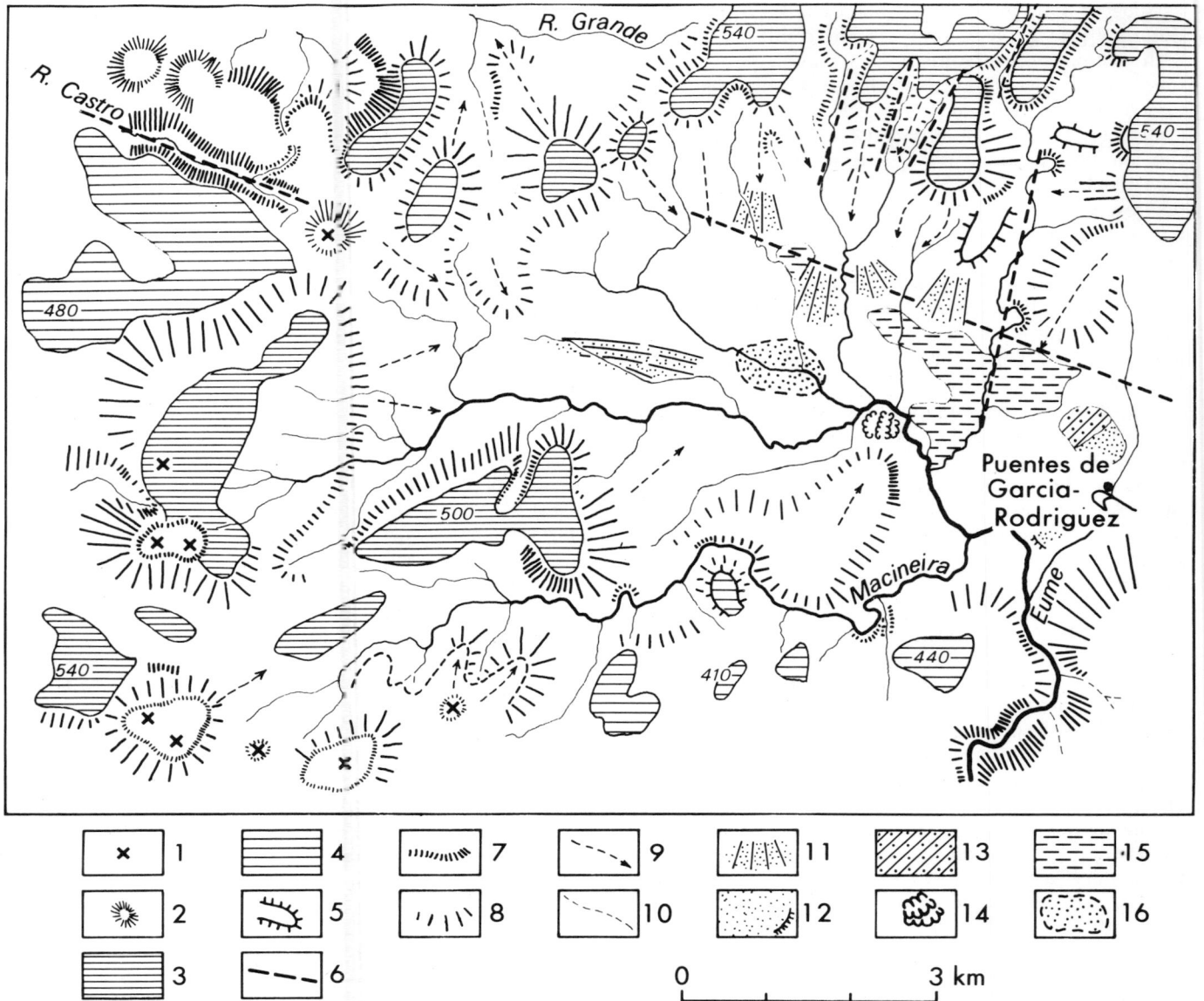

Fig. 12.1 Geomorphology of the tectonic basin of Puentes de García-Rodriguez [after Nonn (1966)]. (1) Tor, (2) residual relief, (3) main upper planation surface, (4) main lower planation surface, (5) bench of uncertain origin, (6) probable fault, (7) slope greater than 16°, (8) slope 8–16°. (9) slope less than 8°, (10) ravine, (11) torrential fan, (12) old river terrace, (13) terrace with sandy cover, (14) scree with limon and gelifluction debris, (15) Quaternary deposit of uncertain age and (16) lignite mine.

established by Nonn (1966), related to the jointing, and the mineral and textural composition. Where the granite is coarse or medium-grained and the jointing closely spaced, the relief consists of gentle small depressions with boulders and tors on the interfluves. A coarse-grained texture and well-spaced jointing tend to produce granite domes. Fine-grained granite with an orthogonal joint system produces castellated tors with piles of angular blocks and boulder-strewn slopes. Microforms, such as karren and weathering pans, are mainly found in coarse-grained rocks with strong granular disintegration.

12.1.5 Quaternary Land-Forming Processes
In many places, the Neogene sediments are covered by a

clearly discordant, more recent deposit, which consists of coarse conglomerates in a red clay matrix. The contrast of these materials with the yellowish colour of the Neogene sediments and with the grey colour of the Palaeozoic rocks is very striking. These deposits and associated landforms have been studied by Nonn (1966) in the rias of Vigo, Arosa and La Coruña, among other valleys. Their general appearance is very similar to the Castilian rañas and they are considered by the author to be a local facies of them, but with differences concerning the size and colour of the materials, still heterogeneous but considerably finer and less rubefied. In the landscape, they form broad plains about 200 m above the river valleys and impart a very special character to the countryside. At lower levels, there are several alluvial terraces following the stream courses (*see* Fig. 12.1).

The presence of coarse deposits should correspond to a notable climatic change with regard to the previous phase, which has already been described, an environment with torrential fluvial transport of debris. Nonn (1966) has also studied the modelling of glacis which should correspond to a more arid climate. The rañas are a type of deposit associated with the Mediterranean pluvials and a Villafranchian age. As in North Africa, the south of France [see Raynal and Tricart (1963)] and north-eastern Spain, the cold periods before the Riss correspond to pluvials, the interpluvials being periods of marked aridity. Nonn and Tricart (1960) have also found forms associated with a late periglacial climate, basically related to lithology and to a northern and coastal position. Head is especially abundant on the cliffs of the north coast, past Cape San Adrián with a substratum dominated by metamorphic rocks. On the granitic areas, the infilling of valley floors with sand and silt deposits has been widespread, but slope deposits are otherwise scarce, showing the importance of surface wash.

12.1.6 Coastal Features

The landforms of the Galician coast are probably one of its most striking features, including some of the best known examples of rias, a name that has become a general term for this type of coastal feature (*see* Fig. 12.2). This coast has been the object of many studies, such as those of Carle (1947), Nonn (1966), Pannekoek (1966), Torre-Enciso (1958) and Mensching (1961).

The main plain of the coast is clearly related to the fundamental tectonics of the Galician massif: a north–south lineation along the west coast and a south-west–north-east lineation along the north-east coast. Into this general setting, the second-order landform of the rias is inserted.

The rias are funnel-shaped indentations on the coast formed by a submergence affecting an area that, in the case of Galicia, can reach from 20 to 35 km inland, and where hills and river valleys meet the sea at right angles. The rias decrease in width and depth as they run inland. The streams that flow into their heads, responsible for eroding the original valleys, are obviously too small for the present sizes of the inlets. In Galicia, the rias are related to a well-incised, sinuous fluvial system cut into a previously peneplained relief and now partially invaded by the sea. In their headwaters, the slopes are clearly fluvial, continuous and with their base level beneath sea level. The submerged floor has not yet been filled up with alluvial materials (*see* Fig. 12.3).

The submergence of this crenulated coast is clearly related to the Flandrian transgression [see Bourcart (1938)]. The shoreline was still at -25 to -30 m in the Vigo ria 8500 years ago [see Margalef (1956a)]. The valleys have different origins. Tectonic lineations, lithology and Tertiary weathering are the main controls leading to a differentiation of three main types of ria: tectonic, fluvial and alveolar [see Nonn (1966)]. The tectonic rias are located on the west coast, between the Miño river and Cape Finisterre, their patterns being related to the south-west–north-east-faulting of the Galician massif. These are the best developed of all the rias (4–6 km wide and 15–35 km deep). These dimensions cannot possibly be related to the fluvial capacity of their corresponding river systems. Vigo, Pontevedra and Noya are also called the low rias (*rias bajas*). Their valley sides, although fault scarps originally, are to a great extent erosion forms that have undergone retreat by pedimentation processes.

In the case of the fluvial or 'funnel' rias, it would have to have been a river of remarkable development for it alone to be responsible for the width and shape of the ria [see Asensio-Amor and Nonn (1964)]. Fluvial excavation has,

Fig. 12.2 Ria landscape of Galicia: Ria de Pontevedra. (Sala)

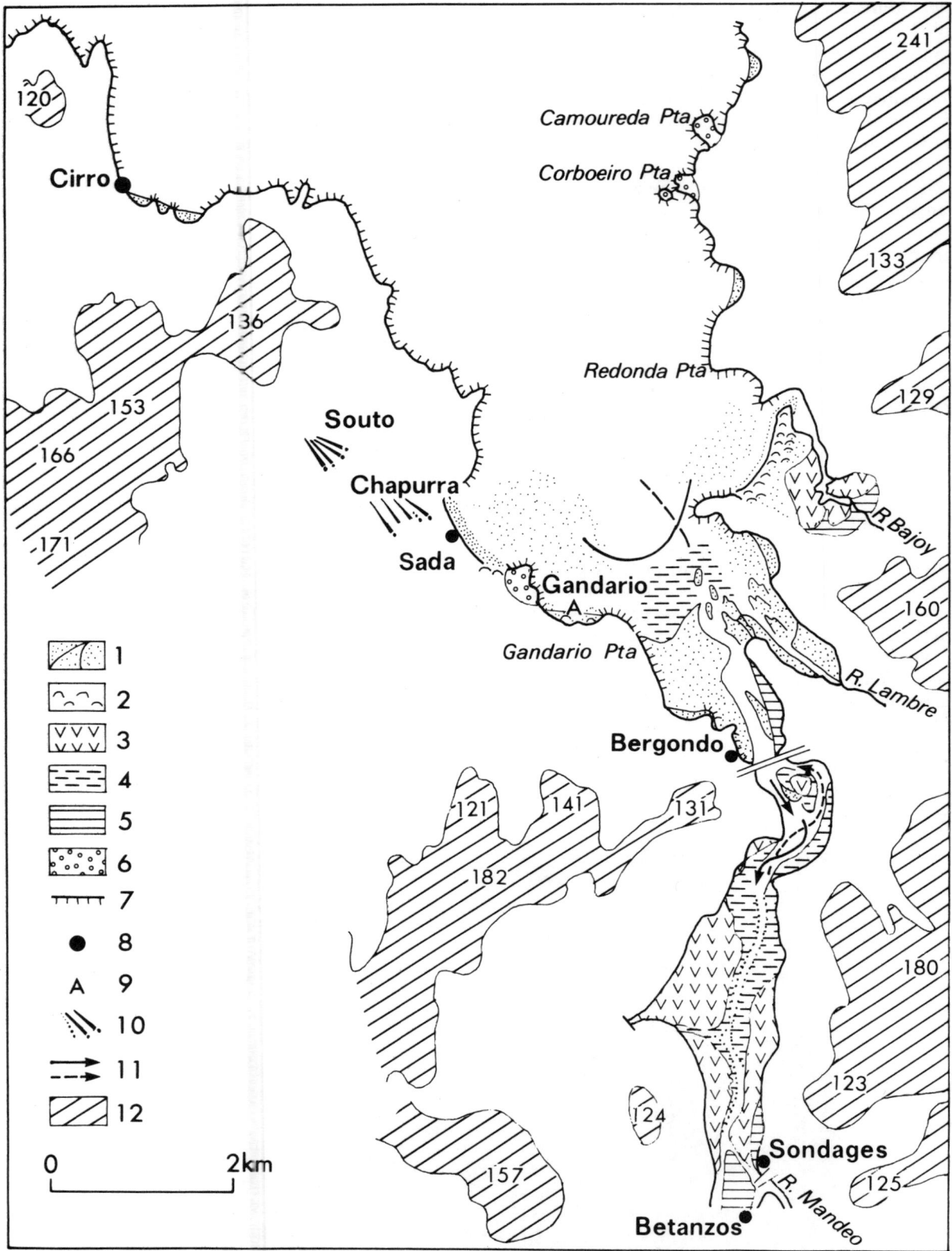

Fig. 12.3 Rias and other geomorphological features of the coastal area near Betanzos, Galicia, north-west Spain [Nonn (1966)]. (1) Sand, (2) dunes, (3) salt marsh, (4) mud flats, (5) cultivated area, (6) rasa at less than 20 m, (7) cliffs, (8) towns, (9) kaolinite clay, (10) older Quaternary alluvial cones, (11) direction of currents and (12) land over 100 m.

however, been facilitated by the existence of lithological differences or abundant jointing, giving areas of low resistance that have been exploited by differential erosion. The rias on the Cantabrian coast or upper rias (*rias altas*) are clear examples, as are many of the central rias (*rias centrales*) such as the Lage ria, where the stripping of a weathered band of schists on a gneissic boss is very apparent.

The alveolar type of ria is represented by those of La Coruña and Arosa, which are broad, circular cuvettes of Tertiary weathering now drowned by the sea. Nevertheless, they can all be considered rias because an important river flows into them and contributes to their formation. Two sectors can be outlined: (1) the headwaters, with a fluvial origin and shape, and (2) the lower section, with an origin related more to the alveolar processes than to fluvial incision.

12.2 Asturian and Leonese Massifs

This high-relief area which extends in an east–west direction adjacent to the northern Spanish coast consists of a well-defined structural unit of Palaeozoic age that owes its present individuality to Alpine tectonics. In the west, the Lugo meseta rises in a fault scarp that separates this crystalline block from the area of Palaeozoic sediments [see Solé-Sabarís and Llopis (1952), Vidal-Box (1941)], outlining and describing an orographic arch that borders the lowlands of Galicia and comprises the Leonese and eastern Galician mountain rim. To the east, the morphotectonic limit is the Saja valley [see Hernández-Pacheco (1932a), Dantin-Cereceda (1912, 1922)] which divides the Palaeozoic high relief of the Picos de Europa and the Liébana graben from the Mesozoic sediments of the Alpine system. The southern and northern borders are also structurally well defined by means of an abrupt fault scarp that sharply separates this mountain range from the Tertiary sediments of the Duero Basin and the Cantabrian Sea. Whereas the southern slopes are long and gentle—the Duero Basin starts at 800 m above sea level—the northern slopes are steep and abrupt, with a rise of approximately 2000 m in 15 km. The coast is rectilinear in its general outline, reflecting its tectonic origin. Its main features are the continuity of the cliffs and a low platform termed the *rasa* that occurs all along the coast.

12.2.1 Structure

Marked changes—a decrease in metamorphism, magmatic activity, folding intensity and age of the sediments—in a west–east direction allow this northern area of the Iberian massif to be divided into two geological areas [see Lotze (1945), Julivert *et al.* (1974)]: (1) the Asturian–Leonese zone and (2) the Cantabrian zone. Altogether the sedimentary succession found in this area shows that, after a stable phase in the Lower Palaeozoic, there followed a phase of growing dynamism culminating in the Hercynian orogenesis. By the Stephanian, sedimentation was already essentially posttectonic, having molasse characteristics. It was in this period that a luxuriant marsh vegetation developed, resulting in the coal deposits of today. The whole area has remained emergent since the Mesozoic and

was fractured into blocks by early mid-Tertiary movements that determine today's relief.

12.2.1.1 Asturian–Leonese zone

This comprises a regional structure in the form of a great arch called the 'Asturian knee', its most important features being—from west to east—an antiform with a porphyritic outcrop called the *ollo de sapo* formation, the Precambrian core of the Mondoñedo overturned fold and the Precambrian core of the Narcea antiform. Particularly noteworthy in this zone are the great thicknesses of the Lower Palaeozoic sediments—the Cambrian is 4 km thick in Narcea and the Ordovician 6 km—a detrital shallow-water facies, and the fact that the Silurian sediments are transgressive over the Ordovician. Some Palaeozoic magmatic activity—both volcanic and plutonic—occurred. The main rock types are slates, schists, granites and quartzites. From a tectonic point of view, important deformation of the Precambrian and Palaeozoic rocks occurred during the Hercynian orogeny accompanied by metamorphism that increased in intensity to the west (i.e. near the Galician shield). The overall structure of the zone is the result of the superimposition of three important episodes of folding. The first phase produced folds overturned towards the east which tend to be of great size, especially in the west, as in Mondoñedo. During the second phase, big shear surfaces gave way to overthrusts. In the third phase, there was a development of folds with subvertical axial planes. All these events took place before the Stephanian sedimentation and before the establishment of plutons.

12.2.1.2 Cantabrian zone

This zone extends from the Narcea antiform to the Mesozoic cover forming the nucleus of the Asturian knee. Although the Palaeozoic succession is not complete, there is at least some representation of each system in an accordant sequence up to the Lower Carboniferous. In this area the Lower Palaeozoic is only 1 km thick and rather incomplete, consisting of platform sediments of a shallow-water regime, a situation that continued during the Devonian. The Lower Carboniferous has a very uniform facies but is only some tens of metres thick. The Upper Carboniferous, in contrast, is very well developed (6 km) both in a detrital facies of shallow-water regime and in a turbidite facies of great depth. Magmatic activity is scarce in this zone, only represented by certain intermittent vulcanism and almost no plutonism. The main rock types are quartzite, slate, sandstone, conglomerate and limestone, this last being the most extensive. The series shows that this was a great area of marine sedimentation during the Carboniferous. In the Westphalian period, a regression began that culminated in the Hercynian folding processes, forming zones of high relief separated by wide continental zones of subsidence in which the coal-bearing deposits accumulated.

Unlike the Asturian–Leonese zone, deformation here was not accompanied by metamorphism, but took place under epidermic conditions in which contrasts in ductility largely controlled the deformation. An earlier tangential phase was followed by the formation of nappe structures by general displacement of the Palaeozoic series over the plastic slippery base of the Lower and Middle Cambrian formation. Such tangential tectonics were responsible for the main structural subdivisions of this zone. Later, a flexure mechanism produced two systems of longitudinal folds with almost vertical axial planes: an arched system that follows the Asturian arc and a radial one that intersects the former.

12.2.2   Relief and Drainage

This is an important area in the Iberian Peninsula because it constitutes the main divide between the large river systems draining south into the Atlantic Ocean and the Mediterranean Sea, and the smaller ones draining north to the Cantabrian Sea. At the same time, this divide is a barrier that, by intercepting the north-westerly winds, creates two very different climatic areas in the Peninsula: the Atlantic, humid northern fringe and the Mediterranean, dry southern lands.

Fracturing into blocks by early and mid-Tertiary movements produced, in addition to the main faults that delimit the whole Asturian block, minor longitudinal faulting that has produced a horst and graben topography [see Llopis (1954)] consisting of two ranges parallel to the coast with an elongated intermediate trough (Oviedo) and, at the eastern end, the massive limestone horst of Picos de Europa and the graben of La Liébana. In the depressions, some remnants of a Mesozoic sedimentary cover are to be found.

Rivers are short and steep, plunging some 2000–2400 m within distances of 15–30 km of the sea. Consequently, erosion is strong and river piracy is frequent. The whole drainage is supposed to be due to a long polycyclic history [see Solé-Sabarís and Llopis (1952), Birot and Solé-Sabarís (1954a)] as in the Galician massif, and the relief of the mountains is derived from a pre-Triassic peneplain. A discordant cover of Permo-Triassic materials shows the existence of a peneplanation phase after the Hercynian folding, giving way to a surface that could have been the point of departure for the morphological evolution of this region, probably even including the heights of the Picos de Europa as these sediments reach Peña Labra (2006 m) and Peña Sacra (2042 m). All in all, there seems to be a succession of planation surfaces, as in Galicia and other areas of the Iberian Peninsula.

Differential erosion has exploited the varied resistance of the underlying rocks in this area, where Devonian and Carboniferous limestones, Silurian quartzites and Devonian sandstones are the resistant elements that form the pronounced headlands and divides, while the valleys are carved into the softer Silurian, Devonian or Carboniferous slates. It is an Appalachian ridge and valley type of relief, with crests aligned perpendicular to the coast.

12.2.3   Glaciation

Small centres of glaciation developed in this area during cold phases of the Pleistocene, especially in the Galician–Leonese area, in the Sierras Segundera, Peña Trevinca, Cabrera and Montes Aquilianos. Stickel (1929) deduced

the existence of 'platform glaciers' developed on the upper erosion surfaces, consisting of ice fields from which glaciers flowed down to levels of about 1000 m. Glacial drift is widespread, and several lakes are held up behind the four arcuate ridges of the terminal moraines of the Tera glacier at the confluence of the Tera, Segundera and Cardena valleys. Stickel (1929) assumed that these features were of Würmian age, which would explain their freshness, based on the relationship between the morainic loops of Sanabria and the fluvial terraces of the Tera valley, where there are three alluvial fills (55, 35 and 10 m above the present river channel). Of these, only the lower ones can be related to the glaciofluvial deposits and the outer moraines of the Sanabria lake.

At the other end of the Asturian range, the Picos de Europa was the most important glacial focus in the north of Spain. Obermaier (1914) described the glaciers of the Andara and Bulnes massifs, where ice originated on upland plateaus and flowed through valleys, forming outlet glaciers. The glacial drift of the Bulnes, Deva and Duje valleys descends to 1400 m, and the glacier in the Duje valley must have been about 7 km in length. In this area, there are also large fields of roches moutonnées and of polished rocks called *lamiares*, among which several lake basins have been carved. Although the terminal moraines might in all probability be related to two different glaciations, the Würmian glaciation seems to be the one that has left a major imprint on the present relief, as is the case in many other Iberian mountain areas.

12.2.4   Karst Landforms

In the Picos de Europa massif, over the features due to the last glaciation several karstic forms have evolved, reflecting to a great extent the high level of the present annual precipitation in this area. Lithologically, it is made of the resistant Lower Carboniferous Limestone that outcrops among the surrounding Carboniferous slates.

The massif is divided by the deep canyons of the Sella, Duje and Deva rivers into three units: from west to east, these are Covadonga (2500 m), Bulnes (2600 m) and Andara (2300 m). Karstification has progressed principally on the upper plains where the previous ice fields existed [see Solé-Sabarís and Llopis (1952)]. Most of the lacustrine basins have been transformed by the formation of dolines and sink holes, down which waters rapidly drain into the massif to reappear at its borders at the contact with the slates. The most important karst feature is the Comerza polje (2.5 km long and 1 km wide). The Comeya plains are also a polje (60 m deep). In the Bulnes massif, there is a great mixture of glacial and karst landforms, with dolines and sink holes that are at the same time permanent nivation hollows extending over a wide platform at 2100–2200 m. In the Covadonga massif, there is a well-developed cave system.

There are also karstic processes at work in other limestone areas of the Asturian massif, but none with the impressiveness of the Picos de Europa, which is probably the only region in Spain with active karst. It is precisely the existence of karstic relief that explains some of the bays in the coast, which are actually drowned dolines.

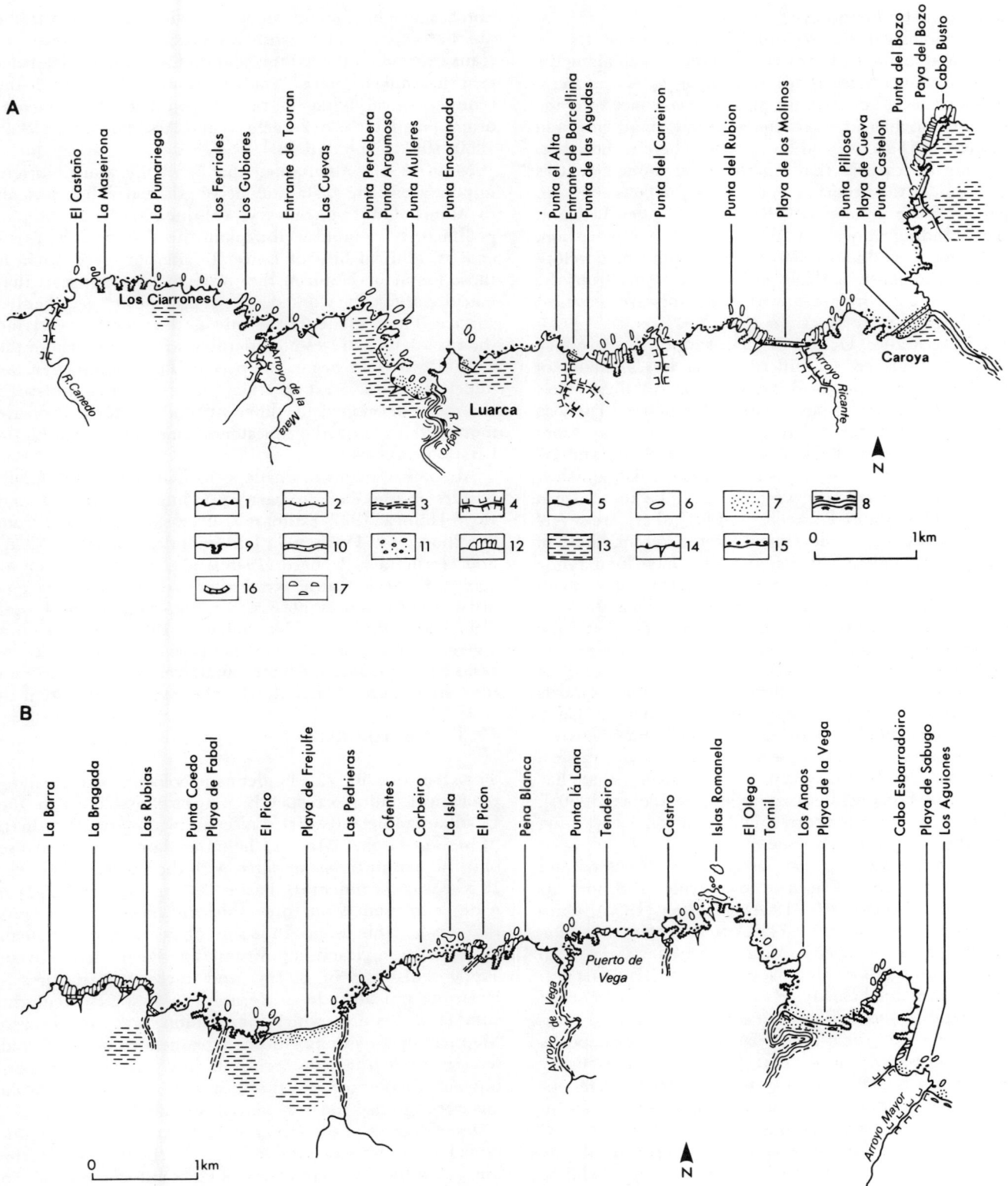

Fig. 12.4 Map of the Asturian coast: (A) from Cabo Busto to Castiecho End; (B) from the Playa de Sabugo to Navia bay [after Asensio-Amor and Martí-Yenderrozos (1981)]. (1) Cliff, (2) stony ground, (3) slightly incised valley, (4) incised valley, (5) low cliff, (6) headlands, (7) sand beaches, (8) river with artificial channel, (9) caves, (10) offshore bar, (11) boulders and shingle, (12) marine abrasion platform (rasa), (13) deposits reworked by the sea, (14) cliff with hanging valley, (15) quartzite outcrops separate from the main cliff, (16) fossil cliff and (17) dunes.

12.2.5   Coastal Features

The morphological features of the Asturian coast are not peculiar to this littoral but are found repeated all along the northern Spanish coast from Foz to Bidassoa. The general appearance is that of a young, sinking coast shown by the abundance of cliffs, the existence of some small rias, and the scarcity of beaches and low coasts. There is, however, one group of features that cannot be explained on this basis, such as the occurrence all along the coast of *rasas*, which are wide erosion benches smoothly bevelled and clearly standing above sea level. The perfect planation, the absence of residual relief, the continuity of development and the gently sloping profile descending from the foot of the coastal ranges towards the cliffs are the most outstanding characteristics (*see* Fig. 12.4).

In earlier studies, Gómez de Llarena and Royo-Gómez (1927) distinguished six different rasa levels, many of them overlain by deposits. But nowhere are all these levels found together. In a later study, Hernández-Pacheco (1949b) reviewed the existing literature [mainly from Hernández-Pacheco (1932a) and Cueto (1930)] and reduced the levels to four. Nonn (1966) distinguished between the eastern and western rasa. In the eastern sector [see Hernández-Pacheco, (1949b)], there are several inserted rasa levels developed over different kinds of rock, but especially on limestones that have undergone karstic solution. There is a general level at 200–230 m, other levels at 120–140 m and 50–70 m, and locally a level at 5 m. In Cabo Peñas, there are only two levels at 200–220 and 120–130 m, and this is the point where the western sector begins, characterized by the existence of only one level but with a decreasing altitude towards Galicia: from 118 m at Cape Vicrias to 5 m at Cape Burela. Nonn (1966) thought that this difference between the two sectors has a structural explanation: the eastern sector corresponds to a zone of fracturing that has produced the horst and graben relief [see Llopis (1950)], while the western sector corresponds to a zone of warping as already described in Galicia.

The origin of the rasas has been amply discussed and successively attributed either to marine abrasion, as by Birot and Solé-Sabarís (1954a), Cotton (1955), Guilcher (1955, 1974), Mary (1971) or to continental erosion as by Hernández-Pacheco (1932a), Hernández-Pacheco (1949b), Llopis (1956) and Nonn (1966). The first hypothesis is sustained chiefly on the basis of the existence of marine deposits overlaying the rasa. However, as discussed by Nonn (1958, 1966), the existence of marine deposits does not necessarily imply that the surface was cut by marine abrasion. Nevertheless, other arguments in favour of marine abrasion are the existence of similar features on the Iberian coast from the Bay of Biscay to the Bay of Cadiz, the fact that the rasa does not penetrate far into the valleys and that the heights of the levels are comparable to those found on other oceanic coasts if the actual levels are interpreted as having been affected by tectonic deformation [see Solé-Sabarís (1978)].

The second hypothesis is based for the most part on the correlation of the higher level with the continental planation surface of this area, on the absence of wave-cut notches at the foot of supposed cliff lines and on the absence of correlation with the classic eustatic levels of transgression on the assumption that no tectonic deformation has taken place. Studies show that most of the sediments overlaying the rasa are, in fact, of continental origin [see Hernández-Pacheco and Asensio-Amor (1959, 1960, 1961), Nonn (1966)].

Nonn (1960) maintained that a wholly marine origin implies a prolonged stability of the coast, which is difficult to admit. On the contrary, continental denudation of pediment type could well explain this feature. The existence of alluvial fans of Lower Quaternary age, such as those found by Nonn in the western rasas, suggests that marine abrasion worked on a previously pedimented surface. The presence of kaolinite in the deposits overlying the rasa demonstrates considerable weathering before the coastal processes began to operate. The resistance of the underlying rock must be taken into account as, although the rasas developed on different kinds of rock, they are more extensive on the limestones that are susceptible to karstic processes.

Absence of fossils in the deposits makes dating difficult. Studies in the eastern part by Gómez de Llarena and Royo-Gómez (1927) estimate a succession of surfaces from the Pliocene to Tyrrhenian I. Hernández-Pacheco (1932a) and Hernández-Pacheco (1949b) also attributed such an age to the rasas. In the western part of the rasas, the age attributed to them is more recent. Solé-Sabarís and Llopis (1952) and Birot and Solé-Sabarís (1954a) agreed on a Tyrrhenian I age, while Nonn (1966) argued that the rasas existed before the Tyrrhenian transgression and that they are deformed after the Tyrrhenian sedimentation.

12.3   Duero Basin

The Duero Basin is a well-defined geological and geomorphological unit corresponding approximately with the Castilian tablelands. It is also known as the Northern Meseta or Higher Meseta, the latter name being because of its higher altitude compared with the Southern Meseta. It is a wide sedimentary basin (200 km long and 250 km wide) surrounded on three sides by mountainous belts that make this area, in spite of its general plateau appearance, a closed depression. The Basin stands at an average altitude of 850 m, and these two features—isolation and altitude—determine to a great extent the climate of this area which, in addition to the generalized Mediterranean summer drought, suffers from long, cold winters and from low precipitation. From a geomorphological point of view, this unit of the Iberian Peninsula has been studied most by García-Fernández (1968).

The contacts of the Basin with the surrounding mountains have different characteristics. In the north-west, the limit with the Asturian massif is clear and corresponds to a fracture line, but in the north-east, although the limit is still sharp, the relief differences are smaller and correspond to the long, narrow synclines of the Cantabrian mountains called *loras*. The eastern part of the Basin is closed by the *parameras* of the Iberian Cordillera, which are high tablelands at the foot of this range rising 100–

Fig. 12.5 Páramo and campiña landscape in the Valladolid sedimentary area, Duero Basin. (Sala)

200 m above the Basin. The southern limit is the Central Cordillera which, in spite of its 500-m difference in elevation from the Basin, does not provide so much of a barrier due to its more gradual contact with the Basin through a pediment several kilometres long. The western limit is only well defined at its northern end, near the Leonese and Galician massifs; farther south, the Duero Basin drops gradually to the peneplain that extends to the Atlantic borders and which is only slightly interrupted by small quartzite lineations.

12.3.1 Geological Structure

The Basin consists of two different units whose dividing line passes approximately north–south through the towns of Leon, Zamora and Salamanca. To the west of this line, Palaeozoic rocks are predominant, while in the east, which is also the more extensive of the two, the Basin is mainly filled with continental sediments of Cenozoic age.

The Palaeozoic rocks are mostly granites and gneiss in the centre of the western region, with alternating bands of Silurian quartzites and Cambrian slates elsewhere. Within the sedimentary area, the central part is filled with a deposit of red clay. To the north, the clay is overlain by quartzite conglomerate in a red matrix, a deposit very similar to that of the rañas. To the east, the clay is covered by a layer of white gypsiferous marl of Vindobonian age, and farther east these strata are in turn overlain by Pontian limestone. In the south, the clay is buried by yellowish-red sands and to the west the sedimentary cover gradually gives way to the Palaeozoic substratum without any topographic change; only the different colours of the two materials—the redness of the sediments and the greyness of the metamorphic rocks—show the structural change.

The tectonic structure of the Duero Basin is simple. The widespread Neogene materials maintain a horizontal bedding over almost the whole area. It is only in the eastern part of the Basin, at the contact between this unit and the Iberian Cordillera, where some Mesozoic and Palaeogene rocks are present, that slight discordances and gently dipping strata can be found. These bear witness to slight tectonic movements that occurred mainly at the end of the Oligocene [see Julivert *et al.* (1974)] at the same time that the principal folding phases occurred in the Iberian Cordillera. In the western Palaeozoic area, the rocks have been folded along a clear north-west–south-east Armorican trend.

12.3.2 Relief and Drainage

The relief of the Duero Basin consists of several horizontal surfaces cut at different levels but always at a great altitude above sea level. Within this vast plain, however, two main types of landform can be observed: (1) the tablelands and lowlands of the sedimentary basin, and (2) the peneplains of the Palaeozoic basement.

12.3.2.1 Sedimentary area

The varying landscapes within the sedimentary sector are the result of differential erosion. Two topographical levels can be found in this wide plain: a higher one or *páramo*, which corresponds to the tablelands that are strongly cut by the drainage net of the rivers on the more resistant materials, and the lower one or *tierra de campos*, which consists of gentle slopes modelled on the soft clays. In the eastern part of the Basin, where the Pontian limestones are found, the proximity of the Iberian Cordillera has given the streams enough potential energy to carve out a deeply incised drainage net, with flat-bottomed dry valleys, steep valley side slopes and broad, flat interfluves (páramos). The páramos, which constitute one of the most typical landscapes of Old Castile, are structural platforms that stand at the general high level of the Basin (850–1000 m), linked to the foothills of the Iberian Cordillera. They stretch for great distances along both sides of the drainage lines, becoming progressively narrower until they finally end, more or less abruptly, in front of the general lowland level. An increase in dissection, mainly because of the thinning and discontinuity of the limestone layer, produces wider valley floors with isolated tabular remnants at different levels which are called, in relation to their decreasing size, *parameras*, *mesas*, *muelas* or *cerros*; when the hills are very small and their tops are conical, they are called *oteros*. In the northern part of the Basin, at the foothills of the Asturian and Leonese massifs, the *raña*—a conglomerate of quartzite cobbles in a red clay matrix—acts as a protector of the underlying clays giving a landscape similar to that of the limestone paramos, but called *raña páramos* in this case.

The central and southern parts of the Basin have been extensively eroded due to the soft clay substratum, resulting in a landscape called *campiña* consisting of a gently undulating plain with small hills and wide valleys. This campiña, which stands at 700–900 m, has been modelled by runoff which has also produced several alluvial fills along the main river branches and a series of corresponding glacis at the foothills of the páramos. There are generally three main terrace levels according to Hernández-Pacheco (1928), which stand at 10, 40 and 70 m above the present-day river course (*see* Fig. 12.6). Variations in these height differences occur in places, related to the incoming of tributaries [see Leguey and Rodriguez (1970), Hoyos *et al.* (1974)]. The modelling of the glacis is attributed to arid periods in the Pleistocene when slope wash was the main geomorphological agent. Whereas at the foothills of the páramos three glacis levels can be distinguished, in the campiña, because slope retreat has on successive occasions reached the weak clay interfluves, only one level is found, which corresponds to the small *lomas* that interrupt the Castilian lowlands. Sometimes the tops of the lomas correspond to remnants of former alluvial terraces that have protected the underlying soft clay.

12.3.2.2   Palaeozoic area

In the west of the Duero Basin, erosion has reduced the Palaeozoic rocks to a more or less perfect peneplain. In the north-west, the bedrock is mainly granite and the peneplain is practically a perfect plain that stands at 700–800 m, with only occasional residuals or monadnocks rising about 200 m above the general level. Only two larger-scale features interrupt the flatness of the peneplain. One is the tectonic depression of Ciudad Rodrigo, which contains 100 m of horizontally bedded sandstones and marls of Eocene age. The interior of the depression has a complex structure with small horsts of Palaeozoic rock 80 m high. The whole area was later fossilized by raña deposits. The other main feature is the *arribes* of the Duero. In its lower course, the Duero falls through about 700 m, thus giving great energy to cut into the peneplain. For approximately 2 km, there are gorges 300–500 m deep forming a deeply incised landscape known as the arribes. Erosion has also exhumed granite boulders resulting from pre-Quaternary deep weathering.

In the south-west (between Salamanca and the Central Cordillera), a peneplain (900–1100 m high) extends for some 60–70 km, cut across slates and quartzites. The difference in the resistance of these two rocks has produced a more undulating landscape than on the other peneplains, with quartzite lineations or monadnocks whose maximum relief is 200 m. As in the case of the other peneplain, several remnants of an Eocene cover can be found, together with widespread raña deposits.

In the Beira district of Portugal (*see* Fig. 12.7), three main morphological regions have been distinguished [see Ferreira (1980)]. From east to west, these comprise the Meseta, the central plateaus and the western mountains. The first is a well-preserved planation surface, whose altitude falls from 950–1000 m at the foot of the Serra da Malcata to 400–450 m near the Duero. A rectilinear scarp (300 m high) running in a general north-north-east–south-south-west direction separates the Meseta surface from the central plateaus, where the main trait of the relief is still the flatness of the interfluves; however, planation surface remnants are scanty and situated at varied altitudes. Contact between these plateaus and the western mountains consists of another scarp trending in the same direction. The relief of these mountains is more rugged than that of the central plateaus. From a study of the deposits on the Meseta and the Mondego platform (central plateaus), Ferreira (1980) concluded that these are polygenetic planation surfaces with a similar evolution. Three distinct erosion levels, probably corresponding to the three generations of deposits preserved there, are recognized. In the Beira district, there is a clear adaptation of the stream pattern to tectonic structure; differential erosion and Quaternary dissection are due to the different resistance of the granites and schists of the area. The granite area presents a steppe relief resulting from successive phases of planation and tectonic movements. Where schists and greywackes are predominant, the erosion levels disappear almost entirely as the rivers have cut down vigorously [*see* Ribeiro (1949)].

Near the mouth of the Duero—in this area known as the Douro—terrace levels are correlated with marine terraces [see Teixeira (1949)]. Some have been dated by fossil remnants and human artefacts. The higher level corresponds to the Calabrian (Upper Pliocene), while the other levels are Sicilian (80–130 m), Milazzian (60–65 m), Tyrrhenian (12–40 m) and Grimaldian (the lowest level).

12.4   Central Cordillera

This is the most outstanding feature of the Iberian massif, forming a natural division between the two high plateaus of Castile and between the drainage basins of the Duero (Douro) and the Tajo (Tejo). It crosses the Peninsula in an east-north-east–west-south-west direction from the Iberian Cordillera to the Atlantic and consists of the following segments: Somosierra, Sierra de Guadarrama, Sierra de Gredos, Sierra de Gata and Serra da Estrela. Overall, it is a geographical unit of striking relief, with heights of up to 2430 m (Peñalara) and 2959 m (Pico de Almanzor), which makes communication between the two plateaus extremely difficult, particularly because of the

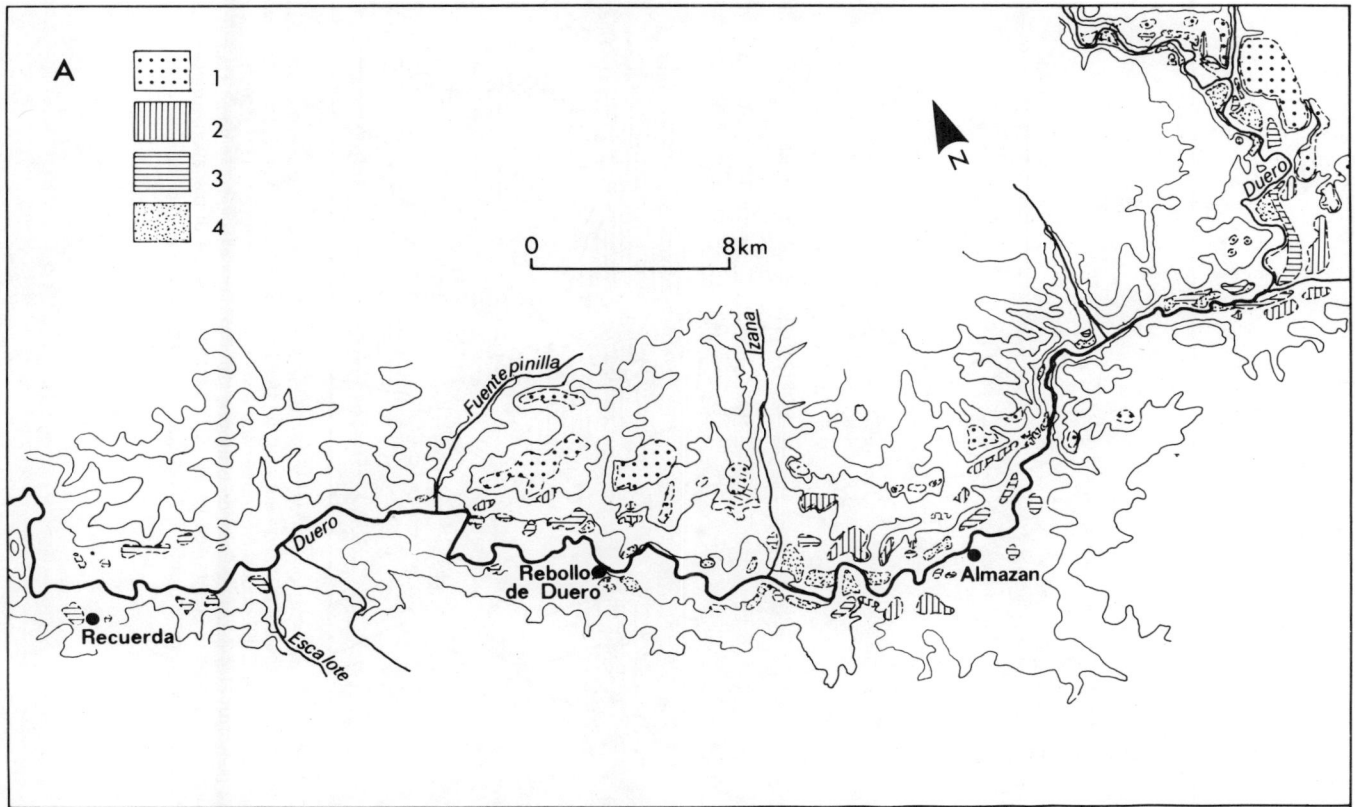

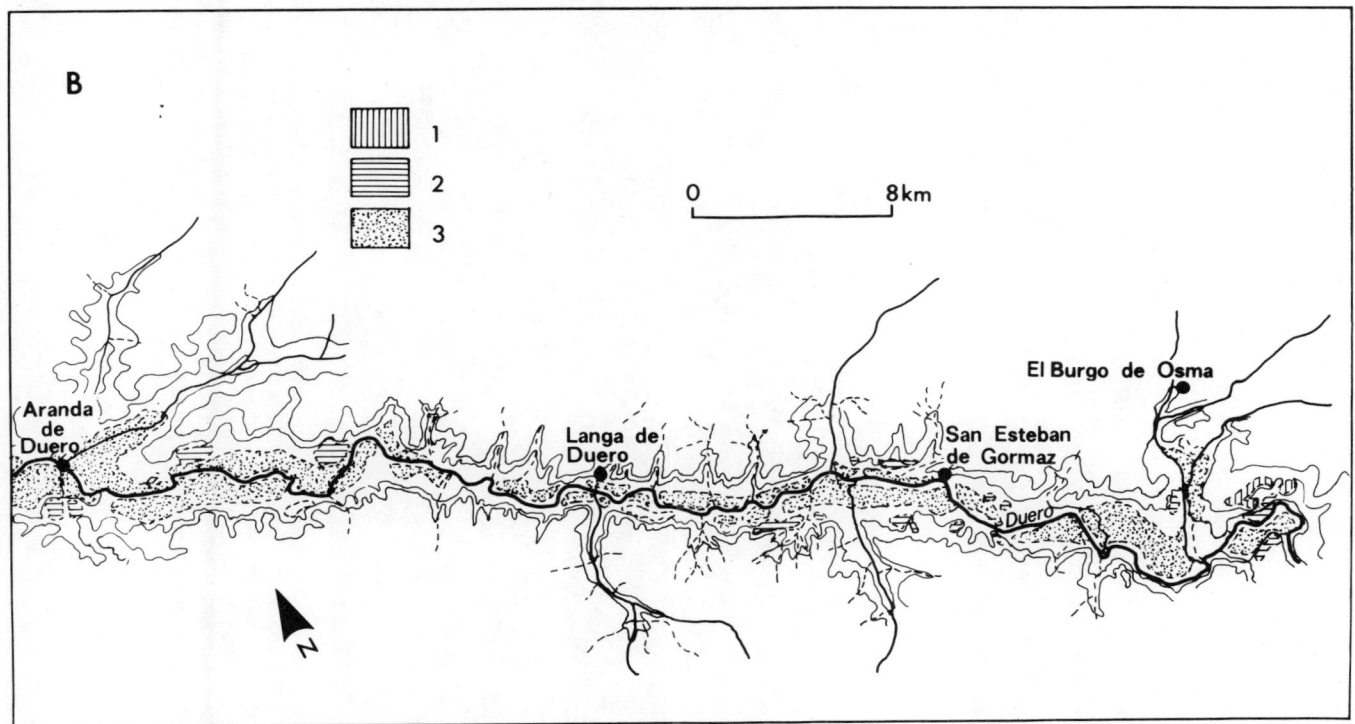

Fig. 12.6 (A) Location of terraces in the upper Duero valley [Hoyos *et al.* (1974)]. (1) Highest terrace, (2) upper middle terrace, (3) lower middle terrace and (4) lowest terrace. (B) Location of terraces in the middle Duero valley [Hoyos *et al.* (1974)]. (1) Upper middle terrace, (2) lower middle terrace and (3) lowest terrace.

Fig. 12.7 Geomorphological map of the northern Beira planation surfaces [after Ferreira (1978)]. *Planation surfaces of the Hercynian Meseta:* (1) fundamental planation surface; (2) levels set into (1). *Surfaces of the Central Planalto:* (3) remnants of the summit planation surface; (4) fundamental planation surface; (5) higher levels of the lower surface; (6) lower levels of the lower surface. *Surfaces in the western mountains:* (7) summit levels; (8) intermediate levels; (9) doubtful intermediate levels; (10) lower levels; (11) doubtful lower levels. *Other planation surfaces:* (12) Mondego platform; (13) coastal platform; (14) Cova da Beira platform; (15) Bacia de Celorico platform; (16) Rio Uíma alluvium. *Tectonic structures, slopes and residual relief:* (17) major fault scarp; (18) minor fault scarp; (19) tectonic alignment; (20) fracture-controlled valley; (21) direction of slope of planation surfaces; (22) mountain edge; (23) convex breaks of slope; (24) concave break of slope; (25) minor canyons; (26) rivers; with knick points; (27) resistant quartzite outcrops; (28) residual relief.

copious snowfalls in winter when the mountain passes of Somosierra, Navacerrada and Guadarrama are usually blocked. The range is about 50 km wide with a marked asymmetry from north to south; gentle slopes meet the northern plateau but there is an abrupt contact with the southern plateau. The cause of this may be structural, including the great fault of the Tajo to the south.

Lithologically, much of the Cordillera consists of crystalline (granitic) rocks together with metamorphosed Palaeozoic sediments, the degree of metamorphism being greater towards the central part and less at the ends of the range. The whole structure is crossed by numerous dykes.

From the tectonic point of view, the essential structural character was determined by Alpine tectonics, as its parallelism with the Baetic Cordillera reveals. Not only was the Central Cordillera raised as a great horst but it was also broken into separate blocks in the form of a series of horsts and grabens (*see* Fig. 12.8).

12.4.1 Geological Structure
The crystalline core of the Central Cordillera has already

been referred to: pre-Ordovician rocks are the main constituents of the range, but within them it is difficult to distinguish the Cambrian from the Precambrian due to the monotonous character of the succession. The Lower Ordovician rocks are strongly transgressive and consist of a layer, several hundred metres thick, of quartzites lying on a base of conglomerates. The Middle and Upper Ordovician consist of slates, sandstones and limestones, while the Silurian is generally constituted by slates with some quartzites, limestones and lidites. Post-Silurian strata are in general poorly represented, the Devonian only being preserved in the cores of certain synclines. Post-Hercynian strata are also rare, restricted to a narrow trench that borders a major strike slip fault parallel to the Hercynian structures; they consist of conglomerates of quartzite, granite intrusives and metamorphic rocks. The Palaeozoic sequences, from the Cambrian to the Lower Devonian, are most completely represented in the Somosierra.

The most striking tectonic feature is the long synform nucleus that, orientated from south-east to north-west,

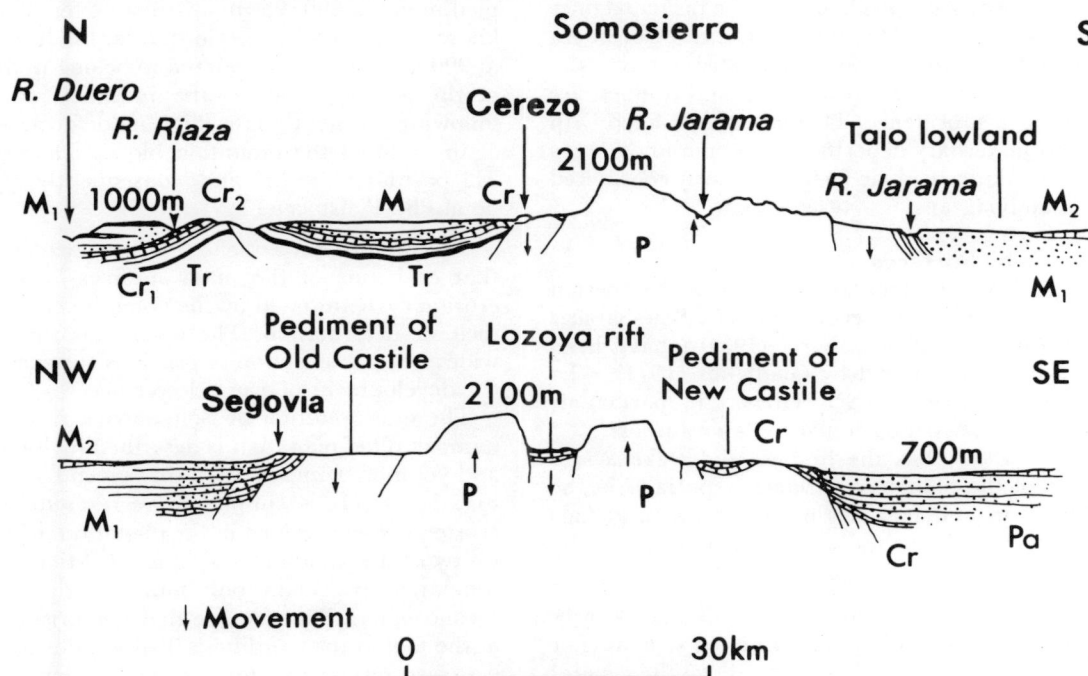

P, Palæozoic; Tr, Triassic; Cr, Cretaceous in general;

Cr$_1$, Conglomerates & sandstones of Cenomanian;

Cr$_2$, Marls & limestones of Upper Cretaceous;

Pa, Continental Palæogene; M, Miocene in general;

M$_1$, Coarse detritic Miocene of the margins of the Cordillers;

M$_2$, Pontian limestones

Fig. 12.8 Two sections through the Central Cordillera, showing eroded fault block structure [Solé-Sabarís and Llopis (1952)]. In the Somosierra, there is a central upstanding block that is tilted towards the south. In the Sierra de Guadarrama, there are various blocks with eroded summits and a central tectonic depression where the Cretaceous has been preserved. The Tertiary depressions of Old and New Castile are limited by important faults. The Pliocene erosion level of the *rampas* (pediment) is at 900 and 1100 m, cutting into the old block and considerably reducing its width.

crosses the range transversely and within which Ordovician and Silurian strata have been preserved.

Three main phases of deformation, which have been described by Babin-Vich (1977), followed one another during the Hercynian orogeny, each identified by its distinctive folding and schistosity characteristics. A late episode of faulting was superimposed on the folding, followed in turn by intrusion of granites, important metamorphism and the division of the massif into fault blocks. The main granite types are adamellite and granodiorite of medium and coarse grain; recent investigations by Aparicio *et al.* (1977) suggest that most intrusive activity was synchronous with tectonism. A series of porphyritic and quartz dykes cross the whole structure in an east–west direction together with a few basic dykes in a north–south direction.

During the Alpine orogeny, this block system was reactivated and the present general structure produced. The main fault lines run in an east-north-east–west-south-west direction and determine the outlines of the whole present-day Cordillera. There are also minor faults running both parallel and perpendicular to the principal ones that are responsible for subdividing the Central Cordillera into horsts and grabens. Several elongated interior depressions, such as those of Lozoya, Tiétar and Ambles, are the result of block subsidence. They have been filled with Neogene and Quaternary deposits that are mainly arkoses with clay intercalations; their facies has been considered as typical of alluvial fans.

12.4.2 Planation Surfaces

Classical studies of the planation surfaces of the Iberian massif by Birot (1933), Schwenzner (1937), Solé-Sabarís (1952), and Birot and Solé-Sabarís (1954b) have been summarized and re-evaluated by Gladfelter (1971), who considers that the views of Schwenzner, in particular, have not been correctly interpreted by later authors.

Schwenzner (1937) was the first to undertake a systematic study of the planation surfaces, especially in the Central Cordillera and its margins. He recognized four surfaces, whose characteristics have been summarized by Gladfelter (1971) (*see* Fig. 12.9).

1) The oldest planation surface (*Dachfläche*) is represented in the modern landscape by isolated remnants or summit surfaces. It stands at 50–100 m above the general elevation of the páramos, and on the ridges and crests of the Cordillera between 1550 and 2200 m. Its origin is supposed to be mid-Oligocene, and its evolution was interrupted by the mid-Tertiary (Savian) orogeny.

2) The Meseta surface (*Mesetafläche*) is divided into three levels which will be referred to, in order of age and height, as M3, M2 and M1.

The M3 surface forms the dominant landscape of the Iberian massif, standing at 1150–1250 m (*see* Fig. 12.10). It was created in a new phase of planation following the Savian orogeny. Later, tectonics again interrupted the cycle and dislocated this surface so that, at present, some of its remnants are found at levels varying from 1300 to 1600 m. Conglomeratic deposits derived from Pontian páramo limestones lying on this surface, and dislocations

by post-Pontian tectonics, are believed to demonstrate that most of the planation must have preceded the Pontian.

3) The M2 surface is extensively developed only among the northern and southern piedmonts of the Central Cordillera, where it is preserved along the margins of the Castilian basins between 1000 and 1150 m, generally on late Tertiary deposits. It is basically a pediment surface, frequently covered by veneers of quartz and quartzite conglomerate, locally increasing in thickness to 30 m. Small inselbergs occasionally rise 20–50 m above the general level. In other areas, the surface is preserved as valley floors cut into the higher M3 surface. A second Tertiary orogeny—the Rhodanian—is the tectonic phase responsible for initiating the erosion cycle which formed this surface. A post-Pontian age is assigned to this bevelling since it cuts across structures in the Pontian limestone.

4) The M1 surface is represented by terraces cut in the Tertiary basins and as valley floors in the Iberian massif. North of Madrid M1 dominates the New Castilian piedmont at 880–950 m, with a cover of fanglomerates known as rañas. In the Cordillera, the head water basins at 990–1110 m are considered to belong to this surface. Its origin is attributed to the initiation of a new cycle following the gentle late Pliocene deformations that once again uplifted the mountain blocks. Development of the M1 postdates the M2 and, consequently, it is thought to be of Plio-Pleistocene age.

The main point attached to Schwenzner's reasoning is that each one of the main surfaces is the result of an erosion cycle initiated by the potential energy provided by tectonic deformation. The result was the dissection and widespread destruction of previous surfaces together with the development of a new lower one.

The interpretation by Solé-Sabarís of the present remnants of relief planation is described by himself as simpler and yet more complex than that given by Schwenzner (*see* Fig. 12.11). It is simpler in relation to the number of erosion cycles involved in the planation, which he reduced to two: the fundamental one of Miocene age, whose remnants are usually only found on the summits, and a second one of Pliocene age that comprises the pediments at the foot of the Cordillera. Erosion during the first cycle was also responsible for the exhumation of older erosion surfaces (pre-Triassic, pre-Cretaceous, pre-Cenomanian, pre-Eocene). The complexity in Solé-Sabarís's scheme comes from tectonics because it is their effect upon the fundamental surface, through deformation, tilting and faulting, that is responsible for the present multi-level appearance. This assertion is thought to be proved through the supposedly easy reconstruction of the outline of the primitive surface and the evidence that the present remnants are not homotaxic levels that correspond to one another on each side of a valley. The age of the fundamental planation is supposed to be Pontian [*see* Solé-Sabarís and Llopis (1952)].

The synthesis that Solé-Sabarís makes is as follows. The Savian (pre-Miocene) and the Styrian (early Miocene) orogenies folded the Mesozoic sediments bordering the

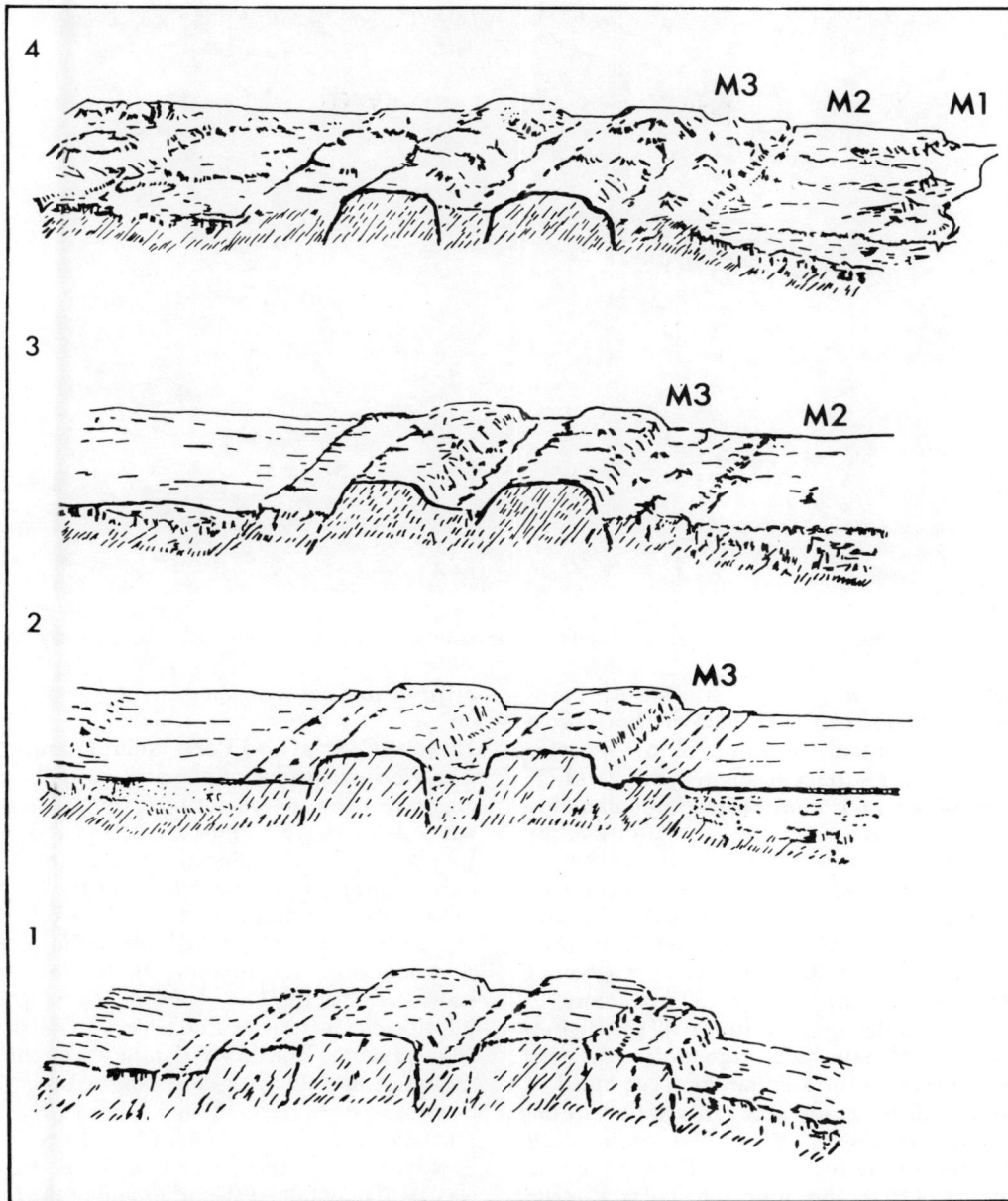

Fig. 12.9 Evolution of planation surfaces according to Schwenzner (1937) and Gladfelter (1971). (1) Interruption of mid-Oligocene to early Miocene planation by the Savian orogeny, (2) formation of M3 surface (mid-Miocene) with residual hills, (3) Rhodanian orogeny followed by planation of the M2 surface during the Pilocene and (4) late Pliocene to early Pleistocene M1 planation and development of river terraces.

Iberian massif and, at the same time, warped and faulted this block creating several units within it: namely, the Central Cordillera, the Montes de Toledo and the two Castilian basins. The development of the Miocene sedimentary cycle was related to the planation of the pre-existing relief and the development of the fundamental erosion surface. This phase terminated in the Pontian. As a consequence of isostatic adjustments, the fundamental erosion surface was deformed and raised to different levels. At the same time, further uplift of the Central Cordillera and the Alpine borders of the massif took place. A new cycle of erosion was then initiated in the Upper Pliocene under an arid climate, the consequence being the development of a series of pediments whose correlative deposits are the rañas. The final events in the evolution of the landscape were the gentle Quaternary deformations and the development of alluvial terraces.

The interpretation of Birot (1933) is even more restrictive than that of Solé-Sabarís in relation to the number of erosion cycles involved, which he reduced to one initiated with the Miocene orogeny and terminating with the Pontian planation. In his opinion, the lower Pliocene erosional levels and the pediments at the foot of the Cordillera are due to a climatic change that occurred at

Fig. 12.10 Main planation surface of the Iberian meseta, broken by broad valley and tributaries in foreground (Guadalajara Province). (Embleton)

the beginning of the Pliocene.

Field mapping by Gladfelter (1971) in the Soria area points to conclusions similar to those of Schwenzner (1937). He identified four surface levels (A, B, C and D) with Miocene, Pliocene and Plio–Pleistocene planation phases associated with the last three. In addition, he correlated basin sedimentation with each planation period: Oligocene deposits with the development of surface A, Tortonian deposits with surface B, and rañas and fluvial deposits with surfaces C and D. In his study area, and particularly in the upper Rio Henares, he suggested that the development of surface D is related more to climate than to tectonic factors. Contrary to the deductions of Birot and of Solé-Sabarís he argued that the very complex fault system which their explanation requires has yet to be recognized in his study area. In relation to the climatic conditions prevailing during the planation periods, study of the correlative deposits of the Oligocene and Miocene indicates dry conditions, particularly in the case of evaporites and limestones. The scarcity of karst features is also counted as a sign of aridity. All this led him to conclude that the surfaces are a result of pediplanation.

Recent work in the Guadalix Basin by Lázaro (1977) deals only with the two more recent surfaces (SI and SII). Their development is related to tectonic readjustments, and their age is equated with Schwenzner's M2 and M1 and with Gladfelter's C and D surfaces, respectively. The raña deposits are correlated with the youngest surface (SII), while the deposit correlated with the SI surface (M2 and C) is called a boulder detrital formation. The difference in terminology of the deposits has no special significance, but only reflects the usage of the word raña in Spanish literature for deposits of quartzite cobbles or pebbles of Villafranchian age [see Hernández-Pacheco (1949c), Solé-Sabarís and Llopis (1952)]. Arid climatic conditions are supposed to have controlled the development of these two surfaces.

12.4.3   Glacial and Periglacial Features

The Pleistocene glacial phenomena in the Central Cordillera have been the object of many studies, a summary of which is given by Vidal-Box (1948) and Solé-Sabarís and Llopis (1952). The glacial features are, however, of only local significance in the context of the overall evolution of the Cordillera, dominated as it is by planation surfaces and fault scarps. The distribution of former glaciers shows a pronounced asymmetry, both along the range and transverse to it. The south-facing slopes were, in the Pleistocene, as now, in the lee of the north-westerly snow-bearing winds from the Atlantic; while the Atlantic end of the range has always received more precipitation than the interior eastern parts. This explains why the most important glacial features are found in the Serra da Estrela, in Portugal. Here, the lower level of permanent glaciation was 1620 m, while at the other end of the Cordillera, in the Somosierra, this limit lay at 1900 m. But nowhere was the relief above such levels extensive enough to be able to support ice accumulation sufficient to nourish major glacier systems, as in the Pyrenees. Most of the glaciers were of the cirque type, sometimes with small tongues but having little erosive power, which explains the lack of U-shaped valleys in the Cordillera.

The pre-Quaternary relief determined the type of glacierization at the outset: small glaciers developed next to the residuals of the planation surfaces and occupied existing river valleys. Structure, especially the fracture lines, also seems to have acted as a major control in the location and evolution of glacial features. Rapid renewal of fluvial action after glaciation caused the washing away of much glacial debris, producing abundant glaciofluvial deposits [see Asensio-Amor and Ontañón (1975)]. Although previously the existence of two systems of moraine

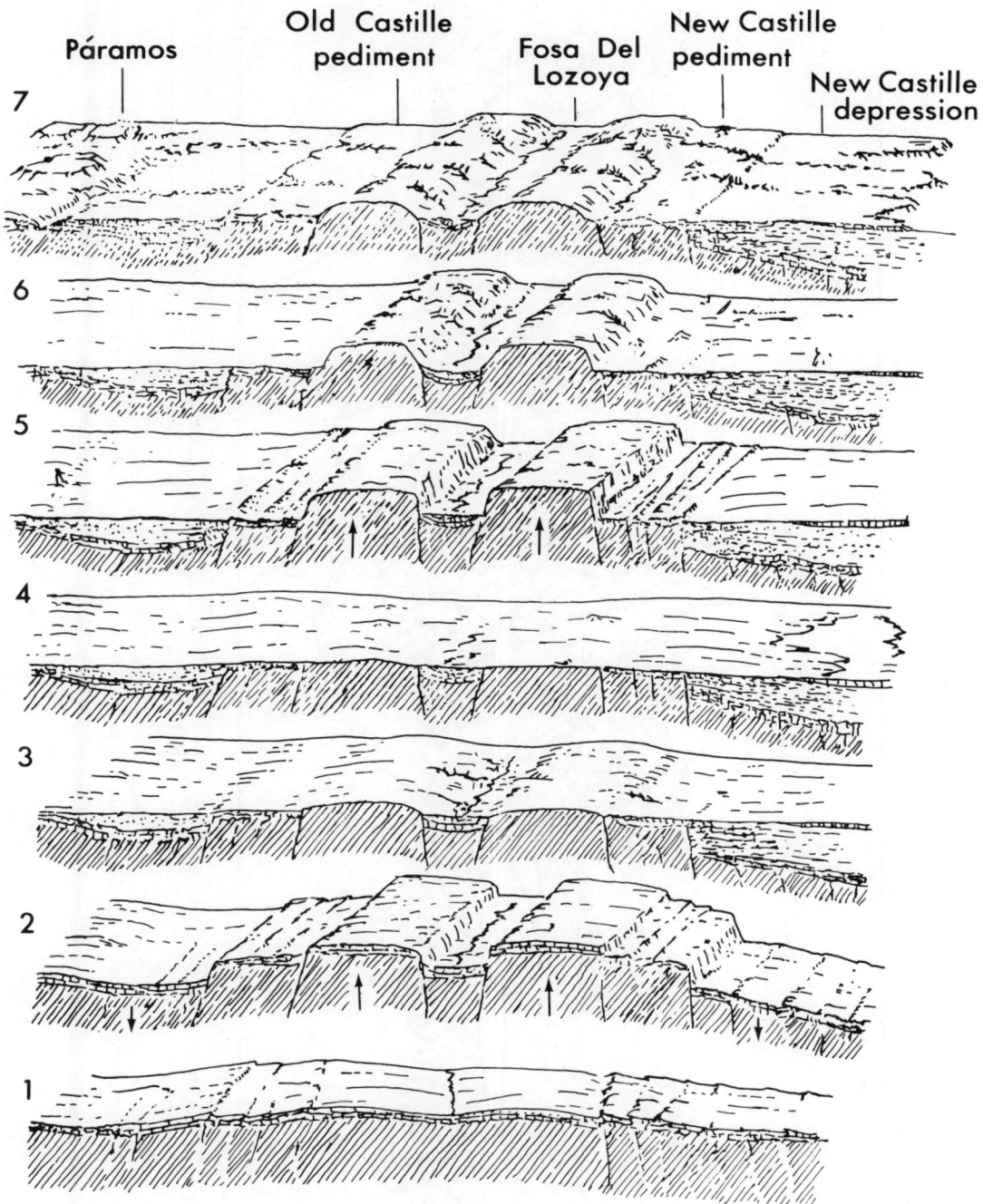

Fig. 12.11 Morphological evolution of the Central Cordillera of Spain [Solé-Sabarís and Llopis (1952)]. (1) At the end of the Oligocene, immediately after the orogeny; folding of the sedimentary cover and beginning of the flexuring of the Central Cordillera and Castilian basins. (2) Fracturing after folding in the Middle Miocene; main initiation of the great fault that divides the Castilian depressions and the Central Cordillera block. (3) Beginning of infilling of the depressions with sediments from the emerging blocks whose relief is being reduced. (4) Pontian peneplain as the result of the ending of the sedimentary cycle; sedimentation of the limestones of the páramos in the lagoons. (5) Rejuvenation of relief as a consequence of the Rhodanian folding; deformation of the peneplain and folding of the Miocene sediments of the margins of the depression. (6) Evolution of the Pliocene peneplain, with a pediment and monadnocks. (7) Post-Pontian rejuvenation and incision of the present drainage net; excavation of the páramos. Diagonal shading: Palaeozoic basement; above this lies a sedimentary cover of folded Cretaceous and Palaeogene; resting discordantly on this is the Miocene, topped by the páramos limestone.

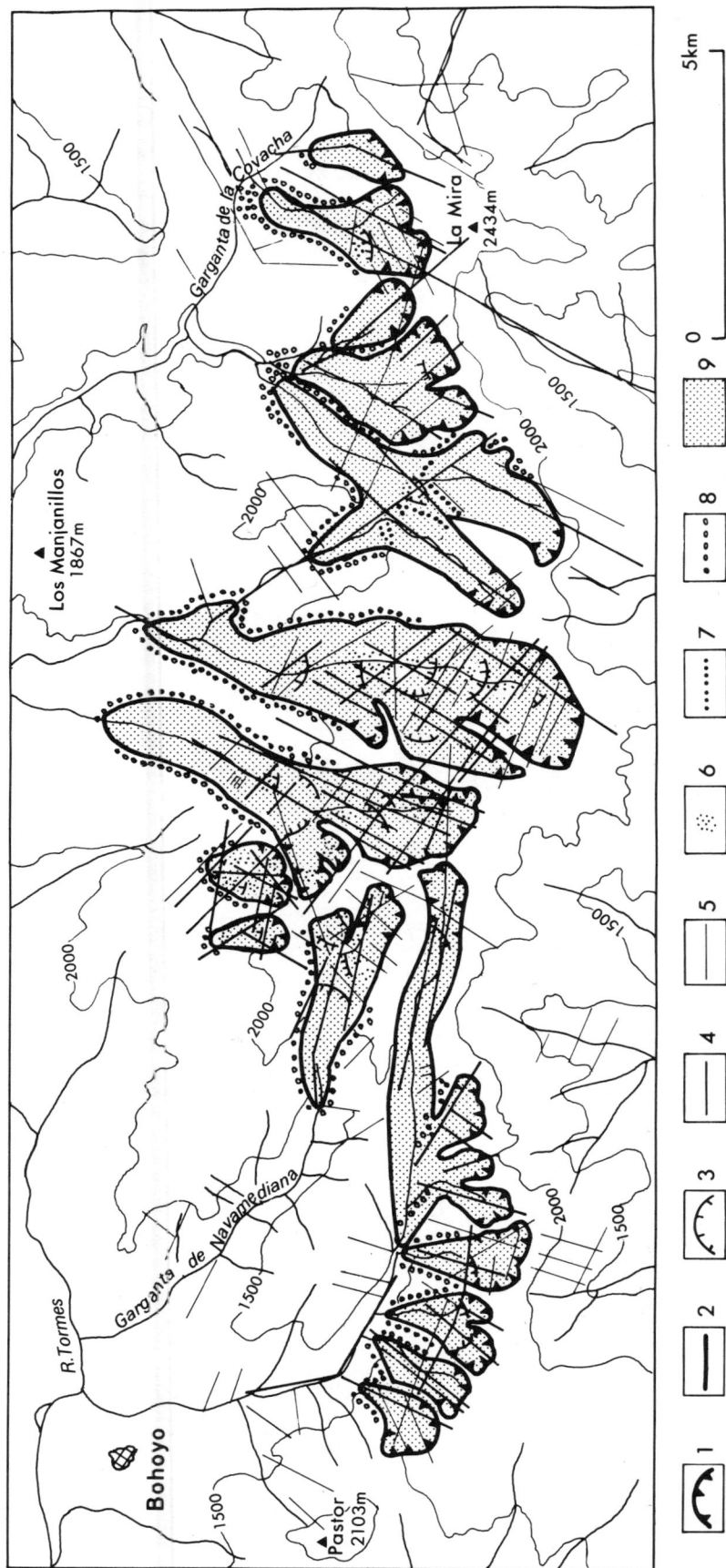

Fig. 12.12 Some glacial geomorphological features of the central Sierra de Gredos [Martínez de Pisón and Muñoz (1972)]. (1) Cirques, (2) edges of glacial troughs and glacially modified valleys, (3) rock bars, (4) faults of major morphological significance, (5) other faults, (6) basins, (7) readvance moraines, (8) major lateral and frontal moraines, and (9) area directly affected by glacial erosion.

loops was interpreted as the result of two glacial periods, the present tendency is to interpret them as stadials of the Würm glaciation.

In the Somosierra, the morphological remnants of glacial action are minimal, consisting only of several small hanging cirques. In the Sierra de Guadarrama, glacial features are mostly located on the southern slopes [see Sanz-Donaire (1978)], the most important glaciated area being located around the summit of Peñalara (2406 m) where there are four main glacial cirques. One of them, the Hoyo de Peñalara, at present contains a lake about 5 m deep. These cirques are approximately 1 km wide, with their exits closed at 1720–2050 m by well-developed morainic loops traditionally attributed to two glacial periods (Riss and Würm). More recent studies by Franzle (1959) and Sanz-Donaire (1978), however, tend to regard them as the result of readvances during a general retreat stage. Location on the southern slopes is believed to be due to more efficient snow accumulation in the lee of the flat summits. On the whole, these glacial features are small in size and relate to short periods of glacial activity.

The Sierra de Gredos is the Spanish part of the Central Cordillera where glaciation was more extensive and intense: consequently, it is the area that has been the object of most investigation [see Hernández-Pacheco (1933), Vidal-Box (1932), Obermaier and Carandell (1916b), Huguet del Villar (1915), Martinez de Pisón and Muñoz (1972)]. The Gredos massif is composed of two horsts: Alto Gredos and Serrota. Three sectors can be differentiated in Alto Gredos [see Martínez de Pisón and Muñoz (1972)]. In the western area, there are nine rather small glacial cirques, supporting short valley glaciers 1–5 km long in the Pleistocene; the maximum depth of ice is believed to have been about 100 m. The central area (*see* Fig. 12.12) is the one where the glaciers were largest but there were only two of them. The Gredos glacier was the biggest; the largest of its three cirques is approximately 2.5 km wide and 8 km long, situated at the foot of the Pico del Moro Almanzor (2592 m). Its glaciated valley ends at Puerto de Roncesvalles (1450 m) and the thickness of the ice is estimated as 350 m. The glacier of El Pinar started at the foot of Risco del Gutre (2568 m); its valley glacier was 6 km long, ended at 1450 m and was fed by three cirques. Several steps along its valley are at present filled with water which produces a lake area known as Circo Lagunas. The ice thickness is estimated to have been 300 m at the terminal moraine. In the eastern area, five small cirque glaciers existed, with glaciated valleys ranging from 1 to 6 km long and with a depth of about 100 m.

To summarize, for the whole of the Alto Gredos, 16 glaciers are spread over an area of 40 km but, with the exception of the two central ones, their incidence and effect on the landscape have been limited. In Serrota, five glacial valleys descend to 1700 m but not all of them come from well-defined cirques.

The Sierra de Béjar (2401 m) has two glacial valleys of 3 km length that descend to 1300 m, and several small cirques and hollows that are now occupied by lakes. The permanent snowline in this area is estimated to have been at 1700 m.

The last glacial vestiges in a western direction are also the most important in the Central Cordillera, being located in the Serra da Estrela (1991 m) in Portugal. The phenomena have been studied by Lautensach (1949) and Daveau (1973). The summit planation surface supported an ice field from which three glacial tongues descended along river valleys and several lobes spread downslope. The Zêzere valley glacier was 13 km long, ending at Manteigas at 600 m above sea level, its ice being 300 m thick. Two more valley glaciers descended in a southerly direction, along the Estrela and Alforfa valleys for distances of 8–9 km, joining at their ends at a height of 700 m. Several smaller glaciers 3–5 km long also existed. The limit of continuous perennial ice was at 1620 m here, but there were also a lacustrine area and several separate cirques below this general level.

Fewer studies have been made of periglacial features than glacial ones, and only recent workers such as Franzle (1959), Gil-Crespo (1964), Asensio-Amor and Ontañón (1972), Martínez de Pisón and Muñoz (1972), Vaudour and Asensio-Amor (1972), Stäblein (1973), Bullón (1978) and Sanz-Herraiz (1978) have carried out such studies. The most widespread periglacial process in the Central Cordillera is frost-wedging, resulting in the development of steep rock faces, pinnacles, scree slopes and block streams. Only in the uplands can remnants of cryoplanation surfaces, gelifluction lobes and terracettes be seen, strongly contrasting with the surrounding scarps. Vaudor and Asensio-Amor (1972) suggested that periglacial processes were most active during the Riss, followed by strong fluvial action and chemical weathering in a temperate climate during the following interglacial. At present, gelifluction is still important. Above 1800 m, soils are frozen until June, but the absence of a thick, widespread mantle of fines limits the operation of mass movements on the slopes.

12.5 Southern Meseta

As in the case of the Northern Meseta, the Southern Meseta can be divided into an eastern and a western unit on the basis of lithology: Palaeozoic crystalline rocks predominate in the west and Tertiary sedimentary rocks in the east. In this area, however, the wider extent of the whole region reflects the greater area underlain by older materials. Unlike the Northern Meseta, the Southern Meseta stands at a lower average altitude (approximately 100 m lower) and for that reason is sometimes called the Lower Meseta. Another difference from the Northern Meseta arises from its lack of geological, morphological and hydrological unity. It is crossed by an important east–west fault system of Neogene–Quaternary age which has produced, from north to south, a tectonic depression at the foot of the Central Cordillera (the Toledo trough), a mountain range (Montes de Toledo), a fault area with volcanic eruptions (Campo de Calatrava) and a great flexure at the southern limit between the Hercynian and Alpine systems (Sierra Morena). The Southern Meseta is drained by two major drainage systems, those of the Tajo and the Guadiana, which have very different hydrological characteristics and consequently have produced different geomorphological landscapes. The lack of hydrological unity is emphasized by the fact that to the east the head waters of two Mediterranean rivers, the Júcar and the Cabril, have captured some of the runoff from the Castilian tablelands, thus depriving the Guadiana. The divergence of runoff between west and east, although the eastern catchment is less extensive, is due to the warping of the Hercynian block. This flexure also determines the individualization of two Tertiary sedimentary basins: a wider and deeper eastern basin, the upper Tajo and Guadiana in the Castilian lands, and to the west, the lower Tajo and Sado Basin on the Portuguese coast. Several sedimentary remnants between them, as in the middle Guadiana near Badajoz, testify to the original continuity of sedimentation, broken only later by erosion processes [see Ribeiro *et al.* (1979)]. The Southern Meseta is not so encircled by a mountain belt as the Northern Meseta and is particularly open towards the Atlantic. The difference in altitude together with the difference in latitude and the lack of a mountainous surround give to this area a milder, more maritime climate than the more extreme continental climate of the Northern Meseta.

12.5.1   Geological Structure

The geological structure of the western, Palaeozoic area is diverse because it embraces three of the main units of the Iberian massif. These units are distributed in broad north-west–south-east bands and comprise the central Iberian zone, the Ossa–Morena zone and the southern Portuguese zone [see Julivert *et al.* (1974), Ribeiro *et al.* (1979)].

1)   The central Iberian zone is a continuation of the structure found in the Northern Meseta and Central Cordillera and has already been described (see p.304).

2)   The Ossa–Morena zone is characterized by extensive Precambrian and Cambrian outcrops, a noteworthy development of intrusive and extrusive igneous activity along long, narrow and distinct bands of Lower Carboniferous age, and a complex and poorly known tectonic structure. The age of the principal Hercynian folding is clearly Lower Carboniferous. The Precambrian rocks consist of gneiss, micaschist, quartzite, slate and grey-wackes. The Cambrian rocks are mainly limestone and dolomite with a certain amount of flysch; several small granite plutons and remnants of gneiss of the *ollo de sapo* type are also found.

3)   The southern Portuguese zone is located in the south-west corner of the Iberian massif and has a less varied lithological composition. The oldest rocks are Upper Devonian and consist of alternating beds of quartzite and slate with intercalations of lava and pyroclastic rocks. There is also a noteworthy pyrite band where some well-known mines are located. On the north-east, there is a thick flysch formation made of greywackes and slates of Lower Carboniferous age and, to the north-west, there is a similar flysch outcrop of the Upper Carboniferous.

A comparison between the palaeogeography and the tectonism of the Ossa–Morena and the southern Portuguese zones makes it evident that the age of the principal deformation becomes younger to the south-west. This south-westerly migration of the orogenic activity corresponds to the north-easterly migration in the northern branch of the old chain, thus imparting a certain degree of symmetry. The relative importance of preorogenic vulcanism is much greater in the southern zones (Ossa–Morena and southern Portuguese) than in the rest of the Iberian massif, while the subsequent vulcanism and plutonism is better represented in the Ossa–Morena zone. Metamorphism is generally of low-pressure type with facies of higher pressure only in certain lineations.

The eastern, Tertiary area has an extremely simple geological structure, comprising mostly horizontal Miocene and Oligocene sediments. The average thickness of the deposit is 300 m, but in the Tajo trough, in front of the Gredos and Guadarrama Sierras, the Miocene can reach 3 km in depth. The rock types include Vindobonian sands and clays, and Pontian gypsum, marls and limestones. There are lateral changes in facies, the finest materials being located at the centre of the basin and the conglomeratic deposits at the foot of the mountain ranges. Evaporites are only found on the eastern side, at the footslopes of the Mesozoic, calcareous Iberian range. The original horizontality of the sediments has remained practically unaffected by tectonism.

In the west, the Tajo and Sado sedimentary basin is a subsident area that has been infilled with Tertiary and Quaternary continental detrital sediments within which lie several marine and brackish water facies that correspond to the maximum of the Miocene transgressions. Their thickness does not exceed 1.4 km. To the east, these series lie directly on the Hercynian basement, while to the west and on the right bank of the Tajo they overlap on the Mesozoic basement of the western border. The structure of the basin is simple, with subhorizontal strata only faulted on the margins of the basin along normal faults that have developed during subsidence [see Ribeiro *et al.* (1979)].

12.5.2   Relief and Landforms

As a whole, the Southern Meseta is a peneplained area within which several distinctive elements are related to tectonic stresses and differential rock resistance.

12.5.2.1   *The highlands*

1)   Montes de Toledo is the name given to an east–west-trending unit that partially divides the Southern Meseta into the Tajo Basin and the Guadiana Basin. The Montes de Toledo stretch for about 100 km, with heights averaging 1200 m, but rising to 1603 m in the Sierra de Guadalupe. They have an asymmetrical cross-profile, their northern slopes being dominated by a fault scarp. On the other margins, they merge gradually into the Extremadura and Alentejo peneplain on the south-west and west, and into the La Mancha sedimentary plain on the south-east and east. They form a massif similar in many ways to the Central Cordillera, but with lower relief and a smaller extent. Several erosion levels have been distinguished by Solé-Sabarís and Llopis (1952) although, in many cases, partly because of tectonic flexuring the differences of elevation are difficult to establish. The

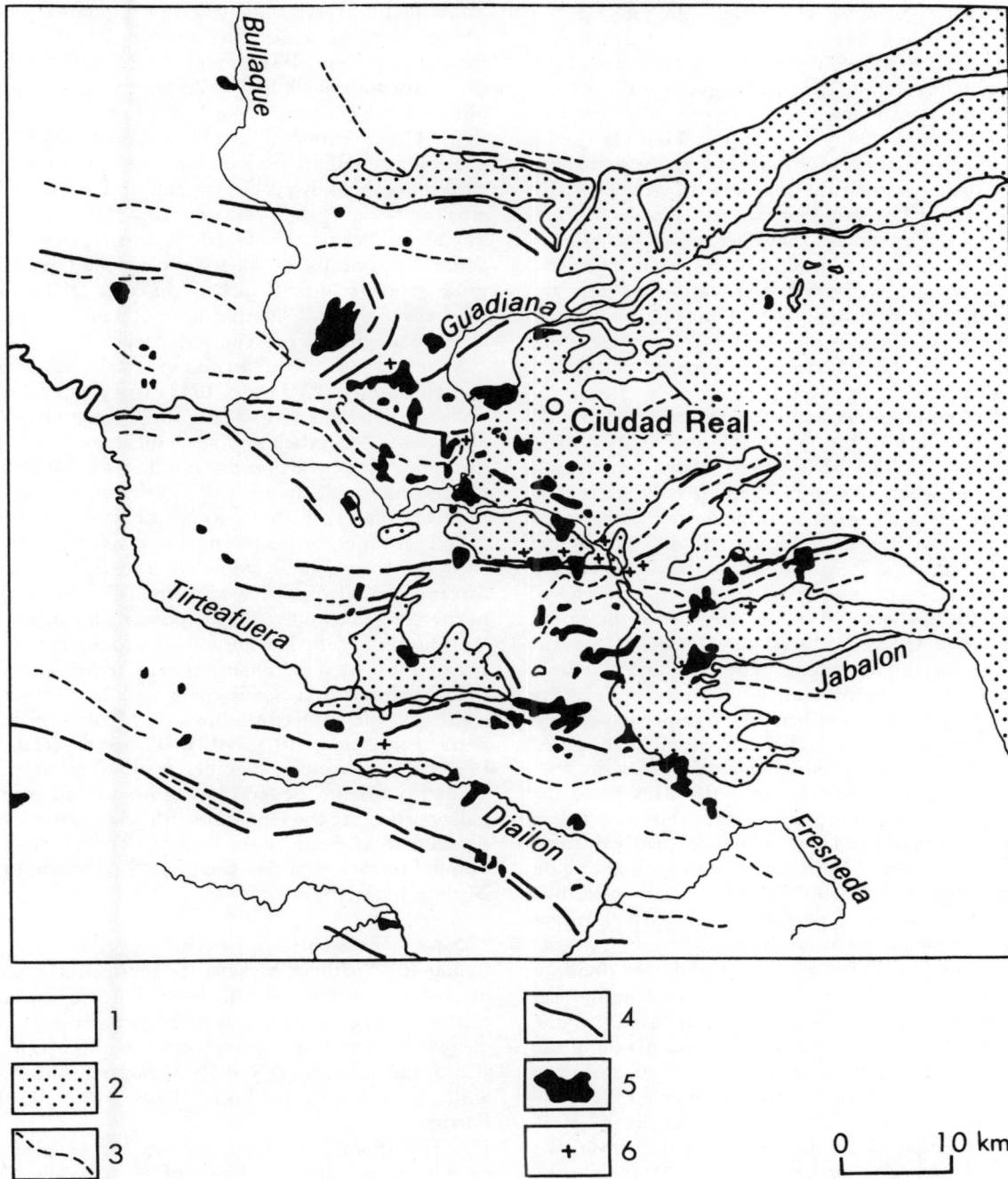

Fig. 12.13 The volcanic region of Campo de Calatrava [after Solé-Sabarís and Llopis (1952)]. (1) Palaeozoic, (2) Miocene, (3) principal anticlinal axes, (4) major faults, (5) volcanic rocks and (6) hot springs.

influence of rock structure is strong, especially in the western sector where the main north-west–south-east trends correspond to quartzite outcrops in the cores of the Hercynian synclines, while the corresponding valleys have been carved in slates. The whole area, especially near Las Villuercas, is a clear example of Appalachian-type relief developed after planation and following rejuvenation by flexuring, causing structural features to reappear by differential erosion [see Solé-Sabarís and Llopis (1952)]. The structural lineations are, however, often interrupted by a system of east–west fractures. As in the Central Cordillera, there are well-developed pediments that are often covered by a raña deposit that makes the contact between the mountain and the plain even smoother and more gradual (*see* Fig. 12.16); no Pleistocene glacial phenomena have been observed.

2) The most remarkable feature of the Sierra Morena, situated in the south of the Lower Meseta, is its north–south asymmetry. While its northern slopes consist of a quite gentle rise when viewed from the Meseta plains, its southern ones are a gigantic, rectilinear wall rising more than 1000 m above the Guadalquivir plain. This slope asymmetry is related to the great flexure that occurs at the junction of the Hercynian and Alpine systems in this area. Associated with this tectonic feature are several rotational faults, and overall it has produced an extremely clear and rectilinear geological boundary between the Palaeozoic rocks of the Sierra Morena and the Tertiary beds of the Guadalquivir Basin. The southern steep slopes have given a strong erosion potential to its streams which have carved deeply into these slopes producing narrow troughs that are the only natural pathways from the Meseta to the Guadalquivir. Headward erosion by these streams has in several cases captured head waters that originally drained to the Guadiana basin. The intensity of erosion acting on the different basement rocks has produced some fine examples of Appalachian-type relief with quartzite ridges protruding in a peneplained landscape.

3) The Campo de Calatrava is an upland area dividing two different geological and geomorphological units. On the western side, the landscape is dominated by the Palaeozoic strata of the peneplained Iberian massif of the Spanish Extremadura lying 300–400 m lower in elevation. On the eastern side, there are the flat-bedded Neogene formations of the broad basin of La Mancha at 100–200 m. The Campo de Calatrava stretches from the Montes de Toledo in the north to the Sierra Morena in the south. Its relief has an important effect on the Guadiana river, representing a barrier that strongly diminishes the power of the upper Guadiana to incise its bed. Its origin is related to neotectonic (Miocene–Quaternary) flexuring of the Meseta block which produced volcanic eruptions and a dense fractured system (even today there are many thermal springs called *hervideros* along the fracture lines). Neotectonic instability has continued until very recently. Molina (1975) has studied the geology and geomorphology of the area in detail. In the Upper Miocene, initial volcanic activity (V_1) was followed by tectonic activity (F_1). In the Lower Pliocene, there was another volcanic phase (V_2) succeeded by a second tectonic phase (F_2).

Subsequent erosion modelled a planation surface on which chemical weathering produced karst features and the development of a red soil. Middle Pliocene tilting of the Hercynian block led to the beginning of a new erosion phase that generated a surface prior to the raña phase (S_{pr}). The evolution of the raña deposits took place during the Villafranchian period. The next erosion phase occurred in the Lower Pleistocene and produced another erosion surface (S_2). During the Pliocene, glacis and fluvial landforms developed. A third phase of volcanic action (V_3) probably started during the formation of the raña deposits and disturbed the alluvial terraces older than the +6-m fill, for the base of some of these is now found beneath the present water level.

Vulcanism in the area (*see* Fig. 12.13) has been studied by Hernández-Pacheco (1932a,b) who, following the classification of Lacroix, describes three eruptive types: Hawaiian, Strombolian and Vulcanian. The first produced basaltic lava that did not flow far out of the crater, generating small domes called *cabezos*; erosion has since reduced many of them to small castellated remnants called *castillejos*, or just to terrain convexities only noticeable by their black colour and called *negrizales*. The Strombolian eruptions were responsible for several composite cone volcanoes; the eruptive rock containing abundant lapilli is called *hormigonera* (meaning concrete mixer) because it is used as construction material. The Vulcanian type produced pyroclastic cones with large craters that in some cases have become the sites of lakes, either permanent or seasonal. All in all, there are 60 relatively well-preserved volcanoes together with a great number of eroded remnants. Several lava flows spread through river valleys affecting the Guadiana drainage and causing badly drained areas. Some of the lava flows lie on top of the 30-m alluvial terrace that has been dated palaeontologically as Middle Quaternary.

12.5.2.2   *The sedimentary lowlands*

Unlike the Northern Meseta, the sedimentary lowlands do not form a continuous unit here. Three main landscapes can be distinguished whose relief is mostly related to the energy of their river systems and to the arrangement of strata: the upper Tajo and La Mancha Basins in the east, and the Ribatejo or lower Tajo and Sado Basin in Portugal.

1) The upper Tajo Basin (*see* Fig. 12.14) is the area that extends from the footslopes of the Iberian Cordillera approximately to the cities of Toledo and Talavera de la Reina. Its landforms are similar to those of the upper Duero Basin described in Section 12.3. Three surface levels can be distinguished: the Alcarria, the High Campiña and the Low Campiña. Alcarria is the local name for páramo, that is the typical Castilian tableland cut in Pontian limestones. As in Old Castile, the tablelands have been strongly dissected; their isolated remnants constitute a complete set of forms of this kind of relief. Karst processes have taken place on this surface producing minor solution forms such as lapiés and small dolines [see Pérez-González *et al.* (1974), Vaudour (1974)]. The High Campiña is a surface modelled on the less-resistant

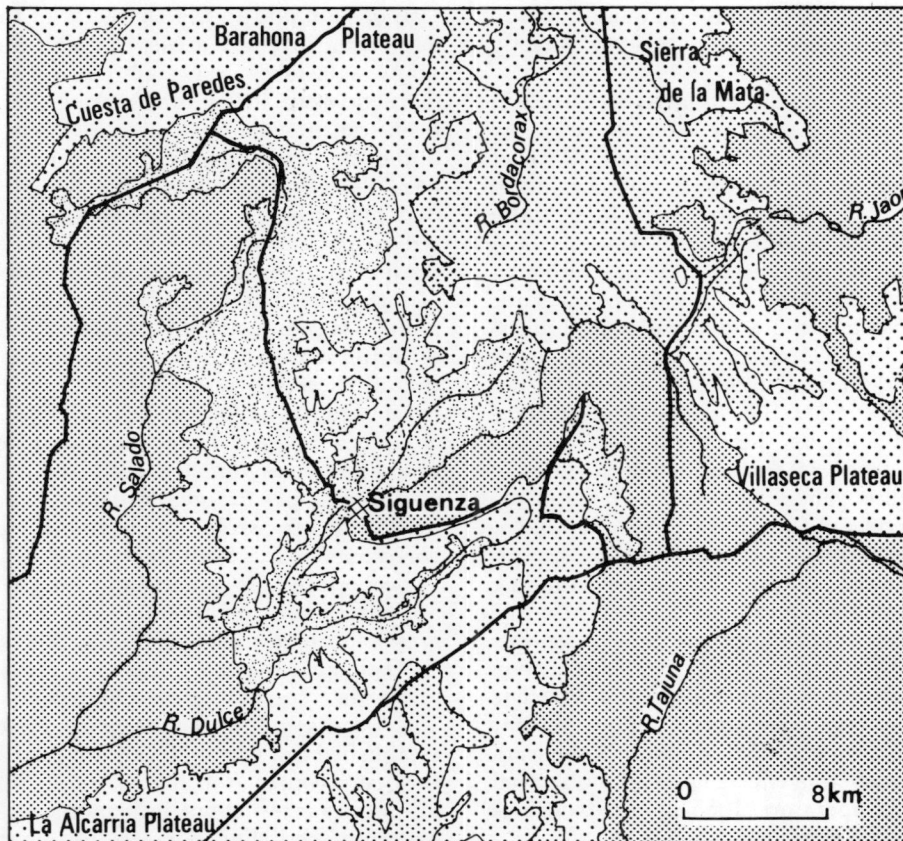

		Relief	Amount of Gentle Slope	Position of Gentle Slope in Profile
Páramo		0 – 40m	>80%	>75% upland
Paramera		40 – 100m	>50%	>50% upland
Serrania		100 – 200m	<50%	< 25% lowland
Campiña		0 – 40m	>80%	>75% lowland

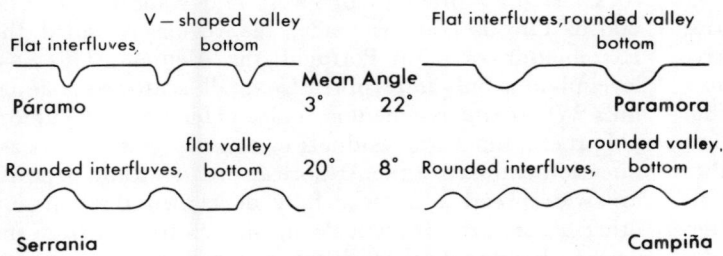

Generalised Interfluve - valley Profiles

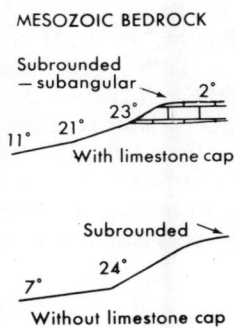

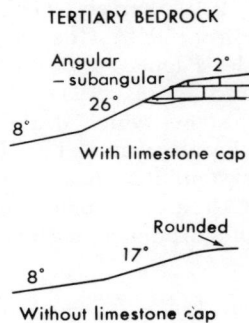

Fig. 12.14 Landforms in the upper Tajo basin [after Gladfelter (1971)]. In the lower part of the figure, characteristic slope angles for different geological situations are given.

Miocene sediments, producing a gently undulating plain. The Low Campiña is an area whose relief is directly related to recent fluvial activity.

The alluvial fills of the Tajo system have been studied by various authors. Traditionally, according to Aranegui (1927), Alia-Medina (1945) and Hernández-Pacheco (1946), four to five terrace levels have been recognized. Modern authors [see Alférez-Delgado (1977), Aleixandre *et al.* (1974), Pérez-González and Asensio-Amor (1973), Vaudour (1979)] have described up to 13 levels, although some of them must be considered as episodes of the same depositional phase. Neotectonic and pseudokarst effects seem probable [see Pérez-González (1971)]. The lower levels are often inserted one into another, while the higher ones are developed *en échelon*. These numerous terrace levels are not paired and, on the opposite side of the valley, the levels are fewer or substituted by a series of glacis. The alluvial fill is generally 3–4 m deep, as in the case of the present floodplain fill, and has sedimentological characteristics that indicate torrential transport, similar to that of oueds (wadis) and ramblas. In the Campiña, erosion has been more effective in the soft, bare Miocene clays than in the areas where the coarse fluvial deposits have protected them, thus often producing relief inversion in which terrace levels stand higher than the Miocene basement. Captures by regressive erosion or by overflow are found near Alcalá; a process also common in the wadi regions. Locally, cryoturbation effects have been observed in the alluvial fill [see Imperatori (1955), Pérez-González (1971)] and on the valley slopes. ·

2) La Mancha is the sedimentary region located in the upper Guadiana Basin. It is a flat, or almost flat, plain underlain by horizontal stata, the most perfect and extensive plain of the Iberian Peninsula and one in which fluvial incision is practically nonexistent. The lack of incision by the Guadiana drainage system can be explained in terms of four main reasons. (a) There is little difference in level between the plain and the head water area. (b) The system is closed in the downstream direction by the Campo de Calatrava threshold which, in addition, is followed by the wide peneplain (200–300 m high) of Extremadura and Alentejo. (c) Water is scarce within the system due to semiarid climatic conditions. (d) There are extensive karst phenomena related to the Keuper and Liassic limestones of the head water area and part of the basement. The fourth reason is probably the most important. Although from a hydrological point of view the main channels of the Guadiana system are on the foothills of the Iberian Cordillera, in the Serranía de Cuenca, its source has been historically placed in the vauclusian springs of Ruidera, at the foot of the Baetic Cordillera, where a series of 17 small lakes stretch for 25 km. Once the drainage reaches the plain, karst sinks and evaporation cause a complete disappearance of this and other branches of the Guadiana. A new resurgence takes place in the area called *Ojos del Guadiana*, but the slight gradient and the Calatrava threshold produce a 30-km stretch of swamps near Daimiel and a wandering system of undefined channels whose waters are easily diverted. Low rainfall (350–400 mm a year) and strong

evaporation add to the character of the Guadiana drainage basin, in many areas, internal drainage systems and many salt lakes, especially near Campo de Criptana. Consequently, it is not surprising that the Guadiana has been traditionally described as the most anomalous river of the Iberian Peninsula, and a constant source of legends from Pliny to Cervantes.

3) The Ribatejo, comprising the lower Tejo and Sado sedimentary basin (*see* Fig. 12.15), has the distinctiveness compared to the basins already described of being an area of subsidence whose borders correspond to active normal faults. Its sediments were affected by only slight warping or tilting so that the strata remain subhorizontal. Denudation of these gently undulating beds has given a typical cuesta landscape [see Ferreira (1980)]. The Tejo Basin is a trough aligned from north-east to south-west with marked slope asymmetry, both facts due to tectonism. The southern slopes are gentle because they are related to a broad flexure while the northern ones are steep and correspond to a fault scarp that was active until the Pliocene [see Solé-Sabarís and Llopis (1952)]. Sediments are alternately continental and marine due to transgressions along the lower Tejo valley. On the left bank of the Tejo near the coast, there are remnants of four terrace levels that correspond to Quaternary sea-level changes. The details of the Tejo terrace system have not been established but certain levels are relatively constant along the river course; the higher ones cut *en échelon* and the lower ones inserted one into another. They then gradually pass into shoreline deposits near the coast. The Sado Basin is a trough trending north-east to south-west, its south-east border corresponding to a semigraben [see Ribeiro *et al.* (1979)].

12.5.2.3   *The crystalline lowlands*

The crystalline lowlands include the area between the Central Cordillera and the Sierra Morena, and from Campo de Calatrava to the Atlantic. The landscape consists of a peneplain cut in the Palaeozoic basement rocks, slightly tilted to the west and standing at 300–600 m. On the Spanish side, the region is called the Extremadura and, in Portugal, the Alentejo. This vast level plain is only interrupted by small, scattered residual hills. Where the basement is made of Hercynian metamorphic rocks, quartzite residuals in a north-west–south-east lineation emphasize the Armorican trends. Raña deposits are widespread and more fully developed than in any other place in the Iberian Peninsula. Protruding from the granitic basement are inselbergs and tors that have always attracted the attention of researchers, both in the Portuguese sector [see Birot (1948), Ribeiro (1940), Zbyszewski (1939), Ferreira (1978)] and in the Spanish sector [see Birot (1945), Chaput (1968), Brinkmann (1932), Hernández-Pacheco (1947), Sanz-Donaire (1979), Solé-Sabarís and Llopis (1952), Vidal-Box (1944)]. The two fluvial systems which drain this area have relatively different characteristics. The Tajo is essentially related to a fault system that has helped it to excavate its valley in the Extremadura peneplain, while in its lower course, because of subsidence, its incision is not as spectacular as that of

Fig. 12.15 Geomorphological outlines of southern Portugal [after Solé-Sabarís and Llopis (1952), Birot (1948), Feio (1949)]. (1) Main planation surface, (2) remnants of higher planation surface, (3) relief rising above the main planation surface, (4) Mesozoic outcrops, (5) Tertiary outcrops, (6) coastal platform on Palaeozoic rocks, (7) coastal platform on Mesozoic rocks, (8) scarps, (9) fracture lines, (10) fault line scarps, (11) monadnocks, (12) quartzite ridges and (13) incised valleys.

the Duero in the Northern Meseta. In contrast, the Guadiana is a meandering river in the Extremadura, only incised 40–50 m, but in the Alentejo it has cut strongly into the peneplain as it flows southwards.

There are certain differences between the peneplains of the Extremadura and the Alentejo. In the Extremadura, the most outstanding features are the existence, in its central, more depressed area along the Guadiana, of remnants of a Tertiary detrital cover in La Serena and Tierra de Barros, which are now the only fertile lands of the area. The Alentejo has been under tectonic stresses which have produced more relief than in any other part of the peneplain, especially near its southern limit which is also the junction of the peneplain with the Algarve region, an elevated block of the Palaeozoic mass partly covered with Mesozoic sediments. One of the most important and widespread features in the entire peneplain is the extensive raña deposits, which will therefore be discussed here in more detail although, as has been noted, they occur in many other places bordering the Hercynian massifs and highlands of the Peninsula.

The rañas have attracted the attention of geomor-

phologists for a considerable time. Gómez de Llarena (1916) first defined these deposits and their related relief forms, and explained them as the result of transport of mountain quartzite fragments by tumultuous streams following torrential rains. Vidal-Box (1944) thought the rañas originated under desert climatic conditions when pediments were also evolving during Miocene and Pliocene times. Later the age of the rañas was restricted to Pliocene by Hernández-Pacheco (1949c). Present research focuses not only on the climatic environment that gave rise to the rañas but also on their age and correlation with pediment formation. Muñoz and Asensio-Amor (1975), following a detailed sedimentological study not only of the rañas but also of the neighbouring slope and fluvial deposits in the footslopes of the Montes de Toledo, concluded that the raña is essentially a deposit related to mountainous relief, originating by frost-weathering and gelifluction of slope materials subsequently washed by torrential stream action and later deposited on the plain below the mountain front. The quartzite fragments are mostly rounded and their mean particle sizes are: 2–6 cm (32–63 percent), 6–12 cm (32–48 percent), 12–24 cm (3–21 percent), 24–60 cm (1–5 percent) and 60–100 cm (0–1 percent). No reference is made to the raña being a type of alluvial fan deposit, although they noticed the gradual diminution in size of the quartz cobbles with distance from the mountain front. Muñoz and Asensio-Amor (1975) accepted the traditional semiarid morphogenesis version, but related the aridity to cold climatic conditions. This morphoclimatic interpretation does not match with Pliocene climatic characteristics but with those of the Quaternary. They have observed, too, that the raña deposit lies on deeply weathered bedrock (up to 20 m) and that it covers indiscriminately the Upper, Middle and Lower Miocene deposits, so its deposition is not post-Pontian but took place after intense erosional activity that affected most of the sediments of the basin. The present fluvial system is cut in the rañas. Molina (1975) defined the raña as a continental detrital formation with a precise morphology and stratigraphic age, a result of a change in tectonic and climatic conditions over the whole Iberian Peninsula. As a landform, he classified it as a piedmont glacis and as an alluvial fan. Stäblein and Gehrenkämper (1977), after a study of the southern slopes of the Sierra de Guadalupe (*see* Fig. 12.16), related the raña deposits and their present mesa-like landforms to piedmont slope processes. The formation of the rañas appears to have taken place in several phases rather than in a single one, and was preceded during the Tertiary period by peneplanation with deep tropical weathering and by uplift of the mountain range. Mechanical weathering in a hot, dry phase provided much of the quartz detritus; a change in climatic conditions at the beginning of the Quaternary increased the transporting capacity of the drainage system. This resulted in the development of glacis penetrating far into the valleys or in the covering of pediments by fanglomerates, as long as there was sufficient supply of pre-existing erosional material. When the hinterland had been stripped of its earlier weathered mantle, dissection of the raña surfaces began, forming valleys which were

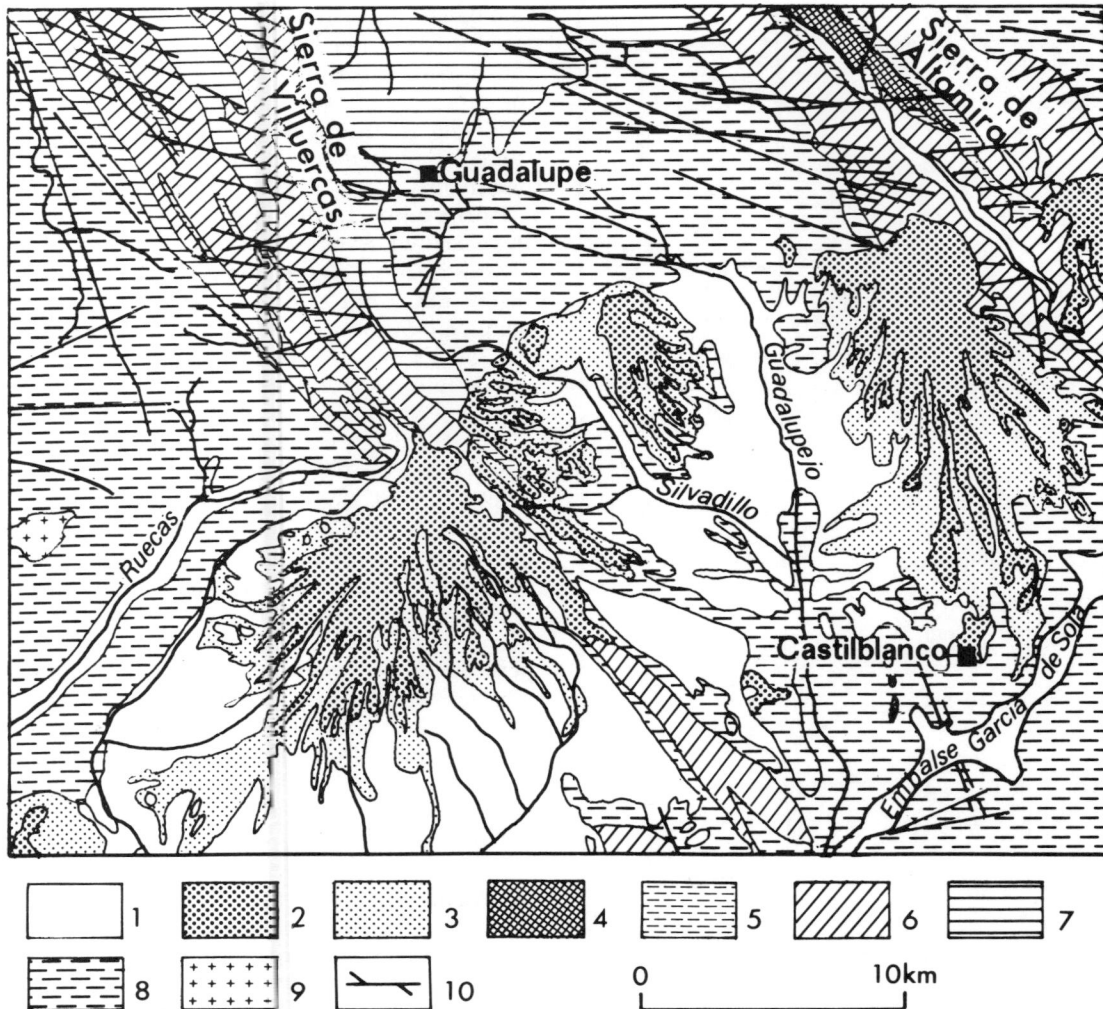

Fig. 12.16 The raña surfaces and geology of the southern foothills of the Sierra de Guadalupe, southern Meseta [after Stäblein and Gehrenkämper (1977)]. (1) Quaternary, (2) rañas, (3) Tertiary, (4) Devonian, (5) Silurian, (6) Ordovician, (7) Cambrian, (8) Precambrian, (9) granite and (10) fractures.

punctuated by terraces. The formation of the rañas is then attributed to a process of adjustment during the transition from one relief generation to the next, under certain epeirogenic and lithological conditions, showing that the rañas are not typical of the morphodynamics of any one particular climatic zone. In mapping the raña deposits, the alluvial fan shapes become very clear, the deposits spreading out from the mountain front.

12.6   Portuguese Extremadura and the Algarve

In Portugal, the epi-Hercynian cover of the Iberian mass occupies two areas on the margins of the massif: the Lusitanian fringe or Extremadura to the west and the Algarve fringe to the south. Various remnants of Mesozoic strata between these two areas permit a reconstruction of the initial continuity of this sedimentary basin bordering the Hercynian mass. The Mesozoic cover of the Palaeozoic platform is, in most cases, tabular or slightly folded and the tectonic style shows the strong influence of

the basement in the deformation of the usually thin cover. It is only in the Serra da Arrábida that the style and intensity of the deformation become more marked and similar to that of the Iberian Cordillera and the Cantabrian mountains. The stratigraphic conditions controlled the tectonic style but at the same time the palaeogeographical evolution was itself controlled by tectonic factors because they determined the extent and depth of these sedimentary margins.

The Extremadura is basically a calcareous platform that appears as a south-western prolongation of the Central Cordillera. It is limited by the same tectonic features and can be considered as its sedimentary replica. It extends along 150 km between the Tajo (Tejo) and Mondêgo rivers and between the Cabo Mondego and the Cabo Roca. It consists of a succession of sedimentary hills and karstic depressions, the most outstanding of the latter being the tectonically determined polje of Mira-Minde. Water circulation is mostly subterranean. At its southern end, there is the small chain of Serra de Arrábida bordering the coast for 30 km. The Algarve is a narrow belt of hilly uplands reaching 500–900 m altitude and 60 km long bordering the Portuguese coast from Cabo São Vicente to the Guadiana estuary. The highest part is in the Serra da Monchique, and everywhere a high degree of dissection of the landscape is typical. To the south, the uplands drop to a narrow coastal strip fringed in part by lagoons and sand bars.

12.6.1 Geological Setting

Julivert *et al.* (1974) and Ribeiro *et al.* (1979) have summarized the main geological features of this area. During the Mesozoic, Extremadura was the setting of a trough elongated in a north-north-east–south-south-west direction. The trough collected sediments from two sources: the Iberian massif to the east and a continental area located to the west, the only remnant of which today is the Berlengas archipelago. This explains the facies distribution and the thickness of sediments: neritic, often reef-type materials on the margins of the trough (0·5–1 km thick) and pelagic facies (up to 5 km thick) in the axis of the basin. Mesozoic sedimentation began with a detrital Keuper complex. Towards the end of the Cretaceous and in the early Eocene, the intrusions of Sintra, Sines and Monchique took place, which were followed by basaltic eruptions in the Lisbon and Leira regions. The general tectonic structure is one of broad folds directly influenced by the underlying basement. Where the sedimentary cover is thicker and the evaporite horizons are missing, the tectonic style can be clearly tabular. In some places the presence of a Liassic evaporite complex has sometimes allowed the cover rocks to behave independently. Overall, three zones with different tectonic styles can be distinguished.

1) A zone of submeridional faults at the junction with the Iberian massif that may be due to reactivation of a late Hercynian fracture belt.
2) West of this zone, there is an area dominated by tectonic structures related to the Baetic folding and by a rigid block structure with broad flexures. The most important raised block is the limestone massif of Extremadura.
3) The Mesozoic trough, with a thick evaporite complex that produces important diapiric phenomena, is related to a zone of normal faults active during subsidence and sedimentation. Although diapirism can be found as far back as the Jurassic, the main activity must have been contemporaneous with fracturing, making possible the extrusion of the basaltic magma of the volcanic complex, probably during the Eocene. As a consequence, domes, dykes and necks intrude the evaporite complex or the materials bordering the diapirs. Diapirism was still more or less active during the Tertiary period; the Pliocene sediments often show deformations.

Within the Baetic zone, the Serra da Arrábida, located south of the Tejo mouth, is much more deformed than the rest of Extremadura. Its structure and tectonic style make it more properly an intermediate type of chain. There the folding style is strongly influenced by the presence of the

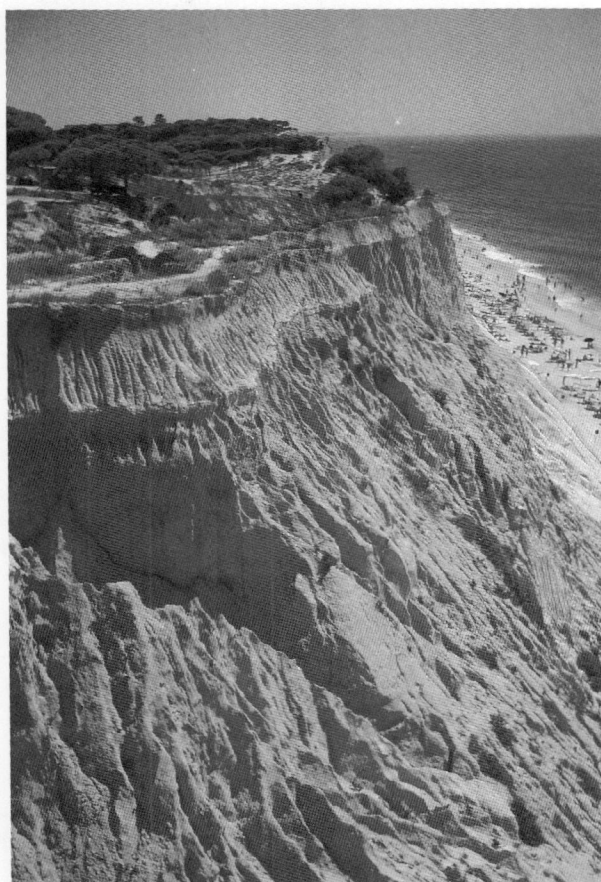

Fig. 12.17 Cliffs near Albufeira, Algarve coast, in soft Plio-Quaternary clays and sands. (Embleton)

evaporite complex that outcrops in the Sesimbra diapir and along some thrustfolds. It is made up of sharp folds arranged *en échelon* in an east-north-east–west-south-west direction and thrust towards the south-south-east; they are cut by reverse slip faults.

The Sintra, Sines and Monchique massifs correspond to subvolcanic annular structures from the Cretaceous–Eocene boundary and are probably aligned along a strike slip fault of north-north-west–south-south-east direction and dextral movement, probably prolonged towards the continental margin. In the Sintra dome, the deformation of the sedimentary cover due to the intrusive processes is very spectacular.

The southern fringe of Algarve (*see* Fig. 12.17) has a palaeogeographical evolution similar to that of the western fringe, the only difference being its much smaller dimension. It has a main flexure line orientated in an east–west direction which separates a zone of platform sediments with dolomitic facies from a zone of deep facies and thicker successions to the south. From a structural point of view, it is possible to distinguish two principal zones. (1) North of the main flexure, the structure is tabular or slightly warped with subvertical faults. (2) To the south there is a band of moderately folded strata with anticlines of east–west trend sometimes slightly overfolded or thrust towards the south. The evaporite complex is injected in the major structures and outcrops in several anticlinal nuclei. Basic volcanic structures are frequent and must be, at least partly, contemporaneous with the volcanic complex of Lisbon.

12.6.2   Lithology and Landforms

The geomorphological map of Portugal (1:500000) offers a synthesis of the general geomorphological setting and the Quaternary evolution of Portugal [see Ferreira (1980)]. The most common lithologies of the Extremadura and part of Arrábida are marls with detrital interbedding and limestones. Various eruptive rocks are found in the Sintra, Sines and Monchique domes (granite, diorite, syenite and gabbro) and a well-developed metamorphic aureole is found in the deformed sedimentary cover. In the Serra da Arrábida, there are conglomerates at the contact with the Hercynian massif followed by sandstones and dolomites with intercalations of gypsum and salt.

Structurally controlled landforms are the most important geomorphological features (*see* Fig. 12.15). Tectonic scarps due to diapiric tectonism are mainly the result of differential erosion between the gypsum beds and salty marls, and the Upper Jurassic limestones, that has produced beautiful examples of relief inversion, of which the Caldas da Rainha is the best. Cuesta landforms are well represented, especially in the Lisbon and Algarve regions, sometimes overriden by basalt lava flows. As a general rule, all resistant beds have been converted to protruding ridges by differential erosion and escarpments are widespread.

Two high platforms of Extremadura are believed to be erosion surfaces that have been carved by karst processes and include multiple depressions containing silica deposits. The summits of the horsts of Serra da Sico, Serra da Montejunto, the basaltic hills around Lisbon and the summits of cuestas are all cut by partly karstified planation surfaces. These surfaces are better preserved in limestones than in marls and sandstones. Solé-Sabarís and Llopis (1952), because of the uniformity in height of the Mesozoic fringe and of some raña deposits, think that the equivalent of the Beira peneplain (i.e. of the Meseta peneplain) may exist here. Most footslopes are cut by erosion glacis.

The outlines of the Portuguese coast have been mainly laid down by tectonics. Marine erosion and deposition have only modified this in detail. The narrow coastal platforms of Extremadura and western Arrábida are surfaces of marine abrasion. In Arrábida, this platform has been uplifted by 300 m. Only at the river outlets has the coast a more open appearance, with consolidated dunes and sandy accumulations, such as can be found in the Setúbal Peninsula, and around Faro and Tavira.

Fluvial landforms in this area are mostly gorge-like, cut in the limestone plateaus. This is particularly common in Extremadura.

12.6.3   Periglacial Features

Observations made by Daveau (1973) in various areas show that periglacial processes were active during the Quaternary, not only in the high mountains but also at lower altitudes. Studies on the western coast indicate that frost action was active in places where snow and ice are practically unknown today. In the calcareous headland of Cabo Mondego, the slopes are covered by a veneer of angular debris imparting a smooth and rectilinear shape to these slopes (*versants reglés*), a modelling considered typical of periglacial slopes in this Mediterranean latitude. Locally, the frost-shattered debris appears interstratified with fossil dune sand on the footslopes that merge directly into the coastal fringe. It seems likely that, during the Würm regression, a vast littoral platform was exposed to wind action and gelifluction. Angular blocks filling the hanging coastal valleys of the Serra de Sintra (Cabo Roca) have traditionally been interpreted as of periglacial origin and their general appearance related to similar features on the Brittany coast [see Guilcher (1950)]. Breuil and Zbyszewski (1942, 1945) have pointed out the existence of a breccia of limestone debris cemented by a resistant red sandstone that contains fragments of shells in the littoral slopes of the Serra da Arrábida in a particularly well-protected site facing south-west and located at 10 m above sea level. In the Algarve, the only typical deposit so far known is located 20 km from the coast in a tributary valley of the Guadiana cut in schist slopes that reach 400 m in altitude.

Daveau (1973) has studied in detail several periglacial deposits on the western slopes of the Serra de Candeiros (Extremadura) at 200–600 m altitude. Daveau found typical grèzes litées (stratified screes) in various places covering slopes of rectilinear form (versants reglés) and nonsorted gelifluction deposits with an abundant reddish fine matrix that seems to have come from gelifluction processes reworking a palaeosol. From these two kinds of deposits, it is deduced that there were alternating dry periods that produced the grèzes litées, and humid episodes characterized by gelifluction. Such climatic oscillations on Atlantic coastlands have also been postulated by Guillien (1962) in Brittany and by Nonn (1966) in Galicia, dry periods corresponding to a steppe landscape with intense runoff that would have produced the stratified deposits and humid periods favouring flows of head.

Chapter 13
Baetic Cordillera and Guadalquivir Basin

The Baetic Cordillera constitutes the main structural unit of the south of the Iberian Peninsula, extending for 600 km from the Strait of Gibraltar to Cabo de la Nao (*see* Fig. 13.1). In its north-eastern part, it merges with the Iberian Cordillera while to the north and north-west it is separated from the Hercynian block by a triangular depression drained by the Guadalquivir. The Balearic Islands mark its prolongation to the east and, past Gibraltar, it is connected to the internal zones of the Rif Atlas in North Africa.

13.1 Baetic Cordillera

Although it possesses peculiarities that differentiate it from other Alpine chains, as in the development and position of preorogenic vulcanism and in the characteristics of its thrust sheets, the Baetic Cordillera harmonizes with the general Alpine style. Solé-Sabarís (1952) described how the general geomorphological features of the Cordillera show a transverse symmetry but that strong lineations are absent; he thinks that this is mainly due to the sedimentological characteristics of the Baetic geosyncline, which was continuous, uniform and of great depth. These features encouraged the deposition of marls with some episodes of a more calcareous nature in certain areas, especially during the Jurassic. In turn, these sediments overlie the gypsiferous marls of Triassic age. This sedimentological arrangement made possible the overthrusting of great masses for long distances, so impeding the development of any relief unity. In addition, only the scattered limestone masses have resisted erosion, while the marls and shales have been worn down and reduced to hillocks. The low intensity of Quaternary glacial processes has also been characteristic for this mountain system.

Lhénaff (1981) noted two geomorphological features that overshadow the Alpine character of the chain, making it of more typical Mediterranean form. In the first place, it is composed of a series of isolated mountains separated by wide basins, and in the second place it lacks a sharp or rugged overall appearance. The main summits are broad and the landscape is dominated by convex hills, access by road always being easy, even to the highest peak [i.e. Veleta (3398 m)]. The monotony of the lithology does not favour differential erosion, and the overturning and thrusting of the folds is not clearly expressed in the relief. In spite of these features, the relief has enough energy to maintain a mountainous appearance; limestones and dolomites produce some impressive cliffs with scree foot-slopes, contrasting with the rounded hills of the footslopes in the schist areas.

13.1.1 Geological Structure

The Baetic Cordillera is a chain whose complex structure has caused much controversy since the beginning of this century. The thrust structure was confirmed after several studies during the first half of the century, especially those by Brouwer (1926), Bemmelen (1927) and Banting (1933), and the systematic work of Staub (1934a). The extreme ideas of Staub on over-thrusting, however, produced a strong reaction. After detailed investigations by Blumenthal (1927), in the western half of the chain, and by Fallot (1931–34), in the eastern part, the thrust structure was definitely confirmed, although the amplitude of the thrusts was considerably reduced. The synthesis of Fallot (1948) is still the best reference work, together with the synthesis written for the tectonic map of the Iberian Peninsula [see Julivert *et al.* (1974)].

In the Baetic Cordillera, three structural units can be distinguished stretching approximately west-south-west to east-north-east: (1) the Pre-Baetic zone, which extends from near Martos (Jaén) to Cabo de la Nao, (2) the Sub-Baetic zone located to the south of the Pre-Baetic zone, and extending west as far as the Gulf of Cádiz and (3) the Baetic zone, also called the Peni-Baetic zone in the older texts, which is located between the Pre-Baetic zone and the Mediterranean, running from Estepona to Cabo de Palos. As in the other Alpine cordilleras, these units can be grouped into two assemblages: (1) the external zones (Pre-Baetic and Sub-Baetic), where the thrust structures of Alpine age only involve the materials of the post-Palaeozoic cover and (2) the internal zones (Baetic) where the Alpine structures affect both the Mesozoic and the Palaeozoic strata. In addition in the internal zone, there has been post-Palaeozoic metamorphism of varying intensity.

In addition to these structural units, which are the fundamental elements of the chain, there are some other independent structural elements: (1) the units of Campo de Gibraltar, consisting of a group of thrust sheets and (2) the interior depressions (Granada, Guadix, Bajo, Segura, etc.) whose individuality is of tectonic origin, but whose development is of later age than that of the fold structures.

13.1.1.1 Pre-Baetic zone
This is only found in the eastern half of the Cordillera. It is distinguished by the absence of a pre-Mesozoic basement and by its relatively simple tectonic structure. The stratigraphy which ranges from the Triassic to the Lower Miocene is more complete and thicker towards the south, and the facies are of continental and shallow marine type;

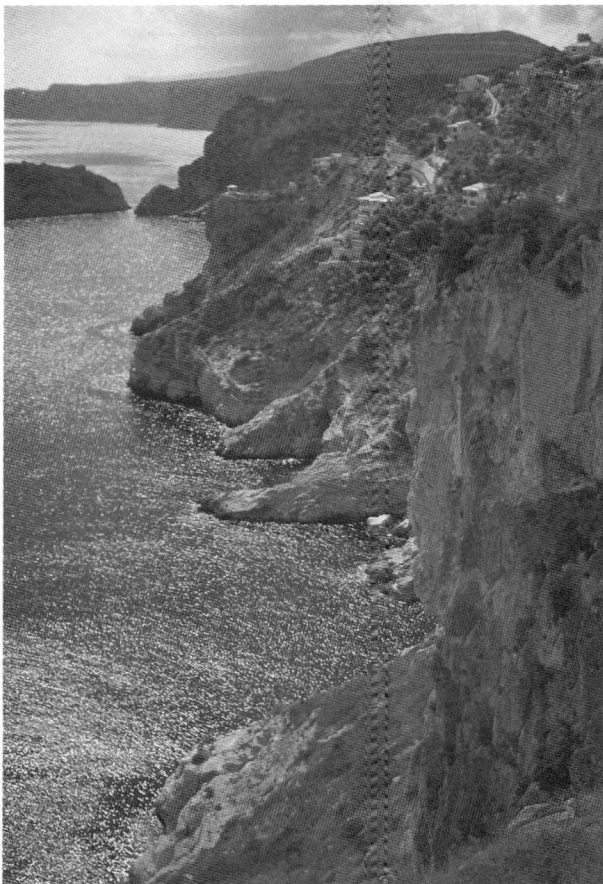

Fig. 13.1 Cabo de la Nao, eastern end of the Baetic Cordillera. (Embleton).

limestones, marls and sandstones are most abundant. As a whole, it shows the distinctive features of a platform palaeogeographical domain. In some places, however, subsidence must have been important because more than 5 km of Jurassic and Lower Cretaceous sediments were deposited [see Busnardo (1960)]; other units of this type have been found by García-Rossell (1973), showing a strong relationship with the Sub-Baetic zone, called intermediate units or internal Pre-Baetic. Magmatic rocks are practically absent in the zone and geosynclinal volcanic rocks are never found. The evaporite rocks of the Triassic have played an important tectonic role, not only by providing a plane of decollement during folding but also because of their halokinetic behaviour. The fundamental structure of the zone is that of a cover more or less separated from its basement and not very intensely folded except for certain sectors (Jaén, Cazorla); the folds show a general inclination towards the exterior of the chain (i.e. towards the north and north-north-west) and adopt various forms related to the local characteristics of the stratigraphic series. In this zone, the main folding took place between the Lower and the Upper Miocene. Prior to the compressive tectonics, the zone had been subjected to normal faulting producing block movements similar in age and significance to those of the Iberian and Cantabrian mountains. Soon after the Alpine folding, but still in the

Upper Miocene, strike-slip faults were produced which cut the folded structure and, in certain cases, even reached the Sub-Baetic zone. These faults have a strong relationship with the changes of direction of the folds in this zone (Sierra de Alcaraz) and also with some diapiric lineations, as in Villena.

13.1.1.2 Sub-Baetic zone

Like the Pre-Baetic, this zone does not have any Palaeozoic outcrops. The Mesozoic comprises a marine sequence from the beginning of the Lias, while the Triassic is similar to that of the other units of the Iberian Peninsula but with a higher relative abundance of marls and clays. In the Jurassic and Cretaceous, pelagic marls and limestones are predominant, with several radiolarian horizons during the Dogger and/or Malm phases. The Upper Cretaceous is made partly of red calcareous marls. The Eocene and Oligocene are represented by flysch deposits in some places and by pelagic deposits in others. The Lower Miocene has mainly organic limestones, marls and diatomites. Between the Upper Cretaceous and the Lower Miocene there exist discontinuities that prove the existence of deformation phases. No outcrops of metamorphic or plutonic rocks are found, but volcanic rocks are widely distributed, especially in Málaga and Guadiana Menor. The tectonic structure is rather complicated, consisting of a broad mass moved to the north or north-west, thrust over most of the internal part of the Pre-Baetic zone and divided in turn into several units whose northern fronts overrun the inner margins of those located immediately to the north [see Fontboté and García-Dueñas (1968)]. The main deformation of this domain was earlier than that of the Pre-Baetic zone, and there is a very important gap between the Lower and the Upper Cretaceous that has to be related to a diastrophic phase in which the nappe movement took place. Some elements may have been displaced far to the north as late as the Lower Miocene, and so it is possible to find Sub-Baetic elements even in the Guadalquivir depression (i.e. farther away from the Sub-Baetic front) in the form of thin gravitational nappes (Jaén–Ubeda–Jodar sector). These gravitationally displaced elements contain Mesozoic areas that had already suffered several deformations in their original settings. After the development of the overthrust structures, which are clearly directed outwards from the Cordillera, the Sub-Baetic zone suffered another phase of deformation in which important reverse faults and overthrusting in the opposite direction (i.e. towards the south or south-south-east) took place. In spite of these detailed complications, the main structural lines were determined chiefly in three phases: Middle Cretaceous, Upper Eocene and Oligocene, and Lower and Middle Miocene. Finally, in the Upper Miocene, in the Pliocene and partly during the Quaternary warping and faulting developed with a growing measure of independence from the previous general features.

13.1.1.3 Baetic zone

Unlike the others, in the Baetic zone, Palaeozoic strata are well represented, mainly by metamorphic rocks and plutonic masses (Serranía de Ronda). In certain places (Málaga complex), the Middle and Upper Palaeozoic is practically nonmetamorphic and has provided fossil remnants. Sediments from the Mesozoic, except for the Málaga complex, are limited to the Triassic; in a considerable part of the zone (i.e. the Alpujárride complex) the Triassic has some peculiarities, the lower part consisting of phyllites with quartzite lenses and the middle and upper zones of a thick limestone and dolomite formation. This is the only zone of the Cordillera where there exists a regional metamorphism of the alpidic cycle involving the Triassic. Volcanic activity was also important and produced a complete set of rocks of this type, while the postorogenic vulcanicity produced broad outcrops related to the fracturing. The tectonic structure of the zone is very complicated, characterized fundamentally by the superposition of thrust nappes which were displaced over several tens of kilometres. As a whole, three great sets of superimposed masses can be distinguished, which are, from the lowest to the highest, (1) the Nevado–Filabre complex, (2) the Alpujárride complex and (3) the Maláguide complex. Each in turn contains several different thrust nappes (but only in some sectors of the Alpujárride). The delimitation and correlation of these nappes is now quite well known, but the age of the major thrusts is still obscure. It seems that important phases of deformation took place during the Cretaceous but it is also evident that there were similar phases in the Palaeogene. As in the rest of the Cordillera, during the Neogene the Baetic zone was influenced by folds and fractures that have strongly contributed to the present-day orographic characteristics [see Solé-Sabarís (1952), Fontboté (1957)].

13.1.1.4 *Units of the Campo de Gibraltar*

These form a set usually called the allochthonous flysch, although these units are not exclusively, nor even mainly, made of flysch, and they are not only found in this area. However, they differ from the Sub-Baetic units and are tectonically superimposed on them. They can be found farther north in smaller outcrops along the limit between the internal and the external zones, and probably even farther to the east, past Granada. It is still more interesting to note that these units have continuity on the other side of the Strait of Gibraltar, thus producing an almost continuous fringe bordering the North African cordillera, and reaching Sicily and southern Italy. It is thought that the marine depths of the Strait of Gibraltar contain a great deal of the flysch. The units of Campo de Gibraltar have several common stratigraphic characteristics: (1) they contain only sediments ranging in age from Lower Cretaceous to Lower Miocene and (2) flysch is important in most parts of the units. As a whole, the Campo de Gibraltar consists of several masses moved by gravitational sliding in noncompetent rocks, which are very thick compared to the distance travelled. The movement was completed by the beginning of the Upper Miocene and in any case it must be younger than the Aquitainian.

13.1.1.5 *The internal depressions*

These are wide synclines affected in varying degree by faulting. Their character is that of intramontane depressions and their tectonic separation happened relatively late (Upper Miocene), when the postorogenic phase in the Baetic Cordillera was already initiated and continued during the Quaternary. They have behaved as sedimentary basins, with continental and marine episodes during which a considerable amount of Neogene and Quaternary sediments have been deposited, with thicknesses of up to 4 km in the north-western margin of the Granada depression and in several areas of the lower Segura. This sedimentation has a clear postorogenic character as in the case of the other sediments of this age in the Cordillera. The structures that developed during this tectonic phase clearly differ from the Alpine ones, and are not to be interpreted as a continuation of the Alpine orogeny. This neotectonic phase has been particularly important in the south and east of the Cordillera, although it is not absent in the western part and it has been proved in the Ronda and Málaga areas [see Lhénaff (1966)]. The best known is the Granada depression, which is limited by normal faults; its floor is also fractured, so that several second-order grabens and horsts can be distinguished within it. Some of the faults have been active during the Quaternary and subsidence has been important. Present-day seismicity proves without doubt that these fractures are still active.

13.1.2 Karst Landforms

In the Baetic Cordillera, karst forms (*see* Fig. 13.2) are abundant and have been the object of several studies, the most recent being those by Fernandez (1964), Lhénaff (1968, 1975), Pezzi (1975a, b, 1977), García-Rossell and Pezzi (1975), and Delgado and Fernandez (1975). The most extensive and detailed work is that of Pezzi (1977), where there is a summary of previous work and a detailed description of the central sector of the Sub-Baetic zone.

Karst phenomena can be found in many areas of the Cordillera—in the Baetic zone (dolomite areas of the Alpujárride complex and in Gran Calar del Mundo; and in the extremities of the Sub-Baetic zone (Serranía de Ronda)—but reach their greatest development in the central part of the Sub-Baetic zone. The reason for this is lithological, related to the occurrence of the Jurassic limestones and dolomites, which outcrop mainly in this zone. Some areas of the zone are underlain by Triassic gypsum and have also been subject to solution processes.

The work of Pezzi (1977) focused on two mountain groups of the external Sub-Baetic zone—Sierra Mágina and Sierra Arana—and on three mountain groups of the internal Sub-Baetic zone—Sierra Cabra, Torcal de Antequera and Sierrra Gorda. The mountains of the external Sub-Baetic zone are of higher altitude than those in the internal zone and, although present karst conditions are of pluvionival type, during the Pleistocene nivoglacial conditions must have prevailed. The Sierra Mágina (2167 m) (*see* Fig. 13.3) is the highest of the mountains studied, explaining the existence of karst features related to the persistent snow on its summits for about 4–6 months each year at present but more permanently during the Pleistocene. There are large dolines developed at the footslopes

Fig. 13.2 Karst landscape in the central Sub-Baetic zone, Baetic Cordillera: Torcal de Antequera with stepped and tabular features. (Sala)

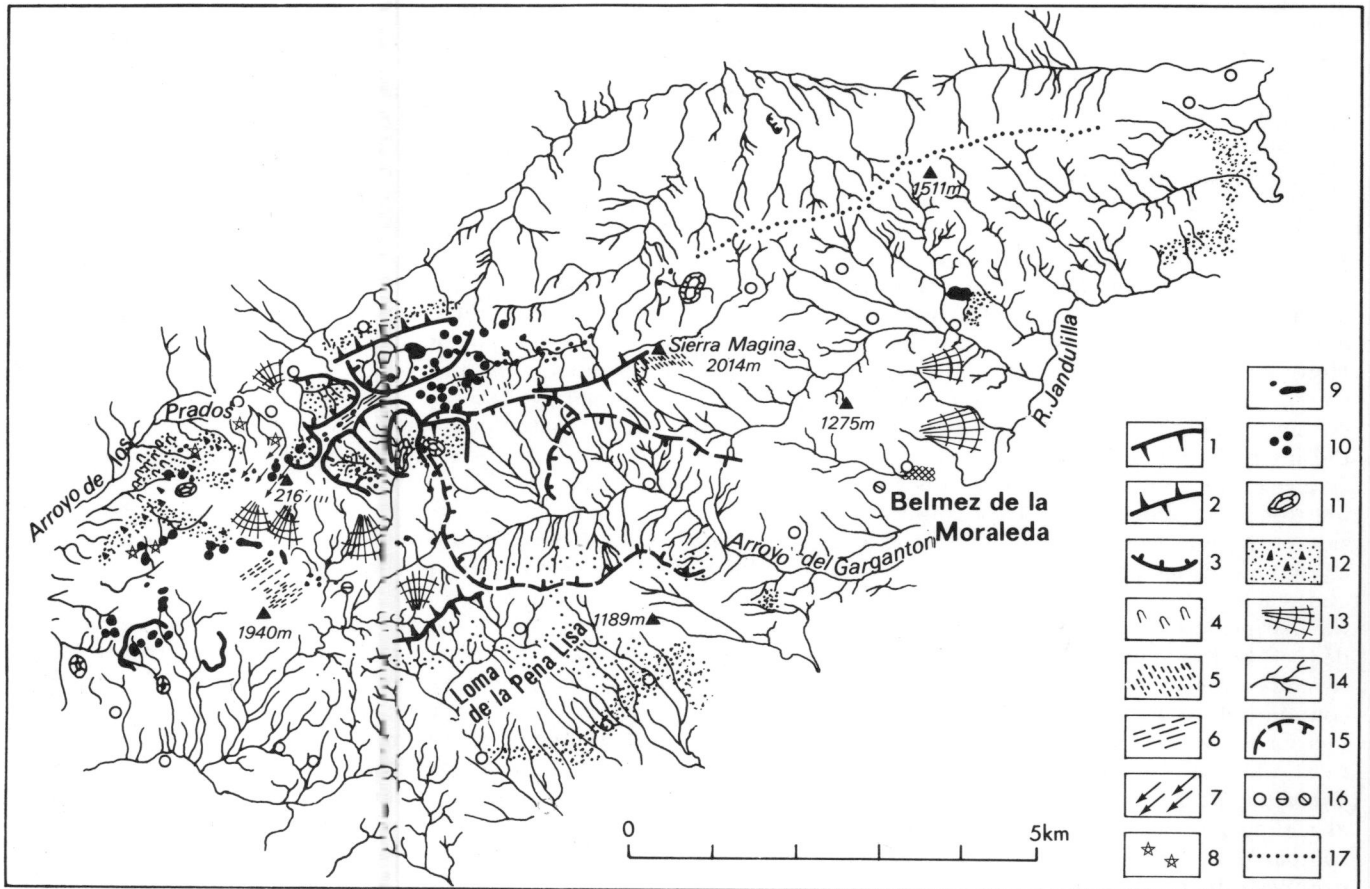

Fig. 13.3 Geomorphological map of the karst landforms of the Sierra Mágina [Pezzi (1977)]. (1) Limestone scarp, (2) other sharp crests or scarps, (3) nivation hollow, (4) gelifluction lobes (5) rill lapiés, (6) small lapiés, (7) large lapiés with sharp crests, (8) fossil lapés, (9) dolines, (10) funnel-shaped dolines, snow wells or *jous*, (11) U-shaped or trough-shaped dolines, (12) scattered debris (13) rectilinear slopes, (14) streams, (15) drainage divides at ground surface, (16) springs, wells, boreholes and (17) limit of limestone material.

of nivation hollows, and asymmetrical dolines related to snow lying on the ubac slopes most of the year. There are also dolines related to structural conditions: fractures, stratification planes or a conjunction of different types of limestones (*see* Fig. 13.4). Frequently dolines appear grouped in large doline fields where an appropriate structural or topographic configuration is present. When located in ancient valleys, the dolines tend to have a smooth, U-shaped appearance, but when along fracture lines, they are steep and funnel-shaped. Lapiés are also well-developed and appear in three essential types: in the summit areas, the grooves have sharp edges, their depth varying from 0.1 to 1 m and they are surrounded by angular debris while, at lower altitudes, lapiés are either scarcely exposed, with a rounded shape, or are of the rill kind. Underground drainage seems to be located mainly at the 1100-m level, with numerous springs.

A. Magina

19.5m

NE ◀

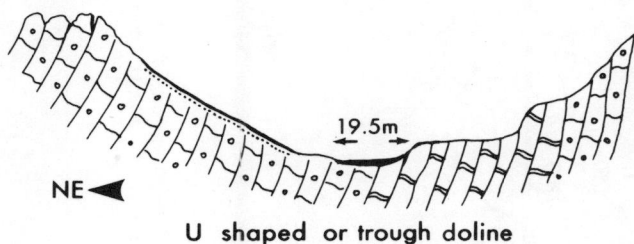

U shaped or trough doline

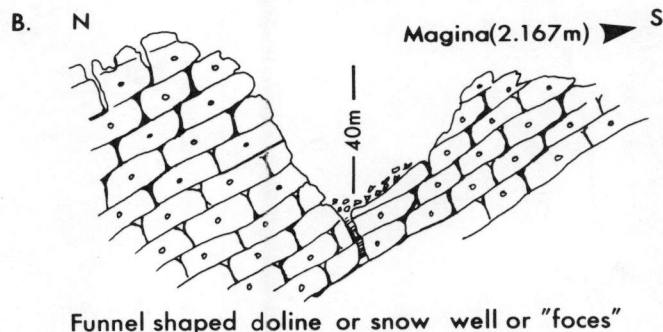

B. N   Magina(2.167m) ▶ S

40 m

Funnel shaped doline or snow well or "foces"

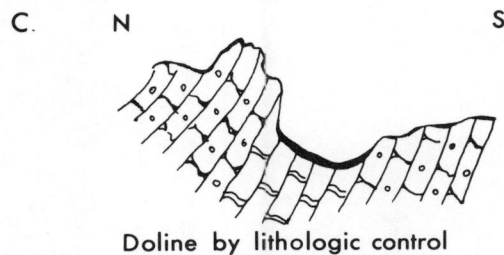

C. N   S

Doline by lithologic control

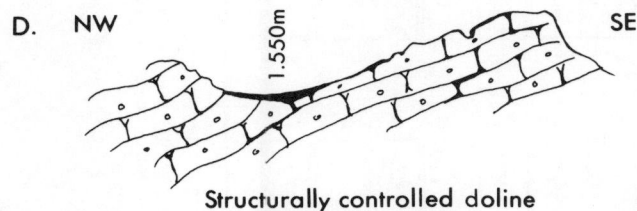

D. NW   1.550m   SE

Structurally controlled doline

Fig. 13.4 Different types of doline [Pezzi (1977)].

In the Sierra Arana (1981 m high), the most widespread features are the lapiés developed along fracture or stratification lines, running for great distances parallel to the anticlinal axis; these fractures may lead into shallow and narrow (0.5 m wide) corridors. Rill lapiés are also present, together with chaotic patterns in the dolomitic areas producing a great deal of debris. Dolines are also very common, with a density of $16 km^{-2}$; this abundance is supposed to be due to the flatness of the relief, the horizontality of stratification, the purity of the limestones (90 percent carbonate), the density of jointing and the frequency of snow.

In the internal Sub-Baetic zone, the most outstanding and best-known karst area is found: the Torcal de Antequera. A very complete set of karst landforms has developed in this thick, horizontal limestone series. The horizontality of the beds has made possible the evolution of depressions where water and snow can remain for a long time, thus favouring intense solution. Corridors are abundant and spectacular, some forming promenades 60 m across and a few centimetres to several metres deep; where two or more corridors meet dolines of irregular form usually appear which in the view of Nicod (1972) are exceptional features for Mediterranean karst. Dolines and uvalas are extraordinarily frequent, usually in neat lines along fractures and with deep, abrupt margins. Sink holes are also numerous and twelve springs are distributed around the mountain, one of them with a continuous water flow. There is one peculiar form in the Torcal, that of the rhythmic tableforms, which is supposed to be due to differential weathering of the limestone beds. In this case, a structure of nodular and brecciated layers is responsible for variations in porosity, the highest being that of the breccia. This contrast is thought to produce, under periglacial conditions, a greater susceptibility to freeze/thaw processes and thus a higher degree of weathering; laboratory tests reported by Pezzi (1975b) support this hypothesis. Pleistocene periglacial action in the area has been confirmed by the existence of typical grèze litées.

In the Sierra Gorda, the most complete, best developed and most typical karst landscape of Andalucía is found (*see* Fig. 13.5). All environmental conditions are favourable: uniformity and thickness of sediments, an extensive anticlinal structure producing a practically horizontal summit, a high degree of jointing, abundance of precipitation (800–1000 mm per year) and the presence of snow for 20–30 days a year. Dolines appear over wide areas with a density of $7.9 km^{-2}$; two types can be found, those with steep walls and rocky floors, and those with smooth sides, filled with terra rossa. Lapiés are also widespread. The most interesting features, however, are the poljes of which there are at least five. The polje of Zamarraya is a prototype and the only functioning one [see Lhénaff (1968), Vera (1969)]. Its shape is irregular (12 km long, 3.5 km wide and 900 m deep), with a well-developed sink hole and several minor swallow holes. In very wet years, the sink holes are incapable of absorbing all the water and the poljes are flooded.

The Sierra Cabra also has poljes, three in total, two of them with a common drainage, and none presently active.

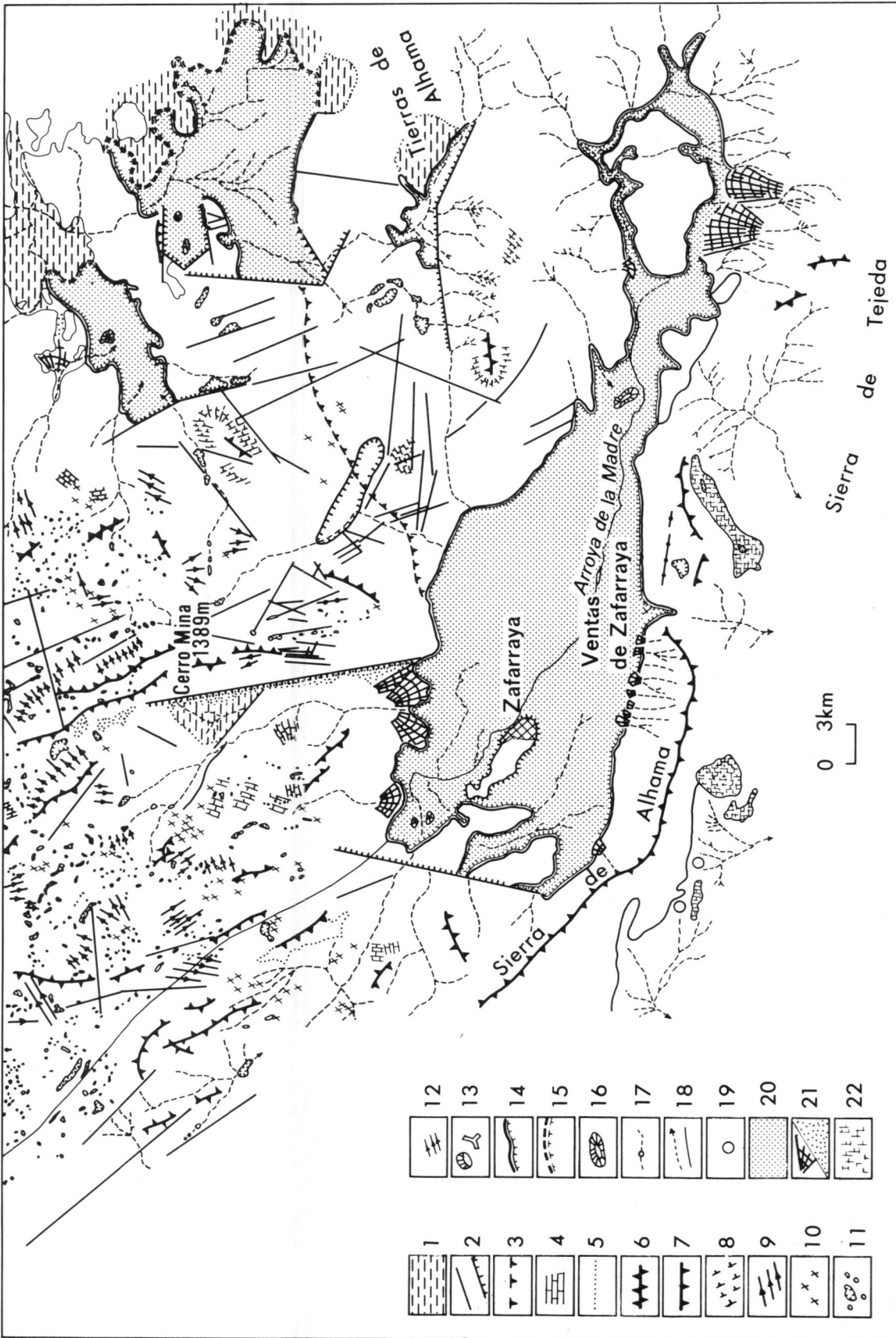

Fig. 13.5 Geomorphological map of karst features in the area between the Sierra Gorda and Sierra de Alhama, Baetic Cordillera [after Pezzi (1977)]. (1) Nonkarst area, (2) fault, (3) reverse fault, (4) joints, (5) lithological boundary, (6) sharp crest, (7) scarp, (8) crossing lapiés, (9) large lapiés, (10) old lapiés, (11) dolines, (12) dolines in rows, (13) asymmetrical dolines and uvalas, (14) edges of poljes, (15) probable polje margins, (16) hums, (17) ponors, (18) temporary and permanent streams, (19) springs, (20) terra rossa, (21) stratified or scattered debris and (22) travertine.

Their well-organized surface drainage and the appearance of marls on the floors limit the possibilities of evolution of these poljes. Dolines may have steep walls (30–40 m high) with rocky floors up to 190 m wide, or smooth sides, terra rossa fillings and a diameter that can reach 800 m. It is also worth noting the existence of lapiés that vary in shape in relation to varying lithology.

In conclusion, the karst landscapes of Andalucía reflect an overall tectonic influence, which is most relevant to the major landforms, and also show the importance of lithology for the general development of solution and microrelief features. Periglacial conditions in the early Quaternary have had a great influence in the area and made possible the development of deep weathering under a niveoglacial régime for the highest areas.

13.1.3 Pre-Quaternary Morphogenesis

Remnants of older landforms in the shape of summit planation surfaces have been preserved in some parts of the Baetic Cordillera, notably in the Sierra Nevada, Filabres and Carrascoy [see Birot and Solé-Sabarís (1959)], as well as in the hills with regular longitudinal profiles cut in the micaschists. Many authors have mentioned these features [e.g., see Sermet (1934), García-Sainz (1943), Solé-Sabarís (1952), Birot (1964)] and have

agreed on their Pontian age considering the generally fine facies of the correlative deposits and their warping. Lhénaff (1981), however, thinks that the relief of this area, although it reached a highly evolved stage of denudation, was never completely planed. In the Sub-Baetic zone, Lhénaff (1981) has described a few small remnants of a possible planation level at 800–1000 m because of the truncated character of the limestone beds, but such an interpretation is doubtful in the case of other accordant surfaces at 1000–1300 m because they could also be karst corrosion surfaces. Although it seems probable that planation surfaces were cut during phases of tectonic stability, they must be the result of short morphogenetic periods because there are no extensive surfaces and they are always located near the margins of the Tortonian sea, probably bearing a certain relationship with it. Solé-Sabarís (1952) was also of the opinion that planation surfaces in the Baetic Cordillera are underdeveloped due to the late tectonic events in the area, which implies a shorter erosion period than that available in the Pyrenees (*see* Fig. 13.6).

On the other hand, the evidence of Villafranchian events and processes is more widely developed and preserved. During that period, traces of older relief forms were largely destroyed and a new set of well-defined

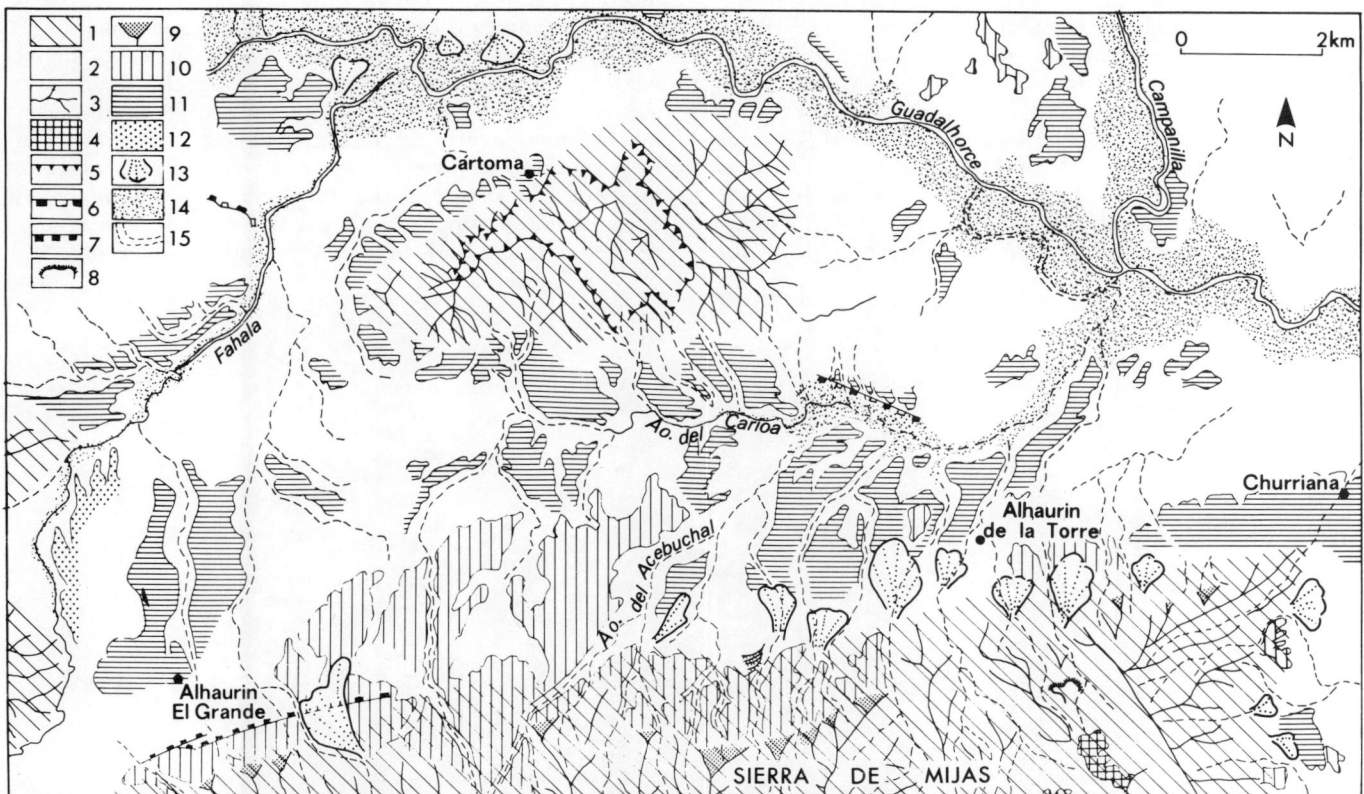

Fig. 13.6 Geomorphology of the piedmont area, Sierra de Mijas and adjacent regions, Málaga Province [Lhénaff (1981)]. (1) Highlands, generalized, (2) less elevated relief, (3) streams (intermittent streams with broken lines), (4) planation surface remnants, (5) ridges, (6) fault line scarp, (7) fault scarp, (8) structural scarps, (9) old alluvial fans, (10) high terrace, (11) middle terrace, (12) low terrace, (13) alluvial cones and fans, (14) holocene alluvium and (15) former river channels.

features was established, especially the deposits and planation surfaces of the mountain foothills. Their location in the landscape and the nature of the correlative deposits make rapid identification possible, although except for some special sites absolute dating is less easy. It is nevertheless clear that the landforming processes took place after the Pleistocene sedimentation, producing a gently sloping surface or glacis covered partially by a veneer of debris of variable thickness that never exceeds 2–3 m. This landscape can be reconstructed in the whole of the Cordillera, the following being the principal criteria for its identification: (1) an elevated position in relation to the present stream courses, (2) the coarse and angular nature of the debris that overlies part of the surface, (3) the reddish colour of the matrix and (4) the generally strong cementation of the deposits, at least at the surface.

The lack of fauna makes exact dating difficult, and confusion with Miocene deposits has sometimes occurred. In the Granada basin, the most precise dating is that of Aguirre (1957, 1961, 1974) who has identified fauna corresponding to the Middle Villafranchian. Comparison with similar deposits in other areas of the Cordillera [see Montenat (1973), Hernández-Pacheco (1932a), Viguier (1974)] and in other western Mediterranean lands [see Raynal and Tricart (1963)] suggests that the age is Middle Villafranchian or Moulouyan.

All researchers agree that the deposit is of torrential type, but the reason for this is sometimes attributed to a semiarid climate [see Birot and Solé-Sabarís (1959), Lhénaff (1981)], especially in view of its calcification, and sometimes to cold conditions according to Raynal (1975). In general, the Villafranchian deposits seem to be due to a process of adjustment during the transition from one relief generation to the next [see Stäblein and Gehrenkämper (1977)]; in the Iberian Peninsula, they are certainly the result of a change in tectonic conditions as well as climate [see Molina (1975)] and must be interpreted in the

context of a mountain front [see Sala (1982)]. They have the same significance as the raña deposits of the Hercynian mountain fronts, the only difference being the deeper previous weathering of the older crystalline masses from which the rana deposits were produced [see Birot and Solé-Sabarís (1959)].

13.1.4   Semiarid Fluvial Landforms (*see* Fig. 13.7)

The major landforms in the Baetic Cordillera are due to fluvial incision, and one of the problems traditionally discussed in this area has been that of stream piracy. Carandell (1935), Sermet (1969) and Solé-Sabarís (1942b) found evidence for river capture by Mediterranean rivers at the expense of the Guadalquivir tributaries in view of some sharp changes of direction of flow, the existence of lakes and peat-bogs upstream from them and the presence of several wind gaps. The steep gradients of the Mediterranean rivers and their proximity to base level would explain the energy necessary for this hypothesis. Lhénaff (1973, 1981), however, has questioned these ideas and has suggested that the changes of direction may be due to a structural adaptation to two sequences of tectonic events: (1) folding and faulting in a north–south direction and (2) warping in an east–west direction. Wind gaps are explained by the existence of flysch outcrops among the massive limestones; the bog areas seem to be precisely located in subsident basins while the lake zones are clearly related to karst processes in the Triassic gypsums.

Fluvial deposition is also important, especially in the western internal depressions and along the main river courses. Lhénaff (1981) has found three post-Villafranchian alluvial levels and, in some cases, a fourth one. The best represented levels are the lower terrace, which is made of grey silts, and the third terrace, which is mostly cemented and topped with a rubified soil. The second level is variable and difficult to identify, sometimes only

Fig. 13.7 Semiarid landscape with badlands, Sierra del Carrascal, north of Alicante. (Embleton)

Fig. 13.8 Tabular residual rising above semiarid pediments, near Mojacar. (Embleton)

represented by discontinuous alluvial fans.

The most typical features of the area, however, are the widespread gently sloping piedmont surfaces resulting from short, but intense rainfall events which are a characteristic feature of the climate of the area. The process of sheet wash is thought to be responsible for these broad piedmont ramps called glacis by French authors working in the Mediterranean area. In the Iberian Peninsula, Solé-Sabarís (1964) has distinguished between an arid type or 'piedmont glacis' and a subarid type or 'terrace glacis'. The piedmont glacis are the most extensive—both temporally and spatially—in south-eastern Spain. They can be found in the Alicante area, but their maximum development is in the area of Murcia, Cartagena and Almería (*see* Fig. 13.8). These glacis have enormous dimensions and are similar to those described in North Africa by Dresch (1938). The entire relief of the lowlands is dominated by the glacis and at least four generations of them can be found, one inset into the other as in a terrace sequence. They appear to be capped by a crust of calcium carbonate which has been the decisive element in their preservation. The second type of glacis consists of those that are related to fluvial terraces in such a way that, in the zone of contact, the angular debris of the glacis is interstratified with the terrace alluvium.

In the lowlands of the Rio Segura, López-Bermúdez (1973) has studied the glacis sequences and found a maximum of five levels, as had been described by Wiche (1961) also in the Segura system, corresponding to terrace levels. These authors think that the four lower levels (at 20, 30, 50 and 70 m) can be related to the four pluvial Quaternary phases of the Mediterranean lands, while the fifth level (at 100–200 m) belongs to the Villafranchian period. In the western part of the Cordillera and in the internal depressions, Lhénaff (1981) has found only small glacis and always related to terrace levels. This supports the hypothesis of Solé-Sabarís that a subhumid climatic

tendency in an area produces a greater development of fluvial processes, and thus of alluvial terraces, to the detriment of glacis development which is optimal in a semiarid environment.

The crusts of calcium carbonate that have developed on the glacis have been studied in detail by Dumas (1967) who defines four different types: (1) pedogenetic crusts (80–150 cm thick) with differentiated horizons, and an upper crust (30–60 cm thick) that progressively thins and loses its compactness, (2) zonal crusts (1–6 cm thick) discontinuous and considered to be the result of subsurface flow, (3) crusts due to diagenetic consolidation following the internal deposition of detrital material, preferably in lenses of coarser material, and (4) crusts developed at the watertable, highly indurated and located between the rock and the debris cover. Pedogenetic crusts and, to a certain extent, zonal crusts are those that can help to identify older surfaces. In the upper parts of the glacis, two sets of crusts can sometimes be found, both due to different genetic processes, but in any case lateral and vertical migration of groundwater is involved. There is a decreasing degree of consolidation from the highest to the lowest crusts.

13.1.5 Glacial and Periglacial Features

Due to its southerly situation and its open relief, the Quaternary glaciations did not produce important phenomena in the Baetic Cordillera in spite of it being the mountain range with the highest altitudes in the Iberian Peninsula. Actual glaciation did occur in the Baetic zone, especially in the Sierra Nevada, and has also been noted in the Sierra Tejeda as reported by Klebelsberg (1928) and Sermet (1934), but in the Sub-Baetic zone only niveo-glacial features have been found [see Pezzi (1977)], in the Sierra Mágina and Sierra Arana. The Quaternary glacial features of the Sierra Nevada were first studied in detail by

Obermaier and Carandell (1916), and later by Dresch (1937), García-Sainz (1943), Paschinger (1954) and Messerli (1965), this last work being the most complete. Glaciation has not greatly altered the relief; cirques are the main glacial landform, often small, although in certain cases they have fed short and shallow valley glaciers. The cirques are located at structural breaks of slope, generally widened by periglacial processes, and the valley moraines are situated in former ravines in which glacial action has not destroyed the previous relief forms. Morainic materials are abundant, however, and, in some cases, are up to 100 m thick due to the fragility of the underlying schists, the importance of periglacial processes and because regressive fluvial erosion has not yet reached them. Both Messerli (1962, 1965) and Lhénaff (1981) have pointed out the difficulty in many cases of making a clear distinction between moraine and gelifluction or debris flow masses.

Obermaier and Carandell (1916) described 12 small glaciers on the southern slopes of the Sierra Nevada, the most important ones being those of Lanjarón and Veleta, the latter being 5 km long and with its lower limit at 2050 m. They also described eight slightly larger glaciers on the northern slopes, the most important being that of Dijar, which was 6 km long and with its terminus at 1780 m. Lhénaff (1981) thinks that aspect is a most important factor, the dominant westerly winds tending to accumulate snow on the eastern slopes.

All glacial landforms were at first interpreted as Würmian due to the fresh appearance of the morainic materials, although the possibility of a Riss glaciation was also considered by García-Sainz (1943) in the Didar valley. At present, remnants of a late glacial phase are also recognized. Paschinger (1954) has suggested that late glacial remnants are to be found between 2800 and 2900 m on slopes protected from the westerly winds and Messerli (1965) has observed extreme situations in which the late glacial features may descend to 2650 m. The average limit of Würm glaciers was established by Obermaier and Carandell (1916) at 2300 m on the southern slopes and at 2000 m on the northern slopes, but Messerli (1965), using modern sedimentary methods instead of only simple topographical mapping, has found lower limits at 1760 and 1680 m, respectively. The lower limit of permanent snow proposed by Messerli is between 2400 and 2300 m, which excludes the possibility of a glaciation in the Sierra Tejeda. Hempel (1958) has pointed out the probable existence of a Riss glaciation due to the presence of several cirques located at a lower altitude than the Würmian snowline; he has distinguished three levels of cirques at 2800–2850 m, 2400–2500 m, and 2100–2200 m, while Butzer and Franzle (1959) believed that the lower concavities are only nivation hollows because of the absence of morainic deposits in them.

As in the case of glacial landforms, study of periglacial features began in the Sierra Nevada [see Hempel (1958b), Messerli (1965)] where gelifluction debris occurs at altitudes ranging from 700 to 1000 m. Lhénaff (1981) has pointed out the difficulty in trying to determine the extent of Quaternary gelifluction processes because they can be easily confused with similar processes active at present; attention should therefore concentrate on cemented and stratified scree slopes or grèzes litées. These deposits are typical of Mediterranean periglacial processes, where frost action and mobilization of debris on the slopes by a combination of gelifluction and slope wash due to snow melt are the dominant processes. Periglacial studies have been undertaken by Pezzi (1977), García-Rossell and Pezzi (1978), Soutadé and Baudière (1970), and Gallegos (1971) not only concerning Quaternary features but also present-day activity.

13.1.6 Coastal Landforms

The Baetic Cordillera lies broadly parallel to the coast and has played a major role in coastal evolution. The range, however, is frequently interrupted due to tectonics and rock composition, and the active Mediterranean streams have been able, in many cases, to build out coastal plains and deltas. In addition, this part of the Mediterranean coast has been subject to tectonic movements since the Pliocene that have often accelerated the sedimentary processes.

From Gibraltar to Cabo de Gata, the coast is predominantly rectilinear due to the proximity of the Baetic zone lineation, but north of the volcanic outcrops of Gata, the Baetic internal depressions and their Neogene sediments allow the development of broad gulfs whose inner parts, especially in the case of the lower Segura river basin, are the sites of well-developed coastal plains with beaches, sand bars and lagoons (e.g., the Mar Menor closed by a sand ridge 21 km long, and the salt lagoons of la Mata and Torrevieja) as described by Sanjaume and Gozálbez (1981), and Mateu and Sanjaume (1981) (*see* Fig. 13.9). Dune fields are also widespread, and it has been possible to establish a Quaternary sequence for them, from Middle Pleistocene to Holocene [see Rosselló and Mateu (1981)]. Littoral sedimentation is also encouraged because this coastal area is somewhat shielded from the north-easterly sea currents. Progradation of about 2 m per year has been observed in the Llanos de Carchuna by Sermet (1964) and in Motril by Bovis (1974) since the end of the eighteenth century; although longshore drift is locally important in the embayments, deposition on this scale indicates considerable erosion of the surrounding uplands.

All along the Baetic coast, former marine levels are numerous, well-known in many cases, as in the Gibraltar and Alicante areas, and well-correlated with the marine levels of other western Mediterranean coastal areas, especially with those of Mallorca [see Solé-Sabarís (1961), Butzer and Cuerda (1962)]. These levels, together with several abrasion surfaces and a good deal of palaeontological data [see Porta and Solé- (1957)], show that there has been considerable uplift along some sectors of this coast. Studies are numerous, the most recent being those of Dumas (1971), Imperatori (1957), Lhénaff (1966), Ovejero and Zaro (1971) and Rosselló (1970). In the synthesis of Solé-Sabarís (1961), the following levels are adduced.

1) At 80 m (possibly Sicilian) found at certain points (e.g., Mojácar, Guardias Viejas, Adra and Nerja) and related to an erosion glacis that has a thick calcareous

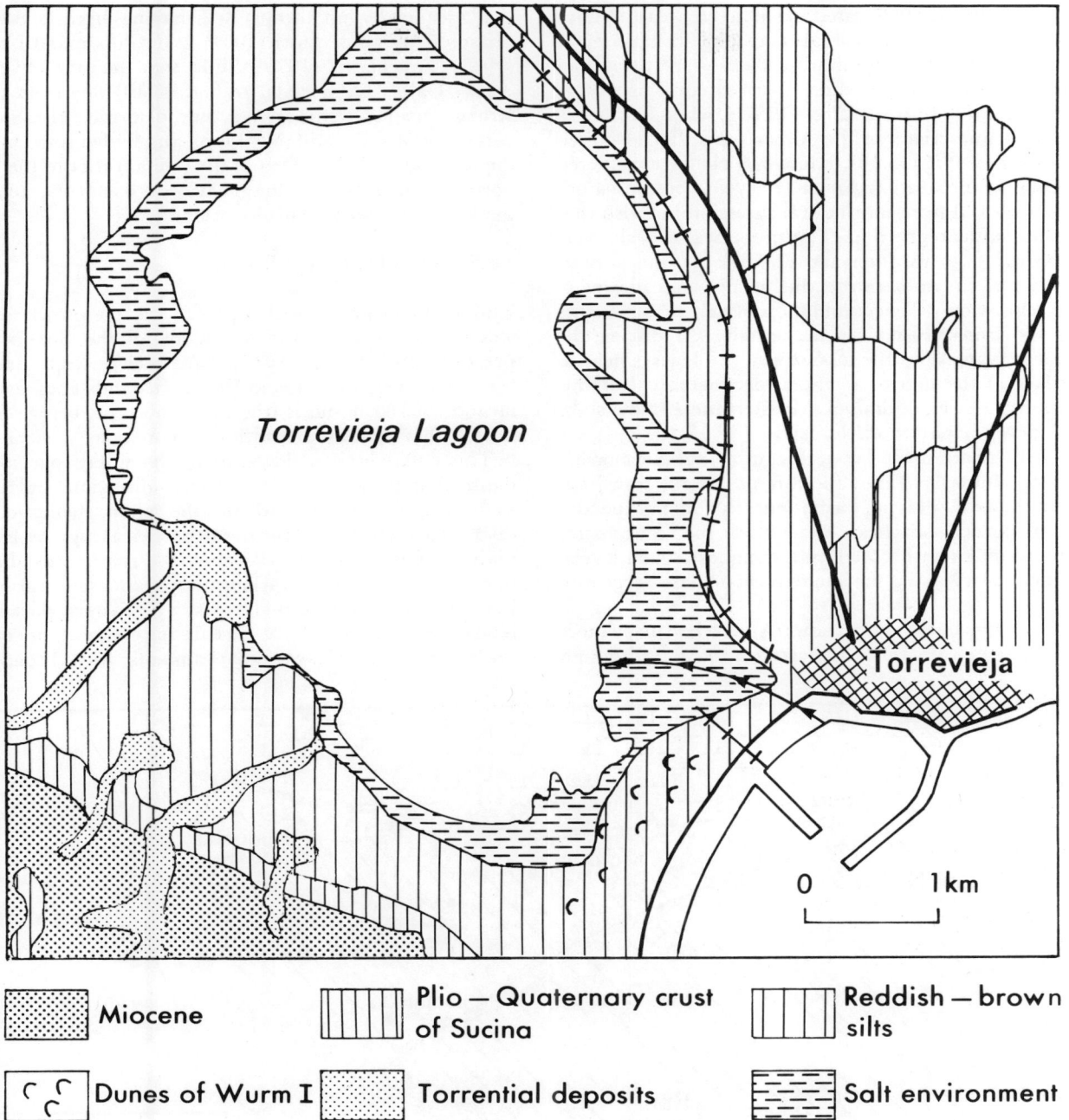

Fig. 13.9 Neogene–Quaternary deposits of the Torreveja coastal area [Mateu and Sanjaume (1981)].

crust. No fauna have been found, but its marine origin is undeniable.

2) At 20–25 m or Tyrrhenian I (Alicante, Villaricos). The marine deposits are covered by strongly cemented continental red silts and, although not very abundant, the fauna of this level are of temperate type, rather similar to the present.

3) At 5–6 m or Tyrrhenian II. This is the best represented level along the whole coast. Continental red silts appear interstratified with the marine deposits and are strongly indurated. Tropical fauna are abundant, with *Strombus bubonius* and other species that are no longer present in the Mediterranean sea.

4) At 2–3 m or Tyrrhenian III. This is partly cut into the previous level. The deposits are less consolidated and only slightly indurated. It is found in the Alicante area and a few other parts of the coast.

5) The Flandrian beaches are not so widespread on this coast as the Tyrrhenian ones and studies of their location and characteristics are not very numerous, but have been undertaken by Ovejero and Zazo (1971), Fourniguet (1975), and Gozalbez and Cuerda (1981). They consist of

small remnants, always located in sheltered areas, of non-cohesive sands and silts of grey colour, containing a mixture of continental and marine shells.

The type area for the three Tyrrhenian episodes is around Alicante [see Imperatori (1957)], while four levels are more characteristic of the Almería area [see Fourniguet (1975)] (*see* Fig. 13.10). Each level can be considered as a complete miniature sedimentary cycle, containing not only the material from marine transgression but also that from continental regression. The processes of each cycle are, according to Brunnacker (1973): (1) mechanical weathering of the substratum during the cold phases of marine regression, (2) accumulation of marine sediments during the transgressive warm periods, (3) during the subsequent regression, the inland watertable rises near to the surface of the marine terraces, producing calcareous indurations; (4) evolution of Mediterranean-type soils; and (5) development of duricrusts.

As for the relationship between continental and marine levels [see Dumas (1971)], there is no lateral transition from continental alluvial formations to the marine deposits of former beaches. Sometimes the continental formations are superimposed on the highest beach levels, sometimes they cut the marine deposits or are inset into them.

At present, growing importance is attached to neotectonic phenomena on the Mediterranean coast, although this idea is by no means new having been previously proposed by Imperatori (1957) and Gaibar and Cuerda (1969). Fourniguet (1975) thinks that the area of Campo de las Dalias (Almería) (*see* Fig. 13.10) has been under strong tectonic stress, which has deformed the Sicilian surface, raised the cliff of la Loma and folded some marine deposits. Dumas (1971) is of the opinion that in the more northern area of the lower Segura basin the tectonic tendency has been a subsident one.

13.2   Guadalquivir Basin

This plain, which is wide open to the sea on the south-west, is the flattest and lowest of all the Spanish depressions and shows very few hard rock outcrops, another feature not very common in Iberia. It stands at an average altitude of 150 m and is triangular in shape, being 330 km long with a maximum width of 200 m at its western end.

The Guadalquivir flows along the upper part of the depression in a roughly east–west direction and in an asymmetrical position within the plain, being clearly displaced towards the northern border along the Hercynian margin. In the Seville area, the river turns sharply towards the south, marking approximately the limit between two differentiated areas. The eastern part has a landscape dominated by undulating relief—the campiña—modelled in Neogene sediments with broad, U-

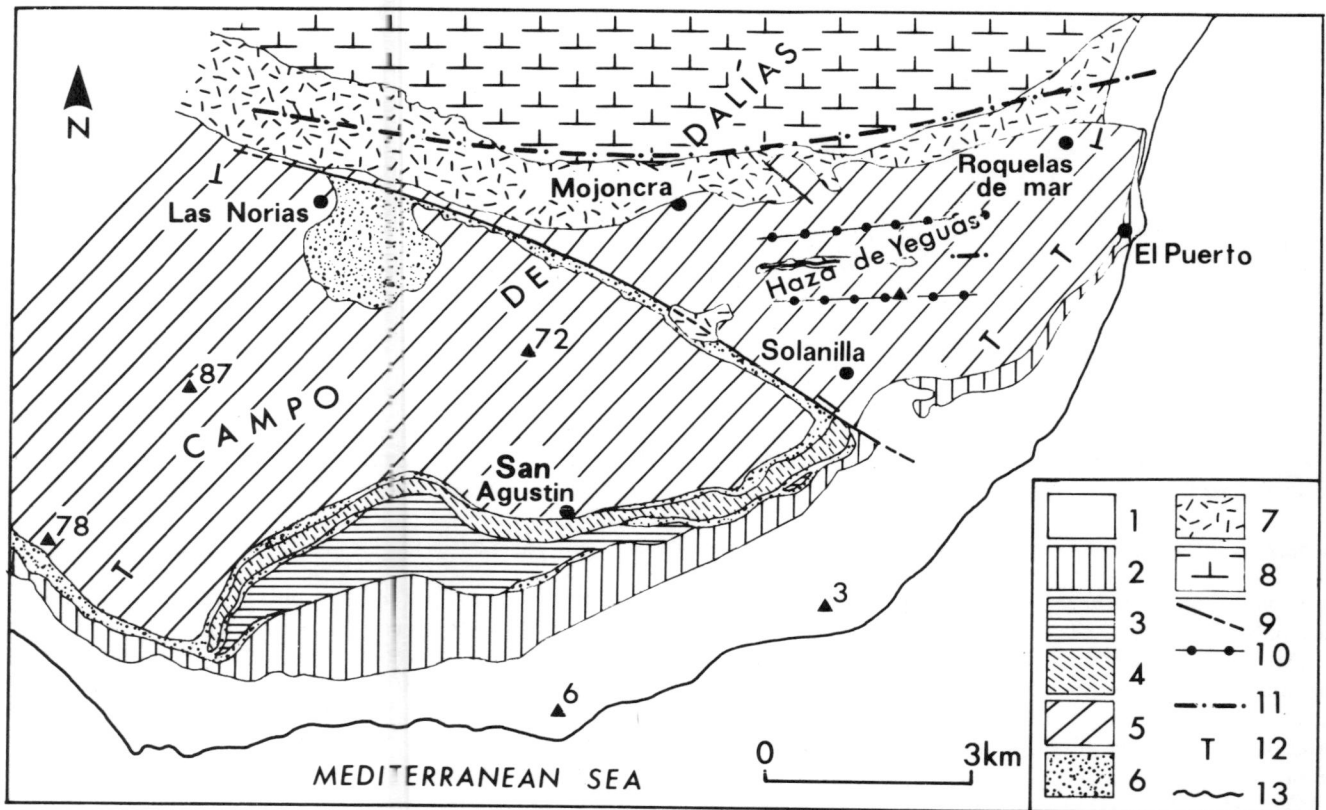

Fig. 13.10 Geological map showing former shore levels in the eastern part of the Campo de Dalias, near Almería, southern Spain [Fourniguet (1977)]. (1) Recent and Flandrian, (2) late Tyrrhenian, (3) middle Tyrrhenian, (4) early Tyrrhenian, (5) Sicilian, (6) Pliocene, (7) limon, (8) alluvial cones, (9) known fault, (10) anticlinal axis, (11) synclinal axis, (12) dip and strike direction of the Sicilian (beds and surface) and (13) fossil shoreline.

shaped valleys bordered by terraces and interrupted in certain places by isolated tablelands cut in somewhat more resistant sediments (*see* Fig. 13.12). West of Seville (i.e. lower Andalucia), the plain becomes flatter and wider, with average altitudes that fall gradually to below 100 m. The depression ends at the sea in a marshy area— Las Marismas—where accumulation processes are dominant, Quaternary silts and sandy clays progressively filling the estuary. This plain was still a lagoon during Roman times and is today a semiaquatic environment that evokes the marshy plains of Flanders.

Like the Ebro Basin, but in an even more remarkable way, the Guadalquivir Basin is a trough related to Alpine tectonics and, in this case, represents the foredeep of the Baetic chain. The great asymmetry between its northern and southern margins is striking: steep and rectilinear to the north, the sediments showing a sharp contact with the Hercynian highlands of the Sierra Morena, but sinuous and discontinuous on the southern, Baetic side, with scattered mountainous inliers interrupting the sedimentary depression as a result of nappes thrust into the depression from the Baetic chain. The depression is filled with Oligocene and Mio-Pliocene sediments of considerable thickness in certain places. The sediments are almost exclusively marine.

13.2.1 Geology

The Guadalquivir plain is a clear example of a piedmont trough between a young chain—the Baetic Cordillera— and the old massif of Iberia. The similarities between this trough and that of the Rhône situated between the Alps and the Massif Central have been often pointed out. The Guadalquivir depression is a geologically recent feature, as shown by the fact that in many places Tertiary sediments lie directly on the Hercynian basement.

The unit has several characteristics intimately related to the geological history of the Baetic Cordillera. Its northern part is marked by the Sierra Morena, the southern end of the Iberian massif, which gradually disappears under the sediments of the depression. The Hercynian basement is covered by a thin representation of Mesozoic sediments, generally reduced to the Trias and sometimes not even the whole of it. It is the Neogene that mostly fills the depression, represented mainly by Miocene sediments whose thickness increases towards the south being 1.5 km at Seville. The Miocene consists mostly of sandstones, limolites, marls and clays, all of marine facies; along the border of the Baetic Cordillera, they have a clear molasse character. The sediments are in general poorly indurated.

One of the most important geological features of the depression is the fact that the Lower Miocene contains marked intercalations of allochthonous elements coming from the Baetic Cordillera [see Perconig (1960–62), Fontboté and García-Dueñas (1968), García-Rossell (1973)]. In the western part of the depression, these materials have an olistostromic character, as in the Carmona nappe, made of greenish clays and marls whose existence was revealed by the oil drillings. In the eastern part, especially east of Jaén, the nappes consist of several

units, not very thick but with a remarkable horizontal continuity rather similar to that found in the Pre-Alpine nappes. The mechanism of movement was gravitational sliding and the order of the nappes is variable: the elements of the eastern part of the depression have similarities with the external Sub-Baetic domains, while the materials of the Carmona nappe have a more internal association. The time of their movement is evidenced by their position within the Lower Miocene series.

Except for the presence of allochthonous elements, the structure of the Guadalquivir Basin is simple. Geophysical investigations by Perconig (1960–62) have shown without any doubt that the northern margin of the depression is not the great fault that was previously supposed. The sinking of the Hercynian massif has been progressive, and it is only locally that some normal faults are found which, nevertheless, may have some influence on the cover. In areas where intercalations of allochthonous elements exist, there may be local complications due to secondary halo-kinetic processes.

13.2.2 Main Relief Features of Upper Andalucia

In the eastern part of the depression or 'upper Andalucia', three types of landscape can be distinguished: (1) the right bank of the Guadalquivir, (2) the left bank and (3) the Guadalquivir alluvial plain.

Along and close to the right bank of the Guadalquivir, the Sierra Morena rises steeply to the north. The drainage system has cut deeply into the old land surface giving the Hercynian mass a more energetic relief than that produced by tectonic flexuring. The mountain foot-slopes are fashioned into a series of ramp-like features, often covered by a detrital mantle of Villafranchian type, which are indistinctly cut into the Miocene sedimentary filling (glacis) and into the crystalline bedrock (pediments) of the Sierra Morena (*see* Fig. 13.11). The glacis appear now in a prominent position because erosion processes have cut into the contact zone with the crystalline basement and carved the El Campo depression. The Aljarafe glacis consists of relatively resistant Miocene sandstones, its foot-slopes often scarred by ancient meanders of the Guadalquivir. The Villafranchian conglomerates, consisting of siliceous pebbles in a reddish, sandy matrix, are not present at the head of the Aljarafe glacis but appear in its lower part with notable thickness and development, spreading out as far as the Ligustine coastline and connecting laterally with the Guadalquivir terraces. On the Aljarafe surface not covered by these conglomerates, red soils have evolved underneath which often appear with calcareous crusts. The Villafranchian surface is found all along the foot-hills of the Sierra Morena as far as the Huelva region [see Pérez-Mateos and Riba (1961), Diaz and Rubio (1981)].

On the left bank of the Guadiana, the landscape is dominated by the undulating relief of the campiña where a succession of hills extends as far as the Baetic uplands. Near Seville, however, there is an outstanding relief feature—the isolated, rectilinear and steep-sided outcrop of Miocene calcarenites known as Los Alcores—rising

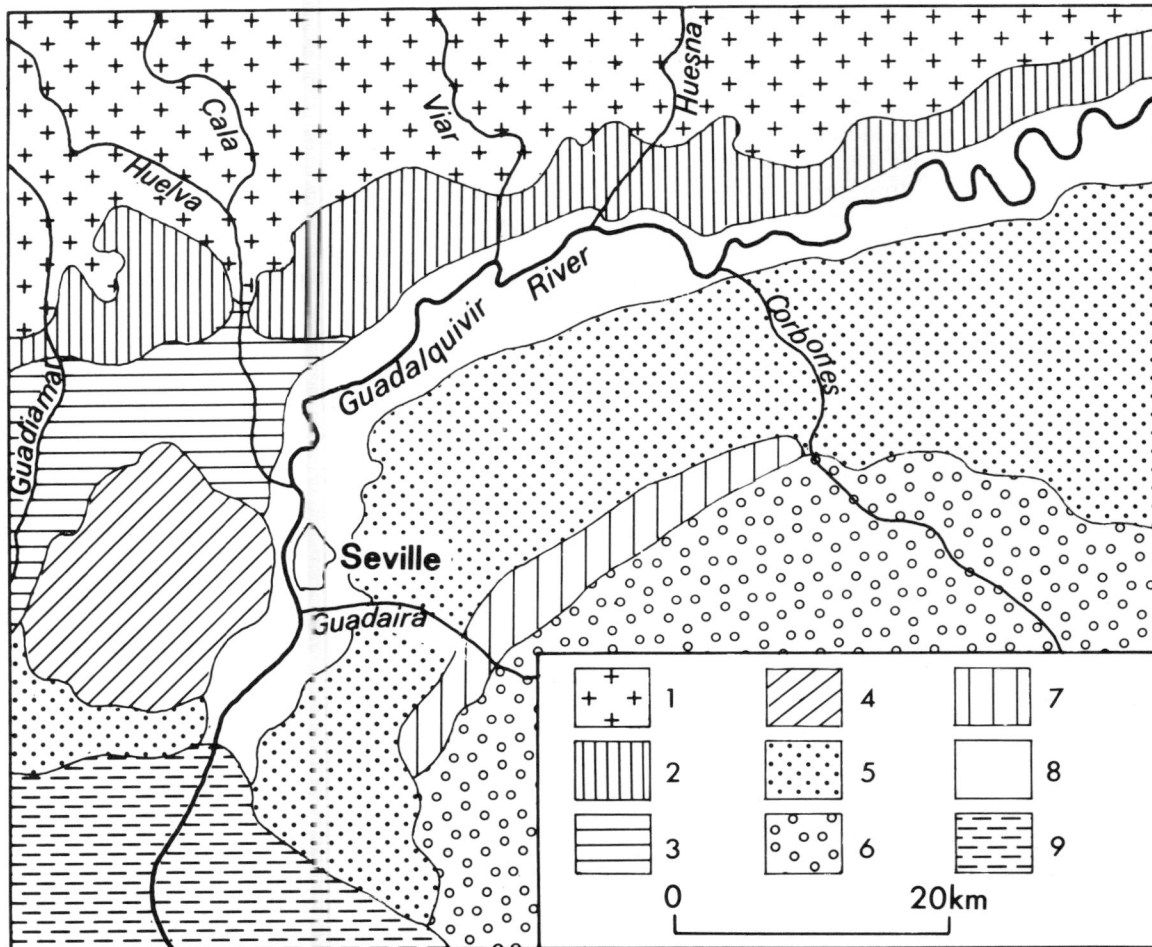

Fig. 13.11 Geomorphological regions of the lower Guadalquivir area [Drain *et al.* (1971)]. (1) Sierra Morena massif, (2) the Miocene fringe and Villafranchian piedmont, (3) El Campo erosional depression, (4) Aljarafe glacis, (5) lower campiña on alluvial fill, (6) upper campiña, (7) Los Alcores, (8) present alluvial plain (la Ribera) and (9) marsh (las Marismas).

150 m above the plain and stretching in a north-east–south-west direction for approximately 30 km. Los Alcores have an asymmetrical profile. The north-west face is gently sloping and merges with the Guadalquivir alluvial terraces, thus looking like a dissected remnant of a Villafranchian glacis. In contrast, the south-east face is cut as a steep slope above the Vega de Carmona, a feature that is thought to be derived from a recently active fault scarp or flexure line, with traces of landslips along it. The Vega de Carmona is a broad, flat plain only interrupted by a few isolated hills that seem to be relics of a Middle Quaternary erosion surface because of their concordance in altitude with the middle terrace of the Corbones river, as described by Drain *et al.* (1971).

13.2.3 Fluvial Processes and Landforms

On the left bank of the Guadalquivir, alluvium is extensive due to the progressive displacement of the river towards the north caused by the greater amount of debris coming from the higher relief of the Baetic Cordillera. Differentiation of alluvial levels resulting from alternating

phases of erosion and sedimentation is difficult. Drain *et al.* (1971) singled out two fundamental levels—a middle and a higher one—the latter having abundant surficial gravels. In a study by IGME (1969), three subdivisions are made: (1) the old Quaternary at 50–170 m, consisting of two fills with no break of slope between, (2) the middle Quaternary, with two levels inset in the older alluvium standing at 10–50 m and (3) the youngest fill made of discontinuous silty deposits. In more detailed work by Clemente and Paneque (1974), in which a comparative study has been made between the pedology [see Guerra *et al.* (1962), Paneque and Mudarra (1966)] and geomorphology, four terrace levels have been distinguished and chronologically correlated with the Quaternary pluvial phases of the western Mediterranean lands [see Raynal and Tricart (1963)] and the corresponding glacial periods. These four terrace levels were earlier described by Cabanás (1957) in Jaén province (*see* Figs. 13.12 and 13.13).

The first phase of fluvial sedimentation was the most extensive one, possibly because the river had not yet clearly carved its bed; as a consequence, a wide, thick

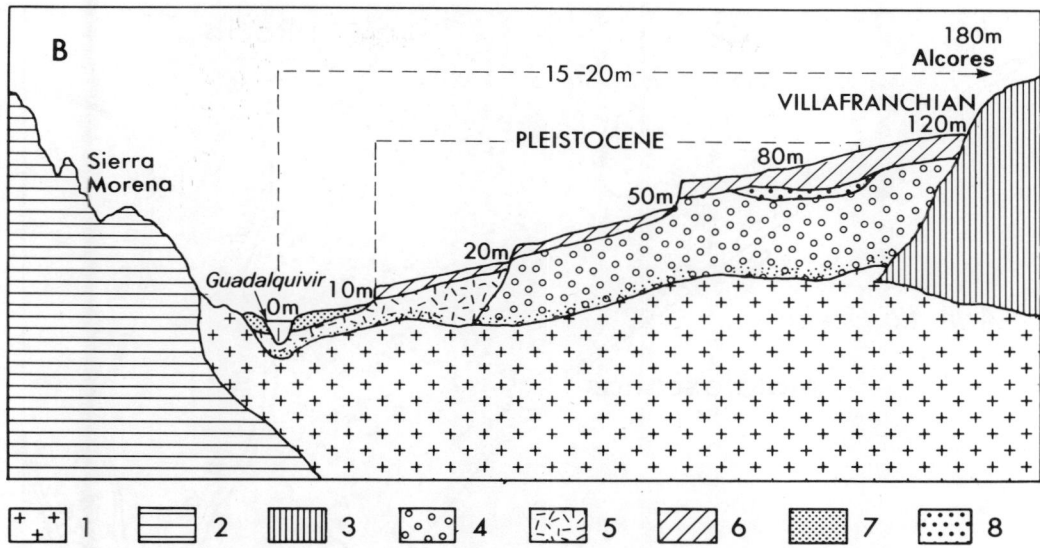

Fig. 13.12 General cross-section (north-south) of the lower Guadalquivir valley [Clemente *et al.* (1981)]. (1) Tertiary marls, (2) Palaeozoic and Mesozoic, (3) calcarenites (Mio-Pliocene), (4) early Quaternary microlitic calcareous sediments, (5) middle Quaternary conglomerates with sands and gravels, (6) middle Quaternary clays, sandy silts, sands and gravels, (7) recent alluvium and (8) early Quaternary calcrete.

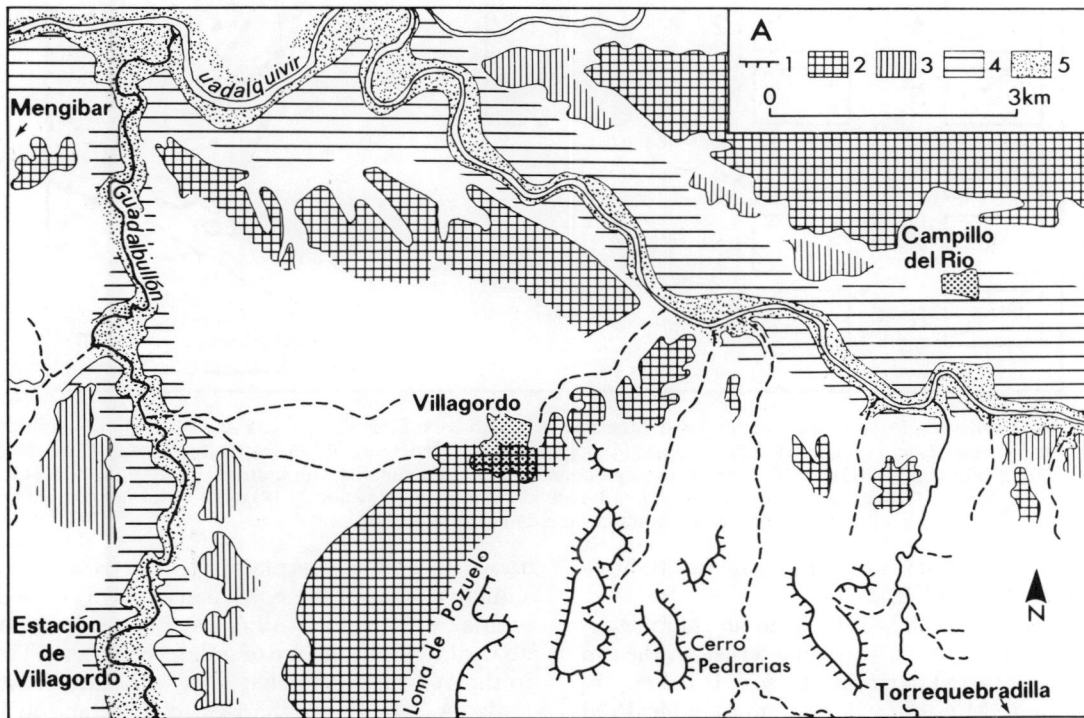

Fig. 13.13 The terraces of the Guadalquivir north of Jaén [after Lhénaff (1981)]. (1) Tabular Miocene residuals, (2) high terrace (Villafranchian), (3) middle terrace (possibly Riss), (4) low terrace (possibly Würm) and (5) recent alluvium.

alluvial cover was laid down in which the later fills have been inset; the soils of this terrace are predominantly siliceous which, together with a lack of fines, is proof of the intensity of leaching processes. The fine matrix of the pedogenetic formation is strongly rubified and the upper horizons are totally decalcified, with only small concretionary remnants. This level stands above 80 m and is connected with the Plio-Villafranchian system of Alcores

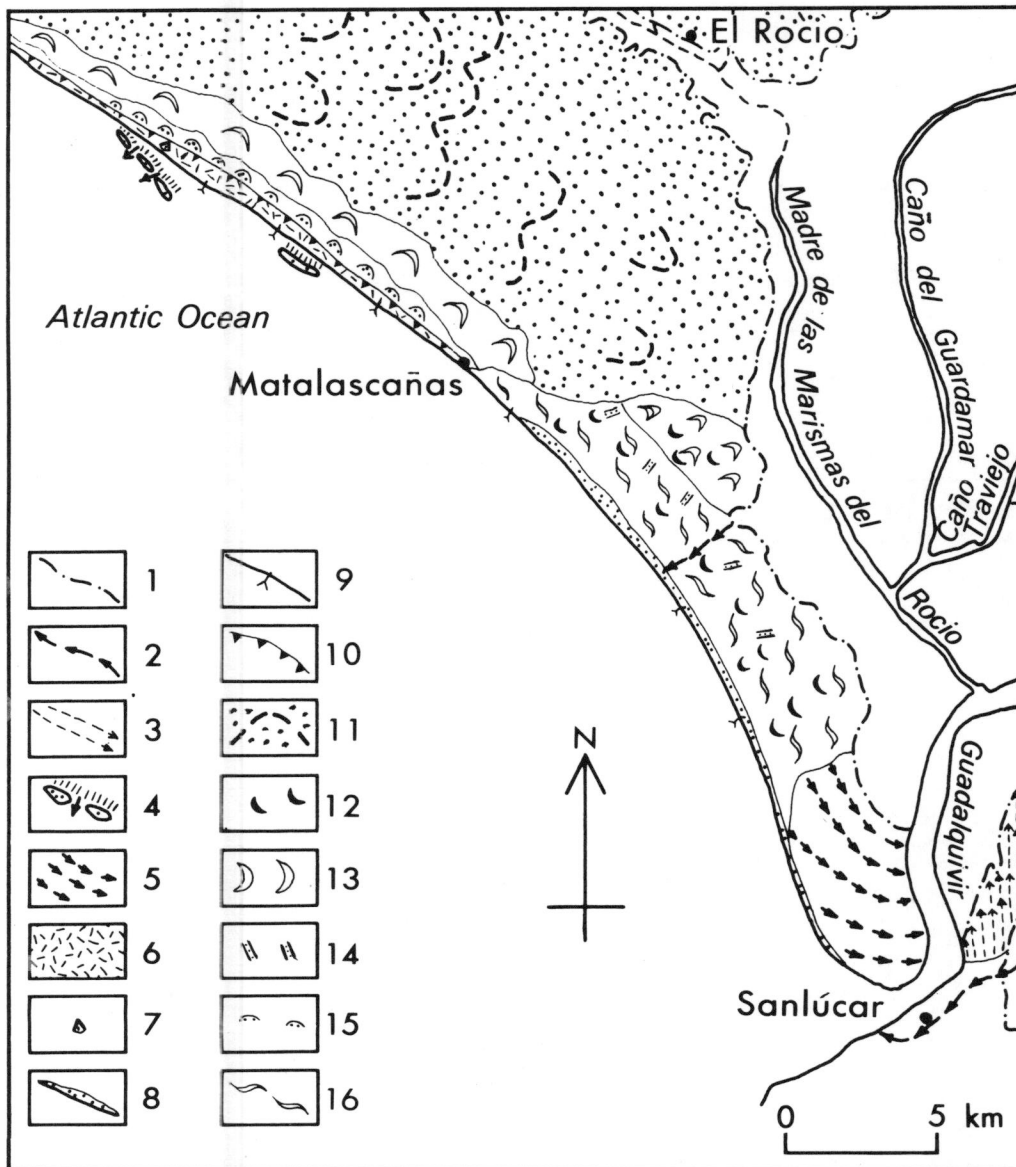

Fig. 13.14 Coastal and estuarine features around the lower Guadalquivir, south-west Spain [after Zazo *et al.* (1981)]. (1) Limit of marsh area, (2) old channels of the Guadalquivir, (3) interdistributary ridges, (4) ridge-and-runnel foreshore, (5) former spit recurves, (6) beach deposits with peat horizons (41 000, 30 000 and 12 260 BP), (7) sandy backshore cones, (8) present sand bar, (9) berms, (10) steep cliff, (11) aeolian mantle with crests of parabolic dunes, (12) barchans (second system), (13) parabolic dunes (third system), (14) arms of former parabolic dunes (of the third system), (15) transverse dunes (fourth system) and (16) transverse dunes (fifth system).

through an erosion glacis. It is dated as Saletian (Günz–Villafranchian) (*see* Fig. 13.13).

The next level is found at 50–80 m, its main characteristic being the development of a travertine crust on the top of which has evolved a red, ferrosiallitic soil. It belongs to the Amirian period (Mindel), when the most typical red soils of the Mediterranean borderlands developed. In the Guadalquivir area, the crust has been eroded in many places prior to the red soil formation. The second level stands between 20 and 50 m and has a hydromorphic soil that is locally replaced by a black soil; this level is correlated with the North African Tensiftian (Riss). On the first level, at 7–20 m, which rises in a well-defined step

above the Holocene plain, the soil is a reddish brown, similar to that of the Soltanian (Würm) episode. Finally, within the present alluvium, two zones are defined according to the degree of soil development. The area next to the Würmian level has soils with clearly differentiated horizons, while near the river course the soil has a great morphological unity.

The present-day alluvial plain is the area called La Ribera. Fluvial landforms are controlled by the enormous variation of water flow, which can be reduced to $10 \, \text{m}^3 \, \text{s}^{-1}$ or less during a dry period, while in an exceptional flood it might rise to $12\,000 \, \text{m}^3 \, \text{s}^{-1}$ and cover the lower terrace. The annual average of $183 \, \text{m}^3 \, \text{s}^{-1}$ is thus meaningless. This is

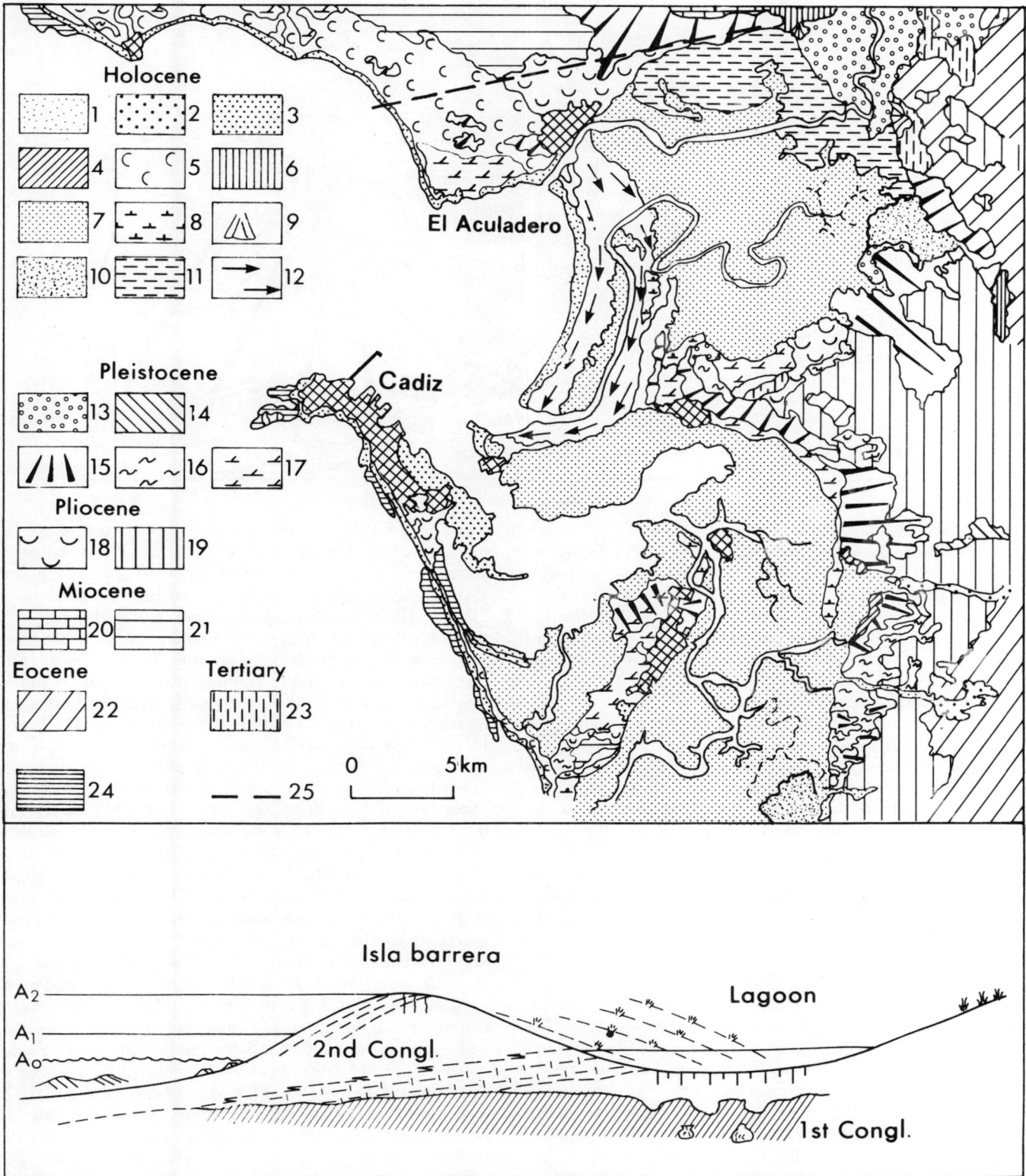

Fig. 13.15 Geomorphology of the area around Cadiz [after Zazo *et al.* (1981)]. (1) Beach, (2) recent alluvium, (3) sand bar, (4) marine level of historical time, (5) coastal dunes, (6) colluvium, (7) mud flats, (8) salt marsh, (9) alluvial cone, (10) fan deposits, (11) floodplain silt, (12) coastal spits, (13) terraces, (14) old alluvium, (15) glacis, (16) red sands, (17) sandstone and marl, (18) conglomerate, (19) limestone, (20) calcarenite, (21) albarizas, (22) Sub-Baetic marls, (23) gypsiferous marls, (24) wave-cut platform and (25) fault.

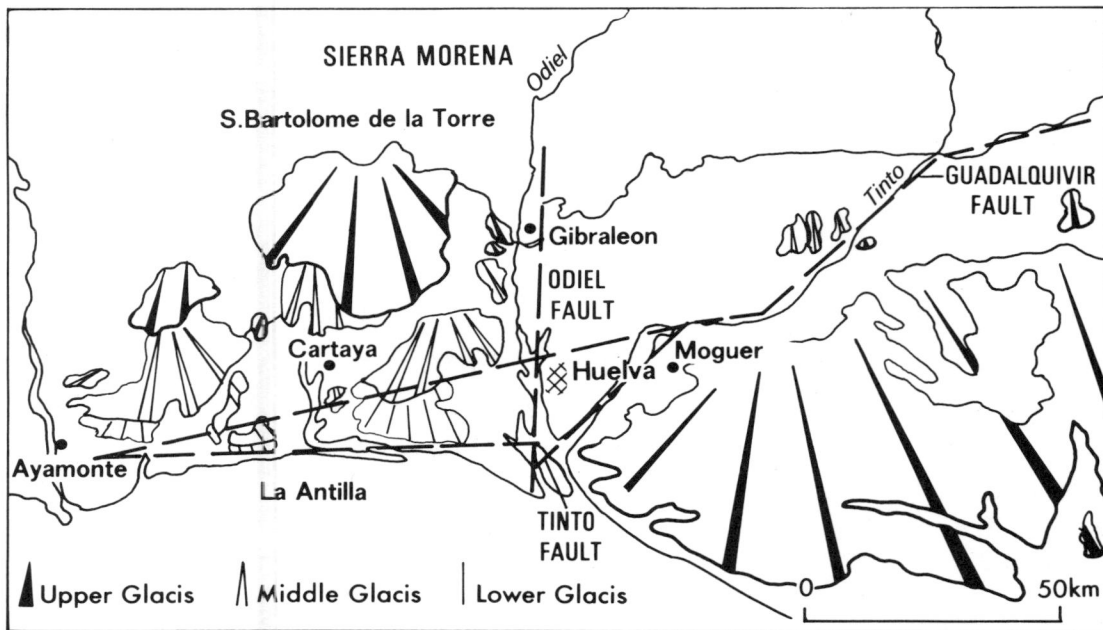

Fig. 13.16 Piedmont glacis in the area around Huelva, southern Spain [Zazo and Goy (1981a)].

the smallest discharge of all the main rivers of the Iberian Peninsula, the Genil in the Baetic Cordillera being its fundamental source. Changes of discharge have created a set of channels, a typical feature of Mediterranean rivers. In this case, there is a minor channel cut within vertical banks. The second channel has a braided pattern, and the flood channel is floored by discontinous silt deposits.

13.2.4   Andalucian Lowlands

The Guadalquivir Basin ends to the west in a marshy area called Las Marismas which extends over an area of 2000 km². This region is the result of the infilling of an ancient lagoon. The marshland is a zone of recent subsidence, which has drawn the Guadalquivir southwards into a former lagoon. The whole of the right bank has remained a marsh and is subject to flooding in winter. A succession of sand ridges that have slowly become consolidated lies between the mouths of the Tinto–Odiel rivers and the Guadalquivir; the development of these bars is still favoured by present-day marine currents and a succession of lagoons in constant regression is found along the littoral (*see* Fig. 13.14).

The last phases of the main marine regression are easy to reconstruct. During Roman times, there was still a great lake (50 km wide) called Ligustinus, which extended from the present coastline to Puebla del Rio. The lagoon has been progressively transformed into marshes, while the old littoral sand bars that closed the lake have developed into the great extension of Arenas Gordas where dunes nearly 30 m high are presently fixed by pine trees. At both ends of the Gulf of Cadiz (*see* Fig. 13.15), remnants of old beaches can be recognized, as at Rota and Tarifa, although a systematic study of the beach levels has not yet been made.

In the lowlands, as in the uplands, there is a set of glacis along the margins of the depression. In the Huelva region

(*see* Fig. 13.16), two sedimentary units have been distinguished by Viguier (1974): (1) basal red sands of estuarine/marine origin and Middle Pleistocene age, and (2) conglomerates of the Villafranchian glacis. West of Huelva, there are three glacis levels but to the east only the oldest level is found, typified by a crust at its base and an Upper Pleistocene aeolian mantle. The basal sands and the glacis, except for the younger one, are cut by faults. The distribution of the red sands is asymmetrical in relation to the Guadalquivir valley, because of a Lower Pleistocene normal fault of submeridional direction, which is responsible for the development of the marshes in the downthrown block partly isolated from the sea by the Doñana spit [see Zazo and Goy (1981a,b)]. A normal fault is also responsible for the Bay of Cadiz; it was probably active during the Upper Miocene and certainly active during the Middle and Lower Pliocene, but movement along it had ceased by the Upper Pliocene. Zazo (1980) has cited archaeological evidence from the lower glacis that the basal red sands in the Cadiz area are fundamentally Quaternary. The Valdelagrana spit stretches along the border of this marshy depressed area.

Along the littoral, a complete set of low coastal features can be found, including a sequence of prograding beaches where shore face and foreshore environments are recognized in the five levels visible along the cliffs, ranging in age from 12 000 to 41 000 BP and in height from 8 to 0.5 m [see Zazo *et al.* (1981)]. Five systems of dunes of different types—parabolic, barchans, transverse—have been described and mapped, the most recent of which are still active and partially invading the marshes (*see* Fig. 13.14).

The present-day river mouths are good examples of estuaries in which the intensity of marine currents prevents the development of a delta. These mouths are frequently flooded and may change their location, as has been proved in the case of the Guadalquivir.

Chapter 14
The Appennines and Sicily

The Appennine chain forms the backbone of the Italian peninsula. It begins, however, in the continental part of Italy, in full continuity with the Alps from which it is difficult to separate exactly and without question. At the other extremity, the chain extends to Sicily—the largest Italian island—separated from the mainland by the Strait of Messina (3–5 km wide).

The Appennine chain is not a single unit, although there is an evident continuity of relief from Liguria to the southernmost point of Calabria. The geomorphological features, whose outlines are controlled by complex tectonic structures, are so different in the various sectors that it is better to speak of a series of ranges or blocks linked to each other. Furthermore, crossing the peninsula from the Tyrrhenian Sea on one side to the Adriatic and Ionian Seas on the other, several different longitudinal belts are clearly distinguishable [see Touring Club Italiano (1957)]. Therefore, each sector will be described in turn, limiting subdivision of the chain to the most obvious units.

It is convenient first to give a short account of the essential geological events, emphasizing three facts important for their morphological consequences. (1) There is great lithological variety, including rocks of very different behaviour in regard to weathering and erosion. (2) The age of the tectonics is recent; besides the main Miocene phase, there was also intense (neotectonic) activity in the late Pliocene–Pleistocene periods, to which the widespread and intense present-day seismicity testifies. (3) Vulcanism has continued through the Pleistocene to the present-day.

Fundamentally, neotectonics is responsible for the great amplitude of the relief (i.e. the large differences of level between summits and neighbouring valleys and plains). In places, it reaches an amount comparable to that in the Alps. In most part of the Appennines, altitudes reach 1000–2000 m, locally exceeding 2500 m, with a relief amplitude generally greater than 500 m, sometimes 1000 and even 1500 m. The Pleistocene uplift has raised Pliocene deposits up to 1000 m above the present sea level [see Ogniben (1975)].

The Hercynian orogenesis is clearly recognizable in some areas of Liguria and Tuscany, in Calabria and especially in Sardinia (see Chapter 9), but it exerts no important control on landforms except in Sardinia. An extended tectonic phase in the Mesozoic gave rise to a deep eugeosyncline in the area extending from Piedmont through Liguria and the western part of Tuscany, to the present Tyrrhenian Sea, in which were laid down marine deposits with flysch facies. To the east, in Tuscany and Umbria, there extended a wide, complex miogeosyncline, where very thick arenaceous, marly and calcareous sedimentation took place, while farther south-east (e.g., in Abruzzi, Latium, Campania) a calcareous and dolomitic sequence of neritic facies was deposited up to the Eocene or Miocene, the so-called 'carbonatic platform' of Italian geologists [see Desio (1973)].

A tectonic compressive phase followed, relating to the Alpine orogenesis, but in the Appennines this is generally younger than in the Alps. Although begun already in the Oligocene, it is chiefly of Miocene age, with important effects also in the Pliocene–Pleistocene. This diastrophism formed a series of folds trending north-west–south-east (i.e. the Appennine trend) or north–south; each fold is generally complicated and magnified, or simply cut by systems of faults, in some cases with overthrusts. Large displacements of huge masses of sediments also took place from west to east, supposedly of gravitational origin. Indeed, vast areas of the Appennines, especially on the Padan–Adriatic slope, consist of irregular sheets of allochthonous deposits, particularly the so-called *argille scagliose* (i.e. scaly clays), which enclose lithic chunks of various ages and sizes (Jurassic to Eocene limestones, silts, sandstones and even ophiolites), up to mountain-size blocks maintaining their own primary stratigraphical connection. Furthermore, according to the latest reconstructions, the whole tectonic structure may be characterized by a series of nappes, at least in the southern Appennines and Sicily.

In the Miocene, an ancient Appenninic chain emerged, forming a peninsula with outlines rather different from those of the present. In the middle and especially the late Miocene, a set of marginal basins was formed, in part of lacustrine or lagoonal character. A general subsidence occurred in the later Cenozoic era and carried the sea again to the interior, on both the Tyrrhenian and the Adriatic/Ionian sides. The sediments laid down in this Pliocene and early Pleistocene (Calabrian) sea are mostly terrigenous and little consolidated. Uplifted by the resumption of tectonic activity, they extend today as hilly belts on both sides of the Appennines.

Pleistocene neotectonics reactivated and raised the previous structures and caused the formation of new folds and faults, so that vigorous erosion followed rejuvenating the relief. Particularly in the interior of the chain, deep intermontane basins were formed, which were later filled in part by lacustrine and fluvial deposits in the early and middle Pleistocene, some of which contain the typical Villafranchian mammalian fauna. Meanwhile, intense volcanic activity took place in some regions, both inland and offshore.

In summary, this is how the essential present-day features of the Italian peninsula have developed, with a mountainous backbone trending generally in the form of an arc, with a concavity facing the Tyrrhenian Sea. Thus it is usual to distinguish an 'internal' slope to the west, and an 'external' one towards the Po plain and the Adriatic and Ionian Seas.

14.1   Northern Appennines

The northern Appennines form a gently curving arc with a general west-north-west–east-south-east trend, starting from the line Savona to Ceva, which marks a good geomorphological boundary with the Alps. Those hills in Piedmont that extend north into the Po plain and reach the Po at Turin belong to the Appennines proper. This area consists of hills and uplands, and is essentially built up from marine deposits, clays, marls, sandstones and conglomerates of Oligocene to late Pliocene age, which geologists have called the Piedmont Tertiary Basin. The geological structure and morphological aspects appear to be the same as those of the Appennines proper.

This area generally assumes the form of a wide basin, with a middle lower belt and two higher marginal belts— the Langhe and the Colline del Po—rising to 600–800 m. The middle zone, which approximates to the Monferrato region, is shaped as hills of Pliocene sands overlying the clays, with ridges uniformly rising to only 100–150 m above the floors of the innumerable small valleys, so that the ridges suggest the existence of an ancient plain of erosion. Relief in the lateral belts has greater amplitude and steeper slopes, owing to the more resistant marly arenaceous rocks of Miocene age and, in some places, hard Oligocene conglomerates. In the southern belt, the uniformity of inclined bedding causes the development, in places, of short cuestas but, in general, sharp ridges predominate because of the intense erosion on both slopes. In some stretches, the streams flow in entrenched mean-

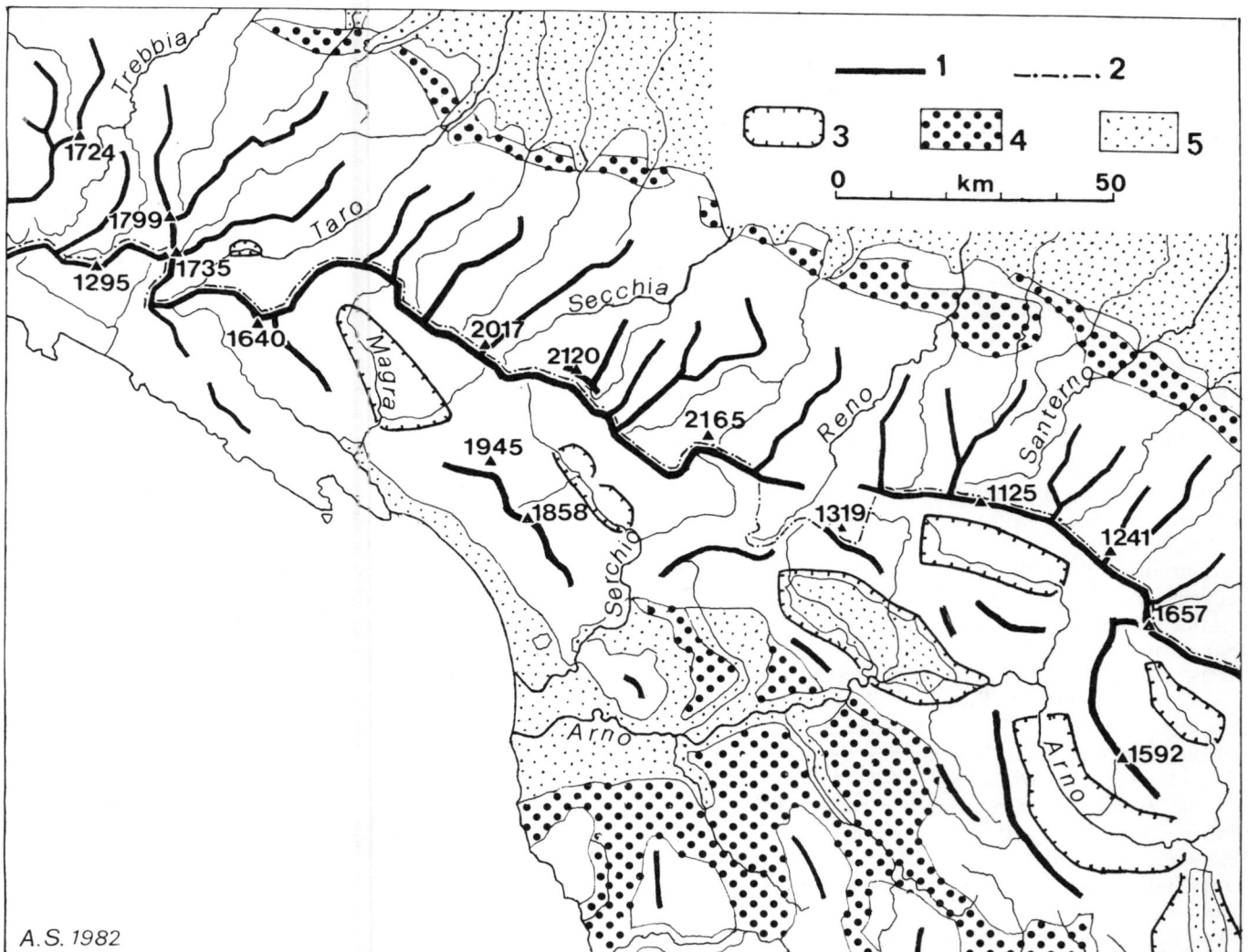

Fig. 14.1 Geomorphological map of the northern Appennines. (Note the asymmetry between the northern and southern slopes.) (1) Ridges, (2) main divide, (3) basins, partly filled with Villafranchian and Pleistocene deposits, (4) Pliocene marine sediments and (5) Holocene alluvial plains.

Fig. 14.2 Earthflow in the *argille scagliose* formation of the Appennines: the Panaro Valley, northern Appennines (Castiglioni).

ders and carry their waters to the Tanaro, an important tributary of the Po which was diverted to the middle belt in Quaternary times.

Between the Giovi Pass (472 m), only 15 km from the sea behind Genoa, and the upper valleys of the Tiber and Metauro, the Appennine chain displays a typical transverse asymmetry, the internal slope being relatively short and characterized by a set of longitudinal broad valleys or basins, elongated in the same general direction as the chain itself. On the other hand, the slope overlooking the Po plain has a more gradual descent, shaped into a series of transverse ridges arranged like a comb, perpendicular to the axis of the chain and its main divide. In consequence, it is only on the internal slope of the chain that the pattern of the relief is related to structure (*see* Fig. 14.1). This consists of several parallel anticline-like 'rises', each much wider and more complex than a common anticline. These were added, one after another, from west to east during the Tertiary orogenesis. Each 'rise', or tectonic ridge, results not simply from folding but from a system of faults, normal faults on the southern or western flanks, reversed on the other flanks, possibly converging at depth (i.e. *cunei composti* or composite wedges).

The structure of the autochthonous masses is complicated and often masked by the overlying sheets of allochthonous type, especially the argille scagliose, present almost everywhere and sometimes reaching the plains beyond the Padan slope. Moreover, local overthrusts, transverse faults and denudation of some ridges make the distribution of the outcrops appear like a very irregular mosaic.

The widespread presence of clays causes frequent *frane* or mass movements, chiefly in the form of earthflows (*see* Fig. 14.2) in the scaly clays [see Almagia (1907, 1910)]. Mass movements also frequently occur in other rocks (e.g., dipping beds of sandstone alternating with shales). In the marginal belt of Pliocene deposits near the Po plain, innumerable gullies of badlands type have developed on the clays (*see* Fig. 14.3); these are locally called *calanchi* and are also present in the argille scagliose [see Panizza (1979)].

The moulding of the relief is essentially fluvial, the erosion being vigorous due to the recent uplift (*see* Fig. 14.2), but it is clearly controlled by the various lithologies, so that selective erosion forms are well marked [see Gonsalvi and Papani (1969)]. The V-shaped valleys are generally narrow in the marly limestones, sandstones and shales of the Ligurian Appennines, locally with entrenched meanders. In contrast, they are broad with alluvial floors in Emilia, owing to the extent of clays and

Fig. 14.3 Badland landscape in the northern Appennines, near Bologna, developed in the *argille scagliose* (Castiglioni).

marls, but occasional defiles occur where the streams meet bars of harder rocks. The ridges are mostly rounded or flat-topped, which is probable evidence of ancient planation surfaces. To the east, in the Romagna region, the ridges are sharper and the valleys narrower, being carved out of a monotonous marly arenaceous formation of Miocene age. A long outcrop of late Miocene gypsum forms a typical longitudinal cuesta.

Glacial landforms are found on the northern slopes at the highest levels, wherever the summits rise above 1700 m. There are many cirques, some troughs and small moraines [see Losacco (1949)].

The post-Pliocene uplift seems to be the cause of the marked discordance between relief and drainage pattern in the Ligurian Appennines. The main divide is near the sea and rises to 1000 m or a little more, while the ridges of the Padan slope reach 1500–1800 m some distance from the watershed.

The intermontane basins on the Tyrrhenian side do not owe their origin to the main tectonic phase of Miocene age but to the resumption of displacements at the end of the Pliocene period. They are sunk deeply between parallel ranges, trending north-west–south-east or north–south, and are drained by the rivers Magra, Serchio, Arno (with three subsequent basins), the tributaries of the Arno—the

Sieve and Chiana—and the upper Tiber. All basins first passed through a lacustrine phase, followed by alluvial sedimentation of sands and gravel, mostly of Villafranchian age. Erosion has removed part of these deposits, but to varying degrees and in different ways, so that all basins display different morphological aspects [see Sestini (1940, 1950)].

The Apuane Alpi range, near the Tyrrhenian Sea, offers some distinctive landforms in its core of Mesozoic limestone, marble, dolomite and crystalline schist. The summits rise to 1500–1950 m with deep, narrow valleys divided by sharp crests. These mountains may be compared with the limestone mountain groups of the Pre-Alps. There were several glaciers in the late Pleistocene, but glacial evidence is not well-marked. Karstic forms are also rather rare, even though underground water circulation is well-developed and there are numerous caves.

South of the Arno, there are some differences in geology and morphology to the extent that previous geologists called this part of Tuscany *Antiappennino*. In fact, the geological formations and structures of the northern Appennines continue, with larger outcrops of Mesozoic limestones and Palaeozoic hard conglomerates and schists, owing to more advanced denudation. Moreover, the marine Pliocene deposits (clays, sands, gravels) assume a broad extension as the transgression reached

Fig. 14.4 Monte Cusna (2354 m) on the main divide of the northern Appennines. [Note the steeper west-facing slope (*see* Fig. 14.1).] Areas at this altitude supported small glaciers in the Pleistocene. (Embleton)

farther into the interior. The mountain pattern is rather confused, elevations are lower (rarely above 1000 m) and the landforms never rugged. The Pliocene deposits have been dissected by innumerable valleys, locally with the development of badlands (*see* Fig. 14.5 A, B). Monte Amiata is an exception, being a complex trachytic volcano that rises to 1738 m, above a sedimentary base at about 1000 m. Scattered remnants of planation surfaces also occur in this Tuscan area.

14.2 Central Appennines

The central Appennines stretch from the Metauro southwards to the Sangro and Volturno valleys. They differ remarkably from the northern Appennines, first because of the large extent of calcareous and dolomitic formations (of Triassic to early Miocene age) and secondly because of a more obvious and younger tectonic structure, with its decisive control on morphology. On the Adriatic side, a wide, uniformly hilly belt of marine Pliocene deposits borders the mountain land. These characteristics are common to the whole central Appennines, but between their north-western and south-eastern parts some remarkable differences occur, which allow an Umbria–Marche section and an Abruzzi section to be distinguished. Indeed, these represent two different geological units, each with its own sedimentary facies. They are narrowly in contact along an important tectonic line that trends nearly north–south. Along the southern part of this line, the Umbria–Marche sequence is thrust over the Abruzzi Appennines.

In the Umbria–Marche sector, where pelagic and marly facies prevail, real folds predominate (anticlines or brachyanticlines) in many cases truncated on one side by longitudinal faults. The direction of the geological struc-

tures, at first north-west–south-east, becomes north–south farther south. The Abruzzi sector is characterized by the predominance of limestone and dolomitic rocks of neritic facies and by tectonics of typical rigid style with intense faulting, expressed in the structure by alternating horsts and grabens and in the relief by huge blocks, mostly rectangular, elongated from north-west to south-east. The cluster of marginal faults is very young (post-Villafranchian and even occasionally post-Rissian). Some faults are easily observed as white bands of mylonite on the slopes of the mountains.

The height of the relief increases southward in the Umbrian Appennines from 1000–1200 m to 1500–2400 m (2457 m in the Monte Vettore in the imposing Sibillini massif, where glacial traces occur) [see Damiani (1975)]. The Abruzzi section bears the highest summits of the whole Italian peninsula, with many crests above 2000 m and a maximum in the Gran Sasso group at 2912 m.

In the Umbria–Marche region, the alignment of the relief in a parallel series of ridges becomes obvious [see Raffey (1980)]. These correspond to anticlines with calcareous cores, limestones of various types, homogeneous and compact, cherty, marly and thin-bedded (*scaglia*) of differing erodibility. From north to south, the anticlines become increasingly stripped of their cover of softer flysch sediments, especially the marly arenaceous formation of Lower to Middle Miocene age. In the northern and westernmost part of the region, some ranges are made entirely of these formations.

The limestone ridges (*see* Fig. 14.6) generally appear as broad swells, rounded or nearly flat-topped—probably remnants of old planation surfaces—with some karstic depressions, underground drainage being well developed. The flanks are commonly steep and continuous, while rugged forms, lined by rocky walls, are characteristic of

Fig. 14.5 (A) Cliff of sands (or sandstone) and (B) badlands developed in the underlying marine clays at Volterra (Tuscany). Both formations are of Pliocene age. The undermining of the cliff is due not only to the action of rain water, but also in part to springs. (Castiglioni)

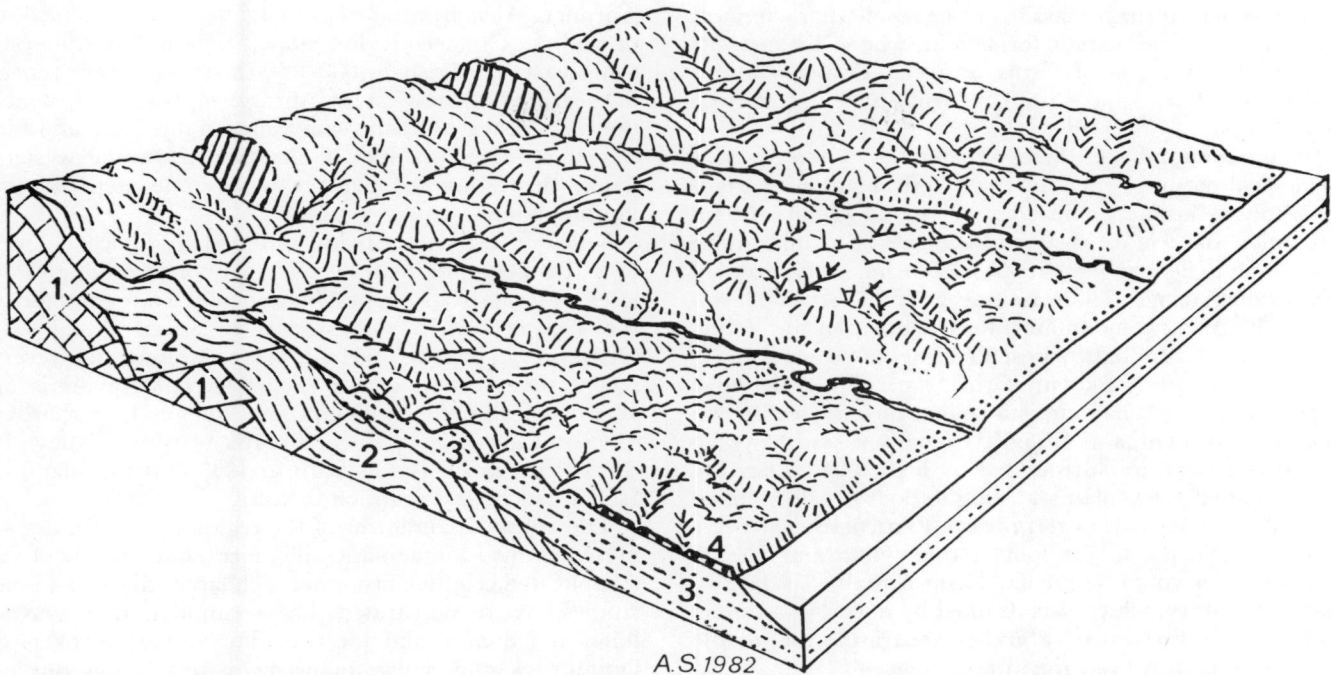

Fig. 14.6 Idealized block diagram of the central Appennines of the Marche, showing longitudinal zones from the interior ridges to the coast. (1) Ridges of Mesozoic limestone, (2) Miocene, chiefly molasse, (3) Pliocene marine sediments and (4) Pleistocene.

some valley stretches where transverse rivers cross the calcareous cores of the anticlines. The gorges of Furlo and Antrodoco are well-known features of this area.

As in Tuscany, there is a marked asymmetry between the internal and external slopes of the chain. In Umbria, the basins are more elongated, but they had a geological history in Plio–Pleistocene times similar to that of Tuscany. The lacustrine and fluvial deposits are commonly preserved at the margins and are covered by an alluvial floor in the centre (e.g., the basins along the Tiber valleys, and the basins of Gubbio, Gualdo Tadino and Norcia). On the Adriatic slope, on the other hand, the rivers follow the tectonic lines only in their upper courses; they then cut through the anticlines and continue on transverse courses as far as the coast, more or less parallel with each other, as in Emilia-Romagna. Before they enter the Pliocene belt, they cross some wide molassic and marly synclines, moulded into hills. Between these, small plains and terraces occur, marking old levels of erosion or alluviation [see Carloni *et al.* (1975)].

The mountainous sector of the Abruzzi is higher, more complex and irregular, owing to its numerous massifs and ranges composed of reef limestones and dolomites of middle Triassic to early Miocene age. The rugged mountains rise from a general base at 500–700 m, so that some speak of an Abruzzi plateau but this name is not strictly appropriate. The amplitude of the relief may exceed 1500 m. Broadly, massifs and ridges are arranged in three longitudinal series, trending generally north-west–south-east. The external series, towards the Adriatic, is made of the Monti della Laga (2455 m), the Gran Sasso group and the Maiella massif (2795 m). The middle zone is a wide

and complex assemblage of massifs and ridges, culminating in the compact Velino massif (2487 m). The third belt is smaller, but again rises to 2000 m in its central sector (i.e. Monti Ernici).

The main divide of the peninsula runs along the ridges of the middle zone, the major rivers rising behind the external massifs. The Tronto rising behind the Monti della Laga, and the Aterno (farther down named the Pescara) and its tributary, the Sagittario, run longitudinally within the basins of L'Aquila and Sulmona and carry their waters through the Popoli defile to the Adriatic, crossing the molassic and Pliocene belts. Here again the discordance between relief and drainage pattern is well displayed. Among the three hypotheses formulated in the past to explain this discordance [see Marinelli (1926)], the most acceptable is antecedence, predating the great Pleistocene uplift.

Each massif has its own special features, but the Laga differs in particular, because it is carved out of a very thick molasse formation of late Miocene age, and is deeply incised by wild, steep valleys. The Gran Sasso—the largest and highest massif—is calcareous and dolomitic, with a long rocky crest notched by cirques, and some summits resembling those of the Alpine Dolomites [see Gentileschi (1967)]. Northwards, it falls abruptly to a landscape of flat-topped molassic mountains, while southwards it breaks down into a series of limestone plateaus and karstic basins of various shapes and sizes, the largest basin being that of Campo Imperatore whose floor is partially interrupted by fans and moraines. Lastly, the Maiella is a huge dome, cut by a set of faults on its western flank, while the eastern flank is incised by some wild ravines and notched by cirques.

The other Abruzzi massifs are less well characterized, but all show clear karstic forms and, where the summits exceed 2000 m, glacial forms as well. Two glaciations (Riss and Würm) have been recognized in the Abruzzi mountains.

In the mountains of the external series, whatever their geological nature, beneath a hypothetical summit surface (possibly Miocene) numerous remnants of an ancient planation surface are clear and widespread. This surface according to Demangeot (1965) is of pediment type and of Villafranchian age. There are traces of other erosion levels in all the Abruzzi mountains: for example, on the Velino massif, much uplifted by post-Pliocene movements.

The abundance of karstic forms, especially of the polje type, is a marked characteristic of the Abruzzi [see Belloni *et al.* (1978), Lehmann (1959)]. In many cases, the poljes of various sizes are surrounded by high relief. They are locally called *piani* (plains), owing to their flat floors filled with alluvial deposits or terra rossa. Particularly extensive is that of Fucino at the foot of the Velino and Sirente massifs. This roughly circular basin contained a lake in the last century, which was drained by means of a tunnel to the upper Liri valley, a graben separating the middle belt of mountains from the internal one.

The longitudinal basins of L'Aquila and Sulmona represent a different type, with a geomorphological history similar to that of the Tuscan–Umbrian basins, and bearing remnants of Pleistocene deposits, mainly conglomerates. Where they leave the basins, the streams tumbling down the limestone or dolomitic outcrops have cut deep gorges: for instance those of Celano (Fucino) and Sagittario.

The Adriatic side of the external high massifs is bordered by a molasse zone of moderate relief. Next lies the Pliocene belt of clays, sands and conglomerates. The upper layers are of early Quaternary age (Calabrian). This belt (normally 20–30 km wide but even wider in northern Marche) is dissected by numerous transverse, more or less parallel, broad valleys between the gentle slopes of the hills. They contain alluvial terraces and have flat floors, while the interfluves show monotonous, undulating profiles or are in part flat-topped, so they appear to continue the Villafranchian planation surface already mentioned. Where, however, the relief is a little high (300–400 m) and the clays attain a greater thickness badlands occur, such as those of Atri. The hills reach the coast in some places, but they are often separated from the sea by a narrow strip of sandy or gravelly plain.

14.3   Latium–Campanian Pre-Appennines

Before describing the southern Appennines, it is necessary to give an account of the belt of heterogeneous relief that stretches parallel to the Tyrrhenian Sea from southern Tuscany to the Gulf of Naples. This belt has some distinctive geomorphological features because of the great extent of two groups of volcanic structures: a wider and more continuous one in Latium and another, more fragmented, in Campania. Between the two groups, however, is a complex alignment of mountains (Lepini,

Aurunci, Ausoni) that represents rigid fault blocks of prevailing Cretaceous limestone, trending north-west–south-east. They rise to 1000–1500 m and generally repeat the features of the western Abruzzi massifs, from which they are separated by the wide valley of the Sacco and Liri rivers, bordered by high fault scarps. On the western flank, the mountain plunges with some steps to the Pontine plain.

Volcanic activity, already initiated in Pliocene times (e.g., La Tolfa mountains) chiefly with acid products and later with more basic types, became much more intense in the Pleistocene. In some areas, it ceased in the late Quaternary, but in others it has continued until very recent times or up to the present-day. Vesuvius is an active volcano, the Solfatara of Pozzuoli is perhaps a quiescent one, the small cone crater of Monte Nuovo in the Phlegrean fields was born in 1538 and the island of Ischia had its last eruption in 1301.

The volcanic landforms of the region are varied not so much because of magmatic differences, but because of the complicated eruptive history of each large edifice as recent studies have demonstrated. The accumulation of lavas as flows and domes, and the spreading of vast blankets of ignimbrites and tuffs, followed by demolition due to explosion or collapse, have led to the total or partial superimposition of several volcanic complexes. Finally, fluvial erosion has contributed to the present-day landscape by carving innumerable furrows that, nevertheless, do not fundamentally alter the general features of the volcanic relief [see Sestini (1969)].

The large volcanic structures of Latium appear on the whole as truncated, rather flat cones, their moderate height contrasting with the large diameter (up to 30 km) of their subcircular bases. The tops of the cones are sometimes marked by real craters, and in others by vast, complex calderas, originating by collapse. The calderas of Bolsena, Vico and Bracciano are now occupied by deep lakes. The Vulcano Laziale (Colli Albani, south-east of Rome) bears a second rim within the more distant walls of an older caldera, and some cones and domes rise in the centre. The western flank is partially breached by explosions, which created the crater lakes of Albano, Nemi and Ariccia. Cones and domes occur elsewhere in Latium, even forming the highest relief [e.g., the Monti Cimini (1051 m) a complex close to the outer wall of the Vico caldera].

Ignimbrite extrusions, outbursts of ash and mud flows have spread out a cover of tuff and formed low plateaus gently sloping in various directions. These are now dissected by numerous small valleys, erosion in some places reaching the underlying Pliocene clays and causing the formation of platforms separated by unstable cliffs, such as Orvieto hill.

In Campania, the Roccamonfina volcano repeats the general features of the Vulcano Laziale: a caldera with interior domes, small parasitic cones and dissected plateaus. Vesuvius—the highest of all, rising to 1277 m— has a more distinct cone shape but is itself accompanied by a partial rim, Monte Somma, the remnant of a caldera that was perhaps breached by the famous eruption of

79 AD which destroyed Pompeii. The hilly district of the Phlegrean fields is an assemblage of cones and craters chiefly made of tuffs. Ischia is a faulted island volcano. Finally, the flat and low tuff plains north of Naples add another distinctive element to the landscape.

14.4 Southern Appennines

Starting from a line from Sangro to Volturno, the southern Appennines stretch southward to the Strait of Messina. The Calabrian Appennines, which form the slender peninsula of Calabria, the southernmost region of peninsular Italy, will be considered separately. This separation is justified by the geological structure, owing to the appearance of a pre-Mesozoic crystalline massif and certain other morphological differences.

The southern Appennines proper end at the vast limestone block of Monte Pollino, which overlooks the plain of the Crati river to the south. This part of the Appennines can be divided into two sections, whose traditional names (Campanian or Neapolitan Appennines, Sannitic Appennines in part and then, farther south, the Lucanian Appennines) are not well defined. This partition, however, is of little significance; more significant is a division into three longitudinal belts, from the internal side facing the Tyrrhenian Sea to the exterior side facing east. These belts correspond to differences in structure, rock type and geomorphology, but the transitions from one belt to another are not simple and each belt is heterogeneous.

According to the recent opinions of many geologists, the structure of the whole southern Appennines, which is controlled by compressive tectonics extending from the Miocene to the Middle Pliocene, results from several stratigraphic units thrust eastwards over one another. Later, strong faulting, uplift or subsidence, continuing

Fig. 14.7 Some elements of the morphology of the southern Appennines. (1) Faulted blocks of Mesozoic limestone with karst phenomena. On the extreme left, monoclinal mountains emerging from the flysch. (2) Uplands of flysch and *argille scagliose*.

until recent times, has been active, recalling the intense seismicity of the whole region.

The first longitudinal zone consists of a series of Triassic–Cretaceous calcareous and dolomitic blocks (*see* Fig. 14.7) These blocks represent the continuation of the Abruzzi massifs related to the Appennine miogeosyncline but are more segmented and more widely separated by depressions filled with arenaceous and marly flysch. Flysch also constitutes most of the Cilento district (a stocky protrusion into the Tyrrhenian Sea). The blocks are horsts controlled by a rigid tectonic style of 'composite wedge' type.

The middle belt is mainly a wide trough where there is a heterogeneous complex of structures, mainly of flysch type, clayey, siliceous, arenaceous or of argille scagliose (*see* Fig. 14.7), the whole being intensely tectonized and sometimes quite chaotic. As in the northern Appennines, the scaly clays include exotic rocks.

The third belt is a Plio-Pleistocene foredeep, called the Fossa Bradanica, filled with marine Pliocene and Calabrian deposits, generally undeformed or with bedding only slightly tilted. This zone continues that already described along the Adriatic coast, but it does not reach the sea everywhere, being partially barred by the Apulian tableland.

Pleistocene movements are largely responsible for the dislocation and altitudes of the calcareous dolomitic massifs. The summits rise usually to 1300–1800 m, with only three massifs exceeding 2000 m—the Matese (2050 m) to the north, Monte Pollino (2248 m) to the south and Monte Sirino (2005 m) in the Lucanian Appennines—all with some glacial forms. The hard, karstic rocks give a rugged landscape, every massif rising abruptly and overlooking a landscape of moderate relief, and in some Campanian areas immediately bordering the plain or the sea. Precipitous rocky scarps—chiefly fault scarps—as well as uneven karstic plateaus are common. The Alburni plateau (1400–1700 m) which is densely dotted with sink holes, is a typical example. Other karst features include several poljes: for example, the elongated basin in the Matese, which is partially occupied by a lake. Underground drainage is well developed and gives rise to powerful springs. In the Lucanian Appennines, there are also some deep basins with normal drainage analogous to those of Tuscany and Umbria, but generally of younger age. Remnants of lacustrine and alluvial deposits are preserved in them.

The main watershed of the peninsula only partially corresponds to the high massifs. In Campania, some of these are located entirely on the Tyrrhenian slope, and again discordance between relief and drainage pattern is apparent. Here, however, the position is reversed, as the divide runs more to the east, on the lower ridges of the second belt.

The middle zone of the southern Appennines is very wide, especially in its northernmost part (in the province of Molise). Monotonous, endless undulating ridges carved out of unresistant flysch or argille scagliose form a moderate relief, even if in some places this surpasses 1500 m. A long stretch around the tributaries of the

Volturno does not even reach 1000 m and some ridges are capped by marine Pliocene conglomerates. Rounded or flat-topped interfluves (probably remnants of old planation surfaces) delimit open valleys where muddy or gravelly streams run. In spite of the gentle slopes, landslides are very common, owing to the widespread clays, and may be as calamitous as in the northern Appennines. A distinctive element of the middle zone is the presence, on its eastern border, of an extinct volcano, Monte Vulture, being the only one on the Adriatic slope of the peninsula.

The eastern belt consists of marine deposits, of Pliocene and early Quaternary (Calabrian) age, clays, sands, conglomerates, but also soft limestone (*tufo calcareo*). The ground rises to a maximum of about 1000 m and is dissected by broad valleys containing terraces at various heights, in part remnants of old erosion levels and in part alluvial. The rounded or flat-topped ridges gradually decline towards the north-east in the Molise region and in northern Apulia, and to the south-east in Lucania, where the rivers flow into the Gulf of Taranto. Also in this zone landslides, mud flows and gullies are frequent, and the lower clayey slopes are in some places moulded into characteristic small domes. Near the Gulf of Taranto, the lowest hills pass into a set of marine terraces of Pleistocene age that drop in steps to the coast, which is bordered by a strip of dunes [see Boenzi *et al.* (1978)].

14.5   Apulian Plateaus

The elongated region in Apulia that extends along the Adriatic coast from the Fortore to the extreme south-east of the Italian peninsula differs remarkably from the mountains and hills of the Appennines. It is a stratigraphic and tectonic unit, separated from the chain by the Fossa Bradanica, and is considered to belong to the foreland. It is not, however, a foreland in the full geological sense because in Mesozoic and Eocene times its geological evolution was similar to that of the Abruzzi and Campanian 'carbonatic platform'. In Apulia, the sequence of limestones and dolomites mostly belongs to the Cretaceous neritic facies.

The landforms match the lithological monotony and the simple tectonic style of very gentle folds with numerous faults, the stratification being commonly near horizontal. The general trend is north-west–south-east and transverse to this. Monte Gargano, a rectangular horst 600–1000 m high, which protrudes into the Adriatic Sea as a stubby promontory, should, however, be distinguished. Here older Mesozoic formations and Eocene nummulitic limestones are also present. The higher part is a rugged karstic plateau, with a few poljes and some areas riddled with sink holes.

The Gargano is separated from the Murge plateau by a wide depression—once a Pleistocene gulf—filled with marine and continental deposits, whose terraces and floor constitute the present plain of Tavoliere. The Murgian plateau is a low tableland (400–600 m high at its centre) to which the name Murge properly pertains. It is gently undulating, but broken longitudinally by two or three abrupt scarps overlooking lower platforms. In the past, these platforms have been interpreted as terraces bevelled by marine erosion, following phases of uplift [see Marinelli (1948)]. Some patches of soft Quaternary limestone (*tufo calcareo*) lie on the platforms, but this is not sufficient to support that opinion. It seems more likely that they represent an old planation surface (one or more) dismembered by faults. The scarps would then be fault scarps, slightly retouched here and there by fluvial erosion. Likewise, the lower plateaus have been incised by a few scattered small furrows (*lame*). Only on the slope towards the Gulf of Taranto have some streams become deeply entrenched in the limestone, forming ravines called *gravine*.

The low efficiency of normal erosion is due to the karstic nature of the whole tableland, being devoid nearly everywhere of surface drainage while underground circulation is well developed. However, external karstic forms, apart from some elongated shallow basins on the highest parts and some scattered sink holes, are rather rare. The Apulian plateaus thus represent a particular type of karst, which has been thought by some to be very old but those present aspects, in fact, are probably not older than early Pleistocene [see Neboit (1975)].

The southernmost part of the tableland—the Salento region or Salentine peninsula, between the Adriatic Sea and the Gulf of Taranto—is an even lower plateau at 50–200 m, but nevertheless marked by cliffs along the Adriatic coast that are higher than elsewhere in Apulia, except the Gargano. A few coastal inlets represent drowned valleys. The Salento platform is covered, in some areas, by layers of Eocene–Miocene deposits (a typical soft limestone) and also by Pliocene sediments. Karst forms are entirely lacking, whereas bare rock outcrops are very common.

14.6 Calabrian Appennines

South of the Pollino massif, the Appennine chain narrows and changes its geological structure and aspect. Essentially, it consists of horsts of crystalline rocks: granites, granodiorites, gneisses and more or less metamorphosed schists up to micaschist and phyllite. These fault blocks have been recently uplifted to different heights; a couple of summits approach 2000 m.

In general, the Calabrian Appennines are an ancient, Palaeozoic crystalline massif, involved in the Hercynian orogeny, when the great granitic intrusion and the related metamorphism took place. The middle to late Carboniferous folding was followed by an erosional phase which led to the peneplanation of the Hercynian mountains. Later, geological events have not been identical everywhere, and partial transgressions left a few marginal patches of Mesozoic limestone. After the dislocation of the ancient massif into various blocks, their peripheral slopes were, in part, covered by Palaeogene and Miocene sediments with flysch facies. These are, at least in part, allochthonous. According to recent interpretations, the entire Calabrian Appennines would result from a series of thrust sheets like the southern Appennines, including the crystalline horsts.

The marine Pliocene is certainly autochthonous. On the Ionian side, it extends as a low hilly belt at the foot of the higher ground, and penetrates up the Crati valley, a graben between the Sila highland and a range along the Tyrrhenian coast named Catena Costiera [see Tomas (1966)]. On the western side, the Pliocene deposits are reduced and scattered, but they pass from one coast to the other in the broad depression of the Catanzaro isthmus between the Gulfs of Sant' Eufemia and Squillace.

Schematically, it is possible to distinguish four main crystalline blocks and one smaller one: the Catena Costiera—narrow but rather high (1000–1800 m)—the Sila plateau, the Serre massif and the Aspromonte group.

The Sila plateau is a broad, massive rectangular block, mostly granitic, with a height of 1000–1400 m. Partly bevelled by a former planation surface, it also bears some rounded ridges (up to 1929 m), where perhaps glacial traces occur, rising above the rolling plateau. The streams run sluggishly across the plateau, plunging at the borders into steep-sided entrenched valleys. These valleys open again where they cross the coastal belt of Tertiary sediments.

South of the Catanzaro isthmus, the Serre massif is even more truncated by a summit surface at 1000–1400 m above sea level, and a set of terraces *en échelon* on the flanks. A similar series of terraces is especially characteristic of the Aspromonte, the mountain group that forms the southern extremity of the Italian peninsula. Such terraces (*see* Fig. 14.8), called *pianalti* in the geomorphological literature, rise step by step up to 1400 m. Arranged around the rugged mountain core, they are dissected by steeply graded streams plunging in all directions. The origin of this gigantic staircase is not yet definitively understood. In the past, it was generally thought that the terraces were cut by marine erosion, as the massif was intermittently uplifted during the Pleistocene (post-Calabrian) times. Others [e.g., Lembke (1931)] have claimed that the pianalti are remnants of one or more planation surfaces, faulted and uplifted to different heights. The intensity of recent uplift is not in doubt, as vigorous neotectonics characterize the whole of Calabria, a region of widespread and strong seismicity. There are also, however, true marine terraces at lower levels, covered with middle and late Pleistocene marine deposits (Sicilian and younger) [see Brückner (1980)].

Another feature of the Calabrian Appennines is the *fiumare*, which are short and violent streams that descend from the mountains, carrying huge amounts of debris in their wide, braided courses during sudden floods. This seems to be in contrast with the crystalline nature of the rocks, such as granite and gneiss, but these have been deeply weathered and the bedrock is often covered by a blanket of incoherent detritus (even if cemented again as arkose) unstable and easily erodible.

14.7 Sicily

This major Italian island shows heterogeneous relief, but appears to be a continuation of the Appennines whose arc assumes here an east–west trend. According to a very

Fig. 14.8 Early Pleistocene marine level at about 400 m, Calabria, dissected by younger valleys. (Embleton)

simplified scheme, the island may be divided into three zones. Part of the northern zone—a narrow belt—is like a chain, even if varied and divided into three distinct sectors. It bounds the deep basin of the Tyrrhenian Sea.

In the north-east corner of the island, the Calabrian crystalline massif continues in the short granitic and gneissic range of the Monti Peloritani (1000–1300 m). Its flanks are sculptured by the deep gorges of the *fiumare*. A higher range follows to the west overlooking the Tyrrhenian Sea, with rounded ridges and moderate slopes due to the predominance of flysch and scaly clays. Lastly, the distinctive calcareous dolomitic massif of the Madonie rises to nearly 2000 m, its highest part being a karstic plateau dotted with sink holes. Westwards, other limestone mountains emerge in striking contrast to the surrounding landscape carved out of less-resistant rocks.

Such a contrast is also found inland in the western part of the middle zone, where several isolated calcareous dolomitic uplands emerge, with small massifs and short ranges. Some of them rise to 1400–1600 m, overlooking undulating ridges or small plateaus of flysch, Miocene molasse and marls. The sharp outlines of these features are not only due to selective erosion, but also to faulting which has raised the limestone features to various altitudes.

Among the Miocene sediments, the youngest Miocene (Messinian) gypsum–sulphur sequence is of particular interest. It is widespread in the central sector of the middle belt and towards the south-west coast. Salt-bearing clays predominate below, while gypsum develops at the top and gives particular features to the local landscape. In some places, it forms sharp ridges and rocky walls emerging from the clays; in others, it has produced rolling plateaus with karstic depressions (sink holes and small blind valleys). The remainder is a tangled assemblage of small plateaus, low mountains and hills with monotonous ridges, crossed by broad valleys with gentle

flanks, but nearly everywhere affected by mass movements.

The Pliocene deposits display similar features. They extend farther and farther eastwards and rise higher, reaching up to 1000 m at Enna in the centre of the island. The Pliocene sequence usually extends to the Calabrian and locally there are also younger sediments. Where the sequence is crowned by conglomerates or shelly limestone, the summits assume tabular forms, surrounded by steep slopes above the valleys which descend in part to the Ionian Sea and in part towards the central Mediterranean.

The relief of south-eastern Sicily shows some distinctive features. On the whole, it consists of a large flattened dome, as demonstrated by the radial drainage pattern. Although it is called Monti Iblei, it is really a tableland, with an average elevation of between 300 and 600 m, rising to nearly 1000 m in the centre, while at the periphery it generally ends with a low step. To the north-east, the plateau is built of flood basalts which erupted when the area was a shallow marine shelf in Miocene times. But the majority of the tableland is made of middle to late Miocene limestone and marls, and of Pliocene soft limestone, with subhorizontal bedding.

The streams have, in places, dissected the tableland, cutting narrow, steep-sided valleys. These possess gorge-like features and are locally called *cave* (no reference to the English meaning of the word) such as that of Ipsica famous for its prehistoric and ancient necropolis.

Mount Etna, a gigantic active volcano, is a morphological region in itself. It is 30–40 km in diameter, with an altitude of 3263 m. On the whole, this volcano appears simple, with a cone that becomes increasingly steep towards the summit, where a crater of moderate size opens. However, with regard to its origin and to the morphological details, Mount Etna is a very complex structure. In Pleistocene times, several volcanoes overlapped, formed by the alternation of lavas and tuffs. A particular characteristic of Etna is the great dent that cuts the eastern flank, known as the Valle del Bove, which is a remnant of an ancient caldera. In addition, about 250 parasitic cones and craters are scattered on the slopes, in some cases arranged in rows. Many of them date from historical times, as do the fresher lava flows extruded from eccentric vents (*see* Fig. 14.9).

Nearly all the small islands around Sicily, including Pantelleria, are of volcanic origin. The Eolie (or Lipari) island group possesses two active volcanoes: Stromboli and Vulcano. Some of the islands are simply a cone or two, others are much more complex. The cones are in part submarine.

Mount Etna is sometimes shaken by local earthquakes, but the whole of eastern Sicily is a region of very high seismicity, with no relation to the volcanic activity, as shown by the catastrophe of Messina in 1908 [see Trevisan (1955)].

14.8 Plains and Coasts

In both the Italian peninsula and Sicily, plains occupy only a relatively small area, the largest being situated in a marginal position near the coasts. The more extensive are the plains of the Tavoliere in Apulia, of the Campanian region and of Catánia in Sicily. They are only in part flat alluvial surfaces. In contrast, the plain of Campania is partially a volcanic ash plain. They are barred towards the sea by chains of low dunes, which form an obstacle to normal drainage, so that reclamation works have been necessary, resulting in the drying-out of some of the coastal lagoons and inland lakes.

Alluvial plains are numerous on the Tyrrhenian side of the peninsula but are never large and are always confined by the hills and mountains. The largest are those of the Tuscan Maremma around Grosseto and the Pontine plain. The remaining plains are generally no more than strips along the coasts, bordered by a simple beach or by dunes whose height rarely exceeds 20–25 m. Smaller alluvial plains are found in the interior, chiefly in the intermontane basins and in the major poljes.

The coasts show a variety of types. There are notable differences between the Adriatic and the Tyrrhenian and Ionian coasts. The former, from the Po plain to the Ofanto River, is a smooth coast with long stretches of straight beaches. In some places, the hills reach the sea forming headlands with cliffs, but these are not prominent features except for the Gargano promontory. The Apulian limestone coast, even though abrupt, is on the whole straight and low, with a few exceptions in the Salento.

The Tyrrhenian coasts, including those of northern Sicily, have generally more characteristic outlines. They present alternating headlands and sandy beaches, the latter forming crescentic or nearly straight lines. These features are found both in the large gulfs and the small embayments. In some cases, the coastal arc is double because of the protruding apices of the deltas of the major rivers. These deltas are all of simple type, being triangular with a wide base inland. They have been built in the last few millennia, several dune ridges (mostly of historical age) marking the advance of the land. Similar features are found on the Ionian side of Calabria and Lucania. This type of coast represents an intermediate stage of shoreline evolution. A transgression, probably caused both by eustasy and subsidence, has taken place in postglacial times, giving rise to large or small embayments, not yet filled by alluvial deposits, while wave erosion of the headlands has been limited.

The Ligurian coast east of Genoa is mostly rocky, but

Fig. 14.9 Part of Mount Etna, Sicily. In the foreground, the road was temporarily blocked by a 1971 lava flow; a little behind are the 1979 lavas. (Embleton)

there are short beaches in several small embayments. Only the Gulf of La Spezia is larger; its indented outline clearly shows the effects of recent submergence.

14.9   Geomorphological Processes and the Age of the Landforms

In addition to the control exerted by the tectonic framework and the distribution of various rock types on the landforms, there is no doubt that endogenic activity has strongly controlled the morphology of the peninsula and the islands, both by neotectonics and by volcanism in some regions. The latter has created some complex volcanic features and tablelands of lavas or tuffs.

The compressive tectonics of the Alpine orogenesis began in the Oligocene in the northern Appennines, but developed over the whole chain mainly in the Miocene, continuing into the early Pliocene. These were responsible for the primary features of the chain, but the major control on present-day morphology derives from the later displacements. The age of these is not exactly the same throughout the Appennines, Sicily included. It should be noticed that the marine Pliocene and Calabrian deposits have been uplifted, both in the north and the south, by up to 1000 m above the present sea level, but the amount of uplift differs greatly from place to place. This has been caused not just by simple epeirogenic movements, but by gentle folding and faulting.

Several neotectonic investigations are now underway in Italy. Their results may show whether the present distribution of altitudes and the discordances between relief and drainage patterns really depend on these late displacements. It is clear that all these recent tectonic events have caused intense fluvial erosion. The Mediterranean climate, together with its Quaternary variations of temperature and rainfall, is favourable to erosion, with alternating dry seasons and longer humid, rainy periods that are especially characterized by the violence of the showers.

Thus, landforms derived from normal erosion predominate and are essentially youthful: the slopes were mostly moulded during the Pleistocene. As a consequence, old planation surfaces, including those of early Quaternary age, are generally poorly preserved, even though they are frequently found. They are reduced to small patches on the ridges or simply represented by rounded summits. The oldest of them are perhaps of late Miocene age, but most appear to date from Pliocene or Villafranchian times.

Traces of high terraces occur in many valleys at various levels (even five or six), but their relationships are not usually clear. On the other hand, lower terraces, up to a few tens of metres above the present valley floors, are in many cases well preserved. Their age seems to be not older than middle Pleistocene and to extend to very recent times. The origin of these terraces is essentially due to Pleistocene uplifts, but climatic oscillations have also influenced the processes of erosion and deposition. Indeed, periods of particularly intense alluviation have been recognized, chiefly in southern Italy.

A not insignificant role in the shaping of the slopes must be attributed to weathering and to mass movements, such as gelifluction and, especially, landsliding. Signs of both old and recent landslides are frequent, mainly where slopes have been carved out of the widespread clays. Badlands are also a common feature, especially in the areas of Pliocene sediments.

Karst is widespread in the Italian peninsula [see Belloni *et al.* (1978)], especially in the central and southern Appennines, owing to the large extent of calcareous formations. Some areas of limestone or gypsum are dotted with sink holes, but basins of polje type controlled by local tectonics are a more characteristic feature. A particular type of karst is that of the Apulian plateau, where surface karst forms are less evident or concentrated in some places only.

Glacial features occur on the northern slopes nearly everywhere that the summits rise above 1600–1700 m in the north and 1800–2000 m in the south. They include cirques, but typical glacial troughs are not numerous and moraines are of moderate or small size. The latter generally relate to the last glaciation—the Alpine Würm—when the snowline was lowered to 1600–1900 m. Traces of the Riss glaciation have been recently recognized. Periglacial forms of different types also occur, chiefly on the limestones, but never at low levels [see Kelletat (1969)].

Marine terraces are found mostly in southern Italy, close to the coast. If it cannot be accepted that the Apulian platforms and the pianalti are terraces flattened by marine erosion, only limited surfaces remain that are really wave-cut benches. Pleistocene marine accumulation terraces distributed at various heights mark some littoral belts. Marine action has also contributed to the formation of the deltas, built mainly in protohistorical and historical times.

Aeolian landforms are limited to the coasts, in the form of dunes of moderate height, often only a few metres high and rarely exceeding 25 m. Some fossilized dunes have also been recognized.

Chapter 15
Carpathian Mountains

The Carpathian Mountains are an epigeosynclinal mountain chain forming the eastern continuation of the Alps. From the Danube Gap near Bratislava (Czechoslovakia) they swing in a wide arc (some 1450 km long) to near the town of Orşova (Romania) in that part of the Danube valley known as the Iron Gate. To the north, north-west, north-east and south, the morphostructures of the Carpathians are bordered by the sub-Carpathian structural depression (*see* Fig. 15.1) which separates the mountains from other morphostructural elements of Europe, especially from the platform areas such as the Bohemian massif (*see* Chapter 9) and the east European platform. Within the arc formed by the Carpathians lies the Pannonian Basin, an area of subsidence consisting of the Little and the Great Central Danubian (Hungarian) Lowland Plains. Separating these two depressions are the Transdanubian block mountains of relatively moderate relief.

Although morphostructurally a counterpart of the Alps, the Carpathians differ considerably from them. Their appearance is less compact and they are divided into a number of mountain blocks separated by basins. The highest peaks—Mount Gerlachovský (2663 m) in the Slovakian western Carpathians or High Tetra and Mont Blanc in the Alps (4807 m)—differ greatly in altitude and overall the elevation of the Carpathian mountain blocks is much lower than that of the Alps. There are further differences in the individual morphostructural elements. The sandstone/shale outcrop known as the flysch zone that flanks the northern margin of the Alps is a narrow strip, but, in the Carpathians, this zone is considerably wider, forming the main component of their central zone. The limestone and dolomitic rocks that form a wide band in the Alps are of secondary importance in the Carpathians. Furthermore, the crystalline and metamorphic rocks, which represent powerfully developed chains in the central part of the Alps, appear in the Carpathians as isolated, smaller blocks surrounded by basins. Another difference is that the Carpathians contain a rugged chain of volcanic uplands.

Differences can also be observed in the relief of these two epigeosynclinal mountain chains, especially in terms of denudational history. The present-day relief of the Alps has been strongly influenced by Quaternary glaciation, affecting most mountain valleys and giving them a distinctive appearance but, in the Carpathians, glaciation has affected only the highest parts and fluvial types of relief are dominant.

Geomorphologically, the Carpathians are subdivided into western, eastern and southern mountain units, the Pannonian Basin, the Transylvanian Alps, and the lower Danube lowland.

15.1 Morphostructure

The Carpathians extend in a system of parallel morphostructural ranges. The outer Carpathians, whose rocks are composed of flysch, run from north-eastern Austria through the central part of Czechoslovakia, along the Polish–Czechoslovak border and through the western Ukraine into Romania, ending in an abrupt bend north of Bucharest. The original morphostructure consisted of nappes belonging to the epigeosynclinal mountains, but today block morphostructures are more important.

The inner Carpathians consist of several separate morphostructures. In the west lies the central Slovakian block, while in the south-east lie the eastern and southern Carpathian blocks. The isolated Bihor block occupies the centre of the Carpathian arc. Among the rock complexes comprising these blocks are ancient crystalline and metamorphic cores over which younger sediments, for the most part Mesozoic limestones and dolomites, have been thrust. The third and innermost range is built of Neogene volcanic rocks, differing in extent in the western and eastern Carpathians. In the former, they form an arc enclosing the central Slovakian block to the south and east; in the latter, they run in an almost straight line from north-west to south-east, following a deep-seated fault zone parallel with this part of the mountains. Between this volcanic zone and the southern Carpathian block lies the Transylvanian Basin, filled with unconsolidated formations of Neogene age.

The Pannonian Basin is surrounded by the Alps, the Carpathians and the Dinaric Alps. Roughly circular in shape, it is morphostructurally a relatively recent form due to subsidence of the Variscan basement in the Tertiary, together with uplift of the surrounding mountains. The Earth's crust in this area is thinner than average: 20–24 km thick beneath the basin compared with 32–60 km under the surrounding mountains. Below the Basin, the Moho surface rises in the form of a dome. Although partial subsidence of the Pannonian Basin began in the Upper Cretaceous, the Basin as a morphostructural unit came into existence only in the late Tertiary, at the time of the greatest extent of the Pannonian Sea. It became dry land during the Upper Pliocene and the Quaternary. Hence, morphostructurally, the Pannonian Basin is a young feature filled with marine

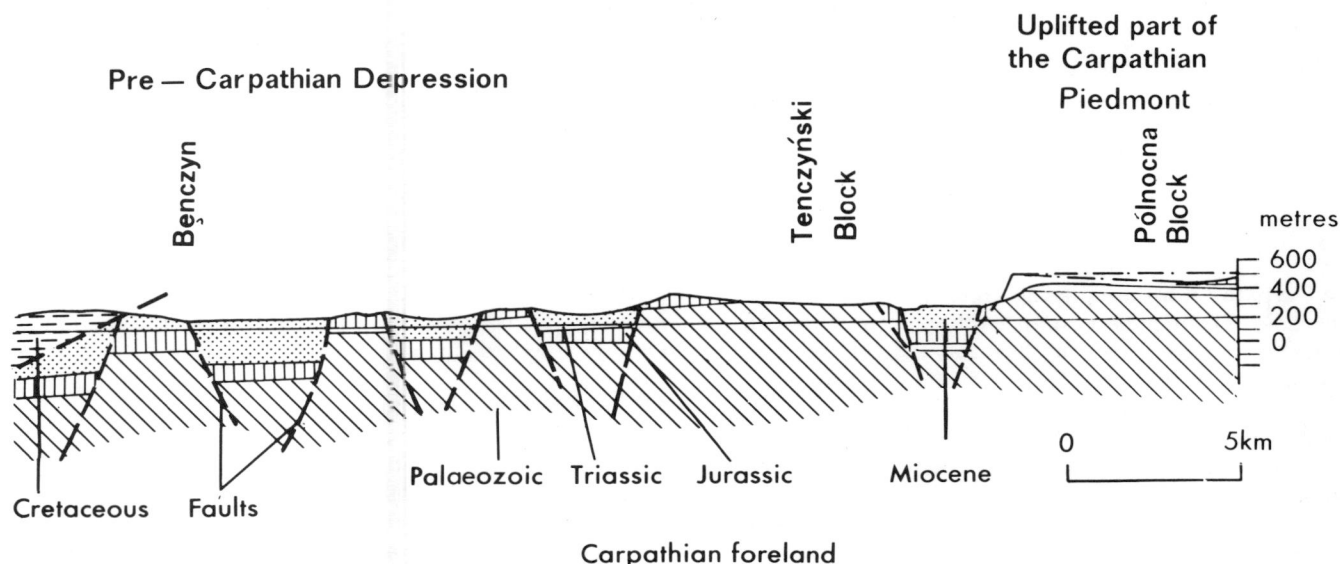

Fig. 15.1 Geological cross-section of the Pre-Carpathian depression and the Carpathian foreland [after Klimaszewski (1972)].

and, subsequently, by fluviolacustrine, fluvial and aeolian deposits. Its subsidence was partly due to the synorogenic (i.e. synchronous uplift and/or subsidence during orogenesis) crustal movements of the Carpathian upfolding and to the volcanic eruptions in the intra-Carpathian volcanic belt [see Pécsi (1970)].

In Romania, there were orogenic movements along the outer flank of the Carpathians up to the end of the Tertiary, producing folding and uplift of the sediments in the sub-Carpathian depression, the result of which was the formation of a foot-hill zone to the Carpathians.

The lower Danube lowland consists of a broad floodplain, with many meanders and oxbow lakes. The gradient of the river is only 0.033 percent and the river channel is 1–1.5 km wide with many sand banks. The formation of numerous alluvial cones spreading out across the fore-Carpathian plain has meant that the river has been pushed to the south against the foot of the northern Bulgarian platform.

15.2 Relief

The relief of the Carpathians has developed mainly during the Neogene. Locally in the inner Carpathians, Palaeogene or even Mesozoic landforms, such as planation surfaces, have survived. Extensive late Mesozoic–early Tertiary planation surfaces on the crystalline and metamorphic rocks are preserved in the Slovakian Ore Mountains. Later block movements have dislocated the Tertiary planation surfaces, leaving a legacy of fragmentary flat-topped relief forms situated at different altitudes and deeply incised valleys on the sides of the block ranges. Some rivers, such as the Váh and Hornád in Czechoslovakia and the Olt in Romania have become superimposed, passing through the mountain ranges in deep gaps.

During the Pleistocene, glaciers developed in the highest ranges of the Carpathians. They never extended more than a few kilometres, the longest being 14 km in the High Tatra of Czechoslovakia where the permanent snowline lay at about 1700 m. Fluvial relief forms are therefore more extensive than glacial forms in the Carpathians.

15.3 Western Carpathians

The morphostructural individuality of the western Carpathians (*see* Fig. 15.2) manifests itself in the form of an extensive, but relatively flat, upwarp of elliptical lay-out [see Mazúr and Stehlík (1965)]. The elliptical outline has a longer axis of about 400 km running from west-south-west to east-north-east and a shorter one of about 250 km. On the inner side, it is sharply delimited, facing the intra-Carpathian and Vienna Basins. On the other side, the western Carpathians face the Hercynian epiplatforms and are confined by a continuous belt of depressions known as the Carpathian foredeep. Only in the north-east is the border of the upwarp less distinctive, being morphostructurally, but not lithologically, marked [see Mazúr (1976)].

A characteristic feature of the upwarp is its inner morphostructural differentiation represented by a mosaic of two contrasted relief types—mountain ranges (*see* Fig. 15.3) and basins—representing differential uplift. This inner morphostructural division is, however, subordinate to a higher-order unification, namely the western Carpathian upwarp. The subordination is reflected in the rise in altitude of both contrasting elements from the fringes of the upwarp to a maximum within the western Carpathians (i.e. the High Tatra). In the case of the upwarp, there are still, however, relative height differences of 1000–1500 m. Thus, it is the core of the upwarp that is

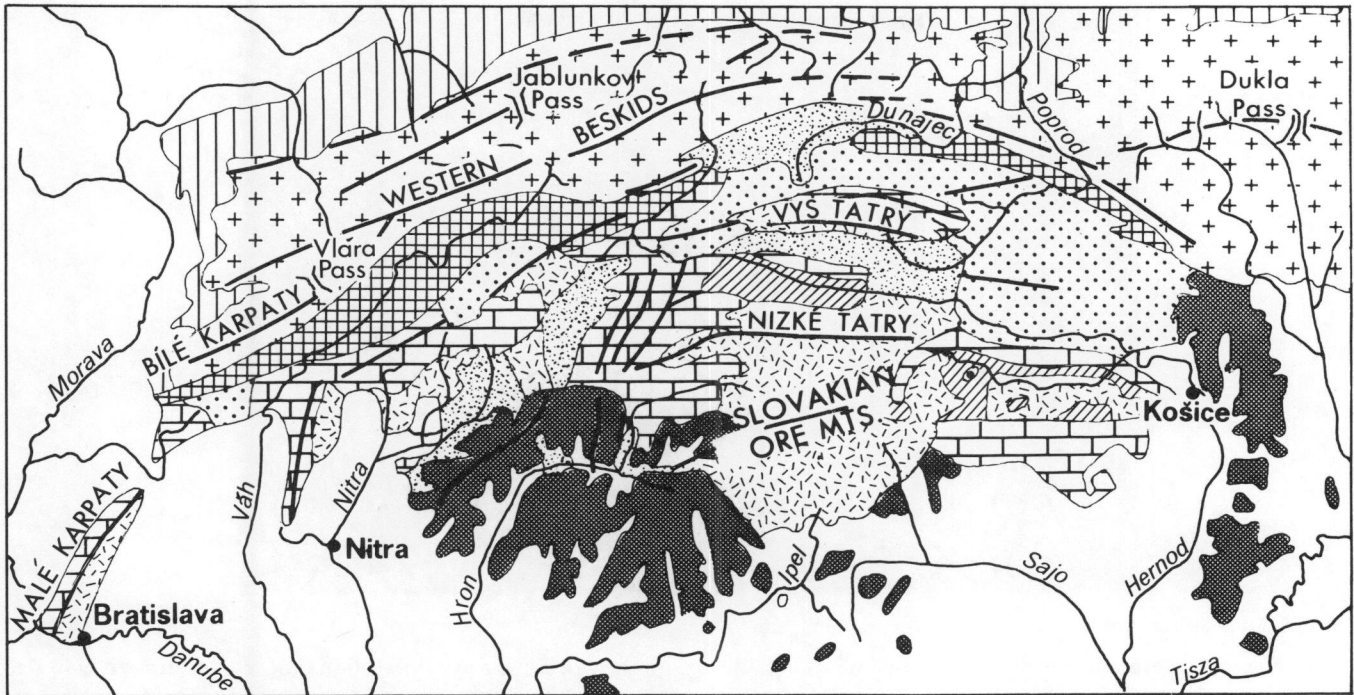

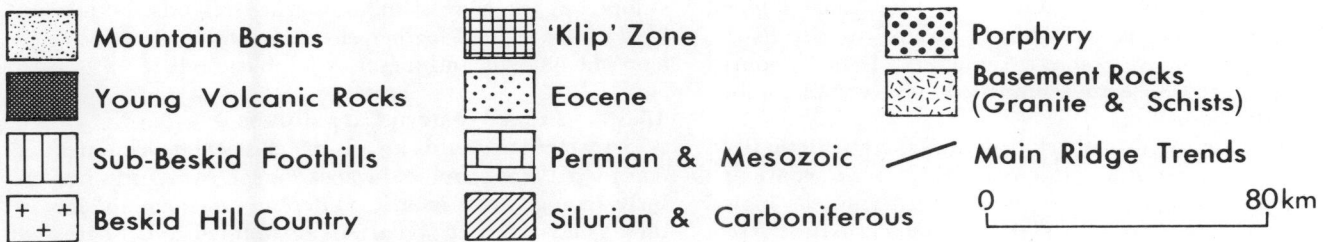

Fig. 15.2 Morphology and geology of the western Carpathians.

most intensively dissected. Another distinctive feature of the dissected upwarp is its asymmetry in both a vertical and a horizontal sense. The inner Carpathians are substantially more dissected than the outer Carpathians. The axis of the upwarp, marked by the Tatra mountains in the north-east, occupies an eccentric position, only about 90 km from the fringe of the western Carpathians. According to plate tectonic theory, this may be explained in terms of a general shift of the morphostructure. Owing to shortening of the space, this structure remains compressed on its outer side [see Mazúr (1976)].

The western Carpathian upwarp is divided into two morphostructural units: the inner and the outer Carpathian blocks, the boundary being represented by a fault encircling the Pieniny mountains [see Zoubek (1960)].

The development of the western Carpathian relief passed through four basic stages. At the beginning, it was associated with the Styrian (Alpine) folding phase which probably conditioned the strongly dissected nature of the relief, as can be assumed on the basis of the coarse Badenian (Miocene) deposits. After the Lower Badenian phase of intensive denudation, the relief became more

subdued in the Sarmatian (Upper Miocene) with relatively flat forms—the so-called summit level of the present-day. On the basis of the general distribution of the remnants of this surface, the Badenian–Sarmatian stage of development is assumed to have been terminated by the uparching of the Carpathians *en bloc* during the Attic (late Alpine) phase. The summit level surface was deformed by differential movements and, during the Pannonian (Pliocene), the lower parts were first modified to a middle mountain surface [see Mazúr and Stehlík (1965)]. The middle mountain planation surface was interrupted by a new general uparching of the western Carpathians in the Rhodanian (Pliocene) phase. These movements were far more marked than those in the Attic phase and imparted the general outlines of the present-day mountain ranges and basins. The partial destruction of the deformed middle mountain planation surface proceeded very irregularly according to lithology. Intense fluvial erosion dissected the uplifted blocks, with the rivers carrying coarse material to form alluvial plains at the foot of the mountain ranges. Towards the end of the Pliocene, extensive valley pediments developed along the river courses. Their de-

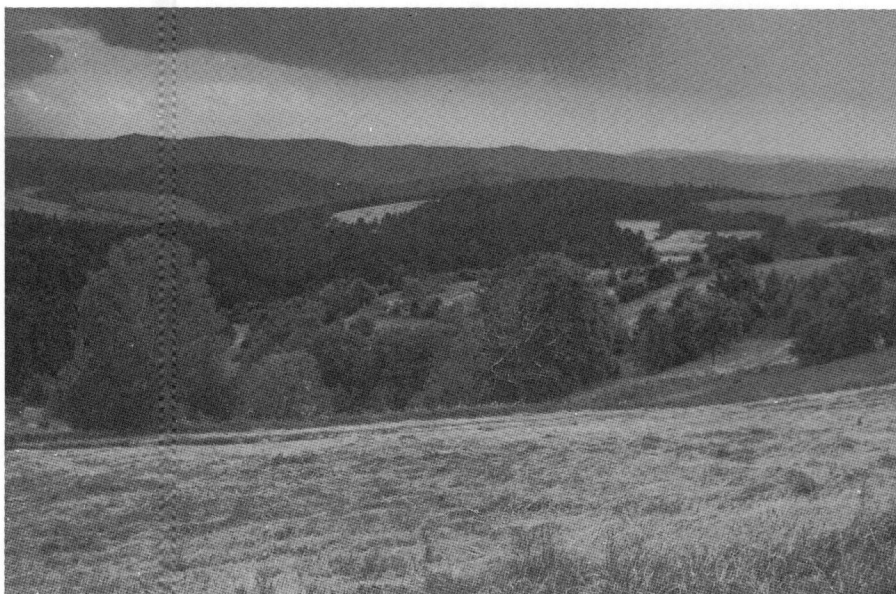

Fig. 15.3 Carpathian mountains near Vsetin, Moravia. (Embleton)

velopment was interrupted by the movements of the Wallachian (latest Alpine) phase.

In the Quaternary, the rapid sequence of climatic changes had a profound influence on the relief. Under periglacial conditions, permafrost formed to depths of about 100 m. On watersheds, remnants of the Tertiary planation surfaces were remodelled and cryoplanation terraces developed.

Cryopediments also developed in the unconsolidated Neogene and Quaternary deposits of the outer Carpathian depressions. They penetrate along valleys into the highlands of the outer western Carpathians. Permafrost created an impermeable layer so that surface runoff was possible even in permeable rocks. Rockslides, landslides and cambering were common on the mountain slopes under periglacial conditions.

The most intense glaciation took place in the Tatra mountains of Czechoslovakia, where traces of at least three glaciations have been found: the Riss glaciation was probably the largest in extent. The number of Würm glaciers on the Czechoslovak side totals about 40, glaciers being from 9 to 14 km long. The Polish side was less severely glaciated [see Klimaszewski (1972)]. In the Low Tatra mountains of Czechoslovakia, there were about 17 glaciers.

15.3.1   Outer Carpathian Depressions

These depressions are the morphostructural expression of the Carpathian foredeep and originated by subsidence of the margin of the epi-Hercynian platform, followed by infilling with marine, lacustrine and fluvial deposits. The basement is formed by old Proterozoic and Palaeozoic rocks forming horsts and grabens. The thickness of the Tertiary and Quaternary deposits ranges from several hundreds to several thousands of metres. The depressions are areas of Quaternary accumulation of colluvial and fluvial deposits due to neotectonic subsidence.

Lowlands form the main element in the relief. The broad present-day floodplains of the main rivers are bordered by flights of river terraces. Extensive piedmont plains consisting of coalescing alluvial cones have developed at the edge of the bordering uplands. Periglacial activity has added further elements (cryopediments, erosion glacis, dells and loess) to the basic relief.

15.3.2   Inner Western Carpathians

A characteristic of this area is the classical development of contrasted block and fold–block morphostructures irregularly arranged in a mosaic. Differential neotectonic movements have formed a pattern of uplifted and depressed morphostructures. The pattern of these neotectonic fault blocks bears little relationship to that of the older structures, which include folded and overthrust Palaeozoic and Mesozoic strata, and even volcanics. The inner western Carpathians are divided into five second-order morphostructures.

1)   The semi-massive structure of the Slovakian Ore Mountains which is the oldest element in the western Carpathians, bearing striking marks of the Hercynian tectogenesis.
2)   The Fatra–Tatra fold–block morphostructure, in which the differential tectonic movements reached their maximum (corresponding to the highest peaks in the High Tatra).
3)   The volcanic block morphostructure of the central Slovakian (Slovenské Stredohorie) mountains, which is built predominantly of volcanic rocks overlying older elements, but including contemporary morphostructural features formed by block mountains.
4)   The block morphostructure of the Lučenec–Košice depression, which is a system of grabens around the southeastern part of the Slovakian Ore Mountains.
5)   The block morphostructure of the Slaná–Mátra arc, which consists of horst structures linked to the inner volcanic arc of the western Carpathians.

A vertical zonation of the various relief features can be clearly observed in the inner western Carpathians. The most uplifted blocks were glaciated during the cold periods of the Pleistocene, but only the highest block—the High Tatra—possesses the typical Alpine scenery of cirques, glacial troughs, moraines and glacial lakes. The other Carpathian mountains exhibit mainly periglacial modification with cryoplanation terraces, frost-riven cliffs, patterned ground, etc. Due to deep dissection, cambering is a common feature on high and steep slopes; cambering was caused by rock creep due to processes connected with the melting of permafrost during the Pleistocene. The lower blocks and intermontane basins are dominated by fluvial processes. Karst features are common in the limestone and dolomite areas, including some extensive cave systems (e.g., Domica Caves in the Slovakian karst connected with the Aggtelek cave system in Hungary).

15.3.3 Outer Western Carpathians

In contrast to the inner western Carpathians, this area is noted both for its zonal relief character and less structural variety. On the basis of age and relief features, it is possible to distinguish between a relatively narrow peri-Pieninian lineament and the flysch zone proper. The peri-Pieninian lineament stretches along the northern circumference of the Malé Karpaty mountains, through the middle Váh valley, along the northern edge of the Fatra and Tatra mountains, and also at a lower level in the Spišsko-Šarišské Medzihorie. The outer flysch zone is marked by a nappe fold structure, which is, however, fragmented to a considerable extent by unequal neotectonic block movements. The development of mountainous horsts and basin-like grabens does not nearly attain the intensity characteristic of the inner western Carpathians. The highest flysch mountains, although, are positively uplifted blocks and not just the product of selective erosion.

Evidence of at least three planation surfaces can be observed in the outer western Carpathians. The upper planation surface shows all the features of a regionally developed planation surface probably cutting, after the Badenian, the whole region of the outer western Carpathians. Neotectonic movements dislocated this surface and raised it to varying altitudes. Under Pleistocene periglacial conditions, the remnants of the post-Badenian planation surface were remodelled by cryogenic processes, producing cryoplanation surfaces. The middle planation surface takes the form of pediments[a] and erosion glacis[b]. They developed in connection with the present-day river pattern and penetrated along valleys from the margins into the inner areas of the flysch highlands and mountains. Tectonic movements caused the break between the formation of the older and younger pediments. On the margins of mountain ranges formed in less resistant rocks, the pediments merge into a pediplain. The lowermost surfaces

developed as cryopediments under permafrost conditions in the Pleistocene. They are related to middle and, in places, to low Quaternary river terraces.

On the slopes, numerous mass movements can be observed. Some of them are large and of catastrophic nature. In the sandstone areas, cambering is noticeable and fissure caves are common. At the foot of the flysch mountains, extensive piedmont lowlands developed due to the merging of alluvial cones. Slope debris has often been washed down to cover the steps of the river terraces in many mountain valleys, the thickness of the debris being 30 m in places.

15.4 Central Danubian Lowland and Transdanubian Mountains

15.4.1 Little Central Danubian (Hungarian) Plain

Situated in western Hungary and crossed by the Danube and one of its main tributaries (the Rába), the Little Plain can be subdivided morphologically into a low floodplain, an alluvial fan in the centre and a dissected older alluvial fan plain on the borders of the basin. The latter alluvial fan grades eastward into a foot-hill surface (*glacis d'érosion*) of the Transdanubian mountains and, towards the west, into the similar features of the Alpine foot-hills.

Structurally, the most important boundary is the Rába line, to the west of which the basement largely consists of crystalline schists, a continuation of the crystalline core of the eastern Alps. East of the Rába line, the basement is composed of downthrown Mesozoic blocks—the continuation of the Transdanubian mountains. The Mesozoic blocks subsided during the early Tertiary, while the crystalline basement began to subside in the Miocene or Pliocene. The trangression of the Pannonian Sea produced more than 1000 m of sediments. The subsidence of the basin continued into the Pliocene and Quaternary, resulting in a thick veneer of fluvial and colluvial deposits.

The main relief features consist of the terraced alluvial fans and flood plains of the Danube, Rába and other rivers. The enormous alluvial fan of the Danube in the Little Plain shows two stages of formation. The younger, floodplain level fan, whose formation is still continuing today, extends from Bratislava to Komárno/Komarom on the Czechoslovak/Hungarian border over a distance of more than 100 km, most of it lying in Czechoslovakia. The deposits laid down by the Danube vary considerably in thickness. Whereas on the uplifted blocks of the basin Neogene rocks come to the surface in the river bed, elsewhere the thickness of recent alluvium can reach 200–250 m, especially in down-faulted basins.

In Austria, west of the Little Plain, the Parndorf plateau is a remnant of the older alluvial fan of the Danube, lying about 50 m above the contemporary floodplain. This older fan merges into the present alluvial fan terrace of the Rába in the Kemeneshát region of Hungary. The contemporary

---

[a] Erosional surface cut across bed rock and backed by a steeper slope with the junction between the pediment and this steeper slope being erosionally, not tectonically, controlled.

[b] Erosion proceeded in softer rocks, with the junction of the glacis and steeper slope behind being usually controlled by lithological differences or tectonics.

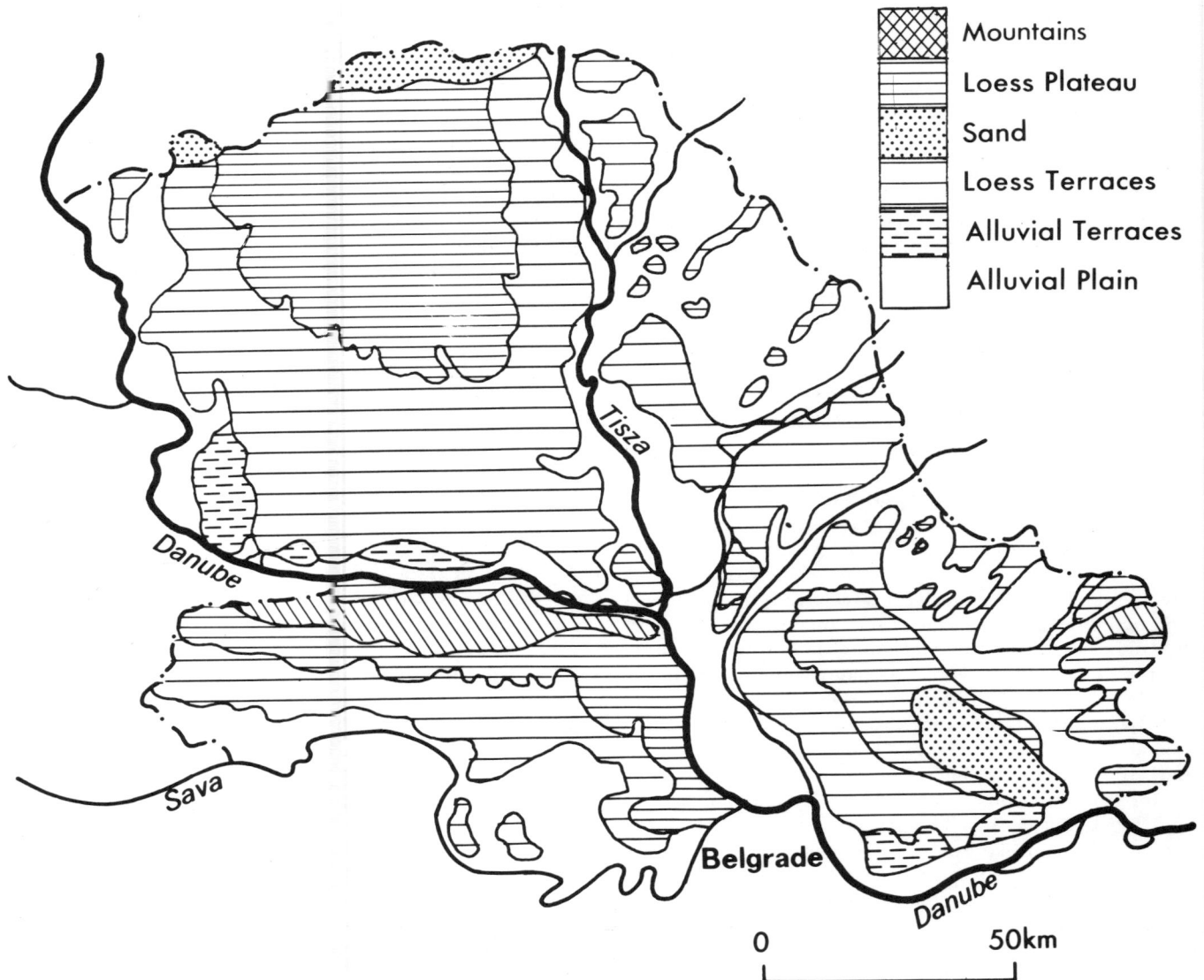

Fig. 15.4 Geomorphology of the Voïvocine area.

Rába alluvial cone merges into the present-day alluvial cone of the Danube. The Sopron–Vas gravel plain is formed by the alluvial cones of rivers coming from the Alps. In the Marcal Basin, the landscape includes some basalt-capped hills, a few of which rise more than 100 m above the basin floor.

15.4.2   Great Central Danubian (Hungarian) Plain

This plain covers almost half of Hungary. In terms of its underlying structure, the great thicknesses of recent unconsolidated sediments rest on a basement composed of a parallel system of north-east-trending faulted ranges of Mesozoic and Palaeozoic rocks. The Palaeozoic formations include gneiss, shales and mica schists, while the Mesozoic strata consist largely of dolomite, limestone and marls. The basement is broken into horsts, small basins and deep furrows formed during several phases of subsidence during the Tertiary. Parts of the upper and middle Tisza Basin had already begun to sink in the Upper Cretaceous. The Pannonian Sea then filled the Basin with

marine deposits (over 3 km thick). Subsidence of the Basin during the Quaternary is the reason why the Quaternary deposits alone are almost 1 km thick.

South of Budapest, the Danube is flanked by a floodplain 20–30 km wide, forming the axis of the Great Plain. The Danube plain is then joined by the wide floodplain of the Drava in Yugoslavia. In the eastern part of the Great Plain, one of the most distinctive features is the floodplain of the Tisza, with its numerous meanders and frequent changes of course.

The alluvial fans of the smaller streams issuing from the Transdanubian mountains coalescing with that of the Danube together constitute the divide between the Danube and Tisza Rivers, which rises by some 50 m above the floodplains of these rivers (*see* Fig. 15.5). Most of this divide is covered with sand dunes. Some loess can be observed on this watershed area, such as in the Gödöllö–Monor Hills east of Budapest, the loessic hills between Kecskemét and Nagykörös, and those of the Bácska. The

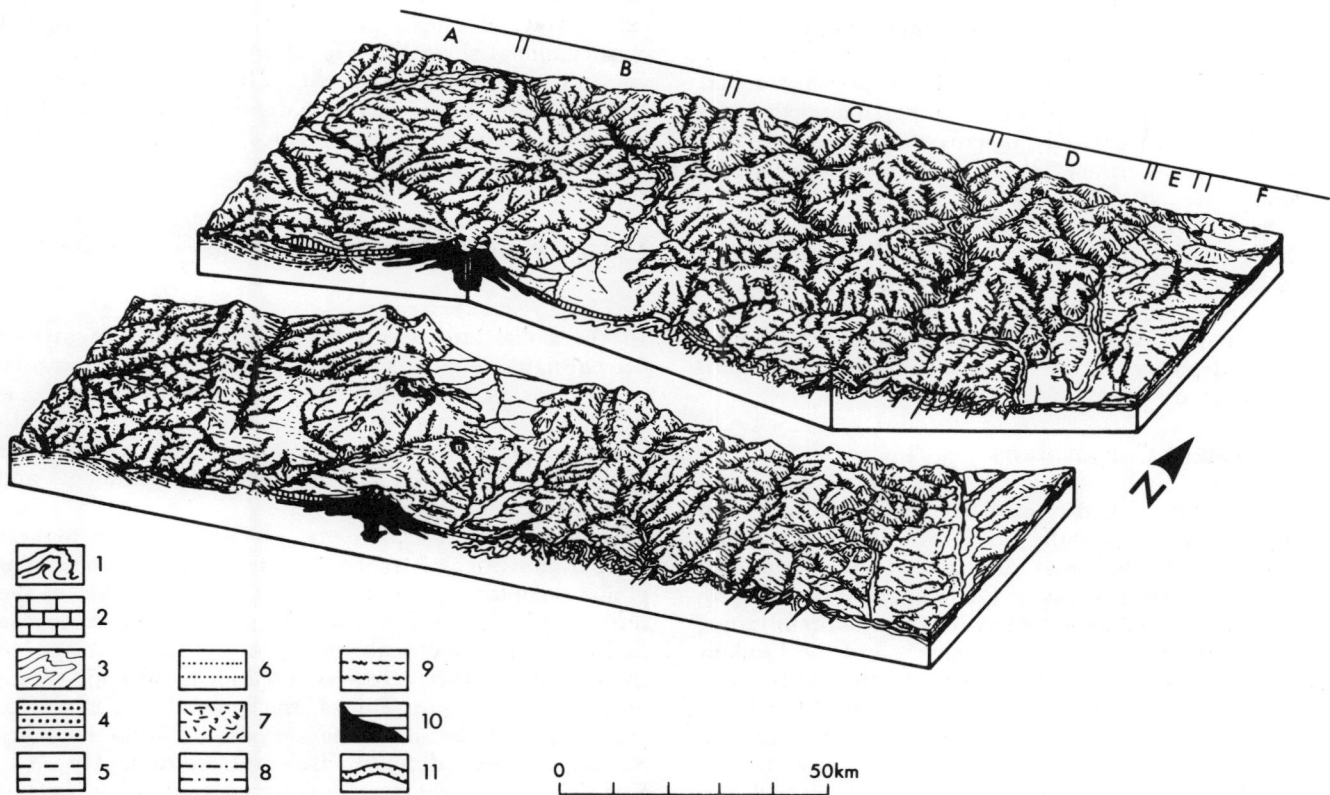

Fig. 15.5 Generalized block diagram of the eastern Carpathians [Academiei Republicii Populare Romine (1960)]. (A) Transylvanian depression, (B) eruptive zone, (C) Mesozoic–crystalline zone, (D) flysch zone, (E) Sub-Carpathian zone and (F) Peri-Carpathian piedmont zone. (1) Crystalline rocks, (2) Triassic and Jurassic, (3) Cretaceous flysch, (4) Eocene, (5) Oligocene, (6) Miocene, (7) salt domes, (8) Sarmatian, (9) Pliocene, (10) andesitic lavas and agglomerates, and (11) Quaternary.

Pest Plain is formed by terraces of the Danube.

Along the northern border of the Great Plain, there is a belt of alluvial cones formed in the Pleistocene by smaller streams. The Nyírség—one of the larger Pleistocene alluvial fans in the north-east part of the Great Plain—was deposited by rivers belonging to the drainage systems of the Tisza–Szamos–Kraszna Rivers and by the Bodrog River. The alluvial cone of the Maros River lies in the south-eastern part of the Great Plain.

15.4.3 Transdanubian Uplands and Mountains
To the south and west of Lake Balaton, down to the broad floodplain of the Mura and Drava Rivers, there is undulating country composed of several more or less independent uplands and low mountains, comprehensively termed the Transdanubian Heights. The Little and Great Central Danubian Plains are separated by these uplands.

The basement structure of the Transdanubian uplands consists of alternating zones of Palaeozoic crystalline and Mesozoic sedimentary rocks, trending more or less parallel to Lake Balaton. The Lake Balaton depression is a broad rift of north-east–south-west trend between the Transdanubian mountains and the Somogy Hills. Its south-western part probably assumed its present form as

early as the Lower Pleistocene, whereas the rest did so only late in the Pleistocene. Generally, however, in contrast to the Little and Great Central Danubian Plains, the Transdanubian uplands rose rather than subsided after the retreat of the Pannonian Sea, which also resulted in more intense dissection of the landscape. The Mecsek mountains are a locally folded and highly faulted uplifted block. In the Upper Cretaceous, volcanic activity occurred at the same time as the folding of the Carpathians, but today, there are only a few remnants left (e.g., dolerite necks).

The Transdanubian mountains are block mountains formed at the end of the Miocene and through the Pliocene. Their present-day altitude of around 500 m, however, is the result of late Pliocene and Pleistocene uplift. Two portions of the Transdanubian mountains— the Bakony Forest and Vértes—consist of blocks of Triassic limestones and dolomites. In the southern forelands of these mountains, Palaeozoic rocks are exposed. South of the Vértes, the basement comprises a batholith rising in the form of the Velencei Hills. In the south-west part of the Bakony Forest and in the Balaton upland, quite extensive outpourings of basalt occurred during and after the Upper Pannonian crustal movements.

Several types of planation surface can be distinguished

in the Transdanubian mountains and uplands. In addition to the summit levels, the blocks carry on their flanks narrow remnants of ancient piedmont surfaces and valley benches. The block mountains as a whole are surrounded by a broad foot-hill surface, made up partly of pediments cut in dolomite and partly of erosion glacis cut across unconsolidated Tertiary deposits. Cryopediments are common.

In the fault blocks of the Transdanubian mountains, largely consisting of limestone and dolomite, fault-controlled karst valleys are quite frequent. On the mountain borders, there are hot springs, some with high discharges—particularly in the Buda mountains—forming travertine deposits. In the Transdanubian uplands periglacial forms are common.

15.5   Eastern Carpathians

The eastern Carpathians consist of mountain ranges in Poland, Czechoslovakia, the USSR and Romania running in a north-west–south-east direction. The junction between the western and eastern Carpathians is conventionally located at the narrowest part of the mountain range, marked by the San River to the north, and the Łupków Pass and the Laborec river valley to the south. The boundary between the eastern and southern Carpathians is marked by the Dîmboviţa river valley. The eastern Carpathians include the flysch zone, which represents the continuation of the outer western Carpathians, and also an inner band of crystalline and volcanic rocks.

15.5.1   Morphostructural Features

The present-day topographical pattern of the eastern Carpathians has resulted essentially from reworking of major fold and nappe structures. Relief elements of various orders are formed in association with deep-seated faults and fractures. Major faults—both longitudinal (the so-called Carpathian trend) and transverse—appear commonly to be related to deep-seated fractures of the basement associated with vertical and horizontal displacements of the separated blocks [see Hofstein (1964)]. The relief of the eastern Carpathians owes its origin much more to block faulting than to folding. Rocks of Neogene age, including volcanic and intrusive rocks as well as the epicentres of Carpathian earthquakes, are confined to the zones of major faults.

Most geomorphologists recognize morphostructurally distinct folded and faulted zones running longitudinally and reflecting the main Carpathian trend (i.e. west-north-west–east-south-east). These zones comprise the fore-Carpathian depression, the outer eastern Carpathians and the inner eastern Carpathians. Compared with the outer western Carpathians, the outer eastern Carpathians are higher and show a more compact banded structure. The highest mountain group is that of Goverla (2061 m) in the USSR.

The inner eastern Carpathians attain their highest altitude in the Rodnei Mountains in Romania. They are built of crystalline rocks and reach 2305 m at Pietrosu. In

this range, there are extensive volcanic plateaus, with extinct volcanoes having kept to some extent their original constructional shape (*see* Fig. 15.5). The highest of these volcanic forms are Căliman (2102 m) and Harghita (1801 m). The fore-Carpathian depression is marked by a narrow zone of foot-hills, which are made up of young folded Tertiary deposits.

15.5.2   Relief and Denudation History

15.5.2.1   Planation surfaces

In the Polish and Czechoslovak sectors of the eastern Carpathians, remnants of three planation surfaces can be observed. The so-called mountain planation surface is developed on the tops of sandstone ridges, while the sub-mountain planation surface developed mostly in the shales of depressions. At the foot of mountains, pediments and cryopediments occur.

In the Soviet Carpathians, planation of the highest Carpathian ridges is more doubtful. Traces of erosional planation of the mountains at various levels seem to be the result of differential arch block tectonics, while the gentle outlines of asymmetrical ridges and the smooth ridges of the Poloniny Ranges are accounted for by tectonic tilting of horsts. The problem of the possible existence of planation surfaces and of the age of uplift of the Poloniny Ranges has been discussed by various authors. In 1925, Rudnicki claimed the existence of one planation surface, but since then other investigators have recognized two, three or even four planation surfaces. Hofstein argued that the so-called Poloniny peneplain involved the uplift of planation surfaces of different age. The data obtained in the course of a more detailed study of the block pattern of the Carpathian relief suggest the existence of local planation surfaces, but these cannot be correlated in view of the fact that there are no deposits or weathered mantles resting on them and that differential uplift has undoubtedly occurred.

In the Romanian Carpathians, the oldest planation surface is the so-called Carpathian pediplain. This is said to have developed under stable tectonic conditions in a warmer, more humid climate, especially between the Cretaceous and Oligocene Epochs. The modelling led to the development of coalescing pediments, finally resulting in a pediplain. At this time, the planation involved almost the whole of Romania. In the eastern Carpathians, the Carpathian pediplain is preserved in the crystalline Rodnei Mountains at heights of about 1800–2000 m—the so-called Nedei surface. The different altitudes at which the remnants of the Carpathian pediplain now lie are the result of neotectonic movements that took place after the Tortonian and that varied in amplitude among the foot-hills, the flysch zone of the outer Carpathians and the crystalline massif. In a fossilized state, the pediplain is also preserved beneath the sediments of the fore-Carpathian depression and in the intra-Carpathian depression; a fact which is also proved by the row of crystalline hills in the north-west of Transylvania [see Posea (1981), Posea *et al.* (1974)].

The lower planation surface is formed by the pediments of the so-called middle Carpathian surface. The beginning of its formation was marked by the Savian movements in the case of most of the crystalline and Mesozoic massifs, and by the Styrian phase in the case of the flysch zone of the eastern Carpathians. The pedimentation stage was generally brought to an end by the Pontian transgression. These pediments apparently developed under a Mediterranean type of climate. In the crystalline massifs of the eastern Carpathians, the pediments lie at 1600–1750 m, which is lower than the remnants of the Carpathian pediplain in the Rodnei Mountains, and at 1300–1700 m in the Bistriţei Mountains, where they occupy a central position. The pediments are well dissected in the Hăsmaşul Mare Mountains and in the Bucegi–Baiu Mountains, occurring mainly at 1500–1700 m. Sporadically,

remnants can be traced in the Moldavian sub-Carpathians.

The next younger Carpathian surface developed after the formation of the Sarmatian piedments, but before the Rhodanian movements (middle Pliocene). In some areas, the surface results from wave abrasion by the Pontian Sea, but it also extends into the upland areas in the form of pediments passing up the valleys. In the outer flysch zone of the eastern Carpathians, the remnants occupy ridges that extend towards the interior of the mountains and result from the dissection of peripheral and valley pediments and glacis. Lower Quaternary valley pediments can also be found. The glacis were extensively formed under Mediterranean climatic conditions in the period from the Miocene to the Villafranchian.

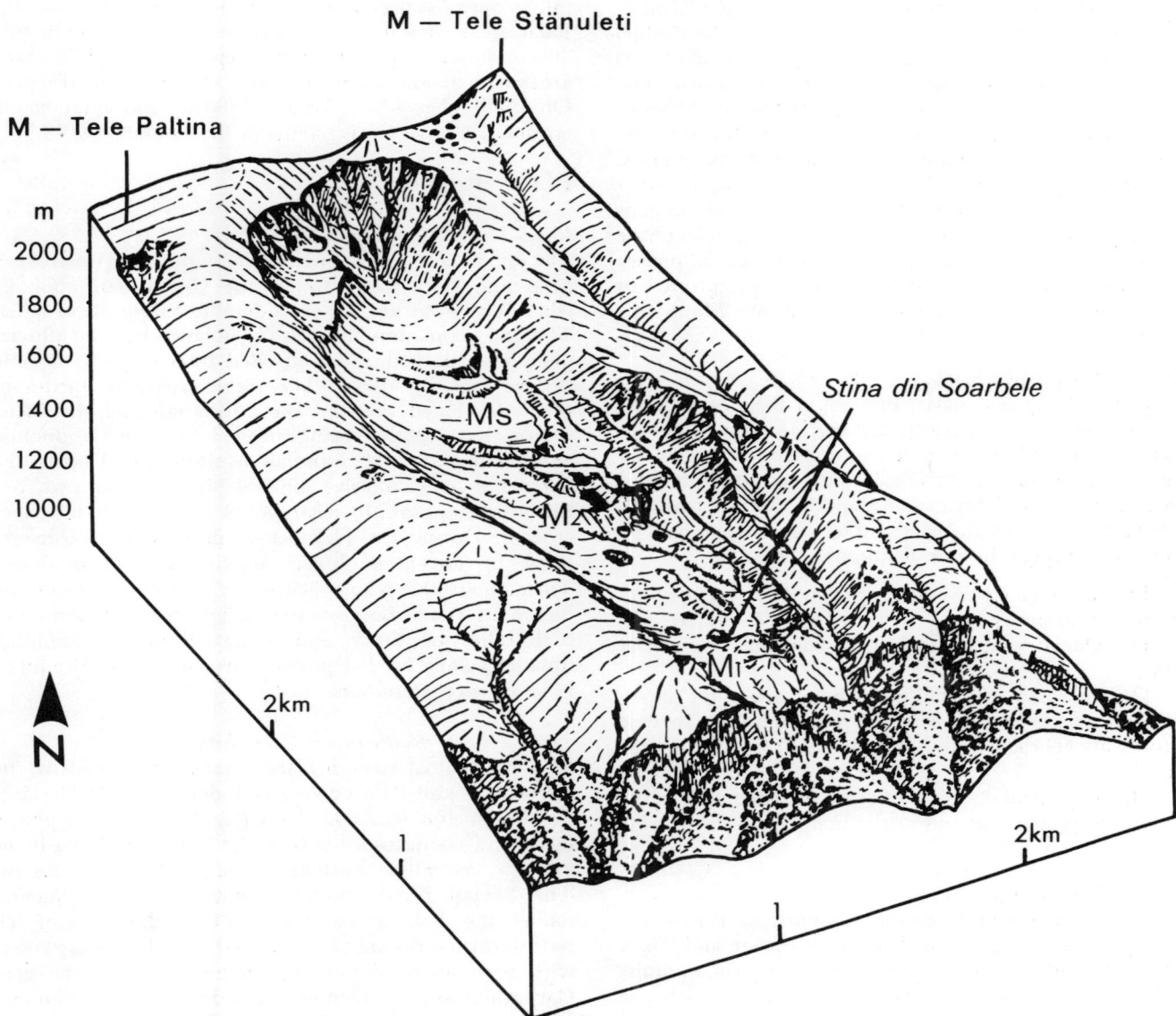

Fig. 15.6 The Soarbele valley, showing cirque head and other glacial features. M1: outer frontal moraines; M2: inner frontal moraines; Ms: moraines of a late glacial readvance [after Niculescu (1965)].

15.5.2.2   Glaciation

In the Soviet Carpathians (in the Svodivets and Chernogor mountain groups), remnants of a single glaciation have been described. The glaciation seems to have been of cirque-type, provided the term cirque is correct in the case of these huge, structurally complex niches. The shape of the glaciers was predetermined by the shape and size of these niches, their step-like relief and their orientation relative to the westerly winds.

In the Romanian Carpathians at the end of the Pleistocene, glaciation affected, in general, only some of the Carpathian massifs (mainly those higher than 1900 m), where topographical conditions allowed the accumulation of snow and maintenance of firn (*see* Fig. 15.6). Analysis of the glacial forms and deposits (mainly three series of end moraines in the Rodnei Mountains and two levels of glaciated valleys in the Bucegi Mountains) shows that two distict glacial phases occurred in the eastern Carpathians: Riss and Würm. The glacier tongues in the Rodnei Mountains descended to 1000 m. In both the Rodnei and Bucegi Mountains, plateau glaciers also developed. The extent of glacial forms is as follows: Maramureş Mountains, one glacial phase with cirque glaciers lying at 1150–1700 m; Rodnei Mountains, two to three glacial phases, especially on the north slope, with many cirques lying between 1500 and 1960 m, from which glacier tongues descended to less than 1000 m; Căliman, one glacial phase with simple cirques; Bucegi mountains, two glacial phases with valley, cirque and plateau glaciers concentrated in the northern part of the massif; the Leaota Mountains with cirque glaciers [see Posea *et al.* (1974)].

15.5.2.3   Periglacial activity

Except for the glacial parts, the rest of the eastern Carpathians experienced a periglacial regime during the cold phases of the Pleistocene. This resulted in periglacial forms (cryoplanation terraces, sharp crests, tors and cryopediments) and periglacial deposits (debris mantles, talus cones, etc.). In the eastern Carpathians, periglacial forms are arranged in four altitudinal zones.

1) The supraglacial cryogenic zone, which developed on the crests and peaks that rose above the ice surface.
2) The glacial zone, with both glacial and periglacial forms.
3) The detrital periglacial zone, lying below the snow-line down to a height of about 700 m, with niveocryogenic features above 1500 m and cryonival features below this altitude.
4) The periglacial margins, lying on the outer edge of the Pleistocene permafrost and at the upper limit of the pine forest.

15.5.2.4   Fluvial features

The block structure of the eastern Carpathians is accentuated by the valley pattern. The main Dnestr and Tisza Rivers are confined to the longitudinal Carpathian fault-guided valleys. The main tributaries of these rivers are controlled by transverse faults. Tributaries of the second and lower orders form an intricate network that corresponds to the fault and minor fracture network. Frequent sharp bends of the river channels reflect the irregular uplift of individual blocks, the more so as locally the rivers flow over bedrock exposed in their channels. Evidence of recent tectonic movements is provided by other features of the river valleys. The Carpathian river beds are generally more stable, variations in their activity being controlled by the succession of narrows and basins along the valleys. Locally, in the depressions, the thickness of floodplain alluvium increases: for example, the alluvium of the Tereblin in the Soviet Carpathians increases from 7 to 20 m in thickness, showing the existence of small local grabens formed within the river valley.

15.5.2.5   Slope processes

Landslides and mudflows are widely developed in the eastern Carpathians, assume the most varied aspects and, sometimes, are the dominant process in the modelling of hill slopes, especially in the foot-hills and in the Palaeogene flysch areas. Torrential runoff frequently affects slopes, especially the steeper ones and in those areas with an annual precipitation of more than 1000 mm. On the higher levels (above 1700 m) torrential runoff erosion affects extensive areas in the summer.

15.5.3   Regions

15.5.3.1   Fore-Carpathian alluvial plain

The plain corresponds structurally to the Carpathian foredeep and developed around the Carpathian uplands, where there was already considerable amplitude of relief. The line of junction between the uplands and the alluvial plain was marked and controlled by a fault or tectonic flexure (*see* Fig. 15.7). The fore-Carpathian piedmont plain developed between the Sarmatian and the Pleistocene, its evolution being affected by periodic climatic and tectonic events, especially subsidence. During the Sarmatian, an extensive alluvial plain was formed, but this was later partially destroyed tectonically and by river erosion. During the Pleistocene, important deposits related to torrential fluvial activity were laid down. Throughout the whole history of the evolution of the piedmont, coarse deposits were associated with periods of high relief amplitude and dissection in the adjoining uplands, while finer deposits represent times of reduced relief or even planation.

15.5.3.2   Fold-faulted eastern Carpathian mountains

The fold-faulted eastern Carpathians are formed by the Cretaceous and Palaeogene flysch deposits. The Cretaceous flysch unit was folded during the Laramian phase, forming a connection between the southern Carpathians and the crystalline nucleus of the eastern Carpathians. The Savian phase of earth movements subsequently folded the Palaeogene flysch. The Palaeogene flysch morphostructures are characterized by sedimentary strata with pronounced Alpine-type folding. During the neo-Carpathian stage, which included the Styrian, Attian and Rhodano–Wallachian phases, the flysch zone was transformed into a fold-faulted morphostructure (*see* Fig. 15.7).

The basic features of the flysch zone are determined by

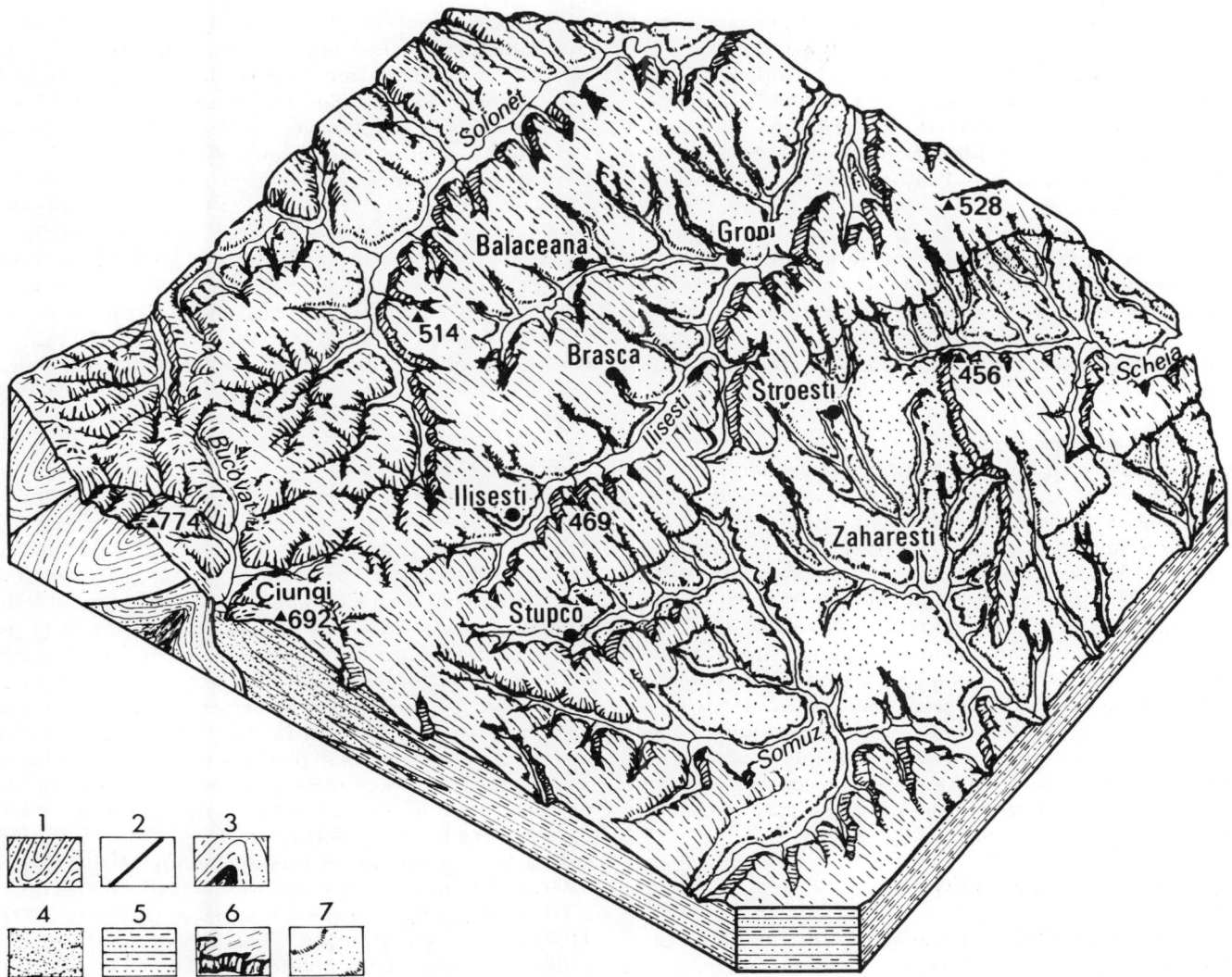

Fig. 15.7 Block diagram showing relationships between geology and relief, eastern Carpathian foothills near Suceava [Academiei Republicii Populare Romine (1960)]. (1) Flysch, (2) overthrust, (3) Helvetian–Tortonian (Miocene), (4) Sarmatian, (5) structural surfaces on sandstones and conglomerates, (6) main cuesta and (7) erosional depressions.

the spreading direction of the flysch layers, the tectonic contact between them, imbricate structures directed towards the east, and by the existence of synclines and anticlines within the nappes. The hogback relief imposed by the flysch layers and imbricate structure is the most distinctive feature of the entire flysch zone. In the Romanian Carpathians especially, hogbacks are typical features in the Stînişoarei, the Goşmanu and the Ciue Mountains. The relief formed on the widespread Palaeogene sediments in the Tibleş Mountains, Bîrgău Mountains, Maramureş and Dorna depressions is characterized by cuestas and various types of structural valley.

15.5.3.3 Crystalline massifs

The crystalline massifs of the eastern Carpathians form the highest parts of the mountain arc. The Marmarosh massif in the Soviet Carpathians represents an uplifted block of probably early Palaeozoic and Precambrian age.

The Rakhov and Chichin massifs of crystalline rocks belong to the basement, as well as the 'cliff' ridges or protruding horsts composed of Jurassic limestones and characterized by summits separated by selective erosion.

In the Romanian Carpathians, the crystalline massifs are composed of Hercynian crystalline rocks, with or without Mesozoic deposits. In the Rodnei Mountains, remnants of the oldest planation surface—the so-called Carpathian pediplain—can be found. Due to neotectonic movements, the crystalline blocks (Maramureş, Rodnei, Bistriţei and Guirgeu Mountains) stand at different heights. On their flanks, two to three levels of the middle Carpathian surface, two to three levels of the border surface and two to three levels of valley pediments can be found [see Posea (1981)]. The massifs have the form of horsts or partial horsts, some of which are tilted (e.g., Rodnei Mountains). The highest crystalline massifs (Maramureş and Rodnei Mountains) were glaciated during the Pleistocene and periglacial forms (cryoplanation terraces, tors, etc.) are common.

Karstic landforms have developed on Mesozoic limestones, especially in the Hǎşmaş and Vîrghis Mountains, where a wide range of karst features is to be found. Specific to the Rodnei Mountains are the caves in the Eocene limestones, such as Tǎuşoarele in Romania, which developed at a great height above the valley floor. On the summits of the Bucegi Mountains, glacial and periglacial erosion processes amplified some forms specific to the karst landscape and degraded others. In the Bucegi Mountains, the cave systems show a complex evolution, while the Guivala–Podu Dîmboviţei region is an example of transitional karst, including areas with exhumed fossil forms.

15.5.3.4   Volcanic mountains
Along deep-seated faults in the miogeosyncline, the volcanic chain of the eastern Carpathians was formed, starting with the Vihorlat Mountains in Czechoslovakia and extending to the Cǎliman–Harghita Mountains in Romania. Volcanic morphology is best preserved in the Cǎliman–Harghita group, with calderas, craters, valleys with angular or radial patterns, and lava plateaus. Structurally two features can be distinguished: a sedimentary–volcanic basic complex, formed in the Pliocene by the accumulation of materials resulting from erosion of the volcanic structures, and an upper complex represented by stratovolcanoes built in the last eruption phase (i.e. approximately 3.5—7 Ma ago). At the contact with the Transylvanian tableland, scarps, depressions, glacis and pediments have developed.

15.6   Southern Carpathians

The southern Carpathian mountains extend from the Dîmboviţa river valley in the east to the Danube river gap in the west. They culminate in the Fǎgǎraş Mountains (the highest point being Moldoveanu at 2543 m) which show Alpine-type relief forms. A longitudinal tectonic depression divides the southern Carpathians into a wider northern and a narrower southern part. In the eastern part, individual mountain groups such as the Fǎgǎraş, the Retezat, and the Parîng Mountains rise to more than 2500 m, while the western part, including the Banat Mountains, does not exceed 1500 m. The boundary is the Timiş valley [see Orghidan (1969), Posea *et al.* (1974), Cernescu (1960), Rosu (1967)].

15.6.1   Morphostructural Features
The typical morphostructures of the southern Carpathians (*see* Fig. 15.8) are associated with crystalline blocks of Hercynian age and with a sedimentary cover. The crystalline basement asserts itself by tectonic/structural control of the massifs, ridges or of some marginal or intermontane corridors (Hideg–Riul Mare, Barusor–the upper Streiu, the upper Dîmboviţa River, etc). The schistosity planes and broader folds are reflected in the details of the relief: small hogbacks, cuestas, structural scarps, obsequent and consequent valleys, and other structurally controlled features. Structural features developed on the sedimentary cover that lies on the crystalline basement follow the alignments of broad synclinoria and synclines (e.g., the zones of Reşiţa–Moldova Nouǎ, Pui, Buila–Vînturariţa, Şviniţa–Şvinecea Mare and Cerna–Cazane). The most conspicuous elements of this relief are the Triassic, Jurassic and Cretaceous limestone morphostructures that dominate the surrounding relief in the form of sharp crests and isolated massifs. Other forms include cuestas, hogbacks, structural surfaces, inverted relief and structurally controlled valleys, which are especially frequent in the Almǎj and Caraş Mountains. In the intramontane depressions (i.e. Brezoi–Titeşti, Petroşani, Bozovici), the structural relief consists of cuestas, structural surfaces and subsequent valleys.

15.6.2   Relief

15.6.2.1   Planation surfaces
The Carpathian pediplain formed between the Cretaceous and the Oligocene as pediments coalesced. Its development in the southern Carpathians was interrupted by the Savian earth movements which led to dislocation of the pediment by fracturing. In the Banat Mountains, pediplain development lasted until the Tortonian when it was interrupted by the Styrian movements. The Carpathian pediplain is now preserved in the southern Carpathians mainly in the form of small plateaus and bevelled hill tops. The highest and most uniform altitudes attained by the pediplain are in the southern Carpathians at 1900–2200 m—the so-called Borǎscu surface—where it is overlooked by summits and crests at various heights (up to 2500 m). In the Banat Mountains, the pediplain occurs at 1200–1400 m—the Semenic surface—and at 800–1100 m—the Almaj surface. The elevated parts of the pediplain, which were uplifted after the Tortonian by neotectonic movements, were dissected by the deepening of valleys, leaving a pattern of residuals along the tops of spurs and watersheds.

The initiation of the younger middle Carpathian surface is marked by the Savian movements in the crystalline Mesozoic massifs and by the Styrian phase in the Banat Mountains. In the southern Carpathians, the higher level of this surface takes the shape of gently sloping watershed areas or sometimes stepped surfaces at altitudes of 1500–1700 m, but rising as high as 1800 m in the Fǎgǎraş Mountains. The lower level occurs on interfluves at heights of 1200–1500 m, with the associated valley pediments penetrating inside the massifs up to 1500–1700 m. In the Banat Mountains, it occupies the main watershed areas situated at lower levels than the remains of the Carpathian pediplain: at 550–800 m for the Almaj and 700–1000 m for the Semenic surface, declining in the west to 350 m where it crosses on to the Pontian formation.

The remnants of the Carpathian border surface form an erosion surface on the flanks of the southern Carpathians and in the Banat Mountains, but sometimes they occur as pediments penetrating into the mountains. The youngest pediments are developed at two levels of which the upper

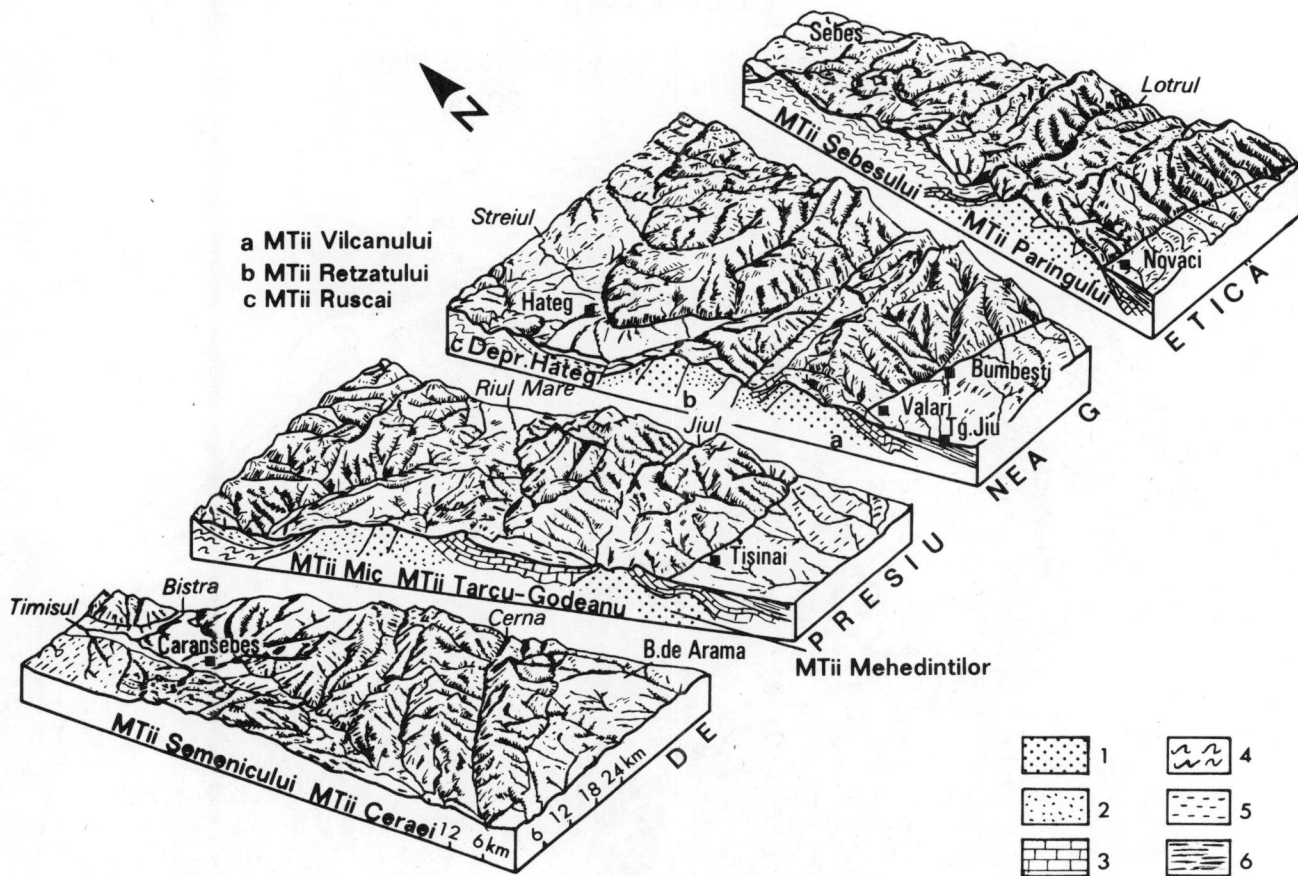

Fig. 15.8 Idealised block-diagram of part of the Southern Carpathians, showing fault-block structure, down-faulted depressions and granitic intrusions [Acad. Republ. Pop. Romine, Bucureşti, (1960)]. (i) intrusive rock, (2) crystalline schists, autochthonous, (3) sedimentary deposits, infra-Getic, (4) crystalline schists, Getic overthrust, (5) sedimentary deposits, Getic overthrust, (6) Neogene sedimentary.

level is better developed.

15.6.2.2 Fluvial landforms
River valleys are mostly controlled structurally and contain several sets of terraces whose ages are generally Quaternary. An important alignment that is not structurally guided is the superimposed valley of the Olt.

15.6.2.3 Glacial landforms
Glacial relief developed over two phases of glaciation in mountains more than 2000 m high (*see* Fig. 15.9). In the southern Carpathians, glacier tongues descended to 1300 m. In the Făgăraş Mountains, there are more than 175 cirques grouped in large complexes facing both south and north, and belonging to two glacial phases. Valley glaciers attained lengths of 4–8 km on the southern slope but only 2–4 km on the northern slopes. In the Paring Mountains, cirques and glacial valley complexes developed in the upper reaches of the Jieţ, Lotru and Latoriţa, with forms that reveal two glacial phases and three stages within the last phase. The most complex and varied glacial forms and deposits developed in the Retezat

Mountains, lying between 1300 and 2200 m. Simple cirques, short valleys and small-sized glacial forms characterize the Godeanu, Iezer, Şureanu, Cindrel and Ţarcu Mountains.

15.6.2.4 Periglacial features
Periglacial forms are similar to those described for the eastern Carpathians. Present-day cryonival processes affect the highest parts of the southern Carpathians above 1700 m with a cold, moist climate (1000–1400 mm precipitation per year, with more than 50 percent as snow), with Alpine and sub-Alpine vegetation. Above 1950–2000 m, against a background of the glacial and periglacial Pleistocene relief, the landscape is dominated by sharp crests, frost-riven scarps, tors, pinnacles, nivation depressions, nivation benches and cryoplanation terraces.

15.6.2.5 Karst landforms
Karst relief in the Banat Mountains centres mainly around two limestone masses stretching north–south (i.e. Reşiţa–Moldova Nouă and Şviniţa–Şvinecea Mare), on which karst plateaus, several cave levels and gorges developed. In the southern Carpathians, the limestones

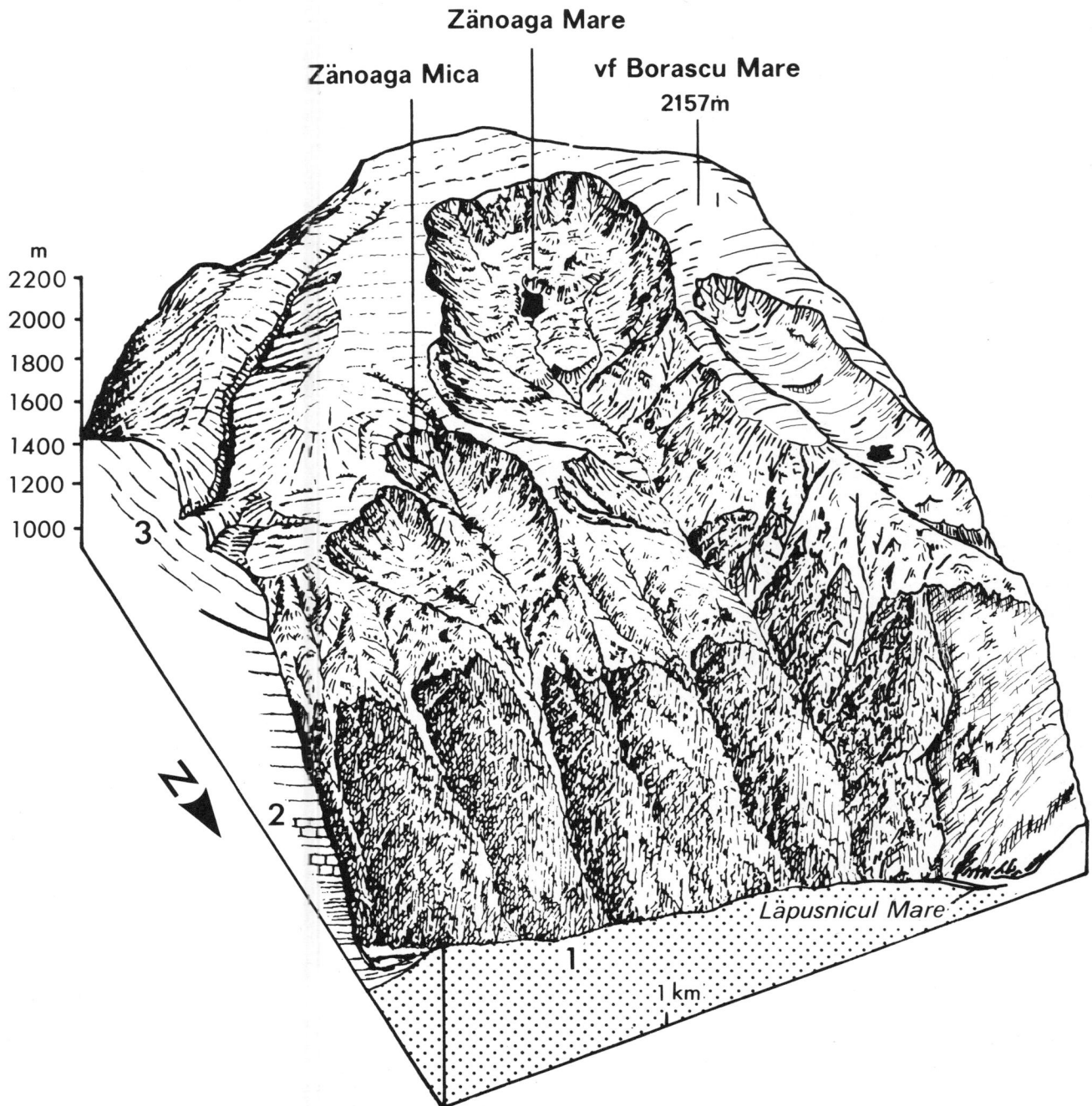

Fig. 15.9 Glacial cirques around Mt Borascu [*after* G. Niculescu, (1965)]. (1) Granodiorite, (2) Autochthonous sediments, (3) Getic overthrust (M) Moraine.

are usually strongly dissected and reduced to a series of isolated summits, crests and peaks lying at different altitudes. These conditions determined the development of two main karst types: (1) bare karst tops situated above 2000 m and (2) bare tops and small plateaus lying below 2000 m. There are also differences determined by the size and thickness of the limestone layers.

15.6.3 Regions

15.6.3.1 *Getic piedmont*

Morphostructurally, the Getic piedmont corresponds partly to the Carpathian foredeep and partly to the subsided Moesian platform. In contrast to the large morphostructural units bordering it [i.e. the mountain ridges of the southern Carpathians (*see* Section 15.6.3.3) to the north and the lower Danube plain to the south], the Getic piedmont represents a transition zone. It is subdivided into two closely related subregions: (1) the Getic sub-Carpathians or hilly piedmont and (2) the Getic tableland or relict piedmont. Within the Getic piedmont, the relief is strongly dissected by valleys in the Getic sub-Carpathians, with the relief amplitude increasing towards the mountain regions (*see* Section 15.6.3.3). It is separated from the mountain ranges by the sub-Carpathian depression. Farther to the south, the relief of the Getic piedmont is formed by broad ridges extending to different distances away from the sub-Carpathian depression. These broad ridges are separated by fewer, but relatively deep, incised valleys. The Getic piedmont evolved between the Sarmatian and Villafranchian stages, after which the area was raised, transformed into a plateau and dissected by valleys [see Velcea (1971)].

15.6.3.2 Southern central piedmont upland

To the east of the geomorphologically important valley of the Dîmboviţa, there is a transition from the mountain ridges, especially the flysch ridges, to the piedmont without the intervention of a sub-Carpathian depression. This region is formed out of folded Tertiary layers. The upland to the east is more dissected and the relief amplitude is higher. Even the youngest Pliocene and Lower Quaternary deposits are folded. Salt domes are common in this region.

15.6.3.3 Ridges and massifs of the central zone

The ridges and massifs of the central zone are built of crystalline rocks dating from the end of the Cretaceous as the large Gettian nappe was formed. Erosion later broke up the nappe into a series of outliers, forming tectonic windows in the Retezat and Parîng massifs. The mountain ridge of Godeanu corresponds morphostructurally to the eastern part of one of the Gettian nappe outliers. The massif is asymmetrical, the northern slopes being steep and the southern ones gentle. The typical structure of the ridges is exemplified by the Godeanu Mountains [see Niculescu (1965)], where an autochthon formed of crystalline rocks and granitic intrusions was covered by Mesozoic rocks. In the east of the massif, the Gettian nappe is composed of crystalline schists and covered by Permian deposits. Due to selective denudation, amphibolite, quartzitic rocks and pegmatites are absent from the higher ground. Slopes formed in gneiss are characterized by cliffs. Among the sedimentary rocks, limestones and conglomerates tend to form ridges.

15.7 Block and Fold-Faulted Mountains of the Bihor Massif (Apuseni Mountains)

The Bihor massif, which occupies an isolated position inside the Carpathian arc, features extensively bevelled summit plains bordered by narrow, deep-cut valleys. The highest peak is Curcubăta (1848 m).

15.7.1 Morphostructural Features

These mountains possess a complex structure, because in this region the crystalline basement, with its Mesozoic sedimentary cover, joins the flysch zone of the Metaliferi Mountains. There are also volcanic structures of various ages. Recent tectonic movement have divided the massif into a system of grabens and horsts. This fragmentation began in the Laramian phase, when banatites were extruded along faults, and continued through the Savian, Styrian and Attian phases into the Rhodano-Wallachian—the main phase of Neogene volcanic activity in the Apuseni Mountains. Tectonic control is the first factor reflected in the morphological configuration of the Bihor massif.

1) The crystalline rocks of the Gilău–Muntele Mare area (mainly granites) form the central nucleus from which other ridges and horsts (e.g., Bihor, Highiş, Codru, Plopiş and Mezes Mountains), separated by basins and grabens, diverge. These rocks control the main river pattern.
2) In the south, the Metaliferi Mountains, which are epigeosynclinal in origin, developed upon Cretaceous flysch rocks with igneous intrusions. They are dominated in their eastern part by massive Mesozoic limestones, displaying complex structural relief.
3) The margins are characterized by depressions aligned along the contacts of different strata and also by epigenetic gorges. The latter were cut discordantly after the stripping of the Tortonian–Sarmatian cover that blanketed the rugged relief on the borders of the Apuseni Mountains (e.g., Crisul Repede at Vad, Crisul Negru at Borz, Crisul Alb at Talagui, Gurahonţ and Cocuiba, the Mureş River at Zam and Şoimuş–Lipova, etc.).
4) The details of the relief are usually related to minor structures, such as are associated with the Mesozoic calcareous deposits, the Cretaceous flysch and the Neogene deposits of intramontane basins. Features include structural scarps related to nappes, structural outliers, structurally controlled valleys, cuestas, hogbacks, and epigenetic and antecedent gorges.

The volcanic morphostructures of the Apuseni Mountains, aligned on fractures in the basement of the Metaliferi Mountains, date from the Neogene. The eruptions were chiefly in the form of stratovolcanoes and, to a smaller extent, volcanoes of Hawaiian or Péléan type. Volcanic activity developed during three phases, only two of which were important in the Bihor massif. During the first phase, Tortonian rhyolites and andesites were extruded, while in the second phase, dacites and andesites dominated the volcanic suites of the Upper Tortonian to the Pannonian. At the end of the Pliocene and in the Pleistocene, basaltic lava flows occurred at several places in the Apuseni Mountains. The volcanic morphostructures of the Apuseni Mountains, situated on top of the older bevelled relief of the basement, are now dissected strongly by erosion. Volcanic activity led to significant changes in the river patterns and also formed volcanic–tectonic depressions.

15.7.2 Relief

15.7.2.1 Planation surfaces

The Carpathian pediplain is well preserved in the Bihor

massif at altitudes of around 1800 m, and in the Gilău massif at 1100–1600 m. The different altitudes of the pediplain remnants are the result of neotectonic movements, which occurred especially after the Tortonian age. Remnants of the middle Carpathian surface of Tortonian (upper level) are situated in the central and eastern part of the Apuseni Mountains on the watersheds at heights of 1000–1100 m (Arieş Basin), 1200 m (Padiş zone), and 800–1200 m (Gilău Mountains). The lower levels of this surface are of Sarmatian–Meotian age in the Apuseni Mountains, frequently in the shape of pediments in high basins at 800–1000 m and declining peripherally to 700–800 m. The surface was partly remodelled by marine abrasion at the time of the Pontian Sea. Marine abrasion also affected the Carpathian border surface in the Apuseni Mountains. The valley pediments developed after the Rhodanian neotectonic phase in the Lower Pleistocene. They may be partly cryopediments. Erosion glacis developed at the abrupt contact of the plains with the Apuseni Mountains, dating from the Pleistocene but also evolving somewhat in the Holocene.

15.7.2.2   Karst processes

In the Apuseni Mountains, karst relief, which developed on thick Mesozoic limestones, has a polycyclic character and displays great complexity of form. In this region, there occur the longest caves in Romania (Peştera Vîntului: 18 km), the most extensive endoreic karst plateau (Padiş–Cetatile Ponorului), typical karstic depressions (rand depressions) developed at lithological contacts, poljes and the only ice caves (Scărişoara, Focul Viu). The most complex karst evolution occurred in the Bihor massif and Pădurea Craiului Mountains, where at least two fossilized karst levels can be distinguished. In the Codrului Mountains, bare karst prevails, developed on a karst planation surface at 700–900 m. In the Trascău Mountains, typical karst relief characterizes the nappe outliers in the Cretaceous flysch mass.

15.8   Transylvanian Plateau

Lying between the Apuseni Mountains and the Carpathian mountain arc, the Transylvanian plateau consists of relatively soft, young Tertiary rocks and is characterized by an unforested, hilly landscape at elevations of 400–700 m. Following neotectonic uplift, the main rivers (Someş, Mureş and Olt) have dissected the landscape to depths of 100–200 m.

15.8.1   Morphostructural Features

Several relief types related to structure can be distinguished.

1)   Structural relief on the Someş plateau is well developed. The predominant features are cuestas on limestones, sandstones and conglomerates, with the rocks, in some areas, being folded or domed (*see* Fig. 15.10). The anticlines have been breached usually by erosion to give central depressions with infacing scarps.
2)   The Tirnava plateau has a monoclinal morphostructure, the strata dipping gently to the north. The strata are,

however, disturbed by several minor anticlines, some of them of diapiric character. The cuestas controlled by this monoclinal morphostructure overlook river valleys with a general east–west direction (Tirnava Mare, Tirnava Mică and Hîrtabaciu Rivers). Around occasional dome structures, smaller cuestas and hogbacks have developed with typical annular river patterns.
3)   In the Transylvanian plain, where clays and marls predominate, structural control is less evident and mass movement features are widespread.
4)   Depressions situated on the southern and western sides of the Transylvanian plateau, at the contact with the mountains, have a definitely subsequent character (the Hoghiz–Veneţia couloir, the Făgăraş, Sibiu, Sălişte depressions, the Mureş couloir between Turzda and Vinţu, etc.).
5)   The sub-Carpathians of Transylvania, situated in the eastern and north-eastern parts of the region, include many complex structurally controlled forms (central depressions in anticlines, structural ridges, synclinal depressions, synclinal ridges, structural surfaces, cuestas and mesas). The morphostructural unit with the most evident sub-Carpathian relief lies between the Olt and Mureş at the contact with the volcanic morphostructures.

Mud volcanoes have developed in the clay–marl region of the Transylvanian plateau, under which gas fields exist under pressure. The typical forms are cones a few metres high and circular hollows full of muddy water through which the gas bubbles out.

15.8.2   Relief

15.8.2.1   Planation surfaces

The Carpathian pediplain passes under the sediments of the Transylvanian plateau, as shown by the row of crystalline hills in the north-west of Transylvania ('the hidden mountains of Transylvania'). Traces of the younger (Sarmatian–Meotian) level of the middle Carpathian surface occur sporadically in the Someş plateau and in the sub-Carpathians. The most important surfaces are the valley pediments formed after the Rhodanian tectonic movements and in the Lower Pleistocene. These surfaces are essentially glacis with marked structural and lithological influence. They developed through lateral erosion of the streams and by parallel retreat of valley slopes. The first pediment level corresponds chronologically and genetically to a surface that bevels the main interfluves of the plateau. A second level extends into the hilly regions, frequently continued as remnants of broad valley floors above the Quaternary river terraces.

15.8.2.1   River terraces and deposits

In the valleys of the Transylvanian plateau, six to seven sets of river terraces can be distinguished. The terraces are most extensive along the larger valleys, such as those of the Olt, Mureş and Someş. In most regions, the terraces are parallel to the direction of the river flow, but at the junctions between the piedmont and the plain or between depressions and the mountain border, they spread out in a fan-like manner. The age of the terraces is Quaternary

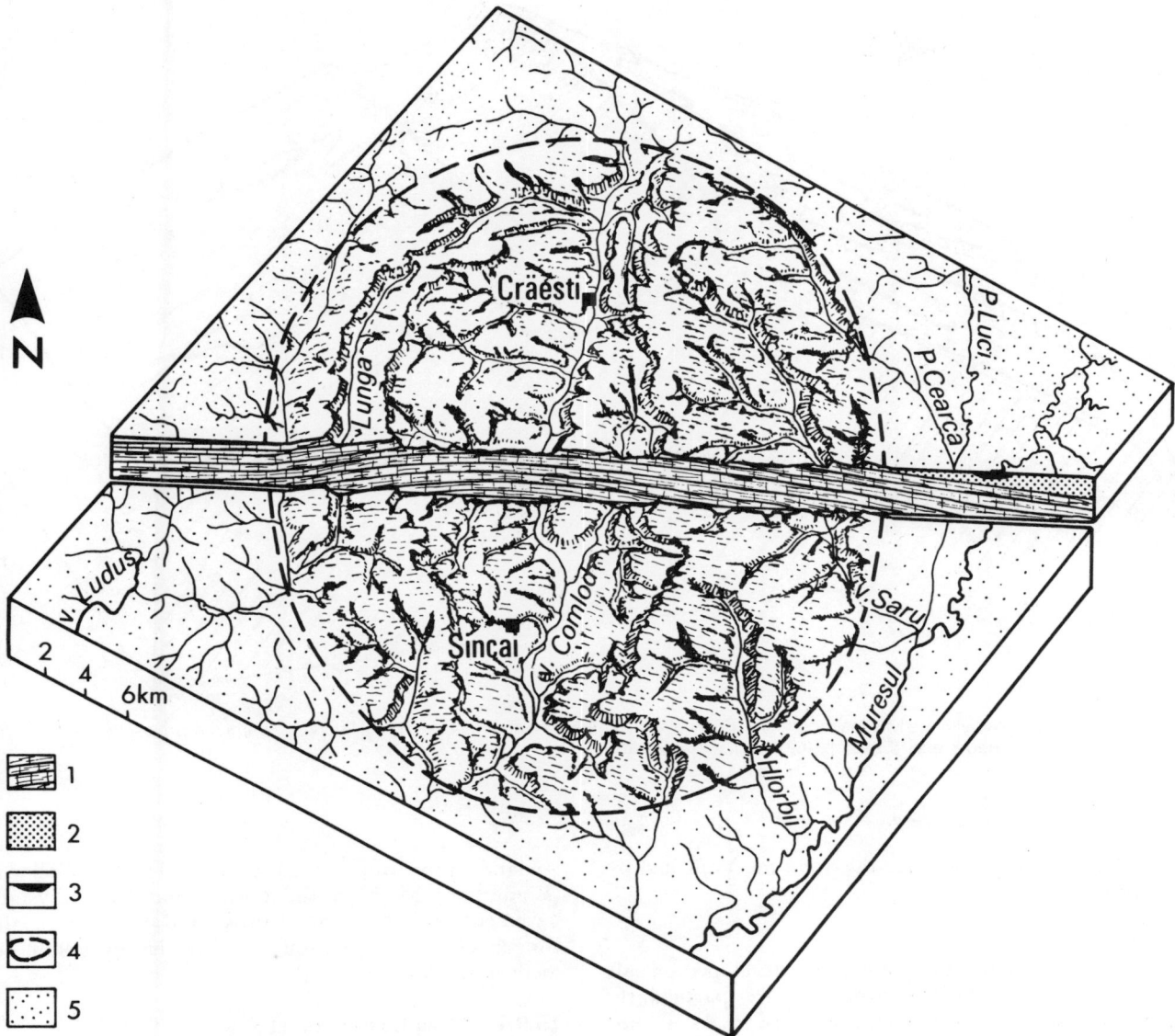

Fig. 15.10 Idealised block-diagram showing morphological characteristics of the Sinca dome in the central zone of the Transylvanian basin, showing eroded anticlinal structure and flanking cuestas [Acad. Republ. Pop. Romine, Bucureşti, (1960)] (1) Sarmatian, (2) Pontian, (3) Quaternary, (4) edges of the dome, (5) area outside the uplift.

generally. Some have been fairly well dated: for example, the 2–4-m terraces on the Someşul Cald are Upper Würm–Holocene, the 26-m terrace on the Someşul Mic is Upper Pleistocene, and the 30–40-m terrace on the Someşul Mare is Würm I. Starting from these landmarks, on the Someş at least two terrace groups can be delimited: (1) the lower terraces at 6–12, 18–22 and 30–40 m, whose deposition belongs to the Würm stadials and (2) the upper group whose age can be dated as Middle and Lower Pleistocene, or even Upper Pleistocene.

On the Transylvanian plateau, the most conspicuous deposits are represented by a series of coalescing proluvial cones, the alluvium of the terraces and valley floors of the Olt, Mureş and Someş, and the thick silty weathered mantle on the watersheds.

15.8.2.3 *Slope forms and processes*

At the present-day, soil erosion due to torrential rain is common on the slopes of the Transylvanian plateau (*see* Fig. 15.11). Rainfall intensities of more than 2.5 mm per minute have been frequently reported. The slopes exposed to the south and west are the more severely eroded— approximately 30 percent more than those facing north or east. The plateau is also an area of high landslide frequency, where old massive landslips are constantly being reactivated, especially in Sarmatian deposits. Frequent superficial slides occur in spring. On the Tîrnava plateau, the cuesta fronts and steep slopes in the catch-

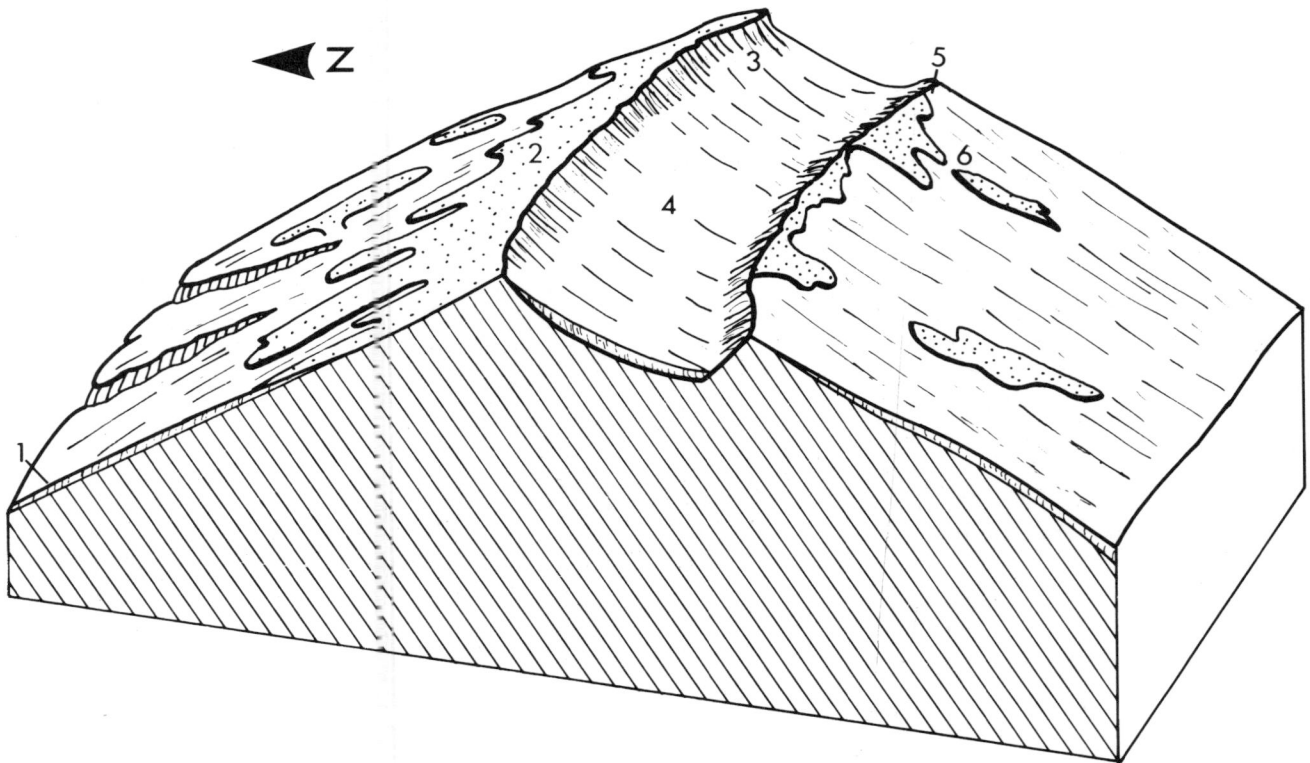

Fig. 15.11 Structural scarps in the Micusa mountains [G. Niculescu, (1965)]. (1) stable slope, (2) solifluction cover, (3) scarp caused by deep creep, (4) depression due to deep creep, (5) scarp, (6) solifluction deposits.

ments of perennial streams are the areas most affected by mass movements.

15.8.2.4 Karst morphology

Karst in the Transylvanian plateau has developed on salt and gypsum deposits. The processes are particularly active, but the area occupied by these rocks is not extensive. Lapiés, rock rills, avens, dolines and a few caves are to be found.

15.8.2.5 Aeolian features

An area of aeolian sand lies between the Someş and Barcău Rivers, covering approximately 31 000 ha, of which 5000 ha are underlain by quicksand and semi-quicksand. The relief consists of an assemblage of main dunes (stretching from south-west to north-east) and secondary dunes with varying orientations. The main dunes are 1–15 km long, 100–350 m wide and 5–15 m high.

15.9 Lower Danube Lowland

This lies between the Carpathian mountains and foot-hills in the north, and the Stara Planina Mountains in the south; the northern Dobrodgea provides a connection with the eastern European plain. The Danube divides the lowland into two unequal parts: (1) the Wallachian (Romanian) plain in the north, where the crystalline basement lies at great depths, and (2) the northern Bulgarian plain in the south, where the crystalline basement is shallower.

15.9.1 Morphostructural Features

The basic morphostructure is the down-faulted Moesian platform. The major Carpathian border fault divides the lowland from the Carpathian mountains, being geomorphologically represented by the depression of the Carpathian foredeep (*see* Section 15.6.3.1). The Carpathian border fault marks the overthrust of the folded Carpathian rocks across the platform cover of the Moesian platform. The southern border is formed by a fault zone stretching from the town of Varna (Bulgaria) in the east to Veliko–Tyrnovo, Lovetch and Mikhaylovgrad in the west. The northern and north-western parts of the lowland are structurally subsiding areas of the Carpathian foredeep, while the south-eastern part is a similar zone of subsidence belonging to the lower Kamtchiya marginal depression. The basement is overlain by three series of deposits: (1) Triassic, (2) Jurassic–Palaeogene and (3) Neogene–Quaternary. The sedimentary cover is up to 8 km thick, but the thickness varies in different parts of the lowland. The intensity of present-day tectonic movements ranges from uplifts of 3–5 mm per year in some areas of Bulgaria to a subsidence of 1 mm per year in the Romanian sector.

15.9.2 Relief and Drainage

The Danube occupies an asymmetrical position in the lowland, the stream having been pressed against the uplands to the south by the alluvial cones of tributaries entering from the north; the Carpathian rivers carry large quantities of sand and gravel forming these extensive alluvial cones. In the west, the Danube enters through the major water gap known as the Iron Gate, at a height of only 36 m above sea level, even though the Danube still has another 1000 km to flow before entering the Black Sea. On the right bank, the river is flanked by a 70–150-m high scarp. To the north of the river, there is an extensive floodplain with many oxbow and river lakes. Its northern boundary is formed by a scarp 20–40 m high; this is the limit of the Wallachian plain. The river bed is 1–1.5 km wide, and its downstream gradient is only 0.033 percent. Below Silistra, the direction of the Danube alters to flow to the north-east. The character of the river also changes, as it separates into two or more channels flowing as much as 20 km apart. The floodplain contains numerous river lakes, oxbow lakes and peaty areas. From Brăila to Tulcea, the river is once again confined to a single large bed. At the head of the delta, it begins to split into branches which occupy an area of 4000 km². Above the floodplain of the Danube, a series of river terraces rises at the following heights above the river: 7–10, 17–22, 27-35, 50–65 and 80–110 m. The development of terraces has been influenced by both tectonics and eustatic movements of the level of the Black Sea. Because of subsidence, the number of terraces decreases from west to east. They are mostly covered by loess, whose thickness is generally 8–15 m, except in the Romanian plain where it reaches 30–40 m. It is of aeolian origin, the material coming from adjacent hilly regions and from the continental shelf of the Black Sea exposed over large areas in the Upper Pleistocene. Piping in the loess cover is an important process, leading to the formation of funnels (4–7 m diameter), avens (4–5 m deep), podis, etc. Gullies with vertical or near-vertical sides are also common in the loess.

Chapter 16
Balkan Peninsula

Geomorphologically, the Balkan peninsula differs from other peninsulas of southern Europe in that it is more widely connected with central Europe. The northern boundary of the peninsula is marked by the Drava river valley, by part of the Danube in the area around the Iron Gate and by the foot of the Stara Planina (Balkan mountains) as far east as Varna on the Black Sea coast. The northern part of the Peninsula is massive and compact, but the southern part is geomorphologically strongly dissected in a system of peninsulas, islands and bays. The southernmost part is formed by the Peloponnese peninsula joined to the mainland only by the narrow isthmus at the head of the Gulf of Corinth.

Morphostructurally, the Balkan peninsula is formed by Hercynian massifs rimmed by Mesozoic and Tertiary epigeosynclinal mountains (*see* Fig. 16.1). The Hercynian massifs are divided into blocks, the largest one of which is represented by the Rila–Rodopi block mountains in Yugoslavia and Bulgaria. The blocks of the crystalline pre-Alpine basement are also found in some eastern Balkan uplands, such as Strandzha, Sakar, central Sredna Gora, the Vitosha mountains, Osogov and Ruyska [see Bognar (1980), Gams (1981), Vaptsarov and Mishev (1978), Zeremski (1972)].

The epigeosynclinal mountains of the Dinaric Alps (Dinarides) form a broad band in the western part of the Balkan peninsula. The southernmost point of the Dinarides is the island of Crete. The epigeosynclinal mountains of the Stara Planina system are of the same age and origin.

The present-day morphostructural patterns are essentially relatively youthful and were formed during several phases in the Neogene and Quaternary Periods. Because of the extent and importance of neotectonic movements, mountain systems are the major geomorphological features of the Balkan peninsula. The main European watershed lies mostly only 60 km from the Adriatic coast, but 500–900 km from the Black Sea.

16.1 Dinaric Alps

The epigeosynclinal mountain range of the Dinaric Alps includes the Outer Dinarides, the Inner Dinarides, and the Pelagonides and Hellenides.

The mountainous part of the Dinarides represents a system of extensive fault-folded block morphostructures. Large parts are formed from Mesozoic limestones. As the intramontane basins and some poljes lagged behind during the general tectonic uplift of the Dinarides in the Neogene, sediments that once formed part of a larger mantle are preserved in them (e.g., basins of Bugojno, Sarajevo, etc.). To the east of Sarajevo, and especially east

of the Drina River, where at present the annual precipitation is less than 700 mm, extensive plateaus have been preserved even in impermeable and semipermeable rocks, representing remnants of former planation surfaces, including Mount Romania (1649 m), Mount Tara (1637 m) and Mount Treskavica (2088 m). The largest remnant of such planation surfaces is the plateau of Stari Vlah, dissected by the deep canyons of the Uvac, Lim and Moravica (a tributary of the western Morava river). Some higher ridges stand above the general planation level [e.g., Mučanj (1543 m)]. In the mountains of Kopaonik (highest elevations, 2017 m) and Sandžak, the morphostructural features show a north and north-north-west trend. The general morphostructural trend in the Yugoslavian Dinarides is, however, north-west to south-west – the so-called Dinaric alignment in the geomorphological literature. Farther south in Albania and Greece, a north-north-west–south-south-east or a north–south trend is more usual. There, along the sea coast, locally extensive Quaternary plains of aggradation appear as a result of the virtual absence of large rivers today in the residual limestone and dolomite (karst) areas of the Outer Dinarides.

In the western part of the Dinarides, outside the limestone regions, the remnants of Pliocene planation surfaces are restricted to the crests of the mountain ridges only. In Slovenian geomorphology, the ridge levels play an important role in the reconstruction of the genesis of the relief. River terraces are virtually absent owing to the strong Pliocene and Quaternary tectonic uplift. The uplift took place in several phases, but the number of phases is still disputed. Bevelled areas of the karst represent remnants of Tertiary landscapes in spite of the continuous lowering of the surface due to karst denudation (mostly by corrosion). No erosional valleys are to be found [see Gams (1981), Lazarevic (1975)].

The highest mountains of the Dinarides are, as a rule, situated along the watershed between the Black Sea or Danube drainage and the Adriatic drainage basins. Alpine mountain characteristics are generally restricted to the highest peaks, such as parts of Mounts Velebit and Durmitor, rising to 2522 m. The main exceptions are parts of the Prokletije mountains [e.g., Mount Jezerces (2694 m)] with crests rising 1500 m above deeply dissected valleys. The main ridges comprising the watershed between the Black, Adriatic and Aegean Seas are aligned toward the east and north-east, possibly reflecting the contact of the lithospheric plates shifting from the south-west (Italian Adriatic plate) and from the south-east (Aegean–Macedonian–Albanian plate) (*see* Fig. 16.2). The parallel alignment of Mount Cukalit in Albania is

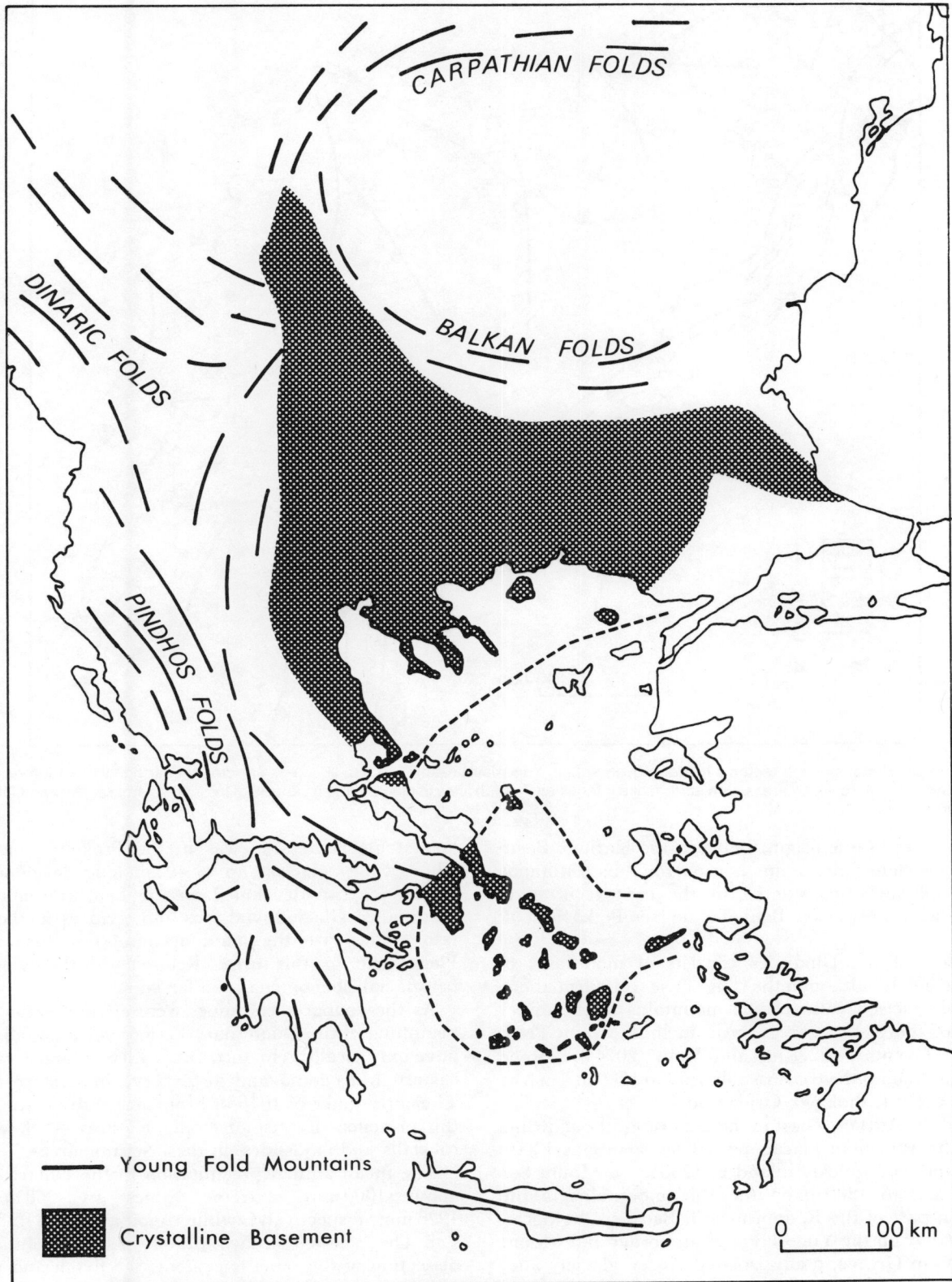

Fig. 16.1 General structural outlines of the Balkans and Hellenides.

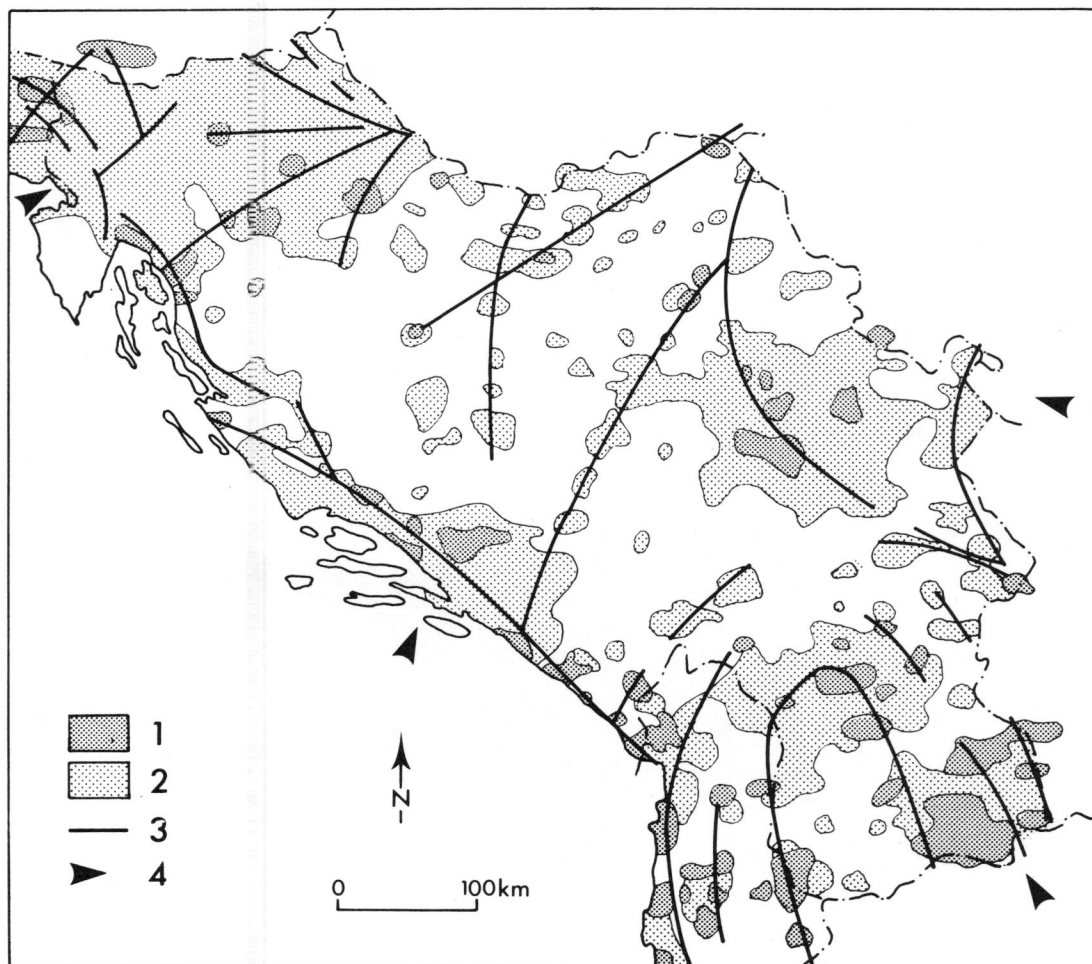

Fig. 16.2 Seismic intensity and tectonic (plate) movements in Yugoslavia [Gams (1981)]. (1) Areas affected by earthquakes of intensity greater than 9 on the Mercalli scale, (2) areas with earthquake intensities 7–9, (3) lines joining the main centres of intense seismic activity and (4) direction of plate movements.

probably of the same origin. Between the Karlova Basin and the Prokletije mountains is a narrow, discontinuous belt of ophiolites, considered in the context of plate tectonics as the Mesozoic Benioff zone [see Petkovic *et al.* (1976)].

The rest of the Dinarides consists of mountains of moderate height, although the ranges rise to 1000 m above the deeply incised valleys. The mountains in the upper reaches of the Neretva river basin include Mount Prenj (2155 m), Čvrsnica (2228 m) and Vran (2074 m). Some mountains behind the coast rise sharply to 1700 m, such as Mounts Velebit, Biokovo, Orjen and Lovčen.

Behind the Adriatic coast in the Dinarides, the snowline during the Würmian glacial period lay at about 1300–1350 m and rose quickly inland to 1740 m on Mount Šar Planina and to 1880 m on the Rila mountains in the Bulgarian part of the Rodopi massif. Between the Soca–Izonco valley on the Yugoslav/Italian border and Mount Grammos in Greece, many isolated cirque glaciers and, exceptionally, a few small ice caps existed in the coastal ranges. Only a few of these glaciers reached the foot-hill zone. Glacial sculpture is therefore very limited. The largest glaciated area was in the Prokletije mountains where valley glaciers up to 40 km long developed. The largest ice cap developed on the Durmitor and Orjen mountains. Glaciofluvial fans and terraces in the mountain valleys are the main accumulation forms of the Pleistocene in this area. Relatively little evidence of periglacial phenomena is so far known.

As the youngest volcanoes were already extinct at the beginning of the Quaternary Period, volcanic relief forms have only locally remained (near Zletovo and Kratovo in eastern Macedonia and at Zvečevo in western Serbia). The earthquake of 1976 in Montenegro drew attention to the presence of seismotectonic features, such as fossil rockfalls and landslides, in these seismic areas.

The mean annual precipitation in the Dinaric Alps is about 1000 mm (extreme values are 258 mm and 4926 mm, respectively), while runoff averages $15.3 \ l \ km^{-2} \ s^{-1}$. The role of surface runoff in the evolution and dissection of the relief is evident; the flysch and molasse hills in eastern Yugoslavia, with annual precipitation of less than 600 mm, are less dissected than are the western parts in Slovenia which receive more rain.

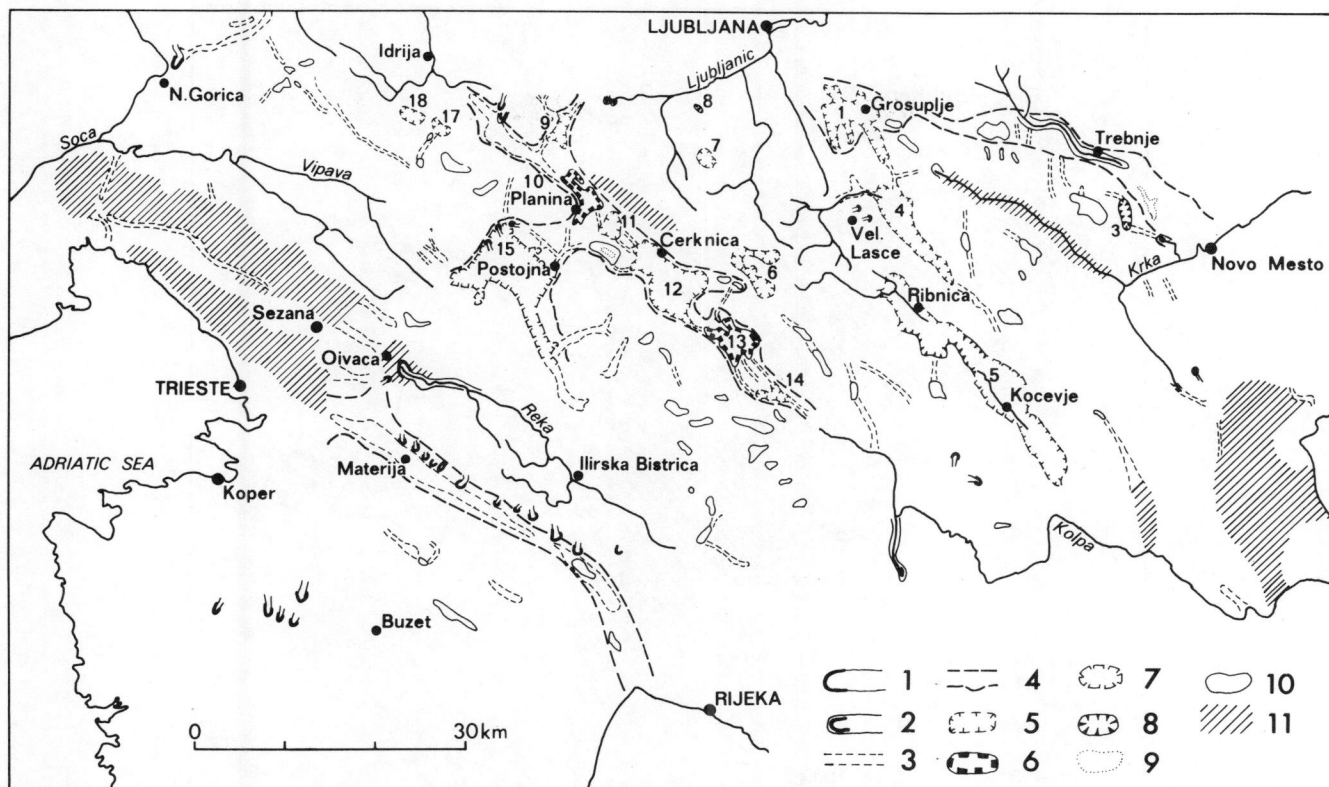

Fig. 16.3 Main surface karst phenomena in the Slovenian Dinaric mountains. Numbers on the map refer to poljes: (1) Grosupeljsko, (3) Globodol; (4) Dobro; (5) Ribniško-kočevsko; (6) Bloke; (7) Rakitna; (8) Ponikve; (9) Logaško; (10) Planinsko; (11) Rakovsko; (12) Cerkniško; (13) Loško; (14) Babno; (15) Postojnsko; (17) Črnovrško; (18) Zadlog. Key: (1) blind valley; (2) pocket valley; (3) dry valley; (4) dry valley system; (5) border polje; (6) overflow polje; (7) peripheral polje; (8) dammed polje; (9) karst doline; (10) uvala; (11) karst plain.

At the beginning of this century, the famous Yugoslavian geomorphologist Cvijić claimed that the bevelled relief of Šumadija in the northern part of Serbia was the result of abrasion by the Neogene Pannonian Sea. Recently, a subaerial erosional origin has been suggested, but Macedonian geomorphologists still acknowledge the role played by wave abrasion and lacustrine deposition in producing the Neogene terraces in the basins of Skopje, Štip, Kavadarcu and Metohia. Of the numerous Neogene lakes, only a few have remained to the present, some of them still retaining their ancient fauna (e.g., Lake Ohrid).

In the part of the Adriatic basin along the Dinarides, north-west of the Ionian trench, block and fault-folded structures covered by Pleistocene fluvial and Holocene marine sediments are typical. Some parts of the coastal belt are less folded and more worn down than the neighbouring Dinarides, so that some geomorphologists and geologists consider that the karst plateaus of Istria and Ravni Kotari near Zadar are a part of the so-called Adriatic mass. These karst plateaus remained at low altitudes due to slight uplift or due to the progressive shifting of the Adriatic basin towards the north-east as a consequence of the drift of the African plate. Another possible explanation is that there has been subsidence of the eastern Adriatic coast. It is also known that the so-called Dalmatian kryptokarst and fossil erosion levels occur on the floor of the Adriatic Sea. The flat floors of the

Bay of Rijeka (50 m deep) and the Bay of Trieste (30 m deep) are examples of fossil subaerial erosion levels now submerged. During the Pleistocene glacials, loess and loess-like sediments also covered the dry floor of the northern Adriatic basin; the island of Susak is built of such sediments. Due to the recent sinking of the eastern Adriatic, the Dalmatian coast has a submergent character; there is extensive evidence of very recent sinking and marine transgression. Because of this, the Adriatic Sea has not been able to cut recent abrasion forms. The morphostructural patterns and the absence of large river valleys have produced a straight coastline, taking into account both the mainland and the island. The famous Bay of Kotor is of tectonic origin and is a graben.

16.1.1 Outer Dinarides

The boundary between the Outer and Inner Dinarides runs from Tolmin (in the valley of Soča–Izonco at the Italian/Yugoslav border) through Žiri, south of the city of Ljubljana, Novo Mesto, Karlovac–Bosanski Novi, Sanski Most, Sarajevo and Kolašin–Škoder, to Albasani in Albania. Mesozoic limestones and dolomites are the principal geological formations. Inland from the Adriatic coast, the Mesozoic sediments become progressively older. Next to the coast, Cretaceous and Palaeocene limestones together with Eocene flysch (the last occurring in the synclines) generally occur, but farther inland Jurassic and

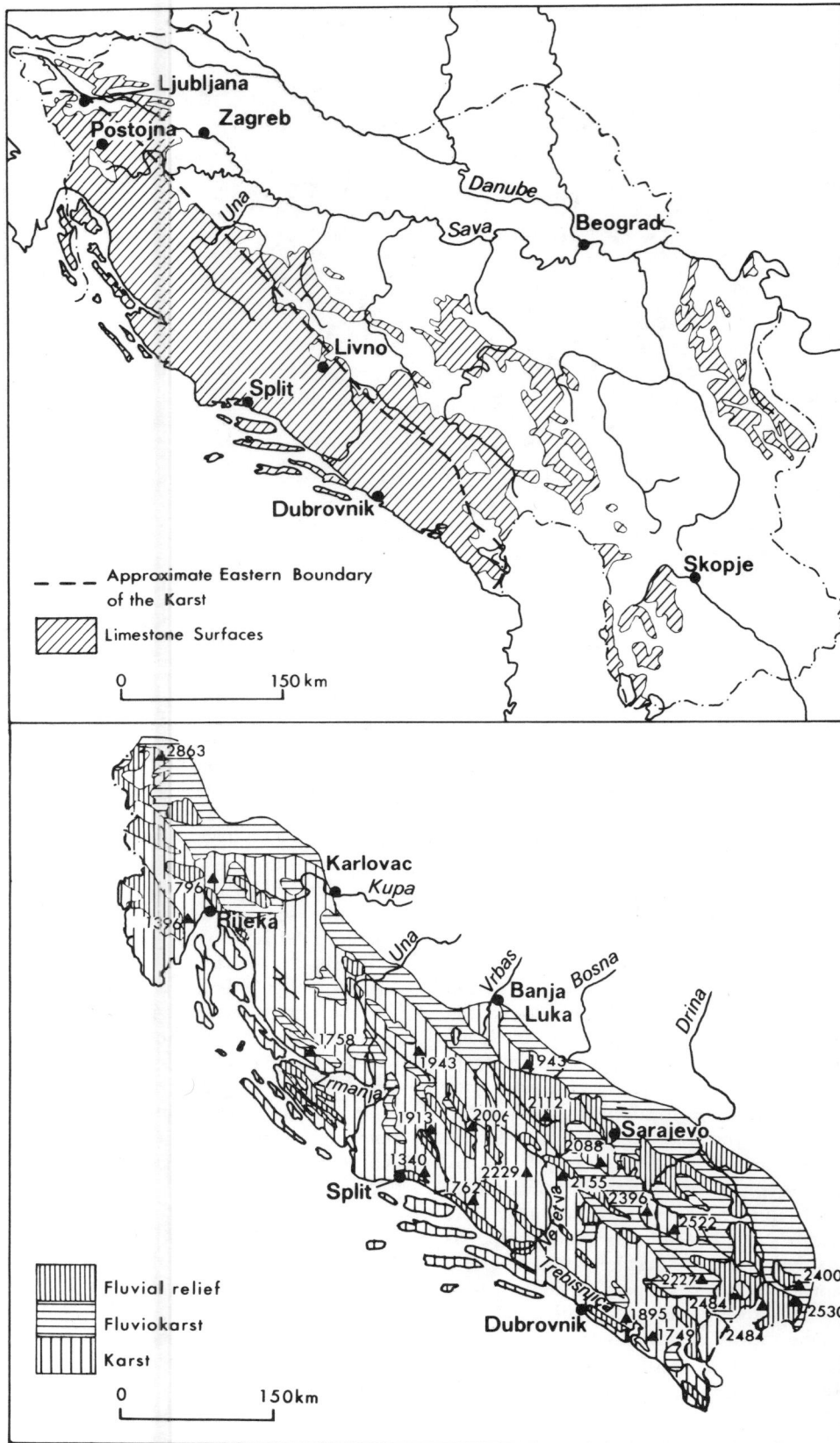

Fig. 16.4 Principal types and areas of karst in Yugoslavia [after Sweeting (1972)].

Fig. 16.5 The problem of definition of a polje [Gams (1974)].

Triassic limestones and dolomites are encountered. The older Mesozoic sediments are usually dolomitized.

In the Mesozoic Era, the Outer Dinaride region consisted of an epiplatform with relative stable tectonic conditions. In the Tethyan Sea, calcareous rocks were laid down, the complete series of the Cretaceous, Jurassic and Triassic amounting to a thickness of 2–3 km in each of these periods.

In the Tertiary, tectonic movements occurred, especially between the Eocene and the Miocene Epochs, and again at the end of the Miocene. Because of this, in many regions along the Adriatic coast, the Palaeocene and Eocene strata are the youngest. It may be concluded that uplift began at first along the western and southern rim of the Outer Dinarides. In the hinterland, up to the middle Pliocene, alternating phases of marine and lacustrine sedimentation dominated. In the Sarmatian and Pannonian periods, the Pannonian Sea was presumably connected with the Mediterranean Sea through western Bosnia. Tectonic movements continued even into the Quaternary Period. Pleistocene overthrusting reached 30 km in some places, such as Trnovski Gozd, the so-called Visoki Krš in Montenegro and Mt Cukalit in Albania. The youngest tectonics along the contact zone with the Pannonian Basin created an upland built of Neogene flysch and molasse, interrupted by isolated mountain groups composed of Palaeozoic schists [e.g., Motajica (652 m)], granites and granodiorites [e.g., Mount Cer (687 m)]. This foot-hill belt is flanked in the south-west by higher mountains of varied geological structure, including Zrinjska Gora (615 m), Kozara (987 m), Majevica (915 m) and Maljen (1103 m) in western Serbia.

The Outer Dinarides contain the largest continuous karst area in Europe (*see* Fig. 16.3). The regional name for the north-westernmost corner of the limestone area of the Outer Dinarides, behind the Bay of Trieste, is *Kras* in Slovenian; this became the word karst in geomorphology. Nearly all karst features are to be found on limestone or dolomite rocks, except for the features developed in gypsum along the Macedonian river Radika [see Gams (1974), Herak and Stringfield (1972)].

The extent of the karst area depends on how one defines the karst phenomena (*see* Fig. 16.4). The area occupied by so-called surface karst amounts to about one-third of Yugoslavia: 67 percent of Montenegro, 45 percent of Bosna and Hercegovina and 27 percent of Slovenia. Nearly the same area again is characterized by fluviokarst where underground karstic drainage persists but where surface karst depressions are absent (e.g., in Slovenia, 42 percent). As well as the subaerial karst on the mainland and islands, submarine karst exists offshore.

Many geomorphological terms have been derived from the Slovenian and Serbo-Croatian languages, such as doline, polje, ponor, uvala and jama. The number of poljes in the Dinarides depends on their definition (*see* Fig. 16.5); some workers claim three times as many as others. In the central and south-eastern part of the Outer Dinarides, the polje floors are mostly underlain by Tertiary sediments, but, in the north-western part, semipermeable dolomites predominate beneath their floors (*see* Fig. 16.6). Along the south-western and north-eastern margins of the Dinarides are the largest karst plains where tectonic uplift has been slow and the groundwater level lies at shallow depths (e.g., the plain behind the town of Karlovac, near Zarad, in Istria). Fig. 16.7 shows examples

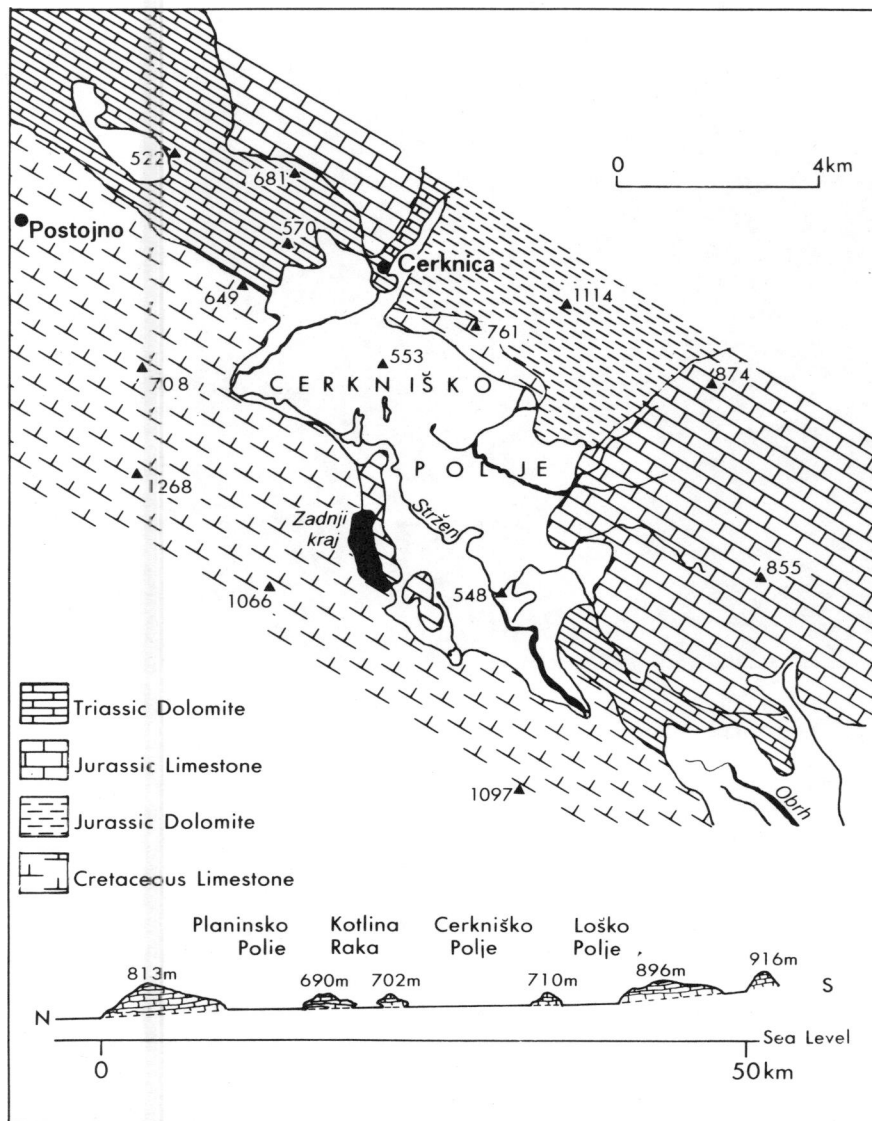

Fig. 16.6 Cerknisko polje [for location, *see* (12), Fig. 16.3]. The section below shows the underground drainage connections [after Sweeting (1972)].

of seasonally flooded poljes, in one of which lies a permanent lake. The rivers have cut some canyons, but the higher uplifted plains are more dissected by karst canyons and dotted by dolines (the so-called Dinaric karst with poljes and dolines) [see Gams (1974)].

16.1.2   Inner Dinarides
The Inner Dinarides are formed by Palaeozoic rocks, Mesozoic (mostly carbonate) sediments, volcanic and metamorphic rocks, and Tertiary molasse. Tectonic phases of Mesozoic and Tertiary age formed the present-day morphostructural pattern (*see* Fig. 16.8).

16.1.3   Pelagonides and Hellenides
The province of Kosovo, together with central and western Macedonia, has a relief in some respects similar to that of the Rila–Rodopi system (*see* Section 16.2) with pre-Mesozoic sediments and large basins. The name of this

region is derived from the basin of Pelagonia (80 km long and 20 km wide) whose floor lies at altitudes of 580–660 m. The same relief type continues into Greece where the same mountain system is known as the Hellenides. In Yugoslavia, the mountain system is built of crystalline schists of Precambrian age and of early Palaeozoic sediments, terrigenous carbonate rocks, volcanics and Mesozoic sediments. The basins are filled mostly by unconsolidated Tertiary deposits. In the lowland of Thessaloniki, in northern Greece, Quaternary fluvial deposits are also widespread. In the older rocks, various types of block mountain are developed. The alignments of mountain crests, valleys and elongated basins mostly conform to the trends of the so-called Macedonian–Albanian morphostructural arch. The northernmost part of this arch is represented by the mountains of Žljeb (2382 m) and Mokra Gora–Suva Planina (1750 m) on the north-west, and by the Šar Planina (Titov vrh, 2748 m) and Korab

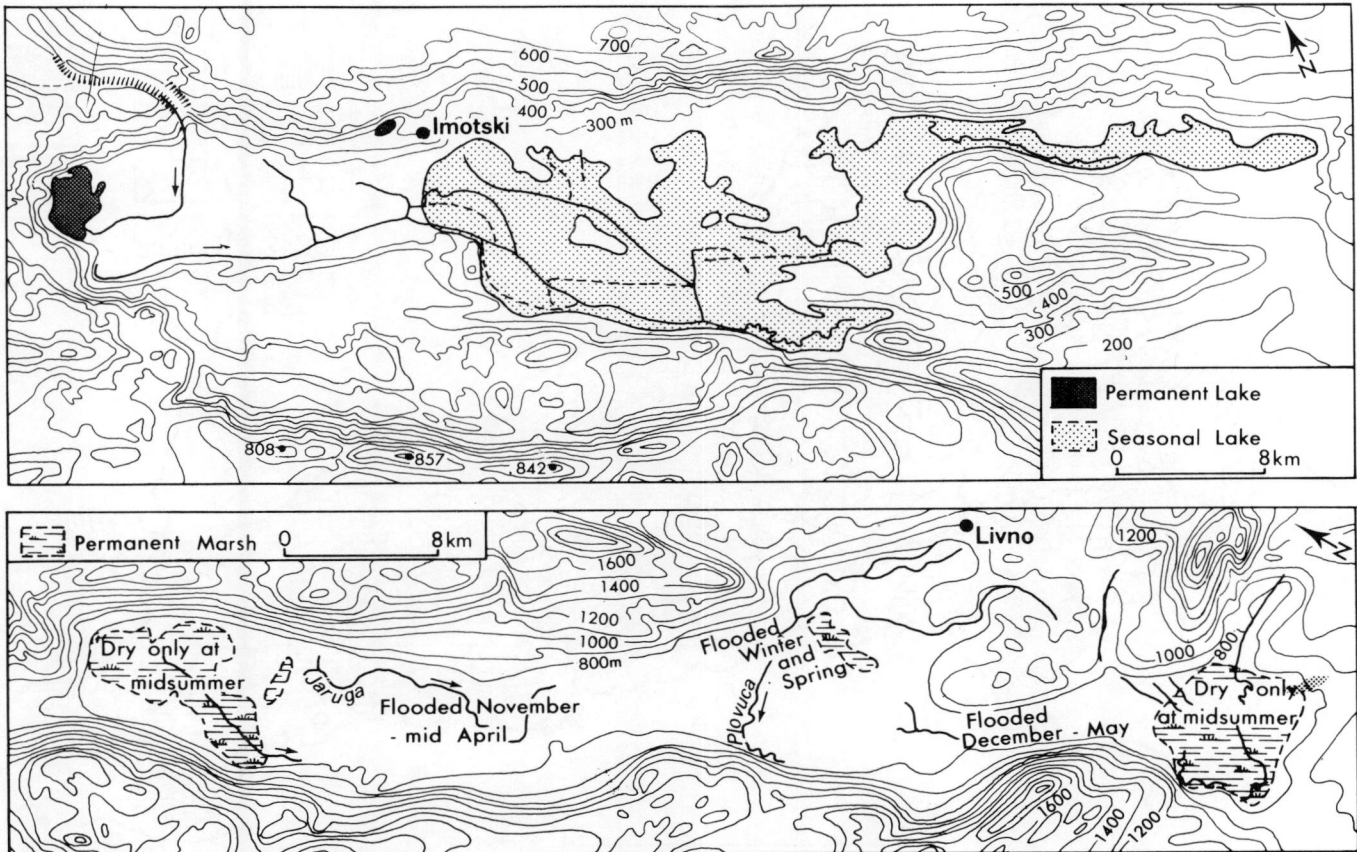

Fig. 16.7 Contour maps of the Livno and Imotski poljes [from 1:100 000 topographic maps, Yugoslavia and Roglić (1938)].

(2764 m) to the south of the large Metohia Basin which measures 80 by 40 km. Farther to the south, there are the Suva Planina (1852 m), Buševa Planina (1751 m), the Galičica mountains between Lake Ohrid and the Prespa Basin, etc. In southern Macedonia and in neighbouring Greece, the mountains are more isolated, examples being the Nidže mountains with Mount Kajmakčalan (2571 m), Kožuf (2171 m) and Vermion (2052 m). Southward in Greece, circular (ring) patterns of morphostructure prevail, some of them with basins and lakes in their centres. Large intramontane basins are typical, such as those of Kosovo, Metohia, Skopje, Polog, Pelagonia, Ohrid, Arnissa, Korce and Florina. In the Miocene and Pliocene, many of these were occupied by lakes. After active tectonic uplift in the Pliocene and Quaternary, the number of lakes was reduced; only those of Ohrid, Prespa, Arnissa and some smaller ones remain today.

The plain of Thessaly consists of a large basin rimmed by mountain ranges, the eastern rim being formed by block mountains built of crystalline rocks and metamorphosed limestones (*see* Fig. 16.9). Three high mountains divided by passes form this range: Olympus in the northern part (2917 m), Ossa (1978 m) in the centre and Pelion (1830 m) in the south. Between Mounts Olympus and Ossa is the deeply incised water gap of the Pinios River. The western rim is formed by the fold mountains of the Ephir mountain system. The morphostructural features are elongated in a north-north-west–south-south-

east direction, forming long broad limestone ridges divided by elongated depressions. Karst features are less developed than in the Yugoslavian karst.

In the central part of Greece, the relief is complex and highly dissected. The main area is formed by the Attica peninsula which comprises a series of massifs of crystalline and carbonate rocks.

The Peloponnese peninsula is also characterized by dissected mountains and an indented coast. The main features are the result of Quaternary tectonic movement which divided the peninsula into a system of grabens and horsts (*see* Fig. 16.9). Block mountains are formed by crystalline schists and Mesozoic limestones; the basins are filled with recent unconsolidated deposits. The island of Crete has mountainous relief built of Jurassic limestones, Triassic shales, and Tertiary marls and conglomerates. Some neovolcanic rocks are also present. The highest points are formed by resistant limestones.

16.2 Rila–Rodopi Massif

Exhibiting a complex morphostructure closely connected with its division into blocks during the early and late Alpine phases of tectogenesis, this massif consists of a series of partially connected or separate units with relief ranging from low to high mountains (*see* Fig. 16.10). The general framework of the massif consists of both west–east (Alpine) and north–south-trending structures. The same

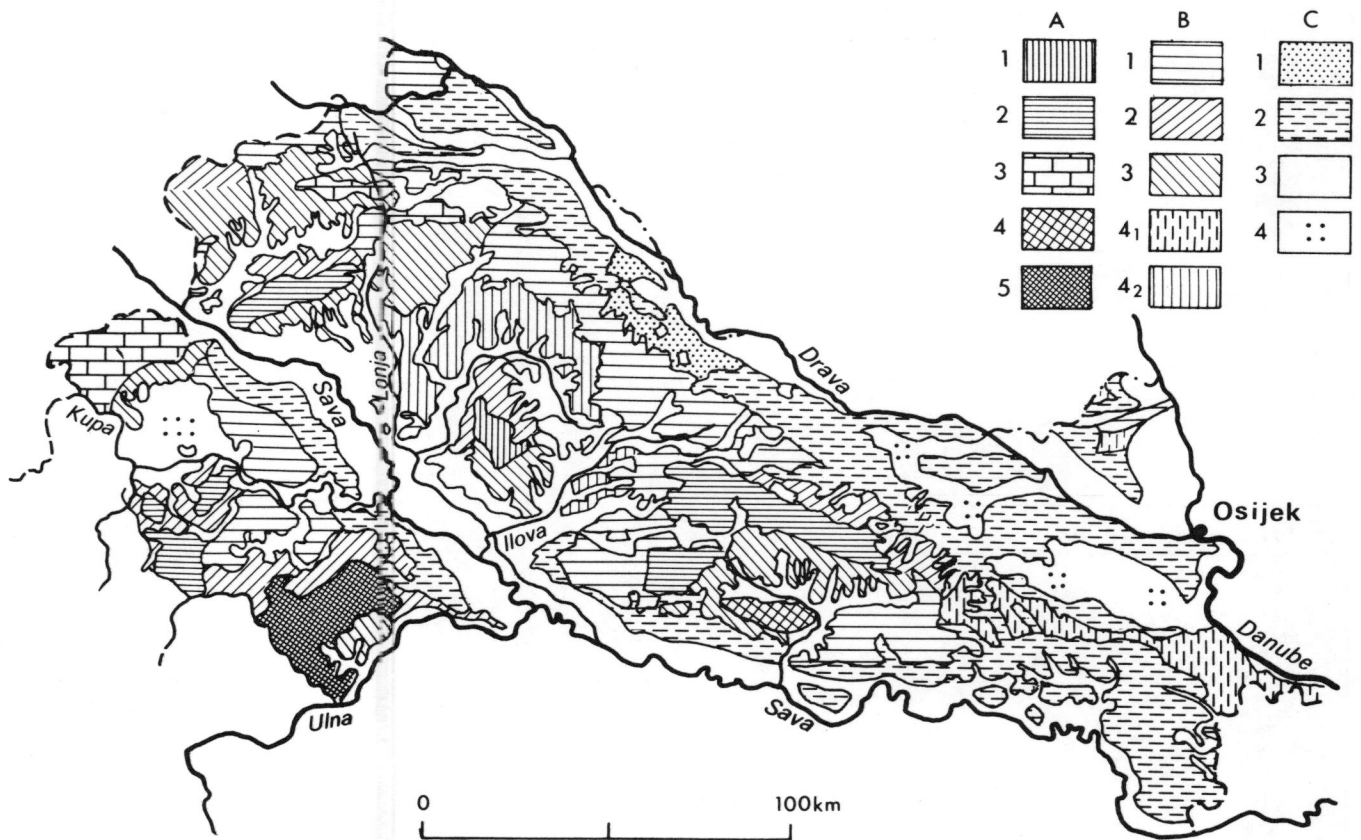

Fig. 16.8 Relief types in Croatia [after Bognar (1979)]. (A) *Denudational–tectonic relief:* (1) faulted and folded Palaeozoic massifs; (2) faulted and folded, reactivated and exhumed Palaeozoic–Mesozoic massifs; (3) faulted and folded, reactivated Mesozoic massifs; (4) faulted and folded, reactivated Mesozoic–Tertiary massifs; (5) faulted and folded, reactivated Palaeozoic–Tertiary massifs. (Note, in each case, the geological age refers to the date of orogenesis.) (B) *Erosional–depositional relief:* (1) uplands; (2) glacis, facing north; (3) glacis, facing south; (4_1) loess plateaus with pseudokarstic relief; (4_2) loess plateaus with linear erosion features. (C) *Depositional–tectonic relief (plains):* (1) fluvial–aeolian plains; (2) plains with river terraces; (3) floodplains; (4) fluvial–lacustrine plains.

orientations can be observed in the peripheral and intramontane basins [see Vaptsarov and Mishev (1978)].

In Yugoslavia (Macedonia and southern Serbia), the massif is built mostly of crystalline rocks (metamorphic, granitoids, porphyrites, dacites, granodiorites). Locally, Neogene deposits are present within basins surrounded by block mountains. Between the Leskovac Basin and the Danube, the relief is somewhat lower. Here, the crystalline basement subsided below the general level during the Neogene period and is today covered mostly by Neogene molasse which fills the valley floors and basins. The pre-Mesozoic basement, which includes some volcanics, rises to the surface in more or less isolated blocks, including the mountains of Jastrebac (1492 m), Rudnik (1132 m) and Avala (511 m). Typical features of the relief are isolated mountains groups, basins and narrow water gaps connecting them [see Zeremski (1972)].

The central part of the massif in Bulgaria is formed by a fold-faulted uplifted zone with rectangular fault-controlled borders. From west to the east, the elevation rises in steps from 1200 to 2200 m. At the higher altitudes, dissection increases and the highest peaks show the distinct features of Pleistocene glacial morphology.

The eastern part of the Rodopi massif is less striking in its relief. It is bounded to the west by the Momchilgrad depression of late Palaeogene age, which developed prior to the end of the Oligocene. At the present-day, on the other hand, slight uplifts are taking place. Planation surfaces, only slightly dissected by river valleys, and standing at altitudes of about 1200 m in the western part and 500–550 m in the eastern part near to the boundary with the lower Thrace depression, are characteristic of the eastern Bulgarian Rodopi uplands.

The Ograshden–Maleshevek uplands in the western part of the massif have similar morphostructural characteristics and relief. The eastern boundary is formed by the deep Struma river graben of submeridional trend, which was actively subsiding during the neotectonic period and filled with thick layers of Pliocene deposits. The eastern and south-eastern periphery of the Rodopi massif consists of Alpine block mountains with sharp relief forms. These include units of early Alpine age (Belasickiy, Pirin, Styrgatschskiy, etc.) or late Alpine age (the Ibredsheckiy massif). They have latitudinal or meridional trends forming a mosaic of relief units divided by deep basins (grabens) with complex structures and striking fault scarps.

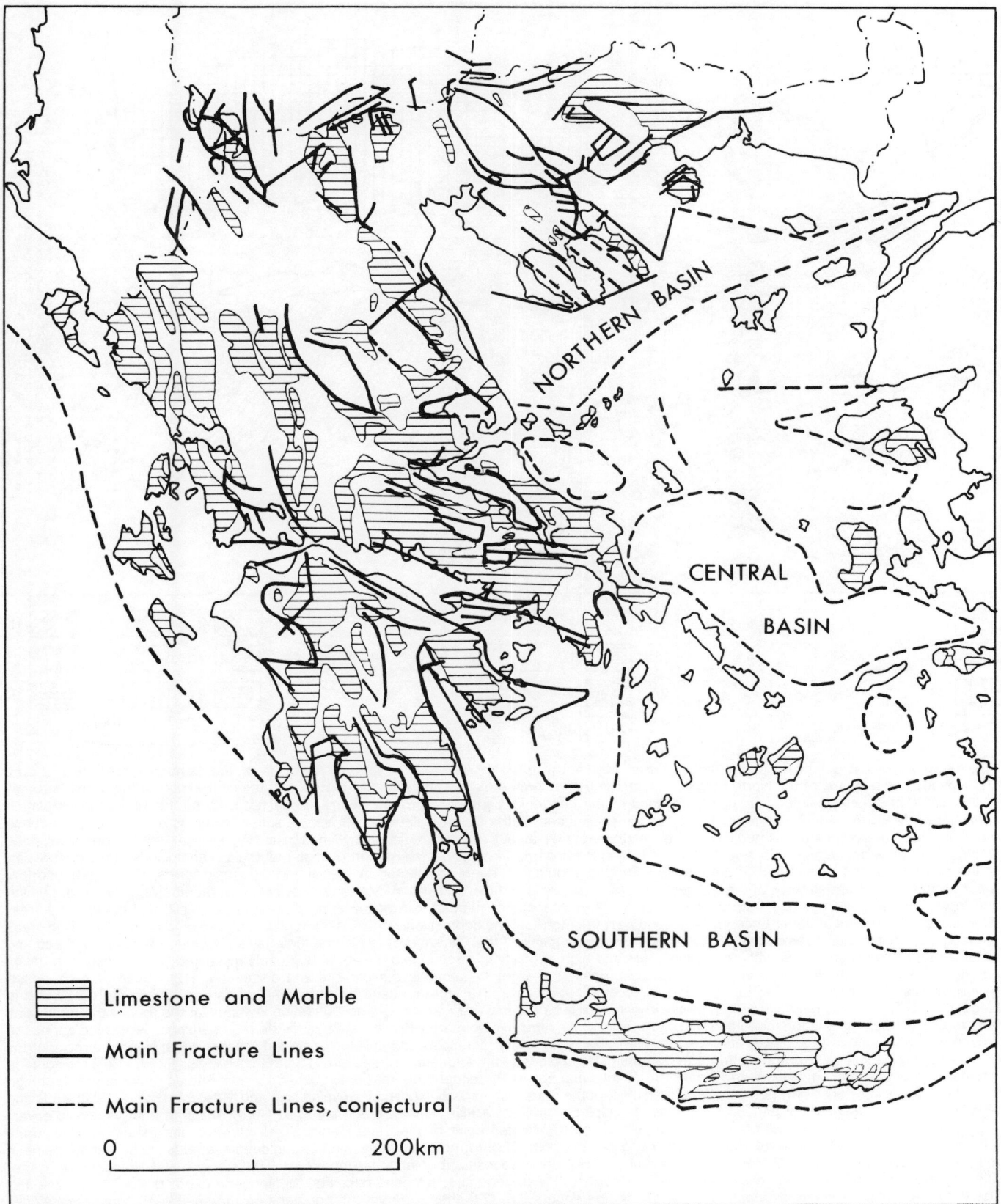

Fig. 16.9 Main fracture lines and areas of limestone in Greece.

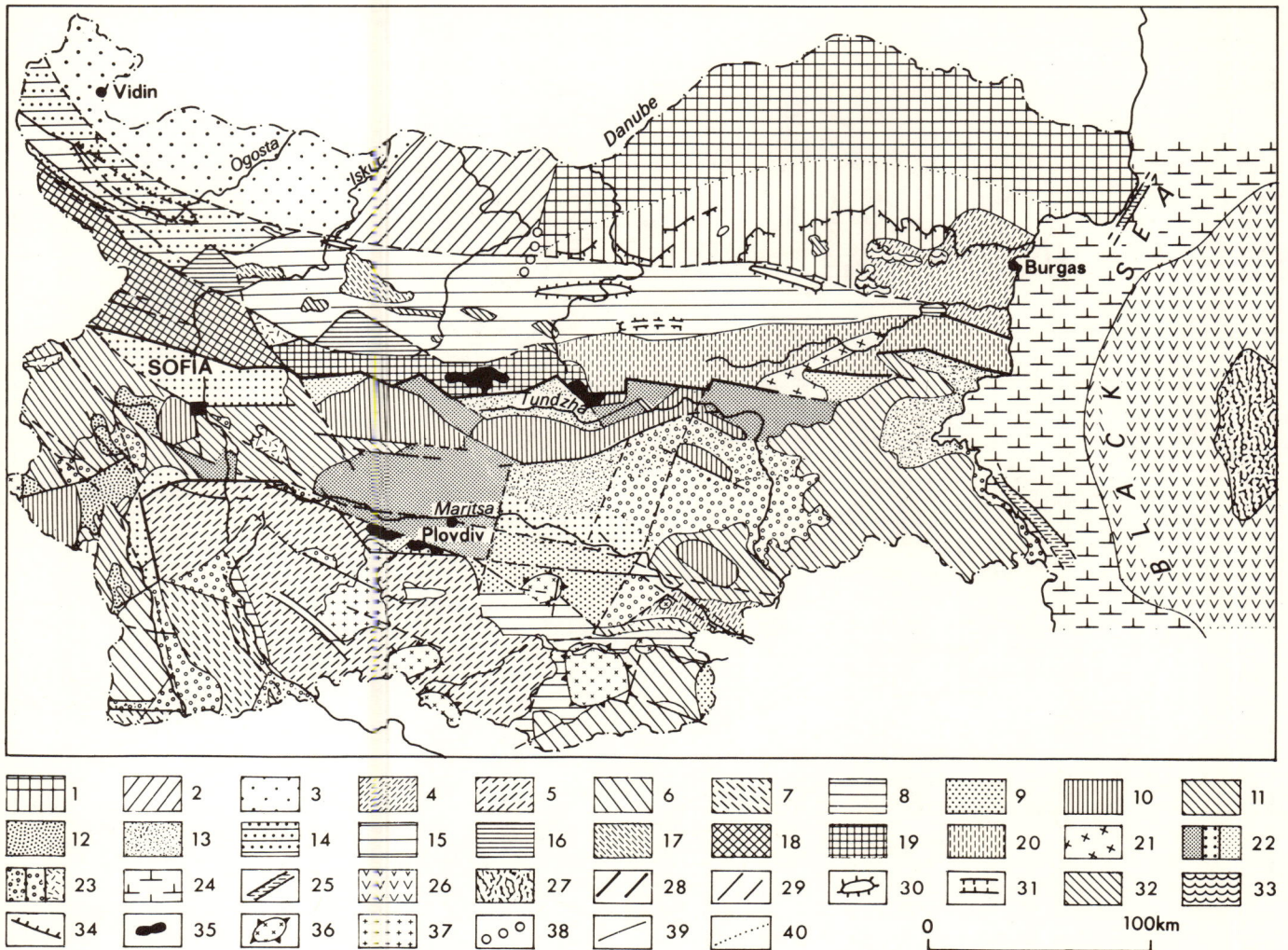

Fig. 16.10 Morphostructural map of Bulgaria [after Vaptsarov, *et al.* (1978)]. (A) *Epiplatform lowlands of the Danube (Moesian) platform:* (1) lowlands and plateaus of the North Bulgarian dome [(1a) structural–erosional lowlands and plateaus of the central part of the dome (strongly dissected), (1b) structural and accumulation lowland of the peripheral parts of the dome (slightly dissected)]; (2) structural and accumulation lowlands on Mesozoic and Neogene rocks of the transitional part of the Danube platform; (3) accumulation lowlands of areas of subsidence (Lomskaya depression), partly uplifted during the Pleistocene; (4) plateaus and accumulation lowlands of late Alpine age (Varna depression), filled by Neogene, Pleistocene and Holocene deposits. (B) *Mountains of the Rhodopi central massif:* (5) high mountains and uplands of the central part of the Rila fold-faulted massif and of the western Rhodopi mountains, deeply dissected by glacial and cryogenic forms; (6) uplands and low mountains in the peripheral block of the western Rhodopi mountains and the Ograshden–Maleshevskaya planina, moderately dissected; (7) high and low mountains, elongated horsts with high or medium intensity of neotectonic movements, dissected by glacial and cryogenic forms; (8) intermontane basins of upper Eocene–Oligocene age; (9) intermontane depressions and/or peripheral basins and piedmont lowlands in areas of Neogene–Pleistocene subsidence. (C) *Epigeosynclinal mountains of the Krayshtidy and Central mountains (Srednegorie):* (10) high and low mountains of fold-faulted areas with crystalline cores and high intensity of neotectonic movements, deeply dissected; (11) planation surfaces cutting uplands on the Hercynian and Mesozoic basement with slight neotectonic movements and dissection; (12) closed or semi-closed exhumed basins related to Upper Eocene and Oligocene depressions; (13) intermontane or peripheral basins and piedmonts developed in Neogene and Quaternary deposits. (D) *Epigeosynclinal mountains of the Stara Planina:* (14) accumulation lowlands at the foot of the mountains with relics of structural relief on the folded Mesozoic rocks of the north-western Stara Planina and the southern Carpathians, depressed during the Neogene and uplifted in the Quaternary (western Pre-Balkans); (15) low mountains and uplands on folded Mesozoic and Neogene rocks of the northern Stara Planina (central and eastern Pre-Balkans); (16) uplands of the fold-faulted areas of the Stara Planina, dissected; (17) intermontane depressions corresponding to exhumed Upper Cretaceous and Lower Palaeogene depressions, uplifted by neotectonic movements; (18) high mountains of the western asymmetrical fold-faulted uplift of the Stara Planina with exposed Hercynian core and Mesozoic cover; (19) uplands and high mountains of the central part of the fold-faulted uplift of the Stara Planina with partly exposed Hercynian core and Mesozoic cover; (20) uplands and high mountains of the eastern part of the fold-faulted uplift of the Stara Planina; (21) intermontane exhumed (structural–lithological) depression. (E) *Morphostructures of young depressions:* (22) intermontane depressions along deep-seated faults bordering the main morphostructures filled by (a) Neogene, Pleistocene and Holocene deposits, (b) Neogene deposits, (c) Pleistocene and Holocene deposits; (23) intermontane and peripheral basins filled by (a) Neogene, Pleistocene and Holocene deposits, (b) Neogene deposits, (c) Pleistocene and Holocene deposits. (F) *Morphostructures of the Black Sea depression:* (24) shelf; (25) youthful grabens on the shelf; (26) continental slope; (27) accumulation plains on the bottom of depressions. (G) *Morphostructural elements of the relief:* (28) fault scarps controlled by deep-seated faults, (a) very distinct, (b) partly eroded; (29) fault scarps of lower-order morphostructures, (a) very distinct, (b) partly eroded; (30) anticlinal ridges; (31) synclinal depressions; (32) synclinal ridges; (33) plateaus; (34) structural slopes, scarps; (35) edge of the overthrust (expressed in the relief); (36) volcanic hills; (37) elongated volcanic bodies; (38) exhumed volcanic forms and volcanic cones of Neogene age; (39) boundaries of higher-order morphostructures; (40) boundaries of lower-order morphostructures.

The major morphostructures of the Rodopi massif are of block-faulted type, but within these major blocks, the forms are differentiated according to lithology. Dominant lithologies include igneous intrusives (including the older Hercynian batholiths), extrusive rocks and sedimentaries that exhibit large anticlinal ridges' and synclinal depressions in the central parts of the Rila—western Rodopi uplifted zone, together with exhumed young Palaeogene grabens. Vulcanism was active in the Oligocene, when the main centres of eruption were located along existing grabens of Palaeogene age and marked by volcanic domes.

16.3 Stara Planina Mountains

This system corresponds to the Alpine epigeosynclinal zone which developed up to the late Palaeogene. It includes the Berkov anticlinorium in the west, the Shipka anticlinorium, the Kotel zone, the Lukokamtchiyskiu synclinorium of the Sytchtinsko—Stara Planina zone, the Pre-Balkan zone and the transitional zone of the Moesian platform. This varied structural base makes it possible to distinguish morphological units as follows: (1) the central mountain zone (main ridge of the Stara Planina) and (2) the Pre-Balkan Mountains (*see* Fig. 16.10).

The central mountain zone of the Stara Planina is formed by a mountain ridge bordered to the south by a steep slope controlled by the Transbalkan major fault, while the northern boundary is formed by a flexure (partly a fault scarp). In the northern part of the central mountain zone, the Vratschanska Planina and Vasilevova Planina represent uplifted parts of the Pre-Balkan morphostructure. Along the northern boundary, the western part of the Pre-Balkan morphostructure, together with part of the southern Carpathians, subsided into the Lomskaya depression. Farther east, the northern boundary of the Pre-Balkan morphostructure meets the lowland of the Moesian platform with a less well-marked and partly planed contact.

The central mountain zone of the Stara Planina has a fault-folded structure. There is a mountain ridge 530 km long from west to east and from 30 to 70 km wide from north to south. The altitudes in the west rise to about 2000 m, but eastward from the Iskyr river gap they fall gradually in the direction of the Black Sea. The western part of the Pre-Balkan morphostructure is formed by the western Balkan fold-faulted asymmetrically uplifted zone following a north-west—south-east trend, corresponding to the Berkov anticlinorium, with its deeply denuded Hercynian core and Mesozoic mantle. The mountain peaks are up to 2000 m high. The eastern part of the morphostructure is deeply dissected by the antecedent gap of the Iskyr. To the south of the Botevgrad depression, the central ridge of the Stara Planina follows an east—west direction. In the area of the Shipka anticlinorium, the central ridge of the Stara Planina possesses a symmetrical structure and is bordered by steep fault scarps. In an eastward direction, the summit altitudes decline from 1500–1700 to 1000–1100 m. An interesting morphostructural element is the Botevyrkh preorogenic (post-middle Eocene) gravitational overthrust of granites and metamorphic rocks which

took place from the direction of the Sredna Gora mountains. This overthrust forms the highest peaks of the central mountain zone, which are more than 2000 m high.

The eastern sector of the central zone of the Stara Planina is morphologically distinctive, including the structural zone of Kotel, the Lukokamtchiyskiy synclinorium and a part of the Sredna Gora zone with its sedimentary and volcanic mantle. These morphostructural features control the development of parallel ridges running in an east—west direction and divided by deeply incised valleys. Maximum heights are 1000 m in the west and 400–500 m in the east. The area is crossed by lower-order morphostructures, particularly the Lulyakov depression of late Eocene age.

The northern morphostructural zone of the Stara Planina—the Pre-Balkan—is a zone of uplands and moderate mountain relief. The southern boundary is formed by a flexure (partly a fault scarp) along the Stara Planina ancient frontal line. The Pre-Balkan is an individual morphostructural unit which corresponds with the zone of folded Mesozoic rocks and is characterized by direct or inverse relationships between geological structure and contemporary relief. Folds are expressed by ridges running perpendicular to subsequent river valleys. The western part of the Pre-Balkan mountains subsided into the Lomskaya depression of Neogene age. Glacis and pediments coalesce with planation surfaces in the Lomskaya depression and appear to be Neogene–Quaternary in age.

16.4 Plains, Plateaus and Highlands of the Lower Danube Basin

The relief of the Danube (Moesian) epiplatform plain is strongly controlled by structure. In the western part—in the Lomskaya depression—the faint low relief shows a slight inclination to the north-east. The watersheds are broad, the valley floors flat with asymmetrical cross-profiles. In the southern part, at the foot of the Stara Planina mountain system, the surface is higher (250–300 m). The boundary of the plain in this part is not well marked due to subsidence of the peripheral parts of the Stara Planina and the southern Carpathians in the neotectonic period. The western and eastern borders of the Danube plain are higher and the river valleys are more deeply incised. The central part of the Danube plain between the valleys of the Rivers Iskar and Yantra is formed by a plateau on Mesozoic and, partly, Neogene deposits. The plateau is more dissected with altitudes of about 400 m. River valleys take the form of asymmetrical canyons with flat floors. The eastern part of the plain is a dome with a radial river pattern (the northern Bulgarian uplift). The central part of this dome rises to about 500 m, while the peripheral parts rise to 200–250 m. The middle of the dome is eroded and divided into plateaus. Around, there are cuestas developed in Mesozoic limestones. The peripheral parts are very flat due to the subhorizontal layers of the epiplatform cover. The north-eastern part is formed by Neogene (Sarmatian) subhorizontal limestones, which continue to the coast of the Black Sea to the north of Varna. To the south, as far as the foot-hills of the Stara Planina mountains, folded morphostructures extend to the active border fault. The Varna–Kamchiya river depression, filled with Palaeogene, Neogene and Quaternary sediments, shows low relief sloping in the direction of the Black Sea, the bed of the Kamchiya river forming the lowest area.

16.5   East Balkan Plains and Uplands

This is a heterogeneous morphostructure including Krayshtidy and Sredna Gora. It is divided from the other main morphostructures by the Transbalkan and Marica deep-seated faults, which are marked by fault scarps (*see* Fig. 16.10). The boundary with the Rodopi Mountains is not clearly marked and is the subject of controversy. It can be followed to the Krupnik fault, dividing the Ograshdenskiy block with its crystalline basement from Krayshtidy with its important Hercynian and Alpine structural stages. The heterogeneous structural base is reflected in the relief. Low, middle and high mountains are present, showing folded and faulted structures, the old cores of the crystalline pre-Alpine basement (Strandzha, Sakar, central Sredna Gora, Vitosha, Oscgov, Ruyska). These mountains have isometric or slightly elongated forms, following a west–east direction and bordered by fault scarps. Isolated inselbergs rise above a broad pedestal that is made from remnants of old planation surfaces. The latter are cut sharply across dislocated rocks. The altitude of the planation surface remnants in the western part is about 1200 m, but in the east they decline to 200–300 m only. The change in level on the boundary between the western (Sofia) Sredna Gora and the central Sredna Gora, and also between the central and the Strandzha Sredna Gora, occurs in a series of steps. River valleys are controlled by faults, folds, overthrusts and flexures. In the Krayshtidy zone, Upper Eocene and Oligocene volcanic morphostructures are present. Sedimentary lowlands and foot-hills of the young basins are important features in the morphostructural differentiation of the area. To the south of the Transbalkan fault zone, rejuvenated or young Alpine basins are marked by border faults, sometimes forming a series of closed or partly closed basins (the Transbalkan group of basins, the Struma river graben complex, etc.) or wide lowlands such as that of upper Thrace. These depressions are elongated along fracture and fault zones. Within the depressions, the relief is complicated by smaller horsts and grabens. Some depressions such as the Thrace lowland developed over a longer time. This lowland originated in a late Alpine phase and developed as a depression during the Miocene and Lower and Upper Pliocene. In some areas, such as the neighbourhoods of Plovdiv and Pasardzhik, it continued also to develop during the Pleistocene and Holocene. The same tendency can be observed in the lowland around the Black Sea (the Varna and Burghaz depressions) and the Mediterranean Sea.

Chapter 17
The Northern Black Sea Lowlands and Crimea

The major part of this region (*see* Fig. 17.1) corresponds to a minor tectonic plate known as the Scythian platform, which lies between the uplifted area of the Dobrogea in the west and the northern part of the Caspian Sea in the east. The contact between the Scythian plate and the south Turanian plate is located beneath the floor of the Caspian Sea. The probable northern boundary of the Scythian plate can be traced from the mouth of the Danube in the east, passing just south of Zmeyniy Island in the Black Sea, through the Perekop Gate to the town of Yeysk on the eastern coast of the Sea of Azov, and then in a south-easterly direction to the Manytch depression. The junction between the Precambrian basement of the eastern European platform and the Palaeozoic (Hercynian) basement of the Scythian plate represents a broad contact zone. In some parts, geophysicists have established the existence of submeridional faults and related marginal depressions. The folded basement of the Scythian platform is of various ages, but generally is composed of Precambrian (mostly Proterozoic) structural blocks of moderate height divided by a system of narrow Palaeozoic folded structures.

The Scythian platform is characterized by bevelled relief, mainly due to the subsidence of the Black Sea depression on the west and of the Caucasian foredeep in the east. The highest parts are the Donets upland and the Stavropol' plateau, which are constituent parts of the meridional uplift zone of the Russian plain.

To the south of the Scythian platform, and occupying the southernmost part of the Crimean peninsula, is an area of very different relief and structure—the Crimean mountains. This region, consisting of both epigeosynclinal mountains and epiplatform ridges, will be described in the final section of this chapter.

17.1 Central Crimea

This denudational surface forms the largest part of the Crimean peninsula. The altitudes range from 30–50 m in the south to 100–150 m in the north. The surface is almost flat, in some parts gently undulating with a slight tilt to the north. The Palaeozoic floor lies at great depths and only in the area of the Stavropol' uplift does it lie within 200–300 m of the surface. The central Crimean lowland is underlain by Pontian limestones and covered almost completely by red clays and silts, with some intercalations of sandstone and conglomerate (the Tauridian layer of middle Pliocene age). The actual surface is covered by loessic loam. Through this cover, a network of dry valleys and gullies can be seen, forming fan-shaped patterns beginning at the foot of the Crimean mountains. Present-day valleys show in their middle courses two to three sets of river terraces. Due to subsidence, these terraces disappear downvalley below the contemporary floodplain. There are typical anthropogenic forms: namely, numerous burial mounds called kurgans.

The denudation chronology of this area is closely contained. The development of the relief started after the regression of the Pontian Sea. Rapid uplift of the Crimean mountains during the Pliocene initiated a phase of erosion that culminated in relief planation. The drainage at that time was directed to the north. In the area of the central Crimean lowland, alluvial cones—the so-called dry deltas—formed on the surface of the red-coloured Tauridian layer. Later during the Pleistocene, a loessic cover was deposited. Present-day fluvial incision is slight.

17.2 Tarkhankut Uplands

In the north-west part of the Crimean peninsula lies the upland known as the Tarkhankut uplands formed by three parallel, flat-topped ridges (75–180 m high) dissected by broad, flat depressions. The ridges are asymmetrical, the slopes exposed to the south being steeper compared with those facing north. Ridges composed of limestones represent eroded brachyanticlinal folds in the platform cover. Structural forms are common; the ridges are usually anticlinal and depressions synclinal. Forms of selective erosion are also common and the watersheds are bevelled.

The main relief features of the Tarkhankut upland began to form during the early Pliocene after the regression of the sea, especially that of the Sarmatian Sea. In the later part of the Pontian and during the Cimmerian (late Pliocene), a planation surface with red-coloured eluvial loams developed. During the later Pliocene, differential neotectonic movements began to encourage erosional activity, which became more marked at the end of the Pliocene and the beginning of the Pleistocene. The planation surface was uplifted and dissected; today only residuals can be found, some on flat-topped ridges. The weathered mantle associated with the surface was also eroded. Substantial gullies were cut into the surface and filled by younger continental deposits of proluvial type. The western part of the upland is most dissected. The eastern part is largely covered by deposits of Plio-Pleistocene age, and erosional forms consist of broad ridges and gentle slopes. The northern, eastern and south-eastern margins, which show undulating plateaus at heights of 100–130 m, are only slightly dissected.

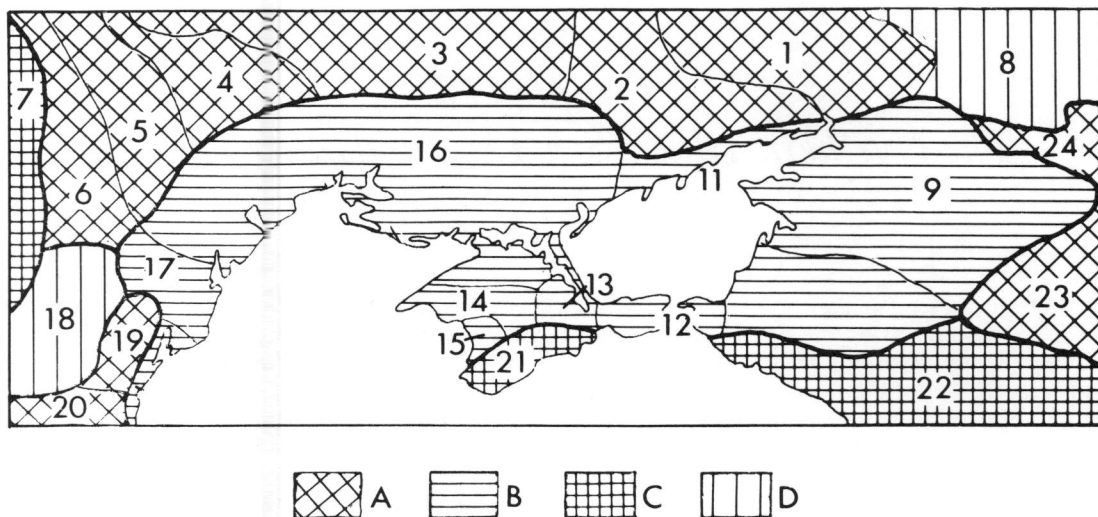

Fig. 17.1 Structural and geomorphological framework of the area around the northern Black Sea. *Megamorphostructures:* (A) zone of uplifted platform; (B) megadepression of the Black Sea lowlands; (C) Alpine orogenic zone; (D) marginal and other depressions. *Morphostructural subdivisions:* (1) Donets upland; (2) Pre-Azov massif (Ukrainian Shield); (3) Pre-Dnepr massif (Ukrainian Shield); (4) southern part of the Volyn'–Podolsk upland; (5) Bessarabian uplands; (6) Moldavian plateau; (7) Carpathian mountains; (8) Don–Sal structural accumulation lowland; (9) Azov–Kuban' foredeep; (10) western Kuban' foredeep; (11) northern Azov foredeep; (12) Kerch–Timan foredeep; (13) Indolskiy foredeep; (14) Tarkhankut–Novoselov uplands; (15) Alma lowland; (16) Odessa–Sivash lowland; (17) Dobrodgea lowland; (18) lower Danube lowland (Moesian platform); (19) Dobrodgea plateaus and uplands; (20) Monoclinal plateau; (21) Crimean mountains; (22) Great Caucasus; (23) Stavropol' plateau; (24) Don–Sal lowland.

The most important present-day processes are marine abrasion and coastal landslides. The coast of the Tarkhankut peninsula is mostly abrasional with active cliffs. Based on archaeological data, cliff recession during the last 2000–2500 years has destroyed large parts of the coast, especially in the north of the peninsula. Landslides are common on the northern coast and are of large dimensions. Rotational slips of large blocks of Miocene limestone have formed a series of steps from the plateau down to the sea. In some cliffs, karst caves can be found, but generally on the plateau the degree of karstification is only slight.

17.3 Donets Ridge

Lying to the west of the North Donets river, this unit represents an old uplift of the Palaeozoic folded basement in the eastern part of the Dnepr–Donets aulacogen. As a relief unit, the highland was formed at the end of the Permo-Triassic period, but later suffered several periods of planation. The present-day relief has been formed mostly during the neotectonic period, but some older structures have been inherited. In the present relief, it is possible to distinguish a central zone with a late Cretaceous and Palaeogene planation surface at altitudes of 250–360 m, and marginal zones dissected by numerous deep valleys and with narrow watersheds.

Many forms are controlled structurally as erosion progressed on different lithologies. Also influential in controlling erosional processes has been a dense network of faults. Planation surfaces are complicated by structural ridges, mosores and monadnocks arranged in zones along the fold directions. Besides the late Cretaceous–Palaeogene regional planation surface, a younger Miocene surface developed in the marginal parts of the highland at altitudes of 180–270 m. Because of intensive dissection, the planation surfaces have been modified into ridges and rows of hills arranged in conformity with the folded structures and forming levels of dissected relief. The river valleys are deep; rivers follow radial and dendritic patterns. Towards the margins, even the smaller valleys are deeply cut with large gradients and steep side slopes. Contemporary erosion is very active, and many gullies have developed on valley slopes.

17.4 Sivash Lowland

The Sivash lowland is a depositional area in the northern part of the Crimea and the southernmost part of the Russian plain. Relief varies between 1–2 and 50 m, with the heights increasing from the centre to the north. To the east lies the Sivash Sea, which is divided from the Sea of Azov by a long bar—the so-called Tongue of Arabat. Morphostructurally, the lowland belongs to the southern marginal part of the Russian platform and to two depressions of the Scythian platform: (1) the Sivash depression and (2) the Indola–Kuban' depression. The Sivash depression is a typical platform depression of Mesozoic–Cenozoic age, while the Indula–Kuban' depression represents a young foredeep of Neogene age. Both depressions are filled by layers of Mesozoic and Cenozoic sediments (7–8 km thick) and, in spite of some differences in development, the neotectonic history was the same for both depressions. This is the reason for the monotonous depositional relief at the present-day. The last marine transgression was in the middle Miocene, although in the eastern Sivash Sea marine bays existed up to the end of the Pliocene. During the Quaternary, loess (30–40 m) accumulated, its base standing near mean sea level.

The lowland surface is mostly extremely flat, although there are slight undulations in some areas. These undulations are due to many gullies filled by loess. The valleys of the Salgira, Burganaka and Indola are very flat, 1–2 km wide near to their mouths and there are no river terraces because of subsidence, although two floodplain levels are apparent. Otherwise, the level surface of the plain is only interrupted by occasional 6–9-m high burial mounds (kurghans) usually found near the rivers and shallow surface depressions (podi) which are the result of suffosion (piping) in the loess.

On the left bank of the lower Dnepr, where the river crosses a pre-delta lowland composed of loams and sands, aeolian processes actively form sand dunes that are only partly covered by vegetation.

17.5 Fore-Caucasian Lowland

17.5.1 Western Section

This consists of the Azov–Kuban' lowland, whose surface towards the coast of the Sea of Azov is very flat. Towards the Stavropol' plateau, the surface becomes gradually more undulating. Altitudes in the west are only 3–5 m, but in the east reach 100 m or more.

Morphostructurally, the lowland represents the northern part of the Indolo–Kuban' foredeep (the ancient delta of the Kuban' River). The basement here is at depths of up to 10 km. Some parts belong morphostructurally also to the Azov–Kuban' marginal platform depression where the basement lies at depths of 2–5 km, and to the wide contact zone of the eastern European platform and the Scythian platform. The lowland is underlain by Mesozoic and Cenozoic deposits, over which is draped a blanket of Quaternary loess. The River Kuban' occupies a broad valley in the southern part of the lowland, within which there are several sets of river terraces. The river valleys in the northern parts are only slightly incised. Suffosion forms, especially flat-floored depressions (i.e. podi), are common. The coast of the Sea of Azov is characterized by estuaries, which developed during the last marine transgression near the river mouths.

During the middle and late Pliocene, the lowland, especially in the south, was flooded several times by the sea (Pontian, Cimmerian, Kulyanickian, Tamanskian and Guriyskian transgressions). The most recent transgression only affected areas close to the present-day coast, but earlier transgressions affected bigger areas. In this way, the development of subaerial relief started earlier in the eastern part of the lowland and gradually extended to the west. The extension of dry land was also affected by the development of the ancient River Kuban' whose present delta is developing rapidly. Offshore bars, with evidence of recent changes in position and growth, characterize the coast.

17.5.2 Eastern Section

The Terek–Kuma lowland in the eastern part of the Fore-Caucasian region extends to the Kuma–Manytch depression in the north, while to the south its limit is formed by the foot-hills of the Great Caucasus. To the west lies the Stavropol' plateau. The altitudes range from 30 m in the eastern part to 150 m in the west, the relief being very varied. The northern part is formed by the Prilumskaya lowland, morphostructurally corresponding to the Scythian platform. The very flat surface is gently inclined to the east. It is underlain by the loamy and silty marine deposits of the Khvalin Sea, by the sandy loamy deposits of the early Khvalin delta (in the west) and by the present-day fluvial deposits of the Kuma and other smaller rivers. The eastern part is typified by numerous salt flats and salt lakes.

In the south-eastern part of the lowland, sand dunes have developed on deltaic deposits in the coastal area next to the Khazar Sea. Aeolian processes are currently very active, and the development of different types of dunes (barkhans, ridges, individual dunes) in this area has been considerably promoted by the activities of man, especially the destruction of the vegetation cover by overgrazing.

The deltas of the Terek and Sulak are of particular interest. That of the Terek has an area of 6000 km², measuring 60 km across and 90 km from its apex to the coast. Farther south lies the delta of the Sulak. Altitudes of the Terek delta vary from 26 m at the coast to 5 m near the town of Kizlyar. The surface shows the typical form of deltas, with many natural levées, oxbow lakes and waterways. The coastline is marked by a zone of dunes (10–12 km wide) formed from reworked marine sand. Bars on the coast are formed by deposits of the Terek and Sulak rivers.

The oldest parts of the lowland are of Khvalin age, while the deltas are still growing.

17.5.3 Stavropol' plateau

This plateau is the watershed between the Azov–Kuban' and Terek–Kuma lowlands. Morphostructurally, it is a part of a zone of recent (post-Sarmatian) uplift of the Russian plain, which extends to the south as far as the Great Caucasus. The Hercynian basement lies at depths of only 1.5–2 km and is covered by slightly faulted Mesozoic and Cenozoic deposits. The actual surface of the plateau, however, is underlain by Palaeogene and Miocene rocks. The whole plateau represents a large dome with maximum altitudes in the centre and south-west, where the typical relief is tabular, with mesas such as Mount Strijamount (832 m). To the north-west, north and east, the surface is slightly inclined. The borders of the plateau on the south and south-west are marked by steep scarps (300–400 m high) controlled by deep-seated faults in the basement or faults or flexures in the platform cover.

Several planation surfaces have been developed on the plateau. The basic and highest surface is of Miocene–Pliocene age (Sarmatian–Ergenian) with altitudes of 500–600 m. Remnants can be found on watersheds, and mesas have been formed by late or middle Sarmatian calcareous sandstones [see Gorelov (1972)]. A lower planation surface is of Akhagyl age, while the lowest one—in the southern part of the plateau only—is Apsheron. The uplift of the dome took place in Meotian–Pontian time, when the radial river pattern originated.

The spatial distribution of gullies varies. Most are concentrated in the south-western and central parts of the plateau, but are especially typical of brachyanticlines still rising at the present time.

17.6   Crimean Mountains

Forming the southernmost part of the Crimean peninsula, these highlands cover about 6000 km². From west-south-west to east-north-east, they stretch for 180 km, the maximum width in the centre being 50 km. They consist of three nearly parallel ridges, the highest—the southern or main ridge—rising to 1500 m. The others, known as the inner and outer ridges, attain heights of about 600 and 300 m, respectively. The ridges are separated by longitudinal depressions. In terms of geological structure, the Crimean mountains represent an asymmetrical anticlinorium, the southernmost part of which is now submerged beneath the Black Sea. The present coastline cuts across the western and, partly also, the eastern end of this anticlinorium.

Three stratigraphic and structural units can be distinguished in the rocks of the Crimean anticlinorium. (1) The lowest consists of intricately folded deposits of Upper Triassic–Liassic age (the Taurid Series). (2) The middle unit consists of much less deformed deposits of Middle and Upper Jurassic and Lower Cretaceous age. (3) The upper one comprises Upper Cretaceous, Palaeogene and Miocene deposits. In some parts of the mountains, Neogene and Quaternary deposits can also be found.

The Crimean mountains represent a very complex system of varied origin and age. Based on morphostructure and age of formation, we can distinguish between (1) the Mountain Crimea or main ridge, an epigeosynclinal mountain belt that has developed since the Jurassic, its main structural features having been formed during the Cimmerian orogeny and (2) the epiplatform piedmont ridges (or inner and outer ridges) formed due to monoclinal uplift of the southern part of the Scythian platform during the Neogene period.

17.6.1   Main Ridge

The main ridge of the Crimean mountains comprises a series of blocks of Upper Jurassic limestone with flat tops. The lower parts of the blocks are composed of shales and sandstones belonging to the Taurid Series. The northern boundary of the ridge is formed by a deep-seated fault, expressed in the relief as the southern longitudinal depression. Transverse faults determine the limits of the individual blocks, known as Yayli. Faults are also thought to control the alignments of many river valleys and the gaps or cols between the blocks: for example the faults marking the eastern and western boundaries of Chatyrdag, Dmerdji and Karabi. The bevelled tops of the individual blocks are remnants of a planation surface of Lower Cretaceous age, best seen in the flattened summits of Babugan, Chatyrdag and Karabi.

The main ridge is an area of intense karstification (*see* Fig. 17.2), favoured by the chemical purity of the Upper Jurassic limestones and by the relatively high precipitation (up to 1000 mm per year in the western part of the ridge). Many varied karst forms can be found, such as lapiés, dolines and poljes; the Chatyrdag and Karabi blocks are particularly rich in these forms and also possess numerous caves. On some summit areas, the planation surfaces are so pock-marked by dolines that they resemble a lunar landscape of craters.

East of Karabi, the main ridge breaks up into a series of short sharp-crested interfluves between deeply incised valleys. The overall altitude decreases from about 1000 to 600 m. In this region, the relief is controlled by the differential resistance of the Middle and Upper Jurassic shales, conglomerates and reef limestones. Precipitation here is much less than in the west—in places as low as 200–300 mm per year—and typical arid landforms of low, intricately carved mountains, especially on the southern side of the ridge where shales of the Taurid Series have developed badlands, make their appearance. Although it is a semiarid area, occasional rainstorms produce flash floods and mudflows.

In the south-eastern area of the main ridge, next to the Black Sea coast, lie the Karadag mountains. Most of this region is built of volcanic rocks of Middle to Upper Jurassic age reaching heights of about 500 m, but some of the ridges are formed in Upper Jurassic limestones, with intervening clay vales. Selective erosion on the slopes and ridge crests has created a distinctive landscape with some interesting forms of weathering and rock removal.

The southern denudational slopes of the main ridge are very steep, overlooking a narrow coastal strip of highly dissected relief next to the Black Sea coast. The towns of Yalta, Gurzuf and Alushta stand on small alluvial river plains. Exhumed Jurassic laccoliths form mountains, such as Castel, Uraga and Ayudag. With the prevalence of steep slopes, the Taurid shales are liable to bulge and deform as they are compressed under the weight of the overlying Jurassic limestones. Also, because this is a seismically active area, mass movements are common. Cambering of the limestone strata may be seen; many hillslopes are unstable because of the occurrence of jointed bedrock. Some low ridges and hills are composed of a breccia of limestone fragments embedded in a red calcareous cement, possibly the cemented debris of ancient slides and scree deposits.

The southern longitudinal depression separating the main and inner ridges is formed in shales, marls and, in parts, conglomerates. Its floor shows an uneven erosional relief consisting of small blocks of highly fissured rock. A former Miocene planation surface has mostly been destroyed and now only remains on the tops of a few interfluves within the depression.

17.6.2   Piedmont Ridges

The inner ridge is a cuesta whose northward-tilted backslope (*see* Fig. 17.3) consists of resistant Upper Cretaceous limestones and Middle Eocene nummulitic limestones. Faults and their associated valleys divide the cuesta into a number of segments. The limestone scarp

Fig. 17.2 Karstic Valley, Crimean mountains. (Wagner)

Fig. 17.3 Structural surface of limestone cuesta, Crimean mountains. (Wagner)

shows the effects of various subaerial denudation processes, but also evidence of wind deflation and honeycomb weathering exists. At its foot, there are erosion glacis developed in Upper Cretaceous or Eocene marls.

The outer ridge is a lower cuesta whose northern dip slope merges gradually into the Crimean lowland. The upper part of the cuesta is composed of Sarmatian conglomerates; it is also bevelled by an Upper Pliocene planation surface with a gravel cover.

17.6.3   Tectonic History and Denudation

The age and importance of the various tectonic phases varies from place to place in the Crimean mountains. The fold structures of the main ridge originated in the early Alpine period of earth movement (the Cimmerian phase).

At the end of the Lower Cretaceous period, mountains 300–500 m high had been formed, and deep erosional–tectonic intermontane depressions filled with Aptian and Albian deposits are important components of the present-day relief. In the Middle Pliocene, the main ridge was uplifted in the form of an asymmetrical dome, emphasizing the features of the Cimmerian morphostructure; the uplift probably amounted to about 1000 m. The inner and outer ridges, together with the southern longitudinal depression, were formed during the Neogene. During uplift, a sequence of five rock-cut terraces formed in the river valleys, while the drainage pattern was modified by successive river captures in the Pleistocene. Most of the excavation of the longitudinal valleys among all three ranges took place by erosion at this time.

Chapter 18
Caucasus Mountains and Armenian Highlands

18.1 Main Relief Features and Morphostructure

The Caucasus mountains and Armenian highlands represent the eastern part of the tectonically young, epigeosynclinal orogenic Alpine zone, lying between the Scythian platform (Hercynian) and the Arabian platform (Precambrian). The Great Caucasus forms the outer ridges of the Alpine orogenic system, while the Armenian highlands form an inner sector, comprising typical block, folded–block and folded systems of ridges, extensive uplands with bevelled surfaces, basins and monoclinal plains; the whole complex assemblage represents a transitional zone to the ancient Arabian platform [see Dumitrashko and Museibov (1977)].

In the eastern part of the Armenian highlands lies the Transcaucasian highland (Zakavkazskoye nagorye) just within the limits of the USSR. In the central part are the Armenian highlands proper; this more restricted definition of the Armenian highlands will be used in the rest of this chapter. Lying a little to the north are the southern Georgian volcanic highlands. The whole group of highlands in Armenia and southern Georgia is flanked, on the north and east, by the ridges of the Lesser Caucasus (see Fig. 18.1). Between the Transcaucasian highland and the Armenian highlands is the intermontane basin of Srednearaksinskiy, whose main axis trends in a north-west–south-east direction. The same direction is followed by the Priaraksinskiy mountain ridges flanking the basin on its eastern side.

Between the Transcaucasian highlands and the Great Caucasus is a large intermontane basin called the Transcaucasian depression, connecting the Black Sea basin with that of the Caspian Sea. At the western end of this depression, the ground falls to the Black Sea coast around Poti; at the eastern end, in Azerbaijan, the floor of the Transcaucasian depression sinks well below global sea level to the level of the Caspian Sea. The structure of the western end includes a crystalline block, the Dzhirulskiy massif, of Baikalian age. Farther west is the Rion depression, geomorphologically known as the Kolkhida lowland. To the east of the Dzhirulskiy massif is the Kura depression, geomorphologically consisting of a series of smaller basins: the Verkhnekartliyskaya, Alazan'–Agricayskaya and Karayazi–Gyandjinskaya depressions. Proluvial lowlands have developed at the foot of the mountains surrounding the basins. The Dzhirulskiy massif, rising to 1900–2000 m, and other smaller massifs represent crystalline basement blocks of Baikalian age, belonging structurally to the eastern part of the Lesser

Caucasus. The southern Georgian volcanic highland, the Armenian volcanic highland, the volcanic area of Elbrus (5633 m) and the Verkhnetschegemskoye highland in the Great Caucasus, together with the laccolitic structures of the Hercynian basement around the town of Mineral'nyye Vody are all parts of the Transcaucasian uplift zone. This zone intersects the Transcaucasian highlands, the Transcaucasian depression, the Great Caucasus and part of the Scythian platform (represented geomorphologically by the Stavropol' plateau: see Section 17.5.3). The southern part of the Scythian platform forms the foredeep of the Great Caucasus.

Morphostructural analysis has shown a very close topographical and genetic connection between the Great Caucasus and the Transcaucasian highland. The axial Alpine ridges of the Great Caucasus show the greatest neotectonic activity (see Fig. 18.2); Neogene and Quaternary uplift probably amounted here to about 4500 m. These high ridges take the form of domes and dome blocks with denudational/structural relief [see Dumitrashko and Museibov (1977)]. Between the ridges, there are grabens in which shales of Lower and Middle Jurassic age have been down-faulted and preserved (e.g., the Arkhizo–Zagedanskaya graben in the central Great Caucasus and the Bezhitinskaya graben in the east.

On the north-western and south-western margins, the main ridges of the Great Caucasus reach heights of 3000–4000 m. The Vodorazhdelnyi, Perodovoy and Bokovoy ridges are surrouneded by erosional mountains and uplands attaining 1800–2500 m. Overall, the Great Caucasus represents a large asymmetrical mega-anticlinorium, the northern slopes occupying a larger area than the southern ones. The latter are characterized by fault scarps and overthrusts, of which the main one runs along the western part of the Vodorazhdelnyi ridge. The western and central parts of the Great Caucasus are composed of Palaeozoic crystalline and metamorphic rocks belonging to the uplifted Hercynian basement, and partly also of Triassic rocks. The eastern Great Caucasus is formed from Liassic shales [see Milanovskiy (1968), Safronov (1969)]. The Svanetian ridge on the southern slope of the Great Caucasus, south of the Elbrus area and lying parallel to the main ridge, has Alpine mountain forms and is cut by the water gap of the Ingur. In the middle section of the Rion river valley, there is a large basin infilled by fluvial deposits in which a series of aggradational river terraces has been developed. South of the Rion valley in western Georgia, the landscape displays a relief strongly controlled by structure, consisting of Upper Jurassic limestones that have been folded and

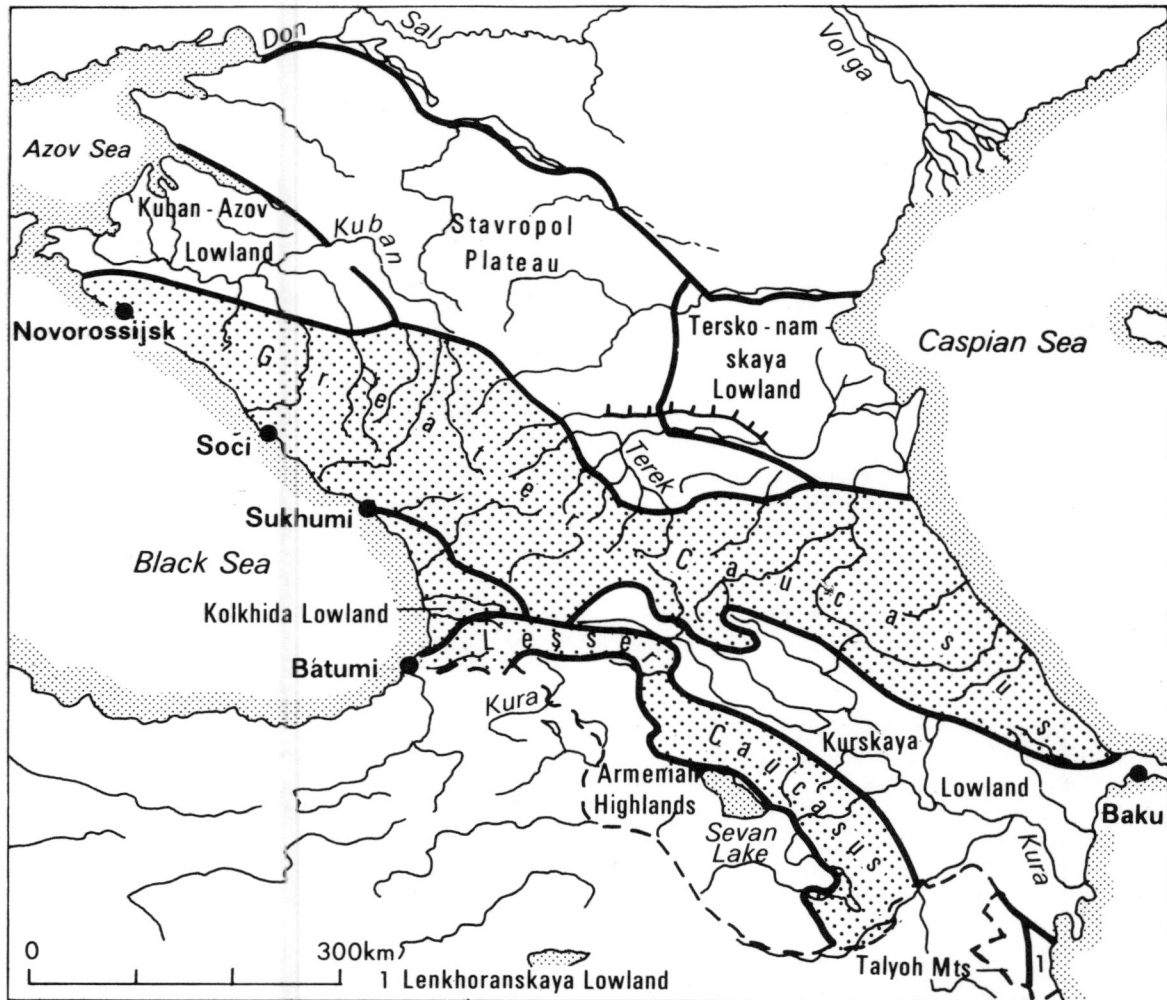

Fig. 18.1 Main geographical regions of the Caucasus and adjacent areas.

faulted. To the north of the axial ridges of the Great Caucasus, there is the broad Severo–Yurskaya depression with a general level of 1800 m and some 25 km in width. Where the zone of the Transcaucasian uplift crosses this depression, the high Betchasyn plateau, with altitudes up to 3000 m, has developed. The depression itself is eroded in gently dipping Jurassic shales and sandstones. To the north, it is limited by the cuestas of the northern Caucasian monocline with altitudes of 1500–3000 m, composed of Upper Jurassic limestones (the so-called Skalistyi or 'rocky' ridge) and of Cretaceous sandstones (the Pastbistchnyi or 'pasture' ridge). The lower northern cuestas consist of Palaeogene rocks, partly covered by Pliocene gravels. The foothills of the eastern Great Caucasus are formed from folded Neogene rocks (*see* Figs. 18.3 and 18.4).

Inner Dagestan is characterized by fold structures with broad anticlinal ridges formed by Upper Jurassic and Cretaceous rocks [see Safronov (1969)]. The south-eastern end of the Great Caucasus is formed by isoclinal folds in Lower Jurassic shales (Lias). In the north-western part of the Caucasus, asymmetrical folds are typical of the Upper Jurassic and Upper Cretaceous flysch [Safronov

(1969)]. The foredeeps of the Great Caucasus—the Kuban' and the Terek–Caspian foredeeps—are filled with Pleistocene, Neogene, Palaeogene and Mesozoic sediments (10–12 km thick) resting on the Hercynian basement of the Scythian platform. The Kuban' foredeep is separated from the Terek–Caspian foredeep by an uplifted part of the Hercynian basement near the town of Mineral'nyye Vody. In this particular area, the Hercynian undermass is covered by Cretaceous and Palaeogene unconsolidated deposits, intruded by laccoliths. Where exhumed, these laccoliths form distinctive inselbergs from 700 to 1400 m in height. In the Terek–Caspian foredeep zone, low anticlinal ridges, such as the Terek and the Sundzha ridges, are composed of Upper Pliocene alluvial, proluvial or glaciofluvial gravels [see Milanovskiy *et al.* (1966)]. The foothills of eastern Dagestan, which were subjected to earlier folding in the Upper Pliocene, show more complex relations between fold structures and relief.

Between the Great Caucasus and the Crimean peninsula lies the Kerch–Taman foredeep, filled with thick Neogene molasse deposits, folded into brachyanticlinal structures that appear clearly in the relief of the Taman peninsula. There are also salt domes and mud volcanoes

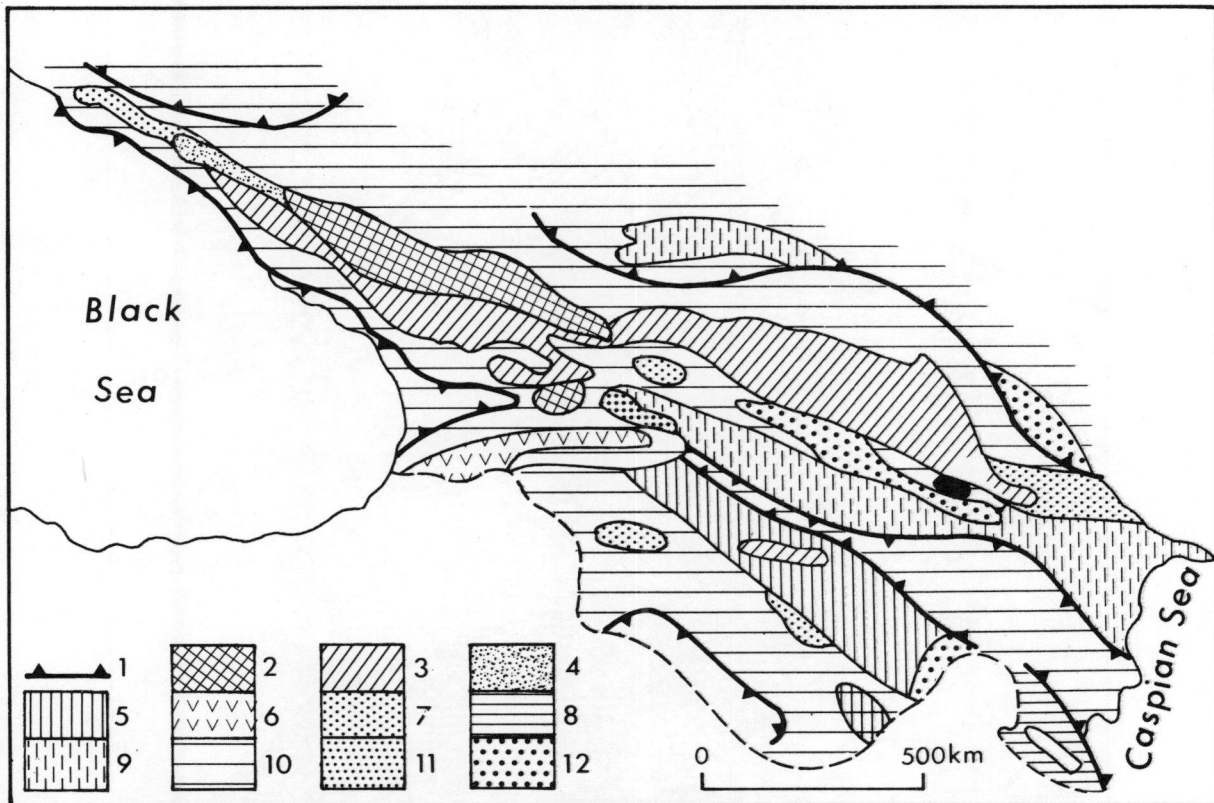

Fig. 18.2 Changes in the direction of tectonic movements in the Caucasus. (1) Regions which were affected by a general inversion: (2) in the Palaeozoic; (3) prior to the late Jurassic; (4) in the late Cretaceous; (5) in the early Cretaceous; (6) in the late Eocene–Oligocene; (7) prior to the Oligocene; (8) in the Miocene; (9) in the Pliocene–Quaternary. (10) and (11) Regions unaffected by tectonic inversion, and (12) superimposed depressions of Pliocene–Quaternary age.

in this area [see Blagovolin (1962)].

The Lesser Caucasus is an area of short fold-faulted ridges, reaching heights of about 3400 m in the north and 3900 m in the south. Neotectonic movements in this area were slightly less intensive than in the Great Caucasus. The main ridges show a denudational/structural type of relief, with ridges corresponding to horsts or anticlines, and basins representing grabens or synclines.

On the margins of the Lesser Caucasus, uplands of moderate height and low mountain ridges have an erosional/structural relief. The more northerly ridges—the Adzhar–Imeretia and Trialetskiy ridges—are composed of calcareous and volcanic rocks of Cretaceous age, and thick Palaeogene flysch and volcanics. The ridges are asymmetrical, their northern slopes being longer. Many overthrusts directed towards the Transcaucasian intermontane depression can be seen on the northern slopes. On the southern slope of the Adzhar–Imeretia ridge, the Aknaltsikhe synclinal basin is filled with Oligocene lacustrine deposits as described by Milanovskiy *et al.* (1966).

The inner eastern ridges of the Lesser Caucasus are formed in the so-called Somkhet–Karabakh zone, representing a system of anticlinoria and synclinoria complicated by faults and overthrusts on their south-western slopes. They are composed of thin Lower Jurassic continental deposits, Upper Jurassic volcanics and calcareous sediments and, in some places, by Lower Cretaceous

deposits. Depressions are filled by thick continental, volcanic and calcareous Lower and Upper Cretaceous deposits. In the Mesozoic, granitic, gabbroid and ultrabasic rocks were intruded, controlled by deep-seated faults [see Milanovskiy (1962)].

South of the Somkhet–Karabakh zone lies the Sevan–Akerinskiy synclinorium, filled with Cretaceous and Palaeogene volcanic rocks and sediments. The western and central part corresponds to a deep-seated fault zone, geomorphologically expressed by the Shiraki, Pambak and Sevan basins; together these form the central basin of the Transcaucasian highland. In the south-western part of the synclinorium, granitic and basic rocks of Upper Eocene and Oligocene age were intruded. The largest intrusion is that of the Ordubad pluton, forming the highest ridge in Armenia, the southern part of the Zangezur ridge.

The mountain ridges of the southern part of Armenia in the Priaraksinskaya zone are part of the Yerevan synclinorium, which is filled by unconsolidated deposits of Upper Cretaceous and Palaeogene age and also by Miocene lacustrine deposits. The Srednearaksinskiy basin is bounded by large faults and flexures, and contains red beds and molasse of Oligocene and Miocene age, covering the Palaeozoic limestones of the central part of the basin.

The Armenian and Jushn–Gruzinskoye volcanic highlands form the central part of the Transcaucasian moun-

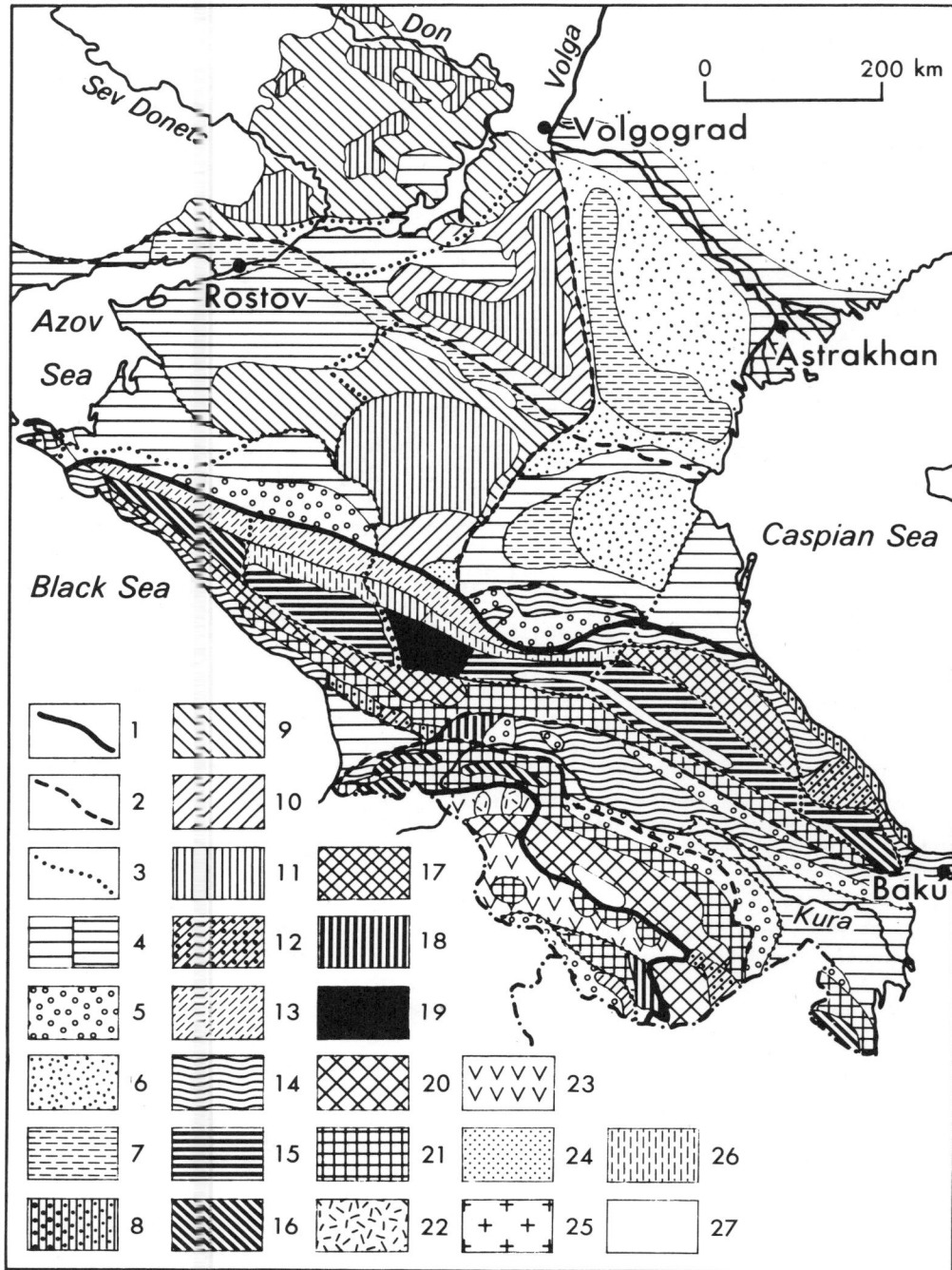

Fig. 18.3 Geomorphological regions of the Caucasus (compiled by N. V. Dumitrashko) (1) Boundaries of major regions. (2) Boundaries of subregions. (3) Minor divisions. *Morphostructures*: (4) to (13) *Lowlands*: (4) alluvial, alluvial–deluvial; (5) alluvial–proluvial lowlands; (6) lowlands due to marine deposition or abrasion; (7) fluvial–marine lowlands; (8) dissected terraced marine lowlands; (9) denudational–depositional lowlands; (10) erosional–denudational lowlands; (11) plateaus; (12) monoclinical plateaus with gently dipping strata; (13) cuestas due to differential erosion. (14) to (21) *Mountains*: (14) anticlinal and synclinal ridges, plateaus and basins; (15) and (16) horst and anticlinal ridges; (17) horst and anticlinal ridges; synclinal massifs with partly inverted relief; (18) horst ridges; (19) faulted domes; (20) fold-faulted ridges; (21) erosional–denudational ridges. (22) to (24) *Volcanic relief*: (22) volcanic massifs with horsts and domes; (23) volcanic plateaus and plains; (24) laccoliths. (25) to (27) *Other features*: (25) mud volcanoes; (26) structural–denudational depressions and (27) graben–synclinal basins (within the area of shading on the map).

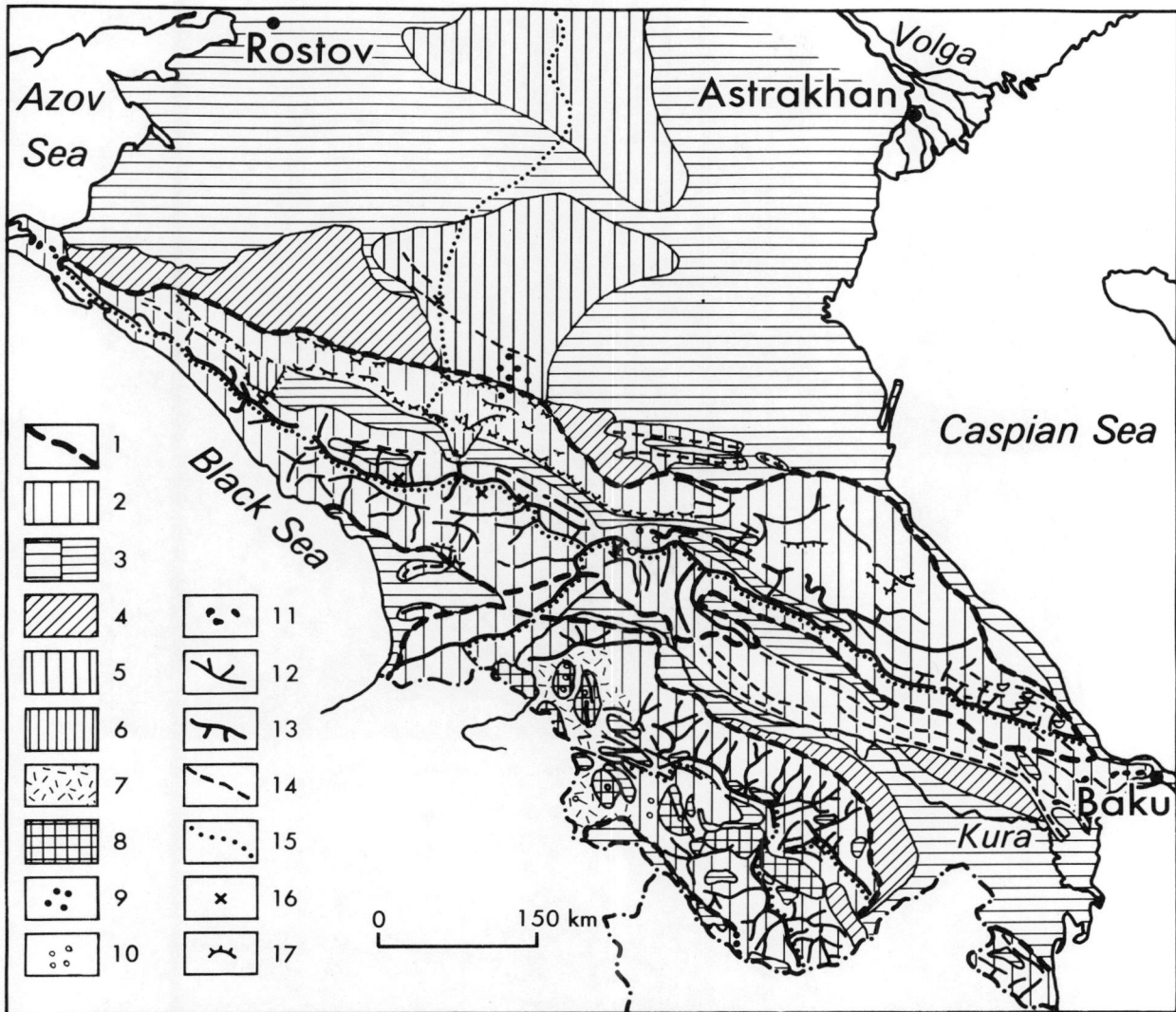

Fig. 18.4 Relief features of the Caucasus (based on work by Antonov, Budagov, Gabrielyan, Gvozdetskiy, Milanovskiy and Khain): (1) Boundaries of mountain areas; (2) mountainous and upland areas; (3) lowlands and plains: (a) intermontane depressions and (b) basins and hollows; (4) sloping submontane plains and plateaus; (5) plateau of the central Pre-Caucasus; (6) mountainous plateau; (7) volcanic plateau; (8) volcanic massifs and highlands; (9) mountains formed by laccoliths; (10) major extinct volcanoes; (11) mud volcanoes and cones; (12) main watershed; (13) secondary watershed ranges of the Great Caucasus and Transcaucasian uplands; (14) main watersheds of the uplands and low mountains of the Pre-Caucasus and Transcaucasus; (15) watersheds of the Black, Azov and Caspian Sea basins; (16) main summits of the ranges; (17) escarpments in areas of cuesta relief.

tain system. These areas comprise a series of high (1500–2000 m) lava plateaus comprising basalt and andesite, dissected by deep canyons and by shield volcanoes [see Balyan (1969)]. The slopes are frequently stepped, the structural terraces representing individual lava flows. The plateaus are sometimes surmounted by volcanic cones. In the Pambak, Lori and Shiraki basins, and around Yerevan, ignimbrites and volcanic tuffs can be found. These basins are volcanotectonic depressions, formed by subsidence caused by the loss of magma during volcanic eruptions. Mount Elbrus in the Great Caucasus was also originally a volcanotectonic depression containing thick layers of volcanic material, but volcanic activity in the Great Caucasus was much less extensive in comparison

with that of the Armenian highlands [see Milanovskiy (1977)].

The south-eastern part of the Transcaucasian highland is formed by the Talysh range, consisting of several ridges with erosional relief forms trending north-west to south-east. They are composed of volcanic–terrigenous rocks and separated by a fault or flexure zone from the narrow strip of the Lenkoran coastal plain. The foot-hills are formed by strongly folded Miocene deposits [see Milanovskiy *et al.* (1966)]. The denudational/structural ridges of the Great Caucasus and the Lesser Caucasus are inherited from Hercynian and Cimmerian structures reactivated by young tectonic movements. The relief of the mountains marginal to the Caucasus uplift was also

Fig. 18.5 Mountains of the eastern Caucasus, with structural forms in limestone. (Demek)

Fig. 18.6 Plantation surface in the eastern Caucasus. The valley sides in the foreground and on the far side of the river show the effects of frequent mass movements. (Demek)

rejuvenated during late orogenic phases: the uplift here probably averaged 1500–2000 m.

In some areas of the Lesser Caucasus, block structures were subsequently developed on previously folded (synclinal) structures (e.g., the Azhar–Imeretia and Trialetskiy ridges) or on other Palaeogene tectonic forms (e.g., the inner ridges around the Sevan and Pambak basins). Inversion of relief, too, has frequently taken place. Neotectonic movements played an important role in the formation of the Armenian volcanic highlands and the Gruzhinskoye highlands, because volcanic activity is closely controlled by recent faults. Late block movements of the crystalline Baikalian and early Alpine basement were responsible for the formation of shield volcanoes, compensational subsidence giving rise to volcanotectonic depressions. Lava plateaus around the depressions often show strongly inverted relief since the lava flows capping them originally occupied erosional and tectonic (synclinal) depressions in the prevolcanic relief [see Antonov *et al.* (1977)].

In the zone of the Transcaucasian depression, occupied by depositional lowlands and low plateaus, the relations between relief and tectonic structure are more direct. In the eastern part of this depression and in the depression next to the southern part of the Caspian Sea, zones of very young (Pliocene–Pleistocene) orogenesis, with anticlinal ridges and synclinal depressions, came into existence. Mud volcanoes (*see* Section 18.4.4) are closely connected with this latest orogenic episode [see Antonov *et al.* (1977)].

In the tectonic structure of the Caucasus and the Transcaucasian highlands, faults are very important, especially young or inherited deep-seated faults. The older ones have in many cases been reactivated many times up to the present-day. These faults mark the boundaries of morphostructures of different orders and represent major features in their dynamic development. Many faults, particularly transcurrent ones, are visible in the relief as fault scarps, dividing various regions of the Caucasus and Transcaucasian highlands. The following are some examples: the Alderskiy fault scarp between the north-west and western Caucasus, the main overthrust scarp along the southern slope of the western Caucasus, and the fault scarp system along the northern slopes of the Alazan'–Agritchayskaya depression in the eastern Caucasus.

Medium-scale relief forms, such as river valleys, their slopes and watersheds, are strongly controlled by lithology. The massive calcareous beds of the Jurassic and Cretaceous (limestones and calcareous sandstones) form the crests of escarpments (*see* Fig. 18.5); synclinal hills and mountains represent fine examples of relief inversions (e.g., the cuestas of the Skalistyi ridge in the Great Caucasus and the limestone area of Dagestan). Limestones also form extensive plateaus. In areas of crystalline (metamorphic or extrusive) rocks, water gaps and deep river canyons are common. The slopes of these deeply incised valleys are often step-like with structural benches as described by Antonov *et al.* (1977). Lavas and other volcanic rocks such as ignimbrites and tuffs form a distinctive landscape with structural terraces and scarps.

In tuffs or agglomerates, selective erosion has sometimes developed forms similar to earth pyramids: for instance, on the Goris plateau in south-western Armenia. The less-resistant rocks of the flysch formation—relatively weakly consolidated sandstones, shales and marls—have usually been eroded into basins, sometimes partly of an inversional character: combes in anticlines, synclinal hills and plateaus in more resistant rocks. The Pliocene molasse in the eastern part of the Transcaucasian depression possesses a youthful folded relief.

18.2 Planation Surfaces and Denudation Chronology

Planation surfaces of subaerial and, in some cases, partly of marine origin are well developed on the watershed ridges of the Great and Lesser Caucasus, and also in the Dzhirulskiy massif, especially in the foot-hill zones. Such surfaces (*see* Fig. 18.6) often consist of accordant round-topped ridges with intervals between successive levels from 100 to 200 m. The older surfaces are poorly preserved. In the area of the main Caucasian ridge, the planation surfaces can be reconstructed only from summit levels, similar to the Gipfelflur in the Alps. For this reason, some geomorphologists [e.g., Safronov (1969)] have expressed doubts about the existence of planation surfaces when they are based solely on such evidence. Much better preserved are the planation surfaces in the foot-hills of the Great Caucasus, especially in Dagestan, the Kusary plateau and in the south-eastern part of the Lesser Caucasus. These planation surfaces are covered by gravels of Baku and Apsheron age, which enable correlations with marine facies in the Caspian Sea basin to be made. Similar correlations are also possible in the molasse area in the zone of the young Plio-Pleistocene folding in the eastern part of the Transcaucasian depression [see Dumitrashko (1974)].

Around Mount Elbrus, remnants of planation surfaces bear weathered mantles on them. Since they are, in turn, covered by Upper Pliocene volcanic rocks and glacial deposits in places, Milanovskiy (1964) has attributed these mantles to Miocene weathering. Near Schakhyurd in the south-eastern Caucasus at a height of 3550 m, Budagov (1966) has located deposits of Upper Sarmatian age, which enabled the high planation surface at 4000–4200 m to be dated as Miocene. This is what Antonov *et al.* (1963) referred to as the Schakhdag surface. In the outer ridges of the eastern part of the Lesser Caucasus, Middle and Upper Miocene marine deposits are exposed, while around Lake Sevan on this planation surface, weathered mantles of Miocene age have been described by Aslanyan (1958) and Milanovskiy (1952, 1968).

It is possible to deduce, then, that the oldest planation surfaces are now represented by the accordant levels of watershed ridges and that these formed during the early stages of the development of the Caucasus and Transcaucasian highlands; moreover, they are in all probability of Palaeogene age. However, the best preserved remnants of planation surfaces on the axial ridges of the Great

Caucasus are evidently of Miocene age. Therefore, most workers prefer to regard the oldest planation surface as being of unspecified early Tertiary age. The formation of the surfaces and their related weathered mantles took place under the warm, humid climatic conditions of the early Tertiary in this area. Subsequently, the mantles were eroded and the basal surface of weathering was exposed.

On the northern and southern margins of the Great Caucasus and in areas of the north-western and south-eastern depressions, remnants of Mio-Pliocene and Lower to Middle Oligocene planation surfaces have been found. The ages of these surfaces have been established by Dumitrashko *et al.* (1961) and Muratov (1964) according to their positions in relation to the older early Tertiary surface and the younger Upper Pliocene surfaces. The remnants of these surfaces also occur in the Lesser Caucasus and around the Araks river [see Budagov (1966), Dumitrashko (1957, 1962)]. In volcanic areas, the surfaces are covered by lavas, and in some places have been subsequently exhumed.

In the foot-hills of the Great and Lesser Caucasus, in the Talysh region and in the zone of Plio-Pleistocene folding in the eastern Kura depression, Upper Pliocene and Lower Pleistocene planation surfaces have been recognized. The age of these has been established by Dumitrashko (1974) by reference to their cover gravels which pass laterally into marine sediments of Apsheron age, as well as in relation to the molasse deposits. These young planation surfaces form quite distinctive elements in the landscape.

Pediments developed under warm, arid climatic conditions in the eastern part of the Caucasus and the Transcaucasian highlands. They are particularly striking in the central and southern parts of the Transcaucasian depression where, in the foot-hills Upper Pliocene pediments can also be seen. It therefore seems likely that, from the Palaeogene to the beginning of the Upper Pliocene, planation processes and especially pedimentation prevailed, and that valley cutting commenced at the end of the Upper Pliocene, continuing into the Pleistocene and Holocene Epochs. In the eastern part of the Caucasus and the Transcaucasian highland, and in adjacent areas near the Caspian Sea, dissection set in earlier during the Middle Pliocene.

The principal reason for river incision in the areas draining to the Caspian Sea was the lowering of the level of the latter during the Middle Pliocene, when the water level fell to 500 m below mean global sea level. The sea then covered only a small southern part of the present-day Caspian depression [see Goreckiy (1964), Kvasov (1964, 1966), Milanovskiy (1968, 1977)]. The causes of this lowering were both climatic and tectonic. Tectonic subsidence formed the deep southern Caspian depression. The climate in the Caspian region became arid and warm during the Middle Pliocene, raising evaporation rates and contributing further to the falling water level. At the end of the Upper Pliocene (the Apsheron phase) rapid uplift contributed to intensive dissection of the relief, increasing relief amplitude and an increase in absolute elevation,

which even brought the higher parts under the influence of glaciation (as on the northern slopes of the central Great Caucasus and the southern part of the Transcaucasian highlands.) Later, meltwater from the glaciers aided further in the process of river incision and valley excavation. During the Upper Pleistocene, an interconnecting network of glaciers developed [see Dumitrashko (1974)] and the glaciers individually became longer than in any previous period of glaciation. The maximum extent of Pleistocene ice was on the northern slopes of the central and western Great Caucasus, where the Teberda glacier attained a length of 77 km, the Terek glacier 72 km, the Baksan 70 km and the Kuban 50 km. At the present-day, glacial ice covers just under 2000 km² overall. Major recession occurred between 1890 and 1930, when, for instance, the glaciers on the drier south-eastern slopes lost up to two-thirds of their surface area. A recent study of marginal moraines in the central Caucasus is that of Serebryanniy and Orlov (1982).

In summary, the most important stages in the development of the Caucasus and the Transcaucasian highlands occurred in the Oligocene, the Miocene–Lower Pliocene and the Upper Pliocene periods. Uplift of the Caucasian region started in the Palaeogene, when a narrow ridge—the precursor of the Great Caucasus and the axis of a mega-anticlinorium—emerged from the sea. By the late Oligocene, the area had taken on the character of a low mountain system. In Armenia, dry land formed to the south of the present-day Lake Sevan [see Antonov *et al* (1977)]. The peripheral parts of the Great Caucasus and nearly the whole of the Lesser Caucasus consisted of erosional lowlands with some flat areas of deposition. Around the present-day Lake Sevan, there was a lacustrine basin in which molasse deposits accumulated. The mountains continued to grow during the Upper Miocene and Lower Pliocene. In the axial part of the Great Caucasus, Alpine ridges surrounded by lower mountains developed. The northern and eastern parts of the Lesser Caucasus contained some further mountain areas, partly with Alpine relief forms. Intensive volcanic activity gave rise to lava plateaus and volcanic mountains in the central part of Armenia and in the southern Georgian highlands at this time. The Caucasian isthmus in the area of the present Transcaucasian depression was very narrow and consisted only of the contemporary Kolkhida lowland in the west and the southern part of the Kura depression and adjacent foot-hills in the east [see Antonov *et al.* (1977)].

The dissection of the relief began essentially in the Middle Pliocene. In the Upper Pliocene, the intensity of erosion was even greater due to rapid uplift. In that period, the main relief features of the Caucasian and Transcaucasian areas, as we now know them, became recognizable. During the Upper Pliocene, the oldest glaciation commenced, witnessed by the glacial and glaciofluvial deposits of this age found in the foot-hills of the Great and Lesser Caucasus [see Dumitrashko (1974)]. Uplift continued through the Pleistocene when three further glaciations are known to have occurred. The forms of the late Pleistocene glaciation are, as expected, the best preserved in the present landscape.

Fig. 18.7 Mountain relief in the eastern Caucasus near Shemakha. (Demek)

Fig. 18.8 Tafoni weathering of sandstone on the shore of the Caspian Sea south of Baku. (Demek)

18.3 Relief Types and Relief Regions

The Caucasian and Transcaucasian highlands are shown on the *International Geomorphological Map of Europe* as young epigeosynclinal orogenic and volcanic mountains. On the basis of morphostructure, they may be divided into several groups.

1) The first group is formed by mountains composed of crystalline and metamorphic rocks: the western Great Caucasus, the central Great Caucasus, the Dzhirulskiy massif and the southern part of the Zangezur ridge in Armenia. These mountains are characterized by Alpine fold-faulted ridges and block massifs. Small crystalline massifs in Georgia and in the north-eastern part of Armenia are shown on the *International Geomorphological Map* as Alpine block ridges with dissected relief.

2) The major basins of the Great Caucasus—those of Arkhiz–Zagedan and Beshitan—are mapped as relief types developed on consolidated rocks. The Seven–Yurs depression is mapped as an intermontane erosional and erosional/structural basin formed in consolidated rocks. East of the River Ardon—a tributary of the Terek—the crystalline and metamorphic rocks of the western and central Caucasus are bordered by a large fault. This fault is also the eastern boundary of the Transcaucasian zone of uplift.

3) The eastern Caucasus (*see* Fig. 18.7), the northern part of the south-eastern Caucasus and the young dissected Kakhetia ridge are Alpine fold-faulted ridges, built of consolidated rocks. The same relief type is represented in the area of the Svanetskiy ridge on the southern slopes of the Great Caucasus.

4) In the Lesser Caucasus, the highest parts of the Adzhar–Imeretia and Trialetskiy ridges have a clear fold-faulted structure. Fold-faulted morphostructures are also characteristic of the inner ridges of Armenia around the Sevan Basin and the Pambak depression, and of the Karabakh ridge on the border of Armenia with Azerbaijan. Generally, the central highest ridges of the Great and Lesser Caucasus consisting of crystalline, metamorphic and consolidated sedimentary rocks have fold-faulted and block morphostructures.

5) The peripheral ridges of the Great and Lesser Caucasus have varied relief types controlled by lithology. In the Great Caucasus, the dissected Goykhtskiy ridge of the north-western sector composed of consolidated rock displays a block structure. The fold-faulted ridges of the Lesser Caucasus and of Talysh, also built of consolidated deposits, are characterized by extensive planation surface remnants. The valleys here are deeply incised.

6) The low mountain block ridges of the north-western Caucasus, the ridges of inner Dagestan and the southern slopes of the Great Caucasus from western Georgia to the south-eastern Caucasus are composed of flysch rocks and shales. In western Georgia, on the edge of the Great Caucasus between Sukhumi and Kutaisi, there is a distinctive zone of fold-faulted limestone ridges, some with Alpine relief forms. The area known as Limestone Dagestan is another region of block ridges. Inversion of relief is widespread with peaks formed in the resistant limestone layers.

7) Limestones and dolomites make up the karst plateau of Lagonak in the western Great Caucasus. A series of limestone and dolomite cuestas extend between the western end of the mountains and the Terek valley. These cuestas gradually merge into folded and monoclinal low mountains on the margin of the eastern Great Caucasus. Limestones also form the Schakhdag, Kizylkaya and other plateaus in the area of the Bokovoy ridge in the south-eastern Caucasus.

8) In the Terek–Caspian foredeep, the Terek and Sundzha ridges are built of unconsolidated sediments and molasse. The mountains around Ordzhonikidze, Machatsch–Kala and Samur, the foot-hills of the south-eastern Caucasus around the Kusary plateau and the Kobystan hills have a folded denudational relief formed prior to the Upper Pliocene.

9) Similar relief types can be found in the western Caucasus along the Black Sea depression, to the north-west of Sukhumi and along the Kolkhida lowland. The latter is surrounded on both the south and the north by young folded foot-hills much dissected by erosion and on the west (west of Kutaisi) by a dissected plateau.

10) Accumulation/denudation plateaus occur along the Srednearaksinskiy depression near Yerevan, around the mouth of the River Veda, north of Nakhichevan' and in part of the Terek–Caspian foredeep of the Great Caucasus.

11) The volcanic relief of the Armenian highlands and of the Yushn–Gruzinskoye highland was formed during the Tertiary and Quaternary, and includes various types of lava plateau, tectonic–volcanic depressions, volcanic foldfaulted highlands and mountains. Smaller areas of volcanic highlands are also developed in the central Caucasus, such as Mount Elbrus and the Kelskoye, Verkhnetchegemskoye and Nizhnetchegemoskoye plateaus.

12) In the Transcaucasian intermontane depression lie the Kolkhida and Kura–Araksinskaya fluvial lowlands. In the Nizhnekurinskaya basin, extensive piedmont lowlands (bajadas) have developed at the foot of the surrounding mountain ridges.

13) Along the coast of the Caspian Sea (*see* Fig. 18.8), from the Samur as far as the frontier of the USSR, there is a narrow coastal plain, the product of both abrasion and deposition. Similar relief characterizes the Apsheron peninsula.

14) In the Kuban and Terek–Caspian foredeeps, fluvial lowlands surrounded by proluvial (piedmont) lowlands extend as a fringe at the foot of the surrounding mountains.

18.4 Present-Day Geomorphological Processes and Natural Hazards

In the Caucasus and Transcaucasian highlands, certain geomorphological processes offer particular hazards to man, especially with regard to settlement, industry, roads, forests, agriculture and recreation. In this connection, such processes as avalanches and mudflows must be considered, but there are also other contemporary processes of a less hazardous nature, such as karst processes and mud volcanoes, that are nevertheless very distinctive for certain areas of the Caucasus.

18.4.1 Mudflows and Avalanches

These are the most dangerous processes in the Caucasus

at the present time because of the high energy inputs due to the amplitude of the relief and because of their widespread distribution.

Mudflows are particularly feared, causing devastation along their tracks and spreading out deposits not only at the foot of the mountains but also farther out on the piedmont areas. They are most common in the central Caucasus, on the southern slopes of the Great Caucasus, in the eastern part of the Lesser Caucasus, in the Armenian volcanic highlands and on the mountain ridges of southern Armenia. Their formation depends on the steepness of the slopes, local concentrations of torrential rain, high relative relief energy and the presence of zones of rock highly fractured by folding or faulting [see Budagov *et al* (1977)]. Present-day tectonic movements, especially earthquakes, often act as triggers setting off such mudflows. Ermakov (1957) and Bugadov (1966) reported that they are common in areas of Alpine relief, but less so in the lower mountains. In the Alpine zone, mudflow development is related to intensive physical rock weathering and accumulations of till or colluvial deposits. Glacier meltwater plays a role, saturating the till and changing it into mud. Rockslides, other mass movement accumulations and gelifluction deposits also provide materials that participate in mudflow formation [see Budagov *et al.* (1977)].

Mudflows may develop frequently in grassed areas, rupturing the turf cover, but are rare in forested areas, partly because of the reduced intensity of mechanical weathering. On the other hand, they are common in dry steppe and semiarid areas that are subject to occasional torrential rains, especially in low mountains where Pliocene and Quaternary clays, conglomerates, loams and loess provide materials particularly susceptible to liquefaction and flow [see Budagov *et al.* (1977)]. The most devastating types are the structural block mudflows, which occur in flysch areas with steep slopes (up to 50° or even 60°) during rainstorms, as on the southern slopes of the eastern and south-eastern Caucasus. In areas of crystalline, metamorphic and volcanic rocks, or of gravel deposits, less blocky material is involved. The volumes of material transported by individual mudflows range from 150000 to 300000 m³. The frequency of such mudflows varies from once every two or three years to once in fifteen years; in some areas, their periodicity is quite regular.

Avalanches are also primarily controlled by the climate and relief of the Caucasus. They form a major source of supply for many glaciers, and like mudflows may also be agents of great destruction. They are most common on slopes of more than 30° and develop as soon as the snow cover reaches 1 or 1.5 m. Under favourable wind and slope conditions, however, the snow pack may build up to 10 m or so before movement is triggered. Avalanche activity usually starts in October or November, as soon as a continuous snow cover forms [see Tushinskiy (1977)]. Akifyeva (1970) and Tushinskiy (1977) have plotted the regional distribution of avalanche risk for the Caucasus and the Transcaucasian highlands. Regions of major avalanche activity comprise the Alpine mountain areas of the Great Caucasus and Armenia. These correspond to areas of present-day and earlier glaciation, and are typified by slopes with many avalanche ravines and chutes active during nearly the whole year. Moderate avalanche activity occurs on the inner ridges of the Lesser Caucasus, while the north-western Caucasus, the mountains and foot-hills of the northern, eastern and south-eastern Caucasus and the northern ridges of the Lesser Caucasus experience only slight avalanche activity.

18.4.2 Karst

Karst forms occur mainly on the northern slope of the Great Caucasus and in the foot-hills, but also in western Georgia. The karst has developed mostly on Upper Jurassic and Cretaceous limestones, rarely on dolomite or gypsum. Karstification probably began in the Cretaceous or Palaeogene periods and is active today in the areas mentioned. Buried and exhumed karst forms can also be found [see Gvozdeckiy and Maruashvili (1977)].

18.4.3 Badlands

In the south-eastern Caucasus, the southern part of the Lesser Caucasus and in the Armenian highlands, extensive badlands have developed under arid and semiarid conditions. Suffosion processes, as well as normal gully erosion, have helped in the badland formation, producing suffosion caves, pits and other collapse phenomena [see Lilienberg (1955a), Lilienberg *et al.* (1977)]. Badlands are relatively useless for agriculture because of the removal of the soil and vegetation cover and because of the steep slopes of the actively eroding gullies.

18.4.4 Mud Volcanoes

Mud volcanoes are concentrated in the periclinal depressions of the Great Caucasus: in the Taman peninsula, the Apsheron peninsula and in neighbouring areas of the south-eastern Caucasus. The greatest number is concentrated in Azerbaijan, especially on the Apsheron peninsula. Their formation is closely connected with oil and gas-bearing strata and their tectonic deformation—the folding and fracturing of brachyanticlinal or dome structures. Also relevant is the existence of thick layers of plastic clays saturated with water. Geomorphologically, it is possible to distinguish cone and shield volcanoes, and fields of small mud volcanoes on the basis of morphology [see Blagovolin *et al.* (1977), Lilienberg (1955b, 1962), Shirinov (1962, 1965)]. Dynamically, they can be divided into active, slightly active and extinct types. Their lifespan may range from about 15 to 65 years; during their lifetime, a periodicity of activity may be apparent. In the period 1810–1950, two slight and two active phases of eruption have been documented. Since 1952, the mud volcanoes of the Apsheron peninsula have been more active than elsewhere. Gorin (1952) argued that their activity is connected with fluctuations in the water level of the Caspian Sea; intensive episodes of eruption can be correlated with a falling water level, while slight eruptions can be associated with a rising water level. This is explained by the changing weight of water masses on the Earth's crust causing compensatory vertical movements. Agabekov and Akhmedbeyli (1956) connected the mud volcanoes with continuing Pliocene–Quaternary folding and faulting, determining the positions of brachyanticlines and domes.

Chapter 19
Ural Mountains

19.1 Introduction

The Ural Mountains, which stretch from north to south for about 3000 km, form a continuous margin on the east of the Russian platform. In many respects, it is a unique mountain system, not just for Europe but in the context of the whole of Eurasia. The main structural forms are elongated and divided by parallel deep-seated faults. Transverse faults, located by geophysical surveys and also expressed in the relief [see Yanshin (1966), Ogarinov (1974)], divide the mountain chain into seven major units, each of which has its own typical pattern of mountain ridges. From north to south, these are Pay-Khoy, Zapolyarnyy, Pripolyarnyy, Polyarnyy, Severnyy, Sredniy, Yuzhnyy Ural and Mugodzhary. The Pay-Khoy ridge trends orthogonally to the main direction of the Ural system, following an alignment from north-west to southeast, through Vaygach Island and extending into Novaya Zemlya (see Fig. 19.1).

The average altitudes of the Urals are around 1000–1300 m, the highest point being Narodnaya peak in the Pripolyarnyy Ural which reaches a height of 1894 m. Other high points are 1467 m in the Pay-Khoy, 1617 m in the Tel'pos-Iz peak of the Severnyy Ural and 1569 m in the Konzhakovskiy Kamen' peak also in the Severnyy Ural. The Sredniy Ural is comparatively low, reaching only a maximum of 746 m, while the Mugodzhary Ural rises to just 675 m. The Yuzhnyy Ural has a number of high points in the Yaman-Tau group reaching about 1640 m.

The geological history of the Ural Mountains is highly complex. The proto-Ural geosyncline first began to evolve in the Precambrian era, the first orogenic phase being termed the Baikalian, which was responsible for the largest structural forms of the Urals. The Caledonian orogeny had a much smaller effect and did not significantly alter the Precambrian structural patterns, but instead served to reinforce them. The closing tectonic phase was Hercynian, at which time the tectonic processes tended also to follow the ancient structural plan, forming a unified orogenic zone.

Because of lithological differences, many ancient structural forms are evident in the present-day physical features of the Urals: for example, resistant Precambrian quartzites mark the cores of denuded horst–anticlinoriums, with less-resistant Palaeozoic rocks being exposed in the synclinoriums. The Mesozoic history of the whole Ural chain is poorly known because of the general lack of deposits of this age. Probably during this time most of the area was undergoing denudation, with the erosion products accumulating outside the mountain area. In this respect, two areas are noteworthy: the down-faulted depressions of Priuralye and Zauralye on the eastern and western flanks, respectively, of the Urals. In these depressions, Jurassic and Cretaceous sediments are preserved, whose position and attitude show that the Urals had by then achieved the general outlines of the present-day mountain chain.

Deep-seated faults are considered to have played a major role in the development of both the ancient and contemporary morphostructural plan of the Urals. These are essentially old faults, subsequently re-activated, the most fundamental one being the Ural-Oman fault system (see Fig. 19.1), also known as a 'superlineament' [see Bush et al. (1980)]. This consists of several meridional or submeridional deep fractures, such as the main Ural fault which divides the Ural Mountains from the Transuralian plains and the eastern Ural fault which forms the easternmost member of the superlineament. Along the eastern Ural fault, the Hercynian basement, which underlies the Transuralian plain in the central southern Urals, has long been subsiding and is now buried by the Mesozoic and Cenozoic deposits of western Siberia. The close connection between the Ural Mountains and the Hercynian basement beneath western Siberia is evidence that the Ural Mountains belong more to Asia than to Europe.

In recent years, the Ural-Oman superlineament has been studied in detail using satellite imagery (and also visually by astronauts). Such studies have revealed that it extends continuously through the whole length of the mountain system to the plains of the Pechora syncline (see Fig. 19.1).

The fact that the Urals span 22° of latitude means that there are substantial differences of climate between north and south. In the area of the Polyarnyy Ural, and in some parts of the northern Severnyy Ural, forms of mountain glaciation appear, including cirque glaciers. However, in the Mugodzhary Ural, arid processes of land sculpture are active.

In summary, the main relief features of the Ural Mountains are related to the following factors.

1) Longitudinal faults (including the Ural–Oman superlineament) which have been active for long periods of the Earth's history.
2) The elongated shape (from north to south) of the main tectonic patterns.
3) Transverse faults, dividing the mountain chain into a series of block morphostructures.
4) The relative resistance of the various rock types, and especially the occurrence of hard rocks in the cores of

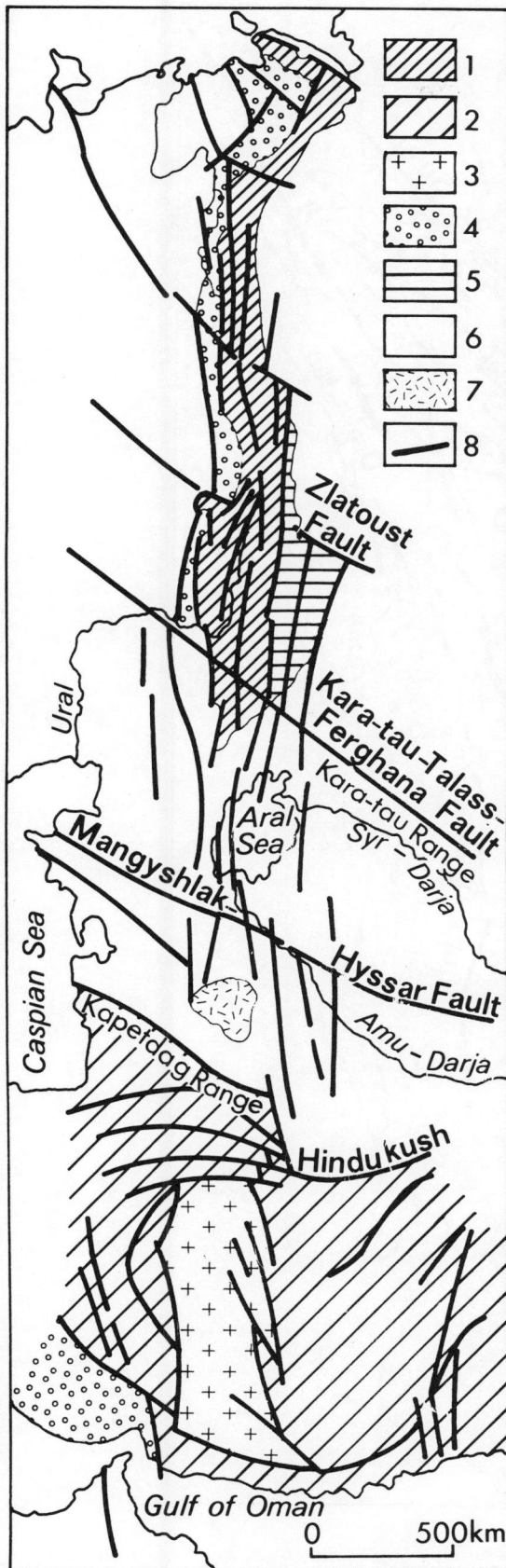

Fig. 19.1 Structure of the Ural–Oman superlineament [after Bush *et al.*, 1980]. (1) Hercynian fold-faulted structures of the Ural basement, (2) Alpine fold structures, (3) crystalline massif involved in (2) (the Lut block), (4) foredeeps and marginal depressions, (5) planation surface of desert lowland on folded basement (so-called *syrts*), (6) platform cover, (7) Karakum dome of the Turanian platform and (8) principal deep-seated faults.

ancient structural forms.

5) Differences of climate affecting the range of erosional processes active at any time.

19.2 Southern (Yuzhnyy) and Central (Sredniy) Urals

19.2.1 General Relief and Structure

From west to east, there are the following main structural zones in the southern central Urals (*see* Fig. 19.2): (1) the Pre-Ural foredeep, (2) the Bashkirian uplift, (3) the Zilairskiy graben–synclinorium, (4) the Ural-Tau uplift (which becomes the central Ural uplift farther north), (5) the Magnitogorsk synclinorium (or 'green rock' synclinorium, continuing in the central and northern Urals as the Tagil synclinorium), (6) the Ural-Tobol anticlinorium and (7) the Kustanay synclinorium. Yanshin (1966) noted that these synclinoria and anticlinoria involve faulting as well as folding.

In addition to the main deep-seated transverse faults which serve to divide the Urals into the seven fundamental units already listed (including the Zlatoustinskiy fault between the southern and central Urals, and the Kara-Tau–Talaso–Ferganskiy fault between the southern Urals and the Mugodzhary mountains: *see* Fig. 19.1), other less important transverse faults delimit a series of complex block morphostructures and are responsible for changes in the direction of mountain ranges, intermontane basins and river valleys. In spite of this, the general longitudinal mountain pattern of the southern Urals is maintained. The legacy of ancient structural patterns is well marked in the mountain areas of the Urals, but later block tectonics, periods of denudation and planation throughout the Mesozoic and Cenozoic eras have also influenced the relief, so that the latter now presents only a skeletal outline of the original structural plan. The largest morphostructures of the southern and central Urals consist of strongly metamorphosed and resistant Precambrian quartzites, forming the massifs of Yaman-Tau and Iremel, and the mountain ranges of Nary, Nurgush, Zigal'ga, Zil'merdag, Mashak, Bakty, Ural-Tau and Uren'ga (*see* Fig. 19.2). All these features are related to two ancient structures: the Bashkirian uplift and the Ural-Tau horst–anticlinorium, divided further into fault blocks.

The differences in altitudes of river terrace sequences in different blocks shows that differential block faulting has continued until recent geological times. Low river terraces, numbering not more than six or seven and of both erosional (rock cut) and aggradational origin, are found in the main valleys of the southern Urals (along the rivers Belaya, Yuryuzan', Bol'shoy Inzer, Malyy Inzer, Sakmara, Ural, etc.) and of the central Urals (rivers Bizhay,

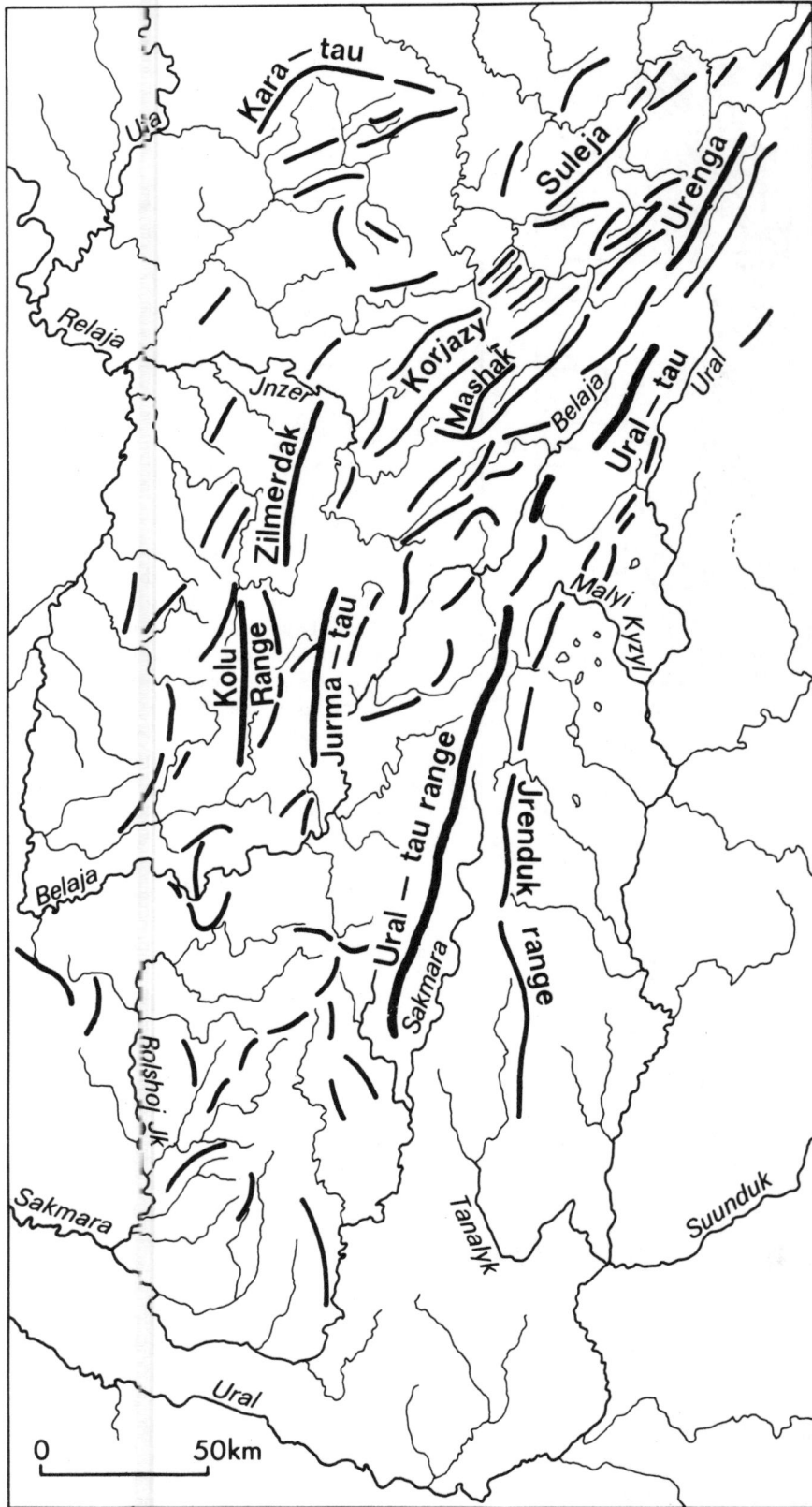

Fig. 19.2 Major ranges and valleys of the southern Urals.

Chusovaya, etc.) (*see* Fig. 19.2). The age of the uppermost terrace is always the same—Oligocene–Miocene—but the relative height and composition of the terraces differ, not just between different valleys but also along the same valley. Such differences can only be ascribed to differential movements of the blocks crossed by the rivers. The terrace systems of the central and southern Urals are now well-known, so that reconstruction of the tectonic history is possible [see Bashenina (1948, 1961), Nikiforova (1946), Vedenskaya and Golubeva (1966)]. In some water gaps, controlled by faults and other dislocations crossing mountain ranges and intermontane basins, river terraces may be absent due to the rapidity of river down-cutting, leaving steep convex slopes, but overall the evidence for neotectonic differential uplift of the Urals is clear.

North of the Zlatoustinskiy fault in the central Urals (*see* Fig. 19.1), the Ural-Tau range has lower altitudes (around 700–750 m), is more bevelled by planation and is structurally shifted to the west. The relief is characterized by low flat-topped or gently convex ridges, the surface cutting across folded and faulted rocks. The same is true of the southern Urals, except that here the heights are less and the degree of dissection is smaller. Farther east lies an upland of folded and faulted Palaeozoic rocks, while south of the Kara-Tau–Talaso–Fergansky fault (*see* Fig. 19.1) the Ural-Tau range ends in the Mugodzhary mountains.

There are no correlative deposits left on the planation surfaces of the central Urals or of the Mugodzhary mountains. The number of river terraces, their ages and composition match those of the southern Urals and, together with the incised river valleys, provide evidence that neotectonic movements have not affected the differences in altitude between the foot-hills of the southern Urals, the central Urals and the Mugodzhary mountains. On the western side, the southern Urals become lower to the south and merge into the Zilairskiy uplands. Both consist of Palaeozoic rocks; the Zilairskiy uplands coincide approximately with the Zilairskiy synclinorium. Farther to the south in the foot-hill zones of the Urals, the foot-hills of the Ural-Tau range and the Zilairskiy uplands are bevelled by an extensive planation surface that cuts across the Proterozoic rocks and is partially buried by marine Cretaceous and Palaeogene deposits up to 40 m thick, consisting of gravels and glauconitic sandstones. The river valleys are incised to depths of 130–135 m and the surface is being actively exhumed. The youthfulness of this incision is shown by the height (135 m) of the fifth Pliocene terrace of the Sakmara river. To the south of where the Ural river flows in an east–west direction and west of the Mugodzhary uplands, another erosion surface cuts across folded and faulted structures.

The Zilairskiy tectonic zone is marked by massive intrusions of Upper Palaeozoic hyperbasic rocks, intruded along the main Ural fault (*see* Fig. 19.1). These intrusions correspond to the Krak and Surtanda massifs of roughly circular outline. Present-day continuing uplift of these massifs has been proved by geomorphological evidence in the form of the steep stepped slopes subject to repeated mass movements and the up-domed river terraces of the valleys crossing these areas. The sixth river terrace of the

Belaya rises to a height of 300 m above the valley floor where it crosses the Zilairskiy tectonic zone. The fifth terrace of the river Ryaza stands at 130 m, but farther downstream at its confluence with the Belaya the terrace height is only 75 m.

The Magnitogorsk or 'green rock' synclinorium on the eastern side of the southern Urals is marked by the Irendyk–Krykty range; this is separated from the Ural-Tau range by an intermontane basin in which the source of the Ural river is to be found. After its confluence with the Mindyak river, the Ural river cuts through the Krykty range in a water gap. Two other rivers, the Malyy Kizil and the Bol'shoy Kizil, draining the southern part of the basin, also traverse the Krykty range in water gaps. The south-eastern part of the basin is comparatively broad and is displaced laterally by a fault which also controls the east–west-trending section of the Ural river. The basin is terminated by the Orskiy graben which lies on the Kara-Tau–Talaso–Ferganskiy fault. The floor of the Orskiy basin consists mostly of an eroded lowland, only the eastern part being infilled to produce an alluvial plain.

The foot-hills east of the Irendyk–Krykty range comprise an upland area known as *melkosopochnik* underlain by folded and faulted Palaeozoic rocks, including intrusions. Farther east, the relief develops into a mosaic of hills and lake basins. The eastern margin is the Chelyabinsk graben within the Ural–Tobol zone, as far as the boundary of the Ural–Oman superlineament. East of the latter are the plateaus of western Kazakhstan and the Turgay lowlands formed from gently dipping sedimentary strata together with various alluvial areas. In the south, east of the Mugodzhary ridge, there are plains of aeolian deposits. All these plains and low plateaus lie in the zone of the Kustanay synclinorium.

The Pre-Uralian plains occupy the area of an ancient foredeep infilled with Permain marine limestones and molasse. The strata are subhorizontal or only gently folded, with altitudes mostly between 200 and 300 m. Salt and gypsum deposits have been squeezed into domes and other structures. The resulting deformation of the surface combined with the effects of solution have produced tectonokarst-like features, with depressions filled with lacustrine and fluvial sediments of Neogene age. Inselbergs known as *schikany*, which are built of reef limestones, rise sharply above the general surface. To the north, the floor of the depression narrows sharply towards the transverse fault, so that the depression is nearly wedged out by the Kara-Tau range, but still farther north it reappears.

19.2.2 Intermontane Depressions of the Southern Urals and the Eastern Slope

The two largest longitudinal depressions of the southern Urals have already been mentioned: the first situated east of the Ural-Tau range and the second being the Chelyabinsk graben which was recognized as a rift valley by Karpinskiy in the nineteenth century. In the foot-hill zone (the melkosopochnik), on the eastern side of the southern and central Urals, several other smaller longitudinally orientated closed depressions occur. Their floors are formed in bedrock, characterized by undulating relief and small lakes; rivers are mostly absent. In a few parts, there are some superficial deposits, mostly fluvial gravels of various ages from Jurassic to Miocene (and Pliocene in the Orskiy graben). The general features are strongly reminiscent of the African rift valleys, although with dimensions one-tenth of the size of the latter [see Bashenina (1977)].

In the southern Urals, depressions are found separating the highest mountain ridges. The ridges and depressions are orientated from north-north-west to south-south-east, and have been laterally displaced by transverse faults. Relative relief amounts to about 700 m, and the slopes of the ridges have concave profiles merging gradually at their lower ends into the basin floors. The depressions are up to 3 or 4 km wide, their floors showing only a thin soil covering over bedrock. Slope deposits are often absent, except for occasional block streams coming from the narrow rocky ridges but usually failing to reach the slope foot. Much of the area lacks any integrated drainage system, but a few rivers rise on the massifs and cross the basins, leaving them by means of water gaps. The courses of the rivers, such as the Yuryuzan coming from the Yaman-Tau massif, appear to occupy ready-made depressions; sometimes there are river terraces with gravel cappings up to 1 m thick. The absence of the slope deposits and any significant floor accumulations in these Ural depressions is puzzling, for as has been pointed out rivers draining the basins are infrequent, so that removal of deposits by this mechanism is unlikely. It seems more probable that the depressions are purely of recent tectonic origin [see Bashenina (1977)].

On the Pre-Uralian plains, grabens are absent. In the central Urals they are relatively rare, but in the northern and Pripolyarnyy Ural they are once again more common.

19.2.3 Fluvial Forms, Deposits and Slope Processes

The major morphostructures of the southern and central Urals had already come into existence by the end of the Mesozoic Era. The main outlines of the relief were similar to those of the present-day: the lowlands of the western slope, the mountain areas of the southern Urals, the hills and uplands of the central Urals and the Transuralian hilly and lowland areas. Each unit underwent further modifications of its relief and further deposition, both related in part to the differential block movements of the neotectonic period. By the Oligocene, this long history of subaerial landscape evolution had allowed the present main elements of the river system in the southern and central Urals to develop as discussed by Bashenina (1948, 1961). The seventh Palaeogene terrace was only locally developed, but the sixth Miocene–Oligocene terrace is present in all the main morphostructural units. The relative heights of this sixth terrace vary considerably from place to place: about 100 m in the Belaya valley of the mountain Urals, 200 m on the western slope and 300 m in the Krak massif. Lower Pliocene fluvial deposits up to 25 m thick are preserved in tectonic and karst depressions, and also on the fifth terrace of valleys on the western slope and in the Transuralian region. In contrast, the thickness of these deposits on the fifth terrace in the mountain Urals

only amounts to 0.5–2.0 m. The relative height of the fifth terrace in different blocks varies from 40 to 130 m.

In the Middle Pliocene, the floors of the depressions suffered further subsidence, accompanied by the accumulation of sediments. The Belskaya graben on the western slope and the depressions in the southern part of the Transuralian region were particularly affected, some depressions being flooded by the sea but later developing as lakes. The valleys draining to the depressions were filled first by marine deposits and later by fluviolacustrine accumulations. Regression of the sea and the subsequent drying out of the lakes caused further incision of valleys into the Pliocene deposits, aided by positive tectonic movements in places. This was the background to the formation of the fourth terrace (mainly of rock cut type) in the mountain areas of the southern central Urals. In some valleys of the Transuralian region and in the Belaya valley on the western slope, the terrace has an aggradational character and the thickness of the Middle Pliocene deposits reaches 120–150 m. This means that in some places their base is below present river level. The relative height of the fourth terrace ranges from 30 to 90 m.

At the end of the Pliocene and in the Lower Pleistocene, the climate became colder, as shown by palaeobotanical evidence. Aggradation by the rivers formed the relatively thin fluvial deposits of the third terrace whose height varies from 10 to about 40 m. The Upper Pliocene–Lower Pleistocene deposits in the Transuralian region are buried by later deposits in some places and the river gravels are mainly slightly rounded, indicating that the rivers flowed only intermittently and that slope processes were operating in a cold, dry environment. Palaeobotanical data indicate that broad-leaf forest had been replaced by coniferous woodland. Pollen analysis shows a predominance of *Pinus*, but there were also *Picea* and *Abies* in the mountain Urals. At the same time, the height of the treeline was lowered in the highlands, and in such areas the physiographic environment became typical of the periglacial zone [see Krasheninnikov (1939)], this change coinciding with the establishment of glacial conditions elsewhere. In the southern central Urals, glaciers failed to develop because of the arid climate, but in the Prilopolyarnyy and Polyarnyy Urals, where there are small cirque glaciers at the present-day, cirque and valley glacier systems developed more extensively in the Pleistocene cold phases.

The Middle Pleistocene is characterized by increased aggradation in the valleys and on the slopes of the southern and central Urals. In the valleys, the second terrace stands 10–15 m above the valley floors. As well as slope wash processes, deflation was also active in the mountains. On the western slope and in the Transuralian region, slope processes caused the infilling of many small valleys and river beds, while a cover of loessic loam was laid down. Many minor changes of river pattern occurred: there were some instances of stream bifurcation, as well as cases of river capture. Some valley pediments, coalescing to form undulating surfaces (*melkosopochnik*), developed at this time according to Bashenina (1948).

In the Upper Pleistocene, the first terrace was developed over the whole area of the southern central Urals. The climate became warmer and more humid, and the tree-line rose. The Holocene saw the formation of the high floodplain, while in the most recent time up to the present-day, there is another lower floodplain forming. Present-day slope processes include wash, rill erosion, creep and, in some high mountain areas, gelifluction. In the northern and Polyarnyy Urals, gelifluction and block stream movements are more active, but on the eastern slopes of the central and especially the southern Urals, the climate is often too dry. In these latter areas, there are still some lakes, which are relics of former wetter conditions, but now they mostly lack outlets, are becoming more saline and are gradually undergoing dessication [see Bashenina (1961)].

19.2.4 Planation Surfaces

There are no planation surface remnants in the mountain areas of the Urals. In the northern and Polyarnyy Ural, however, there are some references in the literature to a pre-Neogene planation surface and to the occurrence of some relics of a weathered mantle of this age. In the extreme south of the Urals, there is the important sub-Cretaceous unconformity beneath the marine Cretaceous and Palaeogene deposits, and in places these are being stripped off to reveal the old surface. This surface was bevelled by subaerial processes before the Cretaceous marine transgression, and only developed around the lower parts of the mountain system [see Bashenina (1967)].

The Transuralian plain—usually termed the Transuralian peneplain—actually comprises a complex series of coalescing pediplains. These appear to have formed during long periods of pedimentation and are often directly related to the seventh, sixth and even the fifth terraces. At lower levels, valley pediments are characteristic.

Above the tree-line, where periglacial conditions were established in the Lower and Middle Pleistocene, intensive cryoplanation occurred [see Bashenina (1948, 1960, 1967), Boch and Krasnov (1951)]. The present-day lower limit of cryoplanation lies much higher, and this group of processes is more active in the higher ranges of the northern and Arctic Urals. The processes give rise to cryoplanation terraces or cryopediments which grow in size by the back-wearing of steep slopes. Cut in bed rock, they are often covered in a veneer of angular debris in their active state, derived from frost-weathering on the scarp slope which stands behind them. Nivation processes are also at work, as the terraces often provide ideal conditions for snow bank accumulation. Gelifluction and wash processes assist the transport of the fragments across the terrace surface. At the foot of the scarps, there is a zone of most intense frost weathering where most moisture tends to collect, aided by snow accumulation. The formation of cryopediments follows the same lines: a zone of higher humidity and more intense weathering is caused by snow accumulation at the foot of steeper slopes, nivation forming the pediment angle or nivation hollows. Birot (1949) was the first to propose this mechanism of cryopediment formation. An important additional factor is the presence of joints and fissures in the bedrock, allowing the penetration of moisture prior to ground freezing.

Cryopediments in the Urals—the so-called goletz terraces—were first studied in 1943–44 by Bashenina and described as long ago as 1946 by Boch. They are seen to be developing on slopes of varied aspect and at a range of altitudes [see Bashenina (1948, 1960), Demek (1969)]. The process of formation appears to begin with the destruction of blockfields and the creation of bedrock terraces. Cryoplanation terraces develop in the next stage, while the final stage is marked by the development of tors. These are gradually destroyed by frost weathering, leaving a heap of blocks which eventually succumb to total frost disintegration. Thus a cryoplain is left; backwearing has ceased because the scarps have been destroyed, and the dominant processes now become gelifluction and deflation. Rates of transport of material across the surfaces vary according to slope, presence of permafrost, availability of moisture or runoff and so on. At the present-day, many different forms of cryoplanation at different stages of development can be observed over a range of altitudes in the mountains of the southern Urals. Active cryoplanation is now restricted to areas above 1350 m; at lower levels, other processes are now tending to destroy the cryoplanation terraces and cryopediments.

19.3 Northern Urals and the Pay-Khoy ridge

19.3.1 General Relief and Structure

The Ural-Tau uplift zone is prolonged northwards as the central Ural uplift zone, with some lateral shifting caused by transverse faults. The relief of the northern Urals shows clearly how the whole mountain system has undergone subsidence to the south. The largest watershed ridges, (e.g., the Zapadnyy Poyasovyy and the Vostochnyy Poyasovyy) and the depressions between them (Verkhne–Pechora, Ilych, etc.) all merge in a southward direction, thus providing evidence of this general subsidence. Both the eastern and western boundaries of the northern Ural mountains are formed by fault scarps, and there is a clear structural control over the altitudes of the various ranges and peaks. The Vostochnyy Poyasovyy range has the highest peaks such as Pas-Ner formed on gabbro-diabase, and Tel'pos-Iz (1617 m) on resistant quartzites. In the Pripolyarnyy Ural, the relief features are more complex, the mountain ranges become broader and higher, and the main trend changes from north–south to north-east–south-west. The highest points occur on the resistant rocks of the Narodo-It'inskiy and Isledovatel'skiy massifs. The general morphostructural features of the central Ural uplift zone and its western slope are continued in the northern part of the Pripolyarnyy Ural. The parallel ridges of Obo-Iz and Zapadnye Saledy represent horst–anticlines etched out by differential erosion and separated by graben–synclines. The altitudes of the ranges and peaks are closely related to rock resistance; peaks such as Menaraga, Narodnaya (1894 m), Kvartzitnaya, and ridges such as Saran'-Khop-Ner, Yuma-Mylk and others correspond to outcrops of resistant quartzites and quartz conglomerates.

The Polyarnyy Ural consists of a relatively narrow (not more than 50 km) monolithic range of north-east–south-west trend. The greater part of the range comprises the Voykar-Syn'inskiy massif, with its peak of Pay-Er (1499 m) built of resistant hyperbasic rocks. Farther north is the Ray-Iz massif, which is roughly circular in shape and bounded by concentric faults.

The Zapolyarnyy Ural is a complicated cluster of mountains. The Bol'shoy Pay-Pudynskiy and the Malyy Pay-Pudynskiy ridges, and the Engane-Pe and Mannta-Nyrd massifs form arcuate patterns whose convex sides face east. From the Sobskiy prokhod pass in the south to the east–west segment of the Longot-Yugan river in the north, the main relief features trend in a north-east–south-west direction. Farther north, the main alignments of the morphostructures alter to approach a north–south or north-west–south-east trend (e.g., the Syum-Keu and Marun-Keu ranges). The western boundary of the Zapolyarnyy Ural is formed by a great fault scarp, being up to 800 m high in places and stretching for over 100 km. This corresponds to the western Ural fault. The relief is characterized by straight valleys and elongated lakes and depressions, picked out by such rivers as the Bol'shaya Pay-Pudyna, Nyarma, Kharbey, Khanmey, and the Bol'shoye Schuche and Maloye Schuche lakes. All these features are controlled by faults of Uralian or Pay-Khoy trend. Valley-side slopes are steep with numerous bedrock exposures, and the whole landform assemblage reflects strong tectonic influence.

The Pay-Khoy coast range is composed of several parallel ridges, conforming to the elongated structures of the Pay-Khoy horst–anticlinorium and following the trend of the Kara Sea coast in general. The Bol'shoy Ezuney ridge (up to 272 m) and the More-Iz peak (467 m) mark the axial zone. Another major feature is the Nizhne–Karskaya annular depression.

It is apparent, then, that the northern Ural Mountains share many characteristics, especially the strong structural control of the relief, with other zones of the Urals. Everywhere, but particularly in areas of more intense neotectonic activity, tectonic movements and lithological composition determine the patterns of ridges, river valleys, depressions and the absolute elevation. Not only are the major relief forms inherited from the Hercynian tectonic plan, but neotectonic movements tended to follow the pre-existing lines of structural and lithological weakness. Thus the Hercynian ranges and ridges were further uplifted, and the Hercynian basins underwent subsidence. Therefore, the present relief is an amalgam of Hercynian and neotectonic features, the whole modified by prolonged periods of subaerial denudation.

19.3.2   Morphostructural Evolution

Throughout the Mesozoic and Palaeogene which was characterized by prolonged tectonic stability (i.e. between the Hercynian and mid-Tertiary orogenic episodes), denudation took over as the main influence on relief formation. Extensive areas of the northern Urals were worn down to surfaces of low relief covered by weathering deposits before the onset of the neotectonic movements.

The age of commencement, the nature and intensity of these movements differed from one part of the region to another, as shown by the palaeogeomorphological reconstructions. During the Upper Oligocene, the northern Urals were planed down to a surface of relatively low relief, with altitudes ranging up to 400 or 500 m. At the present time, in contrast, the mountain blocks stand at heights of 800–1000 m, showing the effects of recent uplift. The most mobile blocks include the Tel'pos-Iz massif and the zone of the eastern Uralian deep-seated fault which divides the mountain Ural from the Transuralian lowland. Along this fault zone, there are deformed Mesozoic, Palaeogene and Quaternary deposits, including Middle Pleistocene glacial and Holocene fluvial deposits.

In the Upper Cretaceous, Palaeogene and Miocene, there were no tectonic movements in the Pripolyarnyy Ural. Here again, this time interval was one in which denudation was paramount, leading to the development of a planation surface which now forms a basic reference datum against which the later neotectonic movements can be judged. Intensive neotectonic movements began during the Miocene and reached their climax during the Pliocene and early Pleistocene. They resulted in radical changes of the relief and the development of the present-day mountain systems. Study of the contemporary deposits has shown that there were particularly intense movements at the Pliocene–Quaternary boundary, witnessed, for example, by the appearance of coarse sediments resting on the lower Pleistocene deposits in the Pechora lowland.

In the Polyarnyy and Zapolyarnyy Urals, neotectonic movements can be shown to have commenced close to the Pliocene–Pleistocene boundary. Again, the basic reference datum is the planation surface evolving up to the late Tertiary, which became modified by differential tectonic movements and denudation into a slightly dissected upland, not unlike the present-day Pay-Khoy upland. Further intensive movements took place at the beginning of the Middle Pleistocene, evidenced by the appearance of coarse sediments of this age in the foot-hill zone and in the deep valleys of the Nemur-Yugan, Sot', Krestovaya and other rivers.

19.3.3   Present-day Relief Features and Processes in the Northern Urals

The most intensive block movements occurred in the central and eastern areas, as evidenced by the extent of valley incision, reaching several hundred metres in places, and by the presence of coarse fluvial deposits in the upper parts of the valleys. Tectonic movements continued through the Pleistocene into the Holocene alongside other processes of glaciation (Alpine valley-type in the Pleistocene and cirque-type in the Holocene), cryoplanation, river erosion and deposition, and changes of relative sea level in the Arctic Ocean. Some river terraces may therefore be of eustatic origin.

The evolution of the river pattern in the northern Urals is far less well-known than in the case of the southern and central Urals. Furthermore, the correlations between the various sets of river terraces have not yet been established. This is due to several factors: inadequate data on the terrain, considerable differences in the number of terraces in different valleys, even in the same mountain block, the influence of Pleistocene valley glaciation on terrace formation and valley deposition, the varying intensity and type of block movements in the different areas, and controversy over the nature of the deposits—fluvial, glaciofluvial or glacial. There is also continuing discussion about the extent of influence of sea-level changes on terrace formation; work by Bashenina (1948) and Kaleckaya (1974), for instance, suggests that these changes only affect the lowermost parts of the valleys, on the other hand, in Pay-Khoy and Novaya Zemlya, marine transgressions are known to have penetrated far inland.

In spite of these uncertainties, an attempt will be made to form a general picture of the terrace sequence. On the western slopes of the northern and Pripolyarnyy Urals, five river terraces can be distinguished. The upper two are rock cut. The ages are probably as follows: fifth terrace, Lower Pliocene; fourth terrace, Middle Pliocene; third terrace, Upper Pliocene; second terrace, Middle Pleistocene; and first terrace, Upper Pleistocene. A similar picture appears on the eastern slopes, but the relative heights of the terraces are different, suggesting the influence of differential tectonic movements. There are traces of older higher terraces, possibly the sixth, seventh and eighth in the series, which are usually dated as pre-Pliocene. It is also agreed that large river systems were already in existence during the Oligocene and that there may well be terraces of such age present too.

In the Pripolyarnyy, Polyarnyy and Zapolyarnyy Urals, river terrace formation has been complicated by Alpine valley glaciation. On the margins of the mountains in these areas, the first and second terraces only can be traced and, farther into the mountains, only the first; both are aggradational terraces. A similar situation is found in the Pay-Khoy valleys [see Kaleckaya (1974)].

The northern Urals supported both valley and cirque glaciers in the Pleistocene, although none of these reached either the Pechora lowlands or the plains of western Siberia. Because of major differences in the amounts of snow accumulation, the glacier systems had very different extents as between the western and the eastern slopes of the Urals. The high mountain ranges of the Pripolyarnyy, Polyarnyy and Zapolyarnyy Urals show fine examples of Alpine glacial relief with sharp rocky peaks. Such glacial forms are also present on the western edge of the northern mountains in the Tel'pos-Iz group. Here, cirques are 300 m deep or more. Many cirques contain lakes, some dammed also by moraines. There are striking examples of glacial troughs, with moraines marking stages in the ice recession. All moraines, and probably most of the erosional forms, are related to the last glaciation; older forms have not survived. Glacial and glaciofluvial sediments have been reworked and incorporated into the fluvial deposits of the lower accumulation terraces. In the Pay-Khoy uplands there are no traces of older glaciations; altitudes of this area were even lower during the Pleistocene. Recency of uplift is shown by the late Pleistocene marine deposits which now stand about 300 m above sea level; prior to this, the elevations cannot have been much more than 150 m. In the Pripolyarnyy, Polyarnyy and Zapolyarnyy Urals today there are 143 glaciers, situated in the axial zone of high mountains and on their western slopes which receive greater precipitation [Kaleckaya, (1974)]. Grosval'd and Kotlyakov (1969) provided some data on mass balance. The island of Novaya Zemlya, which prolongs the Ural structural belt as far north as latitude 77°, has a much more continuous ice cover in its northernmost parts.

A major role in the sculpturing of the upper parts of the mountain ridges is played by cryoplanation. The lower limit of cryoplanation phenomena at the present-day is found at 900 m in the northern Urals, at 700 m in the Pripolyarnyy Ural and at 400–500 m in the Polyarnyy and Zapolyarnyy Urals. Cryoplanation terraces are more extensive in the north where, in places, they completely truncate the mountain ridges. Gelifluction processes are very active, together with frost sorting, producing many types of patterned ground. On the hillslopes, block streams occur in places.

Chapter 20
Submarine Morphology Around Europe

Whereas Chapter 3 was principally concerned with the structural and tectonic framework of the ocean floors around Europe, the present chapter deals with the submarine relief features and their origins. The chapter is organized on a regional basis and deals in turn with the north-western Atlantic margins of Europe, the south-western margins, the Mediterranean Sea, the Black Sea and the Caspian Sea, and finally the more distant parts of the Atlantic Ocean floor as far west as the Mid-Atlantic Ridge.

20.1   North-Western Sector

20.1.1   Southern Barents Sea
The Barents Sea occupies an area of $1438.4 \times 10^3 \, km^2$, its average depth being 186 m, with a maximum depth of up to 600 m. Within the area of the *International Geomorphological Map of Europe*, only its southern part is shown.

The relief of the Barents Sea platform is characterized by a combination of large submarine rises and depressions. The extreme northern part of the map shows the Medvezh'i rise genetically associated with the Spitsbergen fold–block structural feature. To the south there is the Medvezh'i trench (up to 140 km wide), which is clearly marked by a system of submerged valleys cut by glaciofluvial meltwater. South of the trench lies the Kopytov plateau and the Nordkin rise represented by cuesta-shaped rises [see Matishov (1977)] and composed at the surface of Upper Cretaceous rocks [see Dibner (1978)]. The Cretaceous, and locally Permian, rocks are overlain by glacial morainic debris attributed to the penultimate glaciation.

East of the Medvezh'i trench and the Nordkin rise, there is the Demidov rise with banks of the same name in its western part. This rise, according to Matishov (1977) and Dibner (1978) is a structural-denudation elevation composed of Jurassic and Lower Cretaceous deposits (sandstones, marls and carbonaceous clays). The Demidov banks represent, according to Matishov, the expression of diapiric structures.

East of the Demidov rise lies the central depression of the Barents Sea, occupying a syneclise of the same name as described by Klenova and Lavrov (1975) infilled with a sequence of Palaeozoic and Lower Cretaceous sedimentary rocks overlain by Quaternary glacial sediments. The depression borders the Admiralty structural-denudation plain, which is underlain by pre-Quaternary deposits of various ages (Cretaceous to late Palaeozoic) and overlain by glacial sediments. All these relief forms are terminated by a flat coastal plain of denudation that bounds the shores of Novaya Zemlya.

On the south, the Barents Sea is bounded by another series of banks. Some of them, particularly the South Kanin, North Kanin, Gusinaya and Moller banks, represent structural uplifts composed of Upper Palaeozoic rocks. They are overlain by glacial deposits, frequently occurring as morainic ridges, the hollows between them representing glacial and glaciofluvial erosion. A typical feature of the banks is the widespread distribution of sand ridges produced by powerful tidal currents.

A clearly marked structural scarp occurring south of these banks distinctly separates the Pechora plain from the rest of the Barents Sea floor. This plain is distinguished by the flatness of its relief which is due to the great thickness of deposits of which this depressed part of the Russian platform is composed. Exogenic processes have played a vital role in shaping the final details of the bottom relief. Small rises north and west of the island of Kolguyev are related to buried brachyanticlinal structures belonging to the epi-Baikalian platform cover which are only poorly expressed in the relief of the Pechora plain. In general, however, the Pechora plain is a broad offshore flat, much of which represents a bench formed by thermoabrasion. The effect of tidal currents is again manifest in the formation of sand ridges.

As indicated by Dibner (1978), the Kanin Peninsula advancing far into the southern part of the Barents Sea corresponds structurally to the Murmansk–Timan rise, a unusual structural feature of the Russian platform reflecting, in its tectonic regime, the ancient orogenic structures (*see* Fig. 20.1). This Timan uplift appears in the relief of the shelf as the Murmansk rise composed of Permian–Carboniferous rocks and, near the surface, of glacial sediments. The Timan megaswell is joined on the south by the Pre-Timan downwarp [see Dibner (1978)] which is marked on the sea floor by the Nordkin trench (400 m deep).

The narrow coastal strip of the northern Norwegian shore and the Kola Peninsula represents structurally and geomorphologically the flat margin of the Baltic Shield incised by submerged glacial troughs. North of Murmansk, there is the Kildin ridge represented by a narrow horst on the Baltic Shield margin. Fiord-type coasts characterize the entire shoreline.

20.1.2   White Sea
The White Sea covers an area of $90.8 \times 10^3 \, km^2$, its average depth being 49 m with a maximum depth of 340 m. Fault tectonics have determined the main features of its configuration. The northern and western limits of the

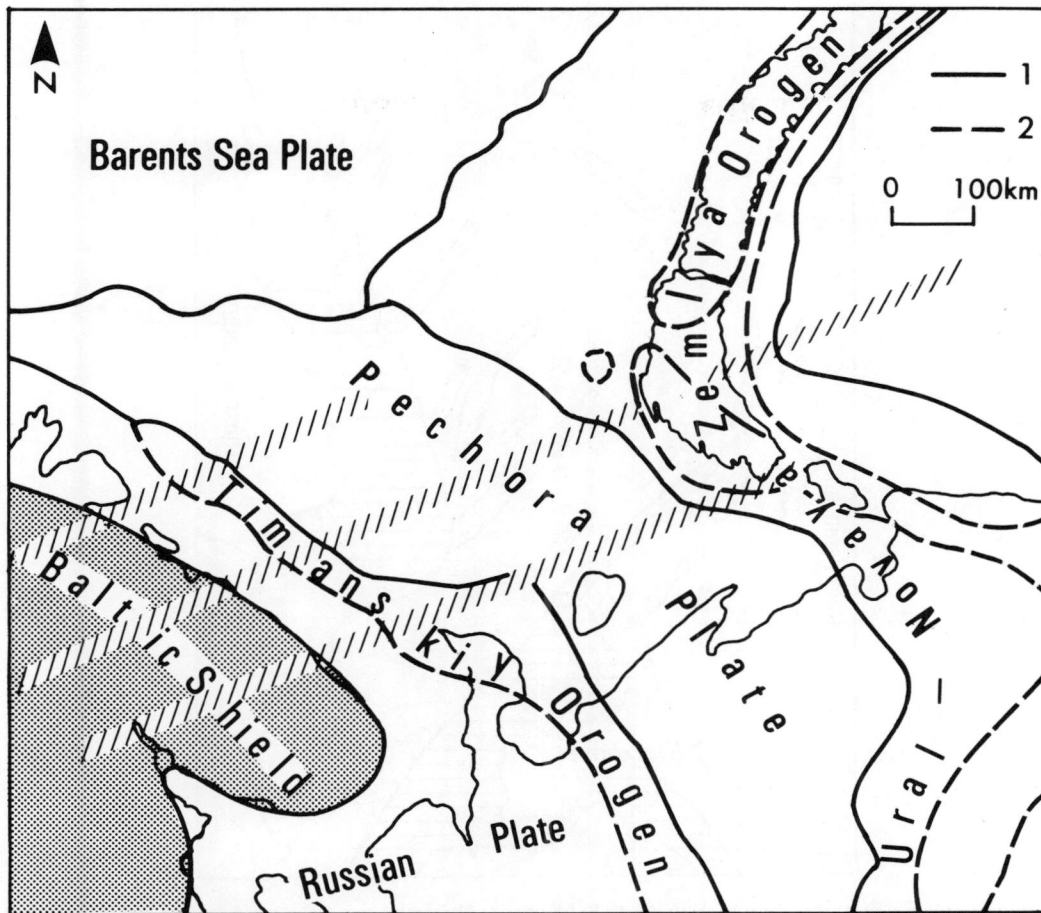

Fig. 20.1 The tectonic framework of the south Barents Sea floor [after Dibner, (1978)]. (1) Boundaries of major tectonic provinces and (2) rises corresponding to the Timan and Ural–Novaya Zemlya orogenic zones.

White Sea coincide closely with the boundary of the crystalline shield. The principal element of the submarine relief is the subbathyal Kandalaksha depression (340 m deep) consisting of a distinctly outlined graben in the zone of the Spitsbergen–Kandalaksha deep-seated fault system. This zone is also connected to the Ingedyupet trench system which separates the Kopytov plateau from the Nordkin rise. The sides of the depression show evidence of submarine landslides, and its floor is level owing to accumulation of sediments. The coasts of most of the Kola Peninsula and Karelia are deeply indented, slightly affected by marine action and are characterized chiefly by fiords and skerries.

The majority of the White Sea floor lies at depths of less than 20 m; being subjected to both tidal and wave action, it is therefore relatively level. The coasts of the Onega and Kuloy peninsulas are being actively modified by wave action. They possess typically graded coasts with features of both abrasion and deposition. Abrasion plays a major role in supplying the White Sea with sediment. Thermo-abrasion causes the coast of Morzhovets Island to retreat at a rate of 20 m per year. In the 'neck' and especially in the 'funnel' of the White Sea, tidal currents play an active relief-forming role, producing sea floor sand

ridges, up to 10–12 m high, extending for tens of kilometres (*see* Fig. 20.2). The ridges follow the direction of the tidal currents and are complicated by smaller cross ridges called sand waves [see Emery and Uchupi (1973), Longinov (1973), Van Veen (1930)]. The heads of the Gulf of Onega and Dvina Bay are distinguished by estuarine depositional processes. The Northern Dvina has built out a substantial delta, but the Mezen' has no delta because of the very strong and high tides (up to 10 m). The mouth of this river is a typical estuary bordered by sand ridges and broad tidal flats.

20.1.3 Norwegian Sea

This covers $1383 \times 10^3 \text{ km}^2$, its maximum depth being 4487 m (*see* Fig. 20.3). The sea floor consists of a shelf, continental slope and rise, and the deep-sea floor. The width of the shelf varies from 15 to 225 km, but is usually narrow and strongly incised by numerous submarine extensions of fiords and intrafiord ridges; it is also marked by skerries. The depth of some submarine valleys exceeds 400–500 m. The flat-topped banks separated by valleys and canyons are considered to be wave-cut surfaces [see O. Holtedahl (1960)], the formation of which is also due to frost weathering and tidal currents. The surfaces have

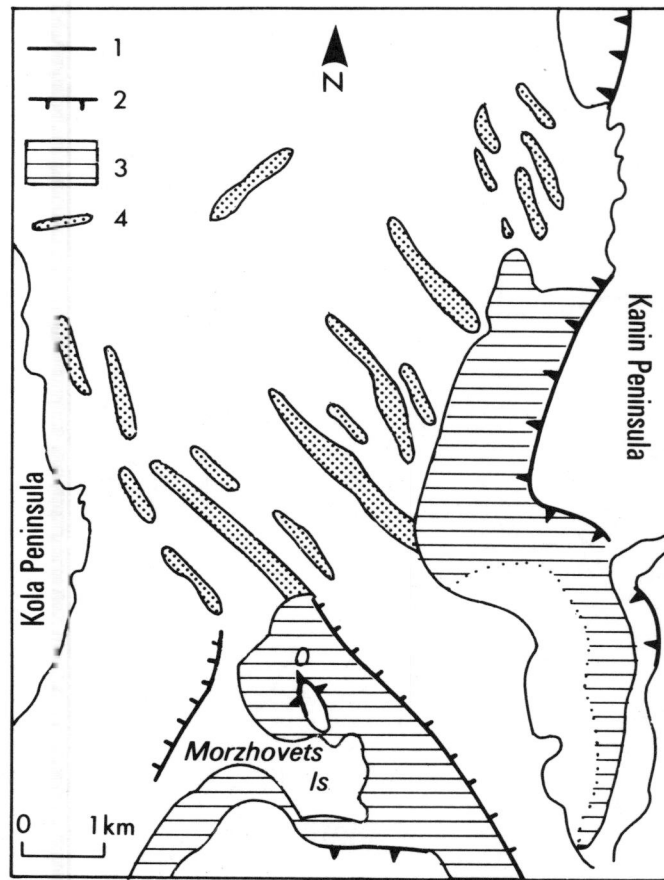

Fig. 20.2 Sandy ridges and benches of the north-eastern White Sea floor. (1) Abrasion shoreline, (2) submarine abrasion cliff, (3) benches and (4) sandy ridges.

been eroded in the dislocated and metamorphosed Caledonian structures (*see* Section 7.1.8).

The continental slope of Norway exhibits a varied structural pattern. Near the Lofoten Islands and to their north-west, it is represented by a simple steep scarp, with a gradient varying from 6° to 20°. Farther south, the middle segment of the Norwegian shelf appears to slope down gently, with a gradient of 1° and pass into a gentle low rise. Below the latter from the 500-m isobath to a depth of 1500–1600 m, there extends the Vöring marginal plateau. This morphostructure is bounded on the west by the Jan–Mayen fracture zone which continues farther out across the sea floor. An asymmetrical horst ridge crosses the north-western part of the Vöring plateau. Irregularities on the plateau surface are associated with diapirs [see Talwani and Eldholm (1974)] and also, probably, with iceberg deposits. The gradient of the plateau is about 2° in the north, but up to 10° in the Jan–Mayen fracture zone. A group of small submarine canyons occurs on the continental slope east of the Vöring plateau.

Seismic data indicate that, under the Vöring plateau and the wider portion of the shelf, there is a thick sedimentary sequence which in the form of a narrow tongue continues farther southwards, merging with the sedimentary basin of the North Sea. Talwani and Eldholm

(1974) reported that the maximum thickness of the shelf sediments in the area of the Halten banks and the Tren banks is 9 km, and in the central part of the Vöring plateau is 7–8 km. Such sedimentary thicknesses suggest the possibility of discovering oil and gas in these areas.

The continental slope south of the Vöring plateau is steep, consists of sedimentary rocks and is incised by a system of closely spaced submarine canyons. At a depth of 800 m, the continental rise is joined by a sloping depositional plain dissected by turbidity current canyons; west of it there is another depositional plain with a surface tilted slightly to the north and extending down to depths of 2500–2900 m. In the south-west, the Norwegian Sea is separated from the North Sea shelf by the deep (1680 m) Faröe–Shetland trench. The trench has steep slopes (up to 3–6°) and a flat floor, inclined to the north-east. Hoshino (1978) considered it to be a relict of the deep-sea trench of the Caledonian orogeny.

The north-eastern boundary of the Norwegian–Greenland basins is represented by a continental slope coinciding morphologically with the Senja fracture zone [see Talwani and Eldholm (1974)] that bounds the western margin of the Barents Sea shelf. The scarp of the slope is incised by submarine valleys, the most prominent of which—the Perseus valley—is localized at the exit of the

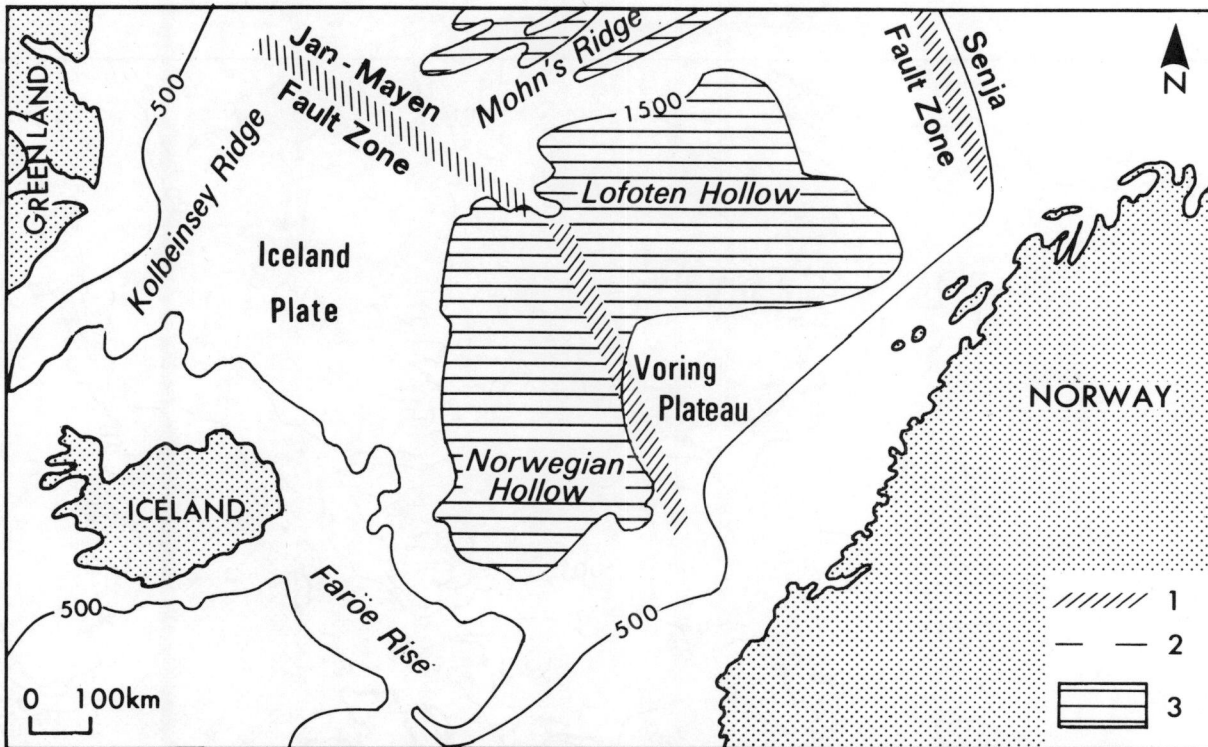

Fig. 20.3 Main topographic units of the Norwegian Sea [after Talwani and Eldholm (1974)]. (1) Fracture zones, (2) 500-fathom (914-m) contour and (3) depths over 1500 fathoms (2743 m).

Medvezh'i trench to the Norwegian Sea. Following the Medvezh'i trench and then farther along the Perseus valley, turbidity currents have transported seawards a great amount of sediment resulting in a deep-sea fan [see Matishov (1977)].

The deep-sea part of the Norwegian Sea is divided by a chain of horsts, which are confined to the Jan–Mayen fracture zone (the Norwegian ridge), into the Norwegian and Lofoten basin. The maximum depth of the Lofoten basin is 3219 m, its floor merging with the flat Dumshaf abyssal plain; the sedimentary thickness exceeds 2 km [see Talwani and Eldholm (1974)]. The submarine relief of the Norwegian basin is more complex. According to seismic data, the bottom of the Norwegian and Lofoten basins represent oceanic-type crust [Neprochov (1979)]. Talvani and Eldholm (1974) believed that the Norwegian ridge is an ancient rift zone of the mid-oceanic ridge which about 40 Ma ago shifted westward forming the Kolbeinsey ridge.

The south-eastern part of the Norwegian basin has undergone aggradation and represents a flat abyssal plain as shown in the *Map of Arctic Regions* (1976). The sedimentary thickness is about 1.5 km. The north-western part of the depression is represented by hilly and low-mountain relief forms.

20.1.4   North Sea and English Channel
The North Sea is 549 km^2 in area, its average depth being 94 m, with a maximum depth of 433 m (*see* Fig. 20.4). As has been already mentioned, the North Sea lies within a large epi-Caledonian depression which has been subsiding from the Palaeozoic Era to the present-day [see Kalinko (1977)]. The post-Caledonian platform cover (about 4 km thick) consists of Upper Palaeozoic, Mesozoic and Palaeogene rocks. The oil-bearing formations belong to the Carboniferous, Permo-Triassic, Jurassic and Cretaceous sequences. The complex pattern of the platform cover is due to faulting and salt tectonics.

With respect to the structural peculiarities of the sea floor, the North Sea can be divided into north-western, north-eastern, central and southern sectors.

In the north-western sector, between the northern edge of the shelf and latitude 58° N, there lies an area of strongly dissected relief, with both banks and submerged valleys. These valleys extend chiefly in a meridional direction and are most clearly marked in the area of the buried Viking graben. The southern limit of this area, well displayed in the relief, corresponds to the former coastline in the Würm and follows approximately the 100-m isobath. Structurally, the greater part of this area corresponds to the eastern Scottish Caledonian platform [see Whiteman *et al.* (1975)].

The central area is characterized by graded relief, a network of valleys and widely distributed submerged peat beds. The former submarine continuations of the Elbe, Rhine, Thames, Humber and other valleys are easily

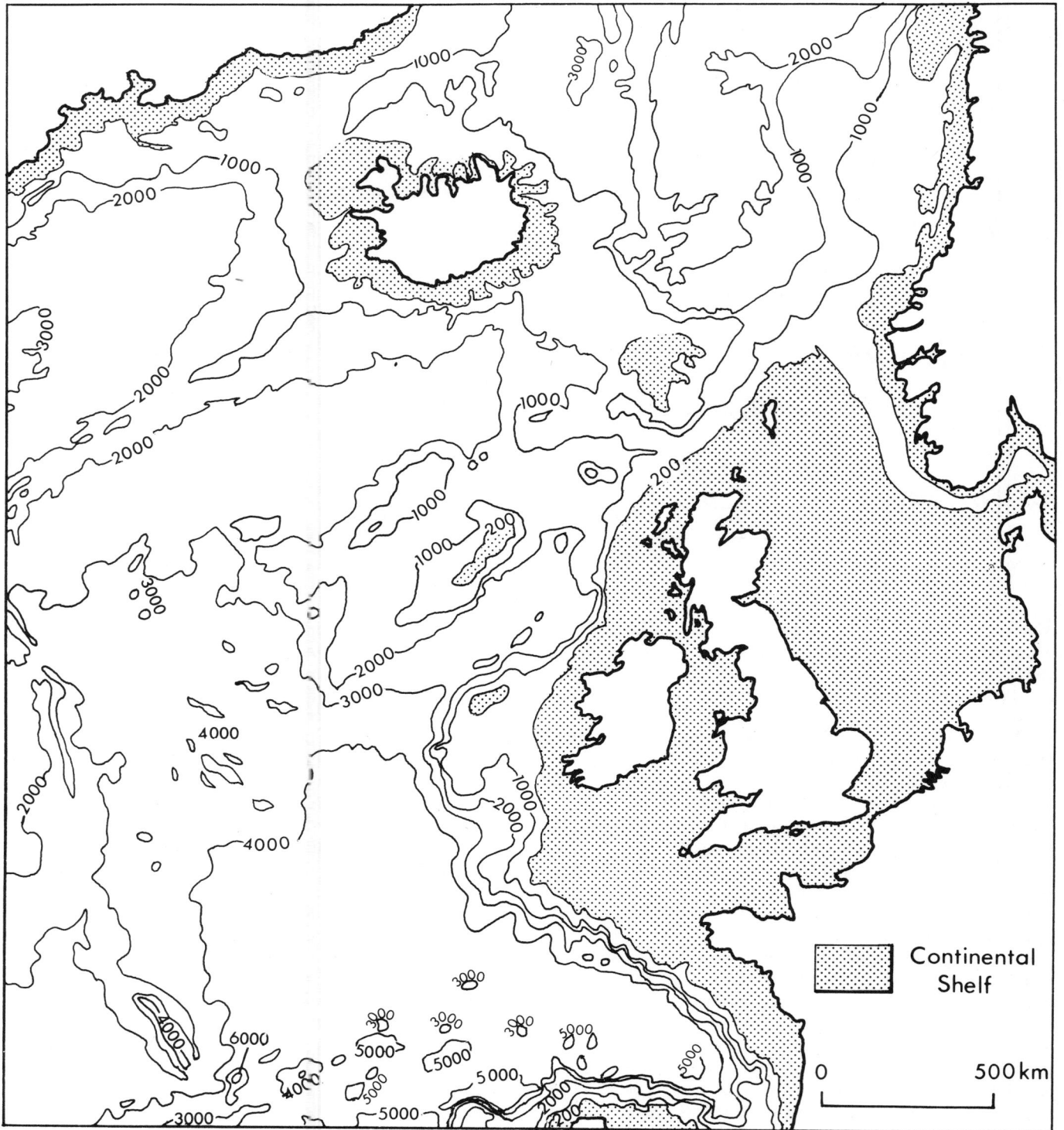

Fig. 20.4 Relief of the Atlantic Ocean and North Sea floors off north-western Europe. Depths are in metres.

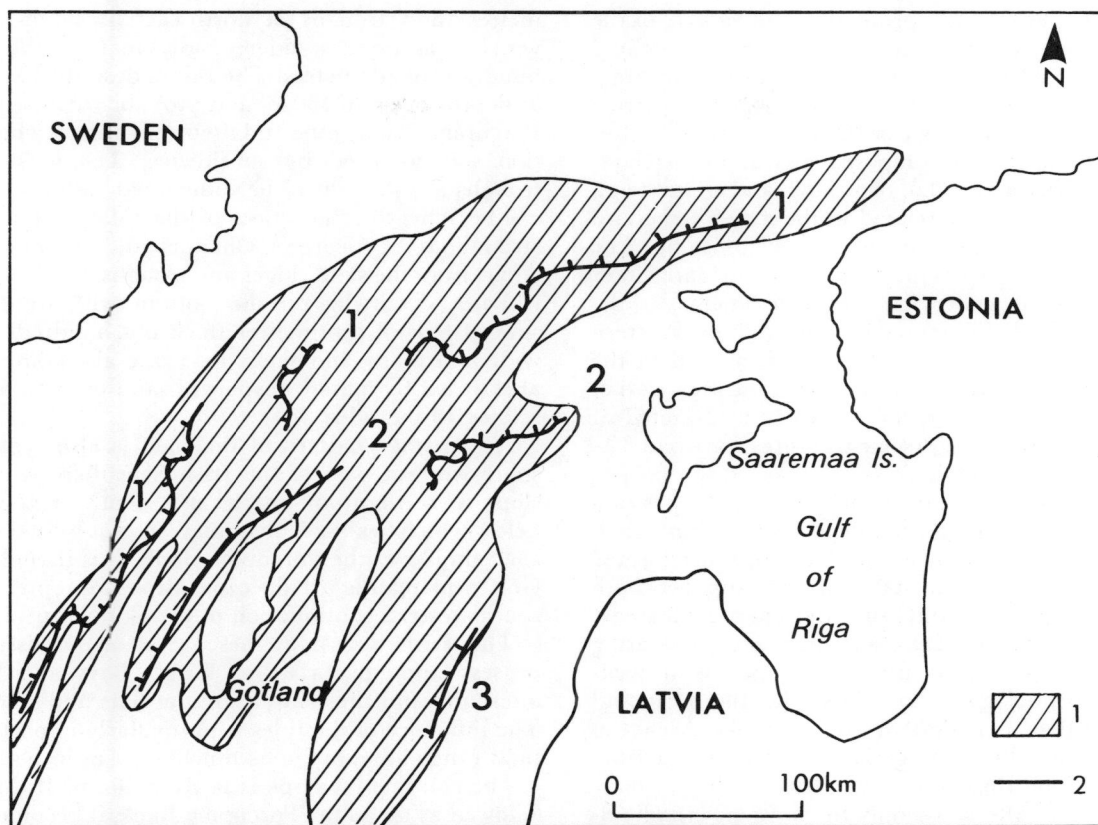

Fig. 20.5 Cuesta relief of the Baltic Sea floor [after Gudelis and Emelianov (1976)]. (1) Main area of cuesta relief and (2) glints. 1: Ordovician; 2: Silurian; 3: Devonian.

traceable. In the southern part of this area, there is the Dogger Bank whose relief features are associated with buried salt domes. The Devil's Hole depression—a semi-buried graben—in which the maximum depth (433 m) of the North Sea is found—lies in the north-western part of the central sector.

The southern sector of the North Sea possesses abundant tidal sand ridges [see McCave (1971), Van Veen (1930)]. In places, these ridges attain 7–10 m in height and tens of kilometres in length. Still larger sand ridges are observed near the shores of south-eastern England and in the Straits of Dover. Also common are sand waves and deep troughs washed out by tidal currents. The tides produce very broad mud flats, particularly prominent in the area of Friesland, in the Wash, and in the Humber and Thames estuaries.

Similar relief features also characterize the English Channel, where strong tidal currents contribute to the formation of sand ridges. The submarine Herd Deep lies along the axis of the channel and most probably represents the submerged continuation of the lower Seine. Extensive mud flats have developed along the coasts (e.g., in St Malo Bay and in the Seine estuary).

The north-eastern sector of the North Sea is marked by the Norwegian trench extending from north to south-east and then to the east, with steep slopes and a wide flat floor. Depths in the trench exceed 200–300 m, the maximum being 908 m (in the east). The trench cuts both Hercynian and younger geological structures and there-

fore should be considered as a superimposed depression brought about by neotectonic development of the Fennoscandian Shield.

The North Sea and English Channel coasts are characterized by a great variety of forms caused by marine action. Numerous beaches, composed of easily eroded Cretaceous and Neogene sediments, alternate with headlands along the shores of eastern and south-eastern England, northern France and Denmark [see Bird (1976), Steers (1946)]. In the north-west of Scotland and on the Norwegian coast, fiords are common. Glacial forms have been almost unaffected by the sea. The largest coastal accumulation form is the Friesland barrier bar extending from Holland to southern Jutland.

20.1.5 Baltic Sea

The submarine features have been extensively described in a monograph on the *Geology of the Baltic Sea* [Gudelis and Emelianov (1976)]. The Baltic Sea is $386 \times 10^3 \, km^2$ in size, its average depth being 86 m and the maximum depth 459 m. The south-western part of the sea floor is shallow and underlain by a blanket of bottom sediments. The coastlines are indented, in places affected by wave abrasion; coastal deposition forms are also present. In the area of The Sound, between Denmark and Sweden, submerged river valleys are clearly evident. Shallow-water low-lying coasts extend along the shorelines. A zone of the sea floor located beyond the reach of wave action is notable for some large scarps trending from north-east to south-west (*see* Fig. 20.5), representing equivalents of the glints that occur on the land caused by monoclinal bedding of the Silurian, Ordovician and Devonian limestones and sandstones [see Martinsson (1960)]. A narrow trough extends along the foot of the most northerly glint coinciding with the area composed of Cambrian clays. Large depressions (e.g., Landsortian) and rises (e.g., Bornholm) are genetically related to the major structural features of the Fennoscandian Shield and its slope. The relief of the southern part, where glacial deposits are either reworked or buried under the recent sediments, is the most evenly graded. Submerged valleys of the Western Dvina, Neman and other rivers are well marked in the Gulf of Riga, the Moonsund straits and in the area west of Klaipeda. Former coastlines have also been detected at depths of 32–47 m (pre-Littorina transgression) and 12–26 m (Littorina transgression).

The present-day coasts are either abrasion/deposition or embayed types. Large coastal depositional forms, such as the Kurshu Neria, Vistula and Hack spits, are formed at the expense of the sedimentary material supplied both directly from the sea floor and from adjacent coastal areas.

The so-called Baltic flexure—a fracture zone separating the Fennoscandian Shield from its slope—is a well-marked geomorphological boundary. To the west and north, the sea floor is underlain exclusively by Archaean rocks and has a hilly or ridge-like relief resulting from glacial erosion and deposition. The floor is dissected by numerous glacial valleys and only the bottoms of individual depressions (grabens) are covered with the young sediments. The coasts are glacially eroded and hardly touched by marine action. Numerous submerged troughs, which are the continuations of fiords, and abundant skerries characterize the offshore areas.

20.1.6   Western Submarine Margins of Europe West of the British Isles

Including the Irish Sea (area about 40 000 km², maximum depth 272 m), this part of the submarine margin is related to the western European structural platform and, in the north-west, to the Eria platform (*see* Fig. 20.4). There are many depressions, troughs and rocky areas on the sea floor. The coasts composed of crystalline rocks display a fiord-type shoreline. The shelf surface west of Ireland is smoother than that north of it; the most uneven surfaces are those interrupted by the submarine valleys that continue out from the rias typical of the western Irish shoreline. The upper step of the shelf is terminated by a distinct edge at depths of 200–250 m. There then follows a gently sloping scarp, at the foot of which lies the second subsided portion of the shelf, the Porcupine Bank. Most of the latter is at depths of 200–400 m, but part lies at only 120 m. The plateau surface is diversified by sand ridges preserved as a result of submergence. The continental slope of the Porcupine Bank is steep and incised by several submarine canyons with evidence of submarine landsliding.

The most characteristic relief forms of the shelf in the area of the southern Irish Sea or Western Approaches are long, high sand ridges extending for many tens of kilo-

metres, and troughs of north-east to south-west trend which separate these ridges. Individual troughs extend for hundreds of kilometres. The ridges described above occur at depths of up to 150 m, and probably represent, like the Porcupine Bank, relict tidal current ridges whose formation was followed by subsidence. The most extensive troughs are thought to be submerged valleys.

The relief-forming effect of the tidal currents is most striking in St George's Channel and in the Irish Sea. Apart from the sand ridges and troughs, the so-called sand ribbons, occurring on the bottom with dimensions of several tens of centimetres thick but hundreds of metres wide and many kilometres long, are also common. They are caused by mass transport of sand by bottom currents related to the tides.

A two-stage pattern of the shelf is also typical of the south-western part of the Irish Sea: there is a break of slope in the bottom profile at a depth of about 200 m, below which lies the Goban plateau, a subsided step of the shelf similar to the Porcupine Bank. It is thought that the Goban plateau is a Palaeozoic structure representing the south-western continuation of the Cornish peninsula.

The Irish Sea shorelines are strongly dissected and possess numerous ria-type forms. Bays and headlands alternate, and there are also tidal flats and salt marshes. The influence of the tides is particularly important as the tidal range amounts to as much as 14 m in places.

The continental slope is fairly uniform. If this slope is followed as far as the Porcupine Bank, it becomes narrow, steep and moderately incised by submarine canyons. North of Ireland, the continental slope is complicated by a step whose depth is 1600–1700 m with a width of up to 70 km. The slope is covered with a sedimentary veneer. Terraces that complicate the structural pattern of the continental slope are probably due, as elsewhere, to step faulting.

The upper limit of the slope near the Porcupine Bank lies at a depth of 1000 m, while in the adjacent bight that is deeply incised into the shelf, the depth of the upper limit of the slope is 2500–2700 m. Where this bight opens to the ocean, there is also a steep (10°) scarp lying at a depth of 3000–4000 m. The continental slope is seen here as being split into upper and lower branches. Near the Goban plateau, the two branches merge together to form a steep, high escarpment. According to seismic surveys carried out on the continental slope, large submarine landslides regularly take place on the slope. A continental rise occurring as a tilted depositional plain within the limits of the area described in this section is not well developed. In all probability, the continental rise includes the Rockall depression which separates the rise of the same name (*see* Section 20.1.7) from the Porcupine Bank and the western Scottish continental slope. The depth of this depression, which appears to open south-westwards, increases in the same direction from 1500 to 4300 m (i.e. until the trough joins the ocean floor). The bottom of the depression occurs as a broad tilted plain complicated in its northern part by the Hebridean, Anton Dorn and Rosemary seamounts which are extinct flat-topped volcanoes [see Ulrich (1964)]. This part of the continental rise is notable for the presence of a large submarine depositional form (the Feny sedimentary ridge.)

A narrow continental rise (up to 100 km wide) extends near the continental slope in close proximity to the Porcupine Bank and Goban plateau. The supply of sediments composing the tilted plain of the rise takes place through submarine canyons incised. into the plateau slopes.

20.1.7 Rockall Rise
This consists of three morphostructural elements: the Hatton plateau in the north-west, the Rockall plateau in the south-east and the Hatton trench between them. The highest peak in the northern part of the plateau is the Rockall cliff, rising to 1–3 m above sea level. The average depths above the Rockall plateau vary from 200 to 500 m, those over the Hatton plateau from 1000 to 1200 m and, in the trench, depths are 1200–1400 m. The two plateaus are characterized by smooth summit surfaces and by an asymmetrical transverse profile: for example, the south-eastern slope of the Rockall plateau dips gradually at an angle of 10–12° north-westwards to the Hatton trench. All three morphostructural features are generally tilted to the south-west, so that the highest parts of the plateau and the greatest depths of the trench are observed in the north-east.

The Rockall uplift also includes the George Blith bank. This is a flat-topped seamount separated by deep saddles from both the plateaus, the lowest depth over it being 400 m.

Roberts (1974), who discussed the formation and development of the Rockall plateau from the standpoint of plate tectonics, concluded that it was involved in considerable horizontal drift in the Mesozoic and Cenozoic Eras. The limited data quoted in his article, however, fail to confirm the correctness of his conclusions. The existence of typical continental crust (about 30 km thick) under the Rockall rise is, though, unquestionable, as is the important role of faulting in its formation. According to the deep-sea drilling data (*Initial Reports, Deep Sea Drilling Project*), in the course of the last 55 Ma, there occurred a slow subsidence of the rise resulting in a change-over from shallow-water sediments to later pelagic varieties.

Landsliding is active on the slopes of the rise. Probably in the recent geological past such landslides served as a source of sedimentary material for the Fenny sedimentary ridge.

20.1.8 Faröe–Iceland Swell (*see* Fig. 20.4)
This consists morphologically of the submarine margins of Iceland and a rise proper which, in turn, includes the Faröe–Iceland plateau, the Faröe plateau and a group of flat-topped banks. It should be remembered that the structural nature of Iceland itself remains controversial.

The Icelandic shelf is dissected by large trenches or valleys, known as *dyupets*, into isolated plateaus tilted towards the ocean; such radial dissection of the shelf is not, moreover, restricted only to the western margin of Iceland. Dyupets pass farther into submarine canyons that dissect the continental slope of Iceland. The edge of the shelf lies at depths of about 200 m. The height of the

continental slope varies: in the south it attains 1800 m and in the east not more than 300 m. The thickness of sediments on the shelf is negligible, represented by pyroclastic sediments resting on the basalts. The shelf surface represents a lava plateau; basaltic lavas also form the continental slope and underlie the continental rise. Westwards, this lava plateau, according to Ilyin (1976), extends as far as the continental rise of Greenland, whereas in the north-east the basalts form the so-called Iceland submarine plateau, which is a relatively uplifted western part of the floor of the Norwegian Sea.

Young basalts are also common on the Faröe–Iceland plateau, which is a relatively depressed block morphostructure with minimum depths of 300–400 m. The north-eastern slope of the Faröe–Iceland swell is steep and dissected by the upper reaches of the east Iceland canyon. South-westwards, the slope gradually declines to a depth of 2000 m at the base of the Iceland continental rise. This slope is also incised by a system of canyons which, north of the Hutton rise, appear to merge into each other, feeding a large abyssal valley named the Mori valley.

The Faröe–Iceland plateau is divided by transverse depressions into four segments, the most northerly of which is the Rosengarten bank. The minimum depth over the bank is about 300 m; the summit surface of the plateau is hilly. The most easterly morphological element of the swell represented by the Faröe plateau, with the islands of the same name in the western part, is a basaltic plateau resting on the ancient crystalline basement. On the shores and shelf of the islands, there are distinct glacial troughs.

The Faröe plateau is bounded by a steep scarp which, on the south, forms the northern side of the Faröe–Shetland trench. Flat-topped banks (e.g., Bailey's Bank) serve as a link between the Faröe–Iceland swell and the Rockall rise. The Rosemary bank is an extinct submarine volcano, while the rest of the banks are small basaltic plateaus formed probably as a result of dissection of a single volcanic massif. The Faröe–Iceland swell (excluding Iceland proper) is generally characterized by mainly negative vertical movements. The position of Iceland astride the mid-ocean ridge points, on the other hand, to predominant uplift. Widespread recent vulcanism in the area of the Rockall rise and the Faröe–Iceland swell is associated with the fault tectonics of the region.

20.2 South-Western Europe and North-West Africa

20.2.1. Bay of Biscay (*see* Fig. 20.4)
A wide shelf plain narrowing towards the south occupies the northern and central parts of the Bay of Biscay; it is termed the Armorican shelf. Its surface is uneven due to bedrock outcrops and the thinness of the cover of superficial deposits. An exception is in the extreme southern portion where the sediment cover is thicker. In the northern parts, there are numerous tidal sand ridges, while submerged river valleys are traceable extending out from the estuaries of the Loire and other rivers.

The continental slope of the northern and central parts of the Bay of Biscay is deeply dissected by a hundred large submarine canyons, some extending for 90–120 km. The continental rise joins the slope and is represented by a tilted depositional plain whose lower limit lies at great depth (up to 4500–4700 m). Channels of turbidity currents incised into the bottom sediments can be locally traced. Magnetic data of Klenova and Lavrov (1975) indicate that the continental crust under the continental rise is strongly downwarped.

A shelf localized west of the Aquitaine coast consists of a depositional plain. The Cap Ferre downwarp, which is a continuation of the Cretaceous Parantis downwarp on land, is buried here under the sediments. South of the Cap Ferre downwarp, the basement of the shelf rises to form the Landes submarine plateau which separates one of the largest submarine canyons–the Cap Breton canyon–from the Aquitaine shelf. In spite of its deep entrenchment (over 1000 m), there are no exposures of ancient rock in the canyon sides and the head of the canyon is cut into unconsolidated turbidities. Montadert *et al.* (1974) also noted that accretion of the shelf edge and continental slope took place exclusively in the Oligocene.

The shelf along the northern coast of the Iberian Peninsula is narrow and bounded by a steep (up to 20°) continental slope, dissected by numerous submarine canyons. Between the Jijon and Santander canyons, the continental slope profile appears to be complicated by the Asturian marginal plateau. The northern edge of the plateau (the Denoit bank) is composed of rocks typical of the Pyrenees rather than of the Cantabrian mountains. Thus, it may be concluded that the Pyrenees extend into the Bay of Biscay up to the Denoit bank. Part of the Pyrenees between the Denoit bank and Cape Machichano has been subjected to subsidence, crushing and destruction. Seismic data point to a great thickness of sediments infilling the depression that forms the structural basement of the continental rise along the northern coast of the Pyrenees. Analysis of the relation between the structure of the Aquitaine sedimentary basin and that of the floor of the Bay of Biscay has led Montadert and his co-workers to conclude that the Bay of Biscay originated by rifting and that the opening of the rift took place in the Albian period. The central part of the Bay had been formed by the end of the Cretaceous, with volcanic activity continuing here until the end of the Eocene. The existence of a group of seamounts (e.g., Cantabria and Gaskon) in the Bay of Biscay can be considered as indirect evidence of former tectonic activity in the region.

20.2.3 Submarine Continental Margin West and South of the Pyrenees

West of the Pyrenees, the submarine margin shows a complex structural pattern. The shelf is narrow, bounded by the 200-m isobath; in the north, the shelf extends out for 1000 m. This part of the shelf—the Ortegal spur—has subsided considerably and is uneven due to bedrock exposures. South of latitude 40° N, large submarine canyons, such as those of Nazaret and Tajo, are incised into the shelf. Seismic survey north of the Nazaret canyon showed a great thickness of sediments (Triassic and Jurassic) and there are also diapiric structures associated

with Triassic–Lower Jurassic evaporites.

Several scarps caused by step faulting are identified on the continental slope. Block tectonics are responsible for the formation of the Galician marginal plateau and seamounts, such as Vigo, Porto, etc (*see* Fig. 20.6). East of the Porto seamounts, seismic profiling has detected diapiric structures.

The scarps of the continental slope are dissected by numerous submarine canyons, the largest one (Nazaret canyon) being initiated 1 km from the shore; it is traceable across the shelf and continental slope, and on the continental rise it passes into an abyssal valley. The total length of the canyon is about 150 km, the depth of its entrenchment into the shelf attains 2000 km and it ends at a depth of about 4500 m. Channel ramparts extend along its lower section. Another large example—the Tajo canyon—starts near the mouth of the Tajo, extends for 60 km and ends at a depth of 4600 m in a large detrital fan.

A unique structural pattern is found in the submarine continental margin south of the Pyrenees and in the Straits of Cadiz. The shelf ends at depths of about 120 m in a gentle convexity, beyond which the floor falls gradually to about 500 m; at greater depths, the relief becomes more rugged down to 3000–3500 m. The floor is dissected by a series of the westerly trending submarine canyons. According to Heezen and Hollister (1972), they owe their origin to the outflow of high-density saline Mediterranean waters through the Straits of Gibraltar into the Atlantic Ocean. The main stream of this bottom current follows a wide ravine within the Straits of Gibraltar and for several tens of kilometres west, with numerous bedrock exposures. Farther, over a distance of about 160 km, the floor of the ravine is underlain by mud, the upper layer of which is eroded. On the west, this ravine is joined by a vast fan of sandy mud ridges resulting from bottom currents. Ridges (3–10 m high, but attaining 20 m on the southern margin of the fan) are common features.

20.2.4   Submarine Margin of North-West Africa

South of the Straits of Cadiz, there extends a shelf plain, several tens of kilometres wide, eroded in the structural continuation of the Atlas mountains. Even now, the shelf undergoes vertical movements, which is indicated by variation in depth from 110 to 200 m. Uplifted parts of the shelf coincide with the continuations of anticlinal folds of the Atlas. The foot of the continental slope also lies at varying depths [see Klenova and Lavrov (1975)]. Numerous diapiric structures associated with Mesozoic–Cenozoic evaporites and submarine canyons have developed on this slope. The well-marked continental rise ends at depths of about 4000 m, its width attaining several hundreds of kilometres.

20.3   Mediterranean Sea (*see* Fig. 20.7)

The Mediterranean Sea occupies an area of $2510 \times 10^3 \, \text{km}^2$, its average depth being 1536 m and the maximum depth 5120 m. The floor of the Mediterranean

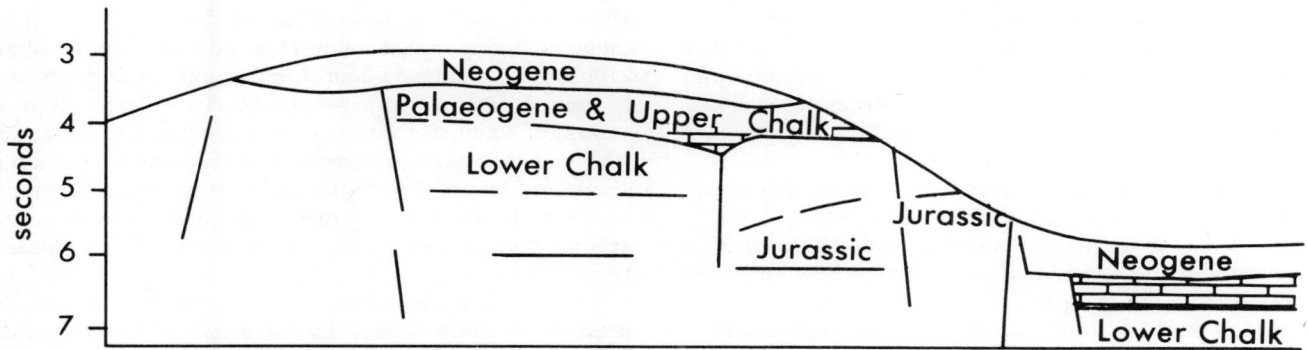

Fig. 20.6 Seismic reflection profile across the Vigo seamount [after Montadert *et al.* (1974)]. Vertical scale is in seconds reflection time (two-way travel).

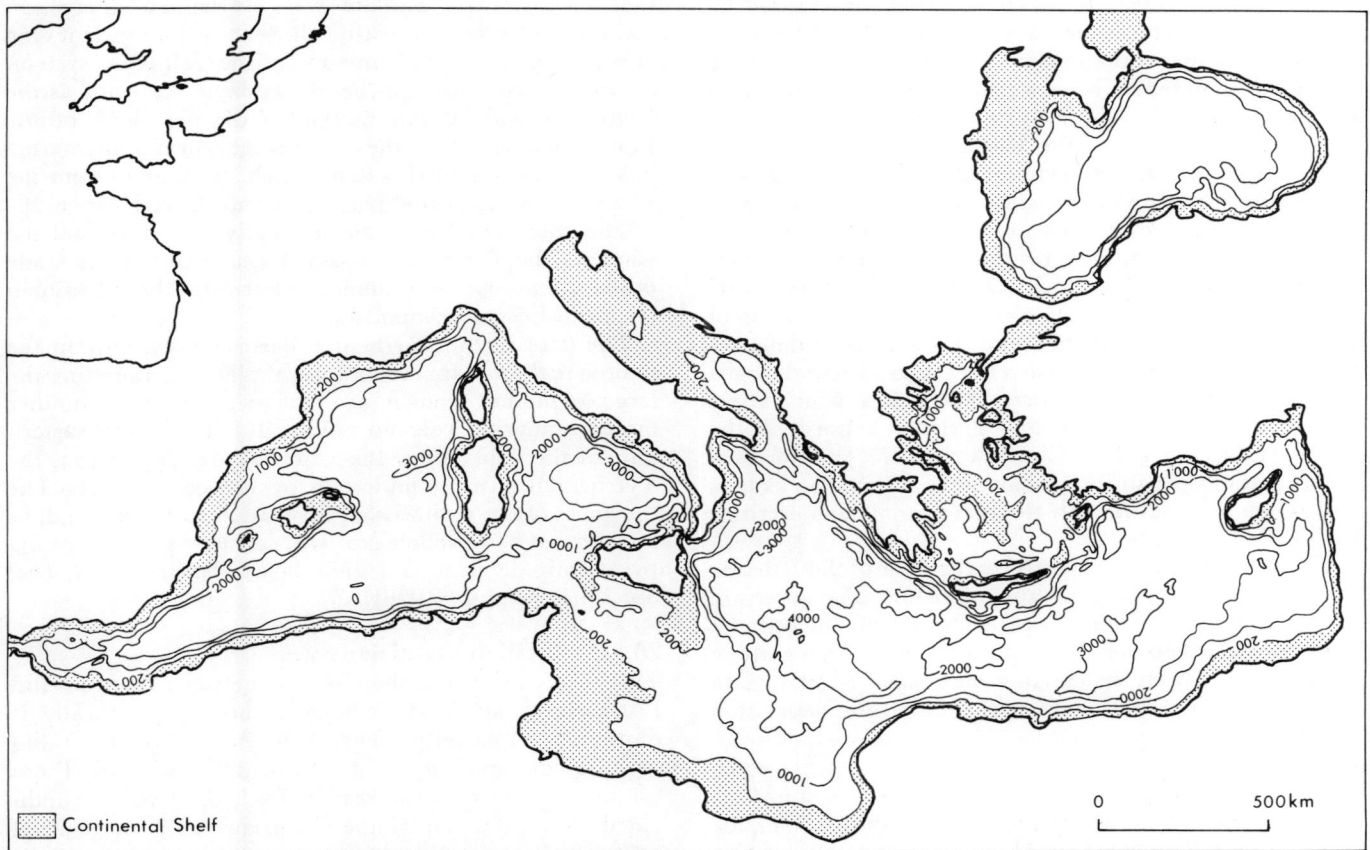

Fig. 20.7 Relief of the Mediterranean and Black Sea floors. Depths are in metres.

Sea is notable for its complex and diverse structural pattern. As has been already mentioned, different views exist with respect to its genesis and structure. The author favours a concept according to which the relief and structure of the Mediterranean Sea are manifestations of the late stage of geosynclinal development [see Leont'ev (1968a,b)].

20.3.1 Western Region

The western region of the Mediterranean Sea comprises three main areas: the Alboran basin, Balearic ridge and the Balearic basin. The Alboran basin occupying the approaches to the Strait of Gibraltar displays complex sea floor relief consisting of a group of troughs and ridges elongated in a sublatitudinal direction. The Alboran ridge and a small volcanic island of the same name is the most important uplift in this area. The maximum depth in the basin is 1570 m. Malovitsky (1976) believed that the Earth's crust under the Alboran basin is of the platform continental type, but the Alboran ridge represents an anticline of Alpine age complicated by volcanic eruptions and fault tectonics.

The Balearic ridge is the termination of the Riff–Baetic

fold system. On the south, the ridge is bounded by a well-marked fracture. The horst–anticline structure of the ridge is clearly expressed in the relief of the continental slope where a steep scarp occurs south of the Balearic Islands at a depth of 500–1500 m, and 200–500 m east of them [see Mikhailov (1965)].

The extensive, deep (2855 m) Balearic or Algerian–Provençal basin occupies a large part of the western region. Off the eastern Pyrenean coast, the shelf is represented by a depositional plain incised by submerged river valleys. The largest of these—the Ebro valley—terminates on the continental slope in a submarine canyon and detrital fan. The entire Gulf of Lions is occupied by vast alluvial fan deposits brought down by the Rhône. The surface of the fan is dissected by numerous turbidity current ravines. The Riviera coastline is characterized by a narrow and rugged shelf and by complex relief on the continental slope and rise. The narrow shelf and strongly dissected continental slope separate the Balearic basin from Africa.

The eastern slope of the basin, which is part of the Corsican–Sardinian (Hercynian) block, is strongly dissected. Numerous submarine canyons off Corsica and Sardinia are direct continuations of rias and, as Shepard and Dill (1966) believed, have resulted from deep subsidence of the margins of this block. On all sides, the Balearic basin is bordered by a wide continental rise, composed of Pliocene–Quaternary turbidites at the surface and dissected by numerous turbidity current ravines. These deposits are underlain by Upper Miocene evaporites, which attest to a lacustrine stage in the Tertiary history of the Mediterranean Sea [see Chumakov (1975)]. Numerous manifestations of salt tectonics appear in the relief as hills and ridges associated with these evaporites as described by Biju-Duval (1974).

Two types of relief characterize the floor of the Balearic basin: abyssal hills and flat abyssal plains. The latter are predominant and result from the burial of the bedrock relief by thick sedimentary deposits, consisting mainly of terrigenous, weakly calcareous materials. Seismic data show that the Earth's crust under the Balearic depression belongs to the suboceanic type.

20.3.2   Tyrrhenian Sea

The sea floor in the Tyrrhenian Sea exhibits a complex relief pattern. The considerable width and strong dissection of the submarine margins of Corsica, Sardinia and the Appennine peninsula allow only tentative boundaries to be drawn between the continental slope and continental rise. In the northern part of the Tyrrhenian Sea, shallow areas (up to 500 m deep) alternate with deep trenches and valleys. A large valley crosses the floor of the Strait of Bonifacio. The continental slope and shelf off the Appennine peninsula are characterized by hilly relief complicated in places by volcanic cones. The lower part of the slope is markedly steeper than the upper part, and the slope and continental rise are dissected by numerous submarine canyons and ravines. The marginal zone is especially complex near the north of Sicily. The shelf is insignificant in width and bounded at a depth of 150–

180 m by a ledge of the continental slope, north of which there is a depression with maximum depths of about 2100 m and an uneven floor. On the north, this depression is limited by the chain of the Lipari volcanic islands, several of which are active (e.g., Stromboli and Vulcano). This volcanic ridge changes in a north-eastern direction to a volcanic belt of the continental rise extending farther to north-west along the Appennines. The volcanic areas around Naples and Rome are branches of this volcanic belt.

The Corsican/Sardinian side of the Tyrrhenian depression is characterized by intricately dissected mountainous relief which, according to Malovitsky (1976), represents a submerged folded Alpine structure. Apart from numerous grabens and depressions of limited extent, the south-western part of the Tyrrhenian depression includes two ridges of east-north-easterly orientation. One of these ridges is a continuation of the Tell Atlas system. Another passes through the Skerky bank and ends as the Ustika volcanic island merging with the chain of the Lipari Islands. Thus, these two ridges form a submarine link in a single folded system which, continuing from the Appennines and Sicily, passes into north-west Africa.

The continental rise can be clearly traced around the whole of the Tyrrhenian basin. Its uneven relief is made up of numerous salt domes related to the Messinian (Upper Miocene) evaporites.

The floor of the Tyrrhenian basin is hilly; only in the middle is there a small flat abyssal plain, surrounding the large seamount known as Vavilov's volcano. Another large submarine volcano (Marsigly) lies in the eastern part of the depression. Like the Balearic depression, the Tyrrhenian basin is underlain by suboceanic crust. The thickness of the sedimentary layer is up to 3–4 km and, as is indicated by Borehole No. 373, consists largely of young pyroclastic deposits and tuffa–lavas [*Initial Reports, Deep Sea Drilling Project* (1978)].

20.3.3   Sicilian–Tunisian Swell

This is an extensive shallow rise between Tunisia and Libya on one side and Sicily on the other. Structurally, it represents an easterly bulge of the African platform and consists correspondingly of continental-type crust. From the description by Evsiukov (1978a), the swell is subdivided into three units: the Tunisian shelf, the central region and the Sicilian margin.

The Tunisian shelf is distinguished by relatively smooth relief and by depths increasing from a few metres near the coastline to 1000–1500 m at the edge of the continental slope, which Evsiukov (1978b) calls the Maltese escarpment. More than half of the Tunisian shelf lies at depths exceeding 200 m. In the eastern, submerged part of the shelf, it is possible to distinguish the Melita bank (minimum depth, 139 m) and the Medina bank (150 m); in the western part there is the Kerkennah bank, with the islands of the same name, which represents an uplift of anticlise type [see Malovitsky (1976)]. At the surface, the shelf is composed of recent and Quaternary oozes. The thickness of Pliocene–Quaternary deposits varies from 30–50 m, in the north-west, to 300 m, in the southern foredeep of the Tunisian shelf.

The central region, which is separated from the Tunisian shelf by the Lampedusa graben and by its eastern continuation, the submarine Heron valley, is distinguished by strongly dissected relief. The submarine Heron valley is marked by alternating ridges (horsts) and depressions (grabens) lying at a depth of up to 1730 m (in the Pantelleria depression), and by a thick blanket of Quaternary and Pliocene sediments occurring on the smooth floors of the depressions. The horst ridges and plateaus are also orientated to the south-east. Some of them rise above sea level forming Malta, Pantelleria and other islands.

The Sicilian marginal zone unites the Maltese and Adventura shelf bulges and the Jolla depression situated between them. At the boundary with the central region, there is a chain of submarine volcanoes. The Jolla depression descends to a depth of 340 m, its floor being covered with a sequence of Pliocene–Quaternary carbonate sediments up to 2 km thick [see Evsiukov (1978a)].

The Sicilian–Tunisian swell is limited on the east by a continental slope—the Maltese escarpment—that starts near the Straits of Messina and dies out south of the Melita bank. Evsiukov (1978b) indicated that the Maltese part of the escarpment is more gentle than that lying north of the Sicilian and Melita areas. North of the Heron valley, the Maltese escarpment is divided into upper and lower steps. The Maltese marginal plateau lies between these two steps. South of the Melita bank, the slope is less distinct and is complicated by a gently dipping marginal plateau. The continental slope described above is dissected by numerous submarine canyons. Evsiukov believes that the Maltese escarpment represents a large flexure complicated by lateral and transverse faults. The flexure dies out towards the south, but Malovitsky thinks that in the deep structure of the Earth's crust it corresponds to the Misuratian deep-seated fault which in turn is related to the western step of the Al Hamra plateau (Libya) on the continent.

20.3.4 Central Basin

The central part of the Mediterranean, which is limited on the east by the Maltese escarpment and on the west by the central Mediterranean ridge, is divided topographically into two areas: the central depression and a southern shallow-water area. In the north lies a narrow shelf that represents an abrasion/denudation step cut in the folded structures of the Appennines and the Calabrian massif. This step is flanked by a strongly dissected continental slope that merges in the south into the broad sloping surface of the continental rise. On the floor of the Bay of Otranto lies a large valley which, according to Malovitsky, was initiated along the fault that passes into a marginal suture separating the Appennines from the foredeep. As has been shown by Evsiukov and Moskalenko, it is along this submarine valley that terrigenous sand and mud are constantly being transported to the deep areas of the Ionian Sea. Southwards, there lies the vast Messina detrital fan. The sedimentary material is supplied to the fan along the bottom of the Sicilian foredeep

and along numerous submarine canyons that cut the Maltese escarpment. The sedimentary deposits of the Messina fan consist mainly of pyroclastic rocks [see Emelianov *et al.* (1979)]. The surface of the fan is dissected by numerous turbidity current ravines that disappear in the abyssal part of the depression. These ravines serve as channels along which sediment is transported to form the Messina abyssal plain. The floor of the central basin (4297 m deep) is largely an undulating abyssal plain; only the extreme south-eastern part is represented by the small Sirte flat abyssal plain [see Ryan *et al.* (1970)].

The central depression is thought by many investigators to be underlain by suboceanic crust. Its northern part is notable for an abundance of minor features associated with salt tectonics.

The southern part of the central depression is sometimes called the Sidra trough. Its depth appears to increase gradually from the Gulf of Sidra toward an abyssal plain (up to 3500 m deep). Apparently, this part of the depression is underlain by continental crust and represents a deeply sunken block of the African platform which, according to Malovitsky (1976), is bounded by two well-marked deep-seated faults (Misuratian and Benghazian). The shelf along most of the Libyan coast is narrow (20–40 km), smooth and characterized by a wide distribution of calcareous oolitic sands. The continental slope is slightly dissected and passes gradually into a broad north-sloping plain—the continental rise—composed of foraminiferal and pteropod oozes. Two small marginal plateaus lie on the north-western and north-eastern extremities of the continental slope of the Sidra depression.

20.3.5 Adriatic Sea (*see* Fig. 20.7)

The floor of the Adriatic is a shelf plain complicated by two basins the central Adriatic (262 m) and the southern Adriatic (1224 m) depressions. The coastal part of the shelf adjoining Yugoslavia is extremely rugged. Its complex topography can be explained by the fact that the sea in the course of successive transgressions covered the partly denuded marginal structures of the Dinaric Alps, preserving their main relief features and trends. Thus the coastline is highly indented and possesses an abundance of islands; the topographic grain runs parallel to the coast which is often described as the Dalmatian type.

In the central part of the Adriatic, there are a few small submerged volcanoes with smooth, flattened summits. Small hills are also observed at the foot of the southern trough side; such hills are probably caused by salt tectonics. Both Adriatic troughs are newly formed features superimposed on the basement structures which are not directly expressed in the sea floor relief. The southern trough is linked with the Ionian Sea by a ravine in the Strait of Otranto along which the turbidity current runs to the northern part of the central basin.

20.3.6 Eastern Basin (*see* Fig. 20.7)

This is the most complex of all the Mediterranean areas in its relief. Only near the Nile delta is the shelf well-developed. Near the coast of Greece and most of Anatolia,

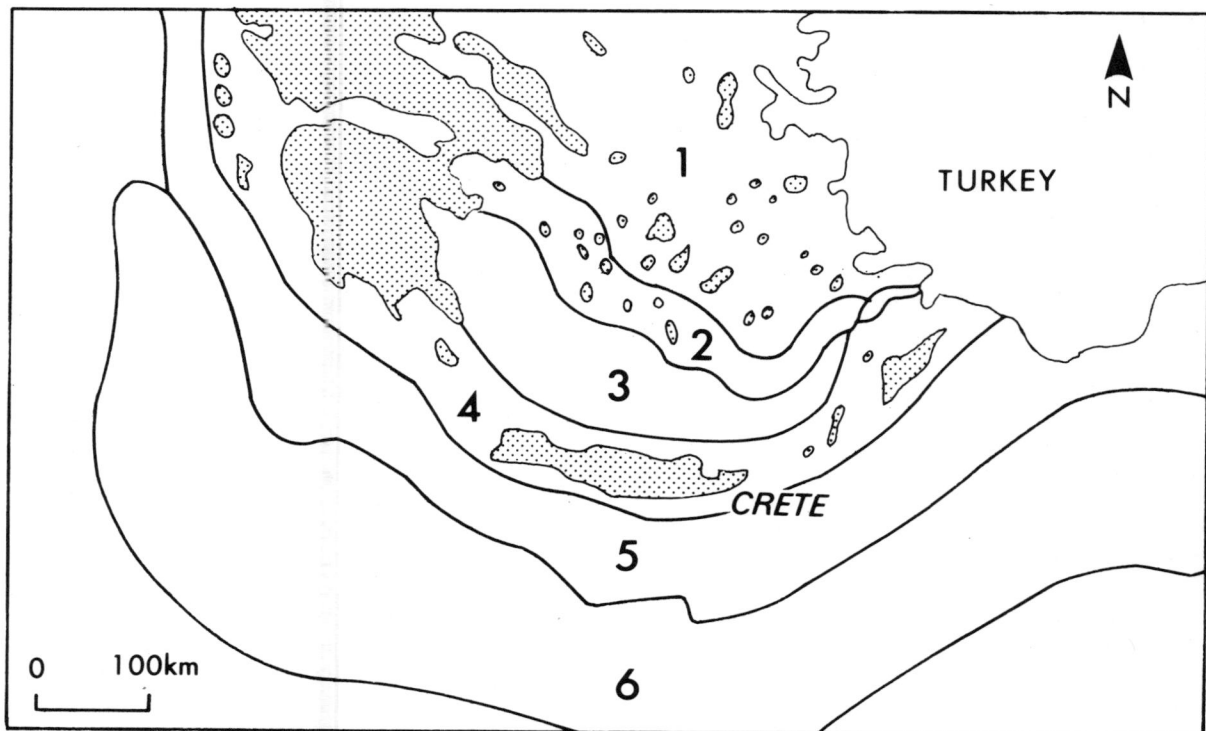

Fig. 20.8 Outline map of major morphostructural units in the Aegean–Crete region. (1) The Aegean Sea foreland, (2) Aegean volcanic ridge, (3) Crete basin, (4) Crete island arc, (5) Hellenic trench: zone of relict deep-sea trenches and rises separating them and (6) east Mediterranean ridge.

it occurs only as a narrow bench several kilometres wide, cut in the Alpine fold structures. The continental slope is highest and steepest at the Anatolian coastline: in the Rhodes trough it descends to 4000 m and is strongly dissected by canyons. The continental slope near the Levantine coast is well-marked but irregular, ending at a depth of 1000–1500 m. The continental slope of Libya and north-west Egypt is moderately wide and dissected by numerous deep submarine canyons. Near the Nile delta, it becomes more gentle and merges with the vast sloping plain of the continental rise. The plain widens eastwards and becomes most extensive north of the Nile delta, where the continental rise is represented by the large Rosetta submarine fan. The latter has been built out by turbidity currents carrying the suspended load of the Nile. North-west of the mouth of the Nile at Damietta is another submarine fan.

Complex relief characterizes the Cretan island arc and the Hellenic deep-sea trench (*see* Fig. 20.8). The Cretan island arc is represented, above sea level, by the chain of the Ionian islands of which the southernmost member is Strofades. Then the chain breaks off and reappears again as Kítira, Crete, Kasos, Karpáthos and Rhodes, passing beyond into the Alpine structures of southern Anatolia. North of Crete lies the deep Crete basin (2591 m), which can be considered as a typical deep-sea trench. It is also bordered by an island arc of active volcanoes, such as Metana, Santorini and Nisiros. The formation of the volcanic arc is probably due to the deep-seated faults that bound the southern edge of the Aegean plate. Santorini,

like other Mediterranean volcanoes, is composed of andesitic lavas [see Neenkovich and Heezen (1965)].

The floor of the Crete Sea occupies an elongated depression with minor structural forms much obscured by the accumulation of sediments at depths exceeding 1600 m. Borehole No. 378 sited in the centre of the depression intersected 343 m of Quaternary calcareous ooze and Pliocene marls, and entered gypsum of the Messina formation. According to the seismic data, the thickness of the crust here is about 30 km and Malovitsky (1976) considered that it is of continental type. Thus, even though it is a deep-sea trench, this depression appears to have been initiated on a continental base. The relief of the eastern part of the depression resembles that of a young mountainous area [see Mikhailov and Goncharov (1962)].

The outstanding feature of the Hellenic deep-sea trench is the range of contrast in its bottom relief. The trench is divided into several isolated narrow troughs (3200–5100 m deep) arranged either *en échelon* or one after the other. The longest one starts with the Zakinthos trough (4200 m) separated by a narrow ridge from the Vavilov trough (5121 m), south of which several other trenches bounding the submarine Menelay plateau are localized. Farther to the south-east, there lies the Crete trench, separated from the shelf around Crete by the Gaudos and Prometheus rises. West of Crete is the Pliny trench (up to 3000 m deep) which is arranged *en échelon* with respect to the Crete trench [see Emery *et al.* (1966)]. South-east of the Crete trench, there are the Strabo and Achilles trenches (up to 3208 m deep) which are terminated by an isometrically shaped trough (4486 m) east of Rhodes.

The submarine rises of Euripid, Sofocle and Democrete are, like those of Menelay, Gaudos and Prometheus, blocks marking the side of the once single Hellenic trench which in the course of its evolution became broken into separate troughs and secondary trenches [see Leont'ev (1980)].

South of the Hellenic trench there is another large rise, extending for more than 1500 km, called the central Mediterranean ridge or swell. It starts in the Ionian Sea, west of the Ionian islands, and ends near Cyprus. The swell is highest in the area between Crete and Cyrenaica, where its depth is 1395 m. With the exception of its more deeply submerged north-western portion, the swell is generally well picked out by the 2500-m isobath. Comparison of the most common depths in the troughs with those in the Hellenic trench that are adjacent to the swell shows that the relative height of the latter varies from 500 to 2500 m. The width of the swell ranges from 180 to 120 km. Its surface relief looks like a hilly plain or low upland (e.g., the Likian plateau or the Anaximandra rise at its north-eastern end). The eastern termination (above the 2000-m isobath) of the central Mediterranean ridge is called the Florence rise [*Initial Reports, Deep Sea Drilling Project* (1978)]. The authors of this work identifying the Cyprus island arc consider this rise to be a western continuation of Cyprus. However, Malovitsky (1976) believes that the Cyprus belt continues, north of the Florence rise, towards Cape Gelidonya on the south coast of Anatolia.

Two boreholes were sunk from the *Glomar Challenger* on the Florence rise. Borehole 378 intersected Pliocene and Miocene deposits, and was completed at a depth of 821 m below the sea floor. The succession is represented by calcareous pelagic muds and marls, and in the Lower Miocene, by pelitomorphic limestones and a 50-m bed of gypsum and dolomite belonging to the Messina formation. Borehole, No. 377 sited in the western part of the central Mediterranean ridge struck calcareous sediments of the Quaternary, Pliocene and Miocene, but failed to detect the Messinian evaporites. Borehole No. 125 drilled on the swell at a depth of 2700 m, however, intersected the Messinian strata. The wide distribution of Messinian rocks has led to the development of numerous salt dome structures and undulating relief.

South of the central Mediterranean ridge lies the abyssal eastern Mediterranean basin, which is narrow in the west and wide in the east. The relief consists of hills and undulating abyssal plains, the narrow, flat abyssal plain of Herodotus that is confined to the axial part of the depression, and several rises in the east. The Herodotus abyssal plain extending intermittently from the Sirta abyssal plain to the continental rise in the area of Cyprus has been formed as a result of accumulation of both fine sediment transported to the eastern Mediterranean depression by turbidity currents and biogenic calcareous matter.

A partially buried uplift with depths above it varying from 1500 to 2000 m divides the ancient submarine fan of the Nile into two parts—the Damietta and Rosetta fans—

and is expressed mainly as a hilly abyssal plain, while in the north it forms the Eratosthena rise. Here Emery *et al.* (1966) identified several hills or seamounts whose relative height attains 700 m. To the north there is the Hekateus rise which is separated from the above uplift by a narrow abyssal plain. Several peaks rise to 300 m above the sea floor, which is strongly dissected by submarine canyons. The rise is bounded on the south by a distinct scarp and therefore could be considered a marginal plateau if it had not been separated from Cyprus by depths greater than those over the submarine slope of this island. According to the map of Malovitsky (1976), it is an oblique horst genetically related to the Cyprus ophiolite belt. The small Mela uplift located near the continental rise south-west of Latakia is also likely to be a horst.

The Levantine depression lies between the seamounts just mentioned and the continental rise. Its central part includes a narrow, ribbon-like, flat abyssal plain of the same name; another small flat abyssal plain also lies west of the Mela uplift. The rest of the bottom topography represent an undulating, locally hilly, abyssal plain.

According to Malovitsky (1976), exposures of the Cretaceous ophiolites in the Cyprus area are related to the ophiolites composing the Amanus and Basit mountain ranges in southern Turkey. As can be seen from the intense magnetic anomalies, the buried ophiolites can be traced north-westward. North-west of Cyprus there is the Antayla depression whose southern part is also included in the Cyprus ophiolite belt. Thus, the Cretaceous Cyprus arc also identified by Biju-Duval and his coworkers is not directly expressed in the relief. North of Cyprus, there extends the Seikhan intermontane basin whose western part is represented by a marginal plateau of the same name, being a part of the Anatolian continental slope.

20.3.6 Aegean Sea (*see* Fig. 20.8)

This represents a sunken and deformed part of the complex Alpine orogenic structures of the Hellenides. In the northern part of the Aegean Sea, ancient median masses predominate while, in the southern part, Alpine folded ridges prevail. In the intensity of its tectonic crushing, the floor of the Aegean Sea represents a typical foreland.

The smooth shallow-water floor areas of the Gulf of Thessaloniki in the north-west of the Aegean Sea, the Gulfs of Mandalya and Izmir, and several other similar coastal localities may be considered as offshore shoals. Most of the northern and central parts of the sea represent shelf plains diversified by rises that form the bases of numerous islands or banks, and by various troughs. The rises are linearly orientated and occur as continuations of the landward orogenic structures, mostly horsts. Depth sounding in the Aegean Sea has detected a subbathyal depression, the north Aegean graben, with a maximum depth of 1575 m, a smooth floor covered with a thick sedimentary layer [see Biju-Dival. (1974)] and with steep sides. The trend of this prominent graben is controlled by the deep-seated northern Anatolian fracture. Subbathyal troughs over 1000 m in depth are also identified in the central and south-eastern parts of the sea. The majority of islands are sited on continuations of horst–anticlinal land structures, and are composed of acid intrusive rocks. Their coasts are subjected to abrasion, but its rate is negligible because of rock resistance, limited exposure or other factors. The fault tectonics of the floor of the Aegean Sea are responsible for the high seismicity of the region and for present-day vulcanism in the southern margins.

20.3.7   Sea of Marmara

The North Anatolian fracture has determined the broad configuration of the northern coastline of the Sea of Marmara and the steep scarp that marks its northern submarine slope. This regional fracture is also associated with narrow depressions or grabens lying along the foot of the slope and looking as though they are incised into the floor of the depression. The maximum depth of the Sea of Marmara (1261 m) is attributed to one of these bottom relief forms. The southern portion of the sea floor represents a smooth shelf surface complicated by a minor rise.

20.4   Black Sea and Caspian Sea

20.4.1   Black Sea (*see* Fig. 20.7)

The floor of the Black Sea can be divided into two unequal parts: the northern, consisting of a vast shelf in the north-west and in the Sea of Azov, and the Black Sea basin proper. Shelves are present also (although to a lesser degree) in other peripheral parts of the Black Sea.

The north-western Black Sea shelf is generally characterized by smooth relief and mainly shallow depths (up to 50–100 m only) and is confined to the structures of the Scythian platform. The coastal strip appears to be the smoothest. Among the large positive relief forms, the Odessa and Karkinitsk banks, which presumably resulted from wave action, may be distinguished, as well as many other smaller rises on the bottom related to the buried platform structures. Old submerged river valleys and, in the south-east, submerged marine terraces have been identified by Varushenko. The same type of shelf is observed farther south along the western coasts. An interesting feature of the Bulgarian shelf is the series of sandy silt submarine ridges which represent either submerged relict bars or forms produced by present-day bottom currents.

The floor of the Sea of Azov is flat and even due to accumulation of sediments; its maximum depth does not exceed 14 m. The shelf area widens somewhat south of the Kerch Strait, but, as has been recently shown, the forms of denuded brachyanticlinal structures can be also faintly detected through the sedimentary blanket. The narrow shelf of the northern Caucasian coast clearly exhibits the structural forms of layers in the flysch sequence that constitutes the shores and submarine slope. In the Pitsunda area, the shelf is dissected by the heads of submarine canyons; the same situation exists farther south at the mouths of the Kadori and Ingur. Along the Anatolian coast, the shelf is formed by a narrow plain cut in young fold structures.

The Black Sea continental slope in the west is relatively wide, being largely represented by a depositional plain.

The most distinctive feature of the relief is a turbidity current ravine that in the past supplied the relict Danube fan. The continental slope of the southern Crimea is complicated by step-wise faults and landslide forms. The Caucasian and Anatolian continental slopes are densely channelled by submarine canyons. The continental rise is well-developed and is widest on the western side of the depression due to the Danube submarine fan. On the lower part of the continental rise in the east, geophysical survey has detected a large rise buried under the sediments and orientated in the same direction as the Caucasus. The continental rise loses its distinctiveness in the central Anatolian area where it is obscured by young horst–anticline ranges.

The boundary of the Black Sea basin floor coincides closely with the 2000-m isobath. The relief of the basin floor consists of a typical smooth abyssal plain (although depths here are less than in the true abyssal zone). Relief gradation is achieved by the accumulation of a thick sedimentary layer. According to seismic data, its total thickness attains 15 km. There are positive Bouguer gravity anomalies over almost all of the depression. From these data, it seems likely that there is no granitic layer under the depression. The bottom is completely aseismic, in contrast to the high seismic activity in the zone of the continental slope.

20.4.2   Caspian Sea

The depression occupied by the Caspian Sea is varied in its character. The northern shallow part of the Caspian lies on a platform structure, and its relief is characterized by relict forms including submerged river valleys and depositional forms such as offshore bars frequently superimposed on buried brachyanticlinal structures. The middle Caspian depression is distinguished by a well-developed offshore shoal (*see* Fig. 20.9) extending from both the western and the eastern sides, and by the deep (over 700 m) Derbent depression. The western slope of the Derbent depression is steep and dissected by landslide processes, while the eastern side is greatly elongated and represents a subsided part of the Scythian platform. The western portion of the Derbent depression lies structurally within the Terek–Caspian marginal foredeep that is part of the north-eastern termination of the Great Caucasus. The sides of the depression include parts of the subsided shelf. The floor of the depression is smooth due to accumulation of sediments.

The middle Caspian depression is bounded on the south by a submarine elevation called the Apsheron rise, composed of a chain of brachyanticlinal structures (Nefty-anyye Kamni, etc.) extending from the Apsheron peninsula in the west to the Cheleken peninsula in the east. The Apsheron rise separates the middle Caspian from the south Caspian. The periphery of the latter is characterized by a well-developed shelf which in the west exhibits structural/denudation relief. Floor irregularities are associated with brachyanticlinal uplifts crossing the shelf in a southerly or south-easterly direction. On the continental slope, these uplifts pass into a submarine ridge that advances far out into the deeper parts of the south

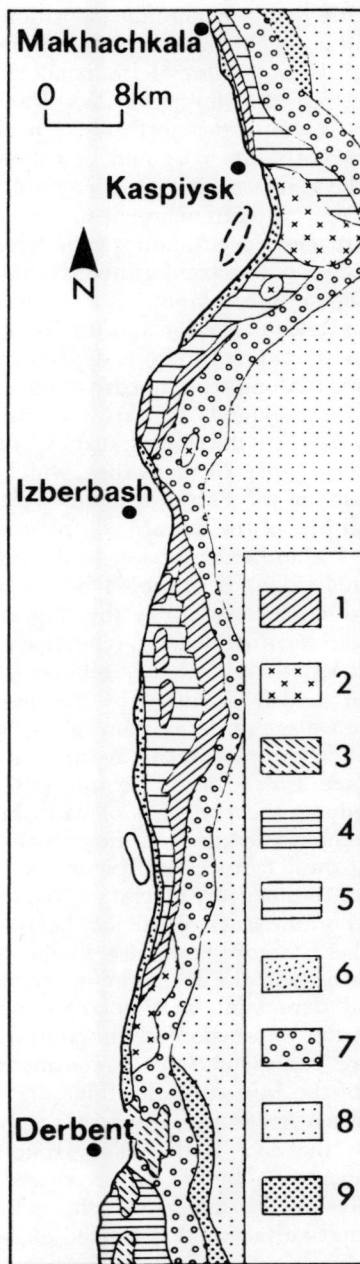

Fig. 20.9 Relief types of the Dagestan shelf of the Caspian Sea [after Leont'ev and Varuschenko]. (1) Bench with ridges, (2) buried bench with ridges, (3) bench with steps, (4) buried bench with steps, (5) buried even bench, (6) nearshore relief of ridges and swales, (7) offshore depositional plain, (8) shelf plain of non-wave accumulation and (9) relict coastal features.

Caspian basin. Along the southern coast the shelf appears to be narrow and flat, and is covered with a thin blanket of shallow-water sediments. The eastern shelf of the south Caspian is distinguished, on the other hand, by its much greater width, consisting of a thick sedimentary layer under which, as is indicated by seismic evidence, occur buried folded structures whose trend is similar to that of Kopet Dagh as reported by Leont'ev (1964).

The continental slope of the southern side of the depression consists of a steep scarp. Elsewhere, it is complicated by the submarine ridges mentioned above. These are a direct expression of the present-day anticlinal structures [see Leont'ev (1964), Maev (1961), Soloviev *et al.* (1952)]. The most prominent ridges (the Abikha, Shatsky, etc.) lie at a relative depth of up to 500 m. Local uplifts are identified on the crests of the ridges; there are many mud volcanoes here which on the western shelf form the archipelago of the Baku islands.

The floor of the south Caspian depression represents a flat accumulation plain with abyssal characteristics, but its depth does not exceed 1000 m.

20.5 Mid-Atlantic Ridge and Adjacent Parts of the Atlantic Ocean Floor (*see* Fig. 20.4)

20.5.1 Mid-Atlantic Ridge

The *International Geomorphological Map of Europe* includes the Kolbeinsey and Mona ridges and the north-eastern part of the Reykjanes ridge. Since the main geological/geophysical characteristics of these ridges are described in Chapter 3, this section will consider only the geomorphological data.

The Kolbeinsey ridge starts on the northern shelf of Iceland as the flat-topped Grimseijar rise, with the basaltic island of Grimsey in its northern part. The ridge consists of two longitudinal chains separated by a median rift valley, the total width being not more than 40–50 km. Its relative elevation above the sea floor in the south is of the order of 500 m, while farther north it attains 1000–1300 m. At latitude 69°N, the ridge is cut by the Spar fracture zone, which corresponds to a distinct transverse trough (400 m deep). North of the trough, the Kolbeinsey ridge changes its direction to north-east and widens markedly, with flanking zones appearing. It is cut across by several large fractures seen as deep troughs and horsts. The ridge is bounded by the large Jan–Mayen transverse fracture zone. The plateau-like rise of Jan–Mayen Island with the Byerenberg active volcano (absolute elevation, 2277 m) lies on the southern side of the trough. Several trachytic cones are also located here. The Byerenberg mountain is composed of trachytes and rhyolites and represents a typical stratovolcano, which most recently erupted in 1818.

The north-eastern continuation of the mid-oceanic ridge appears to have been shifted, along the Jan–Mayen transform fracture, eastward relative to the Kolbeinsey ridge for approximately 180 km. This link in the Mid-Atlantic Ridge is known as the Mohn's ridge and is characterized by the same general features: a distinct rift valley (up to 2000 m deep) and transverse troughs cutting the ridge, associated with transform faults, as well as a block rectangular structural pattern. Farther north, Mohn's ridge is intersected by the Greenland fracture zone. Another link of the Mid-Atlantic Ridge known as the Knipovich ridge lies north of the Greenland fracture zone.

The Icelandic plateau and Norwegian Sea floor have been described in Chapter 3. West of the Kolbeinsey ridge lies part of the Greenland depression whose submarine relief is not fully known. Its southern portion, which forms the Danish Strait, is distinguished by a hilly topography, probably due to accumulation of iceberg debris. The axial part of the Greenland depression includes a well-marked abyssal valley—the Danish channel—whose formation, according to Matishov (1980), is due to the action of turbidity currents, probably subglacial during the Pleistocene glaciations.

The Reykjanes ridge extends from the Reykjanes Peninsula to the south-west for 1580 km. Its width varies from 200 to 300 km and its relative elevation from 1100 to 1800 m. Instead of a rift system, the central and northern parts of the ridge consist of a basaltic horst whose length exceeds 1000 km and whose maximum width is 60 km. The basalts are of Quaternary and Pliocene age. On the Icelandic shelf, this horst is joined to the Reykjanes volcanic zone by a chain of submarine volcanoes and volcanic islands [see Lavrov (1979)]. Lavrov identified crests and ravines trending in sublongitudinal and sublatitudinal directions on the surface of the horst. On the west and east, the sides of the horst are complicated by steps at depths of 1500 and 1750 m. Dissection of the surface attains an amplitude of 500–700 m, but the ravines that are orientated perpendicular to the trend of the horst do not exceed 200 m in depth [see Lavrov (1979)].

The flanks of the Reykjanes ridge consist of ridges and ravines. There are bedrock ridges of sublongitudinal trend and transverse valleys infilled with calcareous sediments [see Ruddiman (1972)]. A slightly dissected, sublongitudinally orientated rise, that of Ulriche was recognized by Matishov (1980) on the eastern flank, and is probably a bottom accumulation form.

The southern part of the Reykjanes ridge south of the transverse fracture (latitude, 57° N) is characterized by predominant longitudinal trends. Moderately deep (1000–1500 m) rift valleys can be seen locally and a young block structure type of relief is common [see Lavrov (1979)]. Individual peaks rise above the 1000-m isobath. The ridge ends at the Gibbs fracture zone that extends across the entire ocean, from the continental rise of Newfoundland to Porcupine Bank. The North Atlantic mid-oceanic ridge south of this fracture zone is displaced along it to the east, relative to the Reykjanes ridge, for approximately 300 km.

20.5.2 Ocean Floor

The north-eastern branch of the Newfoundland basin lying west of the Reykjanes ridge has been called the Irminger basin by Matishov; its maximum depth, within the limits of the map, is 3600 m. The floor of the depression largely comprises an abyssal plain that passes north-westwards into the sloping depositional surface of the continental rise. The most outstanding feature here is the very large abyssal valley of Imarsuak, traceable in a south-westerly direction for more than 1500 km, beyond which it fades out into the north-western abyssal plain. The author suggests that it should be called the Heezen valley after the name of its discoverer and first investigator. The depth of entrenchment of the valley exceeds 100–150 m in places. Several minor valleys join it, starting at the mouths of submarine canyons that dissect the continental slope of Greenland.

Between the Reykjanes ridge and the Rockall rise, there lies the relatively shallow Icelandic depression (maximum depth, 3180 m). Along the north-western foot of the Rockall rise, there extends a narrow abyssal plain that passes northwards into the sloping surface of the continental rise. The large Mory abyssal valley, which starts on the continental slope of the Faröe–Iceland swell, is traceable along the axial line of the flat abyssal plain. The western margin of the Icelandic depression (at its boundary with the mid-oceanic ridge) is occupied by the Garder sedimentary ridge which constitutes the largest abyssal accumulation form showing, locally, relative elevations of up to 1000 m. Its origin is probably related to the bottom abyssal current which is caused by the descent of overcooled oceanic waters to the bottom (as is the case of the western oceanic bottom current in the North American depression) [see Emery and Uchupi (1972), Leont'ev (1975)]. Based on the concepts of Matishov regarding specific glaciofluvial features in the ocean region, it is suggested that these features are due to the evacuation of great masses of sediment by glacial meltwater, since both in Iceland and on the Rockall rise ice sheets existed. One, or both, of these factors gave rise to the Mory valley, which extends beyond the Icelandic depression.

The Iceland depression is a northern branch of the western European depression of the Atlantic Ocean. Its boundaries are represented by a continental rise, the eastern limit of the Mid-Atlantic Ridge and the block of the Azores–Biscay ridge extending in a sublongitudinal direction from the end of the King trough (a narrow graben). The Azores–Biscay ridge, orientated almost perpendicularly to the King trough, exhibits a vast dissected summit surface and is fringed on both sides by deep depressions (up to 5500 m deep). The sharp outer limits of the ridge, in combination with distinct depressions (grabens) that bound it on the north-west and south-east, indicate that this ridge is a horst morphostructure. Its eastern continuation is marked by a few horst uplifts with relatively smooth summit surfaces and abrupt slopes. The most extreme of these rises forms a seamount in the Bay of Biscay, with a minimum depth over it of 2410 m. Farther eastwards, there extends a fracture zone in the Bay of Biscay which is accompanied by the block seamounts of Cantabria and the Gulf of Gascony.

The floor of the west European basin is distinctly divided into western and eastern parts. The western part is characterized by abyssal hilly relief that, in the form of a wide band, bounds the Mid-Atlantic Ridge. The eastern part of the depression consists of a flat abyssal plain. The small Iberian basin (locally over 5600–5800 m deep) lying south of the Azores–Biscay ridge is also characterized by hilly relief in its western part, and by a flat abyssal plain in the east. The southern boundary of this depression is represented by the intricately constructed volcanic uplands of Gorindge, consisting of several ridges and small depressions over 5000 m deep. This area is also characterized by an abundance of guyots, many of which are notable for very small depths of water over their summits: for example, Dasiya (27 m), Amper (40 m), Gettisberg (24 m) and Sen (148 m). The Gorindge upland extends southward and includes the Madeira volcanic area, whereas the Gloria fracture running through this area is related to the volcanic upland of the Azores. Drilling undertaken on the Gorindge upland has indicated the presence of ophiolites. Small depressions within the upland and south of it have flat floors which resemble abyssal plains. A small area of abyssal hills lies south-east of the Sen seamount [see Ilyin (1976)].

References

Aarseth, I. and Mangerud, J. (1974) Younger Dryas end-moraines between Hardangerfjorden and Sognefjorden, western Norway. *Boreas*, **3**, 3.

Abele, G. (1974) Bergstürze in den Alpen. *Wiss. Alpenvereinshefte*, **25**, 1.

Academiei Republicii Populare Romine (1960) *Monografia Geografica a Republicii Populare Romine* (Bucureşti).

Agabekov, M. G. and Akhmedbeyli, F. S. (1956) Concerning the question of studying neotectonics in Azerbaijan. *Izv. Akad. Nauk azerb. SSR*, **7**, 49.

Agip Mineraria (1959) *I Giacimenti Gassiferi dell'Europa Occidentale*. Atti Congr. Milan, 30 November–5 December, 1957 (Accademia Nazionale dei Lincei and ENI, Rome).

Agrell, H. (1979) The Quaternary of Sweden. *Sver. geol. Unders. Afh., Ser. C*, **770**.

Aguirre, E. (1957) Una prueba paleomastológica de la edad cuaternaria de los conglomerados de la Alhambre (Granada). *Estudios geol.*, **13**, 135.

Aguirre, E. (1961) La serie estratigráfica del Neógeno en la depresión de Granada y contribución del género Chlamys a sú caracterización. *Estudios geol.*, **18**, 7.

Aguirre, E. (1974) Depresión de Granada. *Coloquio int. sobre Biostratigrafía continental del Neógeno superior y Cuaternario inferior (Guía)*, **7–10**, 175.

Aguirre, E. *et al.* (1976) Datos paleomastológicos y fases tectónicas en el Neógeno de la Meseta Sur española. *Trab. Neógeno-Cuaternario*, **5**, 7.

Åhman, R. (1977) Palsar i Nordnorge. En studie av palsars morfologi, utbredning och klimatiska förutsättningar i Finnmarks och Troms fylke. *Avh. geogr. Inst. Lunds Univ.*, **78**.

Akifyeva, K. V. (1970) Avalanches in the Caucasus, in *Avalanche Hazard Areas of the Soviet Union*, ed. by G. K. Tushinskiy, p.68 (Moscow State University, Moscow).

Alcaydé, G. *et al.* (1976) Val de Loire (Anjou, Touraine, Orléanais, Berry), in *Guides Géologiques Régionaux*, ed. by C. Pomerol (Masson, Paris).

Aleixandre, T. *et al.* (1974) Guía de Excursión A. Valle alto del Jarama. *Acta Guías Excursiones I Reunión Grp esp. Trab. del Cuaternario*, **2**, 215.

Alexandrowicz, Z. (1977) The origin of sandstone tors in the Polish Western Carpathians. *Bull. Acad. pol. Sci., Sér. Sci. terre*, **24** (2).

Alférez-Delgado, F. (1977) Estudio del sistema de terrazas del Río Tajo al W. de Toledo. *Estudios geol.*, **33**, 223.

Alía-Medina, M. (1945) Notas morfológicas de la región toledana. *Las Ciencias*, **10**, 95.

Alimen, H. (1951) Actions périglaciaires et sols sur le versant nord-Pyrénéen en Bigorre. *Revue Géogr. Pyrénées S.-Ouest*, **22**, 124.

Alimen, H. (1964) Le Quaternaire des Pyrénées de la Bigorre. *Mém. Carte géol. Fr.* (Paris).

Almagia, R. (1907) Studi geografici sulle frane in Italia. *Memorie Soc. geogr. ital.*, **13**.

Almagia, R. (1910) Studi geografici sulle frane in Italia. *Memorie Soc. geogr. ital.*, **14**.

Ampferer, O. (1906) Über das Bewegungsbild von Faltengebirgen. *Jb. geol. Bundesanst. Wien*, **56**, 539.

Andersen, B. (1960) Sørlandet i sen- og postglacial tid. *Norg. geol. Unders. Afh.*, **210**.

Andersen, B. (1965) Glacial chronology of western Troms, north Norway. *Geol. Soc. Am. Spec. Pap.*, **84**, 35.

Annaheim, H. (1946) Studien zur Geomorphogenese der Südalpen zwischen St Gotthard und Alpenrand. *Geographica helv.*, **1**, 65.

Annaheim, H. and Barsch, D. (1963) Geographischer Exkursionsführer der Schweiz. Exkursionsraum Gempenplateau und benachbarte Talregionen. *Geographica helv.*, **18**, 241.

Annaheim, H. *et al.* (1975) Geomorphologie. I. Übersicht, 1:500000, in *Atlas der Schweiz*, Map 8 (Wabern, Bern).

Antonov, B. A. *et al.* (1963) Comparative analysis of the planation surfaces of the eastern Transcaucasian highlands. *Izv. Akad. Nauk georg. SSR*, p.42.

Antonov, B. A. *et al.* (1977) The principal features and stages in the development of the relief of the Caucasus, in *The Geology of the Quaternary Era (Pleistocene)*, p.50 (Akad. Nauk Armenian SSR, Yerevan).

Aparicio, A. *et al.* (1977) Los granitos hercínicos sincinemáticos de la Sierra del Valle (sector oriental de Gredos, Sistema Central español) sus relaciones con las series graníticas post-tectónicas. *Estudios geol.*, **33**, 575.

Aranegui, P. (1927) Las terrazas cuaternarias del Río Tajo entre Aranjuez (Madrid) y Talavera de la Reina (Toledo). *Boln R. Soc. esp. Hist. nat.*, **27**, 285.

Archipov, S. A. and Vdovin V. V. (1970) *West Siberian Lowlands* (Nauka, Moscow).

Argand, E. (1916) Sur l'arc des Alpes Occidentales. *Eclog. geol. Helv.*, **14**, 145.

Arkell, W. J. (1943) The Pleistocene rocks at Trebetherick Point, north Cornwall: their interpretation and correlation. *Proc. Geol. Ass.*, **54**, 141.

Arkhangelskiy, A. D. and Strakhov, N. M. (1938) *Geologicheskoye Strotenie i Istoriya Razvitya Chernogo Morya* (Akad. Nauk, Moscow).

Armand, A. D. and Grave, M. K. (1974) The Baltic Shield, in *Geomorphology of the USSR: Lowlands of the European Part of the USSR*, p.46 (Nauka, Moscow).

Arnberger, E. (1960) *Korsika* (Freytag-Berndt u. Artaria, Vienna).

Arnborg, L. (1959) Nedre Ångermanälven. Del II. Vattenregleringarnas inverkan på morfologi och fluviala processer, särskilt inom det recenta deltaområdet 1947–58. *Avh. geogr. Inst. Univ. Uppsala*, **2**.

Artiushkov, E. V. *et al.* (1980) Global oceanic structures, in *Contemporary Problems of Marine Geology*, vol. 3, p.110 (Institute of Oceanology, Moscow).

Aseev, A. A. (1963) The influence of the climatic rhythms of the Quaternary period on the development of the erosional pattern. *Izv. Akad. Nauk SSSR, ser. Geogr.*, **1**, 8.

Aseev, A. A. (1974) *Ancient Continental Glaciations of Europe* (Nauka, Moscow).

Aseev, A. A. (1978) Common peculiarities of the structure of river valleys of the USSR as indicators of the rhythm of oscillatory movements of the Earth's crust. *Geomorphologiya*, **2**, 3.

Aseev, A. A. and Makkaveyev, A. N. (1976) Glacial geomorphology: results of science and technology. *Geomorphologiya*, **4**.

Aseev, A. A. and Makkaveyev, A. N. (1977) Glacial morphological criteria for the erosion of continental ice-sheets in Europe. *Issled. Materialy glyatsiologicheskih, Chron., Obsuzh.*, **29**.

Asensio-Amor, I. and Martí-Yenderrozos, J. M. (1981) Morfología litoral y acumulaciones detríticas costeras del occidente asturiano (ensenada de Navia-Cabo Busto). *Actas Guías Excursiones V Reunión Grp esp. Trab. Cuaternario*, p.17.

Asensio-Amor, I. and Nonn, H. (1964) Los depósitos de terrazas del Río Eo y de las márgenes de su ría (zona galaico-asturiana). *Estudios geogr.*, **96**, 319.

Asensio-Amor, I. and Ontañón, J. M. (1972) Acumulaciones periglaciares en el valle de las Guarramillas (Vertiente septentrional de la Sierra de Guadarrama). *Estudios geogr.*, **28**, 453.

Asensio-Amor, I. and Ontañón, J. M. (1975) Depósitos torrenciales de aspecto fluvio–glaciar (Alto valle de Lozoya, Sierra de Guadarrama). *Estudios geogr.*, **31**, 365.

Ashauer, H. and Teichmüller, R. (1935) Die variscische und alpidische Gebirgsbildung Kataloniens. *Abh. Ges. Wiss. Göttingen, Math. Phys.*, **3**(16), 16.

Asklund, B. (1923) Bruchspaltenbildungen im süd-östlichen Östergötland nebst einer Übersicht der geologischen Stellung der Bruchspalten Südost–Schwedens. *Geol. För. Stockh. Förh.*, **45**, 249.

Aslanyan, A. T. (1958) *Regional Geology of Armenia* (Aipetrat, Yerevan).

Aubert, D. (1965) Calotte glaciaire et morphologie jurasienne. *Eclog. geol. Helv.*, **58**, 555.

Aumiento, F. *et al.* (1977) Introduction and site reports. *Initial Rep., Deep Sea Drilling Project*, **38**, 1.

Auzende, J. M. and Olivet, J. L. (1974) Structure of the western Mediterranean basin, in *The Geology of the Continental Margins*, ed. by C. A. Burk and C. L. Drake, p.723 (Springer–Verlag, New York).

Babin-Vich, R. B. (1977) Evolución tectónica y posición dentro del Macizo Hespérico del sector occidental de la Sierra de Gredos. *Estudios geol.*, **33**, 251.

Bachmann, R. C. (ed.) (1978) *Gletscher der Alpen* (Hallwag, Bern).

Baker, C. A. and Jones, D. K. C. (1980) Glaciation of the London Basin and its influence on the drainage pattern, in *The Shaping of Southern England*, ed. by D. K. C. Jones, p.131 (Academic Press, London).

Bakker, J. P. and Levelt, T. W. (1964) An enquiry into the probability of a polyclimatic development of peneplains and pediments (etchplains) in Europe during the Senonian and Tertiary periods. *Publs fys.-geogr. Lab. Univ. Amsterdam*, **4**, 27.

Balchin, W. G. V. (1952) The erosion surfaces of Exmoor and adjacent areas. *Geogrl J.*, **118**, 453.

Ball, D. F. (1966) Late-glacial scree in Wales. *Biul. peryglac.*, **15**, 151.

Balyan, S. P. (1969) *Structural Geomorphology of the Armenian Uplands and Surrounding Regions* (Yerevan University, Yerevan).

Banting, A. H. (1933) Der Bau der Betischen Kordillera und ihre Stellung im Mediterranen Orogen. *Geol. Rdsch.*, **24**, 311.

Barbaza, I. (1971) Morphologie des secteurs rocheux du littoral catalan septentrional. *Mém. Docums, Cent. Docum. cartogr. géogr.*, **11**, 1.

Barrère, P. (1951) *La Morphologie des Sierras Oscenses* (Instituto de Estudios Pirenaicos, Zaragoza).

Barrère, P. (1952) Évolution mécanique et nivation sur les versants calcaires de la haute montagne pyrénéenne. *Pirineos*, **24**, 201.

Barrère, P. (1953) Équilibre glaciaire actuel et quaternaire dans l'Ouest des Pyrénées centrales. *Revue Géogr. Pyrénées S.-Ouest*, **24**, 116.

Barrère, P. (1963) La période glaciaire dans l'Ouest des Pyrénées centrales franco-espagnoles. *Bull. Soc. Géol. fr.*, **7**, 516.

Barrère, P. (1966) La morphologie quaternaire dans la région de Biescas et de Sabiñánigo (Haut Aragón). *Bull. Ass. fr. Étud. Quatern.*, **2**, 83.

Barrère, P. (1979) Terrasses et glacis d'érosion en roches tendres dans les montagnes du Haut Aragón. *Mélanges G. Viers*, p.29.

Barrow, G. (1908) The high-level platforms of Bodmin Moor and their relation to the deposits of stream tin and wolfram. *Q. J. geol. Soc. Lond.*, **64**, 384.

Barsch, D. (1968a) Periglaziale Seen in den Karstwannen des Schweizer Juras. *Regio basil.*, **9**, 115.

Barsch, D. (1968b) Die geomorphologische Übersichtskarte 1:250000 der Basler region. *Regio basil.*, **9**, 384.

Barsch, D. (1969) Studien zur Geomorphogenese des zentralen Berner Juras. *Basler Beitr. Geogr.*, **9**, 1.

Barsch, D. (1977a) Alpiner Permafrost: ein Beitrag zur Verbreitung, zum Charakter und zur Ökologie am Beispiel der Schweizer Alpen. *Abh. Akad. Wiss. Göttingen*, **3** (31), 118.

Barsch, D. (1977b) Eine Abschätzung von Schuttproduktion und Schutttransport im Bereich aktiver Blockgletscher der Schweizer Alpen. *Z. Geomorph., Suppl.*, **28**, 148.

Barsch, D. *et al.* (1975) Atlas der Schweiz. Geomorphologie in Übersicht, 1:500000. *Regio basil.*, **16**, 50.

Bartkowski, T. (1967) O formach strefy marginalnej na Nizinie Wielkopolskiej. *Pr. Kom. geogr.-geol., Poznań*, **7**(1).

Bartkowski, T. (1968) O istocie glacitektoniki. *Przegl. Geol.*, **16**, 455.

Basalikas, A. B. (1969) Variety of relief in the glacial accretion area, in *Continental Glaciation and Glacial Morphogenesis*, p.65 (Vilnius).

Bascom, W. N. (1964) *Waves and Beaches* (Doubleday, New York).

Bashenina, N. V. (1948) *Origin of the Relief of the Southern Urals* (Geografgiz, Moscow).

Bashenina, N. V. (1960) Cryoplanation surfaces as zonal types of pediments. *Vestnik mosk. gos. Univ., ser. Geogr.*, **6**, 68.

Bashenina, N. V. (1961) Palaeogeography and Quaternary history of the evolution of the relief and superficial deposits of the Southern Urals. *Materialy sovetschaniya po izucheniyu chatvetichnogo perioda*, **3**, 408.

Bashenina, N. V. (1967) *Formation of the Earth's Present-Day Relief* (Vyshaya Schkola, Moscow).

Bashenina, N. V. (1971) On the glaciation of the Soviet Carpathians and the links between glacial forms and block tectonics. *Izv. vses. geogr. Obschch.*, **2**, 166.

Bashenina, N. V. (1973) An attempt at classifying faults and fractures by their expression in the relief and their influence on various types of morphostructure, in *Methods of Composing Tectonic Maps*, vol. 6 (Nauka, Novosibirsk).

Bashenina, N. V. (1977) The formation of rifts and their expression in the relief. *Vestnik mosk. gos. Univ., ser. Geogr.*, **1**, 45.

Bashenina, N. V. *et al.* (1974a) *Geomorphology of the Elbruz Mountain Area (Caucasus Mountains, USSR)* (Czech Academy of Science, Brno).

Bashenina, N. V. *et al.* (1974b) *A Brief Geomorphological Review of the Pre-Elbruz Area* (MGU, Moscow).

Bashenina, N. V. *et al.* (1977) *Geomorphological Mapping* (Vyshaya Schkola, Moscow).

Bashenina, N. V. *et al.* (1979) On the content of world geomorphological maps (with a morphostructural scheme of the Earth). *Geologiya Geophys.*, **3**, 134.

Baskina, V. A. (1973) Concerning the tectonic position of Iceland. *Geotectonics*, **2**, 24.

Bataller, J. R. and Larragan, A. (1955) *Explicación geológica de la hoja n° 352 (Tabuenca)*, scale 1:500000 (Instituto Geológica y Minero de España, Madrid).

Baulig, H. (1927) La Crau et la glaciation würmienne. *Annls Géogr.*, **36**, 499.

Baulig, H. (1928) *Le Plateau Central de la France et sa Bordure Méditerranéenne. Étude Morphologique* (Armand Colin, Paris).

Bech, J. (1972) Datos sobre la mineralogía de la fracción arena en los regolitos, saprolitos y suelos graníticos del Maresme (Barcelona). *Publnes Inst. Invest. Geol. Dip. Prov.*, **26**, 113.

Beck, P. (1926) *Eine Karte der letzten Vergletscherung der Schweizeralpen* (Geogr. Karten-Verlag, Bern; Kümmerly u. Frey, Bern).

Belderson, R. H. *et al.* (1976) Holocene sediments on the continental slope west of the British Isles. *Rep. Inst. geol. Sci. Lond.*, **70**, 281.

Belloni, S. *et al.* (1978) Karst of Italy, in *Important Karst Regions of the Northern Hemisphere*, p.85 (Elsevier, Amsterdam).

Belousov, V. V. (1954) *Principal Questions of Geotectonics* (Gosgeoltechizdat, Moscow).

Belousov, V. V. (1960) Tectonic map of the Earth. *Geol. Rdsch.*, **50**, 316.

Belousov, V. V. (1975) *Principles of Geotectonics* (Nedra, Moscow).

Belousov, V. V. (1976) *Geotectonics* (MGU, Moscow).

Belousov, V. V. (1978a) *Endogenic Regimes of the Continents* (Nedra, Moscow).

Belousov, V. V. (1978b) Geology of the oceans, in *Tectonosphere of the Earth*, p.100, (Nauka, Moscow).

Belousov, V. V. and Gozjatchev, A. V. (eds.) (1977) *Iceland and the Mid-Oceanic Ridge* (Nauka, Moscow).

Bemmelen, R. W. van (1927) *Bijdrage tot de Geologie der Betische Ketens in de Provincie de Granada*, thesis, University of Delft.

Berglund, B. E. (1979) The deglaciation of southern Sweden, 13500–10000 BP. *Boreas*, **8**, 89.

Bergqvist, E. (1981) *Svenska Inlandsdyner. Översikt och Förslag till Dynreservat* (Arlöv).

Biancotti, A. (1981) Geomorfologia dell'alta Langa (Piemonte meridionale). *Memorie Soc. ital. Sci. nat.*, **22**, 59.

Biju-Duval, B. (1974) Geology of the Mediterranean basin, in *The Geology of the Continental Margins*, ed. by C. A. Burk and C. L. Drake, p.695 (Springer-Verlag, New York).

Bird, E. C. F (1976) *Coasts* (MIT, Cambridge, MS).

Birot, P. (1933) Le relief de la Sierra d'Alto Rey et sa bordure orientale. *Bull. Ass. Géogr. fr.*, **20**, 92.

Birot, P. (1937) *Recherches sur la Morphologie des Pyrénées Orientales Franco-Espagnoles* (Baillière, Paris).

Birot, P. (1945) Sobre la morfología de la Sierra de Guadarrama occidental. *Estudios geogr.*, **6**, 155.

Birot, P. (1948) Notes sur la morphologie du Portugal méridional. *Mélanges géogr. D. Faucher*, **1**, 103.

Birot, P. (1949) Essai sur quelques problèmes de morphologie générale. *Comuns Cent. Estud. Geol. Lisboa*, **176**.

Birot, P. (1960) *Le Cycle d'Érosion sous les Différents Climats* (University of Brazil, Rio de Janeiro).

Birot, P. (1963) Évolution des versants à corniche dans la série miocène au sud de Teruel (Espagne), in *Neue Beiträge zur Internationalen Hangforschung, 3ª Rapport der 'Commission on Slope Evolution'*, ed. by H. Mortensen, p.67 (Vandenhoeck u. Ruprecht, Göttingen).

Birot, P. (1964) La Méditerranée et le Moyen Orient, in *Généralités; Péninsule Ibérique*, vol. 1, 2nd edn (PUF, Paris).

Birot, P. and Solé-Sabarís, L. (1954a) Recherches morphologiques dans le Nord-Ouest de la Péninsule Ibérique. *Mém. Docums Cent. nat. Res. scient.*, **4**, 7.

Birot, P. and Solé-Sabarís, L. (1954b) *Investigaciones sobre morfología de la Cordillera Central española* (CSIC, Inst. Juan Sebastián Elcano).

Birot, P. and Solé-Sabarís, L. (1959) Recherches sur la morphologie du sud-est de l'Espagne. *Revue Géogr. Pyrénées S.-Ouest*, **30**, 119.

Bishop, W. W. (1958) The Pleistocene geology and geomorphology of three gaps in the midland Jurassic escarpment. *Phil. Trans. R. Soc.*, **B254**, 255.

Biskey, G. S. (1959) *Fundamental Questions of the Geomorphology of the Eastern Part of the Baltic Shield* (Moscow).

Björnsson, H. (1979) Glaciers in Iceland. *Jökull*, **29**, 74.

Björnsson, S. (1937) Sommen–Åsunden–området. En geomorfologisk studie. *Meddn Lunds geogr. Instn*, **4**.

Blagovolin, N. S. (1962) *Geomorphology of the Kerch–Taman Region* (Nauka, Moscow).

Blagovolin, N. S. *et al.* (1977) Mud volcanoes, in *General Description and History of the Development of the Relief in the Caucasus*, p.70 (Nauka, Moscow).

Blumenthal, M. M. (1927) Versuch einer tektonischen Gliederung der betischen Cordilleren von Central und Sud-West Andalusien. *Eclog. geol. Helv.*, **20**, 487.

Boch, S. G. (1946) Les névés et l'érosion par la neige dans la partie nord de l'Oural. *Bull. Soc. géogr. USSR*, **78**, 207.

Boch, S. G. and Krasnov, I. I. (1951) The process of cryoplanation and the formation of cryoplanation terraces. *Priroda*, **5**, 25.

Boenzi, F. *et al.* (1976) Caratteri geomorfologici dell'area del Foglio Matera. *Boll. Soc. geol. ital.*, **95**, 527.

Boenzi, F. *et al.* (1978) I terrazzi della valle del Basento (Basilicata). *Riv. geogr. it.*, **85**, 396.

Bogdanov, A. A. (1964) Several general questions concerning the tectonics of ancient platforms. *Sov. Geol.*, **9**.

Bogdanov, A. A. (1972) Europe: geological structures and useful minerals, in *The Great Soviet Encyclopaedia*, vol. 9, p.36.

Bognar, A. (1979) Distribution, properties and types of loess and loess-like sediments in Croatia. *Acta geol. hung.*, **22** (1–4).

Bognar, A. (1980) *Tipovi reljefa kontinentalnog dijela Hrvatske* (Spomen zbornik ob 30. obljetnici GD Hrvatske, Zagreb).

Bomer, B. (1956) Aspects morphologiques du bassin de Calatayud–Daroca et de ses bordures. *Bull. Ass. Géogr. fr.*, **261–2**, 186.

Bondarchuk, V. G. (1949). *Basics of Geomorphology* (Uchpedgiz, Moscow).

Bondesan, M. (1975) Carta geomorfologica del Comune di Ferrara, *Mem. Soc. geol. ital.*, **14**.

Bonjer, K. *et al.* (1975) Geophysical investigations of the Rhine rift system, in *Symposium on Rift Zones of the Earth*, p.77 (Irkutsk).

Bonnifay, E. (1969) *Languedoc–Provence–Côte d'Azur* (International Quaternary Association, Paris).

Borisevitch, D. V. (ed.) (1973) Poverkhnosti vyravnivaniya Evropy, in *Poverkhnosti Vyravnivaniya Evropy, Azii i Afriki*, vol. 3, p.1 (Itogi Nauki i Tekniki, Moscow).

Borisov, A. A. (1966) *Abyssal Structure of the USSR According to Geophysical Data* (Nedra, Moscow).

Borisov, A. A. (1977) *Abyssal Structure of the USSR According to Geophysical Data* (Nedra, Moscow).

Bortolami, C. *et al.* (1977) Land, sea and climate in the northern Adriatic region during the later Pleistocene and Holocene. *Palaeogeogr. Palaeoclimatol. Palaeoecol.*, **21**, 139.

Bortolami, C. *et al.* (1978) Hydrogeological features of the Po valley, northern Italy. *Annls Inst. geol. Publs Hungarici*, **69**.

Bott, M. H. P. (1971) *The Interior of the Earth* (Edward Arnold, London).

Boulton, G. S. *et al.* (1977) A British ice-sheet model and patterns of glacial erosion and deposition in Britain, in *British Quaternary Studies: Recent Advances*, ed. by F. W. Shotton, p.231 (Clarendon Press, Oxford).

Bourcart, J. (1938) Le marge continentale. Essai sur les transgressions et regressions marines. *Bull. Soc. Géol. fr., ser. 8*, **5a**, 393.

Bout, P. and Brousse, R. (1969) *Auvergne–Velay* (International Quaternary Association, Paris).

Bovis, M. J. (1974) Late Quaternary continental deposits of the Motril area, southern Spain. *Z. Geomorph.*, **18**, 426.

Bowen, D. Q. (1971) A re-evaluation of the coastal Pleistocene succession in south-west Britain. *Quaternaria*, **15**, 87.

Bowen, D. Q. (1974) Quaternary of Wales, in *The Upper Palaeozoic and Post-Palaeozoic Rocks of Wales*, ed. by T. R. Owen, p.373 (University of Wales Press, Cardiff).

Bowen, D. Q. (1977) Studies in the Welsh Quaternary: retrospect and prospect. *Cambria*, **4**, 2.

Boyé, M. (1952) Gélivation et cryoturbation dans le massif du Mont Perdu (Pyrénées centrales). *Pirineos*, **23**, 5.

Brand, G. *et al.* (1965) Die lithostratigraphische Unterteilung des marinen Holozäns an der Nordseeküste. *Geol. Jber.*, **82**, 365.

Bravard, J. (1980) Voies nouvelles dans l'étude des Alpes françaises et du piedmont, in *Alpes–Caucase: Problèmes Actuels de la Géographie Constructive de Pays de Montagne*, p.51 (Nauka, Moscow).

Bravard, J. *et al.* (1980) Traits principaux de la rassemblance et de la différence de la nature des Alpes et du Caucase, in *Alpes–Caucase: Problèmes Actuels de la Géographie Constructive de Pays de Montagne*, p.278 (Nauka, Moscow).

Bremner, A. (1942) The origins of the Scottish river system. *Scott. geogr. Mag.*, **58**, 15, 54, 99.

Breuil, H. and Zbyszewski, G. (1942) Contributions à l'étude des industries paléolithiques du Portugal et leurs rapports avec la géologie du Quaternaire. *Comunções Servs geol. Port.*, **13**.

Breuil, H. and Zbyszewski, G. (1945) Contributions à l'étude des industries paléolithiques du Portugal et leurs rapports avec la géologie du Quaternaire. *Comunções Servs geol. Port.*, **26**.

Breyer, F. (1974) Die Entstehungsgeschichte des Südteils des Rheingrabens nach reflexionsseismischen Messungen, geologischen Kartierungen und Tiefbohrungen. *Geol. Jb. Hannover*, **A20**, 1.

Brinkmann, R. (1931) Beticum und Keltibericum in Sudspanien. *Abh. Akad. Wiss. Göttingen, Math. Phys.*, **3**, 749.

Brinkmann, R. (1932) Las montañas fósiles, especialmente en España. *Boln R. Soc. geogr.*, **72**, 387.

Bristow, C. R. and Cox, F. C. (1973) The Gipping till: a re-appraisal of East Anglian glacial stratigraphy. *J. geol. Soc. Lond.*, **129**, 1.

Brouwer, H. A. (1926) Zur Geologie der Sierra Nevada. *Geol. Rdsch.*, **17**, 118.

Brown, E. H. (1952) The 600-foot platform in Wales. *Proc. 8th gen. Assembly & 17th int. Congr., int. geogr. Un.*, p.304.

Brown, E. H. (1960a) *The Relief and Drainage of Wales* (University of Wales Press, Cardiff).

Brown, E. H. (1960b) The building of southern Britain. *Z. Geomorph.*, **4**, 264.

Brückner, H. (1980) Marine Terrassen in Süditalien. Eine quartärmorphologische Studien über das Küstentiefland von Metapont. *Düsseld. geogr. Schr.*, **14**, 1.

Brunet, R. (1956) Un exemple de la régression des glaciaires pyrénéennes. *Pirineos*, **39–42**, 261.

Brunnacker, K. (1973) Observaciones sobre terrazas marinas y formaciones de piedemonte en el Sudeste de España. *Estudios geogr.*, **130**, 133.

Brunsden, D. (1963) The denudation chronology of the River Dart. *Trans. Inst. Br. Geogr.*, **32**, 49.

Brunsden, D. (1968) *Dartmoor* (Geographical Association, Sheffield).

Budagov, B. A. (1966) Planation surfaces in the Azerbaijan region of the Great Caucasus, in *Questions of the Geomorphology and Landscape Formation in Azerbaijan*, p.55 (Akad. Nauk Azerbaijan SSR, Baku).

Budagov, B. A. *et al.* (1977) *General Description and History of the Development of the Relief in the Caucasus*, p.122 (Nauka, Moscow).

Büdel, J. (1948) Das System der klimatischen Geomorphologie. Beiträge zur Geomorphologie der Klimazonen und Vorzeitklimate, V. *Verh. dt. GeogrTags*, **27**, 65.

Büdel, J. (1957) Grundzüge der klimamorphologischen Entwicklung Frankens. *Würzb. geogr. Arb.*, **4–5**, 5.

Büdel, J. (1963a) Klimagenetische Geomorphologie. *Geogr. Rdsch.*, **15**, 269.

Büdel, J. (1963b) Die pliozänen und quartären Kaltzeiten der Sahara. *Eiszeitalter Gegenw.*, **14**, 161.

Büdel, J. (1969) Das System der klimagenetischen Geomorphologie. *Erdkunde*, **23**, 165.

Büdel, J. (1977) *Klima-Geomorphologie* (Bornträger, Berlin).

Büdel, J. (1982) *Climatic Geomorphology* (Princeton University Press, New York).

Bullón, T. (1978) Los fenómenos periglaciares en la Sierra de la Mujer Muerta. *Actas V Coloquio Geogr.*, p.35.

Burillo, F. *et al.* (1981) El cerro del Castillo de Alfambra (Teruel). Estudio interdisciplinar de geomorfología y arqueología. *Kalathos*, **1**, 7.

Burke, M. J. (1969) The Forth valley—an ice-moulded lowland. *Trans. Inst. Br. Geogr.*, **48**, 51.

Bush, V. A. *et al.* (1980) Cosmogeological investigation of the Ural–Oman superlineament. *Issled. Zemli Kosmosa*, **4**, 13.

Busnardo, R. (1960) Aperçu sur le Prébétique de la région de Jaén (Andalousie, Espagne). *Bull. Soc. géol. Fr.*, **7**, 324.

Butzer, K. W. (1964) Pleistocene geomorphology and stratigraphy of the Costa Brava region (Catalonia). *Abh. Akad. Wiss. Lit., Math. Naturw.*, **1**, 3.

Butzer, K. W. and Cuerda, J. (1962) Coastal stratigraphy of southern Mallorca and the implications for the Pleistocene chronology of the Mediterranean Sea. *J. Geol.*, **70**, 398.

Butzer, K. W. and Franzle, O. (1959) Observations on pre-Würm glaciations of the Iberian Peninsula. *Z. Geomorph.*, **3**, 85.

Cabanás, R. (1957) Terrazas cuaternarias del Guadalquivir y sus afluentes en la provincia de Jaén. *Revta Acad. Cienc. exact. fis. nat.*, **51**, 213.

Cadisch, J. (1953) *Geologie der Schweizer Alpen* (Wepf, Basel).

Cailleux, A. and Hupe, P. (1947) Présence de sols polygonaux et striés dans les Pyrénées françaises. *C. r. hebd. Séanc. Acad. Sci. Paris*, **225**, 1353.

Calatayd, P. *et al.* (1980) *Itinerario Geológico y Geomorfológico por el Valle del Najerilla* (Inst. Estudios Riojanos, Logroño).

Caldenius, C. and Lundström, R. (1956) The landslide at Surte on the river Göta: a geological–geotechnical study. *Sver. geol. Unders. Afh.*, **27**.

Calvet, J. (1976) Notas geomorfológicas sobre un sector del contacto Depresión Central Catalana–Cordillera Prelitoral (alrededores de l'Espluga de Francolí). *Acta geol. hisp.*, **11**, 25.

Calvet, J. (1977) *Contribución al Conocimiento Geomorfológico de la Depresión Central Catalana*, thesis, University of Barcelona.

Calvo-Palacios, J. L. (1975) Nota sobre las relaciones de la red fluvial camerana y la tectónica del borde septentrional del Sistema Ibérico. *Berceo*, **88**, 93.

Campredon, R. *et al.* (1975) Alpes maritimes, Maures, Esterel, in *Guides Géologiques Régionaux*, ed. by C. Pomerol (Masson, Paris).

Camus, G. *et al.* (1975) *Volcanologie de la Chaîne des Puys* (Parc national des Volcans d'Auvergne, Clermont Ferrand).

Capote, R. *et al.* (1981) Movimientos recientes en la fosa del Jiloca (Cordillera Ibérica). *Actas V Reunión Grp esp. Trab. Cuaternario*, p.245.

Carandell, J. (1935) Las condiciones del modelado erosivo en la vertiente mediterránea de la Cordillera Bética. *Boln R. Soc. esp. Hist. nat.*, **35**, 39.

Carandell, J. and Gómez de Llarena, J. (1918) El glaciarismo cuaternario en los montes ibéricos. *Trab. Mus. nac. Cienc. nat.*, **22**, 1.

Carbognin, L. *et al.* (1978) Land subsidence of Ravenna and its similarities with the Venice case. *Proc. Engng Fdn Conf. Evaluation Prediction Subsidence, Am. Soc. Civ. Engrs*.

Carle, W. (1941) Karrenbildung im Granit der galicischen Küste bei Vigo (Nordwestspanien). *Geologie Meere Binnengewäss.*, **5**, 55.

Carle, W. (1947). Die westgalicischen Meeresbuchten. *Natur. Völkerk.*, **77**, 5.

Carloni, G. C. *et al.* (1975) Considerazioni geomorfologiche sui terrazzi dell'Aso e del Tenna (Marche meridionali). *Ateneo parmense, Acta Naturalia*, **11**, 649.

Carraro, F. *et al.* (1975) Geomorphological study of the morainic amphitheatre of Ivrea, north-west Italy. *Bull. R. Soc. N. Z.*, **13**.

Castellarin, A. and Vai, G. B. (eds.) (1982) Guida alla geologia del Sudalpino centro-orientale. *Memorie Soc. geol. ital., Suppl. C*, **24**.

Castiglioni, G-B. (1933) Valli sovralluvionate e deviazioni fluviali in Abruzzo e Piceno. *Boll. Soc. geogr. ital.*, **6**, 642.

Castiglioni, G-B. (1935) Ricerche morfologiche nei terreni pliocenici dell'Italia centrale. *Publ. Inst. Geogr. Univ. Roma*, **A4**.

Castiglioni, G-B. (1979) *Geomorfologia* (UTET, Turin).

Castiglioni, G-B. and Pellegrini, G-B. (1980–81) Geomorfologia dell'alveo del Brenta nella pianura tra Bassano e Padova, in *Il Territorio del Brenta* (Provincial Administration of Padua & University of Padua, Padua).

Cattuto, C. (1973) Carta e lineamenti geomorfologici del territorio tra il F. Chiascio e i torrenti Rosina e Saonde. *Geol. Romana*, **12**, 105.

Cavaillé, A. (1965) Les unités morphologiques des basses plaines de la Garonne. *Revue Géogr. Pyrénées S.-Ouest*, **36**, 243.

Cavaillé, A. (1975) Le piémont quaternaire de la Garonne. *Bull. Ass. fr. Étud. Quatern.*, **3–4**, (44–5), 1.

Cepek, A. G. (1967) Stand und Probleme der Quartärstratigraphie im Nordteil DDR. *Ber. dt. Ges. geol. Wiss.*, **A12**.

Čermák, V. (1975) Heat flow and deep temperature distribution in Central and Eastern Europe and their dynamic application, in *Symposium on Rift Zones of the Earth*, p.77 (Irkutsk).

Cernescu, N. (ed.) (1960) *Monografia Geografica a Republicii Populare Romine*, vol. I. *Geografia Fizica* (Academiei Republicii Populare Romine, Bucureşti).

Chabot, G. (1927) Les plateaux du Jura Central: étude morphogénique. *Publ. Fac. Lettres Univ. Strasbourg*, **41**.

Chaput, J. L. (1968) Les inselbergs granitiques de la Meseta ibérique méridionale. *Bull. Ass. Géogr. fr.*, **359–360**, 47.

Chardonnet, J. (1944) *Le Relief des Alpes du Sud*, dissertation, University of Paris.

Charlesworth, J. K. (1929) The South Wales end-moraine. *Q. J. geol. Soc. Lond.*, **85**, 335.

Charlesworth, J. K. (1957) *The Quaternary Era* (Edward Arnold, London).

Charlesworth, J. K. (1963) *Historical Geology of Ireland* (Oliver & Boyd, Edinburgh).

Chebotareva, N. S. and Makaritcheva, I. A. (1974) *Last Glaciation of Europe and its Geochronology* (Nauka, Moscow).

Chikishev, A. G. (1965) Types of karst in the Russian lowlands, in *Tipy Karsta v SSSR*, p.12 (Nauka, Moscow).

Chikishev, A. G. (1978) *Karst of the Russian Lowlands* (Nauka, Moscow).

Cholley, A. (1950) Morphologie structurale et morphologie climatique. *Annls Géogr.*, **49**, 321.

Cholley, A. *et al.* (1956) Carte morphologique du Bassin parisien, 1:400000. *Mém. Docums, Cent. nat. Res. scient.*, **5**.

Chumakov, J. S. (1975) Lake stage of the development of the Mediterranean, in *The History of Lakes*, p.10 (Institute of Lake Research, Leningrad).

Ciabatti, M. (1968) Ricerche sull'evoluzione del Delta Padano. *G. Geol. (Bologna)*, ser. 2, **34**.

Clemente, L. and Paneque, G. (1974) Propiedades, génesis y clasificación de suelos de terrazas del Guadalquivir. I. Factores ecológicos y relaciones edafogeomorfológicas. *An. Edafol. Agrobiol.*, **33**, 215.

Clin, M. (1964) *Études Géologique de la Haute Chaîne des Pyrénées Centrales entre le Cirque de Troumouse et le Cirque du Lys* (Ed. Technip., Paris).

Cloos, H. (1939) Hebung–Spaltung–Vulkanismus. *Geol. Rdsch.*, **30**, 405.

Clozier, R. (1940) *Les Causses du Quercy* (Baillière, Paris).

CNR, Progetto Finalizzato Geodinamica (1978) *Contributi Preliminari alla Realizzazione della Carta Neotectonica d'Italia*, publ. no. 155 (CNR, Rome).

CNR, Progetto Finalizzato Geodinamica (1979) *Nuovi Contributi alla Realizzazione della Carta Neotectonica d'Italia*, publ. no. 251 (CNR, Rome).

Colchen, M. (1974) *Géologie de la Sierra de la Demanda (Burgos, Logroño, España)* (Instituto Geológica y Minero, Madrid).

Collected papers (1960) *Questions of Contemporary Foreign Tectonics* (Inostrannaya Lit., Moscow).

Collected papers (1973) *Planation Surfaces of Europe, Asia and Africa. Results of Science and Technology* (VINITI, Moscow).

Collected papers (1974) *Planation Surfaces and Weathered Mantles of the USSR* (Nedra, Moscow).

Collected papers (1975) *Symposium on Rift Zones of the Earth* (Nauka, Irkutsk).

Collected papers (1977) *Iceland and the Mid-Oceanic Ridge*, ed. by V. V. Belousov and A. V. Gozjatchev, p.196 (Nauka, Moscow).

Collected papers (1980) *Alpes–Caucase. Problèmes Actuels de la Géographie Constructive des Pays de Montagne* (Nauka, Moscow).

Common, R. (1954) The geomorphology of the east Cheviot area. *Scott. geogr. Mag.*, **70**, 124.

Coope, G. R. *et al.* (1961) A late Pleistocene fauna and flora from Upton Warren, Worcestershire. *Phil. Trans. R. Soc.*, **B244**, 379.

Cotton, C. A. (1955) The theory of secular marine planation. *Am. J. Sci.*, **253**, 580.

Cremaschi, M. (1979) The loess of the central-eastern Po valley. *Proc. 15th Mtg, Commn geomorph. Surv. Mapping, int. geogr. Un.*, p. 03.

Cremaschi, M. and Marchesini, A. (1978) L'evoluzione di un tratto della Pianura Padana (Prov. Reggio e Parma) in rapporto agli insediamenti e alla struttura geologica tra il XV sec.a.C. ed il sec.XI d.C. *Archeol. mediev.*, **5**.

Cremaschi, M. and Orombelli, G. (1980) I paleosuoli del Pleistocene medio nel settore centrale della Pianura Padana; il problema del 'ferretto' nella stratigrafia del Quaternario continentale, in *Atti Convegno Ass. it. Quatern., Geogr. fis. din. Quatern*.

Crusafont, M. *et al.* (1964) Contribución al conocimiento de la estratigrafia del Terciarion continental de Navarra y Rioja. *Notas Comun Inst. geol. min. Esp.*, **90**, 1.

Cueto, E. (1930) Nota acerca de las llanuras rasas y sierras planar de la costa de Asturias. *Boln R. Soc. esp. Hist. nat.*, **30**, 241.

Cvijić, J. (1924) *Geomorfologija* (Belgrade).

Cys, P. N. (1966) Review of the main problems of morphogenesis in the Ukrainian Carpathians. *Geogr. polon.*, **10**, 37.

Czudek, T. *et al.* (1965) Tertiary elements in the relief of the Outer Carpathians in Moravia, in *Geomorphological Problems of the Carpathians*, vol. 1. *Evolution of the Relief in the Tertiary*, ed. by E. Mazúr and O. Stehlík, p. 55 (Vydavatelstvo Slovenskej Akad. Vied, Bratislava).

Dahl, R. (1965) Plastically sculptured detail forms on rock surfaces in northern Nordland, Norway. *Geogr. Annlr*, **A47**, 83.

Dainelli, G. (1930) Guida della Excursione alla Penisola Sorrentina; Guida della Escursione al Matese. *Atti XI Congr. geogr. ital.*, vol. 4, pp.59, 101.

Damiani, A. V. (1975) Aspetti morfologici e possibile sistema evolutivo dei Monti Sibillini (Appennino Umbro-Marchi-giano). *Boll Serv. geol. ital.*, **96**, 145.

Dantin-Cereceda, J. (1912) Resumen fisiográfico de la Península Ibérica. *Trab. Mus. nac. Cienc. nat.*, **9**, 1.

Dantin-Cereceda, J. (1922) *Ensayo Acerca de las Regiones Naturales de España*. (Junta Ampl. Estudios, Madrid).

Dantin-Cereceda, J. (1944) Tectónica del macizo galaico. *Estudios geogr.*, **5** 45.

Daveau, S. (1973). Quelques exemples d'évolution quaternaire des versants au Portugal. *Finisterra*, **8**, 5.

Davies, G. L. (1966) Cyclic surfaces in the Roundwood basin, County Wicklow. *Ir. Geogr.*, **5**, 150.

Davies, G. L. and Stephens, N. (1978) *Ireland* (Methuen, London).

Davies, G. L. and Whittow, J. B. (1975) A reconsideration of the drainage pattern of Counties Cork and Waterford. *Ir. Geogr.*, **8**, 24.

Davis, W. M. (1909) Glacial erosion in North Wales. *Q. J. geol. Soc. Lond.*, **65**, 281.

Dedkov, A. P. (1970) *Exogenic relief formation in the Kazan–Ulyanovsk Povolzhje* (Izd. State University, Kazan).

Dedkov, A. P. (1976) Theoretical aspects of contemporary climatic-geomorphological ideas. *Geomorphologiya*, **4**, 3.

Dedkov, A. P. *et al.* (1977) *Climatic Geomorphology of Denuded Lowlands* (Izd. State University, Kazan).

De Geer, G. (1897) Om rullstensåsarnas bildningssätt. *Sver. geol. Unders. Afh.*, ser. C, **173**.

De Geer, G. (1909) Some stationary ice borders of the last glaciation. *Geol. För. Stockh. Förh.*, **31**, 199.

De Geer, G. (1911) Klarälvens serpentinlopp och flodplan. *Sver. geol. Unders. Afh.*, ser. C, **236**.

De Geer, G. (1940) *Geochronologica Suecica Principles* (K. Svenska Vetenskapsakad., Stockholm).

Degens, E. T. and Ross, A. A. (1974) *The Black Sea* (Oklahoma).

Dejou, A. P. *et al.* (1977) *Évolution Superficielle des Roches Cristallines et Cristallophylliennes dans les Régions Tempérées* (INRA, Paris).

Delattre, C. *et al.* (1973) Région du Nord, in *Guides Géologiques Régionaux*, ed. by C. Pomerol (Masson, Paris).

Delgado, S. and Fernandez, R. (1975) Morfología kárstica de las Sierras de Loja y Alhama (Granada). *Cuad. geogr. Univ. Granada*, **1**, 109.

Demangeot, J. (1965) *Géomorphologie des Abruzzes Adriatiques* (CNRS, Paris).

Demarcq, G. *et al.* (1973) Lyonnais, vallée du Rhône, in *Guides Géologiques Régionaux*, ed. by C. Pomerol (Masson, Paris).

Demek, J. (1964a) Castle koppies and tors in the Bohemian Highland (Czechoslovakia). *Biul. peryglac.*, **14**, 195.

Demek, J. (1964b) Slope development in granite areas of the Bohemian massif. *Z. Geomorph., Suppl.*, **5**, 82.

Demek, J. (1969) Cryoplanation terraces, their geographical distribution, genesis and development. *Rozpr. čsl. Akad. Věd*, **79** (4).

Demek, J. (1976a) Pleistocene continental glaciation and its effect on the relief of the north-eastern part of the Bohemian Highlands. *Studia Soc. Sci. torun.* 8C, 63.

Demek, J. (1976b) Planation surfaces and their significance for the morphostructural analysis of the Czech Socialist Republic. *Studia geogr.*, Brno, **54**, 133.

Demek, J. (1976c) Planation surfaces of the Moravian Carpathians, Czechoslovakia. *Sb. čsl. geogr. Spol.*, **81**, 9.

Demek, J. (1977) *The Theory of Systems and the Study of Landscape* (Progress, Moscow).

Demek, J. *et al.* (1965) *Geomorfologie Českych zemí* (Academia, Prague).

De Puydt, F. (1972) De Belgische strand- en duinformaties in het kader van de geomorfologie der zuidoostelijke Noordzeekust. *Verh. K. vlaam. Acad. Wet. Belg.*, **34**, 228.

Desio, A. (ed.) (1973) *Geologia dell'Italia* (UTET, Turin).

Dewey, J. F. (1974) Continental margins and ophiolite obduction: Appalachian Caledonian system, in *The Geology of the Continental Margins*, p.889 (Springer-Verlag, New York).

Dewey, J. F. *et al.* (1973) Plate tectonics and the evolution of the Alpine system. *Bull. geol. Soc. Am.*, **84**, 3137.

Dewolf, Y. (1976) À propos des argiles à silex. Essai de typologie. *Revue Géomorph. dyn.*, **4**, 113.

Dewolf, Y. (1982) *Le Contact Île-de-France – Basse Normandie. Évolution Géodynamique*, thesis, University of Paris.

Diaz, F. *et al.* (1981) Introducción al área litoral y prelitoral del SO español (sector Cádiz–Ayamonte). *Actas Guías Excursiones V Reunión Grp esp. Trab. Cuaternario*, p.309 [see also Diaz, F. and Rubio, J. M., *ibid.*, p.387].

Dibner, V. D. (1978) *Morphostructure of the Shelf of the Barents Sea* (Nedra, Moscow).

Diffre, P. and Pomerol, C. (1979) Paris et environs, in *Guides Géologiques Régionaux*, ed. by C. Pomerol (Masson, Paris).

Dmitriev, L. V. (1975) Deep-sea drilling in the ocean-floor crust. *Nature, Lond.*, **5**, 34.

Dobrovolskiy, V. V. (1969) *Geography and Palaeogeography of the Weathering Crusts of the USSR*. (Mysl., Moscow).

Dobrynin, B. F. (1948) *Physical Geography of Western Europe* (Uchpedgiz., Moscow).

Dobson, M. R. *et al.* (1973) The geology of the south Irish Sea. *Rep. Inst. geol. Sci. Gt Br.*, **73** (11).

Doré, F. *et al.* (1977) Normandie, in *Guides Géologiques Régionaux*, ed. by C. Pomerol (Masson, Paris).

Dorsser, H. J. van (1982) Carte géomorphologique du sud-ouest du Massif du Cantal. *Revue Géomorph. dyn.*, **31**, 1.

Drain, M. *et al.* (1971) *Le Bas Guadalquivir. Introduction Géographique* (Ed. Boccard, Paris).

Drake, C. L. *et al.* (1959) Continental margins and geosynclines: the east coast of North America, north of Cape Hatteras, in *Physics and Chemistry of the Earth*, vol. 3, p.110 (Pergamon, New York).

Dresch, J. (1937) *De la Sierra Nevada au Grand Atlas. Formes Glaciaires et Formes de Nivation*, p.194 (Mélanges Gautier, Tours).

Dresch, J. (1938) Les surfaces de piémont dans les Djebilet et le Massif Central du Grand Atlas. *C.r. Congr. int. Géogr.*, vol. 2, p.135.

Driscoll, E. M. (1958) The denudation chronology of the Vale of Glamorgan. *Trans. Inst. Br. Geogr.*, **25**, 45.

Dubois, M. (1959) *Le Jura Méridional: Étude Morphologique* (Soc. Educ. Enseignement, Paris).

Dumas, B. (1967) Glacis et croûtes calcaires dans le Levant espagnol. *Bull. Ass. Géogr. fr.*, **375–6**, 553.

Dumas, B. (1971) Alternance de niveaux continentaux climatiques et de hauts niveaux marins sur la côte du Levant espagnol. *Quaternaria*, **15**, 161.

Dumitrashko, N. V. (1957) The main stages in the development of the relief of the south-eastern part of the Great Caucasus. *Trans. IV Conf. Geomorph. Caucasus TransCaucasus*, p.13 (Akad. Nauk. armyan. SSR, Yerevan).

Dumitrashko, N. V. (1962) Contemporary river valleys and river terraces, ancient valleys, planation surfaces, in *Geology of the Armenian SSR*, vol. I *Geomorphology*, p.388 (Akad. Nauk armyan. SSR, Yerevan).

Dumitrashko, N. V. (1974) Caucasus, in *Mountain Regions of the European Part of the USSR and the Caucasus*, p.90 (Nauka, Moscow).

Dumitrashko, N. V. and Museibov, M. A. (1977) Comparative description of the relief of the Caucasus and neighbouring mountain regions, in *General Description and History of the Relief of the Caucasus*, p.248 (Nauka, Moscow).

Dumitrashko, N. V. *et al.* (1961) Marine terraces; planation surfaces; ancient valleys (Great Caucasus), in *Geomorphology of the Azerbaijan SSR*, pp.129, 147, 164 (Akad. Nauk Azerb. SSR, Baku).

Durand, S. *et al.* (1977) Bretagne, in *Guides Géologiques Régionaux*, ed. by C. Pomerol (Masson, Paris).

Durand-Delga, M. *et al.* (1978) Corse, in *Guides Géologiques Régionaux*, ed. by C. Pomerol (Masson, Paris).

Dury, G. H. (1959) A contribution to the geomorphology of central Donegal. *Proc. Geol. Ass.*, **70**, 1.

Dury, G. H. (1964) Aspects of the geomorphology of Slieve League peninsula, Donegal. *Proc. Geol. Ass.*, **75**, 445.

Dylik, J. (1956) Coup d'oeil sur la Pologne périglaciaire. *Biul. peryglac.*, **4**, 195.

Edel, J. *et al.* (1975) The Earth's crust and upper mantle of the Rhine rift system from explosion seismic experiments, in *Symposium on Rift Zones of the Earth*, p.77 (Irkutsk).

Eicher, U. and Siegenthäler, U. (1976) Palynological and oxygen isotope investigations on Late-Glacial sediment cores from Swiss lakes. *Boreas*, **5**, 109.

Einarsson, T. (1962) Upper Tertiary and Pleistocene rocks in Iceland. *Rit Vísindaf. ísl.*, **36**.

Einarsson, T. (1973) *Jardfrædi* (Reykjavik).

Eldholm, O. and Ewing, J. (1971) Marine geophysical survey in the south-western Barents Sea. *J. geophys. Res.*, **76**, 3832.

Elhaï, H. (1963) *La Normandie Occidentale entre la Seine et la Golfe Normand–Breton*, thesis, University of Paris.

Elhaï, H. (1967) Colloque sur les argiles à silex du Bassin de Paris. *Mém. Soc. géol. Fr, hors série*, **4**.

Elhaï, H. (1970) Colloque sur la tectonique du Bassin de Paris. *C. r. somm. Séanc. Soc. géol. Fr.*, **7**, 268; *Bull. Ass. Géol. Bassin Paris*, **26**, 319.

Elhaï, H. and Journaux, A. (1969) *Normandie* (International Quaternary Association, Paris).

Embleton, C. (1964) The planation surfaces of Arfon and adjacent parts of Anglesey: a re-examination of their age and origin. *Trans. Inst. Br. Geogr.*, **35**, 17.

Embleton, C. and King, C. A. M. (1975) *Glacial Geomorphology* (Edward Arnold, London).

Emelianov, E. M. *et al.* (1979) Geochemistry of the Mediterranean Sea, in *Scientific Thoughts*, p.132 (Kiev).

Emery, K. O. and Uchupi, E. (1973) Western North Atlantic Ocean: topography, rocks and structure, in *Water, Life and Sediments*, p.488 (Tulsa).

Emery, K. O. *et al.* (1966) Bathymetry of the eastern Mediterranean Sea. *Deep Sea Res.*, **13**, 173.

Emiliani, C. and Flint, R. F. (1963) The Pleistocene record, in *The Sea*, ed. by M. N. Hill, vol. 3, *The Earth beneath the Sea: History*. (Wiley Interscience, New York).

Enjalbert, H. (1960) *Les Pays Aquitains; le Modelé et les Sols* (Bière, Bordeaux).

Ermakov, A. V. (1957) The role of different vertical zones in the formation of planation surfaces. *Izv. Akad. Nauk SSSR, ser. Geogr.*, **6**, 83.

Erzinger, E. (1943) Die Oberflächenformen der Ajoie (Berner Jura). *Mitt. geogr.-ethnol. Ges. Basel*, **6**, 1.

Everard, C. E. (1954) The Solent River: a geomorphological study. *Trans. Inst. Br. Geogr.*, **20**, 41.

Evsiukov, Y. D. (1978a) The relief and sedimentary basins of the Tunisian Straits. *Izv. Akad. Nauk. SSSR*, **240** (1).

Evsiukov, Y. D. (1978b) Geomorphology of the Maltese scarp. *Oceanology*, **6**, 1053.

Fairbridge, R. (1961) Eustatic changes in sea level, in *Physics and Chemistry of the Earth*, ed. by L. H. Ahrens *et al.*, vol. 4, p.99 (Pergamon Press, New York).

Fallot, P. (1931–34) Essais sur la répartition des terrains secondaires et tertiaires dans le domaine des Alpides espagnoles, in *Géologie des Chaînes Bétiques et Subbétiques*, vol. 4, part II.

Fallot, P. (1948) Les cordillères bétiques. *Estudios geol.*, **8**, 83.

Farrington, A. (1966) The last glacial episode in the Wicklow Mountains. *Ir. Nat. J.*, **15**, 226.

Farrington, A. (1968) Suggestions towards a history of the Shannon. *Ir. Geogr.*, **5**, 402.

Feio, M. (1949) *Le Bas Alentejo et l'Algarve*. (International Geographical Congress, Lisbon).

Fénelon, P. (1951) *Le Périgord, Etude Morphologique* (Lahure, Paris).

Fernandez, R. (1964) Contribución al estudio del karst de la Alfaguara (Alpujárrides occidentales). *Boln R. Soc. esp. Hist. nat.*, **62**, 309.

Ferreira, A. de B. (1978) *Planaltos e Montanhas do Norte da Beira. Estudo de Geomorphologia* (Centro de Estudos Geográficos, Lisbon).

Ferreira, D. de B. (1980) Notice explicative. Carte géomorphologique du Portugal. *Memorias Cent. Estud. geogr.*, **5**.

Fezer, F. (1969) Schuttmassen, Blockdecken und Talformen im nördlichen Schwarzwald. *Göttinger geogr. Abh.*, **14**, 45.

Filliol, J. (1955) Aspects physiques de la région des étangs landais d'Arcachon à Soustons. *Revue Géogr. Pyrénées S.-Ouest*, **26**, 28.

Fink, J. (1973) Zur Morphologie des Wiener Raumes. *Z. Geomorph.*, **17**, 91.

Fink, J. (1975) Changes of climate and landforms in the eastern Alps. *An. Acad. brasil. Cienc.*, **47** (Suppl.)

Fink, M. (1969) Beiträge zur Geomorphologie der Voralpen zwischen Erlauf und Traisen. *Geogr. Jber. Öst.*, **32**, 130.

Fischer, F. (1973) Schichtstufenlandschaft und Periglazialklima. *Annls scient. Univ. Besançon, Géol., ser. 3*, **18**, 249.

Fischer, H. (1969) Geologischer Überblick über den südlichen Oberrheingraben und seine weitere Umgebung. *Regio basil.*, **10**, 57.

Fischer, K. (1977) Reliefgenerationen im Gebiet der Montes de Toledo, Zentralspanien. *Würzb. geogr. Arb.*, **45**, 69.

Fleet, H. (1938) Erosion surfaces in the Grampian Highlands of Scotland. *Rapp. Comm cartogr. des Surfaces d'Applanissement Tertiaires, Un. géogr. int.*, p.91 (Paris).

Flint, R. F. (1971) *Glacial and Quaternary Geology* (Wiley, New York).

Fogelberg, P. (1970) Geomorphology and deglaciation at the second Salpausselkä between Vääsky and Vierumäki, southern Finland. *Commentat. physico-math.*, **39**, 1.

Fontboté, J. M. (1954) Sobre la evolución tectónica de la depresión del Vallès-Penedés. *Arrahona* (Publ. Museo Sabadell, Madrid).

Fontboté, J. M. (1957) Tectoniques superposées dans la Sierra Nevada (Cordillères Bétiques, Espagne). *C. r. hebd. Séanc. Acad. Sci. Paris, ser. D.*, **245**, 1324.

Fontboté, J. M. and García-Dueñas, V. (1968) Essai de systématisation des unités subbétiques allochtones dans le tiers central des chaînes bétiques. *C.r. hebd. Séanc. Acad. Sci. Paris, ser. D*, **266**, 186.

Fontboté, J. M. and Guitard, G. (1958) Aperçus sur la tectonique cassante de la zone axiale des Pyrénées orientales entre les bassins de Cerdagne et de l'Ampurdan–Roussillon. *Bull. Soc. géol. Fr., ser. 6*, **8** (8), 884.

Fourniguet, J. (1975) *Néotectonique et Quaternaire Marin sur le Littoral de la Sierra Nevada, Andalousie (Espagne)*, thesis, University of Orléans.

Franceschetti, B. and Masone, G. (1967) Aspetti della degradazione accelerata nei dintorni di Pocapaglia in provincia di Cuneo. *Riv. geogr. ital.*, **74**, 435.

Franzle, O. (1959) Glaziale und periglaziale Formbildung in ostlichen Kastilischen Scheidegebirge (Zentralspanien). *Bonn. geogr. Abh.*, **26**, 1.

Frechen, J. von and Lippolt, H.J. (1965) Kalium–Argon-Daten zum Alter des Laacher Vulkanismus, der Rheinterrassen und die Eiszeiten. *Eiszeitalter Gegenw.*, **16**, 5.

Frödin, G. (1919) Jordskreden och markförskjutningarna i Göta Älvs dalgång mellan Trollhättan och Lilla Edet. *Meddn K. VattenfStyr.*, **19**.

Frutos-Mejias, L. M. (1968) Los glacis del Campo de Zaragoza. *Aportación Española XXI Congr. geogr. int.*, p.423.

Furrer, G. (1978) Fossile Böden in spät- und postglazialen Ablagerungen der Alpen. *Beitr. Quartär- Landschaftsforsch.*, p.177 (J. Fink Festschrift, Vienna).

Gabert, P. (1962) *Les Plaines Occidentales du Po et leurs Piémonts (Piémont, Lombardie Occidentale et Centrale). Étude morphologique* (Gap)

Gabilly, J. *et al.* (1978) Poitou, Vendée, Charentes, in *Guides Géologiques Régionaux*, ed. by C. Pomerol (Masson, Paris).

Gaibar, C. and Cuerda, J. (1969) Las playas del Cuaternario marino levantadas en el Cabo de Santa Pola (Alicante). *Boln Geol. Min.*, **80**, 105

Galibert, G. (1965) *La Haute Montagne Alpine*, thesis, University of Toulouse.

Gallart, F. (1980) *Estudi Geomorfología del Penedès, Sector Anoia i Riudebitlles*, unpublished thesis, University of Barcelona.

Gallegos, J. A. (1971) Una colada de gelivación en calizo-solomías: Sierra Nevada (Cordilleras Béticas). *Cuad. Geol.*, **2**, 31.

Galon, R. (1959) New investigations of inland dunes in Poland. *Przegl. Geogr.*, **31**, 93.

Galon, R. (1960) Problems of geomorphological classification of the Polish coast. *Przegl. Geogr.*, **32**, suppl., 67.

Galon, R. (1965) Some new problems concerning subglacial channels. *Geogr. polon.*, **6**, 19.

Galon, R. (1968) (ed.) *Ostatnie Zlodowacenie Skandynawskie w Polsce* (PWN, Warsaw).

Galon, R. (1982) Altes und Neues zum Problem der Entstehung der Durchbruchstäler im skandinavischen Vereisungsgebiet südlich der Ostsee. *Würzb. geogr. Arb.*, **56**, 159.

Galon, R. and Dylik, J. (1967) *The Quaternary of Poland* (PWN, Warsaw).

Galon, R. and Roszko, L. (1961) Extents of the Scandinavian glaciations and of their recession stages in Poland in the light of an analysis of the marginal forms of the inland ice. *Przegl. Geogr.*, **33**, 347.

Galpezin, E. N. and Cosminskaia, J. P. (eds.) (1964) *The Structure of the Earth's Crust in the Transition Zone from the Continent of Asia to the Pacific Ocean* (Nauka, Moscow).

Gams, I. (1974) *Kras* (Ljubljana).

Gams, I. (1981) Morfografski sistemi u Jugoslaviji. *Glasn. srp. geogr. Društ*, **61** (1).

García-Fernandez, J. (1968) Castilla la Vieja y León, in *Geografía Regional de España*, ed. by M. de Terán, p.100 (Ariel, Esplugues de Llobregat, Barcelona).

García-Rossell, L. (1973) *Estudio Geológico de la Transversal Ubeda–Huelma y Sectores Adyacentes (Cordilleras Béticas)*, unpublished thesis, University of Granada.

García-Rossell, L. and Pezzi, M. C. (1975) Un karst mediterráneo suproforestal en Sierra Mágina (Jaén). Condicionamientos geológicos y geomorfológicos. *Cuad. Geogr. Univ. Granada*, **1**, 19.

García-Rossell, L. and Pezzi, M. C. (1978) Análisis de depósitos periglaciares en el sector central de las Cordilleras Béticas (Andalucía). *Actas V Coloq. Geogr.*, p.99.

García-Ruiz, J. M. (1979) El glaciarismo cuaternario en la Sierra de la Demanda (Prov. de Logroño y Burgos, Espagna). *Cuad. Invest. geogr.*, **5**, 3.

García-Ruiz, J. M. and Arbella-Léon, M. (1981) Modelos de erosión en el piso supraforestal: la degradación de los loess del macizo de Monte Perdido (Pirineo Central español). *Pirineos*, **114**, 1.

García-Ruiz, J. M. and Puig de Fabregas, J. (1982) Formas de erosión en el flysch eoceno surpirenaico. *Cuad. Invest. Geogr. Hist., Logroño*, **7** (in press).

García-Sainz, L. (1943) El glaciarismo cuaternario de Sierra Nevada. *Actas II Reunión Est. Geogr.*, p. 153.

Gareckiy, P. G. (1972) *Tectonics of Young Platforms of Eurasia* (Nauka, Moscow).

Garrido, A. (1968) Sobre la estratigrafía de los conglomerados de Campanué (Sta Liestra) y las formaciones superiores del Eoceno (Extremo Occidental) de la cuenca de Graus-Tremp, Pirineo Central, Provinca de Huesca. *Acta geol. hisp.*, **3**, 39.

Garwood, E. J. (1910) Features of Alpine scenery due to glacial protection. *Geogrl J.*, **36**, 310.

Gaunt, G. D. (1976) The Devensian maximum ice limit in the Vale of York. *Proc. Yorks. geol. Soc.*, **40**, 631.

Gee, D. G. (1979) The Caledonides in Sweden. *Sver. geol. Unders. Afh., ser. C*, **769**.

Geikie, A. (1901) *The Scenery of Scotland Viewed in Connection with its Physical Geology*, 3rd edn. (London).

Gellert, J. F. (1931) Geomorphologische Studien und Probleme in Schwarzwald. *Ber. naturf. Ges. Freiburg*, **31** (1/2).

Gellert, J. F. (1958) *Grundzüge der Physischen Geographie von Deutschland* (Berlin).

Gellert, J. F. (ed.) (1965) *Die Weichsel-Eiszeit im Gebiet der Deutschen Demokratischen Republik* (Berlin).

Gellert, J. F. (1970) Climatomorphology and palaeoclimates of the Central European territory, in *Problems of Relief Planation*, p.107 (Akad. Kiado, Budapest).

Gellert, J. F. (1973) Die geomorphologische Generalkarte der DDR im Massstab 1:1,5 Millionen und deren Beziehung zur Internationalen Geomorphologischen Karte von Europa 1:2,5 Millionen. *Petermanns geogr. Mitt.*, **117**, 76.

Gentileschi, L. (1967) Forme crionivali sul Gran Sasso d'Italia. *Boll. Soc. geogr. ital., ser. IX*, **8**, 34.

Geographical Society of Finland (1960). *Atlas of Finland* (Geographical Society of Finland, Helsinki)

George, T. N. (1932) The Quaternary beaches of Gower. *Proc. Geol. Ass.*, **43**, 291.

George, T. N. (1955) Drainage in the Southern Uplands: Clyde, Nith, Annan. *Trans. geol. Soc. Glasgow*, **22**, 1.

George, T. N. (1961) The Welsh landscape. *Sci. Progr.*, **49**, 242.

George, T. N. (1965) The geological growth of Scotland, in *The Geology of Scotland*, ed. by G. Y. Craig, p.1 (Oliver & Boyd, Edinburgh).

George, T. N. (1966) Geomorphic evolution in Hebridean Scotland. *Scott. J. Geol.*, **2**, 1.

George, T. N. (1974) Prologue to a geomorphology of Britain. *Inst. Br. Geogr. Spec. Publ.*, **7**, 113.

Gerasimov, I. P. (1946) Attempt at a geomorphological interpretation of a general scheme of the geological structure of the USSR. *Izv. Akad. Nauk SSSR*, **12**, 33.

Gerasimov, I. P. (1959) Structural features of the relief of the Earth's crust in the USSR and their origins. *Izv. Akad. Nauk SSSR*, **25**, 99.

Gerasimov, I. P. and Metcherikov, J. A. (eds.) (1967) *Relief of the Earth: Morphostructure and Morphosculpture* (Nauka, Moscow).

Gerasimov, I. P. et al. (1974) *Ravniny Evropeyskoy Chasti SSSR* (Nauka, Moscow).

Gèze, B. (1979) Languedoc méditerranéen, Montagne Noire, in *Guides Géologiques Régionaux*, ed. by C. Pomerol (Masson, Paris).

Gèze, B. and Cavaillé, A. (1977) Aquitaine orientale, in *Guides Géologiques Régionaux*, ed. by C. Pomerol (Masson, Paris).

Ghelardoni, R. (1961) Evoluzione e spostamenti dello spartiacque appenninico tra Monte Fumaiolo e Gualdo Tadino. *Memorie Soc. tosc. Sci. nat.*, **A68**, 57.

Gignoux, M. and Fallot, P. (1927) Contribution à la connaissance des terrains néogènes et quaternaires marins sur les côtes méditerranéennes d'Espagne. *C.r. XIV Congr. Géol. int.*, **2**, 413.

Gigout, M. (1959) À propos du Quaternaire marin sur le littoral de la province de Tarragona (Espagne). *C.r. hebd. Séanc. Acad. Sci., Paris*, **249**, 2531.

Gigout, M. (1960) Cuaternario del litoral de las provincias del levante español. Cuaternario marino. *Notas Comun Inst. geol. min. Esp.*, **57**, 209.

Gigout, M. et al. (1955) Sur le Quaternaire méditerranéen de l'Andalousie. *C. r. Soc. géol. Fr.*, **9–10**, 177.

Gil-Crespo, A. (1964) Periglaciarismo en el macizo central de Gredos. *Boln R. Soc. geogr.*, **100**, 121.

Gill, W. D. (1962) The Variscan fold belt in Ireland, in *Some Aspects of the Variscan Fold Belt*, ed. by K. Coe, p.49 (Manchester University Press, Manchester).

Gillberg, G. (1955) Den glaciala utvecklingen inom Sydsvenska höglandets västra randzon. I. Glacialerosion och moränackumulation. *Geol. För. Stockh. Förh.*, **77** (4), 481.

Gjessing, J. (1960) Isavsmeltningstidens drenering, dens forløp og formdannende virkning i Nordre Atnedalen. Med sammenlignende studier fra Nordre Gudbrandsdalen og Nordre Østerdalen. *Ad Novas, Oslo*, **3**.

Gjessing, J. (1965–66) On plastic scouring and subglacial erosion. *Norsk geogr. Tidsskr.*, **20**, 1.

Gjessing, J. (1966) Some effects of ice erosion on the development of Norwegian valleys and fjords. *Norsk geogr. Tidsskr.*, **20**, 273.

Gjessing, J. (1967) Norway's Paleic surface. *Norsk geogr. Tidsskr.*, **21**, 69.

Gladfelter, G. B. (1971) Meseta and campiña landforms in central Spain; a geomorphology of the Alto Henares basin. *Univ. Chicago, Dept. Geogr. Res. Pap.*, **130**.

Glangeaud, L. et al. (1969) *Massif Central et Bordure Méditerranéenne* (International Quaternary Association, Paris).

Godard, A. (1965) *Recherches de Géomorphologie en Écosse du Nord-Ouest* (Belles Lettres, Paris).

Gómez de Llarena, J. (1916) Bosquejo geográfico–geológico de los Montes de Toledo. *Trab. Mus. Cienc. nat. Barcelona, Geol.*, **15**, 1.

Gómez de Llarena, J. and Royo-Gómez, J. (1927) Las terrazas rasas litorales de Asturias y Santander. *Boln R. Soc. esp. Hist. nat.*, **27**, 19.

Gómez-Ortiz, A. (1981) *Síntesis Geomorfológica del Alto Valle del Segre* (University of Barcelona, Barcelona).

Goncharov, V. P. et al. (1972) *Relief of the Floor and Deep Structures of the Black Sea Plain* (Nauka, Moscow).

Gonsalvi, L. and Papani, G. (1969) Alcune idee sull'evoluzione oro-idrografica dell'Appennino settentrionale. *Atenèo parmense, Acta nat.*, **5**, 29.

Gonzalo-Moreno, A. (1977–78) Los niveles de las terrazas del Ebro en La Rioja. *Geographica, Madrid*, **131**.

Gonzalo-Morena, A. (1979) Capturas y valles muertos en los cursos bajos de los ríos riojanos. *Cuad. Invest. Geogr. Hist., Logroño*, **5**, 27.

Goodchild, J. G. (1888–89) The history of the Eden and some rivers adjacent. *Trans. Cumb. Westmor. Ass.*, **14**, 73.

Goreckiy, G. I. (1958) Periglacial formations. *Izv. Kom. Izuch. Chetvertichn Perioda*, **22**, 3.

Goreckiy, G. I. (1964) *Alluvium of the Major Ancient Rivers of the Russian Lowlands. Ancient Rivers of the Kama Basin* (Nauka, Moscow).

Goreckiy, G. I. (1967) The origin and growth of deep valley-shaped depressions in the relief of the bed of anthropogenic deposits of glacial areas, in *Lower Pleistocene Glacial Regions of the Russian Lowlands*, p.17 (Nauka, Moscow).

Gorelov, S. K. (1972) *Morfostrukturnyi Analiz Neftegazonosnykh Territoriy* (Nauka, Moscow).

Gorin, B. A. (1952) Oscillations of the level of the Caspian Sea and mud volcanoes. *Izv. Akad. Nauk azerb. SSR*, **8**, 119.

Gornung, M. B. and Timofeev, D. A. (1958) Zonal features of the appearance of exogenic relief-forming processes, in *Voprosy Physicheskoi Geographii*, p.74 (Izd. Akad. Nauk, Moscow).

Goskar, K. L. and Trueman, A. E. (1934) The coastal plateaux of South Wales. *Geol. Mag.*, **71**, 468.

Gouvernet, C. et al. (1971) Provence, in *Guides Géologiques Régionaux*, ed. by C. Pomerol (Masson, Paris).

Goy, J. L. and Zazo, C. (1974) Estudio morfotectónico del Cuaternario en el óvalo de Valencia. *Acta I Reunion Grp Esp. Trab. Cuatern.*, p.71.

Gozálbez, V. and Cuerda, J. (1981) Los depósitos flandrienses en el litoral de Torrevieja (Alicante). *Actas V Reunión Grp Esp. Trab. Cuatern.*, p.87.

Granlund, E. (1943) Beskrivning till jordartskarta över Västerbottens län nedanför odlingsgränsen. Karta i skala 1:300000. *Sver. geol. Unders. Afh., ser. C*, **26**.

Grassi, D. (1974) Il carsismo nella Murgia (Puglia) e sue influenze sull'idrologia della Regione. *G. Geol. prat. appl.*, **9**, 119.

Green, C. P. (1974) The summit surface on the Wessex Chalk. *Inst. Br. Geogr. Spec. Publ.*, **7**, 127.

Green, C. P. and MacGregor, D. F. M. (1978) Pleistocene gravel trains of the River Thames. *Proc. Geol. Ass.*, **89**, 143.

Green, J. F. N. (1941) The high platforms of East Devon. *Proc. Geol. Ass.*, **52**, 36.

Greenly, E. (1938) The age of the mountains of Snowdonia. *Q. J. geol. Soc. Lond.*, **94**, 117.

Gregory, J. W. (1915) The Tweed valley and its relation to the Clyde and Solway. *Scott. geogr. Mag.*, **31**, 478.

Gregory, K. J. (1965) Proglacial Lake Eskdale after sixty years. *Trans. Inst. Br. Geogr.*, **36**, 149.

Gresswell, R. K. (1964) The origin of the Mersey and Dee Estuaries. *Geol. J.*, **4**, 77.

Gripp, K. (1952) Die Entstehung der Landschaft Ost-Holsteins. *Meyniana, Kiel*, **1**, 119.

Grosval'd, M. G. and Kotlyakov, V. M. (1969) Present-day glaciers in the USSR and some new data on their mass balance. *J. Glaciol.*, **8**, 9.

Gruppo di Studio delle Universita' Emiliane per la Geomorfologia (1976) Geomorfologia dell'area circostante la Pietra di Bismantova (Appennino Reggiano). *Boll. Serv. geol. ital.*, **97**, 107.

Gudelis, V. K. and Emelianov, E. M. (eds.) (1976) *The Geology of the Baltic Sea* (Nauka, Moscow).

Guerra, A. et al. (1962) *Estudio y Mapa de Suelos* (Cap. del Est. agrobiol. Prov. Sevilla, Seville).

Guidebooks of Excursions (1961) *6th Congr. int. Quatern. Ass., Warsaw* (PWN, Warsaw).

Guilcher, A. (1948) *Le Relief de la Bretagne Méridionale, de la Baie de Douarnenez à Vilaine* (Potier, La Roche-sur-Yon).

Guilcher, A. (1950) Nivation, cryoplanation et solifluction quaternaires dans les collines de Bretagne occidentale et du nord Devonshire. *Revue Géomorph. dyn.*, **1**, 55.

Guilcher, A. (1955) La plage ancienne de la Franca (Asturias). *C. r. hebd. Séanc. Acad. Sci., Paris*, **241**, 1603.

Guilcher, A. (1974) Les rasas, un problème de morphologie littorale générale. *Annls Géogr.*, **83**, 1.

Guilcher, A. and Giot, P. R. (1969) *Bretagne–Anjou* (International Quaternary Association, Paris).

Guilcher, A. *et al.* (1952) Formes de plage et houle sur le littoral des Landes de Gascogne. *Revue Géogr. Pyrénées S.-Ouest*, **23**, 99.

Guillien, Y. (1962) Néoglaciaire et Tardiglaciaire: géochimie, palynologie, préhistoire. *Annls Géogr.*, **71**, 1.

Gutiérrez-Elorza, M. and Peña-Monné, J. L. (1975) Karst y periglaciarismo en la Sierra de Javalambre (Prov. de Teruel). *Boln Geol. Min.*, **86**, 561.

Gutiérrez-Elorza, M. and Peña-Monné, J. L. (1976) Glacis y terrazas en el curso medio del río Alfambra (Prov. de Teruel). *Boln Geol. Min.*, **87**, 561.

Gutiérrez-Elorza, M. and Peña-Monné, J. L. (1977) Las acumulaciones periglaciares del macizo del Tremadal (Sierra de Albarracín). *Boln Geol. Min.*, **88**, 109.

Gutiérrez-Elorza, M. and Peña-Monné, J. L. (1979a) El karst de los llanos de Pozondón (Sierra de Albarracín). *Teruel*, **61–2**, 1.

Gutiérrez-Elorza, M. and Peña-Monné, J. L. (1979b) El karst de Villar del Cobo (Sierra de Albarracín). *Estudios geol.*, **35**, 651.

Gutiérrez-Elorza, M. and Peña-Monné, J. L. (1981) Los glaciares rocosos y el modelado acompanante en el área de la Bonaigua (Pirineo de Lérida). *Boln. Geol. Min.*, **92**, 101.

Gunther, E. (1941) Die quartären Niveauschwankungen im Mittelmeer, unter besonderer Berücksichtigung des Beckens von Alboran. *Jena. Z. Naturw.*, **74**, 1.

Gvozdeckiy, N. A. (1954) *Karst: Questions of General and Regional Karst Studies*, 2nd edn. (Geographgiz, Moscow).

Gvozdeckiy, N. A. and Maruashvili, L. I. (1977) Karst, in *General Description and History of the Development of the Relief of the Caucasus*, p.188 (Nauka, Moscow).

Gwinner, M. P. (1978) *Geologie der Alpen: Stratigraphie, Paläogeographie, Tektonik* (Stuttgart).

Haeberli, W. *et al.* (1976) Der Seewener See: Refraktionsseismische Untersuchung an einem spätglazialen bis frühholozänen Bergsturz-Stausee im Jura. *Regio basil.*, **17**, 133.

Hageman, B. P. (1969) Development of the western part of the Netherlands during the Holocene. *Geologie Mijnb.*, **48**, 373.

Hanne, C. (1930) Stratigraphische und tectonische Untersuchungen in der Provinzen Teruel, Catellon und Tarragona, Spanien. *Z. dt. geol. Ges.*, **82**, 79.

Hanns, C. (1980) Formenschatz und mutmassliches Alter einer spätglazialen Moränenabfolge im inneralpinen Trockengebiet der Vanoise (französische Nordalpen). *Tübinger geogr. Stud.*, **80**, 177.

Hansen, K. (1970) Tunnel valleys in Denmark and northern Germany. *Meddr dansk geol. Foren.*, **20**, 295.

Hansen, S. (1965) The Quaternary of Denmark, in *The Quaternary*, vol. 1, ed. by K. Rankama, p. 1 (Interscience, New York).

Hantke, R. (1978) *Eiszeitalter*, vol. 1. *Die jüngste Erdgeschichte der Schweiz und ihrer Nachbargebiete. Klima, Flora, Fauna, Mensch, Alt- und Mittelpleistozän, Vogesen, Schwarzwald, Schwäbische Alb, Adelegg* (Ott, Thun).

Heezen, B. C. and Hollister, C. D. (1972) *The Face of the Deep* (Oxford University Press, Oxford).

Heezen, B. C. and Johnson, G. L. (1969) Mediterranean undercurrent and microphysiography west of Gibraltar. *Bull. Inst. oceanogr. Monaco*, **67**, 1382.

Heezen, B. C. *et al.* (1959) The floors of the oceans. I. North Atlantic. *Geol. Soc. Am. Spec. Pap.*, **65**, 122.

Heim, A. (1919) *Geologie der Schweiz*, vol. 1, Molasseland und Juragebirge. (Tauchnitz, Leipzig).

Heim, A. (1921) *Die Schweizer Alpen—Erste Hälfte* (Tauchnitz, Leipzig).

Heim, A. (1922) *Die Schweizer Alpen—Zweite Hälfte* (Tauchnitz, Leipzig).

Helland, A. (1879) Über die glazialen Bildungen der nordeuropäischen Ebene. *Z. dt. geol. Ges.*, **31**, 123.

Hempel, L. (1958a) Studien in norddeutschen Buntsandsteinlandschaften. *Forschn dt. Landes- u. Volksk.*, **112**, 1.

Hempel, L. (1958b) Zur geomorphologischen Höhenstufung der Sierra Nevada spaniens. Ein Beitrag zur klimamorphologischen Zonierung der Erde. *Erdkunde*, **12**, 270.

Herak, M. and Stringfield, V. (1972) *Karst. Important Karst Regions of the Northern Hemisphere* (Elsevier, London).

Hernández-Pacheco, E. (1928) Los cinco principales ríos de España y sus terrazas. *Trab. Mus. Cienc. nat. Barcelona, ser. Geol.*

Hernández-Pacheco, E. (1932a) Síntesis fisiográfica y geológica de España. *Trab. Mus. Cienc. nat. Barcelona, ser. Geol.*, **38**.

Hernández-Pacheco, E. (1932b) La región volcánica de Ciudad Real. *Boln R. Soc. Geogr.*, **72**, 131, 145.

Hernández-Pacheco, E. (1933) El glaciarismo cuaternario en la Serrota (Avila). *Boln R. Soc. esp. Hist. nat.*, **33**, 417.

Hernández-Pacheco, E. (1946) Los materiales terciarios y cuaternarios en los alrededores de Toledo. *Estudios geol.*, **23**, 225.

Hernández-Pacheco, E. (1947) Ensayo de morfogénesis de la Extremadura Central. *Notas Comun. Inst. geol. min. Esp.*, **17**.

Hernández-Pacheco, E. (1949a) Geomorfologia de la cuenca media del Sil. *Mems R. Acad. Cienc. exact. fis. nat. Madr.*, **12**.

Hernández-Pacheco, E. (1949b) Las rasas litorales de la costa cantabrica en su segmento asturiano. *C.r. XVI Congr. int. géogr.*, vol. II, p.29.

Hernández-Pacheco, E. (1949c) Las rañas de las sierras centrales de Extremadura. *C. r. Congr. int. géogr.*, vol. II, p.87.

Hernández-Pacheco, E. and Asensio-Amor, I. (1959) Materiales sedimentarios sobre la rasa cantábrica. I. Tramo comprendido entre las rías del Eo y Foz. *Boln R. Soc. esp. Hist. nat.*, **75**.

Hernández-Pacheco, E. and Asensio-Amor, I. (1960) Materiales sedimentarios sobre la rasa cantábrica. II. Tramo comprendido entre la ría de Foz y el casco urbano de Burela. *Boln R. Soc. esp. Hist. nat.*, **73**.

Hernández-Pacheco, E. and Asensio-Amor, I. (1961) Materiales sedimentarios sobre la rasa cantábrica. III. Tramo comprendido entre Santiago de Villapedre (Navia) y Cadaveo (Luarca). *Boln R. Soc. esp. Hist. nat.*, **207**.

Herranz, A. and Carreras, A. (1966) *Observaciones Sobre el Macizo Kárstico de Maboré (Pirineo–Huesca–España)* (Cent. Estud. Hidrográf., Madrid).

Herzog, P. (1956) Die Tektonik des Tafeljura und der Rheintalflexur südöstlich von Basel. *Eclog. geol. Helv.*, **49**, 317.

Hill, M. N. (ed.) (1962, 1963) *The Sea. Ideas and Observations on Progress in the Study of the Seas* (Interscience, New York), 3 vols.

Hilly, J. *et al.* (1979) Lorraine–Champagne, in *Guides Géologiques Régionaux*, ed. by C. Pomerol (Masson, Paris).

Hobbs, W. H. (1911) *Earth Features and Their Meaning* (Macmillan, New York).

Hofstein, I. L (1964) *Neotectonics of the Carpathians* (Akad. Nauk, Kiev).

Högbom, B. (1914) Über die geologische Bedeutung des Frostes. *Bull. geol. Instn Univ. Uppsala*, **12**, 257.

Högbom, B. (1923) Ancient inland dunes of northern and middle Europe. *Geogr. Annlr*, **5**, 113.

Höllermann, P. W. (1977) Zur Verbreitung rezenter periglazialer Kleinformen in den Pyrenäen und Ostalpen. *Göttinger geogr. Abh.*, **40**, 1.

Hollingworth, S. E. (1931) The glaciation of western Edenside and adjoining areas, and the drumlins of Edenside and the Solway Basin. *Q. J. geol. Soc. Lond.*, **87**, 281.

Holmes, A. (1965) *Principles of Physical Geology* (Nelson, London).

Holtedahl, H. (1949) Geomorphology and Quaternary geology of the Opdal-Sunndal area, south-western Norway. *Univ. Bergen Årb., naturvitensk.*, **2**, 1.

Holtedahl, O. (1960) Geology of Norway. *Norg. geol. Unders. Afh.*, **208**.

Hoppe, G. (1948) Isrecessionen från Norrbottens kustland i belysning av de glaciala formelementen. *Geographica*, **20**.

Hoppe, G. (1952) Hummocky moraine regions with special reference to the interior of Norrbotten. *Geogr. Annlr*, **34**, 1.

Hoppe, G. (1959) Några kritiska kommentarer till diskussionen om isfria refugier. *Svensk Naturv.*, p.123.

Hoppe, G. (1967) Case studies of deglaciation patterns. *Geogr. Annlr*, A**49**, 204.

Hoppe, G. (1972) Ice sheets around the Norwegian Sea during the Würm glaciation. *K. svenska Vetensk-Akad. Handl.*, **2**, 25.

Hoppe, G. and Liljequist, G. H. (1956) Det sista nedisningsförloppet i Nordeuropa och dess meteorologiska bakgrund. *Ymer*, **76**, 43.

Hörner, N. (1927) Brattforsheden. Ett värmländskt randdeltakomplex och dess dyner. *Sver. geol. Unders. Afh., ser. C*, **342**.

Hörner, N. G. and Järnefors, B. (1956) Jordartskarta över Uppsalatrakten skala 1:20 000. *Sver. geol. Unders. Afh., ser. B*, **15**.

Hoshino, M. (1978) Origin of trenches. *La Mer*, **16**, 3.

Hoyos, M. A. *et al.* (1974) Las terrazas del río Duero desde Gormaz hasta Peñafiel. *An. Edafol. Agronom.*, **33**, 185.

Hsu, K. (1976) Origin of saline geants: a critical review after the discovery of the Mediterranean evaporites. *Earth Sci. Rev.*, **8**, 371.

Hsu, K. *et al.* (1978) Introduction and explanatory notes. *Initial Reports, Deep Sea Drilling Project*, p.42 (US Government Printing Office, Washington).

Hudson, R. G. S. (1933) The scenery and geology of north-west Yorkshire. *Proc. Geol. Ass.*, **44**, 228.

Huguet del Villar, E. (1915) Los glaciares de Gredos. *Boln R. Soc. esp. Hist. nat.*, **15**, 379.

Ibáñez-Marcellán, M. J. (1975) El endorreismo del sector de la depresión del Ebro. *Cuad. Invest. Geogr. Hist. Logroño*, **1**, 35.

Ibáñez-Marcellán, M. J. (1976) *El Piedemonte Ibérico Bajoaragonés* (Instituto Geografía Aplicada, Zaragoza).

Ibáñez-Marcellán, M. J. and Ménsua, S. (1976) Contribución al estudio de vertientes en condiciones semiáridas: tipos de vertientes sobre yesos en el Valle del Ebro. *Boln R. Soc. geogr.*, **112**, 381.

Ibáñez-Marcellán, M. J. and Ménsua, S. (1977) Los valles asimétricos de la orilla derecha del Ebro. *Actas II Reunión Grp esp. Trab. Cuatern.*, p.113.

IGME (1969) *Mapa Geológico Nacional, 1:200000* (Instituto Geológico y Minero de España, Madrid).

Illies, J. H. (1975) The rift zones of Western Europe and the Alpine system, in *Symposium on Rift Zones of the Earth*, p.75 (Irkutsk).

Ilyin, A. V. (1976) *Geomorphology of the Atlantic Ocean* (Nauka, Moscow).

Imperatori, L. (1955) Documentos para el estudio del Cuaternario madrileño. *Estudios geol.*, **26**, 139.

Imperatori, L. (1957) Documentos para el estudio del Cuaternario alicatino. *Estudios geol.*, **34**, 141.

Initial core descriptions, Deep Sea Drilling Project (1973) leg. 48, 217; leg. 49, 1; leg. 50, 102 (US Government Printing Office, Washington).

Initial Reports, Deep Sea Drilling Project (1976–78), vols. 33 and 42 (US Government Printing Office, Washington).

Ivanov, S. N. *et al.* (1975) Fundamental features in the structure and evolution of the Urals. *Am. J. Sci.*, **275A**, 107.

Jacoby, W. R. (1979) Iceland and the North Atlantic. *Geo. Jnl*, **3**, 3.

Jaffrezo, M. *et al.* (1977) Pyrénées orientales, Corbières, in *Guides Géologiques Régionaux*, ed. by C. Pomerol (Masson, Paris).

Jahn, A. (1980) Main features of the Tertiary relief of the Sudetes Mountains. *Geographica polon.*, **43**, 5.

Jamagne, M. (1973) *Contribution à l'Etude Pédogénétique des Formations Loessiques du Nord de la France*, thesis, University of Gembloux.

Jelgersma, S. *et al.* (1970) The coastal dunes of the western Netherlands: geology, vegetational history and archaeology. *Meded. Rijks geol. Dienst.*, **21**, 93.

Jensen, R. H. and Jensen, K. M. (1976) *Topografisk Atlas Denmark* (K. Danske Geogr. Selsk., Copenhagen).

Jessen, K. *et al.* (1959) The interglacial deposits near Gort, Co. Galway, Ireland. *Proc. R. Ir. Acad.*, **60B**, 1.

John, B. S. (1967) Further evidence for a Middle Würm interstadial and a Main Würm glaciation of south-west Wales. *Geol. Mag.*, **104**, 630.

John, D. T. (1977) The soils and superficial deposits on the North Downs of Surrey. *Kingston Poly. School Geogr. Occas. Pap.*, **4**.

Johnson, R. H. and Rice, R. J. (1961) Denudation chronology of the south-west Pennine upland. *Proc. Geol. Ass.*, **72**, 21.

Johnsson, G. (1956) Glacialmorfologiska studier i södra Sverige. Med särskild hänsyn till glaciala riktningselement och periglaciala frostfenomen. *Meddn Lunds Univ. geogr. Instn*, **31**.

Joly, F. (1939) Le littoral du Cotentin. *Annls Géogr.*, **48**, 225.

Jones, D. K. C. (1974) The influence of the Calabrian transgression on the drainage evolution of south-east England. *Inst. Br. Geogr. Spec. Publ.*, **7**, 139.

Jones, D. K. C. (1981) *South-East and Southern England* (Methuen, London).

Jones, O. T. (1931) Some episodes in the geological history of the Bristol Channel region. *Rep. Br. Ass. Advmt Sci*, p.57.

Jones, O. T. (1951) The drainage system of Wales and the adjacent regions. *Q. J. geol. Soc. Lond.*, **107**, 201.

Journaux, A. (1956) *Les Plaines de la Saône et leurs Bordures Montagneuses. Etude Morphologique* (Caron, Caen).

Jukes, J. B. (1862) On the mode of formation of some of the river-valleys in the south of Ireland. *Q. J. geol. Soc. Lond.*, **18**, 378.

Julivert, M. (1954) Observaciones sobre la tectónica de la depresión de Calatayud. *Arrahona* (Publ. Mus. Sabadell, Madrid).

Julivert, M. *et al.* (1974) *Memoria Explicativa del Mapa Tectónico de la Península Ibérica y Baleares* (Instituto Geológica y Minero de España, Madrid).

Kaitanen, V. (1969) A geographical study of the morphogenesis of northern Lapland. *Fennia*, **99** (5).

Kaleckaya, M. S. (1974) *Mountain Areas of the European Part of the USSR and Caucasus*, p.227 (Nauka, Moscow).

Kalinko, M. G. (1977) *Methods of Comparative Evaluation of the Oil–Gas Bearing Capacities of Water Areas* (Nedra, Moscow).

Karlén, W. (1973) Holocene glacier and climate variations, Kebnekaise mountains, Swedish Lapland. *Geogr. Annlr*, **A55**, 29.

Karlén, W. (1975)) Lichenometrisk datering i norra Skandinavien—metodens tillförlitlighet och regionala tillämpning. *Dept Phys. Geogr. Univ. Stockh. Rep.*, **22**.

Karlén, W. (1976) Lacustrine sediments and tree-ring variations as indicators of Holocene climatic fluctuations in Lappland, northern Sweden. *Geogr. Annlr*, **A58**, 1.

Kelletat, D. (1969) Verbreitung und Vergesellschaftung rezenter Periglazialerscheinung in Appennin. *Göttinger geogr. Abh.*, **48**.

Kendall, P. F. (1902) A system of glacier lakes in the Cleveland Hills. *Q. J. geol. Soc. Lond.*, **58**, 471.

Kent, P. E. (1949) A structure contour map of the surface of the buried pre-Permian rocks of England and Wales. *Proc. Geol. Ass.*, **60**, 87.

Khain, V. E. (1964) *General Geotectonics* (Nedra, Moscow).

Khain, V. E. (1975) Structure and main stages in the tectono-magmatic evolution of the Caucasus: an attempt at geodynamic interpretation. *Am. J. Sci.*, **275A**, 131.

Khain, V. E. (1977a) *Regional Tectonics, Non-Alpine Europe and Western Asia* (Nedra, Moscow).

Khain, V. E. (1977b) The new international tectonic map of Europe and some problems of structure and tectonic history of the continent, in *Europe from Crust to Core*, ed. by D. V. Ager and M. Brooks, p. 19 (John Wiley, London).

Khain, V. E. (1979) *Regional Geotectonics* (Nedra, Moscow).

King, C. A. M. (1959) *Beaches and Coasts* (Edward Arnold, London).

King, C. A. M. (1960) *The Yorkshire Dales* (Geographical Association, Sheffield).

King, C. A. M. and Gage, M. (1961) The extent of glaciation in part of west Kerry. *Ir. Geogr.*, **4**, 202.

King, L. C. (1962) *Morphology of the Earth* (Oliver & Boyd, Edinburgh).

Kinzl, H. (1932) Die grössten nacheiszeitlichen Gletschervorstösse in den Schweizer Alpen und in der Mont Blanc-Gruppe. *Z. Gletscherk.*, **20**, 269.

Kjartansson, G. (1956) Ur sögu bergs og landlags. *Náttúrufraedhingurinn*, **26**, 3.

Kjartansson, G. (1966) Stapakenningin og Surtsey. *Náttúrufraedhingurinn*, **36**, 1.

Klaer, W. (1956) Verwitterungsformen im Granit auf Korsika. *Petermanns geogr. Mitt., Erg.*, **261**, 1.

Klebelsberg, R. von (1928) Beiträge zur Geologie der Sierren zwischen Granada und Malaga (Andalusien). *Z. dt. geol. Ges.*, **80**, 535.

Klein, C. (1974) *Massif Armoricain et Bassin Parisien*, thesis, Paris IV (University of Strasbourg, Fond. Baulig, Strasbourg).

Klenova, M. V. and Lavrov, V. M. (1975) *Geology of the Atlantic Ocean* (Nauka, Moscow).

Klimaszewski, M. (1965) Views on the geomorphological evolution of the Polish West Carpathians in Tertiary times, in *Geomorphological Problems of the Carpathians.* vol. I. *Evolution of the Relief in the Tertiary*, p.91 (Vydavatelstvo Slovenskej Akad. Vied, Bratislava).

Klimaszewski, M. (1972) *Geomorfologia Polski* (PWN, Warsaw).

Kober, L. (1931) *Das Alpine Europa* (Bornträger, Berlin).

Kober, L. (1955) *Bau und Entstehung der Alpen* (Dendicke, Vienna).

Kondracki, J. (1978) *Geografia Fizyczna Polski* (PWN, Warsaw).

Korsakov, O. D. (1979) *Geological Structure of the Norwegian Sea in Connection with the Problems of Oil and Gas Content* (Odessa).

Koutaniemi, L. (1979) Late-glacial and post-glacial development of the valleys of the Oulanka river basin, north-eastern Finland. *Fennia*, **157** (1), 13.

Kozarski, S. (1966–67) Origin of the subglacial channels in the North Polish and North German plain. *Bull. Soc. Amis Sci. Lett. Poznań*, **20**, 21.

Kozarski, S. (1978) Lithologie und Genese des Endomoränen im Gebiet der skandinavischen Vereisungen. *SchrReihe geol. Wiss. Berlin*, **9**, 179.

Krähahn, G. (1964) Die Grossformen des Thüringer Waldes. *Geogr. Ber.*, **9**, 107.

Krasheninnikov, I. M. (1939) Development of the vegetation of the Southern Ural in the Pleistocene and Holocene. *Sov. Bot.*, **6–7**, 67.

Krasniy, L. T. (1977) *Problems of Tectonic Systematization* (Nedra, Moscow).

Kratz, K. O. (1963) Geology of the Karelides of Karelia, in *Studies in the Geology of the Precambrian Period*, p.16 (Izd. Akad. Nauk, Moscow).

Krausse, H. F. (1974) The tectonic evolution of the Western Pyrenees. *Pirineos*, **111**, 69.

Krilov, S. V. *et al.* (1967) Seismic depth research in the area of accretion of the West Siberian plate and the Siberian platform. *Geologiya Geofiz. Novosibirsk*, **2**.

Ksiazkiewicz, M. *et al.* (1965) *Zarys Geologii Polski* (Wydawnictwo Geol., Warsaw).

Kühne, U. (1974) Zur Stratifizierung und Gliederung quartärer Akkumulationen aus dem Bièvre-Valloire, einschliesslich der Schotterkörper zwischen St Rambert d'Albon und der Enge von Vienne. *Heidelb. geogr. Arb.*, **39**, 1.

Kuujansuu, R. (1967) On the deglaciation of western Finnish Lapland. *Bull. Commn. géol. Finl.*, **232**.

Kvasov, D. D. (1964) The hydrology of the Middle Pliocene Caspian Sea. *Izv. Akad. Nauk SSSR*, **158**, 352.

Kvasov, D. D. (1966) The water balance of the Middle Pliocene Caspian Sea. *Byull. Moih. Otd. Geol.*, **12**, 99.

Labhart, T. P. (1977) Aarmassiv und Gotthardmassiv. *Samml. geol. Führ.*, **63**, 1.

Lake, P. (1934) The rivers of Wales and their connection with the Thames. *Sci. Progr.*, **29**, 25.

Larsen, B. T. (1975) Petrological, geochemical and volcanological features of the volcanic rocks in the Permian Oslo palaeorift zone, in *Symposium on Rift Zones of the Earth*, p.78 (Irkutsk).

Larsson, I. (1954) Structure and landscape in western Blekinge, south-east Sweden. *Lund Stud. Geogr.*, **A7**.

Laubscher, H. P. (1961) Die Fernschubhypothese der Jurafaltung. *Eclog. geol. Helv.*, **54**, 221.

Laubscher, H. P. (1962) Die Zweiphasenhypothese der Jurafaltung. *Eclog. geol. Helv.*, **55**, 1.

Laubscher, H. P. (1965) Ein kinematisches Modell der Jurafaltung. *Eclog. geol. Helv.*, **58**, 231.

Laughton, A. S. (1977) Deep drill holes in the ocean floor. *Geogrl J.*, **143**, 235.

Lautensach, H. (1949) Granitische Abtragungsformen auf der Iberischen Halbinsel und in Korea; ein Vergleich. *C. r. Congr. int. géogr.*, p.270.

Lautensach, H. and Mayer, E. (1961) Iberian Meseta und Iberian Masse (Iberische Meseta und Iberische Masse). *Z. Geomorph.*, **5**, 161.

Lavrov, V. M. (1979) *Geology of the Mid-Atlantic Ridge* (Nauka, Moscow).

Lazarevic, R. (1975) *Geomorfologija*, p.451 (Belgrade).

Lazaro, I. (1977) Estudio geomorfológico de la cuenca del río Guadalix (evolución Neógeno-Cuaternaria). *Estudios geol.*, **33**, 101.

Lebedev, L. J. *et al.* (1976) *Geological Structure and Oil–Gas Bearing Platform Region of the Caspian Sea* (Nauka, Moscow).

Leguey, S. and Rodriguez, J. (1970) Estudio de las terrazas y sedimentos de los ríos de la cuenca del Esla. *Boln R. Soc. esp. Hist. nat., ser. Geol.*, **68**, 41.

Lehmann, H. (1959) Studien über Poljen in den Venetianischen Voralpen und in Hochapennin. *Erdkunde*, **13**, 258.

Lembke, H. (1931) Beiträge zur Geomorphologie des Aspromonte (Kalabrien). *Z. Geomorph.*, **6**, 58.

Leonardi, P. (1967) *Le Dolomiti, Geologia dei Monti tra Isarco e Piave* (Manfrini, Rovereto).

Leont'ev, O. K. (1955) *Geomorphology of Sea Shores and Floors* (Moscow State University, Moscow).

Leont'ev, O. K. (1961) *Principles of Geomorphology of Sea Shores* (Moscow State University, Moscow).

Leont'ev, O. K. (1964) Relief and geological structure of the Caspian Sea floor. *Vestnik mosk. gos. Univ., ser. Geogr.*, **5**, 27.

Leont'ev, O. K. (1965) *Geomorphology of the World Ocean Floor* (Moscow State University, Moscow).

Leont'ev, O. K. (1968a) Geomorphological types of transition zone. *Vestnik mosk. gos. Univ., ser. Geogr.*, **2**, 152.

Leont'ev, O. K. (1968b) *The Floor of the Oceans* (Mysl., Moscow).

Leont'ev, O. K. (1971) Types of planetary morphostructure of the Earth and some features of their dynamics in the Cenozoic. *Geomorphologiya*, **3**, 3.

Leont'ev, O. K. (1973) The problem of the origin of the oceans. *Vestnik mosk. gos. Univ., ser. Geogr.*, **6**, 21.

Leont'ev, O. K. (1975) Concerning the peculiarity of the distribution of deep thermal streams in the oceans. *Vestnik mosk. gos. Univ., ser. Geogr.*, **1**, 42.

Leont'ev, O. K. (1977) Relief-forming activity of currents in the deep ocean zones. *Geomorphologiya*, **2**, 3.

Leont'ev, O. K. (1980) On the evolution of deep-sea trenches, in *Contemporary Problems of Marine Geology*, vol. 3, p.147 (Institute of Oceanology, Moscow).

Le Pichon, K. *et al.* (1977) *The Tectonics of Plates* (Mir, Moscow).

Letzer, J. M. (1978) *The Glacial Geomorphology of the Region Bounded by Shap Fell, Stainmore and the Howgill Fells in East Cumbria*, thesis, University of London.

Lewis, C. A. (1970) *The Glaciations of Wales and Adjoining Regions* (Longman, London).

Lewis, C. A. (1974) The glaciations of the Dingle peninsula, County Kerry. *Scient. Proc. R. Dubl. Soc.*, A5, 207.

Lewis, H. C. (1894) *The Glacial Geology of Great Britain and Ireland* (Longman, London).

Lewis, W. V. (1932) The formation of Dungeness Foreland. *Geogrl J.*, **80**, 309.

Lhénaff, R. (1966) Existence d'un haut niveau marin (Pliocène terminal ou Quaternaire ancien) déformé à l'ouest de Malaga (Espagne). *C. r. somm. Séanc. Soc. géol. Fr.*, **10**, 395.

Lhénaff, R. (1968) Le poljé de Zafarraya (Province de Granada). *Mélang. Casa Velazquez*, **4**, 5.

Lhénaff, R. (1973) Estudio geomorfológico del Valle de Lecrín. *Estudios geogr.*, **132–133**, 539.

Lhénaff, R. (1975) Les poljés ouverts de la Sierra de Cabra (Cordillères Bétiques). *Cuad. Geogr., ser. Monogr.*, **1**, 85.

Lhénaff, R. (1981) *Recherches Géomorphologiques sur les Cordillères Bétiques Centro-Occidentales (Espagne)*, thesis, University of Lille.

Lichtenecker, N. (1925) Das Bewegungsbild der Ostalpen. *Naturwissenschaften*, **13**, 739.

Lidmar-Bergström, K. (1982) Pre-Quaternary geomorphological evolution in southern Fennoscandia. *Sver. geol. Unders. Afh.*, C785.

Liedtke, H. (1975) *Die Nordischen Vereisungen in Mitteleuropa. Erläuterungen zu Einer Farbigen Übersichtskarte im Massstab, 1:1000000*. (Bonn, Bad Godesberg).

Lilienberg, D. A. (1955a) Forms related to subterranean erosion (clay pseudo-karst) in the south-eastern Caucasus. *Trudÿ Inst. Geogr. Akad. Nauk SSSR*, **65**; *Information on the Geomorphology and Palaeogeography of the USSR*, vol. 14, p.147 (Akad. Nauk SSSR, Moscow).

Lilienberg, D. A. (1955b) The morphology of mud volcanoes in the southeastern Caucasus. *Trudÿ Inst. Geogr. Akad. Nauk SSSR*, **65**; *Information on the Geomorphology and Palaeogeography of the USSR*, vol. 14, p.173 (Akad. Nauk SSSR, Moscow).

Lilienberg, D. A. (1962) *The Relief of the Southern Slope of the Eastern Part of the Great Caucasus* (Akad. Nauk, Moscow).

Lilienberg, D. A. *et al.* (1977) Arid-denuded relief, in *General Description and History of the Development of the Relief of the Caucasus*, p.196 (Nauka, Moscow).

Lilienberg, D. A. and Shirinov, N. S. (1977) Contemporary tectonic movements, in *General Description and History of the Development of the Relief of the Caucasus*, p.49 (Nauka, Moscow).

Liniger, H. (1966) Das plio-altpleistozäne Flussnetz der Nordschweiz. *Regio basil.*, **7**, 158.

Liniger, H. and Rothpletz, W. (1964) Ein neuer Aufschluss in den Vogesenschotten westlich Delsberg. *Regio basil.*, **5**, 78.

Linton, D. L. (1933) The origin of the Tweed drainage system. *Scott. geogr. Mag.*, **49**, 162.

Linton, D. L. (1934) On the former connection between the Clyde and the Tweed. *Scott. geogr. Mag.*, **50**, 82.

Linton, D. L. (1940) Some aspects of the evolution of the rivers Earn and Tay. *Scott. geogr. Mag.*, **56**, 1, 69.

Linton, D. L. (1951a) Problems of Scottish scenery. *Scott. geogr. Mag.*, **67**, 65.

Linton, D. L. (1951b) Midland drainage: some considerations bearing on its origin. *Advmt Sci.*, **7**, 449.

Linton, D. L. (1951c) Watershed breaching by ice in Scotland. *Trans. Inst. Br. Geogr.*, **15**, 1.

Linton, D. L. (1954) The landforms of Lincolnshire. *Geography*, **39**, 67.

Linton, D. L. (1955) The problem of tors. *Geogrl J.*, **121**, 470.

Linton, D. L. (1956) Geomorphology of the Sheffield region, in *Sheffield and its Region*, ed. by D. L. Linton, p.24 (British Association for the Advancement of Science, London).

Linton, D. L. (1957) Radiating valleys in glaciated lands. *Tijdschr. K. ned. aardrijksk. Genoot.*, **74**, 297.

Linton, D. L. (1959) Morphological contrasts between eastern and western Scotland, in *Geographical Essays in Memory of Alan G. Ogilvie*, ed. by R. Miller and J. W. Watson, p.16 (Nelson, Edinburgh).

Linton, D. L. (1962) Glacial erosion on soft-rock outcrops in central Scotland. *Biul. peryglac.*, **11**, 247.

Linton, D. L. (1963) The forms of glacial erosion. *Trans. Inst. Br. Geogr.*, **33**, 1.

Linton, D. L. (1964) Tertiary landscape evolution, in *The British Isles*, ed. by J. W. Watson and J. B. Sissons, p.110 (Nelson, London).

Linton, D. L. and Moisley, H. A. (1960) The origin of Loch Lomond. *Scott. geogr. Mag.*, **76**, 26.

Lisizyn, A. P. (1978) *Sediment Formation in the Oceans* (Nauka, Moscow).

Ljungner, E. (1927–30) Spaltentektonik und Morphologie der schwedischen Skagerrak-Küste. *Bull. geol. Instn Univ. Upsala*, **21**, 1.

Ljungner, E. (1949) East–west balance of the Quaternary ice caps in Patagonia and Scandinavia. *Bull. geol. Instn Univ. Upsala*, **33**, 11.

Llobet, S. (1975a) Noticia de soliflusión periglaciar en Cataluña. *Estudios geogr.*, **140–141**, 661.

Llobet, S. (1975b) Materiales y depósitos periglaciares en el macizo del Montseny. Antecedentes y resultados. *Revta Geogr.*, **9**, 35.

Llopis, N. (1934) Geologia de la Costa Brava. *Butll. Cent. Estud. Min.*, **139**, 1.

Llopis, N. (1941) El macizo cárstico de La Morella. *Estudios geogr.*, **4**, 413.

Llopis, N. (1947) Contribución al conocimiento de la morfoestructura de los Catalánides. *Publnes Inst. Lucas Mallada, Cons. Sup. Invest. cient.*

Llopis, N. (1950) Rasgos morfológicos y geológicos de la cordillera cántabroastúrica. *Trab. Mems Inst. Geol., Oviedo*, **9**, 51.

Llopis, N. (1954) El relieve de la región central de Asturias. *Estudios geogr.*, **15**, 501.

Llopis, N. (1956) Los depósitos de la costa cantábrica entre los Cabos Busto y Vidio (Asturias). *Speleon*, **6**, 333.

Llopis, N. *et al.* (1953) *Memoria Explicativa Mapa Geológico de España (1:50000)*, no. 366, (Instituto Geológica y Minero de España, Madrid).

Longinov, V. V. (1973) *Studies in the Lithodynamics of the Ocean* (Nauka, Moscow).

López-Bermúdez, F. (1973) La Vega Alta del Segura (clima, hidrología y geomorfología). *Publnes Depto Geogr. Univ. Murcia*.

López-Gomez, A. and Riba, O. (1957) El glaciarismo cuaternario en la sierra de Neila (Prov. de Burgos). *C. r. V Congr. int. Ass. Quatern. Res.*, p.109.

Losacco, U. (1949) La glaciazione quaternaria dell'Appennino Settentrionale. *Riv. geogr. ital.*, **56**, 90, 196.

Lotze, F. (1945) Zur Gliederung der Varisziden der Iberischen Meseta. *Geotekt. Forsch.*, **6**, 78.

Lotze, F. (1962) Pleistozäne Vergletscherungen im Ostteil des Kantabrischen Gebirges (Spanien). *Abh. math.-naturw. Kl. Akad. Wiss. Mainz*, **2**, 1.

Louis, H. (1968) *Allgemeine Geomorphologie* (de Gruyter, Berlin).

Loveday, J. (1962) Plateau deposits of the southern Chiltern Hills. *Proc. Geol. Ass.*, **73**, 83.

Lundqvist, G. (1951) Beskrivning till jordartskarta över Kopparbergs län. Karta i skala 1:250000. *Sver. geol. Unders. Afh., ser. Ca*, **21**.

Lundqvist, G. (1958) Beskrivning till jordartskarta över Sverige. Med en karta i tre blad i skalan 1:1000000. *Sver. geol. Unders. Afh., ser. Ba*, **17**.

Lundqvist, G. (1961) Beskrivning till karta över landisens avsmältning och högsta kustlinjen i Sverige. Med en karta i tre blad i skala 1:1000000. *Sver. geol. Unders. Afh., Ser. Ba*, **18**.

Lundqvist, J. (1957) Övre Klarälvsdalens kvartärgeologi. *Sver. geol. Unders. Afh., Ser. C*, **550**.

Lundqvist, J. (1962) Patterned ground and related frost phenomena in Sweden. *Sver. geol. Unders. Afh., ser. C*, **583**.

Lundqvist, J. (1965) The Quaternary of Sweden, in *The Quaternary*, vol. 1, ed. by K. Rankama, p.139 (Wiley Interscience, New York).

Lundqvist, J. (1969a) Beskrivning till jordartskarta över Jamtlands län. Med fyra planscher samt karta i fyra blad i skala 1:200000. *Sver. geol. Unders. Afh., ser. Ca*, **45**.

Lundqvist, J. (1969b) Problems of the so-called Rogen moraine. *Sver. geol. Unders. Afh., ser. C*, **648**.

Lundqvist, J. (1972) Ice-lake types and deglaciation pattern along the Scandinavian mountain range. *Boreas*, **1**, 27.

Lundqvist, J. (1979) Morphogenetic classification of glaciofluvial deposits. *Sver. geol. Unders. Afh., ser. C*, **767**.

Lundqvist, J. and Lagerbäck, R. (1976) The Pärve fault: a late-glacial fault in the Precambrian of Swedish Lapland. *Geol. För. Stockh. Förh.*, **98**, 45.

Lundqvist, T. (1979) The Precambrian of Sweden. *Sver. geol. Unders. Afh., ser. C*, **768**.

Lunsen, V. H. A. (1970) Geology of the Ara–Cinca region, Spanish Pyrenees, Province of Huesca. *Geologica ultraiect.*, **16**.

Lupia Palmieri, E. and Zuppi, G. M. (1977) Il carsismo degli altipiani di Arcinazzo. *Geol. Romana*, **16**, 309.

Machatchek, F. (1955) *Das Relief der Erde* (Inostrannaya Lit., Moscow).

Mackinder, H. J. (1902) *Britain and the British Seas* (Heinemann, London).

Maev, E. G. (1961) The size of sediments and the rate of sediment formation in the Southern Caspian. *Oceanology*, **11**, 430.

Magnusson, N. H. *et al.* (1962) Description to accompany the map of the pre-Quaternary rocks of Sweden. *Sver. geol. Unders. Afh., ser. Ba*, **16**.

Magnusson, N. H. *et al.* (1963) *Sveriges Geologi*, 4th edn. (Stockholm).

Makkaveyev, A. N. (1975) An attempt at a quantitative evaluation of the relief-forming role of ancient glaciers in the north-western part of European Russia. *Geomorpholiogiya*, **2**, 44.

Makkaveyev, N. I. (1955) *River Channels and Erosion in Their Basins* (Izd. Acad. Nauk, Moscow).

Maldonado, A. and Riba, O. (1966) El delta reciente del río Ebro: descripción de ambiente y evolución. *Acta geol. hisp.*, **1**, 9.

Malovitskiy, Y. P. (1964) *Tectonics of the Water Areas of the Southern USSR and the Possibility of their Oil–Gas Content* (Nauka, Moscow).

Malovitskiy, Y. P. (1976) *Tectonics of the Mediterranean Sea Floor* (Nauka, Moscow).

Mangerud, J. (1970) Late Weichselian vegetation and ice-front oscillations in the Bergen district, western Norway. *Norsk geogr. Tidsskr.*, **24**, 121.

Mangerud, J. *et al.* (1979) Glacial history of western Norway, 15000–10000 BP. *Boreas*, **8**, 179.

Manley, G. (1959) The late-glacial climate of north-west England. *Lpool Manchr geol. J.*, **2**, 188.

Mannerfelt, C. M. (1945) Några glacialmorfologiska formelement och deras vittnesbörd om inlandsisens avsmältningsmekanik i svensk och norsk fjällterräng. *Geogr. Annlr*, **27**, 1.

Map of the Arctic Regions (1976) (American Geographical Society, New York).

Maraga, F. (1980) Morphologie fluviale et migration des cours d'eau dans la haute plaine du Pô, in *Symposium on Holocene Valley Development—Methods and Results*, Project no. 158, (International Quaternary Association).

Marcet, J. (1956) Las formaciones cuaternarias de la región costera del Nordeste de España. *Actas IV Congr. int. Quatern. Ass.*, p.631.

Marcet, J. and Solé-Sabarís, L. (1949) *Memoria esplicativa del Mapa geológico de España, 1:50000, Gerona, no. 334* (Instituto Geológica y Minero de España, Madrid).

Marcinek, J. and Nitz, B. (1973) *Das Tiefland der Deutschen Demokratischen Republik* (Gotha, Leipzig).

Margalef, R. (1956a) Paleoecología post-glacial de la ría de Vigo. *Investigación pesq.*, **5**, 1.

Margalef, R. (1956b) Oscilaciones del clima postglaciar del Noroeste de España registradas en los sedimentos de la Ría de Vigo. *Zephyrys*, **7**, 1.

Marin, A. and San-Miguel, M. (1941) *Memoria esplicativa Mapa geológico de España, 1:50000, Estartit, no. 297* (Instituto Geológica y Minero de España, Madrid).

Marinelli, O. (1926) La maggiore discordanza fra orografia e idrografia nell'Appennino. *Riv. geogr. ital.*, **33**, 65.

Marinelli, O. (1948) *Atlante dei Tipi Geografici Desunti dai Rilievi al 25000 e al 500000 dell'Istituto Geografico Militare*, 2nd edn., ed. by R. Almagia *et al.* (Istituto Geografico Militare, Florence).

Markov, K. K. *et al.* (1968) *The Pleistocene* (Vyshaya Schkola, Moscow).

Marr, J. E. (1906) The influence of the geological structure of the English Lakeland upon its present features. *Q. J. geol. Soc. Lond.*, **62**, 67.

Marres, P. (1935) *Les Grands Causses. Etude de Géographie Physique*, thesis, University of Tours.

Martel, E. A. (1936) *Les Causses Majeurs* (Artières et Maury, Millau).

Martí-Bono, C. (1973) Nota sobre los sedimentos morrénicos del Alto Aragón. *Pirineos*, **107**, 39.

Martí-Bono, C. (1977) Altos valles de los ríos Aragón y Gállego. *Actas II Reunión Grp esp. Trab. Cuatern.*, p.337.

Martí-Bono, C. and González, M. C. (1979) Nota sobre algunos depósitos coluviales del Alto Aragón. *Acta III Reunión Grp esp. Trab. Cuatern.*, p.199.

Martí-Bono, C. and Solé-Sabarís, L. (1971) Nota sobre la geomorfología del Alto Aragón. *Acta geol. hisp.*, **6**, 13.

Martí-Bono, C. *et al.* (1978) Los fenómenos glaciares en la vertiente meridional de los Pirineos. *Actas V Coloq. Geogr. (Granada)*, p.57.

Martin, H. (1939) Die Post-Archäische Tektonik im südlichen Mittelschweden. *Neues Jb. Miner. Geol. Paläont. Abh.*, **82**, 55.

Martínez-Gil, F. J. (1972) Estudio hidrogeológico del Bajo Ampurdán (Gerona). *Mems Inst. geol. min. Esp.*, **84**.

Martínez de Pison, E. and Arenillas M. (1976) La morfología glaciar del Moncayo. *Tecniterrae, Madrid*, **18**.

Martínez de Pison, E. and Muñoz, J. (1972) Observaciones sobre la morfología del Altos Gredos. *Estudios geogr.*, **129**, 3.

Martinis, B. (ed.) (1977) Studio geologico dell'area maggiormente colpita dal terremoto friulano del 1976. *Rivta ital. Paleont. Stratigr.*, **83**.

Martinsson, A. (1958) The submarine morphology of the Baltic. *Bull. geol. Instn Univ. Upsala*, **38**, 11.

Martinsson, A. (1960) The structure and submarine morphology of the Baltic Cambro-Silurian area. *Bull. geol. Instn Univ. Upsala*, **36**, 70.

Martonne, E. de (1942) *La France Physique: Géographie Universelle*, vol. 6 (Colin, Paris).

Mary, G. (1971) Les hautes surfaces d'abrasion marine de la côte asturienne (Espagne), in *Histoire Structurale du Golfe de Gascogne*, vol. 2, p.1 (Technip, Paris).

Mateu, J. F. (1980) Contribución a la morfología litoral del norte valenciano, in *Primer Curso de Geomorfología Litoral Aplicada* (Valencia), p.65.

Mateu, J. F. and Sanjaume, E. (1981) Formaciones continentales y marinas en el cierre de la laguna de Torrevieja. *Actas V Reunión Grp esp. Trab. Cuatern.*, p.69.

Matishov, G. G. (1977) Geomorfologiya dna i problema pleistocenovogo oledeneniya barencevomorskogo shelfa. *Geomorfologiya, Moskva*, p.91.

Matishov, G. G. (1980) *Glacial and Periglacial Relief of the Ocean Floor* (MGU, Moscow).

Mattsson, Å. (1962) Morphologische Studien in Südschweden und auf Bornholm über die nichtglaziale Formenwelt der Felsensculptur. *Meddn Lunds Univ. geogr. Instn*, **29**.

Matveevskaya, A. A. (1975) Western Europe, in *Precambrian Continents*, p.89 (Nauka, Novosibirsk).

Maw, G. (1864) On a supposed deposit of boulder clay in north Devon. *Q. J. geol. Soc. Lond.*, **20**, 445.

Mayr, F. and Heuberger, H. (1968) Type areas of late-glacial and post-glacial deposits in Tyrol, Eastern Alps. *Proc. VII Congr. int. Quatern. Ass.*, p.143.

Mazúr, E. (1976) Morphostructural features of the West Carpathians. *Geogr. Čas.*, **28**, 101.

Mazúr, E. and Stehlík, O. (eds.) (1965) *Geomorphological Problems of the Carpathians* (Vydavatelstvo Slovenskej Akad. Vied, Bratislava).

Mazzanti, R. and Trevisan, L. (1978) Evoluzione della rete idrografica nell'Appennino centro-settentrionale. *Geogr. fis. din. quatern.*, **1**, 55.

McArthur, J. L. (1977) Quaternary erosion in the upper Derwent basin and its bearing on the age of surface features in the southern Pennines. *Trans. Inst. Br. Geogr., new ser.*, **2**, 490.

McCave, J. N. (1971) Sand waves in the North Sea off the coast of Holland. *Mar. Geol.*, **10**, 199.

McConnell, R. B. (1938) Residual erosion surfaces in mountain ranges. *Proc. Yorks. geol. Soc.*, **24**, 76.

Menard, H. W. (1964) *Marine Geology of the Pacific* (McGraw-Hill, New York).

Mensching, H. (1957) Geomorphologie der Hohen Rhön und ihres südlichen Vorlandes. *Würzb. geogr. Arb.*, **4–5**, 47.

Mensching, H. (1961) Die Rias der galicisch–asturischen Küste Spaniens. *Erdkunde*, **15**, 210.

Mensching, H. (1965) Beobachtungen zum Formenschatz des Küstenkarstes in der kantabrischen Küste bei Santander und Llanes (Nordspanien). *Erdkunde*, **19**, 24.

Ménsua, S. and Ibáñez, M. J. (1975) Alvéolos en la depresión del Ebro. *Cuad. Invest. Geogr. Hist. Logroño*, **1**, 3.

Ménsua, S. and Ibáñez, M. J. (1977) Terrazas y glacis del centro de la depresión del Ebro. *Actas III Reunión Grp esp. Trab. Cuatern.*, p.18.

Ménsua, S. and Ibáñez, M. J. (1977–78) Correlación entre glacis de acumulación y terrazas fluviales: las terrazas fosilivadas del Gállego y Cinca. Un nuevo modelo de interpretación. *Geographica, Madrid*, **19–20**, 191.

Ménsua-Fernandez, S. (1964) Sobre la génesis de los glacis del Valle del Ebro y su posterior evolución morfológica. *Aportación esp. XX Congr. geogr. int.*, p.191.

Messerli, B. (1962) Sierra Nevada. *Estudios geogr.*, **86**, 25.

Messerli, B. (1965) *Beiträge zur Geomorphologie der Sierra Nevada (Andalusien)*, thesis, University of Zurich.

Metcherikov, J. A. (1965) *Structural Geomorphology of Plains* (Nauka, Moscow).

Metcherikov, J. A. (1968) Les concepts de morphostructure et de morphosculpture: un nouvel instrument de l'analyse géomorphologique. *Annls Géogr.*, **77**, 539.

Metz, B. (1967) Beiträge zur geomorphologischen Entwicklung dreier Becken im Neuenburger Jura. *Frankf. geogr. Hft.*, **43**.

Meurer, D. (1974) Studien zur Morphologie eines intermontanen Beckens in Subappennin: il Valdarno Superiore. *Frankf. geogr. Hft.*, **50**.

Meyerhoff, A. A. and Hatten, C. W. (1974) Bahamas salient of North America, in *The Geology of the Continental Margins*, ed. by C. A. Burk and C. L. Drake (Springer-Verlag, New York).

Meynier, A. (1947) Influences tectoniques sur le relief de la Bretagne. *Annls Géogr.*, **56**, 170.

Michel, J. P. (1972) *Le Quaternaire de la Région Parisienne*, thesis, Paris VI.

Mihailescu, V. and Niculescu, G. (1967) Les surfaces d'applanissement dans les Carpathes Roumaines, in *Studia geomorph. Carpatho-Balc.*, vol. 1, p.5 (Polska Akad. Nauk, Kraków).

Mikhailov, O. V. (1965) *Relief of the Mediterranean Sea Floor: General Features of Geological Structure, Hydrology and Biology* (Nauka, Moscow).

Mikhailov, O. V. and Goncharov, V. P. (1962) Geomorphology of the Mediterranean sea floor. *Izv. Akad. Nauk SSSR, ser. Geogr.*, **2**, 271.

Milanovskiy, E. E. (1952) New data on the structure of the Neogene and Quaternary deposits in the Lake Sevan basin. *Izv. Akad. Nauk SSSR, ser. Geol.*, **4**, 110.

Milanovskiy, E. E. (1962) The Pambakskaya valley; the Sevan basin; latest tectonics of the Armenian SSR and neighbouring regions of the Trans-Caucasus, in *Geology of the Armenian SSR*, vol. 1, Geomorphology, pp.104, 403 (Akad. Nauk Armyan SSR, Yerevan).

Milanovskiy, E. E. (1964) The Upper Pliocene glaciation of the central Caucasus, in *Information Issue for the International Geophysical Year*, vol. 10, p.9 (Moscow State University, Moscow).

Milanovskiy, E. E. (1968) *Recent Tectonics of the Caucasus* (Nedra, Moscow).

Milanovskiy, E. E. (1976) *Rift Zones of Continents* (Nedra, Moscow).

Milanovskiy, E. E. (1977) Recent tectonics, in *General Description and History of the Development of the Relief of the Caucasus*, pp.31, 206 (Nauka, Moscow).

Milanovskiy, E. E. *et al.* (1966) Geological history and the formation of the relief, in *Caucasus: Natural Environment and Natural Resources of the USSR*, p.35 (Nauka, Moscow).

Miller, A. A. (1937) The 600-foot plateau in Pembrokeshire and Carmarthenshire. *Geogrl J.*, **90**, 148.

Miller, A. A. (1955) The origin of the South Ireland peneplain. *Ir. Geogr.*, **3**, 79.

Miller, R. *et al.* (1954) Stone stripes and other surface features of Tinto Hill. *Geogrl J.*, **120**, 216.

Mitchell, G. F. (1960) The Pleistocene history of the Irish Sea. *Advmt Sci.*, **17**, 313.

Mitchell, G. F. (1965) The St Erth beds—an alternative explanation. *Proc. Geol. Ass.*, **76**, 345.

Mitchell, G. F. (1970) The Quaternary deposits between Fenit and Spa on the north shore of Tralee Bay, County Kerry. *Proc. R. Ir. Acad.*, B**70**, 141.

Mitchell, G. F. (1980) The search for Tertiary Ireland. *Jl Earth Sci. R. Dublin Soc.*, **3**, 13.

Mitchell, G. F. and Orme, A. R. (1967) The Pleistocene deposits of the Isles of Scilly. *Q. J. geol. Soc. Lond.*, **123**, 59.

Mitchell, G. F. *et al.* (1973) A correlation of Quaternary deposits in the British Isles. *Geol. Soc. Lond. Spec. Rep.*, **4**.

Moissenet, E. (1980) Relief et déformations récentes: trois transversales dans les fossés internes des chaînes ibériques orientales. *Revue Géogr. Pyrénées S.-Ouest*, **51**, 315.

Molina, B. (1975) Estudio del Terciario superior y del Cuaternario del Campo de Calatrava (Ciudad Real). *Trab. Neógeno-Cuatern.*, **3**, 1.

Møller, J.J. and Sollid, J. L. (1972) Deglaciation chronology of Lofoten–Vesterålen–Ofoten, north Norway. *Norsk geogr. Tidsskr.*, **26**, 101.

Molodkin, P. F. (1976) *Anthropogenic Morphogenesis of Steppe Plains* (Izd. University Rostov).

Monkhouse, F.J. (1964) *A Regional Geography of Western Europe* (Longman, London).

Montadert, L. *et al.* (1974) Continental margins of Galicia—Portugal and Bay of Biscay, in *The Geology of the Continental Margins*, ed. by C. A. Burk and C. L. Drake, p.323 (Springer-Verlag, New York).

Montadert, L. *et al.* (1978) *Introduction and Explanatory Notes, Initial Core Descriptions*, leg. 48, p.218 (Deep Sea Drilling Project, La Jolla).

Montenat, C. (1973) *Les Formations Néogènes et Quaternaires du Levant Espagnol (Province d'Alicante et de Murcia)*, thesis, University of Paris.

Montijuvent, P. (1980) Glaciation ancienne, in *Alpes–Caucase. Problèmes Actuels de la Géographie Constructive des Pays de Montagne*, p.58 (Nauka, Moscow).

Monturiol, J. (1950) El campo de dolinas del Pla de Campgrós (macizo de Garraf, Barcelona). *Speleon*, **1**, 23.

Morgan, A. V. (1973) The Pleistocene geology of the area north and west of Wolverhampton. *Phil. Trans. R. Soc.*, B**265**, 233.

Mörner, N. A. (1969) The late Quaternary history of the Kattegatt Sea and the Swedish west coast. Deglaciation, shore-level displacement, chronology, isostasy and eustasy. *Sver. geol. Unders. Afh., ser. C*, **640**.

Moscheles, J. (1920) Das Böhmische Mittelgebirge. *Z. Ges. Erdk. Berl.*, **55**, 24, 117.

Mosler, H. (1966) Studien zur Oberflächengestalt des östlichen Hunsrück und seiner Abdachung zur Nähe. *Forsch. dt. Landesk.*, **158**.

Muñoz, J. (1978) Morfología estructural y glaciarismo en la Cordillera Cantábrica, el relieve del Sinclinal de Saliencia (Asturias–León). *Acta V Coloq. Geogr.*, p.57.

Muñoz, J. and Asensio-Amor, I. (1975) Los depósitos de raña en el borde noroccidental de los Montes de Toledo. *Estudios geogr.*, **141**, 779.

Muratov, V. M. (1964) *Neotectonics and the Relief of the North-Western Caucasus*, thesis, Moscow State University.

Muratov, M. V. (1975) *Origin of Continents and Ocean Depressions* (Nauka, Moscow).

Nangeroni, G. (1980) E'ancora valida la ipotesi dell'origine dei grandi laghi prealpini italiani da escavazione glaciale? *Geografia*, **3**, 127.

Nansen, F. (1922) The strandflat and isostasy. *Videskapssellsk Skr.*, **1**; *Mat. naturv.*, **2**.

Neboit, R. (1975) *Plateaux et Collines de Lucanie Orientale et des Pouilles*, thesis, University of Lille.

Neef, E. (1978) *Das Gesicht der Erde* (Brockhaus, Leipzig).

Neenkovich, D. and Heezen, B.C. (1965) Santorini tefra. *Submar. Geol. Geophys., Colston Pap.*, **17**, 413.

Nelson, H. (1910) Om randdeltan och randåsar i mellersta och södra Sverige. *Sver. geol. Unders. Afh., ser. C*, **220**.

Neprochnov, J.N. (ed.) (1979) *Oceanology; Geophysics of the Ocean Floor* (Nauka, Moscow).

Nicod, J. (1972) *Pays et Paysages du Calcaire* (PUF, Paris).

Niculescu, G. (1965) *Muntii Godeanu* (Academiei Republicii Populare Romine, Bucureşti).

Niewiarowski, W. (1965) Conditions of occurrence and distribution of kame landscapes in the Peribalticum within the area of the last glaciation. *Geogr. polon.*, **6**, 7.

Niini, H. (1968) A study of rock fracturing in valleys of Precambrian bedrock. *Fennia*, **97**, 1.

Nikiforova, K. V. (1946) Geological structure and history of development of the drainage pattern in the northern part of the eastern slope of the Central Urals during the Cenozoic. *Byull. mosk. Obshch. Ispyt. Prir., ser. Geol.*, **21**, 57.

Nikolayev, N. I. (1962) Neotectonics and their expression in the structure and relief of the USSR, in *Questions of Regional and Theoretical Neotectonics* (Gosgeoltechizdat, Moscow).

Nikolayev, N. I. (1977) *Map of Modern Tectonics of the USSR and Adjacent Regions* (Moscow).

Nonn, H. (1958) Contribution à l'étude des plages anciennes de Galice (Espagne). *Publs Soc. Océanogr. Fr.*, **3**, 257.

Nonn, H. (1966) *Les Régions Cotières de la Galice (Espagne). Etude Géomorphologique*, thesis, University of Strasbourg.

Nonn, H. and Tricart, J. (1960) Etude d'une formation périglaciaire ancienne en Galice. *Bull. Soc. géol. Fr.*, **2**, 41.

Norrman, J.O. (1964) Lake Vättern. Investigations on shore and bottom morphology. *Geogr. Annlr*, **46**, 1.

Obermaier, H. (1914) Estudio de los glaciares de los Picos de Europa. *Trab. Mus. nac. Cienc. nat., ser. Geol.*, **9**.

Obermaier, H. and Carandell, J. (1916a) Los glaciares cuaternarios de Sierra Nevada. *Trab. Mus. nac. Cienc. nat., ser. Geol.*, **17**.

Obermaier, H. and Carandell, J. (1916b) Contribución al estudio del glaciarismo cuaternario en la Sierra de Gredos. *Publnes Jta Ampl. Estud. Invest. cient., Trab. Mus. nac. Cienc. nat., Madr., ser. Geol.*, **14**.

Obrador, A. *et al.* (1971) Morfología de la costa baja en la provincia de Gerona. *Revta Gerona*, **17**, 29.

Oele, E. *et al.* (eds.) (1979) *The Quaternary History of the North Sea* (Acta University, Uppsala).

Ogarinov, I. S. (1974) *Deep-Seated Structure of the Ural Mountains* (Nauka, Moscow).

Ogniben, L. (ed.) (1975) *Structural Model of Italy, Maps and Explanatory Notes*, p.90 (CNR, Rome).

Orbenok, V. V. (1980) *Physical Principles of the Evolution of the Perisphere of the Earth* (Leningrad State University, Leningrad).

Orghidan, N. (1969) *Vaile Transverstale din Romania* (Academiei Republicii Soc. Romania, Bucureşti).

Orme, A. R. (1964) Planation surfaces in the Drum Hills, County Waterford, and their wider implications. *Ir. Geogr.*, **5**, 48.

Ottmann, F. (1969) *VIII Congr.* (International Quaternary Association, Paris).

Ovejero, G. and Zazo, C. (1971) Niveles marinos pleistocenos en Almería (SE España). *Quaternaria*, **15**, 145.

Ozoray, G. (1972) Structural control of morphology in Alberta. *Albertan Geogr.*, **8**, 35.

Palli, L. (1977) Contribución al estudio morfolitológico del sector costero central del macizo de Begur (Girona). *An. Secc. Cienc. Col. Univ. Gerona*, p.231.

Palli, L. (1978) *Mapa Geológico de Sant Feliu de Guíxols* (Curbet, Girona).

Palli, L. (1980) *Mapa Geológico de Castillo de Aro-Playa de Aro* (Curbet, Girona).

Palmason, G. (1974) Insular margin of Iceland, in *The Geology of the Continental Margins*, ed. by C. A. Burk and C. L. Drake, p.375 (Springer-Verlag, New York).

Palmer, J. and Neilson, R. A. (1962) The origin of granite tors on Dartmoor. *Proc. Yorks. geol. Soc.*, **33**, 315.

Paneque, G. and Mudarra, J. L. (1966) *Morfología Sistemática y Cartografía de los Suelos de Andalucía Occidental*, p.417 (Conf. Suelos Medít., Madrid).

Panizza, M. (1966) Carta ed osservazioni geomorfologiche del territorio di Calopezzati (Calabria). *Riv. geogr. ital.*, **73**, 1.

Panizza, M. (ed.) (1979) *Proceedings of the 15th Plenary Meeting, IGU Commission on Geomorphological Survey and Mapping* (Ist. Geol., Modena).

Pannekoek, A. (1966) The Ría problem. *Tijdschr. K. ned. aardrijksk. Genoot.*, **83**, 289.

Panzer, W. (1926) Talentwicklung und Eiszeitklima im nordöstlichen Spanien. *Abh. senckenb. naturforsch. Ges.*, **39**, 141.

Panzer, W. (1932) Die eiszeitlichen Endmoränen von Puigcerdá (Ostpyrenäen). *Z. Gletscherk.*, **18**, 411.

Parga-Pondal, I. (1958) El relieve geográfico y la erosión diferencial de los granitos en Galícia. *Trab. Lab. Geol. Lage*, **6**, 129.

Parga-Pondal, I. (1963) *Mapa Petrográfico Estructural de Galícia*, (Instituto Geológica y Minero de España, Madrid).

Paschinger, H. (1954) Würmvereisung und Spätglacial in der Sierra Nevada (Spanien). *Z. Gletscherk.*, **3**, 55.

Pasierbski, M. (1979) Remarks on the genesis of subglacial channels in northern Poland. *Eiszeit. Gegenw.*, **29**, 189.

Peach, B.N. and Horne, J. (1930) *Chapters on the Geology of Scotland* (Oxford University Press, Oxford).

Pécsi, M. (1965) Upper Pliocene–post-Pannonian pediments in the middle mountains of Hungary, in *Geomorphological Problems of the Carpathians*, ed. by E. Mazúr and O. Stehlík, p.199 (Vydavatelstvo Slovenskej Akad. Vied, Bratislava).

Pécsi, M. (ed.) (1970) *Geomorphological Regions of Hungary* (Akad. Kiadó, Budapest).

Pécsi, M. and Sarfalvi, B. (1964) *The Geography of Hungary* (Corvina, Budapest).

Peel, R. F. (1949) A study of two Northumbrian spillways. *Trans. Inst. Br. Geogr.*, **15**, 73.

Peguy, C. and Vivian, R. (1980) Particuliarités climatiques et glaciation actuelle, in *Alpes–Caucase. Problèmes Actuels de la Géographie Constructive des Pays de Montagne*, p.69 (Nauka, Moscow).

Peive, A. V. (1956a) General description, classification and spatial distribution of deep faults. *Izv. Akad. Nauk SSR, ser. Geol.*, **1**.

Peive, A. V. (1956b) The linking of sedimentation, folding, magmatism and mineral deposits with deep faults. *Izv. Akad. Nauk SSSR, ser. Geol.*, **3**.

Peive, A. V. (1960) Faults, their role in the construction and development of the Earth's crust. *Rep. Sov. Geol., int. Geol. Congr. (Norden), Akad. Nauk SSSR, Problema*, **18**, 10.

Pellegrini, M. (1969) La pianura del Secchia e del Panaro. *Atti Soc. Nat. Mat. Modena*, **100**.

Pellegrini, M. (1979) The Po valley: methods of study, geological characteristics and examples of geomorphological evolution, in *Proceedings of the 15th Plenary Meeting, IGU Commission on Geomorphological Survey and Mapping*, ed. by M. Panizza, p.83 (Ist. Geol., Modena).

Pellegrini, M. and Vezzani, L. (1978) Faglie attive in superficie nella Pianura Padana presso Correggio (Reggio Emilia) e Massa Finalese (Modena). *Geogr. fis. din. Quatern.*, **1**, 63.

Pellegrini, M. *et al.* (1979) *La Situazione Morfologica degli Alvei degli Affluenti Emiliani del Po* (Ministero dei Lavori Pubblici, Convegro di Idraulica Padana, Parma).

Pelletier, J. (1960) *Le Relief de la Sardaigne* (Institut des Études Rhodaniennes, Lyon).

Pellicer, F. (1980) El periglaciarismo del Moncayo. *Geographicalia*, Zaragoza, **7–8**, 3.

Penck, A. (1883) Die Eiszeit in den Pyrenäen. *Mitt. Ver. Erdk. Lpz.*, **11**, 1.

Penck, A. and Brückner, E. (1909) *Die Alpen im Eiszeitalter* (Tauchnitz, Leipzig).

Penck, W. (1924) *Die Morphologische Analyse. Ein Kapital der Physikalischen Geologie* (J. Engelhorns Nachf., Stuttgart).

Pennington, W. (1973) The recent sediments of Windermere. *Freshwater Biol.*, **3**, 363.

Perconig, E. (1960–62) Sur la constitution géologique de l'Andalousie occidentale, en particulier du bassin du Guadalquivir (Espagne meridionale), in *Livre Mémoire Prof. Paul Fallot*, p.199 (Société Géologique de France, Paris).

Pérez-Gonzalez, A. (1971) Estudio de los procesos de hundimiento en el valle del río Jarama y sus terrazas (nota preliminar). *Estudios geol.*, **27**, 317.

Pérez-Gonzalez, A. and Asensio-Amor, I. (1973) Rasgos sedimentológicos y geomorfológicos del sistema de terrazas del río Henares-Azuqueca (Nota preliminar). *Boln Geol. Min.*, **84**, 15.

Pérez-Gonzalez, A. *et al.* (1974) Guia de Excursión B: valle del Henares-Jarama. *Actas I Reunión Grp esp. Trab. Cuatern., Trab. Neógeno-Cuatern.*, **2**, 222.

Pérez-Lorente, F. (1979) Niveles de erosión y acumulación en La Rioja Central y Oriental. *Cuad. Invest. Geogr. Hist. Logroño*, **5**, 27.

Pérez-Mateos, A. and Riba, O. (1961) Estudios de los sedimentos pliocénicos y cuaternarios de Huelva. *Actas II Reunión Sedimentología, CSIC*, p.87.

Peterlongo, J. M. *et al.* (1972) Massif Central: Limousin, Auvergne, Velay, in *Guides Géologiques Régionaux*, ed. by C. Pomerol (Masson, Paris).

Petković, K. *et al.* (1976) Tektonika Jugoslavije kroz evoluciju misli domaćih i stranih istraživača. *Zborn. Rad. rud.-metal. Inst. bakar, Borž*, **19**.

Petrucci, F. (1968) Studio geomorfologico dei terreni pleistocenici fra il fiume Taro e il torrente Baganza. *Ateneo parmense, Acta nat.*, **4**, 93.

Pezzi, M. C. (1975a) Le Torcal d'Antequera (Andalusie): un karst structural retouché par le périglaciaire. *Méditerranée*, **21**, 23.

Pezzi, M. C. (1975b) Algunas observaciones sobre sistemas morfoclimáticos y karst en las Cordilleras Béticas. *Cuad. geogr. Univ. Granada*, **1**, 59.

Pezzi, M. C. (1977) Morfología kárstica del sector central de la Cordillera Subbética. *Cuad. geogr. Univ. Granada*, **2**, 1.

Piirola, J. (1967) Die glaziale Oberflächenformen und die Entwicklung der Täler. Auf den Fjelden Marastotunturit und Viipustunturit in Finnisch-Lappland. *Suomal. Tiedeakat. Toim. (Helsinki), ser. Geol.-Geogr., A. (III)*, **92**.

Pinchemel, P. (1954) *Les Plaines de Craie du Nord-Ouest du Bassin Parisien et du Sud-Est du Bassin de Londres et leur Bordures. Étude de Géomorphologie* (Colin, Paris).

Piotrovskiy, M. V. (1972) Methodology of mapping arch-block morphotectonics, in *Modern Tectonics, Deposition and Man* (MGU, Moscow).

Pissart, A. (1963) Les traces de 'Pingos' du Pays de Galles (Grande-Bretagne) et du Plateau des Hautes Fagnes (Belgique). *Z. Geomorph.*, **7**, 147.

Pomerol, C. and Feugueur, L. (1968) Bassin de Paris, Île-de-France, pays de Bray, in *Guides Géologiques Régionaux*, ed. by C. Pomerol (Masson, Paris).

Poole, E. G. and Whiteman, A. J. (1961) The glacial drifts of the southern part of the Shropshire–Cheshire basin. *Q. J. geol. Soc. Lond.*, **117**, 91.

Porta, J. (1957) Observaciones sobre las terrazas marinas de Cataluña. *Résumés Communs V Congr. int. Quatern. Ass.*, p.148.

Porta, J. and Solé, N. (1957) La fauna cuaternaria marina del litoral mediterráneo español. *Résumés Communs V Congr. int. Quatern. Ass.*, p.149.

Portmann, J. P. (1977) Variations glaciaires, historiques et préhistoriques dans les Alpes suisses. *Les Alpes*, **53**, 145.

Posea, G. (1981) Types de montagnes en Roumanie. *Revue roum. Géol. Géophys. Géogr.*, **25**, 13.

Posea, G. *et al.* (1974) *Relieful Romaniei* (Editura Stiintifica, Bucureşti).

Price, R. J. (1960) Glacial meltwater channels in the upper Tweed drainage basin. *Geogrl J.*, **126**, 485.

Puig de Fàbregas, C. (1975) *La Sedimentación Molásica en la Cuenca de Jaca* (Instituto Estudios Pirenaicos, Jaca).

Puig de Fàbregas, J. *et al.* (1982) *Consideraciones Sobre el Balance Hídrico de la Laguna de Sariñena (Huesca)*.

Quirantes, J. (1965) Nota sobre las lagunas de Bujaraloz–Sástago. *Geographica*, Zaragoza, **12**, 30.

Quirantes, J. and Riba, O. (1973) Materiales pirenaicos depositados en la depresión terciaria del Ebro. *Pirineos*, **107**, 13.

Quitzow, H. W. (1974) Das Rheintal und seine Entstehung. Bestandsaufnahme und Versuch einer Synthese, in *L'Évolution Quaternaire des Bassins Fluviaux de la Mer du Nord Méridionale*, p.53 (Centenaire de la Société Géologique de Belgique, Liège).

Radlowska, C. (1970) Morfotektonika w rozwazaniach nad rozwojem rzeźby. *Przegl. geogr.*, **42** (3).

Raffey, J. (1980) Les marges occidentales de l'Appennin Ombrien. *Bull. Soc. languedocienne Géogr.*, **3**, 23.

Raitt, R. W. (1963) The crustal rocks, in *The Sea*, vol. 3, *The Earth beneath the Sea: History*, ed. by M. N. Hill, p.85 (Wiley Interscience, New York).

Ramsay, A. C. (1866) The geology of North Wales. *Mem. Geol. Surv. Gt Br.*

Ramsli, G. (1951) Snø og snøskred. *Landbruksdepartementets småskrifter* (Oslo), vol. 96. (Universitetsforlaget, Oslo).

Rankama, K. (ed.) (1967) The Quaternary, in *The Geologic Systems* (Wiley Interscience, New York).

Rapp, A. (1961) Recent development of mountain slopes in Kärkevagge and surroundings, northern Scandinavia. *Geogr. Annlr*, **42**, 71.

Rapp, A. (1965) Studies of mass wasting in the Arctic and the Tropics, in *Mass Wasting*, ed. by E. Yatsu *et al.*, p.79. (Geo-Abstracts, Norwich).

Rapp, A. and Clark, G. M. (1971) Large non-sorted polygons in Padjelanta national park, Swedish Lappland. *Geogr. Annlr*, A**53**, 71.

Rapp, A. and Rudberg, S. (1960) Recent periglacial phenomena in Sweden. *Biul. peryglac.*, **8**, 143.

Rapp, A. and Strömqvist, L. (1976) Slope erosion due to extreme rainfall in the Scandinavian mountains. *Geogr. Annlr*, A**58**, 193.

Rapp, A. *et al.* (1962) Iskilar i Padjelanta? *Ymer*, **82**, 188.

Raskatov, G. I. (1969) *Geomorphology and Neotectonics of the Voronezh Anteclise* (Voronezh State University, Voronezh).

Rat, P. (1972) Bourgogne, Morvan, in *Guides Géologiques Régionaux*, ed. by C. Pomerol (Masson, Paris).

Raynal, R. (1973) Quelques vues d'ensemble à propos du périglaciaire pleistocène des régions riveraines de la méditerranée occidentale. *Biul. peryglac.*, **22**, 249.

Raynal, R. (1975) Essai de synthèse de quelques données acquises sur le Quaternaire ancien des marges humides du domaine méditerranéen. *Étud. géogr. Mélang. Viers, Toulouse*, **2**, 467.

Raynal, R. and Tricart, J. (1963) Comparisons des grandes étapes morphogénétiques du Quaternaire dans le Midi méditerranéen français et au Maroc. *Bull. Soc. géol. Fr.*, **7**, 587.

Reffay, A. (1972) *Les Montagnes de l'Irlande Septentrionale: Contribution à la Géographie Physique de la Montagne Atlantique* (Grenoble).

Reinhard, H. (1965) Das Pommersche Stadium südlich des Mecklenburgischen Grenztales, in *Die Weichsel–Eiszeit im Gebiet der Deutschen Demokratischen Republik*, ed. by J. F. Gellert, p.89 (Akademie Verlag, Berlin).

Riba, O. (1955) Sur le type de sedimentation du Tertiaire continental de la partie ouest du bassin de l'Ebre. *Geol. Rdsch.*, **43**, 363.

Riba, O. (1959) *Estudio Geológico de la Sierra de Albarracín* (Instituto Lucas Mallada, Madrid).

Riba, O. (1964) Estructura sedimentaria del Terciario continental de la depresión del Ebro en su parte riojana y navarra. *Aportación esp. XX Congr. geogr. int.*, p.127.

Riba, O. (1971) *Comentario a la hoja no. 32 (Zaragoza)* (Mapa Geológico de España, Madrid).

Riba, O. and Rios, J. M. (1960–62) Observations sur la structure du secteur sud-ouest de la chaîne ibérique (Espagne), in *Livre Mémoire Prof. Paul Fallot*, p.275 (Société Géologique de France, Paris).

Ribeiro, A. (1970) Position structurale des Massifs de Morais et Bragança (Trás-os-Montes). *Comunções Servs geol. Port.*, **54**, 115.

Ribeiro, A. *et al.* (1979) *Introduction à la Géologie Générale du Portugal* (Servs Geologico Portugal, Lisbon).

Ribeiro, O. (1940) Problemas Morfológicos do Maciço Hispérico Português. *Las Ciencias*, **6**, 315.

Ribeiro, O. (1949) *Le Portugal Central* (International Geographical Congress, Lisbon).

Ribeiro, O. (1955) Portugal, in *Geografía de España y Portugal*, ed. by M. de Teran, p. 5 (Madrid).

Richter, A. (1956) Las cadenas ibéricas entre el valle del Jalón y la Demanda. *Publnes extranj. Geol. Esp., Madrid*, **9**.

Richter, D. (1974) *Grundriss der Geologie der Alpen* (de Gruyter, Berlin).

Richter, G. D. (1936) Orographic regions of the Kola peninsula, in *Studies of the Institute of Physical Geography*, vol. 19 (Izd. Akad. Nauk, Moscow).

Richter, G. and Teichmüller, R. (1933) Die Entwicklung der Keltiberischen Ketten. *Abh. Ges. Wiss. Göttingen, Math. Phys., ser. 3*, **7**, 1.

Richter, H. (1963) Das Vorland des Erzgebirges. *Wiss. Veröff. dt. Inst. Länderk.*, **19–20**.

Richter, K. (1937) *Die Eiszeit in Norddeutschland* (Berlin).

Riedl, H. (1966) Neue Beiträge zum Problem der Raxlandschaft–Augenstein-landschaft. *Mitt. öst. geogr. Ges.*, **108**, 96.

Riedl, H. (1977) Die Problematik der Altflächen am Ostsporn der Alpen. *Würzb. geogr. Arb.*, **45**, 131.

Roberts, D. G. (1972) Slumping on the eastern margins of the Rockall Bank, north Atlantic Ocean. *Mar. Geol.*, **13**, 225.

Roberts, D. G. (1974) Structural development of the British Isles, the continental margin and the Rockall Plateau, in *The Geology of the Continental Margins*, ed. by C. A. Burk and C. L. Drake, p.343 (Springer-Verlag, New York).

Roeleveld, W. (1975) Le vieux bac (Ardres), in *Coastal Excursion Guide 1975* (Holocene Committee and Northwest European Shoreline sub-committee, International Quaternary Association).

Roglić, J. (1938) *Imotski Polje* (Eds. Spec. Soc. Géogr., Belgrade).

Rondeau, A. (1961) *Recherches Géomorphologiques en Corse*, thesis, University of Paris.

Rose, J. and Allen, P. (1977) Middle Pleistocene stratigraphy in south-east Suffolk. *J. geol. Soc. Lond.*, **133**, 83.

Rosselló, V. (1963) Notas preliminares de la morfología litoral del norte de Valencia. *Saitabi, Valencia*, p.105.

Rosselló, V. (1969) *El Litoral Valencia* (L'Estel, Valencia).

Rosselló, V. (1970) Clima y morfología pleistocena en el litoral mediterráneo español. *Pap. Depto Geogr. Univ. Murcia*, **2**, 79.

Rosselló, V. (1971) Notas sobre la geomorfología litoral del Sur de Valencia. *Quaternaria*, **15**, 121.

Rosselló, V. and Mateu, J. F. (1981) Formaciones dunares en los alrededores de Torrevieja (litoral subvalenciano). *Actas V Reunión Grp esp. Trab. Cuatern.*, p.40.

Rosu, A. (1967) *Subcarpatii Olteiei dintre Motru si Gilort* (Editura Academei Republicii Soc. Romania, Bucureşti).

Rouire, J. and Rousset, C. (1973) Causses, Cévennes, Aubrac, in *Guides Géologiques Régionaux*, ed. by C. Pomerol (Masson, Paris).

Rozdestvenskiy, A. P. (1971) *Recent Tectonics and the Development of the Relief of the Pre-Urals* (Nauka, Moscow).

Rózycki, S. Z. (1965) Traits principaux de la stratigraphie et de la paleogéomorphologie de la Pologne pendant la Quaternaire. *Rep. VII Congr. int. Quatern. Ass.*, **1**, 123.

Rubina, E. A. *et al.* (1977) *Geomorphological Mapping* (Vyshaya Schkola, Moscow).

Rudberg, S. (1944) Enköpingsåsen mellan Mälaren och Dalälven. En geomorfologisk studie, in Geografiska studier tillägnade John Frödin den 16 April 1944. *Geographica, Uppsala*, **15**, 152.

Rudberg, S. (1949) Kursudalur i Norrbotten. En preliminär översikt. *Geol. För. Stockh. Förh.*, **71**, 442.

Rudberg, S. (1950) Ett par fall av skred och ravinbildning i Västerbottens fjälltrakter. *Geol. För Stockh. Förh.*, **72**, 139.

Rudberg, S. (1952) Some observations on the geology and morphology of the Paltsa area. *Bot. Notiser*, **3** (2).

Rudberg, S. (1954) Västerbottens berggrundmorfologi. Ett försök till rekonstruktion av periglaciala erosionsgenerationer i Sverige. *Geographica, Uppsala*, **25**.

Rudberg, S. (1960) Geology and morphology, in *A Geography of Norden*, ed. by A. Sømme, p.27 (J. W. Cappelens, Oslo).

Rudberg, S. (1962) A report on some field observations concerning periglacial geomorphology and mass movement on slopes in Sweden. *Biul. peryglac*, **11**, 311.

Rudberg, S. (1964) Slow mass movement processes and slope development in the Norra Storfjäll area, southern Swedish Lappland. *Z. Geomorph., suppl.*, **5**, 192.

Rudberg, S. (1967a) Det kala bergets utbredning i Fennoskandia—en problemdiskussion, in *Teknik och Natur. Festskrift tillägn. G. Beskow*, ed. by K. G. Eriksson (Gothenburg).

Rudberg, S. (1967b) The cliff coast of Gotland and the rate of cliff retreat, in Landscape and processes. Essays in geomorphology dedicated to Filip Hjulström, 6 October 1967. *Geogr. Annlr*, A**2**, 283.

Rudberg, S. (1968) Wind erosion—preparation of maps showing the direction of eroding winds. *Biul. peryglac.*, **17**.

Rudberg, S. (1970a) The sub-Cambrian peneplain in Sweden and its slope gradient. *Z. Geomorph., suppl.*, **9**, 157.

Rudberg, S. (1970b) The areas of bare rock in Fennoskandia. *Acta geogr. Univ. todz.*, **24**.

Rudberg, S. (1970c) Geomorphology, in *Atlas över Sverige*, sheets 5 and 6 (Stockholm).

Rudberg, S. (1973) Glacial erosion forms of medium size—a discussion based on four Swedish case studies. *Z. Geomorph., suppl.*, **17**, 23.

Rudberg, S. (1974) Some observations concerning nivation and snow melt in Swedish Lapland, *Abh. Akad. Wiss. Göttingen, Math. Phys.*, p.263.

Rudberg, S. (1976) River valley anomalies—one approach to the study of Fennoscandian bedrock relief. *Norsk geogr. Tidsskr.*, **30**, 81.

Rudberg, S. (1977) Periglacial zonation in Scandinavia, *Abh. Akad. Wiss. Göttingen, Math. Phys.*, **3**, 92.

Ruddiman, W. F. (1972) Sediment redistribution on the Reykjanes ridge: seismic evidence. *Bull. geol. Soc. Am.*, **83**, 2039.

Rukhin, L. B. (1959) *Principles of General Paleogeography* (Gostoptechizdat, Moscow).

Russell, R. J. (1942) Geomorphology of the Rhône delta. *Ann. Ass. Am. Geogr.*, **32**, 149.

Ruuhijärvi, R. (1960) Über die regionale Einteilung der Nordfinnischen Moore. *Suomal. eläin- ja kasvit. Seur. van. Julk.*, **31**.

Ryan, W. *et al.* (1970) The tectonics of the Mediterranean Sea, in *The Sea*, vol. 4, ed. by M. N. Hill, p.387 (Interscience, New York).

Safronov, I. N. (1969) *Geomorphology of the Northern Caucasus* (University of Rostov, Rostov).

Sala, M. (1978) Présence de formes et formations périglaciaires dans le massif du Montnegre (chaîne côtière catalane). *Actes Coloq. périglac. Médit.*, p.161.

Sala, M. (1979a) *La cuenca del Tordera. Estudio geomorfológico*, thesis, University of Barcelona.

Sala, M. (1979b) Alluvial depoits in the Tordera basin, in *Proceedings of the 15th Plenary Meeting, IGU Commission on Geomorphological Survey and Mapping*, ed. by M. Panizza, p.185 (Ist. Geol., Modena).

Sala, M. (1979c) *L'Organización de l'Espai Natural a les Gavarres* (Fundació Salvador Vives i Casajuana, Barcelona).

Sala, M. (1982) Glacis d'épandage dans la chaîne côtière catalane (Espagne). *Actes Coloq. Villafranquien Médit.*, p.219.

Salathé, R. H. (1961) Die stadiale Gliederung des Gletscherrückganges in den Schweizer Alpen und ihre morphologische Bedeutung. *Verh. naturf. Ges. Basel*, **72**, 137.

Salop, L. J. (1978) *General Stratigraphy of the Precambrian Era* (Nedra, Leningrad).

Sanjaume, E. and Gozálbez, V. (1981) Cuaternario continental en la albufera de Elx. *Actas V Reunión Grp esp. Trab. Cuatern*, p.53.

San Miguel, M. (1931) Resumen geológico-geognóstico de la Costa Brava. *Ass. esp. prof. Cienc., Congr.*, p.1.

Sanz-Donaire, J. (1978) El glaciarismo en la cara Sur del macizo de El Barco de Avila. *Actas V Coloq. Geogr.*, p.41.

Sanz-Donaire, J. (1979) Acera de los montes-isla españoles. *Actas III Reunión Grp esp. Trab. Cuatern.*, p.187.

Sanz-Herraiz, C. (1978) Morfología glaciar en la Sierra de Guadarrama: el modelado de las áreas glaciares y periglaciares (Peñalara—los Pelados). *Actas V Coloq. Geogr.*, p. 49.

Sauro, U. (1981) Morphogenetic and chronological aspects of some karst areas in the Italian Alps. *Proc. VIII int. Congr. Speleology*, ed. by F. Beck, p.556 (Georgia Southwestern College, Americus).

Sauro, U. and Meneghel, M. (1980) Dati preliminari sulla neotettonica dei fogli 21 (Trento), 35 (Riva), 36 (Schio), 49 (Verona). *Contributi preliminari Carta Neotettonica d'Italia, C.N.R., Progretto Finalizzato Geodinamica*, Publ. no. 356, p.331.

Schatskiy, N. S. (1946) The main features of the structure and development of the East European platform. Comparative tectonics of ancient platforms. *Izv. Akad. Nauk SSSR, ser. Geol.*, **1**.

Schatskiy, N. S. (1948) Deep dislocations, including platforms and folded areas. Comparative tectonics of ancient platforms. *Izv. Akad. Nauk SSSR, ser. Geol.*, **5**.

Schou, A. (1949) The landscapes, in *Atlas of Denmark*, ed. by N. Nielsen (Copenhagen).

Schriel, W. (1929) Der geologische Bau der katalonischen Küstengebirge zwischen Ebromündung und Ampurdan. *Abh. Ges. Wiss. Göttingen, Math. Phys.*, **1**, 62.

Schunke, E. (1968) Die Schichtstufenhänge im Leine–Weser-Bergland in Abhängigkeit vom geologischen Bau und Klima. *Göttinger geogr. Abh.*, **43**, 1.

Schutzbach, W. (1976) *Island—Feuerinsel am Polarkreis* (Bonn).

Schwenzner, J. (1937) Zur morphologie des zentralspanischen Hochlandes. *Geogr. Abh.*, **3**, 1.

Schytt, V. (1974) Inland ice-sheets—recent and Pleistocene. *Geol. För. Stockh. Förh.*, **96**, 299.

Seefeldner, E. (1926) Zur Morphologie der Salzburger Alpen. *Geogr. Jber. Öst.*, **13**, 107.

Segre, A. G. (1948) I fenomeni carsici e la speleologia del Lazio. *Pubbl. Ist. geogr. Univ. Roma, ser. A*, **7**.

Seguret, M. (1967) Mise en évidence sur le versant sud des Pyrénées centrales d'une nappe à matériel crétace deversé au sud: la nappe du Cotiella. *C. r. hebd. Séanc. Acad. Sci. Paris, ser. D*, **265**, 1448.

Seguret, M. (1969) Contribution à l'etude de la tectonique sud-pyrénéenne. *C. r. hebd. Séanc. Acad. Sci. Paris, ser. D*, **268**, 907.

Selli, R. and Ciabatti, M. (1977) L'abbassamento del suolo della zona litoranea ravennate. *G. Geol. (Bologna), ser. 2*, **42**.

Senftl, E. and Exner, G. (1973) Rezente Hebung der Hohen Tauern und geologische Interpretation. *Verh. geol. Bundesanst. Wien*, **2**, 209.

Seppälä, M. (1966) Recent ice-wedge polygons in eastern Enontekiö, northernmost Finland. *Publtiones Inst. geogr. Univ. Turkuens*, **42**, 274.

Seppälä, M. (1971) Evolution of eolian relief of the Kaamasjoki–Kiellajoki river basin in Finnish Lapland. *Fennia*, **104**, 3.

Seppälä, M. (1972a) Pingo-like remnants in the Peltojärvi area of Finnish Lapland. *Geogr. Annlr, A*, **54**, 38.

Seppälä, M. (1972b) Location, morphology and orientation of inland dunes in northern Sweden. *Geogr. Annlr*, A**54**, 85.

Serebryanniy, L. R. (1978) *Dynamics of Ice-Sheet Glaciation and Glacio-Eustasy in the Late Quaternary* (Nauka, Moscow).

Serebryanniy, L. R. and Orlov, A. V. (1982) Genesis of marginal moraines in the Caucasus. *Boreas*, **11**, 279.

Sermet, J. (1934) Sobre unos vestigios de topografía glaciar en la Sierra Tejeda (Andalucía). *Boln R. Soc. esp. Hist. nat.*, **34**, 187.

Sermet, J. (1943) Sierra Nevada. *Actas II Reunión Est. Geogr.*, p.65.

Sermet, J. (1950) Réflexions sur la morphologie de la zona axiale des Pyrénées. *Pirineos*, **17–18**, 323.

Sermet, J. (1964) Un cas de morphologie litorale heritée sur la côte de Granada. *Boln R. Soc. esp. Hist. nat.*, p.619.

Sermet, J. (1969) *L'Andalousie de la Méditerranée. Région Géographique Espagnole*, thesis, University of Toulouse.

Serrat, D. (1974) Nuevos datos sobre el glaciarismo en el Pirineo oriental. *Actas I Reunión Grp esp. Trab. Cuatern.*, p.175.

Serrat, D. (1977) Nota sobre unos derrubios estratificados en el Pirineo oriental español. *Actas II Reunión Grp esp. Trab. Cuatern.*, p.275.

Serrat, D. (1979) Rock glacier morainic deposits of the eastern Pyrenees, in *Moraines and Varves*, ed. by C. Schlüchter, p.93 (Balkema, Rotterdam).

Serrat, D. (1980) Estudio geomorfológico del Pirineo Oriental (Puigmal, Costabona). *Notas Geogr. fís. Barcelona*, **2**, 39.

Sestini, A. (1940) Evoluzione morfologica della Toscana costiera. *Relazioni XXVIII Riunione Soc. ital. Progr. Sci.*, p.415.

Sestini, A. (1950) Sull'origine della rete idrografica e dei bacini intermontani dell'Appennino settentrionale. *Riv. geogr. ital.*, **57**, 249.

Sestini, A. (1969) La morfologia vulcanica in Italia. *Atti XX Congr. geogr. ital.*, vol. 2, p.7.

Sgavetti, M. (1974) Le caratteristiche di riflettivita spettrale di una zona della Pianura Padana su immagini ERTS-1. *Boll. Geol.*, **33**.

Shepard, F. P. (1973) *Submarine Geology* (Harper & Row, New York).

Shepard, F. P. and Dill, R. (1966) *Submarine Canyons and Other Sea Valleys* (McNally, Chicago).

Sheridan, R. E. (1974) Atlantic continental margin of North America, in *The Geology of Continental Margins*, ed. by C. L. Burt and C. A. Drake, p.391 (Springer-Verlag, New York).

Shirinov, N. S. (1962) Geomorphological characteristics of the mud volcanoes of the Apsheron peninsula. *Izv. Akad. Nauk azerb. SSR, ser. Geol. Geogr.*, **5**, 105.

Shirinov, N. S. (1965) *Geomorphology of the Apsheron Oil-Bearing Region* (Akad. Nauk Azerb. SSR, Baku).

Shotton, F. W. (1953) Pleistocene deposits of the area between Coventry, Rugby and Leamington. *Phil. Trans. R. Soc.*, **B237**, 209.

Shotton, F. W. and Osborne, P. J. (1965) The fauna of the Hoxnian interglacial deposits of Nechells, Birmingham. *Phil. Trans. R. Soc.*, **B248**, 353.

Shukhin, I. S. (1954) Morphology of unconsolidated formations in highlands with a continental climate. *Vop. Geogr.*, **35**, 24.

Sidorenko, A. V. (1958) *The Preglacial Weathering Mantle of the Kola Peninsula* (Izd. Akad. Nauk, Moscow).

Simony, F. (1851) Beobachtungen über das Vorkommen von Urgebirgsgeschieben auf dem Dachsteingebirge. *Geol. Jb. Reichsanstalt*, **2**, 159.

Simpson, J. B. (1933) The late-glacial readvance moraines of the Highland border west of the River Tay. *Trans. R. Soc. Edinb.*, **57**, 633.

Sissons, J. B. (1958) Subglacial stream erosion in southern Northumberland. *Scott. geogr. Mag.*, **74**, 163.

Sissons, J. B. (1960) Erosion surfaces, cyclic slopes and drainage systems in southern Scotland and northern England. *Trans. Inst. Br. Geogr.*, **28**, 23.

Sissons, J. B. (1967) *The Evolution of Scotland's Scenery* (Oliver & Boyd, Edinburgh).

Sissons, J. B. (1974a) The Quaternary in Scotland: a review. *Scott. J. Geol.*, **10**, 311.

Sissons, J. B. (1974b) A late-glacial ice cap in the central Grampians, Scotland. *Trans. Inst. Br. Geogr.*, **62**, 95.

Sissons, J. B. (1976) *Scotland* (Methuen, London).

Sissons, J. B. *et al.* (1966) The vertical displacement of shorelines in Highland Britain. *Trans. Inst. Br. Geogr.*, **39**.

Skvor, V. (1975) The geological development of the Ohže River rift (Middle Europe), in *Symposium on Rift Zones of the Earth*, p.80 (Irkutsk).

Smith, C. (1744) *The Ancient and Present State of the County of Down* (Dublin).

Smith, W. E. (1961) The detrital mineralogy of the Cretaceous rocks of south-east Devon, with particular reference to the Cenomanian. *Proc. Geol. Ass.*, **72**, 303.

Soergel, W. (1919) *Lösse, Eiszeiten und Paläolithische Kulturen* (Jena).

Solar, F. (1964) Zur Kenntnis der Böden auf den Raxplateaus. *Mitt. öst. Boden Ges.*, **8**, 1.

Solé, N. (1957) Algunas terrazas marinas a 4–6 m en el litoral del SE y S de España. *Résumés Communs V Congr. int. Quatern. Ass.*, p.176.

Solé, N. and Porta, J. de (1957) El cuaternario marino de los alrededores de Alicante. *Résumés Communs V Congr. int. Quatern. Ass.*, p.176.

Soler, M. and Garrido, A. (1970) La terminación occidental del manto de Cotiella. *Pirineos*, **98**, 5.

Soler, M. and Puig de Fàbregas, C. (1970) Líneas generales de la geología del alto Aragón occidental. *Pirineos*, **96**, 5.

Solé-Sabarís, L. (1940) Superficies de erosión en las Cordilleras Litorales Catalanes. *Mems Comuns, An. Univ. Barcelona*, **1**, 145.

Solé-Sabarís, L. (1942a) La canal de Berdún. *Estudios geogr.*, **7**, 271.

Solé-Sabarís, L. (1942b) Estado actual de nuestros conocimientos sobre los Alpides españoles. *Boln Univ. Granada*, **14**, 425.

Solé-Sabarís, L. (1951) *Los Pirineos; el Medio y el Hombre* (Martin, Barcelona).

Solé-Sabarís, L. (1952) Morfologia comparada de los Pirineos y las Cordilleras Béticas. *Mems. R. Acad. Cienc. Artes Barcelona, ser 3.*, **31** (1), 1.

Solé-Sabarís, L. (1961) *Oscilaciones del Mediterráneo Español durante el Cuaternario* (CSIC, Barcelona).

Solé-Sabarís, L. (1962) Observaciones sobre la edad del volcanismo gerundense. *Mems R. Acad. Cienc. Artes Barcelona*, **34**, 359.

Solé-Sabarís, L. (1963) Ensayo de interpretación del cuaternario barcelonés. *Miscelánea barcilonensis*, **3**, 7.

Solé-Sabarís, L. (1964) Las rampas o glacis de erosión de la península ibérica. *Aportación esp. XX Congr. geogr. int.*, p.13.

Solé-Sabarís, L. (1972) *Memoria explicativa del mapa geológico de España 1:200000, Barcelona*, no. 35 (Instituto Geológico y Minero de España, Madrid).

Solé-Sabarís, L. (1978) Las costas españolas, in *Geografía General de España*, ed. by M. de Terán and L. Solé-Sabarís, p.125 (Ariel, Esplugues de Llobregat, Barcelona).

Solé-Sabarís, L. and Llopis, N. (1947) *Explicación de la Hoja no. 216 (Bellver) del Mapa Geológico de España* (Instituto Geológico y Minero de España, Madrid).

Solé-Sabarís, L. and Llopis, N. (1952) Geografía física: el relieve, in *La Peninsula Iberica*, ed. by M. de Teran, p.1 (Muntaner y Simón, Barcelona).

Solé-Sabarís, L. and Llobet, S. (1957) Formations quaternaires du Vallés et du Besós (Barcelona). *Résumés Communs V. Congr. int. Quatern. Ass.*, p.177.

Solé-Sabarís, L. and Porta, J. de (1955) Las formaciones tirrenienses del Cabo de Salou (Tarragona). *Mems Comun. Inst. Geol. prov.*, **12**, 5.

Solé-Sabarís, L. and Riba, O. (1952) El relieve de la sierra de Albarracín y zonas limítrofes de la cordillera ibérica. *Teruel*, **7**, 7.

Solé-Sabarís, L. *et al.* (1957a) Levante y Mallorca. *Livret Guide V Congr. int. Quatern. Ass.*

Solé-Sabarís, L. *et al.* (1957b) Environs de Barcelona et Montserrat. *Livret Guide V Congr. int. Quatern. Ass.*

Solé Sugranyes, L. (1973) Nota sobre el límite inferior de los derrubios estratificados de vertiente en el sector de St Llorenç de Morunys (Prepirineo Oriental, Lérida). *Acta geol. hisp.*, **8**, 167.

Sollid, J. L. and Kristiansen, K. (1982) *Hedmark Fylke, Kvartdergeologi og Geomorfologi 1:250000* (Institute of Geography, University of Oslo, Drammen).

Sollid, J. L. and Sørbel, L. (1974) Palsa bogs at Haugtjørn in Dovrefjell, south Norway. *Norsk geogr. Tidsskr.*, **28**, 53.

Sollid, J. L. and Sørbel, L. (1975) Younger Dryas ice-marginal deposits in Trøndelag, central Norway. *Norsk geogr. Tidsskr.*, **29**, 1.

Sollid, J. L. *et al.* (1973) Deglaciation of Finnmark, north Norway. *Norsk geogr. Tidsskr.*, **27**, 233.

Soloviev, V. F. *et al.* (1952) Mountain ridges on the floor of the southern Caspian Sea. *Nature, Lond.*, **8**, 80.

Sommé, J. (1977) *Les Plaines du Nord de la France et leur Bordure. Étude Géomorphologique*, thesis, Paris I.

Sommé, J. (1980) *Synthèse Géologique du Bassin de Paris*, vol. 1, *Stratigraphie et Paléogéographie*; vol. 2, *Atlas*; vol. 3, *Lexique des Noms de Formation*.

Sonder, R. A. (1956) *Mechanik der Erde. Elemente und Studien zur Tektonischen Erdgeschichte* (Schweitzerbart'sche Verlag, Stuttgart).

Soons, J. M. (1960) The sub-drift surface of the lower Devon valley. *Trans. geol. Soc. Glasgow*, **24**, 1.

Sorochtin, O. G. (1976) *Global Evolution of the Earth* (Nauka, Moscow).

Souquet, P. (1967) *Le Crétacé Sudpyrénéen en Catalogne, Aragon et Navarre*, thesis, University of Toulouse.

Soutadé, G. (1970) Exhumation de sols polygonaux et dégradation de la pelouse d'altitude sur le Pla de Gorra Blanc (2450 m), massif du Puigmal, Pyrénées méditerranéennes. *Bull. Ass. Géogr. fr.*, **384**, 259.

Soutadé, G. (1973) Aspects du modelé périglaciaire supra-forestier des Pyrénées Orientales. *Bull. Ass. fr. Étud. Quatern.*, **4**, 239.

Soutadé, G. (1980) *Modelé et Dynamique Actuelle des Versants Supra-Forestiers des Pyrénées Orientales* (Imp. Coop. Sud-Ouest, Albi).

Soutadé, G. and Baudière, A. (1970) Végétation et modelé des hauts versants septentrionaux de la Sierra Nevada. *Annls Géogr.*, **79**, 709.

Sparks, B. W. (1971) *Rocks and Relief* (Longman, London).

Sparks, B. W. *et al.* (1972) Presumed ground-ice depressions in East Anglia. *Proc. R. Soc.*, **A327**, 329.

Spiridonov, A. I. (1978) *Geomorphology of the European Part of the USSR* (Vyshaya Schkola, Moscow).

Spönemann, J. (1966) Geomorphologische Untersuchungen an Schichtkämmen des Niedersächsischen Berglandes. *Göttinger geogr. Abh.*, **36**, 1.

Spreitzer, H. (1961a) Hochstand und Rückzug des eiszeitlichen Murgletschers in Kärnten und deren zeitliche Einreihung. *Carinthia I*, **151**, 351.

Spreitzer, H. (1961b) Der eiszeitliche Murgletscher in Steiermark und Kärnten. *Geogr. Jber. Öst*, **28**, 1.

Staalduinen, C. J. *et al.* (1979) The geology of the Netherlands. *Meded. Rijks geol. Dienst*, **31–32**, 9.

Stäblein, G. (1968) Reliefgenerationen der Vorderpfalz. *Würzb. geogr. Arb.*, **23**.

Stäblein, G. (1973) Rezente und fossile Spuren der morphodynamik in Gebirgsrandzonen des Kastilischen Scheidegebirges. *Z. Geomorph., suppl.* **17**, 177.

Stäblein, G. and Gehrenkämper, H. (1977) Rañas der Sierra de Guadalupe: Untersuchungen zu Gebirgsrandformationen. *Z. Geomorph.*, **21**, 411.

Starkel, L. (1969) The age of the stages of the development of the relief of the Polish Carpathians in the light of the most recent geological investigations. *Studia geomorph. Carpatho-Balcan.*, **3**, 33.

Staub, R. (1924) Der Bau der Alpen. *Beitr. geol. Karte Schweiz*, **52**, 1.

Staub, R. (1934a) Der Deckenbau Südspaniens in der Betischen Cordilleren. *Vjschr. naturf. Ges. Zürich*, **79**, 271.

Staub, R. (1934b) *Grundzüge und Probleme Alpiner Morphologie* (Gebrüder Fretz, Zürich).

Steers, J. A. (1946) *The Coastline of England and Wales* (Cambridge University Press, Cambridge).

Steers, J. A. (1958) Physiography of salt marshes. *Izv. Akad. Nauk SSSR, ser. Geogr.*, **6**, 18.

Stephens, N. (1966) Some Pleistocene deposits in north Devon. *Biul. peryglac.*, **15**, 103.

Stephens, N. and Synge, F. M. (1965) Late Pleistocene shorelines and drift limits in north Donegal. *Proc. R. Ir. Acad.*, **B64**, 131.

Stephens, N. and Synge, F. M. (1966) Pleistocene shorelines, in *Essays in Geomorphology*, ed. by G. H. Dury, p. 1 (Heinemann, London).

Stephens, N. *et al.* (1975) The late Pleistocene period in north-eastern Ireland: an assessment 1975. *Ir. Geogr.*, **8**, 1.

Stickel, R. (1929) Observaciones de morfología glaciar en el NW de España. *Boln R. Soc. esp. Hist. nat.*, **29**, 297.

Stille, H. (1924) *Grundfragen der Vergleichenden Tektonik* (Bornträger, Berlin).

Stille, H. (1936) The present tectonic state of the Earth. *Bull. Am. Ass. Petrol. Geol.*, **20**, 849.

Stillman, C. J. (1968) The post-glacial change in sea level in south-western Ireland: new evidence from freshwater deposits on the floor of Bantry Bay. *Scient. Proc. R. Dubl. Soc.*, **A3**, 125.

Stoddart, D. R. (1969) Climatic geomorphology: review and re-assessment. *Progr. Geogr.*, **1**, 160.

Stride, A. H. (1965) Periodic and occasional sand transport in the North Sea, in *La Revue Petrolière*, vol. 1, *La Petrole et la Mer*, p.111.

Strøm, K. M. (1949) The geomorphology of Norway. *Geogrl J.*, **114**, 19.

Strömqvist, L. (1973) Geomorphological studies of blockfields in northern Scandinavia. *Meddn Upps. naturgeogr. Instn*, **22**.

Sundborg, Å. (1956) The river Klarälven. A study of fluvial processes. *Geogr. Annlr*, **38**, 127.

Sundborg, Å. (1977) *Älv—kraft—miljö, vattenkraftutbyggnadens miljöeffekter*. (Rapp. St. Naturvårdsverk, Stockholm).

Sundborg, Å. and Norrman, J. (1963) Göta Älv. Hydrologi och morfologi med särskild hänsyn till erosionsprocesserna. *Sver. geol. Unders. Afh., ser. Ca*, **43**.

Svensson, H. (1961) Några iakttagelser från palsområden. *Norsk geogr. Tidsskr.*, **18**, 212.

Svensson, H. (1962) Ice wedges in fossil tundra polygons on the Varanger peninsula. *Svensk geogr. Årsb.*, **38**, 185.

Svensson, H. (1963) Tundra polygons. Photographic interpretation and field studies in north Norwegian polygon areas. *Norg. geol. Unders.*, **223**, 298.

Svensson, H. (1964a) Traces of pingo-like frost mounds. *Lund Stud. Geogr., ser. A*, **30**, 93.

Svensson, H. (1964b) Fossil tundramark på Laholmsslätten. *Sver. geol. Unders. Afh., ser. C*, **598**.

Svensson, H. (1969) Deflationsytor vid Varangerfjorden. *Svensk geogr. Årsb.* **45**, 150.

Sweeting, M. M. (1950) Erosion cycles and limestone caverns in the Ingleborough district. *Geogrl J.*, **115**, 63.

Sweeting, M. M. (1955) The landforms of north-west County Clare, Ireland. *Trans. Inst. Br. Geogr.*, **21**, 33.

Sweeting, M. M. (1972) *Karst Landforms* (Macmillan, London).

Synge, F. M. (1950) The glacial deposits around Trim, County Meath. *Proc. R. Ir. Acad.*, **B53**, 99.

Synge, F. M. (1968) The glaciation of west Mayo. *Ir. Geogr.*, **5**, 372.

Synge, F. M. (1970) The Irish Quaternary: current views 1969, in *Irish Geographical Studies in Honour of E. Estyn Evans*, ed. by N. Stephens and R. E. Glasscock, p. 34 (Queen's University, Belfast).

Szafer, W. (1954) Pliocenska flora okolic Czorsztyna i jej stosunek do plejstocenu. *Pr. Inst. geol. Warsz.*, **2**, 1.

Taillefer, F. (1951) *Le Piémont des Pyrénées Françaises* (Privat, Toulouse).

Taillefer, F. (1957) Glaciaire pyrénéen: versant nord et versant sud. *Revue Géogr. Pyrénées S.-Ouest*, **28**, 221.

Taillefer, F. (1969) Les glaciations des Pyrénées, in *Études Françaises Quaternaires*, vol. 1, p.19 (International Quaternary Association, Paris).

Talwani, M. and Eldholm, O. (1974) Margins of the Norwegian–Greenland Sea, in *The Geology of the Continental Margins*, ed. by C. A. Burk and C. L. Drake, p.361 (Springer-Verlag, New York).

Talwani, M. *et al.* (1971) Reykjanes ridge crest: a detailed geophysical study. *J. Geophys.*, **70**, 473.

Tanner, V. (1938) Die Oberflächengestaltung Finlands. Eine übersichtliche Darstellung der Morphographie und Morphologie sowie der Morphogenie in chronologischer Beziehung. *Bidr. Känn. Finl. Nat. Folk*, **86**.

Tavernier, R. and Ameryckx, J. (1970) Kust, Duinen, Polders, in *Atlas van België*, sheet 17 (Brussels).

Tavernier, R. and De Moor, G. (1974) L'évolution du bassin de l'Escaut, in *L'Évolution quaternaire des bassins fluviaux de la mer du nord méridionale*, p.159 (Société Géologique de Belgique, Liège).

Teixeira, C. (1944) Tectonica plio-pleistocena do noroeste peninsular. *Bolm Soc. geol. Port.*, **1–2**, 19.

Teixeira, C. (1949) Plages anciennes et terrasses fluviatiles du littoral du nord-ouest de la Péninsule ibérique. *Bolm Mus. Lab. miner. geol. Univ. Lisboa*, **17**, 1.

Temple, P. H. (1965) Some aspects of cirque distribution in the west-central Lake District, northern England. *Geogr. Annlr*, **47**, 185.

Te Punga, M. T. (1957) Periglaciation in southern England. *Tijdschr. K. ned. aardrijksk. Genoot.*, **64**, 401.

Terán, M. de (1968) Galícia, in *Geografía Regional de España*, ed. by M. de Terán, p.31 (Ariel, Esplugues de Llobregat, Barcelona).

Ters, M. and Pinot, J. P. (1969) *Littoral Atlantique* (Livret-Guide de l'excursion A10, VIII Congr. int. Quatern. Ass., Paris).

Tex, E. D. and Floor, P. (1971) A synopsis of the geology of western Galicia, in *Histoire Structurale du Golfe de Gascogne*, vol. 1, p.1.3–1 (Institut Français Pétrole, Paris).

Thomas, G. S. P. (1976) The Quaternary stratigraphy of the Isle of Man. *Proc. Geol. Ass.*, **87**, 307.

Thorarinsson, S. (1944) Tephrokronologiska studier på Island. Thjórsárdalur och dess förödelse. *Geogr. Annlr*, **26**, 1.

Thorarinsson, S. (1960a) Die Vulkane Islands. *Naturw. Rdsch., Stuttg.*, **13**.

Thorarinsson, S. (1960b) Der Jökulsá–Canyon und Ásbyrgi. *Petermanns geogr. Mitt.* **104**.

Thorarinsson, S. (1960c) Iceland, in *A Geography of Norden*, ed. by A. Sømme, p. 203 (J. W. Cappelens, Oslo).

Thorarinsson, S. (1961) Uppblástur á Íslandi ljósi öskulagarannsókna (Wind erosion in Iceland. A tephrochronological study). *Ársr. Skógraektarf. Ísl.*, p.17.

Thorarinsson, S. (1975) Geology and physical geography in, *Iceland, 874–1974 AD*. (Reykjavik).

Thornes, J. B. (1968) Glacial and periglacial features in the Urbión mountains, Spain. *Estudios geol.*, **24**, 249.

Tischer, G. (1966a) El delta weáldico de las montañas ibéricas occidentales y sus enlaces tectónicos. *Notas Comun Inst. geol. min. Esp.*, **81**, 53.

Tischer, G. (1966b) Datos geomorfológicos sobre la cuenca superior del río Alhama. *Notas Comun Inst. geol. min. Esp.*, **84**, 55.

Tollmann, A. (1963) *Ostalpensynthese*, vol. 8 (Deuticke, Vienna).

Tollmann, A. (1968) Die paläogeographische, paläomorphologische und morphologische Entwicklung der Ostalpen. *Mitt. öst. geogr. Ges.*, **110**, 224.

Tomas, F. (1966) La formation d'un fossé méditerranéen: la vallée du Crati. *Revue Géogr. Lyon*, **41**, 155.

Tomlinson, M. E. (1925) River terraces of the lower valley of the Warwickshire Avon. *Q. J. geol. Soc. Lond.*, **81**, 137.

Tooley, M. J. (1974) Sea-level changes during the last 9000 years in north-west England. *Geogrl J.*, **140**, 18.

Torre-Enciso, E. (1958) Estado actual del conocimiento de las rías gallegas, in *Libro en Homaxe a Ramón Otero Pedrayo*, p.237 (Trab. Lab. Geol. Lage).

Touring Club Italiano (1957) *L'Italia Fisica* (Touring Club Italiano, Milan).

Tozer, E. T. and Schenk, P. E. (eds.) (1978) *Caledonian–Appalachian Orogen of the North Atlantic Region*, project 27 (Geological Survey of Canada, Ottawa).

Trevisan, L. (1955) Les mouvements tectoniques récents en Sicile, hypothèses et problèmes. *Geol. Rdsch.*, **43**, 1.

Tricart, J. (1949) *La Partie Orientale du Bassin de Paris*, vol. 1, *La Genèse du Bassin*; vol. 2, *L'Évolution Morphologique Quaternaire*, (SEDES, Paris).

Tricart, J. (1965) *Principes et Méthodes de la Géomorphologie* (Masson, Paris).

Tricart, J. (1966) Quelques aspects des phénomènes périglaciaires quaternaires dans la Péninsule ibérique. *Biul. peryglac.*, **15**, 313.

Tricart, J. (1974) *Structural Geomorphology* (Longman, London).

Tricart, J. and Cailleux, A. (1965) *Introduction à la Géomorphologie Climatique* (SEDES, Paris).

Trotter, F. M. (1929) The Tertiary uplift and resultant drainage of the Alston block and adjacent areas. *Proc. Yorks. geol. Soc.*, **21**, 161.

Tushinskiy, G. K. (1977) Avalanches, in *General Description and History of the Development of the Relief of the Caucasus*, p.114 (Nauka, Moscow).

Udintsev, G. B. (1959) Research into the relief of the sea and ocean floors, in *Scientific Results and Achievements of Oceanology*, vol. 3, p.6 (Institute of Oceanology, Moscow).

Udintsev, G. B. (1972) *Geomorphology and Tectonics of the Pacific Ocean Floor* (Nauka, Moscow).

Ulrich, J. (1964) The variety of the forms of the sea floor, in *Relief and Geology of the Floors of the Oceans*, p.240 (Progress, Moscow).

Ushakov, S. A. and Galushkin, U. I. (1978) *Physics of the Earth*, vol. 3 (Viniti, Moscow).

Valentin, H. (1953) Present vertical movements of the British Isles. *Geogrl J.*, **119**, 299.

Van Veen, T. J. (1930) *Onderzoeklingen in de Hoofden* (Vanwege het Ministerie van Waterstaat, The Hague).

Van Zuidam, R. (1980) Un levantamiento geomorfológico de la región de Zaragoza. *Geographicalia, Zaragoza*, **6**, 103.

Vaptsarov, I. and Mishev, K. (1978) Osnovnye zakonomernosti formirovaniya morfostruktur Bolgarii. *Geomorfologiya*, no. 1, 16.

Vaptsarov, S. *et al.* (1969) Morphostructural analysis of the relief of the Vasilovska mountains and neighbouring parts of the old mountain chain and central Pre-Balkans. *Izv. geogr. Inst. bŭlg. Akad. Nauk*, p.12.

Vardabasso, S. (1935) Origine ed evoluzione del rilievo del massiccio sardo–corso. *Atti XII Congr. geogr. ital.*, p.123.

Vardanyanz, L. A. (1932) Geotectonics and seismic activity of Daryal as a main cause of catastrophic rock falls. *Izv. vses. geogr. Obshch. voronezh.*, **65**.

Vardanyanz, L. A. (1933) The Quaternary history of the Caucasus. *Izv. gos. russ. geogr. Obshch.*, **65**.

Vardanyanz, L. A. (1935) Seismo-tectonics of the Caucasus. *Izv. Akad. Nauk SSSR, Seismol. Inst.*, **64**.

Varutshenko, S. J. (1975) Analysis of the late Pleistocene and Holocene history of the development of the natural environment of the north-west part of the Black Sea shelf, in *Oscillations of World Sea Level and Problems of Marine Geomorphology*, p.50 (Nauka, Moscow).

Vasilkovskiy, N. P. (1974) On the subsidence of the ocean floors, in *Problems of Geology and Geophysics in the North-Western Part of the Pacific Ocean*, p.11 (Vladivostok).

Vaudour, J. (1974) Recherches sur la Terra Rossa de la Alcarria (Nouvelle Castille). *Méms Docums, Cent. nat. Res. scient.*, vol. 15, *Phénomènes Karstiques*, part 2, p.49.

Vaudour, J. (1979) *La Région de Madrid. Alterations, sols et Paléosols* (Ophrys, Paris).

Vaudour, J. and Asensio-Amor, I. (1972) Los depósitos periglaciares del alto valle del río Navacerrada (vertiente meridional de la Sierra de Guadarrama). *Estudios geogr.*, **28**, 77.

Vedenskaya, I. E. (1963) The main types of morphosculpture in the USSR. *Izv. Akad. Nauk SSSR, ser. Geogr.*, **5**, 64.

Vedenskaya, I. E. (1969) Morphoclimatic zonation of the USSR. *Izv. Akad. Nauk SSSR, ser. Geogr.*, **3**, 12.

Vedenskaya, N. V. and Golubeva, I. I. (1966) Ancient and present-day relief of the Vizhaya and Koyvy valleys. *Sb. nauch. Trud. Permskogo politekh. Inst.*, **20**.

Velcea, I. (ed.) (1971) *Piemontul Getic* (Editura Academiei Republucii Soc. Romania, București).

Velitchko, A. A. (1965) Problems of the geochronology of loess in Europe. *Izv. Akad. Nauk SSSR, ser. Geogr.*, **4**, 19.

Velitchko, A. A. (1968) Main climatic divisions and stages of the Pleistocene. *Izv. Akad. Nauk SSSR, ser. Geogr.*, **3**, 5.

Vera, J. A. (1969) *Estudio Geológico de la Zona Subbética en la Transversal de Loja y Sectores Adyacentes*, thesis, University of Granada. *Mems Inst. geol. min. Esp.*, **72**.

Veyret, P. (1980) Relief et structure géologique, in *Alpes–Caucase. Problèmes Actuels de la Géographie Constructive des Pays de Montagne*, p.13 (Nauka, Moscow).

Vidal Box, C. (1932) Morfología glaciar cuaternaria del macizo oriental de la Sierra de Gredos. *Boln R. Soc. esp. Hist. nat.*, **32**, 117.

Vidal Box, C. (1941) Contribución al conocimiento morfológico de las cuencas de los ríos Sil y Miño. *Boln R. Soc. esp. Hist. nat.*, **39**, 121.

Vidal Box, C. (1944) La edad de la superficie de erosión de Toledo y el problema de los montes-isla. *Las Ciencias*, **9**, 83.

Vidal Box, C. (1948) Nuevas aportaciones al conocimiento geomorfológico de la Cordillera Central. *Estudios geogr.*, **30**, 5.

Viera, L. I. and Torres, J. A. (1979) El Weáldico de la zona de Enciso (Sierra de los Cameros) y su fauna de grandes reptiles. *Munibe, San Sebastián*, **31**, 141.

Viers, G. (1961) Le glaciaire du massif du Carlit (Pyrénées Orientales) et ses enseignements. *Revue Géogr. Pyrénées S.-Ouest*, **32** 5.

Viers, G. (1962) *Les Pyrénées* (PUF, Paris).

Vigneaux, M. (1975) Aquitaine occidentale, in *Guides Géologiques Régionaux*, ed. by C. Pomerol (Masson, Paris).

Viguier, C. (1974) *Le Néogène de l'Andalusie Nord Occidentale (Espagne). Histoire Géologique du Bas-Guadalquivir*, thesis, University of Bordeaux.

Vilborg, L. (1977) The cirque forms of Swedish Lapland. *Geogr. Annlr*, **A59**, 89.

Virgili, C. and Zamarreño, I. (1957) Los depósitos continentales del interglaciar Riss–Würm del litoral catalan. *Résumés Commun. V. Congr. int. Quatern. Ass.*, p.194.

Vita-Finzi, C. (1969) *The Mediterranean Valleys. Geological Changes in Historical Time* (Cambridge University Press, Cambridge).

Vittorini, S,. (1977) Osservazioni sull'origine e sul ruolo di due forme d'erosione nelle argille: calanchi e biancane. *Boll. Soc. geog. ital.*, **10**, 25.

Vossler, P. (1931) Eiszeitstudien im NW Spanien. *Z. Gletscherk.*, **19**, 89.

Wach, H. (1969) *Erdkunde in Stichworten, vol. III, Deutschland* (Hirt, Kiel).

Wager, L. R. (1953) The extent of glaciation in the island of St Kilda. *Geol. Mag.*, **90**, 177.

Walsh, P. T. (1966) Cretaceous outliers in south-west Ireland and their implications for Cretaceous palaeogeography. *Q. J. geol. Soc. Lond.*, **122**, 63.

Walsh, P. T. and Brown, E. H. (1971) Solution subsidence outliers containing probable Tertiary sediments in north-east Wales. *Geol. J.*, **7**, 299.

Walsh, P. T. *et al.* (1972) The preservation of the Neogene Brassington Formation of the southern Pennines and its bearing on the evolution of Upland Britain. *Q. J. geol. Soc. Lond.*, **128**, 519.

Waters, R. S. (1960) The bearing of superficial deposits on the age and origin of the Upland Plain of east Devon, west Dorset and south Somerset. *Trans. Inst. Br. Geogr.*, **28**, 89.

Waters, R. S. (1965) The geomorphological significance of Pleistocene frost action in south-west England, in *Essays in Geography for Austin Miller*, ed. by J.B. Whittow and P. D. Wood, p.39 (University of Reading, Reading).

Watson, E. (1966) Two nivation cirques near Aberystwyth, Wales. *Biul. peryglac.*, **15**, 79.

Watson, E. (1972) Pingos of Cardiganshire and the latest ice limit. *Nature, Lond.*, **236**, 343.

Watson, E. and Watson, S. (1967) The periglacial origin of the drifts at Morfa Bychan, near Aberystwyth. *Geol. J.*, **5**, 419.

Watts, W. A. (1967) Interglacial deposits in Kilcromin townland, near Herbertstown. Co Limerick. *Proc. R. Ir. Acad.*, **B65**, 339.

Watts, W. A. (1970) Tertiary and interglacial floras in Ireland, in *Irish Geographical Studies in Honour of E. Estyn Evans*, ed. by N. Stephens and R. E. Glasscock, p.17 (Queen's University, Belfast).

West, R. G. (1961) Vegetational history of the early Pleistocene of the Royal Society borehole at Ludham, Norfolk. *Proc. R. Soc.*, **A155**, 437.

West, R. G. (1968) *Pleistocene Geology and Biology* (Longman, London).

West, R. G. and Sparks, B. W. (1960) Coastal interglacial deposits of the English Channel. *Phil. Trans. R. Soc.*, **B243**, 95.

Whiteman, A. *et al.* (1975) *North Sea Troughs*.

Wiche, K (1961) Beiträge zur Formenentwicklung der Sierren am unteren Segura (Südostspanien). *Mitt. dt. Ges.*, **103**, 125.

Wienberg-Rasmussen, H. (1968) *Danmarks Geologi* (Gjellerups liniebøger for seminarier og gymnasier, Copenhagen).

Wilhelmy, H. (1958) *Klimamorphologie der Massengesteine* (G. Westermann Verlag, Braunschweig).

Wilhelmy, H. (1974) Zur Genese der Blockmeere, Blockströme und Felsburgen in den deutschen Mittelgebirgen. *Ber. dt. Landesk.*, **48**, 17.

Williams, P. W. (1963) An initial estimate of the speed of limestone solution in county Clare. *Ir. Geogr.*, **4**, 432.

Williams, P. W. (1966) Limestone pavements with special reference to western Ireland. *Trans. Inst. Br. Geogr.*, **40**, 155.

Wills, L. J. (1924) The development of the Severn valley in the neighbourhood of Ironbridge and Bridgnorth. *Q. J. geol. Soc. Lond.*, **80**, 274.

Wills, L. J. (1937) The Pleistocene development of the Severn from Bridgnorth to the sea. *Q. J. geol. Soc. Lond.*, **94**, 161.

Winkler-Hermaden, A. (1950) Zum Entstehungsproblem und zur Altersfrage der ostalpin Oberflächenformen. *Mitt. geogr. Ges. Wien*, **92**, 171.

Winkler-Hermaden, A. (1955) Ergebnisse und Probleme der quartären Entwicklungsgeschichte am östlichen Alpenraum ausserhalb der Vereisungsgebiete. *Denkrschr. öst. Akad. Wiss., Math. Nat. Kl.*, **110**, 180.

Winkler-Hermaden, A. (1957) *Geologisches Kräftespiel und Landformung* (Springer-Verlag, Vienna).

Winterhalter, B. (1972) On the geology of the Bothnian Sea, an epeiric sea that has undergone Pleistocene glaciation. *Geol. Surv. Finl.*, **258**.

Woldstedt, P. (1950) *Norddeutschland und Angrenzende Gebiete im Eiszeitalter* (K. F. Köhler, Stuttgart).

Woldstedt, P. (1955) Die Probleme der Terrassenbildung. *Eiszeitalter Gegenw.*, **5**.

Woldsetdt, P. (1961) *Das Eiszeitalter* (Ferdinand Enke, Stuttgart).

Woldstedt, P. (1967) The Quaternary of Germany, in *The Quaternary*, ed. by K. Rankama, p.239 (Wiley Interscience, New York).

Woldstedt, P. and Duphorn, K. (1974) *Norddeutschland Angrenzende Gebiete im Eiszeitalter* (K. F. Köhler, Stuttgart).

Wolff, J. J. (1929) The dunes of Sabloney near Arcachon. *Geogrl J.*, **73**, 453.

Wood, A. and Woodland, A. W. (1968) Borehole at Mochras, west of Llanbedr, Merionethshire. *Nature, Lond.*, **219**, 1352.

Wooldridge, S. W. (1954) The physique of the South West. *Geography*, **39**, 231.

Wooldridge, S. W. and Linton, D. L. (1955) *Structure, Surface and Drainage in South-East England* (George Philip & Son, London).

Worsley, P. (1966) Some Weichselian fossil frost wedges from east Cheshire. *Mercian Geol.*, **1**, 357.

Worzel, J. L. (1974) Standard oceanic and continental structure, in *The Geology of the Continental Margins*. ed. by C. L. Burt and C. A. Drake, p.50 (Springer-Verlag, New York).

Worzel, J. L. and Shurbet, G. L. (1955) Gravity interpretation from standard oceanic and continental section. *Geol. Soc. Am. Spec. Pap.*, **62**.

Wramner, P. (1973) Studies of palsa bogs in Taavavuoma and the Laiva valley, Swedish Lapland. *Dept. phys. Geogr., Göteborgs Univ., Occas. Pap.*

Yanshin, A. L. (ed.) (1966) *Tectonics of Eurasia* (Nedra, Moscow).

Yates, E. M. (1963) The development of the Rhine. *Trans. Inst. Br. Geogr.*, **32**, 65.

Yates, E. M. and Moseley, F. (1967) A contribution to the glacial geomorphology of the Cheshire Plain. *Trans. Inst. Br. Geogr.*, **42**, 107.

Zabrodskaya, M. P. (1961) Some questions concerning the geomorphology of the Dinarsky uplands. *Izv. vses. geogr. Obshch. voronezh. Otd.*, **93**, 537.

Zagwijn, W. H. (1960) Aspects of the Pliocene and early Pleistocene vegetation in the Netherlands. *Meded. geol. Sticht.*, *C-III*, **1** (5).

Zagwijn, W. H. (1974) Palaeogeographical evolution of the Netherlands during the Quaternary. *Geol. Mijnb.*, **53**, 369.

Zagwijn, W. H. and Doppert, J. W. C. (1978) Upper Cenozoic of the southern North Sea basin: palaeoclimatic and palaeogeographical evolution. *Geol. Mijnb.*, **57**, 577.

Zanferrari, A. *et al.* (1982) Evoluzione neotettonica dell'Italia nordorientale. *Memorie Sci. geol. Padova.* **35**.

Zapletal, L. (1969) *Úvod do Antropogenní Geomorfologie* (Palackého University Olomouc, přírodov. Fak.).

Zavarichiy, A P. (1946) Some new facts required in the study of tectonic structures. *Izv. Akad. Nauk SSSR, ser. Geol.*, **2**.

Zazo, C. (1980) *El Cuaternario Marino-Continental del Litoral de las Provincias de Càdiz y Huelva*, thesis University of Madrid.

Zazo, C. and Goy, J. L. (1981a) Litoral de Huelva. *Guía Excursión 1, V Reunión Grp esp. Trab. Cuatern.*, p.354.

Zazo, C. and Goy, J. L. (1981b) Litoral de Càdiz. *Guía Excursión 2, V Reunión Grp esp. Trab. Cuatern.*, p.376.

Zazo, C. *et al.* (1981) Litoral de Huelva. *Guía Excursión 1, V Reunión Grp esp. Trab. Cuatern.*, p.357.

Zbyszewski, G. (1939) Observations sur la structure et la morphologie du Bas Alentejo et de l'Algarve. *Bull. Étud. Port., Inst. Français au Portugal*.

Zeese, R. (1978) Die Reculées des mittleren französischen Plateaujura. *Erdkunde*, **32**, 258.

Zenkovich, V. P. (1946) *Coastal Dynamics and Morphology* (Mozskoi Transport, Moscow).

Zenkovich, V. P. (1962) *Principles of the Study of Coastal Evolution* (Akad. Nauk SSSR, Moscow).

Zeremski, M. (1972) Morfostrukturna podela reljega Jugoslavije saglasna novoj geotektonskih podeli. *Glasn. srp. geogr. Društ*, **53** (2).

Ziegler, W. H. (1975a) Outline of the geological history of the North Sea. Western Europe, in *Petroleum and the Continental Shelf of North-Western Europe*, ed. by A. W. Woodland, p.165 (John Wiley, London).

Ziegler, W. H. (1975b) Geological evolution of the North Sea and its tectonic framework. *Bull. Am. Ass. Petrol. Geol.*, **59**, 1073.

Zienert, A. (1957) Die Grossformen des Odenwaldes. *Heidelb. geogr. Arb.*, **2**.

Zienert, A. (1961) *Die Grossformen des Schwarzwaldes* (Bundesanstalt für Landes-kunde und Raumforschung, Bad Godesberg).

Zimmermann, H. W. (1963) *Die Eiszeit im westlichen zentralen Mittelland (Schweiz)* (Naturforschende Gesellschaft, Solothurn).

Zoller, H. (1966) Postglaziale Gletscherstände und Klimaschwankungen im Gotthardmassiv und Vorderrheingebiet. Geologisch-glazialogische Untersuchungen von C. Schindler und H. Röthlisberger. *Verh. naturf. Ges. Basel*, **77**, 97.

Zoubek, V. (1960) *Tectonic Development of Czechoslovakia* (NČSAV, Prague).

Zunica, M. (1971) *Le Spiagge del Veneto* (Tipogr. Antoniana, Padua).

INDEX